Advances in Flow Analysis

Edited by
Marek Trojanowicz

Related Titles

Gauglitz, G., Vo-Dinh, T. Moore, D. S. (eds.)

Handbook of Spectroscopy

Second Edition

2010
ISBN: 978-3-527-32150-6

Sauer, M., Hofkens, J., Enderlein, J.

Handbook of Fluorescence Spectroscopy and Imaging

From Ensemble to Single Molecules

2008
ISBN: 978-3-527-31669-4

Guo, J. (ed)

X-Rays in Nanoscience

Spectroscopy, Spectromicroscopy, and Scattering Techniques

2009
ISBN: 978-3-527-32288-4

Yan, B

Analysis and Purification Methods in Combinatorial Chemistry

2004
E-Book
ISBN: 978-0-471-53199-9

Günzler, H., Williams, A. (eds.)

Handbook of Analytical Techniques

2001
ISBN: 978-3-527-30165-2

Advances in Flow Analysis

Edited by
Marek Trojanowicz

WILEY-VCH

WILEY-VCH Verlag GmbH & Co. KGaA

The Editor

Professor Marek Trojanowicz
Warsaw University
Faculty of Chemistry
Pasteura 1
02-093 Warsaw
Poland
and:
Institute of Nuclear
Chemistry and Technology
Dorodna 16
03–195 Warsaw
Poland

Library of Congress Card No.: applied for

British Library Cataloguing-in-Publication Data
A catalogue record for this book is available from the British Library.

Bibliographic information published by the Deutsche Nationalbibliothek
Die Deutsche Nationalbibliothek lists this publication in the Deutsche Nationalbibliografie; detailed bibliographic data are available in the Internet at http://dnb.d-nb.de.

© 2008 WILEY-VCH Verlag GmbH & Co. KGaA, Weinheim

Typesetting Thomson Digital, Noida, India
Printing Strauss GmbH, Mörlenbach
Binding Litges & Dopf GmbH, Heppenheim
Cover WMX Design GmbH Heidelberg

Printed in the Federal Republic of Germany
Printed on acid-free paper

ISBN: 978-3-527-31830-8

Contents

Introduction   *XIX*
List of Contributors   *XXV*

I   **Methodologies and Instrumentation**   *1*

1   **Theoretical Aspects of Flow Analysis**   *3*
          Fernando A. Iñón and Mabel B. Tudino
1.1   Introduction   *3*
1.2   Classification of Flow Systems. Fundamentals   *4*
1.2.1   Continuous Flow Analysis   *4*
1.2.2   Flow Injection Analysis   *4*
1.2.3   Sequential Injection Analysis   *4*
1.2.4   Multicommutation in Flow Injection Analysis   *5*
1.2.5   Stopped Flow   *6*
1.2.6   Batch Flow Injection Analysis (BFA)   *6*
1.3   Dispersion in Flow Injection Analysis: From the Movement of Fluids in Open Tubes to Controlled Dispersion   *7*
1.3.1   Transport of Fluids   *7*
1.3.1.1   Viscosity   *7*
1.3.1.2   Thermal Conductivity   *8*
1.3.1.3   Diffusivity   *8*
1.3.1.4   Diffusion   *8*
1.3.2   The Diffusion–Convection Equation in Open Conduits   *9*
1.3.3   The Distribution of Times of Residence   *10*
1.3.3.1   Characterization and Experimental Domain of Flow Systems: Dimensionless Numbers and Their Meaning   *11*
1.3.4   From the RTD Curve to the Generation of Signals in Flow Injection Systems   *12*
1.3.4.1   The Dispersion Process   *12*
1.3.4.2   The Concept of Controlled Dispersion and Analytical Implications   *13*
1.3.4.3   The Transient Profile   *14*

Advances in Flow Analysis. Edited by Marek Trojanowicz
Copyright © 2008 WILEY-VCH Verlag GmbH & Co. KGaA, Weinheim
ISBN: 978-3-527-31830-8

1.4   The Measurement of Dispersion   *15*
1.4.1   The Coefficient "D"   *16*
1.4.2   Peak Width and Time of Appearance   *17*
1.4.3   Peak Variance and Theoretical Plate Height   *18*
1.4.4   Degree and Intensity of Axial Dispersion   *19*
1.4.4.1   Degree of Axial Dispersion   *19*
1.4.4.2   Intensity of the Radial Dispersion   *20*
1.4.5   Other Approaches to the Measurement of Dispersion   *21*
1.5   Contribution of the Different Components of a Flow System to
           Dispersion   *21*
1.5.1   Injection   *23*
1.5.2   Detection   *24*
1.5.3   Transport: Different Models   *25*
1.5.3.1   Descriptive Models or "Black Boxes"   *25*
1.5.3.2   Deterministic Models: Dispersive Models and Tank-in-Series Model   *25*
1.5.4   Probabilistic Models   *28*
1.5.4.1   Random Walk   *28*
1.6   Design Equations   *29*
1.6.1   Influence of the Different System Variables   *29*
1.6.1.1   Reactor Length   *29*
1.6.1.2   Geometric Configuration   *29*
1.6.1.3   Flow Rate   *30*
1.6.1.4   Tube Radius   *32*
1.6.1.5   Injection Volume   *32*
1.6.2   Optimization of Flow Systems   *34*
1.7   Concluding Remarks   *38*
           References   *39*

2   **Injection Techniques in Flow Analysis**   *43*
           Víctor Cerdà and José Manuel Estela
2.1   Introduction   *43*
2.2   Continuous Flow Analysis (CFA)   *44*
2.3   Segmented Flow Analysis (SFA)   *45*
2.4   Flow Injection Analysis (FIA)   *46*
2.4.1   Syringe-based Injection   *46*
2.4.2   Injection with Rotary Valves   *47*
2.4.3   Proportional Injection   *48*
2.4.4   Merged Injection   *48*
2.4.5   Injection Following a Prior Flow Sample Processing   *49*
2.4.5.1   Multiparametric Determination   *49*
2.4.5.2   Dialysis   *50*
2.4.5.3   Gas Diffusion   *50*
2.4.5.4   Pervaporation   *52*
2.4.6   Hydrodynamic Injection   *53*
2.5   Sequential Injection Analysis (SIA)   *53*

2.5.1   Original Procedures   53
2.5.2   Conventional Injection   54
2.5.3   Controlled Variable Volume Injection   54
2.5.4   Cumulative Injection   55
2.5.5   The Sandwich Technique   55
2.5.6   Multiparametric Analysis   57
2.5.7   Gas Diffusion   57
2.5.8   Dialysis   59
2.5.9   Mixing Chamber-Based Injection   60
2.5.10   Bead Injection   60
2.5.11   Hydrodynamic Injection   61
2.6   Multicommutated Flow Injection Analysis (MCFIA)   61
2.6.1   Time-based Injection   61
2.6.2   Tandem Streams   62
2.7   Multisyringe Flow Injection Analysis (MSFIA)   63
2.7.1   Injection with Rotary Valves   64
2.7.2   Time-based Injection   65
2.7.3   Controlled Variable Volume Injection   66
2.7.4   Synchronous Injection with Two Valves   66
2.7.5   Synchronous Injection   67
2.7.6   Injection with a Dual Module   68
2.7.7   Bead Injection   69
2.8   Multipumping Flow Systems (MPFS)   69
2.8.1   Pulse-based Injection   70
2.8.2   Merged Injection   70
2.9   Combined Injection Methods   70
2.9.1   Dual Sequential Injection Analysis   71
2.9.2   Sequential Injection-multisyringe Flow Injection Systems   71
2.9.3   Multisyringe Flow Injection-multipumping Flow Systems   72
2.10   Concluding Remarks   73
           References   74

3   **Application of Moveable Suspensions of Solid Particles in Flow
           Analysis**   79
           Marek Trojanowicz
3.1   Introduction   79
3.2   Solid Microparticles Used in Suspensions in Flow Analysis   81
3.3   Handling Suspensions of Particles in Flow Systems   82
3.4   Detection Methods Employed in Flow Systems with Particle
           Suspensions   86
3.5   Renewable Columns in Flow Systems   91
3.6   Microfluidics with Processing of Particle Suspensions   94
3.7   Nanoparticles in Flow Systems   98
3.8   Conclusions   100
           References   101

4 **Batch Injection Analysis** *107*
Christopher M.A. Brett
4.1 Introduction *107*
4.2 Theory of Batch Injection Analysis *108*
4.3 Experimental Aspects – Cell Design and Detection Strategies *110*
4.4 Applications of Batch Injection Analysis *114*
4.4.1 Stripping Voltammetry of Trace Metal Ions and Applications in Environmental Monitoring *115*
4.4.2 Other BIA Environmental Monitoring Applications *116*
4.4.3 Analysis of Foodstuffs, Pharmaceuticals and Clinical Analysis *117*
4.4.4 Capillary Batch Injection Analysis *119*
4.4.5 Non-Electrochemical Methodologies and Detection Strategies *119*
4.5 Comparison of BIA with Flow Injection Techniques *120*
4.6 Future Perspectives *121*
References *121*

5 **Electroosmosis-Driven Flow Analysis** *127*
Petr Kubáň, Shaorong Liu, and Purnendu K. Dasgupta
5.1 Introduction *127*
5.2 Pumping Systems *128*
5.2.1 Pressure Pumping Systems *128*
5.2.2 Electroosmosis-Driven Pumping Systems *129*
5.2.2.1 Theoretical Considerations *129*
5.2.2.2 Theoretical Considerations for an EOF-Induced Pumping System *130*
5.3 EOF Pumping Systems *133*
5.3.1 EOF Pumping Without Electric Isolation *133*
5.3.2 Electrically Isolated EOF Pumps *136*
5.3.2.1 Open Tubular Electrically Isolated EOF Pumps *136*
5.3.2.2 Microchannel-based Electrically Isolated EOF Pumps *138*
5.3.2.3 Electrically Isolated EOF Pumps Based on Packed Columns *139*
5.3.2.4 Electrically Isolated EOF Pumps Based on Porous Media and Monoliths *141*
5.4 EOF Injection Methods Utilized in FIA, SIA and Micro Total Analysis Systems (μ-TAS) *142*
5.5 Applications of EOF-Driven Pumping in Flow Analysis *143*
5.6 Future Prospects *145*
References *145*

6 **Flow Analysis in Microfluidic Devices** *149*
Manabu Tokeshi and Takehiko Kitamori
6.1 Introduction *149*
6.2 Continuous Flow Chemical Processing in Microfluidic Devices *149*
6.2.1 Integration of Heavy Metal Analysis System *149*
6.2.2 Integration of Multi-Ion Sensing System *152*
6.2.3 Integration of a Bioassay System *155*

6.2.4   Real-time Monitoring of a Chemical Reaction   *157*
6.3   Flow Injection Analysis in Microfluidic Devices   *159*
6.3.1   Flow Injection Analysis System: On-Chip PDMS Valve   *159*
6.3.2   Flow Injection Analysis System: Chip-based Slide Valve   *160*
6.3.3   Flow Injection Analysis System: Immunoassay   *162*
6.4   Perspectives   *163*
         References   *165*

7   The Concept of Multi-commutation in Flow Analysis   *167*
         Mário A. Feres, Elias A.G. Zagatto, João L.M. Santos, and José L.F.C. Lima
7.1   Introduction   *167*
7.2   Concepts   *169*
7.3   The Discretely Operated Devices   *171*
7.3.1   Passively Operated Devices   *171*
7.3.2   Actively Operated Devices   *172*
7.4   System Design   *172*
7.5   Tandem Streams   *174*
7.6   Processes Involving Multi-Commutation   *175*
7.6.1   Sample Introduction   *175*
7.6.2   Sample Dispersion   *176*
7.6.3   Solution Additions to the Sample   *178*
7.6.4   Sample Incubation   *178*
7.6.5   Analyte Separation/Concentration   *179*
7.7   Applications   *180*
7.7.1   Selection Criteria   *180*
7.7.2   Selected Applications   *180*
7.8   Conclusions   *181*
7.9   Trends   *181*
         References   *192*

8   Advanced Calibration Methods in Flow Injection Analysis   *203*
         Paweł Kościelniak
8.1   Introduction   *203*
8.2   Advanced Calibration Procedures   *205*
8.2.1   Preparation Stage   *205*
8.2.2   Reconstruction Stage   *208*
8.2.3   Transformation Stage   *211*
8.2.4   Critical View   *212*
8.3   Advanced Calibration Concepts   *214*
8.3.1   Integrated Calibration   *214*
8.3.2   Gradient Dilution Calibration   *216*
8.3.3   Reaction-based Calibration   *217*
8.3.4   Multi-Component FIA Calibration   *218*
8.4   Trends and Perspectives   *219*
         References   *221*

9   **Multicomponent Flow Injection Analysis**   *227*
         Javier Saurina
9.1   Introduction   *227*
9.2   Principal Strategies for Multicomponent Analysis   *229*
9.2.1   Optimization of FI Multicomponent Methods   *229*
9.2.1.1   Types of Objective Functions in Optimization   *230*
9.2.1.2   Univariate versus Multivariate Optimization   *231*
9.2.1.3   A Practical Case   *232*
9.2.2   Reagents   *234*
9.2.3   Manifolds for Multicomponent Determinations   *239*
9.2.3.1   Sequential Configurations   *240*
9.2.3.2   Parallel Configurations   *241*
9.2.3.3   Serial Configurations   *242*
9.2.4   Detectors   *242*
9.2.5   Chemometric Techniques for Data Analysis   *245*
9.2.5.1   Data Sets from Flow Systems   *245*
9.2.5.2   Calibration Methods Applied to Flow Data   *247*
9.2.5.3   A Practical Example of Second-Order Calibration   *252*
9.3   Trends and Perspectives   *254*
         References   *256*

10   **Flow Processing Devices Coupled to Discrete Sample Introduction Instruments**   *265*
         M. Valcárcel, S. Cárdenas, B.M. Simonet, and R. Lucena
10.1   Introduction: The Problem of Sample Treatment   *265*
10.2   Roles of Flow Processing Devices   *265*
10.3   Ways of Coupling Flow Processing Devices to Discrete Sample Introduction Instruments   *266*
10.4   Coupling Flow Processing Devices to Gas Chromatographs   *267*
10.4.1   Interfaces and Types of Coupling   *267*
10.4.2   Analytical Uses   *268*
10.4.3   Critical Discussion   *270*
10.5   Coupling Flow Processing Devices to Liquid Chromatographs   *271*
10.5.1   Interfaces and Types of Coupling   *271*
10.5.2   Analytical Uses   *274*
10.5.3   Critical Discussion   *274*
10.6   Coupling Flow Processing Devices to Capillary Electrophoresis Equipment   *274*
10.6.1   Interfaces and Types of Coupling   *275*
10.6.2   Analytical Uses   *280*
10.6.3   Critical Discussion   *281*
10.7   Future Prospects   *283*
         References   *283*

11   **On-line Sample Processing Methods in Flow Analysis** *291*
           Manuel Miró and Elo Harald Hansen
11.1   Introduction *291*
11.2   On-line Sample Pretreatment Protocols for Aqueous and
           Air Samples *292*
11.2.1   On-line Dilution *292*
11.2.2   Derivatization Reactions *293*
11.2.3   Solvent Extraction *294*
11.2.4   Sorbent Extraction *297*
11.2.5   Precipitation/Co-Precipitation *300*
11.2.6   Gas–Liquid Separation *301*
11.2.7   Membrane-based Separation *302*
11.2.7.1   Gas Diffusion *302*
11.2.7.2   Dialysis *304*
11.2.7.3   Pervaporation *305*
11.2.8   Digestion Protocols *305*
11.3   On-line Processing of Solid Samples: Leaching/Extraction
           Methods *306*
11.4   Trends and Perspectives *307*
           References *311*

12   **Flow Analysis and the Internet – Databases, Instrumentation,**
           and Resources *321*
           Stuart J. Chalk
12.1   Introduction *321*
12.2   Databases *321*
12.2.1   Dr. Elo Hansen's Flow Bibliography *322*
12.2.2   Google Scholar Beta *322*
12.2.3   The Flow Analysis Database *323*
12.3   Journals *323*
12.3.1   Analytica Chimica Acta (Elsevier) *326*
12.3.2   Talanta (Elsevier) *326*
12.3.3   Analytical Chemistry (American Chemical Society – ACS) *327*
12.3.4   The Analyst (Royal Society of Chemistry – RSC) *327*
12.3.5   Journal of Flow Injection Analysis (Japanese Association of Flow
           Injection Analysis) *327*
12.3.6   Journal of Automatic Methods and Management
           in Chemistry (Hindawi) *327*
12.3.7   Analytical and Bioanalytical Chemistry (Springer) *327*
12.3.8   Analytical Sciences (Japanese Society for Analytical Chemistry) *328*
12.3.9   Analytical Letters (Taylor and Francis) *328*
12.3.10   Journal of Chromatography A (Elsevier) *328*
12.3.11   Electroanalysis (Wiley) *328*
12.3.12   Journal of Analytical Atomic Spectrometry (RSC) *328*
12.3.13   Fenxi Huaxue (Wanfang Data) *329*

12.3.14   Bunseki Kagaku (Japanese Society for Analytical Chemistry)   *329*
12.3.15   Clinical Chemistry (American Association for Clinical Chemistry)   *329*
12.4   Instrumentation   *330*
12.5   Standard Methods   *330*
12.5.1   International Standards Organization   *330*
12.5.2   US Environmental Protection Agency   *332*
12.5.3   American Society for Testing and Materials   *334*
12.5.4   APHA/AWWA/WEF Standard Methods   *334*
12.5.5   US Geological Survey Standard Methods   *336*
12.6   Other Useful Websites   *337*
12.6.1   Tutorials   *337*
12.6.2   Books (Chronological Order)   *337*
12.6.3   Webpages of Prominent Researchers (Alphabetical)   *338*
12.6.4   Other   *339*
12.7   Future Directions   *339*
12.7.1   The Semantic Web   *339*
12.7.2   Extensible Markup Language (XML)   *340*
           References   *341*

II   **Advances in Detection Methods in Flow Analysis**   *343*

13   **Luminescence Detection in Flow Analysis**   *345*
       Antonio Molina-Díaz and Juan Francisco García-Reyes
13.1   Introduction   *345*
13.2   Luminescence Detection in Continuous Flow Systems   *346*
13.2.1   Fluorescence   *346*
13.2.1.1   Introduction   *346*
13.2.1.2   Fluorescence Detection in Flow Methods   *347*
13.2.2   Phosphorescence   *355*
13.2.2.1   Introduction   *355*
13.2.2.2   Room Temperature Phosphorescence in Ordered Media   *356*
13.2.2.3   Non-Protected Room Temperature Phosphorescence   *356*
13.2.3   Chemiluminescence   *357*
13.2.3.1   Introduction   *357*
13.2.3.2   CL Flow Methods   *359*
13.2.3.3   Bioluminescence   *365*
13.2.4   Solid-phase Luminescence Based Detection in Flowing Streams   *366*
13.2.4.1   Solid-phase Fluorescence Detection   *366*
13.2.4.2   RTP-Based Optosensing   *371*
13.2.4.3   Solid-phase Chemiluminescence Detection   *372*
13.3   Recent Trends and Perspectives   *374*
13.3.1   Miniaturization   *374*
13.3.2   Molecularly Imprinted Materials   *375*
13.3.3   Quantum Dots   *375*
           References   *377*

14   **Enzymes in Flow Injection Analysis**   *395*
            Robert Koncki, Łukasz Tymecki, and Beata Rozum
14.1   Introduction   *395*
14.2   Enzyme Substrates as Analytes   *396*
14.2.1   Biosensor-based FIA Systems   *397*
14.2.2   Bioreactor-based FIA Systems   *403*
14.2.3   Additional Benefits Offered by FIA   *403*
14.2.4   Analytical Applications. FIA Systems
            as Monitors   *411*
14.3   Methods Based on Enzyme Activity
            Measurements   *412*
14.3.1   Enzyme Activity Detection   *413*
14.3.2   Enzyme Inhibitor Detection   *415*
14.3.3   Enzyme Cofactor Detection   *416*
14.4   Conclusions   *416*
            References   *417*

15   **Flow Potentiometry**   *425*
            M. Conceição B.S.M. Montenegro and Alberto N. Araújo
15.1   Introduction   *425*
15.2   Background Concepts   *425*
15.3   Electrode Developments and Detector Cell Designs   *428*
15.4   Flow Analytical Techniques Based on Potentiometry   *436*
15.5   Trends and Future Prospects   *443*
            References   *443*

16   **Flow Voltammetry**   *455*
            Ivano G.R. Gutz, Lúcio Angnes, and Andrea Cavicchioli
16.1   Introduction   *455*
16.2   Voltammetric/Amperometric Flow Analysis   *456*
16.2.1   Principles and Techniques   *456*
16.2.2   Electrode Materials   *460*
16.2.3   Commercial Flow Cells   *461*
16.2.4   Adaptors for Commercial Batch Cells   *463*
16.2.5   Specially Designed Flow Cells and Systems   *465*
16.3   Strategies for Improving Selectivity, Sensitivity,
            and Durability   *467*
16.3.1   Preconcentration   *467*
16.3.2   Medium Exchange   *468*
16.3.3   Oxygen Removal   *469*
16.3.4   Catalytic Electrode Processes   *469*
16.3.5   Spectroelectrochemistry   *470*
16.3.6   Modified Electrodes   *471*
16.4   Trends and Perspectives   *472*
            References   *473*

17 **Affinity Interaction Profiling of Protein–Protein and Protein–Ligand Interactions Using Flow Analysis** *483*
J. Kool, N.P.E. Vermeulen, H. Lingeman, R.J.E. Derks, and H. Irth
17.1 Introduction *483*
17.2 Profiling of Noncovalent Protein–Protein and Protein–Ligand Interactions Based on Mass Spectrometry Flow Analysis *485*
17.2.1 Flow Injection and Continuous Infusion Mass Spectrometry of Noncovalent Complexes Using Electrospray Ionization *485*
17.2.1.1 Introduction *485*
17.2.1.2 Electrospray Ionization Mass Spectrometry of Noncovalent Complexes *485*
17.2.1.3 Limitations of Electrospray Ionization Mass Spectrometry for the Study of Noncovalent Complexes *486*
17.2.2 Flow Injection Mass Spectrometry Assays Using Reporter Molecules *487*
17.2.2.1 Introduction *487*
17.2.2.2 Requirements for Mass Spectrometry-Based Biochemical Assays *488*
17.2.2.3 Flow Injection Ligand-Binding Assays Using Mass Spectrometry as Readout *488*
17.2.2.4 Flow Injection Enzyme Assays Using Mass Spectrometry as Readout *488*
17.2.3 Reporter-Free Assays after Dissociation of Protein–Ligand Complexes *489*
17.3 Integration of Flow Analysis and High-Performance Liquid Chromatography for the Bioaffinity Screening of Mixtures *490*
17.3.1 Introduction *490*
17.3.2 On-line Coupling of Flow Biochemical Assays to HPLC *491*
17.3.2.1 General Principles *492*
17.3.2.2 On-line Ligand-Binding Flow Assays Coupled to HPLC *493*
17.3.2.3 On-line Enzyme Flow Assays Coupled to HPLC *498*
17.4 Conclusions *502*
References *503*

18 **Atomic Spectroscopy in Flow Analysis** *511*
José L. Burguera and Marcela Burguera
18.1 Introduction *511*
18.2 Flame Atomic Absorption Spectrometry *512*
18.2.1 Preconcentration/Separation Systems Using FAAS Detection *512*
18.2.1.1 Chemical Vapor Generation *512*
18.2.1.2 Preconcentration of Trace Elements by Solid-phase Extraction *514*
18.2.1.3 Other Preconcentration Systems for FAAS Detection *514*
18.2.1.4 Indirect Determinations *519*
18.3 Sample Dilution *519*
18.4 Electrothermal Atomic Absorption Spectrometry *520*

18.4.1   Analyte Preconcentration for ET AAS Detection   520
18.4.1.1   Precipitation or Coprecipitation/Dissolution Reactions   520
18.4.1.2   Sorption on Columns   521
18.4.1.3   Solvent Extraction   523
18.4.2   Analyte Separation Prior to ET AAS Detection   524
18.4.3   Miscellaneous   526
18.5   Atomic Fluorescence Spectrometry   527
18.5.1   Preconcentration of the Analyte for AFS Detection   527
18.5.2   Hyphenated Techniques for Speciation Studies with AFS
            Detection   529
18.6   Conclusions and Further Developments   531
            References   532

19   Flow Injection Mass Spectrometry   545
            Maria Fernanda Giné
19.1   Introduction   545
19.1.1   The Role and Importance of Flow Injection Analysis Mass
            Spectrometry   545
19.1.2   Mass Spectrometry   545
19.1.2.1   Quadrupole Mass Spectrometers (QMS)   547
19.1.2.2   Sector Field Mass Spectrometers   547
19.1.2.3   Multicollector Mass Spectrometer   548
19.1.2.4   Time of Flight Mass Spectrometers (TOFMS)   549
19.2   FIA-MS Sample Introduction Devices   550
19.2.1   Transient Sample Introduction into the MS Ionization Chamber   551
19.2.2   FI Sample Introduction to External Ionization Sources – MS   553
19.3   Flow Systems Coupled to External Ionization Sources – MS   553
19.3.1   FIA-ICP-MS   553
19.3.1.1   The ICP Source   554
19.3.1.2   FIA-ICP-MS for Reducing Matrix Effects   555
19.3.1.3   FIA Systems to Perform Isotope Dilution (ID)   556
19.3.1.4   FIA Applications of ID-ICP-MS   560
19.3.1.5   FIA Coupled with Hyphenated Techniques to ICP-MS   561
19.3.2   FIA-Thermospray MS   563
19.3.3   Electrospray Ionization (ESI – MS)   564
19.3.3.1   Electrospray Ionization   564
19.3.3.2   FIA-ESI-MS   564
19.4   Conclusions   566
            References   566

III   Applications   575

20   Environmental Applications of Flow Analysis   577
            Shoji Motomizu
20.1   Introduction   577

20.2   Analysis of the Aquatic Environment by Flow Methods   *577*
20.2.1   Substances Related to Eutrophication   *577*
20.2.1.1   Nitrogen Compounds   *578*
20.2.1.2   Phosphorus Compounds   *582*
20.2.2   Organic Compounds Related to Water Pollution   *584*
20.2.2.1   Surfactants   *584*
20.2.2.2   Chemical Oxygen Demand (COD)   *585*
20.2.3   Organic Compounds Related to Toxic/Hazardous Problems   *586*
20.2.4   Metals and Metal Compounds Related to Water Pollution and
            Toxic/Hazardous Problems   *587*
20.2.4.1   Spectroscopic Detection   *587*
20.2.4.2   Other Detection Methods   *590*
20.3   Analysis of an Atmospheric Environment by Flow Methods   *590*
20.3.1   Denuder (DN) and Gas Diffusion Scrubber (GDS)   *590*
20.3.2   Chromatomembrane Cell (CMC)   *591*
20.3.3   Simple Batchwise Collection/Concentration Method for Substances
            in Air   *593*
20.4   Analysis of the Geosphere Environment by Flow Methods   *594*
20.5   Future of Environmental Analysis   *595*
            References   *596*

21   Flow Methods in Pharmaceutical Analysis   *601*
            J. Martínez Calatayud and J.R. Albert-García
21.1   Introduction   *601*
21.2   Analysis of Pharmaceutical Formulations   *602*
21.2.1   Spectrophotometry (UV–vis and IR). Homogeneous Systems   *602*
21.2.1.1   Applications of Flow-Based Molecular Absorption Spectrophotometry
            to the Determination of Drugs   *602*
21.2.1.2   UV–vis Heterogeneous Systems (Turbidimetry, Solid–Liquid and
            Liquid–Liquid)   *606*
21.2.1.3   Infrared Absorption   *607*
21.2.1.4   Flame Atomic Absorption Spectrometry   *608*
21.2.1.5   Liquid–Liquid (Extraction) Systems Involving the Formation of
            Ion-Pairs or Neutral Chelates with the Analyte   *609*
21.2.2   Luminescence   *610*
21.2.2.1   Flow-Fluorimetry in Drug Analysis   *611*
21.2.2.2   Phosphorimetry   *614*
21.2.2.3   Chemiluminescence   *614*
21.2.3   Electrochemistry   *619*
21.2.3.1   Conductimetry   *619*
21.2.3.2   Potentiometry   *619*
21.2.3.3   Polarography   *621*
21.2.3.4   Amperometry   *622*
21.2.3.5   Continuous Flow Voltammetry   *624*
21.2.3.6   Continuous Flow Amperometry   *625*

21.3	Flow Process Analyzers in the Pharmaceutical Industry	*626*
21.3.1	Process Analysis in Pharmaceutical Production	*626*
21.3.2	Automated Drug Dissolution and Drug Release Testing	*628*
21.3.3	Membrane Diffusion	*631*
21.3.4	Functional Cellular Assays for Screening Potential Drugs	*633*
	References	*635*
22	**Industrial and Environmental Applications of Continuous Flow Analysis**	*639*
	Kees Hollaar and Bram Neele	
22.1	Introduction	*639*
22.2	Overview of Environmental and Industrial Fields	*640*
22.2.1	Environmental Applications	*640*
22.2.2	Plant and Soil Applications	*641*
22.2.3	Pharmaceutical Applications	*642*
22.2.4	Beer and Wine Applications	*643*
22.2.5	Tobacco Applications	*644*
22.2.6	Food Applications	*645*
22.3	Applications and Their Ranges	*645*
22.3.1	Total Nitrogen	*647*
22.3.2	Total Phosphate	*649*
22.3.3	Cyanides	*650*
22.3.3.1	Total Cyanide	*651*
22.3.3.2	Free Cyanide	*651*
22.3.4	Phenol Index	*651*
22.3.5	Total Reducing Sugars	*653*
22.3.6	Bitterness	*656*
22.4	Development of Flow Analysis Applications	*658*
22.5	Trends in Continuous Flow Analysis	*659*
	References	*661*
	Index	*663*

Introduction

Chemical analysis is an indispensable element in all areas of contemporary life. Together with progress in science and technological development, and also with constantly increasing demand, one can observe progress in analytical chemistry as a scientific discipline, and methods and techniques of chemical analysis for practical application purposes. The increasing demand for analytical determinations results from the necessity to analyze an increasing number of different samples, as well as from the need to design analytical instruments and methods which can be employed directly by an end-user without the need for the services of specialized analytical laboratories. This increasing demand includes also the need for improvement of the quality of analytical determinations. Depending on the area of application, there may be a need to shorten the time of analysis, minimize the amount of sample needed for analysis, achieve lower limits of detection or better selectivity (resolution) in multi-component determinations, or obtain better precision and/or accuracy of the determination.

Progress in the development of analytical methods occurs in various ways and is a combination of various factors. There are as many factors affecting it as there are different parameters that affect the results of an analytical determination. There are factors resulting from progress in natural science, material science, electronics and informatics, as well as from progress in engineering of materials and devices, and their utilization in analytical procedures. Human inventiveness and the urge for discovery, which is a driving force for fundamental scientific research, is practically unlimited, hence no stage of the development of science or technology should be treated as definitive. This obviously also concerns progress in analytical chemistry and the methods and techniques of chemical analysis.

The carrying out of analytical determinations in flow conditions can, at first glance, be treated as a simplification of the conventional non-flow procedure by omitting a sampling step. Probably such measurements in the area recognized nowadays as process analysis were, in the 1930s or 1940s, the first examples of analytical measurements in flowing conditions, for example, conductivity measurements in process streams. Then, with the development of detection methods and progress in the construction of measuring instruments, measurements of redox potential, pH, turbidity, absorbance at a given wavelength, to mention just the most

Advances in Flow Analysis. Edited by Marek Trojanowicz
Copyright © 2008 WILEY-VCH Verlag GmbH & Co. KGaA, Weinheim
ISBN: 978-3-527-31830-8

common ones, were carried out in the same way in flow mode. These methods are commonly used in modern process analysis. This separate area of chemical analysis, very well developed in the last half century, with a large arsenal of specially designed measuring instruments, has numerous specific problems to cope with. There is a wide literature on this field of chemical analysis and this area of flow analysis will be not discussed in this book.

The area of chemical analysis that is covered by this book is laboratory flow analysis. It is associated with different environmental and technical conditions, and different scales of processes and devices. In this case it is much easier to identify a commonly recognized author and the background of the idea that analytical measurement can be carried out during sample flow through the detector. The background to this invention was the urgent demand from clinical analytical laboratories in large hospitals in the 1950s, which were overloaded with a huge amount of samples, to find a solution as to how to speed up analytical procedures. The inventor of the first laboratory flow analysis system was the American biochemist Leonard J. Skeggs Jr., known also as the co-inventor of the modern artificial kidney, from the hospital associated with Case Western Reserve University Medical School. He designed the first laboratory flow system for determination of blood urea nitrogen with photometric detection (L.T. Skeggs Jr, *Am. J. Clin. Pathol.*, **1957**,*28*, 311). Based on the intuition that such a concept of analytical measurements could significantly improve laboratory analysis, the author of this idea quickly patented the new instrument and, in three years, it was launched on the market by Technicon Co with great success. Already the first prototypes contained several breakthrough instrumental solutions, not only a flow-through photometer and strip-chart recorder for continuous signal recording, which determined the success of the concept. I will mention a few of them to illustrate the many inventions and pioneering solutions that were involved in the construction of this system. It was constructed for analysis of blood samples, hence it was necessary to design a rotary sampler for aspiration of samples from vials. The sample, aspirated into the tubing, could disperse during the flow, but the extent of this dispersion could be limited by segmentation of the liquid stream with air bubbles. The determination of urea required removal of proteins, hence it was necessary to design a flow-through membrane dialyzer. The continuous detection in a flowing stream was already known from liquid chromatography, which was being developed with various detections earlier. The developed air-segmented flow analyzing system allowed the mechanization of numerous operations (sample introduction, addition of reagents, incubation, dialysis), and this was the most essential breakthrough in laboratory analysis.

For about the next 20 years instruments based on this concept of mechanization of chemical laboratory analysis predominated in large (and rich) clinical laboratories, but then numerous other ideas of mechanization of analytical procedures became more and more competitive. They include centrifugal analyzers, devices employing solid-reagent strips and, especially, various designs of discrete analyzers, which in the last 20 years have completely replaced clinical flow analyzers. They were more efficient and versatile, designed to perform several tens of assays from one sample. The air-segmented flow analyzers are still quite widely used, however, in routine analytical laboratories for environmental protection, agriculture analyses and food control.

A crucial new impulse was given to further development of laboratory flow analysis in the mid-1970s by the invention of flow measurements with injection of a small volume of sample into the stream of a flowing carrier or directly into a reagent flowing solution. Even some years earlier one can find in the literature reports on flow measurements with the introduction of a smaller sample volume than needed to achieve a steady-state equilibrium signal in the detector in air-segmented systems, with conclusions that the transient signal obtained in such systems can obviously also be used for analysis with the advantageous possibility of increasing the sampling rate. Based on the existing literature one can notice that the concept of flow injection measurements came simultaneously from different branches of analytical instrumentation. In one case it can be considered as an evolution of the earlier developed air-segmented systems by elimination of segmentation of the flowing stream and injection of a small volume of sample by an injection port instead of by continuous aspiration. This led to obtaining a transient signal and, with simultaneous reduction of the diameter of the tubings, such a system provided attractive fast analytical signals. Alternatively, the same concept of measurements originated from the application of commercial instrumentation for liquid column chromatography, and its utilization in flow measurements without a separation column. The selectivity of the determination of a particular analyte can be achieved by application of appropriate chemical conditions specific for the given analyte.

The rapid increase in interest in this methodology of analytical measurements in the next years (almost exponential, if it is measured by the number of published papers in analytical journals) has to be assigned, to a great extent, to a tandem of authors J. Ruzicka and E.H. Hansen, and their research group in the Technical University of Denmark, who in numerous pioneering publications demonstrated that, for academic laboratory research, flow injection measuring systems can be built easily with low-cost, simple components in almost every analytical laboratory, without big instrumental investment. This can be a way to realize various ideas of technical design, to carry out in such systems various chemical reactions and sample treatment operations, and to employ various detection methods. It is then a very attractive way of mechanization of analytical procedures in flow systems, but it has to be admitted that this is not a way to the automation of measurements, as this term is very commonly misused. Performing analytical measurement in flow conditions does not mean automation of measurements as, according to automation theory, and also following the IUPAC terminology recommendations, the automated system has to be equipped with an intelligent control system which, with the use of a feed-back loop mechanism, can control and regulate conditions of measurements without the participation of a human operator.

The flow injection methodology of analytical determinations, being developed since the 1970s, has gained already very many technical modifications such as the most commonly known flow systems: with sequential injection of sample and reagents into a single line system (called *sequential injection analysis* – SIA), flow measurements in tubeless systems with direct injection to the detector sensing surface (called *batch injection analysis* – BIA), or application in flow injection systems with moveable solid particles, called *bead-injection analysis*, with the same abbreviation BIA. Another aspect of the evolution of flow injection measuring systems is the rapidly progressing miniaturization of particular modules of the flow system, as well as their integration, for example, by incorporation of some modules into the injection valve (named generally as the *lab-on-valve* concept), or their miniaturization down to microfluidic format.

Generally, it seems that flow analytical systems can be described as analytical measuring devices in which all operations of sample pretreatment and detection of analyte are carried out in flowing streams. This seems to be a very common understanding of flow analysis, but at the same time one can find several inaccuracies or problems with such a description. Can one include as flow analysis simple AAS measurements with flame atomization, where the sample is aspirated, nebulized and then transported to the flame for optical detection? Can we talk about flow analysis in the case of mass spectrometry measurements with direct sample injection, where the injected liquid sample is evaporated, the analytes are ionized (and also can be fragmented), separated and then transported to the detector? And the most difficult problem to solve, namely the differentiation of liquid column chromatography and flow analysis. In column chromatography analytes present in an injected sample are separated on the column, then they can also sometimes be derivatized and then transported to a flow-through detector. A similar situation arises with capillary electrophoresis. From the point of view of tradition and history of development, and also, much more importantly, the role in analytical chemistry, it does not seem to be appropriate to include column chromatography in flow analysis. On the other hand, in typical flow systems of any kind (air-segmented continuous systems, flow injection systems etc.), packed reactors are very commonly used for sample clean-up, or preconcentration, and the operations carried out are chromatographic ones if we follow the common mechanism of chromatography. So, where to draw the line? For the sake of a framework for the subject of this book, as flow analysis is meant analysis in measuring systems where all operations of sample treatment and detection are carried out in flowing solutions but without multicomponent chromatographic or electrophoretic separation. Most often it is a single component method with mechanized sample pretreatment, while multicomponent analysis in flow systems is carried out in a system with more complex manifolds or by employing detectors that are multicomponent in the mechanism of their sensing. The dynamic properties of flow measurements are widely employed in sample processing, in many cases for improvement of the parameters of some detection methods, however, still very little is done on the design of multicomponent determinations.

The principal intention in the preparation of this joint work was to present the achievements of flow analysis in recent years that may be helpful in the determination of its position in modern chemical analysis. In spite of many thousands of papers published during the 60 years since the pioneering invention of Skeggs, this methodology of analytical measurements seems to be underestimated in various fields of routine chemical analysis. Certainly, the spectacular success in the 1960s and 1970s was the application of commercial flow analyzers with air-segmentation in clinical laboratories. Long years of development, numerous published papers, some commercial instruments for flow injection methods, have not introduced flow injection methods sufficiently into routine analytical laboratories. Nowadays, if some flow analyzers are used in routine analytical laboratories, they are mostly continuous flow analyzers with air-segmented flow and recording of a steady-state equilibrium signal.

The selection of subjects for all chapters was my subjective choice, as I am convinced that in these areas of flow analysis the largest progress has been made in recent years. I express thanks to our Publisher for acceptance of my choice. My special thanks I address to all the authors who accepted my invitation to contribute to this joint work. I am convinced they all share my hope that this book will be fruitful for the further development of flow analysis and the promotion of these methods of chemical analysis.

I would also address my thanks to all colleagues who accepted my invitation for collaboration, and who reviewed some chapters, namely Professor Diane Beauchemin of Queen's University, Kingston, Canada, Professor Ursula Bilitewski of Helmholtz-Centre for Infection Research, Braunschweig, Germany, Professor Ari Ivaska of Abo Akademi University, Finland, Professor Bo Karlberg of Stockholm University, Sweden, Professor Pawel Koscielniak of Jagiellonian University, Cracow, Poland, Professor Petr Kuban of the Mendel University of Agriculture and Forestry, Brno, Czech Republic, Professor Mark E. Meyerhoff of the University of Michigan, Ann Arbor, USA, Professor Boaventura Reis of CENA, University of Sao Paulo, Piracicaba, Brazil, Professor Petr Solich of Charles University, Hradec Kralove, Czech Republic, Professor Julian Tyson of the University of Massachusetts, Amherst, USA, Dr Bogdan Szostek of DuPont, Wilmington, USA, and Professor Paul Worsfold of the University of Plymouth, UK. I am grateful for their valuable help in giving a final shape to the reviewed chapters. I thank also all the staff members of Wiley-VCH Verlag who took part in this project, particularly Dr Manfred Kőhl, Dr Waltraud Wüst and Ms Claudia Nussbeck for their collaboration.

Marek Trojanowicz

List of Contributors

J.R. Albert-García
University of Valencia
Department of Analytical Chemistry
Dr. Moliner, 50
46100 Burjassot
València
Spain

Lúcio Angnes
Universidade de São Paulo
Instituto de Química
Av. Prof. Lineu Prestes 748
05508-900 São Paulo
Brazil

Alberto N. Araújo
Universidade do Porto
REQUIMTE
Departamento de Química Física
Faculdade de Farmácia
Rua Aníbal Cunha 164
4050 Porto
Portugal

Christopher M.A. Brett
Universidade de Coimbra
Departamento de Química
3004-535 Coimbra
Portugal

José L. Burguera
Los Andes University
Faculty of Sciences
Department of Chemistry
P.O. Box 542
Mérida 5101-A
Venezuela

Marcela Burguera
Los Andes University
Faculty of Sciences
Department of Chemistry
P.O. Box 542
Mérida 5101-A
Venezuela

J. Martínez Calatayud
University of Valencia
Department of Analytical Chemistry
Dr. Moliner, 50
46100 Burjassot
València
Spain

S. Cárdenas
Universidad de Córdoba
Department of Analytical Chemistry
Campus de Rabanales
14071 Córdoba
Spain

Advances in Flow Analysis. Edited by Marek Trojanowicz
Copyright © 2008 WILEY-VCH Verlag GmbH & Co. KGaA, Weinheim
ISBN: 978-3-527-31830-8

Andrea Cavicchioli
Universidade de São Paulo
Escola de Artes, Ciências e Humanidades
Rua Arlindo Béttio
1000 Ermelino Matarazzo
03828-000 São Paulo
Brazil

Víctor Cerdà
University of the Balearic Islands
Automation and Environment
Analytical Chemistry Group
Department of Chemistry
07122 Palma de Mallorca
Spain

Stuart J. Chalk
University of North Florida
Department of Chemistry and Physics
1 UNF Drive
Jacksonville
FL 32224
USA

Purnendu K. Dasgupta
University of Texas at Arlington
Department of Chemistry and
Biochemistry
Arlington
TX 76019-0065
USA

R.J.E. Derks
Vrije Universiteit Amsterdam
Division of Analytical Chemistry &
Applied Spectroscopy
De Boelelaan 1083
1081 HV Amsterdam
The Netherlands

José Manuel Estela
University of the Balearic Islands
Automation and Environment
Analytical Chemistry Group
Department of Chemistry
07122 Palma de Mallorca
Spain

Mário A. Feres Jr
Universidade de São Paulo
Centro de Energia Nuclear
na Agricultura
Avenida Centenario, 303
Caixa Postal 96
13400-970 Piracicaba SP
Brazil

Juan Francisco García-Reyes
University of Jaén
Department of Physical and Analytical
Chemistry
Paraje Las Lagunillas S/N
23008 Jaén
Spain

Maria Fernanda Giné
University of São Paulo
Centro de Energia Nuclear
na Agricultura CENA
Avenida Centenario, 303
Caixa Postal 96
13400-970 Piracicaba SP
Brazil

Ivano G.R. Gutz
Universidade de São Paulo
Instituto de Química
Av. Prof. Lineu Prestes 748
05508-900 São Paulo
Brazil

Elo Harald Hansen
Technical University of Denmark
Department of Chemistry
Kemitorvet
Building 207
2800 Kgs. Lyngby
Denmark

Kees Hollaar
Skalar Analytical B.V.
Tinstraat 12
4823 AA Breda
The Netherlands

Fernando A. Iñón
University of Buenos Aires
Facultad de Ciencias Exactas y Naturales
Pabellón 2
1428 Buenos Aires
Argentina

Hubertus Irth
Vrije Universiteit Amsterdam
Department of Analytical Chemistry &
Applied Spectroscopy
De Boelelaan 1083
1081 HV Amsterdam
The Netherlands

Takehiko Kitamori
The University of Tokyo
Department of Applied Chemistry
Graduate School of Engineering
7-3-1 Hongo
Bunkyo-ku
Tokyo 113-8656
Japan

Robert Koncki
University of Warsaw
Department of Chemistry
Pasteura 1
02-093 Warsaw
Poland

Jeroen Kool
Vrije Universiteit Amsterdam
Department of Analytical Chemistry &
Applied Spectroscopy
De Boelelaan 1083
1081 HV Amsterdam
The Netherlands

Paweł Kościelniak
Jagiellonian University
Faculty of Chemistry
R. Ingardena Str. 3
Kraków, 30-060
Poland

Petr Kubáň
Mendel University
Department of Chemistry
and Biochemistry
Zemědělská
Brno 61300
Czech Republic

José L.F.C. Lima
Universidade do Porto
REQUIMTE
Departamento de Química-Física
Faculdade de Farmácia
Rua Aníbal Cunha 164
4050 Porto
Portugal

Henk Lingeman
Vrije Universiteit Amsterdam
Department of Analytical Chemistry &
Applied Spectroscopy
De Boelelaan 1083
1081 HV Amsterdam
The Netherlands

Shaorong Liu
Texas Tech University
Department of Chemistry
and Biochemistry
Lubbock
TX 79409-1061
USA

R. Lucena
Universidad de Córdoba
Department of Analytical Chemistry
Campus de Rabanales
14071 Córdoba
Spain

Manuel Miró
University of the Balearic Islands
Department of Chemistry
Faculty of Sciences
Carretera de Valldemossa, km 7.5
07122-Palma de Mallorca
Spain

Antonio Molina-Díaz
University of Jaén
Department of Physical and Analytical
Chemistry
Paraje Las Lagunillas S/N
23008 Jaén
Spain

M. Conceição B.S.M. Montenegro
Universidade do Porto
REQUIMTE
Departamento de Química Física
Faculdade de Farmácia
Rua Aníbal Cunha 164
4050 Porto
Portugal

Shoji Motomizu
Okayama University
Department of Chemistry
Tsushimanaka
Okayama 700-8530
Japan

Bram Neele
Skalar Analytical B.V.
Tinstraat 12
4823 AA Breda
The Netherlands

Beata Rozum
University of Warsaw
Department of Chemistry
Pasteura 1
02-093 Warsaw
Poland

João L.M. Santos
Universidade do Porto
REQUIMTE
Departamento de Química-Física
Faculdade de Farmácia
Rua Aníbal Cunha 164
4050 Porto
Portugal

Javier Saurina
University of Barcelona
Department of Analytical Chemistry
Martí i Franquès 1-11
08028 Barcelona
Spain

B.M. Simonet
Universidad de Córdoba
Department of Analytical Chemistry
Campus de Rabanales
14071 Córdoba
Spain

Marek Trojanowicz
University of Warsaw
Department of Chemistry
Pasteura 1
02-093 Warsaw
Poland
and
Institute of Nuclear Chemistry and
Technology
Dorodna 16
03-195 Warsaw
Poland

Manabu Tokeshi
Nagoya University
Department of Applied Chemistry
Graduate School of Engineering
Furo-cho
Chikusa-ku
Nagoya 464-8603
Japan

Mabel B. Tudino
University of Buenos Aires
Facultad de Ciencias Exactas y Naturales
Pabellon 2
1428 Buenos Aires
Argentinia

Łukasz Tymecki
University of Warsaw
Department of Chemistry
Pasteura 1
02-093 Warsaw
Poland

Miguel Valcárcel
Universidad de Córdoba
Department of Analytical Chemistry
Campus de Rabanales
14071 Córdoba
Spain

N.P.E. Vermeulen
Vrije Universiteit Amsterdam
Division of Molecular Toxicology
De Boelelaan 1083
1081 HV Amsterdam
The Netherlands

Elias A.G. Zagatto
Universidade de São Paulo
Centro de Energia Nuclear
na Agricultura
Avenida Centenario, 303
Caixa Postal 96
13400- 970 Piracicaba SP
Brazil

I
Methodologies and Instrumentation

Advances in Flow Analysis. Edited by Marek Trojanowicz
Copyright © 2008 WILEY-VCH Verlag GmbH & Co. KGaA, Weinheim
ISBN: 978-3-527-31830-8

1
Theoretical Aspects of Flow Analysis

Fernando A. Iñón and Mabel B. Tudino

1.1
Introduction

Flow injection analysis (FIA) arises as a consequence of the growing trend towards automation in chemical analysis, and as a natural evolution of the so-called "continuous flow analysis" (CFA) which has revolutionized the conception of chemical analysis, especially in the field of clinical analysis and sample manipulation.

FIA belongs to a family of analytical methods based on the injection of a sample (containing the analyte or its reaction products) into a non-segmented carrier stream, which in turn carries it through a chemical or physical modulator towards the detector. To this category of methods belong also the different chromatographies and capillary electrophoresis. Figure 1.1a depicts the block diagram common to all these methods. Methods are thus differentiated by the nature of the "modulator" which transforms the square wave (the injection) into a chromatogram, electroferogram, or just a FIA peak (fiagram) (Figure 1.1b).

Thus, the differences between these techniques may be better understood in terms of the modulator function and the prevailing forces (or properties) interacting with it.

The primary difference between FIA and chromatography lies in the mass transfer between two phases that exists in the latter, but the diversity of FIA systems published to date makes it difficult to draw a dividing line between the two techniques.

The main difference between chromatography and FIA lies in the fact that in the former separations are achieved through a repetition of interactions which modulate the differential migration rate of the species through the system, while FIA exploits chemical reactions that transform the analytes into species which may be quantified by a detector.

Advances in Flow Analysis. Edited by Marek Trojanowicz
Copyright © 2008 WILEY-VCH Verlag GmbH & Co. KGaA, Weinheim
ISBN: 978-3-527-31830-8

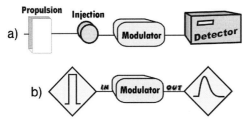

Figure 1.1 Stimulus-response in flow systems.

1.2
Classification of Flow Systems. Fundamentals

1.2.1
Continuous Flow Analysis

CFA, introduced by Skeggs [1], relies on the fact that the sample and the reactive react while traveling in the tube that conducts them to the detector. Air bubbles separate the portions of fluid (reactive) in order to diminish the longitudinal dispersion of the sample and the interaction between samples (carry over), which in turn favor mixing and increase the analysis frequency.

1.2.2
Flow Injection Analysis

In 1975 Růžička and Hansen introduced FIA [2] (calling it initially *non-segmented continuous flow analysis*) showing that bubble separation was unnecessary to prevent carry over and that it also introduced uncertainty in the reproducibility of the sample residence time inside the system. With the non-segmented technique the integrity of the sample pulse is kept by careful control of the hydrodynamic conditions of the system. The analytical consequence is important: in CFA the detection should be performed once the reaction is finished (once equilibrium has been reached) while in FIA this condition is unnecessary. Although in CFA equilibrium is usually attained, and thus the sensitivity is the same as for "batch" techniques, it is not possible to exploit chemical kinetics and mixing kinetics (gradient techniques) as variables to adjust the analytical response. This fact should be added to the difficulty posed by the presence of air bubbles in certain detection systems.

1.2.3
Sequential Injection Analysis

Sinusoidal flow systems [3] avoid fast delivery of the sample to the detector (shortening also the distance to be covered by the sample) in order to allow better mixing and more complete reaction between the sample and the reagents. This is accomplished by a fast change in the carrier flow direction (back and forward) until the desired degree of reaction is reached. This technique results in a drastic reduction in consumption of reagents but at the same time reduces the analytical frequency to a half (typically). Sequential injection analysis (SIA) is based on this principle [4–6].

SIA appeared as a response to two essential problems found in FIA: the increased complexity of the manifolds employed, and the need for its use in process control. The former is a consequence of the addition of steps to a given system, giving rise to many working channels, which in turn leads to higher reagent consumption and very difficult "tuning" of the system. The latter, requires systems that are robust, reliable and stable in the long term and which could proceed unmanned and with low reagent consumption.

FIA, being a continuous flow system, presents several disadvantages such as high consumption of samples and reagents, a need for constant supervision of the peristaltic pumps, frequent recalibration and manual adjustment of the system. These facts are not a problem on the laboratory scale, but become a major issue when dealing with industrial process control.

SIA tackles these difficulties by simplifying the system: a single piston pump (which allows precise flow control in both directions), a single flow channel and only one selecting (instead of injection) valve, are used. Through this valve accurately measured volumes of carrier, sample and reagents are introduced into the reaction coil. Once this step is accomplished, the valve is changed to direct the flow towards the detector. During this step mixing is accomplished through back and forth movements, and finally in one direction to transport the reaction coil load to the detector.

On the other hand, the complexity of multiple channel FIA systems is reduced in SIA systems by using mostly single lines around a selecting valve. Operational parameters in SIA are: volume taken of each solution, flow rate, and the time in which the flow is stopped or changes its direction. Occasionally more pumps could be needed (one for each line) which makes controlling the system very difficult. On-line dilution is still an issue in SIA systems as the sample pulse is greatly dispersed and hence its signal profile becomes much distorted [7–9].

1.2.4
Multicommutation in Flow Injection Analysis

Multi-commuting arose as a FIA modification directed towards increasing the versatility of flow systems, diminishing reagent consumption, improving mixing and facilitating automation. This concept was first mentioned in the 1980s [10, 11], but was only formally introduced in 1994 [12] since when numerous papers have shown its potential [13, 14]. One of the main advantages of this technique lies in exploiting successive injections of sample and carrier (tandem stream [12]) in which the volumes of each section are easily selected.

In this technique three-way solenoid valves are employed and controlled temporally through a microprocessor. System designs are usually simple and it is easy to imagine that, depending on the set-up, the advantages of the technique may be fully exploited and compared with FIA or SIA systems. (See Figure 1.2.)

Multi-commuting appears to be the technique which has the advantages of FIA together with those of SIA and overcomes all their disadvantages [13].

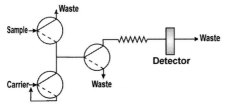

Figure 1.2 Multicommuted system.

Definitely, the key to optimizing a flow technique lies in balancing the dispersion of the injected pulse and reagent mixing, taking into account the time required to reach the optimal degree of conversion of reagents into detectable species.

1.2.5
Stopped Flow

Amongst modern techniques derived from FIA, it seems that *"stopped flow"* techniques are the easiest to implement and the most versatile [15]. In these techniques, the sample is injected as in conventional FIA but the carrier flow is stopped when the sample and the reagents reach the flow cell (at the detector) giving rise to several advantages [15, 16]. While physical dispersion is minimized, as the driving force is diffusion only, chemical reaction proceeds (uncoupling both processes) increasing sensitivity as there is a longer reaction time and lower dispersion of the products. Thus, reagent consumption and waste production diminish and, by controlling the stopping time, the concentration range can be selected, which is of the utmost importance for kinetic determinations at different wavelengths. However, stopped flow has the disadvantage of requiring a very robust and reproducible time control in order not to affect precision.

1.2.6
Batch Flow Injection Analysis (BFA)

Batch flow injection analysis was proposed by Honorato *et al.* [17] as an alternative way to carry out titrations in flow systems. Sampling and signal processing are done as in usual flow systems, whereas chemical reactions take place in a reaction chamber similar to those found in batch systems. The approach is then a flow–batch hybrid system combining the intrinsic advantages of flow systems, such as high sampling rate, low sample and reagent consumption, low cost, ease of automation, and so on, with the wide application range inherent to batch systems.

The method is based upon the use of three three-way solenoid valves to transmit the fluids inside the reaction chamber and to the reactor where a magnetic stirrer ensures instantaneous homogenization of the cell. Reproducible selection of the added aliquots is found for ON/OFF valve intervals controlled by a personal computer.

1.3
Dispersion in Flow Injection Analysis: From the Movement of Fluids in Open Tubes to Controlled Dispersion

A fluid is any kind of substance that can support a shear force when it is not in motion and undergoes a continuous change when it is stressed. A fluid is ideal when, in motion, it cannot transfer or transport heat from one portion to another, nor produce friction with the inner surface of the tube. This can also be defined as an ideal fluid has no viscosity. Obviously, all fluids used in analytical chemistry have viscosity, that is, inner friction. If viscosity is independent of time for any applied shear force, the fluid is known as Newtonian (such as water). The viscosity of non-Newtonian fluids depends on shear force modulus and time (even for a constant shear force). Most carriers used in flow systems are aqueous solutions that can be considered Newtonian, therefore only Newtonian fluids are normally considered.

Fluidynamics is the science devoted to the study of fluids and has a plethora of applications in other fields: biology, chemistry, engineering, medicine, and so on. It is based on the conservative laws (such as mass, movement, energy and thermodynamics) and constituent laws (i.e., Fick's first law that establishes the dependence of mass transfer on concentration gradients). With this set of laws a series of differential equations can be written for any process, from which it may be possible to infer, predict or describe the whole process.

Nevertheless, mathematical solutions to these equations are rarely found for real systems. On the other hand, many constituent laws are empirical [18], and their coefficients (viscosity, diffusion, etc.) are only known under certain experimental conditions and in many cases cannot be extrapolated to real systems.

Transport phenomena is the area of fluidynamics devoted to the study of different types of transport under different experimental conditions (i.e., dimensions of the system, geometry, etc., type of fluids, etc.). One of their objectives is to provide equations applied to reduce systems, that is, systems whose actual size is normalized by some parameter which requires the definition of adimensional numbers (i.e., Reynolds, Schmidt, Peclet numbers, etc.). The transport process in a real-size system can be explained by considering the transport process in its reduced dimension.

1.3.1
Transport of Fluids

It is worth stressing the difference between the transport process and transport phenomena: the process is the phenomena integrated in a given dimension (spatial or temporal). Transport laws are based on the two most relevant properties of fluids: viscosity and thermal conductivity and, in the case of solutions, diffusivity. Transport properties, in contrast to equilibrium, are related to the velocity of a given process.

1.3.1.1 Viscosity
Viscosity (η) is the capacity of a fluid to resist deformation and depends on the composition of the fluid and its temperature. For Newtonian fluids, η is the

proportional coefficient between the tangential shear force and the generated velocity gradient normal to this force.

For aqueous solutions, viscosity can be calculated using Equation (1) [19].

$$\ln\left(\frac{\eta(20\,^\circ\text{C})}{\eta(T)}\right) = \frac{1.37023 \cdot (T - 20) + 8.36 \cdot 10^{-4}(T - 20)^2}{109 + T} \tag{1}$$

where $\eta(20\,^\circ\text{C}) = 1\,\text{cp}$ (centi-poise, $1 \times 10^{-3}\,\text{Pa s}$). Diffusion coefficients, or at least their order of magnitude, can be predicted knowing η.

Kinematic viscosity is the ratio between the viscosity and the density (δ) of a fluid, and its units are square length divided by time ($1 \times 10^{-6}\,\text{m}^2\,\text{s}^{-1}$ for water at $20\,^\circ\text{C}$). Kinematic viscosity is useful when comparing mass transfer ratios by normalizing diffusion coefficients.

From the analytical point of view, the viscosity of different samples must be kept as constant as possible because the concentration gradient profile inside a flow system, and thus its instrumental response, depends drastically on this property, as was studied by Brooks *et al.* [20].

Moreover, when there is a sharp difference between sample and carrier solutions, there is a spurious signal in most detectors, known as Schlieren effect, that drastically affects detection limits. In order to minimize this error source, Betteridge and Růžička [21] suggest that the viscosity of standard and sample solutions must be identical. Another possibility is to include a mixing chamber [22] with a magnetic stirrer when it is not possible to match viscosity. This strategy is mainly used in BFA.

1.3.1.2 Thermal Conductivity

This is the capacity of the fluid to transport heat by conduction and depends on the fluid composition and its temperature. Some methods require heating of some parts of the flow system and this property allows one to dimension the heat.

1.3.1.3 Diffusivity

As viscosity refers to moment transfer and heat conductivity to heat transfer, diffusivity refers to mass transfer. It is related to the velocity of molecules that travel in a "sea" of other molecules (actually, the 'sea' can also be the same substance and in this case the term is called autodiffusivity). It depends on temperature, molecule/particle size and the strength of intermolecular forces. Diffusivity in the solid state is extremely low, and in liquids it is lower than in gases. Differently from the diffusivity in the gas phase, where the diffusivity of gas A in gas B is nearly equal to the diffusivity of gas B in gas A, the diffusivity of solute A in solvent B almost always is very different to the diffusivity of solute B in solvent A.

1.3.1.4 Diffusion

Diffusion is a transport phenomenon that refers to the movement of solutes in the absence of any convective contribution and it is a consequence of a difference in chemical potential, due to a difference in the activity of the solute, in a given direction of the fluid. Any difference in activity (concentration in dilute solutions) tends to be homogenized by the diffusion process, even in the absence of advection.

The net flux of mass in a given direction depends on the concentration gradient in the same direction. This diffusive flux can be represented as a vector with components J_x, J_y, J_z defined in Equation (2):

$$J_x = -k\frac{\partial C}{\partial x} \quad J_y = -k\frac{\partial C}{\partial y} \quad J_x = -k\frac{\partial C}{\partial z} \tag{2}$$

where C is the concentration of the solute, and x, y, and z the axis directions. These equations are known as Fick's first law and k is known as the diffusion coefficient (D_m) of that solute in this medium. This law means that the Brownian movement causes, on average, a net flux of solute from a more concentrated to a less concentrated region of the fluid.

The magnitude of the diffusion coefficient in gases is around 10^{-5} to $10^{-4}\,\mathrm{m^2\,s^{-1}}$ (at atmospheric pressure), whereas in liquids it is typically around 10^{-10} to $10^{-9}\,\mathrm{m^2\,s^{-1}}$. Diffusion coefficients increase with temperature and their magnitude depends on changes in molar density, pressure and viscosity. Diffusion in flow systems plays a very important role and has enormous analytical implications, as will be discussed in this chapter. Nevertheless, for working with flow injection systems, the convective mass transport needs to be taken into account.

1.3.2
The Diffusion–Convection Equation in Open Conduits

Mass transport inside a fluid travelling inside a pipe can be iso, endo or exothermic. Non-isothermic changes will produce changes in fluid properties such as viscosity [18, 23]. For Newtonian, uncompressible (constant density), constant viscosity (isothermic change) fluids flowing inside a cylindrical pipe it is possible to derive Equation (3). It considers convective transport based on a linear velocity (u) independent of the z coordinate, but dependent on the radial coordinate (Figure 1.3). This is only strictly true in perfectly straight open tubes with constant inner diameter, without any curvature, perturbations and splices as they will alter velocity profile.

$$\frac{\partial C}{\partial t} + u(r) \cdot \frac{\partial C}{\partial z} = D_m \cdot \left(\frac{1}{r}\frac{\partial C}{\partial r} + \frac{\partial^2 C}{\partial r^2} + \frac{\partial^2 C}{\partial^2 z} \right) \tag{3}$$

The concentration as function of time depends on convection, represented by the second left-hand term, and the diffusive transport, represented by the right-hand side of the equation. It can be clearly noted that diffusion influences both radial and axial mass transport.

Convection, coaxial to the flow movement, together with diffusion are the main causes of axial dispersion in tubes. Taylor [24, 25] and Levenspiel [26] are pioneers

Figure 1.3 Coordinates definition.

in the theoretical study of transport and dispersion inside tubes and their works set the basis for theoretical studies and modeling of dispersion in the field of flow analysis systems.

Particularly for FIA, Růžička and Hansen [27] have provided the first theoretical aspects for describing processes inside these systems. It must be taken into account that the differential Equation (3) has no general solution and its applicability in any system is based on several assumptions.

1.3.3
The Distribution of Times of Residence

The residence time of the substance, related to the residence time of each fluid element, inside a system is one of the key factors in FI and it depends on the design and operational parameters of the system (size, flow rate, geometry, etc.). This task requires knowledge of the residence time distribution (RTD) of the different portions of injected substance inside the system plus all balances related to mass, composition of chemical reactants and heat transfers.

Experimentally, RTD curves are obtained by introducing a tracer (i.e., a colored substance) and recording the concentration of that tracer in the outlet of the system (i.e., by measuring the absorbance at the characteristic absorption wavelength of the substance), see Levenspiel [26]. RTD curves can be normalized using Equation (4): the area under the curve is equal to 1, supposing that all tracer elements leave the system.

$$\int_{t=0}^{\infty} E(t)\partial t = 1 \tag{4}$$

$$t_m = \int_{t=0}^{\infty} t \cdot E(t)\partial t \tag{5}$$

where E is the fraction of fluid with an age t. $E(t)$ is characteristic of a given flow pattern. Knowledge of the RTD profile may help one to know the flow profile inside the system and to develop a quantitative expression to describe system dispersion. The mean time of residence (t_m) can be calculated using Equation (5).

A common variable reduction for comparing different systems is by using the ratio of the time (t) and the mean time of residence (t_m). This ratio is expressed by the adimensional number θ equal to $t \cdot t_m^{-1}$.

There are different ways in which RTD curves can be displayed. "*C-curves*," are the equivalent to the E curve displayed in Figure 1.4 and are calculated by dividing the concentration of the tracer at the output of the system ($C(t)$) by the area under the curve. "*F-curves*" are those obtained when $C(t) \cdot C_0^{-1}$ is calculated as a function of time, and can be associated to the integral of the C-curve. Figure 1.5 shows these curves for systems under three different flow regimes: plug-flow, ideally mixed and arbitrary flow. The dispersion of a given system is inversely related to the slope of its F-curve and directly related to the width of its C-curve.

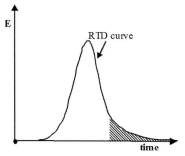

Figure 1.4 Residence times curve.

Taylor [24] has shown that with dimensions similar to an analytical flow system, diffusion has a key role. Although axial diffusion may be negligible compared to convection, radial diffusion is always important and, for small i.d. tubes and low flow rates, it may represent the principal component of dispersion.

1.3.3.1 Characterization and Experimental Domain of Flow Systems: Dimensionless Numbers and Their Meaning

Many adimensional numbers have been defined in order to characterize hydraulics in flow systems, based upon linear flow rate (u), diffusion coefficient (D_m), time (t), reactor length (L), tube radii (a), and so on. The main objective of these adimensional variables is the integration of differential equations, such as Equation (3), for a group of similar systems and not for each particular one. Painton and Mottola [28, 29] have been pioneers of the use of adimensional numbers in FIA.

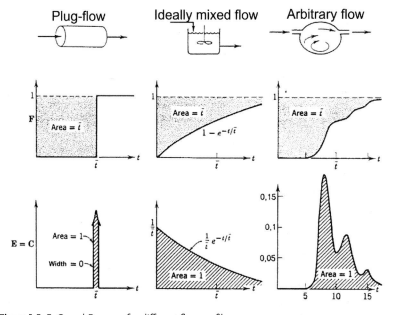

Figure 1.5 E, C, and F curves for different flow profiles.

Table 1.1 Definition of adimensional numbers.

Reynolds (Re)	Peclet (Pe$_L$)	Peclet radial (Pe$_r$)	Fourier (τ)	Schmidt (Sc)
$2 \cdot \bar{u} \cdot a \cdot \left(\dfrac{\eta}{\delta}\right)^{-1}$	$\dfrac{\bar{u} \cdot L}{D_m}$	$\dfrac{\bar{u} \cdot a}{D_m}$	$\dfrac{D_m \cdot t_m}{a^2}$	$\dfrac{\eta}{\delta} \cdot \dfrac{1}{D_m}$

The most common reduced numbers used in analytical flow systems are: Peclet, Schmidt, Fourier and Reynolds, and are defined as shown in Table 1.1.

The Reynolds number is a balance between inertial forces ($\delta \cdot \bar{u}^2/a$, transport moment by convection) and viscous forces ($\eta \cdot \bar{u}/a^2$, transport moment by diffusion) and indicates the flow regimen inside a tube. For straight tubes, Re lower than 2000 indicates laminar flow, which means that the flow lines of the fluid elements are parallel to the direction of the flow. There is not a given Re number from which the flow pattern becomes turbulent and there is a transition zone typified by a mixed flow regime. The end of the transition zone is characterized by a critical Reynolds number (Re$_c$) that, for straight tubes, is around 2300. Coiled tubes stabilize laminar flow, incrementing Re$_c$ as the ratio between the tube i.d. and coil diameter increases.

Radial Peclet, or reduced rate [30], relates mass transport by convection ($\bar{u} \cdot \Delta C/a$) to mass transport by diffusion ($D \cdot \Delta C/a^2$). The axial Peclet number (Pe$_L$) compares axial mass transport by convection to that of diffusion. A similar, but not the same, ratio is more commonly used in analytical flow systems and is known as the reduced distance $\left(\dfrac{D_m \cdot L}{2\bar{u}\cdot a^2}\right)$. Gunn and Pryce (see [31]), have shown that for relatively small Re numbers (0.02–420) there is a dependence between radial and axial Pe numbers with Re number.

The inverse of Pe$_L$ has been defined in the chemical engineering field as *"reactor dispersion number"* (D_N) and has been used to describe RTD variance, which is a dispersion estimator (see next section) [26]. At higher Pe$_L$ convection dominates transport, whereas small values indicate greater influence of diffusion.

The Fourier number (τ) does a scale reduction similar to the Peclet number but in time coordinates, comparing residence time with molecular diffusion. High τ values indicate a higher diffusion contribution to mass transport, whereas small values are related to a higher contribution from convection.

The Schmidt number relates solute and fluid properties instead of flow characteristics. High Sc values (moment diffusion higher than molecular diffusion) are common in liquids (Sc values ranging between 100 and 10 000 are found in aqueous solutions) and shows that convection prevails over diffusion. Pe can be calculated as the product between Re and Sc.

1.3.4
From the RTD Curve to the Generation of Signals in Flow Injection Systems

1.3.4.1 The Dispersion Process
Taylor [24, 25] has published the first experimental and theoretical contributions devoted to understanding the dispersive process inside small i.d. tubes. He studied

different stages related to the introduction of a chemically-inert solute into the fully developed laminar flow pattern of a solvent flowing inside a cylindrical pipe.

It was observed [24] that for short times (high linear velocity), axial convection is the main transport and thus the main contribution to the increase in dispersion of the solute zone. For longer times, radial molecular diffusion contributes to the dispersion process. The net effect of molecular diffusion is to reduce axial stretching of the solute zone, thus decreasing axial dispersion. For even longer times, molecular diffusion continues to limit axial dispersion, generating a more uniform concentration of the solute zone. At longer times, axial and radial diffusion contributes to dispersion, and the solute zone can be regarded as a plug in which the concentration profile in the axial directions tends to be Gaussian. It is important to notice that this does not mean the absence of a radial velocity profile. These conditions are never reached in FIA or multicommutation systems.

Analytical resolution of the convective-diffusion equation (Equation (3)) is only available under two extreme boundary conditions: (i) mass transport is only due to convective transport (short residence times) and (ii) diffusive transport is the main transport phenomenon contributing to dispersion (long residence times). Under these boundary conditions, analytical flow systems are not well represented.

1.3.4.2 The Concept of Controlled Dispersion and Analytical Implications

A consequence of dispersion is the loss in sensitivity and limit of detection. So, from the analytical point of view, the dispersion is a determinant factor in the performance of the system of analysis. In flow systems the term dispersion was associated directly with the loss of sensitivity in comparison with that obtained when the stationary state is reached. Even though this is a consequence, it does not describe the dispersion phenomenon which involves physical and chemical dispersion.

Physical dispersion is the result of the processes of spatial redistribution of mass that undergoes the pulse injected into a carrier in the absence of chemical reaction. This redistribution is carried out due to the parabolic linear velocity profile in non-segmented flow systems. A consequence of this profile is the appearance of a radial concentration gradient that tends to be homogenized by diffusion and, when they exist, by the presence of secondary flows (in directions other than that of the principal flow).

When a chemical reaction is involved, the term chemical dispersion is considered and it is related to the degree of completion of the reaction at the moment of detection (general case in FIA). Thus, chemical dispersion is always accompanied by physical dispersion, since it depends on both the mixing of reagents and the redistribution of the formed product. The concentration gradient affects the kinetics of the reaction and this in turn modifies the gradient [28]. It should be understood that, from the theoretical point of view as opposed to the experimental one, the study of the chemical dispersion is highly complex and needs numerous simplifications to approach it.

"Dilution" is a term that many authors [16, 20, 32–34] erroneously associate with the term "dispersion." However, dilution can be understood as a particular case of dispersion where the function of redistribution of mass is homogeneous in time and space, which is not the situation in FI where the term dispersion is more suitable. So, in FIA it is only possible to talk of an average dilution of the sample in a transverse section of fluid, but this average dilution is not indicative of the way in which the dispersive process was carried out.

The process of dispersion has its origin in the inhomogeneity that is imposed on a substance when it is submitted to diverse types of gradients. In flow systems, the injection of a pulse of substance in a carrier that flows submits the substance to two types of gradient: one of mass and one of velocities (or in general of moments). The diverse ways in which the dispersion takes place is a result of the homogenization of these gradients. These complex processes involve the study of redistribution of mass and energy from their original distribution to that obtained in the outlet of the systems. However, as the dispersion process can be controlled, trustworthy analytical results can be obtained.

1.3.4.3 The Transient Profile

When a sample/tracer is injected into a flow system, its spatial distribution changes and this change depends upon different dimensional and operational variables of each part of the flow system. The signal profile obtained by locating a detector at the system outlet is directly related to the spatial and temporal distribution of the sample plug. Taylor [24, 25] and Levenspiel [35, 36] have studied these cause/effect relationships.

From the analytical point of view, the information required to relate the concentration domain to the domain of a measurable variable (i.e., voltage) is obtained through the response signal. In flow systems, this response can be of two kinds: (i) the steady state signals obtained when sample, reactants and carrier are homogenized and all possible reactions reach their equilibrium and (ii) dynamic signals obtained when a limited portion of the sample undergoes a dispersion process in a given carrier and the possible chemical reaction may or not reach equilibrium.

In the last case, the concentration profile must only depend on analyte concentration and the reproducibility of this profile is essential.

In CFA systems, it is important to reach a *plateau*. Under this condition it is possible to improve precision by averaging the response signal n at the top of the plateau. Moreover, as equilibrium is reached, the sensitivity is usually higher than in other flow systems, but the sample throughput is decreased and sample and reactants consumption increases [37].

The asymmetric peak observed in FIA systems is due to the influence of the injection mode and the contribution of convection and diffusion to mass transport.

A common approach is to try to fit a response signal by a given model and, in the case of peaks, the Gaussian model is one of the simplest to apply. In FIA systems, signals obtained under high dispersion conditions can be fitted by a Gaussian model. If dispersion is medium to high, it can be described as an exponential modified Gaussian (EMG) [38]. When dispersion is low or medium to low, there is no general model to fit a FIA peak.

The interdispersion of sample plug and carrier solution is mandatory in FIA. It is expected that if the analytical response is due to the concentration of a given analyte in the sample, its maximum will be located at the maximum concentration of the analyte in the sample plug. Nevertheless, if a product of the reaction between the analyte and the carrier is monitored this assumption is usually false as it depends on the ratio analyte:reactant which optimizes the chemical reaction.

The operational and design parameters that have the greatest influence on the carrier/sample interdispersion are: tube radius, spatial and geometrical disposition of the reactor/manifold, flow rate, injection and reactor volume. The effect of each parameter on the response signal will be discussed below.

The physical properties of carrier and sample solution are also important. The effect of temperature and viscosity can be summarized by the diffusion coefficient. An increment of its value will be visualized as sharper and narrow peaks, due to a better radial mass distribution. The limit case, infinite radial diffusion, is the basis of the ideal plug-flow pattern flow.

Peak shape is also notoriously influenced by the detector type and the dimension and shape of the detector cell [39–42]. Kinetics phenomena in the flow systems will determine how temporal and spatial concentration gradients enter the flow cell, but the kinetic effects of detection may completely alter these gradients.

Regarding radial mixing, its improvement allows one to obtain a flow pattern more similar to that of plug flow reported by Johnson *et al.* [32] when studying the use of packed loops with different spatial configuration (i.e., "Serpentine II") as a strategy for reducing FIA peak width.

Peaks with shoulders (or humped peaks) have been observed in FIA systems in the absence of chemical reaction, but distortions in the laminar flow pattern tend to smooth them [43]. Nevertheless, in systems where a chemical reaction takes place humped peaks are obtained if the reagent concentration is not in excess with regard to the sample. A limit case of the humped peak is the double peak. In this case the reactant cannot reach the center of the sample plug and the obtained (double) signal is due to the product formation on the head and tail of the sample plug.

Regarding the detection, Růžička and Hansen [16] have suggested that whilst the instantaneous signal of the detector behaves linearly with concentration, there is no difference with the property of the signal used (height or area). Nevertheless, strictly speaking, as also happens in chromatography, only the area of the signal is directly related to the total amount of the analyte injected. Peak height may be used only when the peak profile is not deformed by changing any physical property of the sample.

1.4
The Measurement of Dispersion

The transient signal can be described through different parameters such as peak height (h), time of appearance (t_a), baseline-to-baseline time or peak width (Δt_b), time elapsed between the injection and that of the maximum of the signal (t_r), signal area, and so on. Mostly these parameters cannot by themselves describe the dispersion phenomena and the obtained values are compared with the "ideal" ones. So, these parameters must not be an absolute magnitude since they refer to changes in the mass distribution (redistribution) when compared to the initial condition.

In the next sections, the different parameters able to evaluate dispersion through the characteristics of the transient signal profile will be presented. The employment of these easily calculable parameters is interesting as they allow one to optimize analytical systems in a very practical way.

1.4.1
The Coefficient "D"

The Růžička coefficient of dispersion D is the most popular experimental parameter able to evaluate dispersion in flow systems [27, 44]. Since it is easily calculable, D has been accepted as the way of measuring the degree of dilution of the sample. However, this criterion has been objected to since D involves a punctual definition of dispersion and the main characteristic of the injected pulse is mass redistribution. So, dispersion studies need to be based on a comparison of parameters able to evaluate this distribution. In this way, several papers have focused on study of the peak width rather than D [29, 45] for evaluating dispersion. This will be discussed in the next section.

However, D constitutes a very practical way of comparing different flow systems in relation to their operational variables. Růžička and Hansen proposed a classification of the different flow systems based on D values. This approach allows one to adequate the FI variables to the figures of merit that the analyst wants to optimize.

Růžička and Hansen [27] related the coefficient of dispersion D and the time of residence t_R with the different variables of the system as follows:

$$D = 3.303 \cdot a^{0.496} \cdot L^{0.167} \cdot q^{-0.0206} \tag{6}$$

$$t_R = 1.349 \cdot a^{0.683} \cdot L^{0.801} \cdot q^{-0.977} \tag{7}$$

Vanderslice *et al.* [43] studied the shape of the concentration profiles inside the tube for different elapsed times. Based on these profiles the authors showed that the dispersion coefficient increases with the tube radius. Moreover, it was demonstrated that the asymmetric initial profile becomes quasi-Gaussian at reduced times (τ) close to 1. However, no analytical equation able to show this dependence was presented by the authors.

Narusawa and Miyamae [46–48] employed ZCFIA (zone circulating FIA) and simulation techniques to correlate the system variables with dispersion under conditions of no chemical reaction. The relationships presented by the authors are given in Equation (8):

$$\begin{aligned}
D_A &= 0.045 \cdot L^{1.01} \cdot q^{-1.46} \qquad & D_r &= 0.349 \cdot L^{0.559} \cdot q^{-0.222} \\
D_A &= 0.076 \cdot t^{1.02} \cdot q^{-0.42} \qquad & D_r &= 0.449 \cdot t^{0.564} \cdot q^{0.335} \\
\frac{D_r - 1}{D_A - \theta_t} &= 7.5 \cdot L^{-0.45} \cdot q^{1.24} \qquad & D_r &= 0.388 \cdot L^{0.351} \cdot t^{0.206} \\
\end{aligned} \tag{8}$$

$$\frac{D_r - 1}{D_A - \theta_t} = 6.7 \cdot t^{0.75} \cdot q^{-0.48}$$

where D_A represents the axial dispersion and D_r is the real radial dispersion (note that it is different from D since the component of radial diffusion is not included in the latter).

Nevertheless, some problems with the experimental set-up, mainly related to the overlap between the head and the coil of the sample plug during the "round trip" of the sample, invalidated the conclusions, as was reported by the same authors later [49].

1.4.2
Peak Width and Time of Appearance

The time of appearance (t_a) is not an estimate of dispersion but it is able to complement the useful information given by the peak width. This last parameter symbolized as Δt_b by Vanderslice *et al.* [50], can be obtained from the transient signal observed in FIA. In some cases, it is useful to employ the peak width expressed as a volume $(\Delta t_b V_b = \Delta t_b \times q)$ in order to eliminate the influence of the volumetric flow rate on the residence time of the fluid inside the manifold.

Vanderslice *et al.* reported the dependence of Δt_b on the FI variables. From the numerical integration of the convective-diffusive equation under the conditions imposed by Ananthakrishnan *et al.* [30], the relationships of t_a and Δt_b with the operational variables were obtained as shown in the following equations

$$t_a = \frac{109 \cdot a^2 \cdot D^{0.025}}{f} \left(\frac{L}{\bar{u}}\right)^{1.025} \tag{9}$$

$$\Delta t_b = \frac{35.4 \cdot a^2 \cdot f}{D^{0.36}} \left(\frac{L}{\bar{u}}\right)^{0.64} \tag{10}$$

where f is an adjustment factor which needs to be included since the conditions imposed by Ananthakrishnan *et al.* for solving the convective-diffusive equation are not applicable to the conventional FI systems. It should be noticed that t_a and Δt_b are independent of the loop length "l." Even when this is understandable for t_a, since the head of the sample will arrive at the detector at the same time regardless of the length of the injected pulse, it is not the same for Δt_b. In another work [51], the same authors reported the conditions under which the equations presented above are valid.

Regarding the adjustment factor f, Gómez Nieto *et al.* [52] reported that it depends on the flow rate, the length of the manifold and the tubing diameter. Several works were then devoted to the employment of multiple regressions in order to find the experimental coefficients able to fit the experimental variables of the systems with Δt_b or t_a:

$$t_a = 0.898 \cdot a^{0.950} \cdot L^{0.850} \cdot q^{-0.850} \tag{11}$$

$$\Delta t_b = 69.47 \cdot a^{0.293} \cdot L^{0.107} \cdot q^{-1.057} \tag{12}$$

Kempster *et al.* [53] reported the following correlations:

$$\Delta t_b = 48.43 \cdot a^{0.444} \cdot L^{0.282} \cdot q^{-0.893} \tag{13}$$

$$\Delta t_b = 32.33 \cdot a^{0.504} \cdot L^{0.367} \cdot q^{-0.888} \tag{14}$$

Equation (13) was obtained for FI systems with spectrophotometric detection employing a flow cell of 30 μl. Equation (14) was obtained using ICP-OES with a mixing chamber of 300 μl.

Korenaga [45] reported several experiments to correlate t_a and Δt_b with the variables of the system, taking special care to avoid perturbations in the flow pattern. In the case of tube radius (r), the author found a linear relationship between Δt_b and r^2 for $r > 0.33$ mm. For lower values, the relationship was linear with $d^{0.7}$. However, it is difficult to find the reason for these differences since the other experimental variables are not given.

For the reactor length (L), Korenga showed a linear relationship between Δt_b and $L^{0.64}$ which is in agreement with Vanderslice *et al.* (Equation (10)). The same agreement was found for the relationship between Δt_b and q and Δt_b and D (see Eq. (6); Růžička and Hansen [27]).

However, the peak width fails as an estimator of dispersion since dispersion is not only influenced by the length or volume of the injected sample but also by the width of the original sample plug and its deformation while traveling towards the detector. A useful alternative should be the inclusion of the quotient between the final peak width and that of the original sample plug.

1.4.3
Peak Variance and Theoretical Plate Height

According to Tijseen [54], the definitions of dispersion parameters for flow systems are those developed in the field of chromatography for the evaluation of peak broadening. In this way, the variance (σ_{tt}) is proposed as an estimator of dispersion as it is related to the distribution of times of residence of the injected pulse. The peak variance (σ_{tt}) and the theoretical peak height (H), both estimators of dispersion in chromatography, are related by: $H = L \, (\sigma_{tt}/t_m)^2$ (where L is the reactor length and t_m is the time at the peak maximum). When working with Gaussian profiles, such as those obtained in chromatography and in Tijssen's experiments, these estimators of dispersion are proportional to $\Delta t_{1/2}$.

Painton and Mottola [29] reported the relationship between the number of dispersion ($D_N = Pe_L^{-1}$) and the variance of the curves C vs. t. These relationships are based on the resolution of the convection-diffusion equation under different conditions. The authors found that the variance is proportional to D_N^2.

By employing the exponential modification of a Gaussian function (EMG) [43, 55], Brooks *et al.* [20] evaluated the dependence of the variance on the volumetric flow rate q in a FI system. The peak adjustment through the employment of the equation is only acceptable in systems where a good radial mixture is obtained (use of coiled reactors, low flow rates, long manifolds, etc.).

As stated before, σ_τ and Δt_b fail at the moment of describing dispersion since the size of the initial injected pulse is not taken into account. In other words, if the injection volume is increased under conditions where the injection contribution is not negligible, the variance of the obtained signal will be increased without increasing

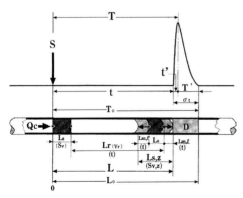

Figure 1.6 Li and Gao Model (Ref. [31]). S = injection point, D = detector, t = time of appearance (t_a), t' = time for the peak ascent, T' = time elapsed between the maximum and the return to the baseline, T_o = total residence time, Q_c = volumetric flow rate of the carrier, Sv = volume of the injected sample, Sv, z = volume of the sample zone, L = reactor length, Ls = sample loop length, Lo = length of the whole manifold, Ls = original length of the injected sample, Ls,z = length of the zone occupied by the injected sample, Lm,f = forward distance in the axial direction (purely dispersive), Lm,b = backward distance in the axial direction (purely dispersive), θ_t = peak width of the injected sample in time units.

dispersion. So, it is convenient to evaluate the different contributions to dispersion in a separate way.

1.4.4
Degree and Intensity of Axial Dispersion

Li and Gao [31] reported two new parameters for the evaluation of dispersion in flow systems. The flow model employed by the authors is shown in Figure 1.6.

The model describes all the parameters that are relevant for a flow system in a very complete way. However, it employs an average of the flow rates developed inside the tube which is not precise since the profile of flow rates in straight tubes is parabolic and the elements of fluid in the central axis travel at twice the average flow rate.

1.4.4.1 Degree of Axial Dispersion
The degree of axial dispersion was defined by Li and Gao as A_D:

$$A_D = \frac{S_{v,z}}{S_v} = \frac{L_{s,z}}{L_s} \simeq \frac{\sigma_t}{\theta_t} \tag{15}$$

This adimensional parameter is defined as the quotient between the volume of the sample zone $S_{v,z}$, and the injected volume S_v. $S_{v,z}$, is estimated by assuming that the sample occupies the whole segment $L_{s,z}$. The temporal peak width is obtained by the product of $L_{s,z}$ and the average flow rate, which is not correct. In the case of σ_t (the width of the dispersed sample zone), the flow rate is related to the carrier flow rate and the "dispersive molecular flow", which is negligible. Regarding the determination of distances in the center of the tube and their conversion into volume, it is not clear. With respect to the relationship with the system variables, they are shown through Equation (16).

$$A_D - 1 = 2\pi \cdot a^2 \cdot t^{0.5} \cdot A_f^{0.5} \cdot S_V^{-1}$$
$$A_D - 1 = k_7 \cdot S_V^{-1} \cdot L_R^{\mu_1} \cdot Q_c^{\mu_2} \, (\mu_1 > 1, \mu_2 < 0)$$
$$A_D - 1 = k_8 \cdot S_V^{-1} \cdot t^{\mu_3} \cdot Q_c^{\mu_4} \, (\mu_3 > 1, \mu_4 > 0)$$

(16)

where A_f is the coefficient for axial dispersion which derives from the concept of model of dispersive axial flow. This model will be presented below. The coefficients k and μ need to be experimentally fitted and their values depend on the tube length and radius, the flow rate and the system configuration.

1.4.4.2 Intensity of the Radial Dispersion

The intensity of the radial dispersion is defined as J_f:

$$J_f = \frac{D}{A_D} = \frac{S_V}{S_{V,z}} \frac{C^0}{C^{max}}$$

(17)

where D is the Růžička coefficient of dispersion and A_D the degree of axial dispersion. According to Li and Gao, the larger the J_f value, the higher the radial dispersion of the sample zone and the closer to a square shape. In terms of the signal variance and the signal height, Equation (17) can be written as follows:

$$J_f = \frac{\theta_t}{\sigma_t} \cdot \frac{h_0}{h_{max}}$$

(18)

This parameter describes the combined action of sample stretching which is contrasted by the radial mass transfer.

However, when relating J_f with the system variables (Equation (19)), the approach of Li and Gao presents some problems. As an example, the parameter J_f' ("pure intensity of radial dispersion") defined as $J_f' = (D-1) \cdot (A_D - 1)^{-1}$ needs to be equal to J_f^{-1} in order to find consistent relationships with the system variables. These two expressions are not equivalent.

$$J_f' = \frac{D-1}{A_D - 1} \cdot S_V \cdot Q_c^{-1}$$

$$J_f' = k_9 \cdot S_V \cdot L_r^{\mu_5} \cdot Q_c^{\mu_6} \, (\mu_5 < 0, \mu_6 > 0)$$

$$J_f' = k_8 \cdot S_V \cdot t^{\mu_7} \cdot Q_c^{\mu_8} \, (\mu_7 < 0, \mu_8 < 0)$$

(19)

In conclusion, it can be said that even when the parameters for estimating dispersion such as D, peak width and variance fail when performing an exact description of dispersion, they are easily attainable through experimental data and are of great help in developing new analytical methodologies.

Other Approaches to the Measurement of Dispersion

In order to advance in the study of dispersion, Andrade *et al.* [56] reported a new methodology: the integrated conductimetric method (*ICM*) which allows one to follow the mass redistribution process of the injected pulse as a function of time by simply monitoring the electrical conductance (*G*) of a whole single line system as time elapses. *G* was measured by placing platinum electrodes in both ends of the manifold. The authors demonstrated that the characteristic profile *G* vs. time can be assigned to the mass redistribution of the injected pulse along the tube. Later, the same authors [57] proposed a model that fitted the experimental curves with great precision and allowed correlation of the model parameters with the typical FI variables. From these curves it was possible to define a new dispersion descriptor *IDQ* (integrated dispersion quotient) [57] which was closely related to the Růžička and Hansen dispersion coefficient *D*. *IDQ* presented several advantages with respect to *D* that are summarized as follows: *IDQ* can be followed continuously with time (*D* can be evaluated only as a function of t_R), *IDQ* allows one to perform calculations on a single response curve and, no detector contribution (zero dead volume detection cell) is observed in the measured conductimetric values.

However, as in the case of *D*, *IDQ* only takes into account the radial contribution to the dispersion. Another limitation is that, to date, this methodology has only been tested for single line manifolds where no chemical reactions are involved.

1.5
Contribution of the Different Components of a Flow System to Dispersion

A FIA system block diagram is usually composed of four basic elements: a propulsion module, an injection module, the reactor and the detection module. So far, several attempts to study the individual contribution of each module to the overall dispersion have been reported, among these the papers by Johnson *et al.* [32] and Spence and Crouch [34] appear to be the most relevant.

Golay and Atwood [58, 59] applied chromatographic models to FIA and proposed the addition of the variances produced by each section of the FIA system to obtain the global variance of the system. This assumption may be applicable if the peaks are symmetric which, according to the authors, could be reached with at least 30 theoretical plates. However, they demonstrated that such a sum is hard to achieve as any perturbation in the laminar flow means that the sections of the system could not be considered independent of each other (which is the condition for adding the variances), furthermore, in linear systems symmetrical peaks are rarely achieved.

Poppe [60] also showed similarities between chromatography and FIA in terms of peak broadening. It was postulated [55, 61] that the overall broadening may be estimated from the sum of the individual variances due to injection, transport and detection

$$\sigma_{Total}^2 = \sigma_{Injection}^2 + \sigma_{Transport}^2 + \sigma_{Detection}^2 \qquad (20)$$

Note that Equation (20) does not take into account the effects produced by chemical reactions. Detection variance is by far the easiest to minimize and accounts for no more than 10% of the other variances [32]. The importance of these variances lies in the fact that the maximum analytical frequency (f_A) attainable, depends on these variances. Generally speaking, f_A is inversely related to peak broadening [60]. Total variances for Gaussian peaks are easily obtained experimentally, but those for non-Gaussian peaks need to be calculated, as will be mentioned in the next section.

Several studies have been done in order to relate the dispersion coefficient (D) with the individual dispersion coefficients. On one hand Valcárcel and Luque de Castro [62] maintain that D is the addition of all the individual dispersion coefficients of the system, that is, injection, transport and detection, while, on the other hand, Růžička and Hansen [16] propose that D is the result of the product of all individual dispersion coefficients. The Růžička and Hansen approach is based on the idea that the signal decays a certain degree while passing through each section of the system, hence the overall decay should be modeled as the product of all individual decays. Valcárcel and Luque de Castro do not explain the fundamentals of their approach.

Spence and Crouch [34], based on the theory of the independent contribution of the individual variances to the global variance described by Equation (20), analyzed the contribution to dispersion arising from the different components of a conventional FIA system and a capillary FIA system by varying the dimensions of the detection cell, the length of the FIA conduit and the injection volumes. They were able to obtain estimates of the contribution of each component for a given linear flow rate. Linear regressions between the global variance and reactor length, injection volume and flow cell volume were established, and the authors assumed a linear relationship between these variables and the variances of each system. As was shown in the previous section, these relationships are not linear (i.e., detection cell variance is proportional to the volume of the cell squared), but the regression coefficients presented were very close to 1000. The authors concluded that, for a conventional FIA system, the contribution to the global variance due to the detection is about 6%, while for a capillary FIA this contribution rises to 40%. Similarly, the injector contributes 40 and 28%, respectively, and the reactor contributes 60 and 32%, respectively. The authors consider that the main contribution is due to the flow cell and this assumption may not be correct, as was mentioned previously.

A deeper analysis of the data presented in the paper shows several facts not considered by the authors:

1. The slope of the regression of the global variance vs. cell volume for the capillary system is four times larger than for the conventional FIA system, indicating the importance of the cell volume in capillary systems. As the linear flow rate (u) was kept constant in both systems, this enhanced dependence of the variance on the cell volume should undoubtedly be related to the lower volumetric flow rate (q) in capillary systems, which in turn produces a larger residence time inside the flow cell, and hence a larger variance.

2. The analysis of the dependence of the variance on the reactor length (L) reveals that the slope for capillary systems is half that of a conventional system, indicating a

Table 1.2 Constants for Equation (21).

	Capillary system	Conventional system
k_1 [s^2 cm^{-1}]	0.60	1.39
k_2 [s^2 μl^{-1}]	6.39	8.39
k_3 [s^2 μl^{-1}]	80.9	23.1

smaller variance for the former system or, which is the same, that the sample "stretches" less in capillary systems.

3. The contribution of the injector to global dispersion is more relevant in conventional systems than in capillary systems. This may be explained by considering that in both cases the injection is done "in time" and the contribution of the injection is influenced by the transport type, and hence the dispersion is smaller as the tube radius is diminished.

4. The global variance for capillary systems is only 75% of the global variance of conventional systems, meaning an increase in peak height (lowering the detection limit) and a decrease in peak width (increasing the analysis frequency).

The expression for the global variance as a result of the regression analysis may be written as:

$$\sigma^2_{peak} = k_1 \cdot L + k_2 \cdot S_V + k_3 \cdot V_{cell} \tag{21}$$

where each term represents the variance of each component in Equation (20). In Table 1.2, the values for each constant are the slopes of the regression graphs and the units should be coherent with the units used at the time the data are adjusted.

Equation (21) may be used to evaluate the result of the global variance produced modifying some of the operational variables of the system. It must be emphasized that the injection variance cannot be minimized as this would imply diminishing the injection volume which in turn would affect the signal height if the cell volume is not reduced consequently.

Unfortunately it is not possible to estimate the error in Equation (21) as the authors have not reported the regression errors for the slopes (they do report intercept errors). However, as the regression coefficients reported are close to 1, it would be expected that this equation should be valid for the range of values explored for each variable: L (50–400 cm), S_V (0.72–1.25 μl), V_{cell} (0.325–1.14 μl).

1.5.1
Injection

The injection contribution and its variance have been studied by several authors and it was concluded that the way in which the injection is carried out has a great influence on the dispersive process and on the profile obtained and that this influence tends to grow as the injection volume increases [32, 35, 36, 61, 63].

As was mentioned before, the most common injection technique consists in emptying a loop filled with sample into a carrier stream. The usual FIA conditions impose a parabolic and laminar flow profile that provokes a deformation of the injected plug which, in turn, leads to an increase in dispersion. Johnson *et al.* [32] have reported some alternatives to diminish the deformation of the injected plug by using packed reactors and "serpentine" reactors to obtain a plug type injection.

Time injection, in which the loop is not emptied completely, is an alternative technique that allows reduction of the peak width [63] as only the first portion of the sample is injected, eliminating "tailing." One of the drawbacks of this type of injection lies in the pressure changes that might not be reproducible, thus affecting the overall reproducibility of the system.

Injection variances obtained for time injection and loop injection were analyzed by Reijn *et al.* [61]. They found that the values of the variance obtained in each case were related to the injection ratio squared.

1.5.2
Detection

The contribution to the dispersion arising from the detection system may be described as the result of several individual contributions associated with minor components of the detection system: connecting conduits, geometry of the cell, response time of the detection cell [42], and so on. Poppe [60] analyzed the detection variance, expressed in volume terms, as:

$$\sigma^2_{\text{Vol,detection}} = \sigma^2_{\text{Vol,transp}} + \sigma^2_{\text{Vol,cell}} + \sigma^2_{\text{Vol,resp.time}} \tag{22}$$

As it is not possible to know the flow pattern generated inside a flow cell, the contribution of cell geometry to the variance is very difficult to predict. Poppe estimated that for a flow cell of V_{cell} volume, the volumetric standard deviation ($\sigma_{\text{Vol,cell}}$) may vary between 0.29 and 1 times the cell volume.

$$0.29^2 S^2_V < \sigma^2_{\text{Vol,cell}} < S^2_V \tag{23}$$

This estimate arises after considering two extreme cases: (i) the cell behaves as an ideal mixing chamber whose washing follows an exponential decay (in this case the standard deviation could be deduced to be 0.29 V_{cell}) or (ii) the flow is "plug type" inside the cell and in this case the standard deviation is equal to the cell volume. It follows that when the injected volume is similar to the cell volume, one could expect some errors in the evaluation of the effects of the different variables on the dispersion, as the cell would act as a mixing chamber.

A dispersion model should be used to estimate the contribution due to transport phenomena (see below) as this model adjusts best to the environmental conditions of system conduits. Volumetric variance due to response time is equal to the flow rate multiplied by the time constant of the detector [60]. Taking into account these variables, it may be deduced that volume peak broadening due to detection increases with flow rate as, excluding the variance due the cell geometry, the other variances depend on "*u*."

1.5.3
Transport: Different Models

Several models have been generated in order to find correlations among the different system variables. The purpose of this section is to give a global view of those models and their scope, as well as an evaluation of their advantages and drawbacks. If the reader is interested in this field, we recommend some reviews on the subject [64, 65].

1.5.3.1 Descriptive Models or "Black Boxes"

Experimental results may be very well represented by descriptive models, but these models usually do not provide a better understanding of the chemical and physical mechanisms which gave rise to such results. In this sense, systems are considered as "black boxes" where the model tries to correlate the results to the entrance parameters.

The tank-in-series model is applied in chemical engineering as a deterministic model for dispersion in big reactors, and although FIA conditions differ from these environmental variables, Růžička and Hansen [27] used this model when dealing with recommendations for optimizing FIA systems. This approach has been fully explained and analyzed by the authors and will not be discussed here.

On the other hand, several authors have depicted the FIA peak as a Gaussian peak modified by an exponential function (EMG). This was first introduced by Foley and Dorsey [38] and applied to a number of FIA systems [66]. Although these models do not describe the dispersive processes, they are useful for estimating peak parameters which is necessary when comparing and evaluating different systems. Later, peak parameters, known as statistical momenta, have gained importance after the studies of Ramsing *et al.* [67] and Reijn *et al.* [68], as they could be used to evaluate any profile.

Regression analysis has been widely used in FIA, and the dependence of the dispersion coefficient D, peak width, appearance time, and so on, on the different FIA system variables has been reported [64] using least squares methods. The validity of these equations is restricted to the systems for which they were deduced as the influence of the component characteristics on the responses makes generalization impossible [20].

Artificial learning models based on neuronal networks have been used recently to model FIA systems [69]. This method is based on generating a network of processing units (neurons), each neuron receives a certain number of input signals and, according to a previously established priority order, the neuron is caused to emit an output signal. The priority is established through a set of experiments in which the neurons "learn" how to behave. Once the learning is complete the system might be used in a predictive way, the capacity of prediction being proportional to the number of neurons. There is a natural limit to the number of neurons which is set by the increase in the capacity of the neurons for learning "noise." The major drawback of this approach is the need for large processing capacity.

1.5.3.2 Deterministic Models: Dispersive Models and Tank-in-Series Model

Deterministic models are those models in which the global behavior of a system is followed through numeric relationships between different parameters. The value assigned to each parameter is imposed by the environmental condition of the system. Most of the systems analyzed consider that the dispersion only takes place in the reactor.

Dispersive models take into account the variations in behavior of the system in each region, considering both diffusive and convective contributions as a function of the environmental conditions of the system. They are mainly based on the diffusion-convection equation or on phenomenological modifications of it.

Applied to cylindrical tubes, the dispersive model is depicted by the differential equation which shows the changes in concentration of a substance as a function of time along the system length:

$$\frac{\partial C}{\partial t} + u(r) \cdot \frac{\partial C}{\partial z} = \frac{1}{r} \cdot \frac{\partial}{\partial r}\left(r \cdot D_r \cdot \frac{\partial C}{\partial r}\right) + \frac{\partial}{\partial z}\left(D_L \cdot \frac{\partial C}{\partial z}\right) + s \tag{24}$$

where $u(r)$ is the velocity profile in the tube, D_L the axial dispersion coefficient (which in turn depends on r), D_r the radial dispersion coefficient (which is a function of z) and s accounts for the kinetic contribution due to the appearance or consumption of substance as a result of a chemical reaction.

This model has been used only twice for FIA: the uniform dispersion model and the axially dispersed plug flow model. In the former the following assumptions are made: (i) a parabolic velocity profile is developed, (ii) phenomenological dispersion coefficients (axial D_L and radial D_r) are equal and are also equal to D_m the diffusion coefficient of the substance. In the second model, a uniform velocity profile is assumed, and hence it is assumed that only axial diffusion is possible.

Uniform Dispersion Model   This model is based on Equation (24), taking into account the actual velocity profile inside the conduit, which depends on the geometry of the manifold. It is assumed that axial and radial dispersion coefficients are constant along the system and equal (or proportional) to the molecular diffusion coefficient. Equation (24) can then be written as:

$$\frac{\partial C}{\partial t} = \frac{D_r}{r} \frac{\partial}{\partial r}\left(r \frac{\partial C}{\partial r}\right) + D_L \frac{\partial^2 C}{\partial z^2} - u(r) \cdot \frac{\partial C}{\partial z} + s \tag{25}$$

As was mentioned before, in the conditions usually employed in FIA, the flow pattern inside the tubes is laminar (Re < 2300), the velocity profile for straight tubes is parabolic and this situation is known as Poiseuille–Hagen flow. In this particular case Equation (25) can be written as:

$$\frac{\partial C}{\partial t} = D_m\left(\frac{1}{r}\frac{\partial C}{\partial r} + \frac{\partial^2 C}{\partial r^2} + \frac{\partial^2 C}{\partial z^2}\right) - 2\bar{u}\left(1 - \frac{r^2}{a^2}\right)\frac{\partial C}{\partial z} + s \tag{26}$$

where $\bar{u}$ is the fluid mean velocity and a is the tube radius.

Axially Dispersed Plug Flow Model   If a uniform velocity profile is assumed, and hence there is no radial variation of velocities:

$$\frac{\partial C}{\partial t} = \frac{D_r}{r} \frac{\partial}{\partial r}\left(r \frac{\partial C}{\partial r}\right) + D_L \frac{\partial^2 C}{\partial z^2} - u \frac{\partial C}{\partial z} + s \tag{27}$$

where u is independent of r and equal to the mean fluid velocity ($\bar{u}$). No radial concentration gradient should be created under this assumption and the term which accounts for this diffusion is logically negligible. This leads to the model known as axially dispersed plug flow.

$$\frac{\partial C}{\partial t} = D_L \frac{\partial^2 C}{\partial z^2} - \bar{u} \frac{\partial C}{\partial z} + s \tag{28}$$

Equation (28) may also be written as:

$$\frac{\partial C}{\partial t} + \frac{\bar{u}}{L} \cdot \left(\frac{\partial C}{\partial z'}\right) = \frac{D_m}{L^2} \cdot \left(\frac{\partial^2 C}{\partial z'^2}\right) \tag{29}$$

where z' is a reduced dimensionless expression for the distance, $z' = z \cdot L^{-1}$.

Considering a delta injection, waterproof tube walls and an endless tube, the variance of the concentration profile at time t and distance L from the origin is:

$$\sigma_t^2 = \frac{2 \cdot D_L \cdot L}{\bar{u}^3} \tag{30}$$

Although this model has been used in several fields, such as chemical engineering, chromatography and even FIA, its application is valid only in those systems where the flow pattern does not generate a radial concentration gradient, which is not the case for FIA where the parabolic flow profile generates a radial gradient which is not negligible.

Tanks-in-Series Model The tanks-in-series model [27, 70] assumes that point to point variations within the FIA system are very small, and hence the system may be considered as homogeneous regarding its properties and dependent variables. In this case, a complete mix between the injector and the detector is considered, being either hypothetical (Tyson and Iris [71]) or obtained through the use of some kind of mixing chamber (Pungor *et al.* [22]). This model is similar to that used in chromatography (ETPH) as it assumes N ideal mixing stages. Nevertheless, it is necessary to stress, that chromatographic peak broadening is the result of dynamic processes caused by the column packing which are absent in most FIA applications.

The equations used are based on residence time curves which may be found in most Chemical Engineering books [26] and will not be discussed here. The model description and the main conclusions are reported elsewhere (see [72]).

However, it should be noticed that experimental peak profiles, when compared to predicted profiles, show a good match in the ascending and descending portions of the curve but not in the maximum nor in the shape. It might be concluded that this model, although simple, does not allow a deep knowledge of the processes occurring, and hence its use for modeling systems without a mixing chamber is limited. On the contrary, this model predicts very well those systems in which a mixing chamber is used, or if single bead string reactors or gradient chambers [64], such as those used in FIA titrations, are present.

1.5.4
Probabilistic Models

1.5.4.1 Random Walk

Random walk has been widely employed in physicochemical studies in order to explain and predict the effect of diffusion of the solutes [19]. The model involves multiple steps, the direction of each step being independent of the direction of the previous stage. Generally, an origin is fixed and then steps of a given length are given in the x and y directions:

$$(\Delta x_1, \Delta y_1), (\Delta x_2, \Delta y_2), (\Delta x_3, \Delta y_3), \ldots, (\Delta x_N, \Delta y_N) \tag{31}$$

where N is the total number of steps and the distance traveled since the initial point (R) is given by:

$$
\begin{aligned}
R^2 &= (\Delta x_1 + \Delta x_2 + \ldots + \Delta x_n)^2 + (\Delta y_1 + \Delta y_2 + \ldots + \Delta y_n)^2 \\
R^2 &= \Delta x_1^2 + \Delta x_2^2 + \ldots + \Delta x_N^2 + 2\Delta x_1 \Delta x_2 + 2\Delta x_1 \Delta x_3 + \ldots
\end{aligned}
\tag{32}
$$

The expression above is general and independent of the direction of the *walk* since the chance for moving back and forward is the same. In this way, on average, for a great number of steps the crossed terms of Equation (32) are cancelled giving the following expression:

$$
\begin{aligned}
R^2 &\cong \Delta x_1^2 + \Delta x_2^2 + \ldots + \Delta x_N^2 + \Delta y_1^2 + \Delta y_2^2 + \ldots + \Delta y_N^2 \\
R^2 &\cong N\langle r^2 \rangle \\
R &\cong \sqrt{N} r_{\text{rms}}
\end{aligned}
\tag{33}
$$

where r_{rms} is the root mean squared step size. This result can be extended to an axis of three coordinates with the same assumption. In agreement with the equation above, even if the average distance of the total walk is $N \cdot r_{\text{rms}}$, the distance from the initial point is $r_{\text{rms}} \cdot N^{0.5}$.

Betteridge *et al.* [73] were pioneers in the application of this model since they studied the effect of different variables on the signal given by a single line FI system where the injection of a discrete number of molecules was simulated. This approach was then extended to confluence FI systems [74] and to SIA [4]. Later, Wentzell *et al.* [75] studied different FI systems without chemical reaction where different flow profiles and conditions of infinite radial diffusion were incorporated.

Since this model is based on the fate of each molecule, it is easy to simulate the effect of sample size, physical dispersion and chemical reaction. The obtained results show good agreement with reported experimental studies in systems *without chemical reaction*.

The random walk simulation has a great capacity for predicting the shape of the curve that should be obtained by monitoring the whole FI system in a longitudinal way. The equations employed in the simulation will not be presented here. However, if the reader needs a deeper knowledge of this issue, several papers can be consulted [19, 73–77].

Regarding the advantages of the random walk modem, they were reported by Kolev [64] and can be summarized as follows: (i) ease of observation of the influence of different flow types, (ii) mathematical and computational simplicity, (iii) good graphic visualization of the results. Amongst the disadvantages, the method has shown certain problems with quantitative prediction which can be improved by a longer computational time.

1.6
Design Equations

Reijn *et al.* [78] proposed design equations to maximize the frequency of analysis and minimize the consumption of reactants for linear systems in coiled and single bead string reactors. Tijssen [54] proposed design equations for systems with coiled reactors, as shown previously. Other design equations based on the tanks-in-series model were provided by Růžička *et al.* These design equations fail for low dispersion cases [79]. Other reported equations will not be presented here as they only work in very specific systems.

The following section aims to provide evidence for the effect of different operational variables on the obtained signal. In order to do this, a $CoSO_4$ solution with an absorbance value equal to 0.341 of absorbance was prepared. Double deionized water (DDW) was used as carrier, and the absorbance was monitored at the maximum of the cobalt–water complex ($\lambda_{máx} = 525$ nm). Different flow rates and tubes of different radius, length and geometry were used [77, 80].

1.6.1
Influence of the Different System Variables

1.6.1.1 Reactor Length
In all cases, and for any reactor configuration, an increase in the traveled distance, providing that all the other variables stay unchanged, means an increase in the dispersion of the sample (peaks tend to be shorter and wider, Figure 1.7).

1.6.1.2 Geometric Configuration
The influence of the spatial configuration of the reactor depends on the other system variables. The more the contribution of the dispersion due to transport to the total dispersion and the more efficient the secondary profiles production, the more notorious will be the differences on dispersion found between straight and coiled reactors.

Regarding this last factor, the arising of secondary flows will be more evident as reactor length and carrier linear rate increase. This behavior is found also in other types of reactor that want to maximize secondary flow production, like knotted reactors in which the flow direction changes successively, favoring the mass radial transfer. The more often this event occurs, the more uniform the rate distribution will be, as well as the similarity with the plug-type flow [16, 81]. A major efficiency in the minimization of the dispersion is reached when knotted reactors (periodic or random

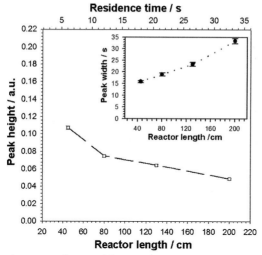

Figure 1.7 Influence of the reactor length on dispersion.

knots) are used instead of coils. This is due to the frequent changes in the direction of flow: in the knotted reactors this change is periodic, whereas reactors with irregular knots produce an odd disturbance of the flow profile [16].

1.6.1.3 Flow Rate

The dependence of the different descriptors, mainly D, on the flow rate seems to be unclear. Li and Ma [82] studied this dependence by employing coiled reactors in the system but, unfortunately, in this kind of geometry, the capability to reduce dispersion is, precisely, a function of the flow rate. The authors found a maximum in the D vs. q plot with medium flow rates. This maximum was found to be independent of the length and diameter of the tube, as well as of the injection volume. However, it showed some relation to the substance diffusion coefficient (the value of q in the maximum is smaller as this coefficient increases).

On the other hand, in a previous work, Stone and Tyson [83] analyzed the different factors that affect dispersion in flow systems, including flow rate, and found that, for short tubes, dispersion decreases monotoneously as the flow rate increases. Nevertheless, as tube length increases, a maximum in the D vs. q plot can be seen for low flow rates. The q value for which this maximum appears depends on reactor dimensions, in contrast with Li and Ma's observations [82] for exactly the same systems. In spite of this, it is worth analyzing the trend of the D vs. q plot.

The differences found in classic bibliography about the dependence of D on q can be explained in terms of the point at which the maximum appears. As an example, the experimental conditions used by Růžička and Hansen [16] who support that D increases as q increases, are located before this maximum, while those used by Valcárcel and Luque de Castro [62], who support that D decreases as q increases, are located after it.

From the simplest point of view, if the sample goes through a plug-type flow regime, it would have the same mass radial distribution as that of injection at any time during its travel. These conditions can be approached in FIA at very (impractical) low flow rates. Then, as the flow rate increases, the mass radial distribution contributes to the increase in the coefficient D. Nevertheless, at greater flow rates, where the main dispersion source is convection, the coefficient D decreases again (assuming that the instrumental time of response is null). In this way, the observations made by the authors are valid, and the difference in their conclusions can be ascribed to a misinterpretation of the different experimental conditions.

Stone and Tyson, and Li and Ma, were correct when they realized that q and L are factors that affect the residence time and, therefore, comparison of the different systems becomes difficult when q and L are modified due to the different contribution of the convective and diffusive processes. For example, for short manifolds and high rates, the contribution is only convective. Nevertheless, if, keeping q constant, L increases, the influence of the diffusive process increases. If a plot of this, as a function of the residence time, is made (Figure 1.8), the result is an increase in D as t_R increases. This profile is much influenced by the flow cell used, in certain cases, D and q are found to be independent for some values of q [83, 84], depending on the length of the tube and the volume of the flow cell [83].

The maximum value on the D vs. q chart is due, then, to the existing dependence between the residence time of the solute and the flow rate. The residence time is inversely proportional to q and it is commonly calculated [16] as the ratio of the reactor length and the average flow rate $(t_R \approx L \times q^{-1})$. Nevertheless, Li and Ma found that actually t_R is proportional to q^{-k}, and that k is less than 1 for low flow rates and greater than 1 for high rates. According to the theoretical expression of D given by Růžička and Hansen [16], it can be said that D is proportional to $(t_R \cdot q)^{1/2}$. Therefore, if k is less than 1, the t_R value is a little bigger than expected if q^{-1} is taken into account and D increases as q increases. The opposite is observed if k is greater than 1.

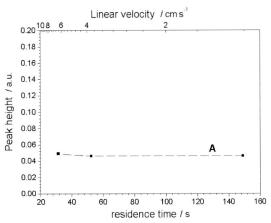

Figure 1.8 Effect of the flow rate on dispersion.

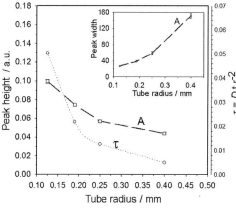

Figure 1.9 Effect of the tube radius on dispersion.

For the case of the peak width, decreasing hyperbolas described by negative exponents of q are obtained as a function of q. The coefficients that describe these hyperbolas depend on the reactor spatial configuration, being smaller in absolute value for the case of coils than for straight tubes. As long as the flow rate decreases, this difference is less noticeable.

1.6.1.4 Tube Radius

As shown in Figure 1.9 and the different published works on this issue [45, 82, 83], as the tube radius decreases, D and the peak width also decrease, at a constant linear flow rate (u). However, it is necessary to say that certain studies, such as Li and Ma's [82], were performed keeping the volumetric flow (q) constant instead of the linear flow rate (u). Thus, the decrease in dispersion is mainly due to a lower t_R for the tubes of smaller diameter, the contribution of the radial mass being less significant in these cases. This contribution can be observed in terms of τ, where the greater diffusive contribution is shown as an increase in the reduced time value.

Figure 1.9 (inset) shows the obtained peak width as a function of the tube radius. There is an important factor to be considered when these plots are compared keeping linear flow rates (u) constant, using different tube radius. In these cases the employed volumetric flow rate (q) changes, and so does the flow rate inside the flow cell, making the comparison unfair. Thus, the real peak width value should be corrected by the average residence time inside the cell, which is a factor of very low importance for high rates and tubes of large radius, but is relevant for low rates and tubes of small radius. As can be seen in Figure 1.9, the peak width increases with the tube radius (a). When a decreases, the signal gets taller and the peak gets narrower, which implies a relatively greater analysis frequency gain rather than a gain in sensitivity.

1.6.1.5 Injection Volume

The dependence of the transient signal on the injection volume S_V has been shown by Růžička and Hansen [16]. This dependence is key to the control of the dispersion, since it affects significantly the signal, the residence time and, for large injection

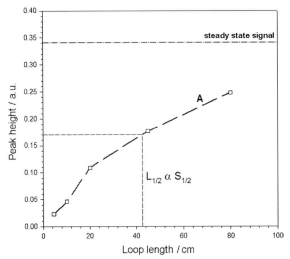

Figure 1.10 Effect of loop length on dispersion.

volumes, the peak width. When the signal height as a function of the injection volume (or of the length of the loop) is plotted, the curves shown in Figure 1.10 are obtained. This dependence has been described [16] through the following phenomenological equation:

$$\frac{1}{D^{\max}} = \frac{C^{\max}}{C^0} = 1 - e^{-K \cdot S_V} = 1 - e^{-0.693 \frac{S_V}{S_{1/2}}} \tag{34}$$

The parameter $S_{1/2}$, (the required sample volume to reach a half of the signal of the stationary state or $D = 2$) is useful to define the performance of a system as it depends on different variables of the system. The coefficient K is inversely related to $S_{1/2}$ and indicates how steep is the slope of the curve shown in Figure 1.10.

In the ideal case of a system without dispersion, the injection volume does not influence the signal height, but has a direct relation to its width. This is true only when the injection volume is bigger than the cell volume. In this ideal situation, K should tend to infinity and D, as expected, should be equal to 1.

Equation (34) can be linearized [83], obtaining:

$$\ln\left(1 - \frac{1}{D^{\max}}\right) = -\frac{0.693}{S_{1/2}} S_V \tag{35}$$

With this, $S_{1/2}$ can be obtained through the slope of the curve that better fits this dependence.

When evaluating this parameter in real systems, care must be taken, especially with small tube radius where the influence of any dead volume inside the system is notorious. For the case of injection volumes that generate peak heights statistically non-different from the signal of the stationary condition, fitting of Equation (35) will not be good as it predicts that the signal is reachable only for S_V tending to infinite.

Stone and Tyson [83] show that D has similar values in systems that keep the relation between the injection volume and the reactor volume constant, without considering the tube radius. In contrast, there is a strong reduction in $S_{1/2}$ if the tube's radius gets smaller.

Moreover, these authors find the major symmetry of the transient signal with temporal injections rather than slug-type ones. They also show the influence of the injection "mode" which becomes more relevant as long as the influence of the injection on the global dispersion increases (i.e., systems where the loop length is close to the reactor length). In these cases, the dispersion due to transport is not relevant.

Regarding the dependence of the peak width on the injection volume, it has been discussed that it is minimal if the length of the loop is less than 20% of the reactor length. This is because, in these cases, the transport contribution to dispersion is greater than the injection contribution. Nevertheless, the general trend is that when the injected volume increases, the width of the obtained peak also increases.

Taking into account Stone and Tyson's considerations about the equality in D values observed in those systems where the relationship between the injection volume and the reactor volume is the same, Equation (34) can be re-written as follows:

$$\frac{1}{D^{\max}} = 1 - e^{-K\frac{S_V}{V_{\text{reactor}}}} = 1 - e^{-\left[\frac{0.693}{\left(\frac{l}{L}\right)_{1/2}} \cdot \frac{l}{L}\right]} \tag{36}$$

where $\left(\frac{l}{L}\right)_{1/2}$ is the length ratio for D equal to 2, and it should be independent of tube radius in order to fulfill the presumptions of Stone and Tyson. This expression may be used as a design equation: the loop to reactor length ratio is chosen so as to obtain the most wanted D value (i.e., based on the desired sensitivity) and the flow rate is chosen to obtain a given residence time (i.e., based on the reaction time).

1.6.2
Optimization of Flow Systems

This way of optimization looks for the maximum utilization of the flow secondary profiles, considering the set of instruments commonly available in FIA. It is based on the work of Reijn *et al.* [85] and Tijssen [54].

The use of a peristaltic pump in FIA imposes limitations to the range of flows that can be used and to the system back pressure that the pump can support. The minimum residence time in a system regarding these limitations can be calculated as follows:

$$t_p = \frac{8 \cdot L^2 \eta}{\Delta p \cdot a^2} \tag{37}$$

t_p is known as "limited time by pressure." The maximum backpressure (Δp) for a peristaltic pump is about 5 bar, while for a plunger pump in liquid chromatography it is about 400 bar. In coiled tubes, the existence of secondary flows causes an increase in the axial pressure, although for De less than 25 the excess of pressure is less than 10%, compared with a straight tube. Although it can be thought that the sophistication of the pump system is a way to increase the maximum backpressure, it should be noticed that for an increase of 10 in Δp, the minimum radius decreases [85] by a factor of 1.47, which is not a great improvement considering the increment in the instrumental costs.

Another limitation due to the maximum flow that the propulsion system can provide is known as "limitation by flow." For this case the minimum obtainable residence time is calculated by:

$$t_F = \frac{L \cdot \pi \cdot a^2}{q_{max}} \tag{38}$$

Considering a peristaltic pump, the maximum flow that can be supplied is about $30 \, cm^3 \, min^{-1}$.

The analytical throughput (f_A) is determined by the peak width which, in Taylor's conditions, can be written as:

$$f_A = \frac{60}{6 \cdot \sigma_t} = 10\sqrt{\frac{24 \cdot D_m}{a^2 \cdot t}} = \frac{49}{a}\sqrt{\frac{D_m}{t}} (in \, min^{-1}) \tag{39}$$

where t is the residence time in the system. As can be seen, the frequency of analysis should be independent of variables such as flow (q), pressure fall (Δp) and reactor length (L), although these variables impose limits to the working a values and the reachable t values.

The carrier consumption per peak (F, in cm^3) [54] can be obtained as follows:

$$F = q \cdot 6\sigma_t = 6\pi a^2 \bar{u}\sqrt{\frac{a^2 t}{24 \cdot D_m}} = \frac{3\pi}{4}a^4\sqrt{\frac{\Delta p}{\eta D_m}} \qquad \left(\frac{\Delta p}{L} = \frac{8\bar{u}\eta}{a^2}\right) \tag{40}$$

Regarding the system optimization, the data obtained by Tijssen can be presented in a practical way in Figure 1.11. Taking into account the desired analytical frequency and

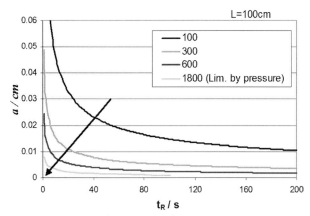

Figure 1.11 Analytical frequency as a function of t_R in FIA conditions.

the residence time, the required diameter of the tube can be obtained. With this, for a given analytical frequency, the greater the desired residence time, the smaller should be the tube radius.

Regarding the pressure fall that the peristaltic bomb can support (about 4 bar) the maximum reachable frequency of analysis in FIA is, theoretically, about 1800 peaks per hour.

For coiled tubes, the decrease in the theoretical plate height improves the analytical frequency, and the carrier consumption per peak decreases. Nevertheless, high flow rates are necessary to obtain big differences. As F, f_A, and H are direct functions of σ_t, it can be seen that the improvement in the height of the theoretical plate for coils compared to straight tubes (H/H_0, where the subscript "0" is used for straight tubes) is directly related to F and f_A through:

$$\sqrt{\frac{H}{H_o}} = \frac{S_o}{S} = \frac{f_A}{f_{A_o}} \tag{41}$$

For example, to double the analytical frequency, $(H/H_0)^{1/2}$ should be equal to $^1/_2$, which is obtainable with a De^2Sc greater than 10^4. In practice, values of De^2Sc close to 10^8 can be obtained, which implies a frequency of about 100 times greater in coil tubes. Nevertheless, this is valid only when transport is the main contributor to the total dispersion.

It is now time to think how to proceed to choose the optimal conditions for a system when a certain residence time is required. If systems with a chemical reaction are analyzed, the residence time only modifies the degree of reaction.

The idea is, then, to calculate the maximum value of De^2Sc to maximize the effects of the secondary flow, once the reaction time is selected. To do this it is necessary to work with flow and pressure fall as large as possible, and so it is only left to set the residence time and take into account the limitations of the maximum pressure fall and the flow rate obtainable to determine the maximum De^2Sc. It is obvious that the desired residence time should be reachable by considering the maximum pressure fall and the maximum flow. So, the following equations can be derived:

$$t_R = t_p = \frac{8 \cdot L^2 \eta}{\Delta p \cdot a^2} = t_F = \frac{L \cdot \pi \cdot a^2}{q_{max}}$$

$$L_{optimo} = \sqrt[3]{\frac{\Delta p_{max} \cdot q_{max} \cdot t_R^2}{8 \cdot \eta \cdot \pi}} \tag{42}$$

$$a_{optimo} = \sqrt{\frac{q_{max} \cdot t_R}{\pi \cdot L_{optimo}}}$$

Once these values are obtained, H/H_0 can be evaluated calculating De^2Sc (for a given λ) through the formula given above. Figure 1.12 shows an experimental adjustment done with the data presented by Tijssen in order to make this calculation easier and to obtain an estimate of the obtainable improvement in coiled tubes using peristaltic pumps instead of plunger ones as Tijssen suggested.

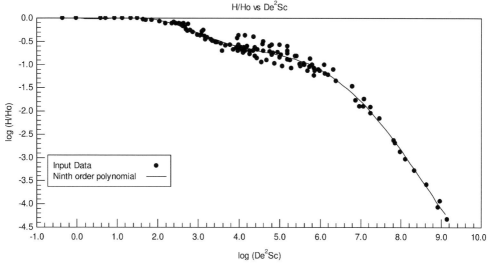

Figure 1.12 H/H_0 relation adjustment as a function of De^2Sc.

Once H/H_0 is known, it is possible to calculate f_A and F. Table 1.3 shows the calculated values with Tijssen's conditions and under other proposed conditions.

It can be concluded from these data that the best results are those where the maximum flow and pressure fall were used. The calculation for the reachable conditions using conventional FIA instrumentation shows a great improvement in the frequency of analysis (about 5 times greater), but the improvement in reactants consumption per peak was of less importance, about only 3 times less.

As the product De^2Sc increases when the tube length increases or the radius decreases, a compromise is reached, in which maximum values for reactor length (L) and tubes not too wide (0.1–0.5 mm i.d.) have to be obtained. One reason for this is

Table 1.3 Optimum operational conditions for different systems.

p_{max}	bar	400	400	80	4
$q_{máx}$	$cm^3 s^{-1}$	0.5	0.5	0.1	0.5
t_R	s	10	100	10	60
L	Cm	4301	19 965	1471	3060
A	Cm	0.0192	0.0282	0.0147	0.0559
De^2Sc		1.96×10^8	9.08×10^7	1.34×10^7	2.32×10^7
H/H_0		6.36×10^{-4}	1.62×10^{-3}	1.25×10^{-2}	7.32×10^{-3}
$(H/H_0)^{-1/2}$ (coil/ recto improvement)		39.7	24.8	9.4	11.7
f_{A_0}	peaks min^{-1}	3.12	0.67	4.08	0.44
f_A	peaks min^{-1}	124	17	37	5.1
F_0	cm^3 peak^{-1}	9.6	44.6	1.5	68.4
F	cm^3 peak^{-1}	0.24	1.80	0.16	5.85

$D_m = 1.5 \times 10^{-5} cm^2 s^{-1}$, $\lambda/v = 0.07 cm^{-2} s$.

that the minimum obtainable residence time due to the backpressure and the maximum flow depend more on the radius (squared power) than on the length.

The same analytical frequency can be achieved in straight tubes by reducing the tube radius. As an example, if the coil reactor increases in the analytical frequency fivefold, a straight tube of radius five times less would have the same analytical frequency. In turn, the carrier consumption can be drastically reduced; this is because F depends linearly on a^4. In the previous example, if a straight reactor is used, it will require a volume per peak of 625 times less.

This optimization only takes into account the transport contribution to dispersion. Nevertheless, it shows clearly, in a system optimization, how all the variables should be included and their influence on the different parameters that "make" the analytical performance.

Equation (36) shows a suitable way to select the variables of a system from which a certain analytical performance is required. It is worth mentioning that this equation is not a real optimization, as it minimizes the $\left(\frac{l}{L}\right)_{1/2}$ value as a function of the flow rate and/or the residence time. If this relation decreases, the absolute slope of the D vs. S_V curve increases, minimizing D and the peak width.

1.7
Concluding Remarks

In this chapter, we have presented a critical review of the different factors that are able to describe the dispersion process in flow systems. Mainly, the review was biased towards flow injection systems as these systems have been extensively studied from a theoretical point of view. Special emphasis was placed on the dispersion descriptors and on the relevance of finding global descriptors that allow the analyst to predict the best working conditions in order to maximize analytical frequency together with optimal sensitivity. Based on this, the relevance of knowing the influence of the different components of the flow system on the global dispersion was demonstrated by analyzing the different models reported to date. It was observed that, mainly, the different approaches pay no attention to the influence of detection and injection on global dispersion. Thus, it is important to stress that the different parts of a system can influence the obtained results, which obliges one to develop configurations where the contribution of the detector is negligible. Moreover, two different ways of selecting and optimizing a flow system in terms of the desired dispersion degree were shown.

It is clear that the analytical strength of the technique rests in controlling the diffusive processes of mixing, keeping the integrity of the injected sample plug. This needs to be attained through knowledge of the transport phenomena and the factors that influence them. In this way, the development of an equation able to relate global dispersion to the typical variables of FIA should be of great interest when predicting the behaviour of systems. However, the different theoretical models and experimental arrangements reported in the literature have failed in monitoring the mass redistribution and thus, dispersion, in a continuous way as time elapses.

Acknowledgments

The authors thank Dr. Osvaldo Troccoli for his contribution to this chapter.

References

1 Skeggs, L.J. (1957) An automatic method for colorimetric analysis. *American Journal of Clinical Pathology*, **28**, 311–322.

2 Ruzicka, J. and Hansen, E.H. (1975) Flow Injection Analysis. Part I. A new concept of fast continuous flow analysis. *Analytica Chimica Acta*, **78**, 145–157.

3 Ruzicka, J., Marshall, G.D. and Christian, G.D. (1990) Variable Flow Rates and Sinusoidal Flow Pump Flow Injection Analysis. *Analytical Chemistry*, **62**, 1861–1866.

4 Ruzicka, J. and Marshall, G.D. (1990) Sequential injection: a new concept for chemical sensors, process analysis and laboratory assays. *Analytica Chimica Acta*, **237**, 329–343.

5 Gubeli, T., Christian, G.D. and Ruzicka, J. (1991) Fundamentals of sinusoidal flow sequential injection spectrophotometry. *Analytical Chemistry*, **63**, 1861–1866.

6 Gubeli, T., Christian, G.D. and Ruzicka, J. (1991) Principles of stopped-flow sequential injection analysis and its application to the kinetic determination of traces of a proteolytic enzyme. *Analytical Chemistry*, **63**, 1680–1685.

7 Baron, A., Guzman, M., Ruzicka, J. and Christian, G.D. (1992) Novel single-standard calibration and dilution method performed by the sequential injection technique. *Analyst*, **117**, 1839–1844.

8 Masini, J.C., Baxter, P.J., Detwiler, K.R. and Christian, G.D. (1995) Online spectrophotometric determination of phosphate in bioprocesses by sequential injection. *Analyst*, **120**, 1583–1587.

9 van Staden, J.F. and Taljaard, R.E. (1997) Online dilution with sequential injection analysis: a system for monitoring sulfate in industrial effluents. *Fresenius' Journal of Analytical Chemistry*, **357**, 577–581.

10 Giné, M.F., Bergamin Filho, H. and Zagatto, E.A.G. (1980) Simultaneous determination of nitrate and nitrite by flow injection analysis. *Analytica Chimica Acta*, **114**, 191–197.

11 Krug, F.J., Bergamin Filho, H. and Zagatto, E.A.G. (1986) Commutation in flow injection analysis. *Analytica Chimica Acta*, **179**, 103–118.

12 Reis, B.F., Giné, M.F., Zagatto, E.A.G., Lima, J.L.F.C. and Lapa, R.A.S. (1994) Multicommutation in flow analysis. 1. Binary sampling: concepts, instrumentation and spectrophotometric determination of iron in plant digests. *Analytica Chimica Acta*, **293**, 129–138.

13 Reis, B.F., Morales-Rubio, A. and de la Guardia, M. (1999) Environmentally friendly analytical chemistry through automation: comparative study of strategies for carbaryl determination with p-aminophenol. *Analytica Chimica Acta*, **392**, 265–272.

14 Zagatto, E.A.G., Reis, B.F., Oliveira, C.C., Sartini, R.P. and Arruda, M.A.Z. (1999) Evolution of the commutation concept associated with the development of flow analysis. *Analytica Chimica Acta*, **400**, 249–256.

15 Christian, G.D. and Ruzicka, J. (1992) Exploiting stopped-flow injection methods for quantitative chemical analysis. *Analytica Chimica Acta*, **261**, 11–21.

16 Ruzicka, J. and Hansen, E.H. (1988) *Flow Injection Analysis*, Wiley, New York.

17 Honorato, R.S., Araujo, M.C.U., Lima, R.A.C., Zagatto, E.A.G., Lapa, R.A.S. and Costa Lima, J.L.F. (1999) A flow-batch titrator exploiting a one-dimensional

optimisation algorithm for end point search. *Analytica Chimica Acta*, **396**, 91–97.

18 Probstein, R.F. (1994) *Physicochemical Hydrodynamics: An Introduction*, John Wiley & Sons, New York.

19 Atkins, P.W. (1978) *Cap 26, 27 and Appendix Vol*, Addison-Wesley Iberoamericana, New York.

20 Brooks, S.H., Leff, D.V., Torres, M.H. and Dorsey, J.G. (1988) Dispersion coefficient and moment analysis of flow injection analysis peaks. *Analytical Chemistry*, **60**, 2737–2744.

21 Betteridge, D. and Ruzicka, J. (1976) Determination of glycerol in water by flow injection analysis – a novel way of measuring viscosity. *Talanta*, **23**, 409–410.

22 Pungor, E., Feher, Z., Nagy, G., Toth, K., Horvai, G. and Gratzl, M. (1979) Injection techniques in dynamic flow-through analysis with electrothermal analysis sensors. *Analytica Chimica Acta*, **109**, 1.

23 Streeter, V.L. and Wylie, E.G. (1981) *Mecánica de Fluídos*, McGraw-Hill, Bogot.

24 Taylor, G. (1953) Dispersion of soluble matter in solvent flowing through a tube. *Proceedings of the Royal Society of London, Series A*, **219**, 186–203.

25 Taylor, G. (1954) Conditions under which dispersion of a solute in a stream of solvent can be used to measure molecular difusion. *Proceedings of the Royal Society of London, Series A*, **225**, 473–477.

26 Levenspiel, O. (1962) *Chemical Reaction Engineering*, Wiley, New York.

27 Ruzicka, J. and Hansen, E.H. (1978) Flow injection analysis. Part X. Theory, Techniques and Trends. *Analytica Chimica Acta*, **99**, 37–76.

28 Painton, C.C. and Mottola, H.A. (1984) Kinetics in continuous flow sample processing. *Analytica Chimica Acta*, **158**, 67.

29 Painton, C.C. and Mottola, H.A. (1983) Dispersion in continuous-flow sample processing. *Analytica Chimica Acta*, **154**, 1–16.

30 Ananthakrishnan, V., Gill, W.N. and Barduhn, A.J. (1965) Laminar dispersion in capillaries. 1. Mathematical analysis. *American Institute of Chemical Engineers Journal*, **11**, 1063.

31 Li, Y.S. and Gao, X.F. (1996) Two new parameters: Axial dispersion degree and radial dispersion intensity of sample zone injected in flow injection analysis systems. *Laboratory Robotics and Automation*, **8**, 351.

32 Johnson, B.F., Malick, R.E. and Dorsey, J.G. (1992) Reduction of injection variance in flow injection analyisis. *Talanta*, **39**, 35–44.

33 Brooks, S.H. and Rullo, G. (1990) Minimal dispersion flow injection analysis systems for automated sample introduction. *Analytical Chemistry*, **62**, 2059–2062.

34 Spence, D.M. and Crouch, S.R. (1997) Factors affecting zone variance in a capillary flow injection system. *Analytical Chemistry*, **69**, 165–169.

35 Levenspiel, O., Lai, B.W. and Chatlynne, C.Y. (1970) Tracer curves and residence time distribution. *Chemical Engineering Science*, **25**, 1611–1613.

36 Levenspiel, O. and Turner, J.C.R. (1970) The interpretation of residence-time experiments. *Chemical Engineering Science*, **25**, 1605–1609.

37 Snyder, L.R. (1980) Continuous-flow analysis: present and future. *Analytica Chimica Acta*, **114**, 3–18.

38 Foley, J.P. and Dorsey, J.G. (1983) Equations for calculation of chromatographic figures of merit for ideal and skewed peaks. *Analytical Chemistry*, **55**, 730–737.

39 Betteridge, D., Cheng, W.C., Dagless, E.L., David, P., Goad, T.B., Deans, D.R., Newton, D.A. and Pierce, T.B. (1983) An automated viscometer based on high-precision flow injection analysis. 2. Measurement of viscosity and diffusion-coefficients. *Analyst*, **108**, 17.

40 Stone, D.C. and Tyson, J.F. (1986) Flow cell and diffusion-coefficient effects in flow injection analysis. *Analytica Chimica Acta*, **179**, 427.

41 van Staden, J.F. (1990) Effect of coated open-tubular inorganic-based solid-state ion-selective electrodes on dispersion in flow injection. *Analyst*, **115**, 581–585.

42 van Staden, J.F. (1992) Response-time phenomena of coated open-tubular solid-state silver-halide selective electrodes and their influence on sample dispersion in flow injection analysis. *Analytica Chimica Acta*, **261**, 381.

43 Vanderslice, J.T., Rosenfeld, A.G. and Beecher, G.R. (1986) Laminar-flow bolus shapes in flow injection analysis. *Analytica Chimica Acta*, **179**, 119–129.

44 Ruzicka, J., Hansen, E.H. and Zagatto, E.A. (1977) Flow injection analysis. 7. Use of ion-selective electrodes for rapid analysis of soil extracts and blood serum. Determination of potassium, sodium and nitrate. *Analytica Chimica Acta*, **88**, 1–16.

45 Korenaga, T. (1992) Aspects of sample dispersion for optimizing flow injection analysis systems. *Analytica Chimica Acta*, **261**, 539.

46 Narusawa, Y. and Miyamae, Y. (1994) Zone circulating flow injection analysis: theory. *Analytica Chimica Acta*, **289**, 355–364.

47 Narusawa, Y. and Miyamae, Y. (1994) Radial dispersion by computer-aided simulation with data from zone circulating flow injection analysis. *Analytica Chimica Acta*, **296**, 129–140.

48 Narusawa, Y. and Miyamae, Y. (1995) Evidence of axial diffusion accompanied by axial dispersion with zone circulating flow injection analysis data. *Analytica Chimica Acta*, **309**, 227–239.

49 Narusawa, Y. and Miyamae, Y. (1998) Decisive problems of zone-circulating flow injection analysis and its solution. *Talanta*, **45**, 519.

50 Vanderslice, J.T., Stewart, K.K., Rosenfeld, A.G. and Higgs, D.H. (1981) Laminar dispersion in flow injection analysis. *Talanta*, **28**, 11–18.

51 Vanderslice, J.T., Beecher, G.R. and Rosenfeld, A.G. (1984) Dispersion and diffusion coefficients in flow injection analysis. *Analytica Chimica Acta*, **56**, 292–293.

52 Gomez-Nieto, M.A., Luque de Castro, M.D., Martin, A. and Valcarcel, M. (1985) Prediction of the behavior of a single flow injection manifold. *Talanta*, **32**, 319–324.

53 Kempster, P.L., van Vliet, H.R. and Staden, J.F. (1989) Prediction of FIA peak width for a flow injection manifold with spectrophotometric or ICP detection. *Talanta*, **36**, 969.

54 Tijssen, R. (1980) Axial dispersion and flow phenomena in helically coiled tubular reactors for flow analysis and chromatography. *Analytica Chimica Acta*, **114**, 71–89.

55 Reijn, J.M., van der Linden, W.E. and Poppe, H. (1981) Transport phenomena in flow injection analysis without chemical reaction. *Analytica Chimica Acta*, **126**, 1.

56 Andrade, F.J., Iñón, F.A., Tudino, M.B. and Troccoli, O.E. (1999) Integrated conductimetric detection: mass distribution in a dynamic sample zone inside a flow injection manifold. *Analytica Chimica Acta*, **379**, 99–106.

57 Iñón, F.A., Andrade, F.J. and Tudino, M.B. (2003) Mass distribution in a dynamic sample zone inside a flow injection manifold: modelling integrated conductimetric profiles. *Analytica Chimica Acta*, **477**, 59–71.

58 Golay, M.J.E. and Atwood, J.G. (1979) Early phases of the dispersion of a sample injected in poiseuille flow. *Journal of Chromatography*, **186**, 353–370.

59 Atwood, J.G. and Golay, M.J.E. (1981) Dispersion of peaks by short straight open tubes in liquid-chromatography systems. *Journal of Chromatography*, **218**, 97–122.

60 Poppe, H. (1980) Characterization and design of liquid phase flow-through detector system. *Analytica Chimica Acta*, **114**, 59–70.

61 Reijn, J.M., van der Linden, W.E. and Poppe, H. (1980) Some theroretical aspects of flow injection analysis. *Analytica Chimica Acta*, **114**, 105–118.

62 Valcárcel, M. and Luque de Castro, M.D. (1987) *Flow injection analysis: Principles and Applications*, Ellis Horwood, Chichester.

63 Coq, B., Cretier, G., Rocca, J.L. and Porthault, M. (1981) Open or packed sampling loops in liquid-chromatography. *Journal of Chromatographic Science*, **19**, 12–112.

64 Kolev, S.D. (1995) Mathematical modelling of flow inject systems. *Analytica Chimica Acta*, **308**, 36–66.

65 DeLon Hull, R., Malic, R.E. and Dorsey, J.G. (1992) Dispersion phenomena in flow injection systems. *Analytica Chimica Acta*, **267**, 1–24.

66 Brooks, S.H. and Dorsey, J.G. (1990) Moment analysis for evaluation of flow injection manifolds. *Analytica Chimica Acta*, **229**, 35.

67 Ramsing, A.U., Ruzicka, J. and Hansen, E.H. (1981) The principles and theory of high-speed titrations by flow injection analysis. *Analytica Chimica Acta*, **129**, 1.

68 Reijn, J.M., Poppe, H. and van der Linden, W.E. (1984) Kinetics in a single bead string reactor for flow injection analysis. *Analytical Chemistry*, **56**, 943–948.

69 Hartnett, M., Diamond, D. and Barker, P.G. (1993) Neural network-based recognition of flow injection patterns. *Analyst*, **118**, 347–354.

70 Hungerford, J.M. and Christian, G.D. (1987) Chemical kinetics with reagent dispersion in single-line flow injection systems. *Analytica Chimica Acta*, **200**, 1.

71 Tyson, J.F. and Idris, A.B. (1981) Flow injection sample introduction for atomic absorption spectrometry. Applications of a simplified model for dispersion. *Analyst*, **106**, 1125.

72 Burguera, J.L. (1989) *Flow Injection Atomic Spectroscopy*, Marcel Dekker, New York.

73 Betteridge, D., Marczewski, C.Z. and Wade, A.P. (1984) A random walk simulation of flow injection analysis. *Analytica Chimica Acta*, **165**, 227–236.

74 Crowe, C.D., Levin, H.W., Betteridge, D. and Wade, A.P. (1987) A random-walk simulation of flow injection sytems with

merging zones. *Analytica Chimica Acta*, **194**, 49.

75 Wentzell, P.D., Bowdridge, M.R., Taylor, E.L. and Macdonald, C. (1993) Random-walk simulation of flow injection analysis — evaluation of dispersion profiles. *Analytica Chimica Acta*, **278**, 293–306.

76 Levine, I.N.(1988) *Chapter 16*, McGraw-Hill, Madrid.

77 Iñón, F.A. (2001) in Un nuevo enfoque en el estudio del proceso de dispersión en FIA: el método conductimétro integral (ICM) y modelado matemático de las curvas ICM Vol. Ph. D. Universidad de Buenos Aires, Buenos Aires, p. 464.

78 Reijn, J.M., Poppe, H. and van der Linden, W.E. (1983) A possible approach to the optimization of flow injection analysis. *Analytica Chimica Acta*, **145**, 59.

79 Betteridge, D. (1978) Flow injection analysis. *Analytical Chemistry*, **50**, 832A–846.

80 Andrade, F.J. (2001) in Estudio de dispersión en sistemas de análisis por inyección en flujos (FIA) y su aplicación al análisis de vestigios Vol. Ph. D. Universidad de Buenos Aires, Buenos Aires, p. 381.

81 Leclerc, D.F., Bloxham, P.A. and Toren, E.C.J. (1986) Axial dispersion in coiled tubular reactors. *Analytica Chimica Acta*, **184**, 173.

82 Li, Y.H. and Ma, H.C. (1995) Two trends of sample dispersion variation with carrier flow rate in a single flow injection manifold. *Talanta*, **42**, 2033.

83 Stone, D.C. and Tyson, J.F. (1987) Models for dispersion in flow injection analysis. 1. basic requirements and study of factors affecting dispersion. *Analyst*, **112**, 515–521.

84 Fang, Z. (1995) *Flow Injection Atomic Absorption Spectrometry*, John Wiley & Son Ltd, New York.

85 Reijn, J.M., van der Linden, W.E. and Poppe, H. (1980) Dispersion phenomena in reactors for flow analysis. *Analytica Chimica Acta*, **114**, 91–104.

2
Injection Techniques in Flow Analysis

Víctor Cerdà and José Manuel Estela

2.1
Introduction

Sample injection in flow analysis involves the reproducible insertion of a carefully selected volume of sample into a carrier stream. Depending on the way such a volume is measured, injection can be volume-based or time-based.

In its most simple version, volume-based injection involves filling a sampling loop with sample and discarding the excess, the loop volume dictating the sample volume that can be injected. Loop-based injection is especially effective with small sample volumes. As the injected volume increases, however, the increased analytical path length resulting from the insertion of the sampling loop in the carrier stream causes a marked decrease in throughput.

Time-based injection techniques use a commutation device to switch between the streams to be processed. Thus, with the switch in one position the sample is driven to the detector and the carrier discarded. After the time required to insert the selected sample volume elapses, the switch is actuated so that the carrier solution is driven through the analytical path in order to sweep the sample and flush the system.

Time-based injection is particularly attractive for inserting large volumes as it uses samples more sparingly and results in less dispersion in the analytical path than does volume-based injection. With small sample volumes, however, the efficiency of time-based injection can be limited by flow rate fluctuations and the reproducibility of inserted volumes suffers when short injection times or high flow rates are used. Some flow analysis techniques are better suited to time-based injection, whereas others are more efficient with volume-based injection and still others are amenable to both, depending on their intrinsic characteristics and on available technological developments. A given injection technique can also be used in different ways with the same flow technique in order to develop new variants, facilitating the analytical processing of samples. The sensitivity, selectivity, accuracy, precision, throughput, sample and reagent consumption, waste production, robustness, monitoring capabilities, automatability, miniaturizability and ability to perform specific analyses or

Advances in Flow Analysis. Edited by Marek Trojanowicz
Copyright © 2008 WILEY-VCH Verlag GmbH & Co. KGaA, Weinheim
ISBN: 978-3-527-31830-8

multideterminations of a given flow technique depend largely on the injection mode it uses. The following sections describe some of the more common injection techniques implemented in the major flow analysis techniques.

2.2
Continuous Flow Analysis (CFA)

In continuous flow analysis, the sample is continuously inserted into the manifold via a tube of the propelling system for continuous monitoring of its temporal evolution. The CFA technique can be implemented in both open and closed configurations (Figure 2.1).

In open CFA configurations, the sample is discarded after passing through the detection system. This allows the incorporation of several channels holding appropriate reagents for merging with the sample in order to facilitate detection [1]. Usually, only one sample at a time is monitored for temporal changes. Using a single

Open system

Evolving system   **Peristaltic pump**

Detector

Waste

Reagent

Closed system

Evolving system   **Peristaltic pump**

Detector

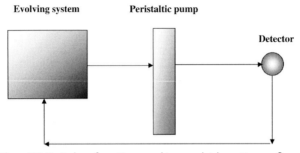

Figure 2.1 Typical configurations used in completely continuous flow analysis.

sample avoids the need for flushing cycles and makes these configurations especially suitable for monitoring waters or industrial effluents, where the sample is typically abundant and inexpensive, and requires continuous control.

Closed CFA configurations are also used to monitor a single sample at a time; however, the sample is not discarded on passage through the detector, but rather returned to the monitored system (i.e., recirculated). This precludes the use of additional reagent channels in order not to contaminate or disturb the monitored evolving system. However, it is commonplace to use two phases and monitor one of them following passage through an appropriate phase separator [2].

2.3
Segmented Flow Analysis (SFA)

Segmented flow analysis is an automatic continuous technique developed by Skeggs in 1957 [3]. An SFA system typically includes a peristaltic pump to continuously aspirate the sample and reagents, a variable number of tubes constituting a manifold to circulate liquids and a detector (Figure 2.2). Aspirated samples are segmented by injecting air bubbles that should be removed before they can reach the detector. The insertion of air bubbles was initially intended to prevent sample carry-over and dispersion of the sample plug, and also to help establish a turbulent flow regime in order to facilitate homogenous mixing of the sample and reagents in between each pair of bubbles. From the beginning, however, bubbles were found not to completely avoid carry-over, which necessitated the insertion of an intermediate washing solution. The sample, air and the intermediate washing solution were inserted into the flow system by means of an aspirating tip in such a way that each sample plug was sandwiched between two air bubbles; these were in turn separated by the intermediate washing solution used to flush any residues of the previous sample from the tube walls. Once the air bubbles were removed, each segment was separated by its neighboring washing solutions. A square-shaped signal results for equilibrium type measurements, where sufficient time is allowed to obtain a steady level of signal. The

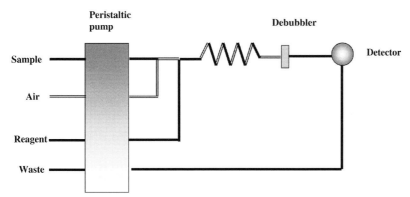

Figure 2.2 Typical segmented flow analyser.

height of the rectangle is proportional to the analyte concentration provided that the reagents are always present in overstoichiometric amounts. This multisegmented system contrasts with the so-called monosegmented (MSFA) system [4], a rather exceptional and not typical one. From the chemical point of view, monosegmented injection allows determinations to be performed under non-steady state conditions, ensured by the high timing reproducibility found in the monosegmented system due to the presence of only two gas bubbles.

2.4
Flow Injection Analysis (FIA)

2.4.1
Syringe-based Injection

The term "flow injection analysis" was coined in 1975 by Ruzicka and Hansen, the developers of this technique [5]. While, initially, it resembled segmented flow analysis both conceptually and practically, it has provided rather different results. Thus, FIA is based on a combination of three different principles, namely: sample injection, controlled dispersion of the injected sample zone and reproducible timing. Therefore, the chemical reaction takes place while the sample material is dispersing within the reagent (i.e., while the concentration gradient of the sample zone is being formed by the effect of dispersion). Otherwise, the components of an FIA system are essentially the same as in SFA and include a peristaltic pump to propel the sample and reagents, an array of tubes to drive the liquids to the detector and a device for injecting or inserting samples (Figure 2.3). Unlike SFA, where the sample is aspirated in a continuous manner, in FIA a constant volume of sample is injected into a liquid stream (the carrier) and successively merged with the reagents needed by the particular analytical method. The reaction time is dictated by the tube length and

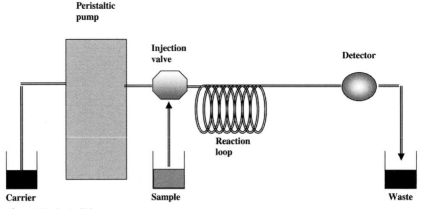

Figure 2.3 Basic FIA system.

the rotation speed of the peristaltic pump. If a long reaction time is needed for kinetic reasons, a very long, usually coiled, tube is inserted in order to prolong the residence time as required. Unlike SFA, FIA operates under a laminar rather than turbulent flow regime; moreover, FIA requires no air bubbles to separate the samples, so the flow is unsegmented and no physical or chemical equilibrium need be reached by the time the measurements are made. In early FIA work, samples were in fact injected into a carrier stream by using a syringe furnished with a hypodermic needle – hence the word "injected" continues to be widely used in connection with FIA, even though "insertion" is certainly more accurate given the types of devices now used to introduce samples.

2.4.2
Injection with Rotary Valves

The rotary valve, also known as a "six-way valve" or "rotary hexagonal valve," is among the most widely used devices, in combination with a peristaltic pump, to introduce samples into FIA systems. A rotary injection valve consists of six ports internally connected in pairs, three acting as inlets and three as outlets. The ports are externally connected to the manifold tubes holding the sample and carrier. The valve can be switched between two positions usually designated "filling" (or "fill") and "injection" (or "inject") (see top of Figure 2.4).

In the filling position, the sample is circulated in a continuous manner through a channel of the peristaltic pump connected to an inlet port of the rotary valve and passed through a piece of tubing of short, known length (the *sample loop*, which is

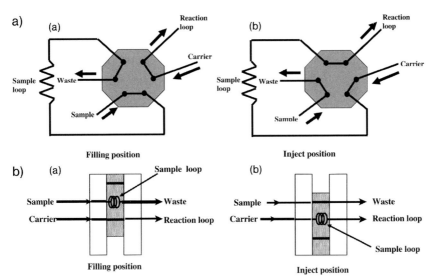

Figure 2.4 (A) Rotary injection valve in the filling (a) and injection positions (b). (B) Proportional injection valve in the filling (a) and injection position (b).

used to connect an inlet port and an outlet port of the valve) and then evacuated via another piece of tubing connected to an outlet port of the valve. Simultaneously, another channel of the peristaltic pump, connected to a valve inlet port, continuously circulates the carrier via another tube fitted to an outlet port (the *reaction coil*) and leading to the detector flow-cell and on to waste.

In the injection position, the valve ports are internally connected in a different manner. Thus, the carrier flows through the sample loop and sweeps the volume sample held in it to the detector via the reaction loop. Simultaneously, the channel holding the sample is directly connected to the outlet port and its contents are sent to waste without passing through the sample loop.

This device provides a simple, expeditious, highly reproducible means of injecting an appropriate volume of sample into an FIA system. It also facilitates automatic injection. However, changing the injected sample volume requires the use of several sample loops of different volumes, switching between them when needed. In addition, special care must be taken to efficiently flush the system and remove any residue of the previous sample before the next is injected in order to avoid inserting air bubbles.

2.4.3
Proportional Injection

Proportional injection was developed by Krug *et al.* [6] While similar to a rotary valve, it has been less widely used and only in connection with the so-called "zone-trapping" mode in FIA [7] and with monosegmented flow analysis [4, 8].

A proportional injector (see bottom of Figure 2.4) consists of three drilled polyethylene or Perspex blocks of which two are fixed and the other can be placed in two different positions for filling and injection, respectively. In the filling position, a channel of the propulsion system (usually a peristaltic pump) continuously circulates the sample through a sample loop and a piece of tubing leading to waste; simultaneously, the other channel is used to circulate the carrier via another loop and a piece of tubing leading to the detector flow cell and subsequently to waste. In the injection position, the sample loop is placed in the position of the carrier loop and the sample inserted into the system for transfer to the detector; simul-taneously, a conduit present in the mobile block is placed in the position of the sample loop in order to drive the sample stream to waste. Proportional valves are prone to leakage at the contact surfaces between blocks.

2.4.4
Merged Injection

This way of introducing the sample and reagents into a flow system allows their economical use and is therefore especially attractive with expensive or polluting reagents, and also with any types of substances to be used sparingly or when producing large amounts of waste is to be avoided. A merged injection system typically consists of two injection valves that are used to load the reagent and sample, respectively; in this way, the two are simultaneously injected into two carrier channels for subsequent merging. This facilitates mixing and reaction development, and saves sample and reagent [9, 10].

2.4.5
Injection Following a Prior Flow Sample Processing

This injection mode, by which the sample is treated in some way before it is inserted into the flow system, has enabled the development of FIA multiparametric determination systems operating in a sequential manner and requiring no alteration in order to perform the planned analyses. It is also employed to increase selectivity, dilute the analyte in concentrated samples and for expanding field analysis to gas samples.

2.4.5.1 Multiparametric Determination

This approach has been used to determine nitrates, nitrites and total nitrogen in waste water [11]. The experimental set-up consists of two peristaltic pumps located in two distinct parts of the configuration (Figure 2.5). The upper part is used to select the way the sample is inserted, whereas the lower part is used for measurement. The sample is inserted by actuating the switching valve VS1, then it is passed through a C_{18} resin cartridge in order to remove potentially interfering colored organics. The injection valve (VI) is filled with sample. When the sample is injected and VS2 is in its H_2O position, the sample is mixed first with water and then with Shinn's reagent (R3), the nitrite ions form a dye with Shinn's reagent which can be spectrophotometrically monitored in order to quantify the nitrites. If the process is repeated in such a way that VS2 is now used to aspirate hydrazine (R2) instead of water, the nitrates are reduced to nitrites by the hydrazine, the reaction being accelerated by the

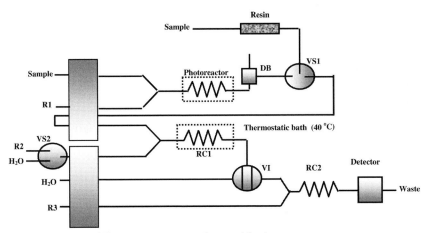

Figure 2.5 Injection following reaction as implemented for the determination of nitrates, nitrites and total nitrogen in waste water. DB, debubbler; VS1 and VS2, switching valves; IV, injection valve; R1, persulfate reagent; R2, hydrazine reagent; R3, Shinn's reagent; RC1 and RC2, reaction coils.

use of a thermostatic bath. Nitrite ions originally present in the sample and those resulting from the reduction of nitrates react with Shinn's reagent to form the dye, the concentration of which will be proportional to the combined nitrate and nitrite contents of the sample. Finally, if the sample is aspirated by switching VS1 to the other position, then it will be mixed with persulfate reagent (R2) and the nitrite ions it contains are photooxidized to nitrate ions by the action of the lamp. The debubbler removes bubbles formed in the decomposition of persulfate ion and the mineralized sample is used to fill the loop of the injection valve. The total nitrogen content of the sample can be determined by injecting the treated sample with VS2 in its hydrazine aspiration position.

2.4.5.2 Dialysis

Dialysis is the process by which a semi-permeable membrane is used to separate chemical species between two solutions. Whether a species can cross the membrane depends on its molecular size and on the pore size of the membrane. Initially, this sample insertion system was commonly used in clinical applications of segmented flow analysis; now, however, dialysis units are used as cells or probes in a number of FIA applications, the latter affording direct sampling by immersion in the medium to be analyzed. In these FIA systems, the sample is circulated through a channel connected to the dialyser, which can be placed in different parts of the manifold, depending on the particular purpose. The dialysis cell comprises two blocks, each of which is engraved with a semi-tubular channel. The channel in one of the blocks is the mirror image of that in the other block, so the two form a single, tubular channel when the two blocks are joined. The membrane, which can be of various types (neutral, ion-exchange), is sandwiched between the two blocks in order to hinder passage of some compounds from one cavity (the donor channel) to the other (the acceptor channel) and the blocks are tightly screwed together. Once the analyte has passed through the membrane and been collected by the solution in the acceptor channel – which can be still or in motion– it is carried through the manifold for analytical processing. This analyte introduction system is usually employed to increase selectivity by isolating the analyte from potential interferents present in the sample; also, the low efficiency of the dialysis process can be used to dilute the analyte in highly concentrated samples.

Figure 2.6 illustrates the use of this introduction system to feed the sample loop of an injection valve in an FIA system for the analysis of nitrates and nitrites using the Griess reagent to form a dye and hydrazine to reduce nitrates to nitrites.

A special dialyser design has been used by Koropchak and Allen [12] for Donnan dialysis through ion-exchange membranes. By replacing the sample injection loop in an injection valve with a Nafion 811 cation-exchange tubular membrane containing the acceptor stream, they succeeded in preconcentrating samples for detection by atomic absorption spectrophotometry (AAS) in an FIA system.

2.4.5.3 Gas Diffusion

Samples can be introduced into a flow system by having volatile species diffuse across permeable membranes. This has facilitated the determination of analytes in gases

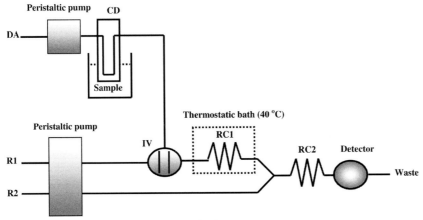

Figure 2.6 FIA set-up using a dialysis probe for the spectrophotometric determination of nitrates and nitrites with the Griess reagent. R1, hydrazine reagent; R2, Griess reagent; RC1 and RC2, reaction coils; DA, acceptor solution; CD, donor solution (sample); IV, injection valve.

and solutions by using separation units similar to those described in the previous section.

Analytes contained in gaseous samples pass through the membrane and are absorbed by the acceptor solution. Frenzel *et al.* [13] have used a modified gas separation unit for the *in situ* spectrophotometric determination of atmospheric nitrogen dioxide. The sensor uses a microporous polypropylene membrane which affords analyte collection and trapping in a temporarily halted liquid absorber and *in situ* spectrophotometric detection with fiber optics of the product of the reaction between NO_2 and a mixed reagent containing sulfanilamide and N-(1-naphthyl)-ethylenediamine.

A modified separation unit intended to facilitate detection simultaneously with separation has also been used for the determination of ammonia in the gas and liquid phase by diffuse reflectance spectroscopy, using a Celgard 2500 hydrophobic membrane and an integrated fiber optic detector [14]. Similarly, chromatomembranes produced from porous hydrophobic PTFE have been used for the spectrophotometric determination of SO_2, as have tubular silicone membranes accommodated in the sample loop of an injection valve to monitor HCN traces in process gas streams [15].

With liquid samples, the analyte released or on-line transformed into a volatile species (e.g., CO_2, NH_3, SO_2, sulfide, O_3, HCN, NO, iodide, bromide, chlorine, acetone, ethanol, hydride-forming elements) passes through the membrane and is collected by the acceptor solution, where it can be reacted with an appropriate reagent in order to facilitate detection. Most often, the detector is a spectrophotometer [16, 17], but other detectors such as conductimetric [18], potentiometric [19, 20], amperometric [21, 22], chemiluminescence [23, 24] or cold vapor atomic absorption spectrophotometric (CV-AAS) may also be used [25]. The optimum choice of detector in each case is dictated by the nature of the analyte, and the sensitivity and selectivity levels required.

2.4.5.4 **Pervaporation**

Pervaporation combines evaporation of the analyte or a volatile reaction product from the sample matrix, whether solid or liquid, with gas diffusion through a membrane to a still or flowing acceptor solution [26–28]. Pervaporation coupled with flow injection has been used for the direct measurement of volatile and semi-volatile analytes in samples that may damage the gas-permeable membranes used in FI gas-diffusion systems [29]. The use of an air gap between the donor solution (sample) and the membrane avoids contamination or deterioration of the latter in FI pervaporation systems.

The earliest pervaporation unit was made of polystyrol and developed by Prinzing *et al.* [30]. and was subsequently superseded by one made of aluminum and furnished with a temperature controller [31]. Some, more recent units are made of methacrylate [32]. Figure 2.7 shows a pervaporation unit consisting of a chamber for the donor solution, to which the sample is driven either continuously or by injection into a carrier if it is liquid and by direct weighing if it is solid; an upper chamber where an appropriate acceptor solution for the analytes is continuously circulated or halted; a membrane support and spacers of variable thickness to adjust the chamber volumes. The body of components is usually held between two blocks of aluminum, steel or another appropriate material which are tightly fitted by means of screws, clamps or some other fastening device. With slight modifications, these units have been used for various purposes such as monitoring evolving systems [33], facilitating handling of corrosive samples or reagents [34, 35], integrating detection with a view to monitoring the kinetics of the separation process [36], affording the analysis of solid samples [37–39], allowing the use of microwaves to improve performance and/or facilitate the sequential determination of species of disparate volatility [40], and also as an alternative to headspace sampling [41, 42].

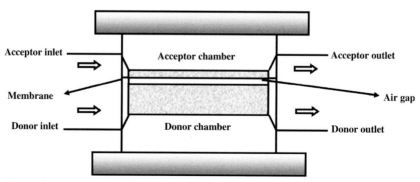

Figure 2.7 Conventional pervaporation module.

2.4.6
Hydrodynamic Injection

Hydrodynamic injection, which was developed by Ruzicka and Hansen in 1983 [43], has been used for various purposes the most interesting of which is probably introducing samples into capillary electrophoresis (CE) systems [44]. Briefly, a small fraction of sample was hydrodynamically injected into the capillary as closing the outflow of an interface channel resulted in temporary overpressure. This will be discussed in a later chapter.

**2.5
Sequential Injection Analysis (SIA)**

The SIA technique was developed as an alternative to FIA by Ruzicka and Marshall in 1990 [45]. Time has shown that SIA has a rather different scope and that it circumvents the two greatest shortcomings of FIA, namely: the need to use a different manifold to implement each analytical method and the need to use peristaltic pumps, which are expensive to maintain. In fact, SIA is a simple, powerful, robust, flexible, easily automated tool that is especially useful for handling liquids. Its advantages have been discussed in several reviews, especially interesting among which is that by Barnett *et al.* [46]. SIA has been widely used in combination with UV–Vis spectrophotometric, electrochemical, spectroscopic and luminescence detection for purposes such as bioprocess monitoring, immunoassay, and environmental, food and drink, drug, industrial and chemometric analysis.

2.5.1
Original Procedures

The earliest experiments towards the development of the SIA technique were conducted at the Analytical Chemistry Processing Center (ACPC) of the University of Washington using a disposable syringe mounted on a round cam, the body of which accommodated fiber optics and operated as both a pump and a detection cell. Each stroke introduced fresh reagents and sample into the syringe that were mixed by the resulting turbulence and monitored via absorbance measurements. The initial placement of the detector within the syringe was soon abandoned as loading the reagents from small-bore (0.8 mm) tubes was found to be more effective than using the syringe bodies for this purpose. A sinusoidal flow pump was used instead, the detection cell being placed between the pump and the switching valve (SV), which were connected with a long enough piece of tubing in order to prevent the reagents from reaching the pump. By sequentially aspirating the reagents through the detection cell, a first signal was obtained that was supplemented by a second, broader, shorter signal obtained as the flow direction was reversed in order to send the sample to waste.

2.5.2
Conventional Injection

The results obtained with the earliest procedures led researchers to move the detection cell to the other side of the switching valve in order to obtain a single peak. Although, initially, research focused on the liquid driver, the switching valve soon revealed itself as the truly innovative element. In fact, connecting a central port of the valve to some of its side ports allowed the sample to be injected and mixed with the reagents in a manner differing from the original procedure. Thus, sample and reagent volumes were stacked in a specific sequence within a piece of tubing (the holding coil, HC) connecting the liquid driver with the central port of the SV and mixed to give detectable products by the effect of diffusion and reversal of the flow. Figure 2.8 depicts a typical SIA system. The side ports of the valve were connected to vessels containing the reagents needed for implementation of the particular analytical methods, the sample and the measuring detector. The side ports can also be connected to waste or to a device such as microwave oven, photooxidation unit or mixing chamber. This sample introduction system is obviously time-based and, although it initially used a piston pump, it has also employed peristaltic pumps [47].

2.5.3
Controlled Variable Volume Injection

This way of introducing samples into an SI system was first implemented by Cladera *et al.* [48]. It involves using computer-controlled burettes equipped with step motors

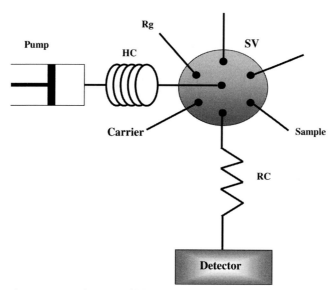

Figure 2.8 Typical SIA manifold including a piston pump, six ports switching valve (SV), holding coil (HC), reaction coil (RC), reagent channel (Rg), sample channel, carrier channel, two free channels and detector.

to actuate the syringe pistons as two-way liquid drivers. The burettes are usually employed in automatic titrations. Appropriate volumes are aspirated into the holding coil through one of the ports of the switching valve by advancing the syringe piston a given number of steps – the aspirated volume is a function of the number of steps and the syringe capacity. Use of a step motor additionally allows flow rates spanning very wide ranges from that required to completely unload the syringe in 20 s (75 mL min^{-1} with a 25 mL syringe) to extremely low values to be controlled and programmed. At present, burette syringes can be unloaded in up to 40 000 steps (i.e., $10 000/40 000 = 0.25\ \mu L$ at a time for a 10 mL syringe and $500/40 000 = 0.0125\ \mu L$ at a time for a 0.5 mL syringe). This sample injection approach provides an advantageous alternative to typical SIA liquid drivers such as piston pumps that are difficult to operate or exhibit a non-linear flow pattern, and also to peristaltic pumps. By contrast, peristaltic pumps are more commonplace in the laboratory and their sampling cycle is substantially shorter than that of piston pumps as they require no periodic replenishment; however, the fragility of their tubing hinders the use of aggressive fluids and causes oscillations in flow rates that call for frequent recalibration in order to ensure accurate handling of volumes.

2.5.4
Cumulative Injection

This injection mode allows reaction products to be preconcentrated by using large sample volumes in combination with the optimum experimental conditions required for maximal development of a reaction in a much lower volume. The approach was originally conceived by Miró *et al.* [49] for the spectrophotometric determination of nitrites in natural water using on-line solid-phase extraction and preconcentration. The method involves the on-line formation of an azo dye and its subsequent extraction on a solid phase consisting of monofunctional C_{18} material which is held in a glass column incorporated into the system (Figure 2.9). A large sample volume (1 mL) is sequentially segmented with sulfanilamide and N(-naphthyl)ethylenediamine dihydrochloride by using an iterative method. The resulting azo dye is propelled to the C_{18} mini-column for retention. The segmentation-retention process is repeated up to 10 times (i.e., up to 10 sample injections are accumulated on the mini-column). Following cumulative retention, the azo dye is eluted with a small volume of 80% methanol for detection with a diode array spectrophotometer. The principal shortcoming of this method is that the preconcentrated volume cannot exceed 10 mL owing to the pre-elution effect (analyte breakthrough) present.

2.5.5
The Sandwich Technique

The sandwich technique is an attractive, convenient operational mode with a view to determining two parameters simultaneously by FIA [50, 51] or SIA. The sample is sandwiched between two reagents each of which is used to determine a different parameter. Usually, the sample is large enough to prevent the reagents on each end

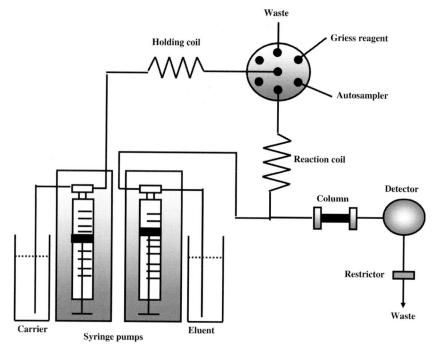

Figure 2.9 SIA set-up for cumulative injection.

from interfering with their mutual determinations. This allows two parameters to be simultaneously determined with a single injection.

One typical example of SIA sandwich systems is that used in the simultaneous determination of Fe(III) and nitrites in water [52]. The reaction ingredients are aspirated in the following sequence: o-phenanthroline (a chromogenic reagent for iron), sample (in a large volume) and the Griess reagent (which forms a dye with nitrite ion). Then, the contents of the loading coil are driven to a photometric detector. A CCD-based detector allows each species to be measured at its optimum wavelength plus another, allowing the Schlieren effect to be corrected. The nitrite peak is very sharp as a result of its product being the first to reach the detector. By contrast, the large sample volume used results in increased dispersion in the Fe(III) peak, which delays overlap with the nitrite peak.

One other example is the SIA determination of nitrate and nitrite in water [53] using a copperized cadmium column (Figure 2.10). By inserting the column in the loading coil and placing the coil at a long enough distance from the switching valve, the system can be used to aspirate the Griess reagent first and a large volume of sample next in order to ensure that one end will penetrate the reducing column while the other remains outside. Finally, the Griess reagent is aspirated and the whole contents of the loading coil are driven to the detector. In this way, the portion of sample not entering the column gives a peak due to the dye formed with nitrite ions in the sample, followed by another resulting from the reaction of

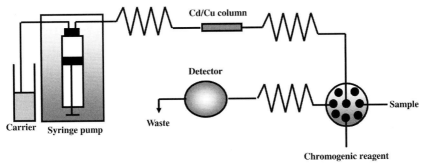

Figure 2.10 SIA system for the determination of nitrates and nitrites using the sandwich technique.

nitrite ions in the sample and those formed by reduction of nitrate ions entering the column.

2.5.6
Multiparametric Analysis

By virtue of its operational foundation, SIA is one of the most suitable flow techniques for multiparametric analysis. In fact, a switching valve furnished with an adequate number of side ports can be used to aspirate the sample and as many reagents as are required to sequentially determine several parameters. Figure 2.11 shows an SIA system for monitoring waste water [54] that affords the sequential determination of up to 12 different parameters with a view to characterizing the incoming and outgoing flows of a water purifier. First, the sample is aspirated via a port of the upper switching valve and driven to a diode array detector connected to a port of the lower valve; following application of a multivariate analysis method, the values of seven major parameters, including BOD, detergents, nitrate and TPS can be determined without the need for a chemical reagent. Subsequently, previously optimized individual methods are implemented by using appropriate reagents delivered via the switching valves in order to determine other major parameters such as the ammonium content (with a gas diffusion cell), nitrite (using the modified Griess reagent), total nitrogen (by photooxidation with persulfate), orthophosphate (by the Molybdenum Blue reaction) and total phosphorus (by photooxidation with persulfate and formation of Molybdenum Blue).

2.5.7
Gas Diffusion

Gas diffusion in SIA is implemented by using units similar to those described in Section 2.4.5 for FIA. In FIA, however, the donor and acceptor solutions are simultaneously propelled to the opposite sides of the membrane, so the pressure exerted by each solution on one side is cancelled by the other. By contrast, gas diffusion in SIA is usually accomplished by using a single propulsion channel, so the

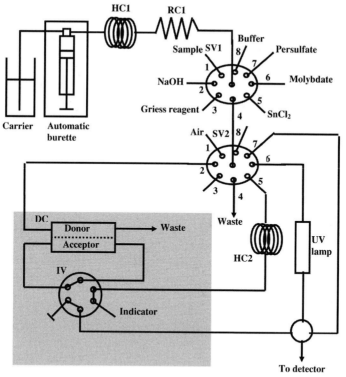

Figure 2.11 SIA multiparametric system for monitoring an urban waste water treatment plant. The system can determine the following parameters in the water inflow and outflow of the treatment plant in cycles of 20 min: (i) BOD, COD, TOC, detergents and total suspended particulates; and (ii) ammonium ion, nitrates, nitrites, total nitrogen, orthophosphate and total phosphorus. The former group of parameters can be determined in a direct manner by deconvoluting the spectrum for the waste water and the latter by using appropriate reagents. IV, injection valve; DC, dialysis cell; SV1 and SV2, switching valves; HC1 and HC2, holding coils; RC1 and RC2, reaction coils.

acceptor stream must remain still while the donor solution is propelled on the other side of the membrane. This generates overpressure on the side of the donor solution and, by virtue of the membrane flexibility, can cause the channel carrying the acceptor solution to empty unless an effective measure is taken. Alternatively, a wide outlet tube can be fitted to the donor channel in order to reduce overpressure in it and a narrow-bore one used for the acceptor channel so as to prevent it from emptying.

The highlighted part of Figure 2.11 shows a different solution involving the use of a rotary valve in combination with colorimetric detection of an indicator. The process is started by aspirating the indicator (Bromothymol Blue) in its acid form; this involves switching the injection valve, as shown in Figure 2.11, and connecting the central port of the switching valve to its side port no. 5. Then, the injection valve is actuated in order to fill the acceptor channel with indicator by delivering carrier from the burette. Once the acceptor stream is conditioned, the switching valve is actuated in order to

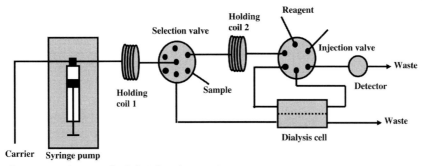

Figure 2.12 SIA system for dialysis-based separation.

aspirate NaOH and sample, which are circulated through the donor channel – where ammonium ion will have already been converted into ammonia and transferred to the acceptor stream. The fact that the SIA system is computer controlled facilitates the conduct of additional operations leading to improved sensitivity. Thus, after the sample is propelled through the donor channel, the flow can be reversed in order to push it back through the cell; in this way, the sample is passed three times across the diffusion membrane rather than once, which increases the amount of ammonia diffusing through it – more than two flow reversals result in no substantial further improvement in sensitivity, however.

2.5.8
Dialysis

Figure 2.12 illustrates a simple, practical, interesting use of SIA in combination with dialysis. The sample is dialysed on-line to remove polymeric substances in order to avoid their interference with subsequent UV spectrophotometric measurements. The system consists of an autoburette, a valve module including an injection valve and a switching valve, and a spectrophotometer. The manifold uses two loading coils, two waste outlets and a dialysis cell comprising two PTFE blocks which have symmetric winding channels and hold an intervening membrane in place. Optimum operation of this system relies on appropriate selection of the selection valve (SV) ports, the fill and inject positions of the injection valve (IV) and the movements of the autoburette piston. The acceptor channel of the dialysis cell constitutes the sample loop of IV. With this valve in its filling position, the sample loop is loaded with acceptor solution held in a channel connected to one of the IV ports following aspiration through an SV port connected to holding coil 2 (HC2). Then, IV is switched to its inject position in order to propel a volume of sample previously aspirated into HC1 via the corresponding SV port through that connected to the donor channel of the dialysis cell; in this way, the analyte is transferred across the dialysis membrane. Once the sample volume has been delivered, SV is switched to the port connected to HC2 in order to drive the acceptor solution to the detector.

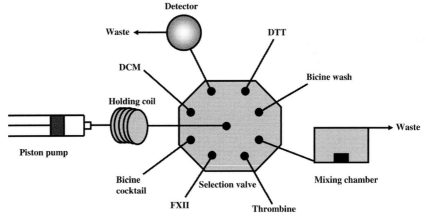

Figure 2.13 Single pump–single valve SI configuration for mixing chamber-based injection used to determine factor thirteen.

2.5.9
Mixing Chamber-Based Injection

This injection mode was developed by Guzman *et al.* [55] for the automation of complex analytical procedures by SIA. The fluorimetric assay of factor thirteen (FXIII), an important enzyme which plays a major role during the final stages of blood coagulation, requires the injection of six different zones into the holding coil and subsequent mixing in a chamber at specific stages of the analysis sequence (Figure 2.13). In the absence of a mixing chamber, it would have been difficult and impractical to stack so many reaction zones in the holding coil and still ensure proper mixing with the sample. Connecting a mixing chamber to one of the ports is the most direct solution. The wash, sample and appropriate reagent solutions are sequentially aspirated in a specific sequence into the holding coil and then transferred to the mixing chamber at the required time for each analysis. In order to facilitate reaction, the contents of the mixing chamber must be gently mixed. After the fluorescence product starts to form, a number of aliquots are withdrawn from the chamber at preset times and propelled through the detector in order to obtain quantitative and kinetic information about the rate of formation of the product.

2.5.10
Bead Injection

This technique involves injecting microbeads into a SIA manifold bearing an appropriate interactant immobilized on their surface in order to construct a micro-column within a flow cell (usually a jet ring cell) that is used for both sample injection and detection. Because the microcolumn can be renewed after each analysis, the use of beads as a growth support does not compromise the integrity of the results [56, 57]. The inception of Lab-on-Valve (LOV) systems [58] facilitated the miniaturization of

SIA and the expansion of bead injection applications. LOV systems are SIA systems where detection can be performed inside the manifold [59, 60] (on valve mode) or outside [61–63] (off valve mode). Bead injection analysis is the subject of a later chapter.

2.5.11
Hydrodynamic Injection

This injection mode has also been implemented in SIA-LOV systems [64], by coupling a Lab-on-Valve to capillary electrophoresis equipment (LOV-CE). The chamber constituting the FIA-CE interface described previously is replaced with a multipurpose cell. The overpressure needed for injection is obtained by using a configuration where the flow outlet is plugged by means of an injection valve. Buckhard *et al.* have proposed an SIA-CE system for the analysis of nitrophenol isomers [65].

2.6
Multicommutated Flow Injection Analysis (MCFIA)

The MCFIA technique, devised by Reis *et al.* [66], relies on the use of fast-switching three-way solenoid valves. Its conceptual foundation, applications and trends are examined in an interesting review elsewhere [67]. The earliest MCFIA systems used a single-channel propulsion device to aspirate the liquids to be employed via individual valves. Because aspiration devices tend to insert air bubbles or degas liquids in the system, it is preferable to use liquid propulsion devices such as peristaltic or piston pumps instead. One major shortcoming of solenoid valves is the adverse effect of heat released by solenoid coils when valves are kept ON for a long time. The resulting temperature rise can deform the Teflon inner membranes of the valves and render them unusable. Overheating here can be avoided by using an effective electronic protection system. The MCFIA technique shares some advantages of FIA including increased throughput – which is higher than that for SIA – and reduced consumption of reagents – a result of unused sample and reagents being returned to their reservoirs for reuse. On the other hand, MCFIA also shares one major disadvantage with FIA (viz. the vulnerability of pump tubing to aggressive reagents and, especially, solvents).

2.6.1
Time-based Injection

This is the most common way of introducing measured volumes of sample and reagents into an MCFIA system, and relies on the main operational feature of this technique: multicommutation. The volumes to be inserted are introduced at a preset flow rate by using appropriate valves [68]. Figure 2.14 depicts an MCFIA manifold consisting of three switching valves (V_1–V_3). Each valve can be switched between two positions connecting it to the analytical processing path (solid lines) and the fluid

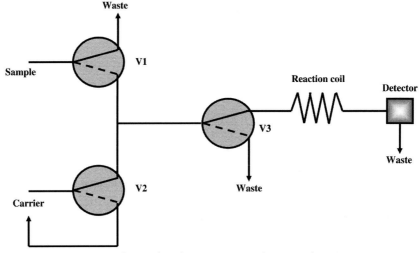

Figure 2.14 MCFIA set-up for time-based injection. V_1–V_3, three-way valves.

recycling or discarding path (dotted lines). By simultaneously switching V_1 and V_2 to the position defined by the dotted line, the sample is introduced into the analytical path and the carrier solution recycled. After a preset interval which determines the amount of sample injected, V_1 and V_2 are switched back to their initial position and as a result the sample is propelled to the detector by its carrier stream.

2.6.2
Tandem Streams

Tandem streams have been used for time-based injection in a number of ways which, according to Rocha *et al.* [67], have been given names such as "tandem injection," "binary sampling," "multi-insertion principle" or "tandem flow." The sample and reagent are inserted as aliquots by rapidly and sequentially switching the commutators (usually computer-controlled valves in a manifold such as that of Figure 2.15). This unique stream can be seen as a set of neighboring solution slugs that undergo

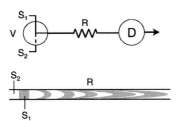

Figure 2.15 Establishment of a tandem stream. S_1 and S_2, miscible solutions; V, three-way valve; R, reaction coil; D, detector.

fast mixing while being transported through the analytical path. With a constant total volume of sample and reagent, mixing can be boosted by decreasing the aliquot volumes and increasing the number of slugs. This approach was first used to develop a single-channel manifold for the spectrophotometric determination of several parameters in natural waters [69]. Subsequently, it has been used for sample dilution prior to sequential ICP-OES or ICP-MS [70], and also to improve sample–reagent mixing in spectrophotometric analysis [71] with a view to minimizing the effects of differences in refractive index [72].

2.7
Multisyringe Flow Injection Analysis (MSFIA)

This variant of FIA was developed in 1999 [73]. An MSFIA system (Figure 2.16) consists of a conventional automatic titration burette adapted in such a way that the motor can simultaneously move the pistons of four syringes, thereby avoiding the need to have separate burettes operated in parallel. This is accomplished by using a metal bar that is moved by the step motor of the burette, the bar accommodating the four syringes and each syringe head containing a fast-switching solenoid valve. The motor moves the pistons of the four syringes at once, which is equivalent to using a multichannel peristaltic pump in FIA but avoids the disadvantages of its fragile tubing. The flow rate ratio between channels can be changed by using

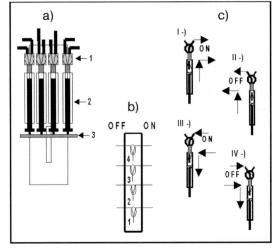

Figure 2.16 Photograph of a multisyringe burette and (A) 1, three-way solenoid valve; 2, syringes; 3, piston driver bar. (B) Schematic depiction of the delivery system and multisyringe burette. (C) Flow directions. ON connection with manifold fines; OFF connection with reservoirs or waste.

syringes of appropriate cross-sectional dimensions similarly to tubing diameters in FIA.

MSFIA systems combine some of the advantages of the above-described flow techniques. Thus, incorporating the sample and reagents in parallel results in improved mixing efficiency, a high throughput, increased robustness, reduced sample and reagent consumption, a high flexibility in designing manifolds and the ability to use MCFIA solenoid valves – which can be actuated without the need to stop their pistons.

2.7.1
Injection with Rotary Valves

Injecting the desired volumes in MSFIA entails using a module including a conventional FIA rotary injection valve, which allows the delivered volume to be adjusted via the dimensions of its loop. The process leading to the obtainment of an analyte peak (Figure 2.17) involves two steps. In the first, the syringe pistons are lowered in order to aspirate the sample and reagents. While the solenoid valve for syringe 1 (the leftmost one), which is used to deliver the sample, is ON (left), the others are OFF (right). While the injection valve is in its fill position, syringe 1 aspirates the sample through the loop defining the volume to be injected. Simultaneously, the other syringes aspirate the carrier and reagents. Once the two are loaded, the second step is started by reversing the direction of the pistons. First, the injection valve is switched ON to drive excess sample to waste and the valve of syringe 2 is switched ON to have the carrier flush the sample held in the loop into the manifold.

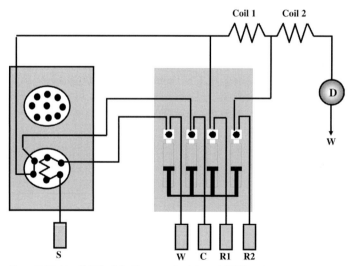

Figure 2.17 Manifold for injecting samples into an MSFIA system by means of a rotary valve. S, sample; C, carrier; R1 and R2, reagents; D, detector; W, waste.

The valves of syringes 3 and 4 are kept OFF in order to return reagents R_1 and R_2 to their respective containers. Immediately before the sample merges with R_1, the valve of syringe 3 is switched ON in order to have the two liquids mix. Such a valve is switched OFF after the sample passes by the merging point and the procedure is repeated with R_2.

2.7.2
Time-based Injection

This injection mode has been made possible by the ability to implement multi-commutation with a multisyringe module, whether with valves mounted on syringe heads and/or with independent valves controlled via the module. Figure 2.18 shows a simple MSFIA manifold using a switching valve (SV) and the steps involved in

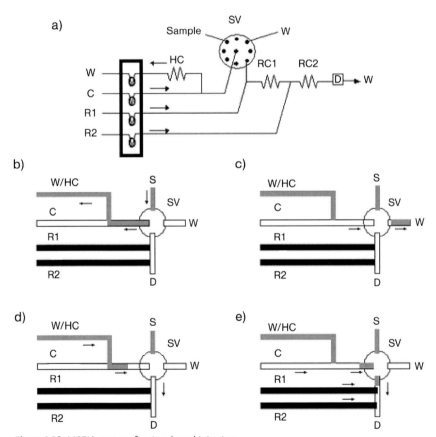

Figure 2.18 MSFIA step-up for time-based injection.
(A) Manifold. (B)–(E) Injection steps. C, carrier; W/HC, waste/holding coil; RC1 and RC2, reaction coils; SV, switching valve; S, sample; R1 and R2, reagents; D, detector; W, waste.

time-based injection with it. The process starts by aspirating the sample from one of the ports of the switching valve (HC). Then, excess sample is discarded by sending it to a waste channel connected to another port of SV. For injection, SV is connected to the detector port in order to deliver the carrier via HC; with the valve of the syringe (C) being ON and (W) OFF, the two are switched over a preset time after which the initial connections of the valves are restored [74].

One other interesting way of performing time-based injection is with a MSFIA manifold which uses an independent commutation valve affording injection and sequencing of samples and reagents. The sample is loaded into a holding coil via an independent commutation valve. Then, a four-way cross-type connector is used to effect time-based injection by switching ON and OFF the valves mounted on the syringe heads, which hold the carrier and reagent, and also the independent valve, in the following sequence: sample/sample + reagent/carrier delivered through a reaction coil. Another two time-based injections are performed uninterruptedly with a single stroke of the piston and are propelled to the detector by the carrier. This MSFIA system affords not only time-based injection, but also a high throughput [75].

2.7.3
Controlled Variable Volume Injection

As noted earlier, multisyringe burettes are adaptations of the burettes typically used to implement SIA; therefore, they also allow accurately measured variable volumes of samples and reagents to be delivered in strokes of the syringe pistons controlled by step motors, similarly to the technique described in Section 2.5.3.

2.7.4
Synchronous Injection with Two Valves

Multisyringe burettes currently include four-outlet connecting strips supplying 12 V at up to 300 mA each in order to facilitate the control of single, double and triple solenoid valves via the burettes themselves. The strip can also be used to govern other devices (e.g., relays, pumps) operating at the same voltage. Provided the maximum nominal current is not exceeded, each outlet can be used to connect several devices for synchronous operation (e.g., the pair of single solenoid valves needed for injection). The incorporation of two fast-switching valves, independent of the syringes but controlled by the same burette, has allowed the injection valve module to be suppressed. The modified system allows accurately known volumes of liquids to be injected, as illustrated in Figure 2.19. The syringes are filled by switching solenoid valves V_5 and V_6 (see dashed lines in the figure); in this way, syringe 2 aspirates a preset volume of sample, and syringes 1, 3, and 4 aspirate the carrier and reagents, respectively. Measurements are made by reversing the flow direction, which is accomplished by switching valves V_5 and V_6 OFF (solid line in the figure). In response, syringe 1, with its valve ON, propels the sample volume held in the loop into the manifold. Simultaneously, the valve of

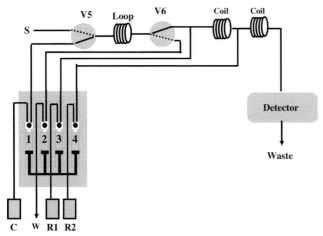

Figure 2.19 Typical flow manifold including two additional commutation valves. The independent solenoid valves V5 and V6 operate as an injection valve. S, sample; C, carrier; R1 and R2, reagents.

burette 2 is switched OFF to have excess aspirated sample sent to waste. Because valves V_5 and V_6 operate synchronously during injection, they can be replaced with one double valve actuated by a single solenoid. This allows the burette to actuate a double valve for injection and frees another valve output for some other purpose.

2.7.5
Synchronous Injection

In MSFIA systems, the sample and reagents are only injected when needed; otherwise, they are returned to their corresponding containers, which provide substantial savings in both. This is especially important with samples and/or reagents that are scant, expensive or hazardous. The manifold of Figure 2.20 was used for the chemiluminescence-based determination of glucose by on-line conversion in an oxidase packed-bed reactor and post-column reaction of the resulting oxidizing species with luminol (3-aminophthalhydrazide) in an alkaline medium containing a dissolved metal (Co^{2+}) or an organic catalyst (horseradish peroxidase, HRP) [76]. In conventional FIA, both the catalyst and the reagents are continuously circulated through the manifold; this results in massive consumption and a high cost per analysis which can only be reduced by using the merging-zones mode or immobilizing the catalyst as well. In MSFIA, however, the sample is only injected (via an appropriate three-, four- or five-way connector and synchronously with the reagent) when it crosses a preset manifold region (points T1 or T2 in Figure 2.20); otherwise, the sample and reagent are returned to their respective reservoirs by switching the respective valves on the syringe heads.

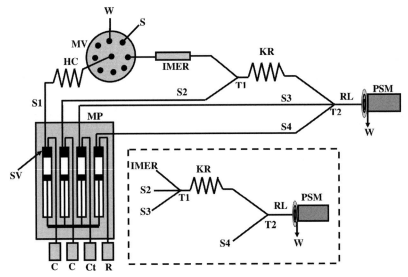

Figure 2.20 Multisyringe flow injection manifolds for the chemiluminescence-based determination of glucose using a Co(II) or HRP homogeneous catalyst. The modifications required to accommodate the HRP-mediated oxidation of luminol are depicted at the bottom. C, carrier; Ct, catalyst; R, chemiluminogenic reagent; MP, multisyringe pump; S1–S4, syringe channels; SV, solenoid valve; MV, multi-position valve; HC, holding coil; KR, knotted reactor; RL, reaction line; S, sample; IMER, immobilized enzyme reactor; PSM, photosensor module; T1 and T2, T-connector; W, waste.

2.7.6
Injection with a Dual Module

The use of an independent burette, as shown in Figure 2.21 results in substantially increased throughput. In the first step of the injection process, the multisyringe is

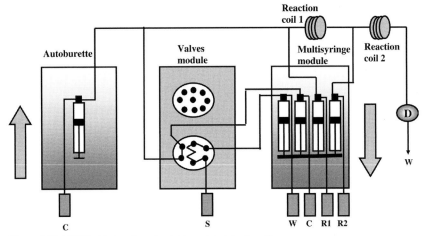

Figure 2.21 MSFIA aided system for dual module injection. S, sample; C, carrier; R1 and R2, reagents; D, detector; W, waste.

lowered in order to fill the sample loop and reagent syringes; simultaneously, the burette on the left propels the carrier to the detector. In the second step, the burette on the right is raised to inject the sample and, after an appropriate interval, the reagents as well; meanwhile, the burette on the left is filled with carrier. The carrier is not propelled to the detector by the multisyringe; rather, beyond the point of merging with the second reagent, the first step is restarted and the carrier volume delivered by the burette on the left propels the sample to the detector while the multisyringe refills the injection loop with sample. This alternate use of the two burettes allows the throughput to be doubled (up to 200 injections h^{-1}, which can hardly be matched by any other flow technique).

2.7.7
Bead Injection

This injection mode has also been implemented in MSFIA. Thus, Quintana *et al.* [77] have used the MSFIA-LOV system, in its off-valve mode, coupled to a liquid chromatograph for the determination of a mixture of acid pharmaceuticals including non-steroidal anti-inflammatory drugs and lipid regulators. The overall sample processing cycle involves four steps, namely: system preconditioning and column packing, sample loading, analyte elution and transportation into the MSFI-LC interface, and bead withdrawal. Unretained matrix species are removed from the sorbent material by rinsing the microcolumn with carrier solution. In order to prevent dispersion of the organic eluent plug into the carrier solution, an air segment is loaded into the holding coil before the eluent is aspirated. The resulting analyte-containing plug is delivered downstream with carrier solution and collected in the injection loop of the rotary valve, the valve being switched to its injection position and the LC protocol initiated. Finally, the bead-packed column is easily removed from the LOV microconduits by moistening with eluent and sent to waste with carrier solution. In this way, the system is made ready for a new analysis cycle with a fresh portion of beads and any risk of carry-over between consecutive runs is avoided.

2.8
Multipumping Flow Systems (MPFS)

Multipumping flow systems, which were developed in 2002 [78], use solenoid piston micropumps where each stroke propels a preset volume of liquid at a flow rate adjusted via the stroke frequency. Having the pump operate as both a liquid propeller and a valve results in a high flexibility, robustness and cost-effectiveness. In fact, an MPFS uses samples and reagents very sparingly; also, analyte peaks are taller than those provided by other flow techniques, which can be ascribed to the pump piston strokes causing turbulence that facilitates mixing of the sample and reagents. The simplicity and economy of MPFS should facilitate the development of portable equipment for field measurements.

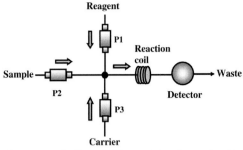

Figure 2.22 MPFS manifold comprising three solenoid micropumps (P$_1$–P$_3$) and merging point (●).

2.8.1
Pulse-based Injection

Figure 2.22 shows an extremely simple MPFS manifold which allows samples and reagents to be introduced in a variety of ways by using three solenoid micropumps (P$_1$–P$_3$) for each of the individual solutions (reagent, sample and carrier). Micropump P$_2$ is used to insert the sample solution and P$_1$ the reagent solution – which also serves as the carrier solution. Initially, the reagent-carrier solution is used to establish the baseline by having P$_1$ operate at a fixed pulse frequency which defines the flow rate for a given stroke volume. The sample is inserted at a preset number of pulses which define the sample volume. The pulse frequency also determines the flow rate during sample insertion. Pulsed flow is an interesting feature of MPFS which allows the flowing stream to be viewed as a continuous string of very small segments corresponding to the volume of each pulse. The sample zone is propelled to the detector by actuating P$_1$. The third solenoid pump in the figure, P$_3$, can be used to insert an inert carrier solution in order to economize reagent and transfer the reacting plug to the detector.

2.8.2
Merged Injection

In the manifold of Figure 2.22, the carrier, sample and reagent solutions are driven to the analytical path while the corresponding pumps are ON. This allows several sample–reagent plugs to be added by repeated rapid switching of P$_2$ and P$_1$. By having both pumps operate simultaneously, the sample can be mixed with the reagent similarly to the merging zones mode and a pulsed flow established inside the main channel [79]. Pump P$_3$ can be used to propel the reacting plug to the detector.

2.9
Combined Injection Methods

The above-described flow techniques can be combined in various ways in order to boost their advantages and alleviate their shortcomings. Selected examples of such combinations are briefly described below.

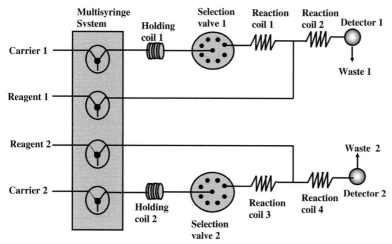

Figure 2.23 Dual SIA system.

2.9.1
Dual Sequential Injection Analysis

The multisyringe burettes typically used in MSFIA can be employed in combination with a module consisting of two SIA switching valves in order to construct a dual SIA system such as that of Figure 2.23. By programming the combined system in an appropriate manner, one can simultaneously determine several target parameters with a high throughput. The system in the figure was designed with a view to monitoring several parameters of interest (viz. conductivity, pH, iron, ammonium ion, and hydrazine) to the control of corrosion in a closed system used to cogenerate electrical energy at an urban solid waste incineration plant [80].

2.9.2
Sequential Injection-multisyringe Flow Injection Systems

SIA has been used in combination with MSFIA for the potentiometric determination of several species with selective electrodes [80]. Constructing calibration curves in the form of potential vs. logarithmic analyte concentration plots requires the use of ionic strength buffering solutions (ISA, TISAB) of composition compatible with the target analyte. Thus, some buffers are compatible with certain analytes but cannot be used with others.

The use of a single-syringe burette and a valve module in the SIA system of Figure 2.24 facilitates preliminary manipulation of the sample, after which ionic strength adjustors are dispensed with a multisyringe burette that propels a different buffering solution with each syringe in order to drive the contents of each channel to a selective electrode compatible with the buffer composition. The three-way solenoid valves controlled by the digital outputs located at the back of the multisyringe burette are used to process the controlled-volume injection performed in each of the four

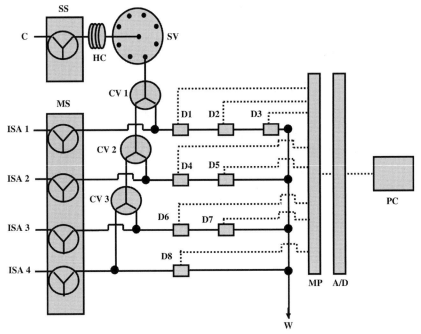

Figure 2.24 SIA-MSFIA system for the simultaneous potentiometric determination of several species with selective electrodes. SS, single-syringe burette; MS, multisyringe burette; HC, holding coil; SV, switching valve; CV1–CV3, commutation valves; D1–D8, potentiometric detectors; AD, analog-to-digital converter; MP, multiplexer; C, carrier; W, waste; ISA (1–4), ionic strength adjustors; PC, personal computer; (– –) electric lines, (– – –) flow lines.

channels in the SIA system. The potentials generated by the selective electrodes are acquired by a multiplexer.

2.9.3
Multisyringe Flow Injection-multipumping Flow Systems

Many of the above-described applications testify to the high flexibility of multi-commutated flow techniques for preparing samples prior to their analysis with an appropriate type of on-line detector. However, it is not always possible to perform the required measurements in this way. One case in point is when determining radioactive isotopes in environmental samples. Because such isotopes are usually present at very low levels, their determination takes a long counting time (hours or even days). In this situation, pretreating the sample by using a suitable flow technique can facilitate off-line counting of isotopes with an appropriate detector.

By way of example, Figure 2.25 shows an MSFIA-MPFS system for the determination of radium in waters [81] based on its adsorption on MnO_2. The procedure involves preparing the adsorbent in an appropriate form. To this end, a piece of cotton is placed in a column through which $KMnO_4$ is subsequently passed. Cellulose in the cotton reduces the permanganate to manganese dioxide, which is retained on its

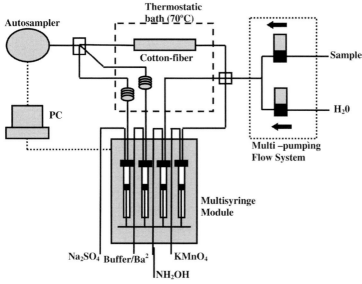

Figure 2.25 Off-line system for sample preparation as coupled to an MSFIA-MPFS set-up for the determination of radium in water.

surface. With the aid of solenoid micropumps, the cotton is cleaned with water and a few milliliters of sample propelled through the column, any radium present being retained by MnO_2. Next, an adequate amount of hydroxylamine is propelled with a multisyringe to redissolve manganese dioxide as soluble Mn(II), radium being released and collected in the sampler. A barium and sodium sulfate salt is then added to cause the coprecipitation of radium with barium sulfate. Finally, the sample is filtered and measured by alpha spectrometry, using a low-background proportional counter.

2.10
Concluding Remarks

The two basic methods for the introduction of a desired volume of sample and reagents into a flow system (time-based and volume-based modes) may be easily implemented in most of the different flows. The injection procedure has to be carefully evaluated in every analytical system, since the success and analytical performance of a method may be critically dependent on this selection. Currently, in the more modern flow techniques, the control of the introduction of sample and reagents, as well as the data treatment, is completely automated by means of microprocessors or computers. Indeed, some flow techniques cannot be implemented without the use of these microprocessors or computers. These requirements have provided both some advantages and some drawbacks. On the one hand the analytical methods have gained in automation, flexibility, precision speed and selectivity. On the other hand their development has been slow due to the need for better knowledge and skill in electronics and informatics. One challenge for the flow analytical developers is to find appropriate collaboration with experts on electronics and informatics and together develop powerful and flexible instruments based on the use of friendly software.

References

1 Gotto, M. (1983) Monitoring of environmental water using continuous flow. *Trends in Analytical Chemistry*, **2**, 92–94.

2 Watari, H., Cunningham, L. and Freiser, H. (1982) Automated system for solvent extraction kinetic studies. *Analytical Chemistry*, **54**, 2390–2392.

3 Skeggs, L. (1957) An automatic method for colorimetric analysis. *American Journal of Clinical Pathology*, **28**, 311–322.

4 Pasquín, C. and Oliveira, W.A. (1985) Monosegmented system for continuous flow analysis. Spectrophotometric determination of chromium(VI), ammonia and phosphorus. *Analytical Chemistry*, **57**, 2575–2579.

5 Ruzicka, J. and Hansen, E.H. (1975) Flow injection analyses: Part I. A new concept of fast continuous flow analysis. *Analytica Chimica Acta*, **78**, 145–157.

6 Bergamin, H.F., Zagatto, E.A.G., Krug, F.J. and Reis, B.F. (1978) Merging zones in flow injection analysis: Part 1. Double proportional injector and reagent consumption. *Analytica Chimica Acta*, **101**, 17–23.

7 Krug, F.J., Reis, B.F., Gine, M.F., Zagatto, E.A.G., Ferreira, J.R. and Jacinto, A.O. (1983) Zone trapping in flow injection analysis: Spectrophotometric determination of low levels of ammonium ion in natural waters. *Analytica Chimica Acta*, **151**, 39–48.

8 Teixeira Diniz, M.C., Fatibello Filho, O., Vidal de Aquino, E. and Rohwedder, J.J.R. (2004) Determination of phosphate in natural water employing a monosegmented flow system with

simultaneous multiple injection. *Talanta*, **62**, 469–475.

9 Zagatto, E.A.G., Krug, F.J., Bergamin, H.F., Jorgensen, S.A. and Reis, B.F. (1979) Merging zones in flow injection analysis: Part 2. Determination of calcium, magnesium and potassium in plant material by continuous flow injection atomic absorption and flame emission spectrometry. *Analytica Chimica Acta*, **104**, 279–284.

10 Ruzicka, J. and Hansen, E.H. (1979) Stopped flow and merging zones – a new approach to enzymatic assay by flow injection analysis. *Analytica Chimica Acta*, **106**, 207–224.

11 Cerdà, A., Oms, M.T., Forteza, R. and Cerdà, V. (1996) Speciation of nitrogen in wastewater by flow injection. *Analyst*, **121**, 13–17.

12 Koropchak, J.A. and Allen, L. (1989) Flow-injection Donnan dialysis preconcentration of cations for flame atomic absorption spectrophotometry. *Analytical Chemistry*, **61**, 1410–1414.

13 Schepers, D., Schulze, G. and Frenzel, W. (1995) Spectrophotometric flow-through gas sensor for the determination of atmospheric nitrogen dioxide. *Analytica Chimica Acta*, **308**, 109–114.

14 Baxter, P.J., Ruzicka, J., Christian, G.D. and Olson, D.O. (1994) An apparatus for the determination of volatile analytes by stopped-flow injection analysis using an integrated fiber optic detector. *Talanta*, **41**, 347–354.

15 Olson, D.C., Bysouth, S.R., Dasgupta, P.K. and Kuban, V. (1994) A new flow injection analyzer for monitoring trace HCN in

process gas streams. *Process Control and Quality*, **5**, 259–265.

16 Frenzel, W. (1990) Gas-diffusion separation and flow injection potentiometry. A fruitful alliance of analytical methods. *Fresenius' Journal of Analytical Chemistry*, **336**, 21–28.

17 Motomizu, S. and Yoden, T. (1992) Porous membrane permeation of halogens and its application to the determination of halide ions and residual chlorine by flow injection analysis. *Analytica Chimica Acta*, **261**, 461–469.

18 Hauser, P.C. and Zhang, Z.P. (1996) Flow-injection determination of lead by hydride generation and conductometric detection. *Fresenius' Journal of Analytical Chemistry*, **355**, 141–143.

19 Frenzel, W., Liu, O.Y. and Oleksy-Frenzel, J. (1990) Enhancement of sensors selectivity by gas-diffusion separation: Part 1. Flow-injection potentiometric determination of cyanide with a metallic silver-wire electrode. *Analytica Chimica Acta*, **233**, 77–84.

20 Ohura, H., Imato, T., Asano, Y., Yamasaki, S. and Ishibashi, N. (1990) Potentiometric determination of ethanol in alcoholic beverages using a flow injection analysis system equipped with a gas diffusion unit with a microporous poly(tetrafluoroethylene) membrane. *Analytical Sciences*, **6**, 541–545.

21 Bartrolí, J., Escalada, M., Jonquera, C.J. and Alonso, J. (1991) Determination of total and free sulfur dioxide in wine by flow injection analysis and gas-diffusion using *p*-aminoazobenzene as the colorimetric reagent. *Analytical Chemistry*, **63**, 2532–2535.

22 Milosavljevic, E.B., Solujic, L. and Hendrix, J.L. (1995) Rapid distillationless "free cyanide" determination by a flow injection ligand exchange method. *Environmental Science and Technology*, **29**, 426–430.

23 Aoki, T. and Wakabayashi, M. (1995) Simultaneous flow injection determination of nitrate and nitrite in water by gas-phase chemiluminescence. *Analytica Chimica Acta*, **308**, 308–312.

24 Gord, J.R., Gordon, G. and Pacey, G.E. (1988) Selective chlorine determination by gas-diffusion flow injection analysis with chemiluminescent detection. *Analytical Chemistry*, **60**, 2–4.

25 Fang, Z., Zhu, Z., Zhang, S., Xu, S., Guo, L. and Sun, L. (1988) On-line separation and preconcentration in flow injection analysis. *Analytica Chimica Acta*, **214**, 41–55.

26 Mattos, I.L. and Luque de Castro, M.D. (1994) Study of mass-transfer efficiency in pervaporation processes. *Analytica Chimica Acta*, **298**, 159–165.

27 Papaefstathiou, I., Tena, M.T. and Luque de Castro, M.D. (1995) On-line pervaporation separation process for the potentiometric determination of fluoride in "dirty" samples. *Analytica Chimica Acta*, **308**, 246–252.

28 Papaefstathiou, I. and Luque de Castro, M.D. (1995) Integrated pervaporation/detection: continuous and discontinuous approaches for treatment/determination of fluoride in liquid and solid samples. *Analytical Chemistry*, **67**, 3916–3921.

29 Wang, L., Cardwell, T.J., Cattrall, R.W., Luque de Castro, M.D. and Kolev, S.D. (2000) Pervaporation-flow injection determination of ammonia in the presence of surfactants. *Analytica Chimica Acta*, **416**, 177–184.

30 Prinzing, U., Ogbomo, I., Lehn, C. and Schmidt, H.L. (1990) Fermentation control with biosensors in flow injection systems: problems and progress. *Sensors and Actuators*, **B1**, 542–545.

31 Ogbomo, I., Steffl, A., Schuhmann, W., Prinzing, U. and Schmidt, H.L. (1993) On-line determination of ethanol in bioprocesses based on sample extraction by continuous pervaporation. *Journal of Biotechnology*, **31**, 317–325.

32 Luque de Castro, M.D. and Papaefstathiou, I. (1998) Analytical pervaporation: a new separation technique. *Trends in Analytical Chemistry*, **17**, 41–49.

33 Papaefstathiou, I., Bilitewski, U. and Luque de Castro, M.D. (1996) Pervaporation: an interface between fermentors and monitoring. *Analytica Chimica Acta*, **330**, 265–272.

34 Papaefstathiou, I. and Luque de Castro, M.D. (1997) Simultaneous determination of chemical oxygen demand/inorganic carbon by flow injection-pervaporation. *International Journal of Environmental Analytical Chemistry*, **66**, 107–117.

35 Papaefstathiou, I., Luque de Castro, M.D. and Valcárcel, M. (1996) Flow-injection/ pervaporation coupling for the determination of sulphide in Kraft liquors. *Fresenius Journal of Analytical Chemistry*, **354**, 442–446.

36 Papaefstathiou, I. and Luque de Castro, M.D. (1997) Nitrogen speciation analysis in solid samples by integrated pervaporation and detection. *Analytica Chimica Acta*, **354**, 135–142.

37 Papaefstathiou, I., Bilitewski, U. and Luque de Castro, M.D. (1997) Determination of acetaldehyde in liquid, solid and semi-solid food after pervaporation-derivatizacion. *Fresenius' Journal of Analytical Chemistry*, **357**, 1168–1173.

38 Belitz, H.D. and Grosch, W. (1987) *Food Chemistry*, Springer, Berlin.

39 Bryce, D.W., Izquierdo, A. and Luque de Castro, M.D. (1996) Continuous microwave assisted pervaporation/atomic fluorescence detection: an approach for speciation in solid samples. *Analytica Chimica Acta*, **324**, 69–75.

40 Burguesa, M. and Burguesa, J.L. (1998) Microwave-assisted sample decomposition in flow analysis. *Analytica Chimica Acta*, **366**, 63–80.

41 Bryce, D.W., Izquierdo, A. and Luque de Castro, M.D. (1997) Pervaporation as an alternative to headspace. *Analytical Chemistry*, **69**, 844–847.

42 Papaefstathiou, I. and Luque de Castro, M.D. (1997) Hyphenated pervaporation-solid-phase preconcentration-gas chromatography for the determination of volatile organic compounds in solid samples. *Journal of Chromatography A*, **779**, 352–359.

43 Ruzicka, J. and Hansen, H. (1983) Recent developments in flow injection analysis: gradient techniques and hydrodynamic injection. *Analytica Chimica Acta*, **145**, 1–15.

44 Kuban, P., Pirmohammadi, R. and Karlberg, B. (1999) Flow injection analysis–capillary electrophoresis system with hydrodynamic injection. *Analytica Chimica Acta*, **378**, 55–62.

45 Ruzicka, J. and Marshall, G.D. (1990) Sequential injection: a new concept for chemical sensors, process analysis and laboratory assays. *Analytica Chimica Acta*, **237**, 329–343.

46 Lenehan, C.E., Barnett, N.W. and Lewis, S.W. (2002) Sequential injection analysis. *Analyst*, **127**, 997–1020.

47 Ivaska, A. and Ruzicka, J. (1993) From flow injection to sequential injection: comparison of methodologies and selection of liquid drives. *Analyst*, **118**, 885–889.

48 Cladera, A., Tomás, C., Gómez, E., Estela, J.M. and Cerdà, V. (1995) A new instrumental implementation of sequential injection analysis. *Analytica Chimica Acta*, **302**, 297–308.

49 Miró, M., Cladera, A., Estela, J.M. and Cerdà, V. (2000) Sequential injection spectrophotometric analysis of nitrite in natural waters using an on-line solid-phase extraction and preconcentration method. *Analyst*, **125**, 943–948.

50 Alonso, J., Bartrolí, J., del Valle, M., Escalada, M. and Barber, R. (1987) Sandwich techniques in flow injection analysis: Part 1. Continuous recalibration techniques for process control. *Analytica Chimica Acta*, **199**, 191–196.

51 Araujo, A.N., Lima, J.L.F.C., Rangel, A.O.S.S., Alonso, J., Bartroli, J. and Barber, R. (1989) Simultaneous determination of total iron and chromium(VI) in wastewater using a flow injection system based on the sandwich technique. *Analyst*, **114**, 1465–1468.

52 Estela, J.M., Cladera, A., Muñoz, A. and Cerdà, V. (1996) Simutaneous determination of ionic species by sequential injection analysis using a sandwich technique with large sample volumes. *International Journal of Environmental Analytical Chemistry*, **64**, 205–215.

53 Cerdà, A., Oms, M.T., Forteza, R. and Cerdà, V. (1998) Sequential injection sandwich technique for the simultaneous determination of nitrate and nitrite. *Analytica Chimica Acta*, **371**, 63–71.

54 Thomas, O., Theraulaz, F., Cerdà, V., Constant, D. and Quevauviller, Ph. (1997) Wastewater quality monitoring. *Trends in Analytical Chemistry*, **16**, 419–424.

55 Guzman, M., Pollema, C., Ruzicka, J. and Christian, G.D. (1993) Sequential injection technique for automation of complex analytical procedures: fluorometric assay of factor thirteen. *Talanta*, **40**, 81–87.

56 Ruzicka, J. and Ivaska, A. (1997) Bioligand interaction assay by flow injection absorptiometry. *Analytical Chemistry*, **69**, 5024–5030.

57 Lahdesmaki, I. and Ruzicka, J. (2000) Novel flow injection methods for drug-receptor interaction studies, based on probing cell metabolism. *Analyst*, **125**, 1889–1895.

58 Ruzicka, J. (2000) Lab-on-valve: universal microflow analyzer based on sequential and bead injection. *Analyst*, **125**, 1053–1060.

59 Wu, C.H., Scampavia, L., Ruzicka, J. and Zamost, B. (2001) Micro sequential injection: fermentation monitoring of ammonia, glycerol, glucose, and free iron using the novel lab-on-valve system. *Analyst*, **126**, 291–297.

60 Wu, C.H. and Ruzicka, J. (2001) Micro sequential injection: environmental monitoring of nitrogen and phosphate in water using a "Lab-on-valve" system furnished with a microcolumn. *Analyst*, **126**, 1947–1952.

61 Wang, J. and Hansen, E.H. (2000) Coupling on-line preconcentration by ion-exchange with ETAAS: a novel flow injection approach based on the use of a renewable microcolumn as demonstrated for the determination of nickel in environmental and biological samples. *Analytica Chimica Acta*, **424**, 223–232.

62 Wang, J. and Hansen, E.H. (2001) Interfacing sequential injection on-line preconcentration using a renewable micro-column incorporated in a "lab-on-valve" system with direct injection nebulization inductively coupled plasma mass spectrometry. *Journal of Analytical Atomic Spectrometry*, **16**, 1349–1355.

63 Wang, J. and Hansen, E.H. (2001) Coupling sequential injection on-line preconcentration by means of a renewable microcolumn with ion-exchange beads with detection by electrothermal atomic absorption spectrometry. Comparing the performance of eluting the loaded beads with transporting them directly into the graphite tube, as demonstrated for the determination of nickel in environmental and biological samples. *Analytica Chimica Acta*, **435**, 331–342.

64 Wu, C.H., Scampavia, L. and Ruzicka, J. (2002) Microsequential injection: anion separation using "Lab-on-Valve" coupled with capillary electrophoresis. *Analyst*, **127**, 898–905.

65 Hostkotte, B., Elshols, O. and Cerdà, V. (2007) Development of a capillary electrophoresis system coupled to sequential injection analysis and characterization by the analysis of nitrophenols. *International Journal of Environmental Analytical Chemistry*, **87**, 797–811.

66 Reis, B.F., Giné, M.F., Zagatto, E.A.G., Lima, J.L.F.C. and Lapa, R.A. (1994) Multicommutation in flow analysis. Part 1. Binary sampling: concepts, instrumentation and spectrophotometric determination of iron in plant digests. *Analytica Chimica Acta*, **293**, 129–138.

67 Rocha, F.R.P., Reis, B.F., Zagatto, E.A.G., Lima, J.L.F.C. and Lapa, R.A. (2002) Multicommutation in flow analysis: concepts, applications and trends. *Analytica Chimica Acta*, **468**, 119–131.

68 Zagatto, E.A.G., Reis, B.F., Oliveira, C.C., Sartini, R.P. and Arruda, M.A.Z. (1999) Evolution of the commutation concept associated with the development of flow analysis. *Analytica Chimica Acta*, **400**, 249–256.

69 Malcome-Lawes, D.J. and Pasquín, C. (1988) A novel approach to non-segmented flow analysis. Part 2. A prototype high-performance analyzer. *Journal of Automatic Chemistry*, **10**, 25–30.

70 Israel, Y., Lasztity, A. and Barnes, R.M. (1989) On-line dilution, steady-state concentrations for inductively coupled plasma atomic emission and mass spectrometry achieved by tandem injection and merging-stream flow injection. *Analyst*, **114**, 1259–1265.

71 Kronka, E.A.M., Borges, P.R., Latanze, R., Paim, A.P.S. and Reis, B.F. (2001) Multicommutated flow system for glycerol determination in alcoholic fermentation juice using enzymatic reaction and spectrophotometry. *Journal of Flow Injection Analysis*, **18**, 132–138.

72 Comitre, A.L.D. and Reis, B.F. (2000) Automatic multicommutated flow system for ethanol determination in alcoholic beverages by spectrophotometry. *Laboratory Robotics and Automation*, **12**, 31–36.

73 Cerdà, V., Estela, J.M., Forteza, R., Cladera, A., Becerra, E., Altamira, P. and Sitjar, P. (1999) Flow techniques in water análisis. *Talanta*, **50**, 695–705.

74 Albertús, F., Cladera, A., Becerra, E. and Cerdà, V. (2001) A robust multi-syringe

system for process flow analysis. Part 3. Time based injection applied to the spectrophotometric determination of nickel(II) and iron speciation. *Analyst*, **126**, 903–910.

75 De Armas, G., Miró, M., Cladera, A., Estela, J.M. and Cerdà, V. (2002) Time-based multisyringe flow injection system for the spectrofluorimetric determination of aluminium. *Analytica Chimica Acta*, **455**, 149–157.

76 Manera, M., Miró, M., Estela, J.M. and Cerdà, V. (2004) A multisyringe flow injection system with immobilized glucose oxidase based on homogeneous chemiluminescence detection. *Analytica Chimica Acta*, **508**, 23–30.

77 Quintana, J.B., Miró, M., Estela, J.M. and Cerdà, V. (2006) Automated on-line renewable solid-phase extraction-liquid chromatography exploiting multisyringe flow Injection-bead injection lab-on-valve analysis. *Analytical Chemistry*, **78**, 2832–2840.

78 Lapa, R.A.S., Lima, J.L.F.C., Reis, B.F., Santos, J.L.M. and Zagatto, E.A.G. (2002) Multi-pumping in flow analysis: concepts, instrumentation, potentialities. *Analytica Chimica Acta*, **466**, 125–132.

79 Carneiro, J.M.T., Zagatto, E.A.G., Santos, J.L.M. and Lima, J.L.F.C. (2002) Spectrophotometric determination of phytic acid in plant extracts using a multi-pumping flow system. *Analytica Chimica Acta*, **474**, 161–166.

80 Cerdà, V. (2006) Introducción a los Métodos de Análisis en Flujo, SCIWARE, S.L. Editions, Palma de Mallorca.

81 Fajardo, Y., Gómez, E., Garcías, F., Cerdà, V. and Casas, M. (2007) Development of an MSFIA-MPFS pre-treatment method for radium determination in water samples. *Talanta*, **71**, 1172–1179.

3
Application of Moveable Suspensions of Solid Particles in Flow Analysis

Marek Trojanowicz

3.1
Introduction

The main advantages of analytical measurements in flow conditions, where detection is carried out during the flow of solution through the flow detector, are the possibility of manipulation of transient signals, utilization of kinetic effects and facilitation of numerous operations of sample processing. These effects allow one to obtain with flow measurements several advantages over conventional non-flow measurements, carried out with various detection methods. In some cases one can gain improved sensitivity and selectivity of determinations, with some detection methods an improved detection limit can be obtained, and – as is usual with mechanization and automation of analytical determinations – a better precision and larger sampling rate.

Solid particles of micrometer size are commonly used in many analytical methods, offering, first, a large ratio of surface area to mass compared to other forms of similar materials. The most common applications of solid microparticles are liquid and gas column chromatography, thin-layer chromatography and capillary electrochromatography. Conducting particles are employed in electrically modulated chromatography where polarization of the stationary-phase allows one to modify the selectivity of retention. The second large field of applications of microparticles is solid-phase extraction methods, from both gaseous and liquid phases, which are very widely used for sample clean-up and analyte preconcentration prior to analytical measurements [1]. In the case of sorption from liquids a particle bed can be packed into a flow-through column or a small amount of particles can be added to a large volume of solution. Another area of application of microparticles in non-flow conditions is their use in diagnostic tests and assays [2]. This can include, for example, enzyme-linked immunosorbent tests and assays, agglutination tests and assays and chromatographic strip tests. In immunological methods, especially, coated magnetic particles are often used, for example in immunoassays for pesticides (including organohalogen pesticides [3] or triazine ones [4]) or bacterial magnetic particles (BMP) with immobilized antibodies, where the amount of antibody coupling with the BMPs

Advances in Flow Analysis. Edited by Marek Trojanowicz
Copyright © 2008 WILEY-VCH Verlag GmbH & Co. KGaA, Weinheim
ISBN: 978-3-527-31830-8

is higher than that with artificial magnetic particles of the same size [5]. An example is the chemiluminescence enzyme assay for immunoglobulin G [6] or model food allergen lysozyme [7]. Solid particles have been used as tags to identify cells or cell surface antigens or microscopic slides and, in recent years, in high throughput screening for drug discovery based on flow cytometry (which in fact means application of solid particles in flow conditions) [8]. Another method of sandwich immuno-assay is a bead suspension array developed for the simultaneous detection of multiple pathogens [9].

Solid particles in flow analytical measurements are employed principally in packed flow-through reactors, providing a large surface of contact with solution, facilitating better efficiency of the heterogeneous processes of sorption, chemical exchange or reaction with solid material in flow-through reactors. When they are used as reactive beds they have to be periodically exchanged after use, whereas when used in sorption or clean-up they can be regenerated in the elution process. Microbeads can also be immobilized for the same purpose on the walls of open columns, which gives the advantage of small flow resistance.

Micrometer and submicrometer solid particles can also be transported as a suspension and for more than 30 years this has been successfully employed in various variants of flow analysis. The main aim of using a moveable sorbent bed is the possibility of performing a reaction on the surface of the solid material, which can be transported between different parts of the measuring system. This can be employed for carrying out many heterogeneous reactions and operations and for modification of various detection methods. The use of renewable beds enables elimination of numerous problems associated with the use of packed bed reactors (progressively tighter packing, malfunctioning of active sites or surface deactivation). Under continuous flow conditions one can carry out immunoassays with antibodies immobilized on magnetic particles [10]. In flow cytometry, microparticles have been used for many years, mainly as an inexpensive means of optimizing surface conjugation. Microbeads are used for the calibration of flow cytometers. Their advantages compared to the use of microwell plates have been demonstrated, for instance, in indirect quantitation of human immunodeficiency virus (HIV), based on the use of derivatized beads to capture single nucleic-acid targets [11]. Beads containing eight levels of fluorescence improved the specificity and class determination of human leukocyte antibodies by flow cytometry [12].

A pronounced impetus was given to various new applications of a moveable suspension of particles in flow analysis in the mid-1990s by the works of Ruzicka and his coworkers from the University of Washington in Seattle, on their use in various flow injection systems. Their first studies dealt with the formation of a renewable surface by a bed of particles, which can be probed spectroscopically [13], and a renewable surface for immunoassays with fluorescence detection [14]. The term bead-injection analysis (BIA) is used quite broadly to describe such systems, although it can be confusing as the same abbreviation can be found in the analytical literature for many other measuring systems and methods (e.g., bioimmunoassay, batch-injection analysis, bioligand interaction assay, or bioelectric impedance analysis). The use of moveable particles in flow injection systems has been the subject of several reviews [15, 16], and is well presented in a free access tutorial file available on the Internet [17].

3.2
Solid Microparticles Used in Suspensions in Flow Analysis

The size of solid particles used in suspensions and their chemical properties depend on the type of flow system used, the detection method and the particular analytical application. Generally, in flow measurements with particle suspensions a good precision is obtained with regular spherically shaped rigid particles of uniform size. Small size is also favored as this allows the formation of concentrated and stable suspensions, although the optimum size can be affected by their transport in the measuring system and their temporary immobilization in the system. The most commonly used particles have diameters in the range 10–200 μm. In particular cases such as flow cytometry, the particles are of a fraction to a few micrometers in diameter, but this type of measurement will not be discussed here in detail, as it is a special type of method and instrumentation used in biochemistry, biology and medicine with wide specialized literature, (see e.g., [18]). It is also worth mentioning here the early-developed flow measuring system for the continuous size separation of particles suspended in a liquid, utilizing a laminar profile in a pinched microchannel [19]. Such a system can be used for size-dependent separation of various kinds of particles, cells, gels and nanometer-size particles.

The chemical nature of the particles used in suspensions depends on the particular chemical application in the flow system. Various commercially available beads can be used; these were developed for gel filtration, cell culture or solid-phase extraction. Special care has to be taken when using fragile elastic beads in order to avoid their disintegration. Particles of ion-exchanger have been used for the determination of inorganic analytes, trace element preconcentration and selective capture of biomolecules. The cation-exchange resin SP sulfopropyl Sephaedex C-25 was used in the determination of Ni in biological and environmental samples with electrothermal atomic absorption spectrometry (ETAAS) detection [20], and Ni and Bi with inductively coupled plasma-mass spectrometry (ICP-MS) detection [21]. For the determination of radionuclides ^{90}Sr, ^{99}Tc, and ^{241}Am specially developed selective ion-exchange resins have been used [22]. An anion-exchange resin with adsorbed silver sols was applied as the active surface with nicotinic acid as a model analyte in a flow analysis-based surface-enhanced Raman spectroscopy (SERS) study [23]. In a sequential injection analysis (SIA), a flow system with the same detection for the determination of 9-aminoacridine, cation-exchange beads were used with an active layer surface formed on the bead surface [24]. The preconcentration of trace metal ions can be carried out by sorption of their chelate complexes on non-polar sorbents. The commercial bead material N-vinylpyrrolidone-divinyl benzene (Oasis HLB from Waters) was used for the determination of Ni as the complex with dimethylglyoxime in the SIA system with ETAAS detection [25]. The same sorbent was successfully employed in a flow extraction system hyphenated with high-performance liquid chromatography (HPLC) for determination of selected acidic pharmaceutical residues [26].

As suspensions in flow systems, controlled-pore glass beads and acrylic beads were used to develop a renewable electrochemical sensor system for use in flow injection measurements [27]. Spherical glass beads, polystyrene, acrylamide and sepharose were used in the SIA measurements in nucleic acid isolation and purification [28]. In bioligand interaction assays by flow injection analysis (FIA), Sepharose 4B gel and various functionalized agaroses, and also avidin Sepharose G were used [29]. Sepharose agarose beads were employed for affinity capture-release of biotin-containing conjugates with mass spectrometry (MS) detection [30]. Numerous biochemical applications have also been reported for the use of Protein A and Protein G Sepharose 4B conjugates [31–33]. Collagen-modified Cytodex 3 beads were used for non-covalent attachment of a phosphorescent oxygen probe (Pt-porphyrin complex) for detection of the oxygen consumption of adherent cell cultures [34]. A description of many various microbeads together with their preparation, properties and applications can be found in the current chemical literature, for example in Ref. [35].

Magnetic microparticles have been employed in flow analyzers since the 1970s [10]. In FIA determinations of the organophosphorus pesticide methamidophos by inhibition of acetylcholinesterase (AChE) the enzyme was covalently immobilized on amino-terminated magnetic particles [36]. In another example, for the immuno-chemical SIA determination of the phospholipoprotein vitellogenin, which is a biomarker for evaluating environmental risk caused by endocrine disruptors, horseradish peroxidase (HRP)-labeled antibody was immobilized on magnetic beads coated with agarose gel [37]. A schematic diagram of the measuring system, is shown in Figure 3.1. The detection was based on the chemiluminescent reaction of HRP with hydrogen peroxide and p-iodophenol in a luminol solution. The magnetic particles of relatively small diameter (2.8 μm), activated with toxyl groups to immobilize the protein, were used to design a magnetoimmunosensor for operation in a FIA system [38]. In a renewable amperometric sensor used in FIA measurements a glassy carbon spherical powder and graphite powder were employed [27].

3.3
Handling Suspensions of Particles in Flow Systems

As was mentioned above, in most applications the beads used in suspensions have diameters ranging from 10 to 200 μm. In most cases a suspension solution volume of 2–10 μL is used, containing 2000–10 000 beads. For SIA determinations with magnetic beads, a much larger suspension volume, 150 μL, is employed with concentration 25 μg mL^{-1} [37].

Determinations carried out in flow systems with microbead suspensions require the formation of a stable suspension, its transfer within a measuring system, and temporary immobilization in order to detect the measured signal. Stabilization of the suspension can be achieved, for example, by adding surfactant to the solution [39] or by mechanical agitation, for instance, by purging air through the suspension [24],

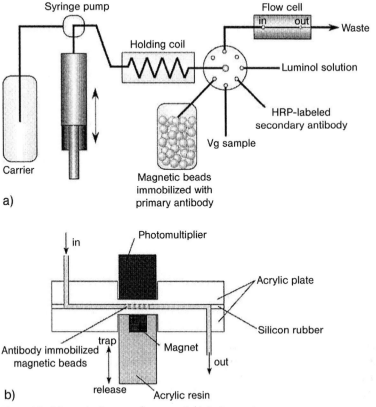

Figure 3.1 Schematic diagram of a sequential injection system employing magnetic microbeads (a) and a flow cell for chemilumienscence detection (b) developed for the flow immunoassay of vitellogenin, a biomarker of the degree of pollution with endocrine disrupting chemicals [37].

mixing with a magnetic stirrer [27], or by keeping the suspension in a rotating flask [13, 27].

Most often movement of the suspension in a flow system involves flow driven by pressure with the use of a syringe or peristaltic pumps. Other methods involve electroosmosis [40], electric field or gravity.

Especially crucial for the functioning of such systems is the temporary immobilization of the particles in a flow cell or renewable column. A review of applied designs for this purpose has been published [41], and the discussed designs are reproduced in Figure 3.2. From the point of view of ease of construction the simplest designs seem to be those with a magnet for stopping the magnetic particles in a flow-through detector, see, for example, Refs. [10, 36–38]. In micromagnetic systems, a microfabricated circuit rather than permanent magnets or an electromagnet is used to generate a local magnetic field [39]. In the system fabricated with soft lithography, current-carrying microcircuits generate a strong magnetic field and the paths can be made simply by changing the current flowing dynamic configuration.

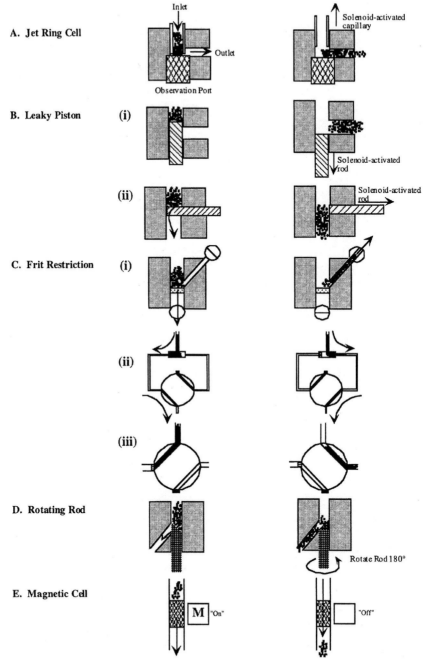

Figure 3.2 Schematic presentation of different designs of renewable columns for flow measurements with a suspension of microparticles [41]. See detailed description in the text.

In the jet ring cell (Figure 3.2A) a moveable capillary end is placed against a faceplate, such that the beads are retained while the solution flows radically outward [13]. The bed of particles forms a renewable surface, which can be probed by reflectance, fluorescence or chemiluminescence using a microscope. After measurement, the beads are removed by retracking the capillary or reversing the fluid flow. This design was used in SIA determinations of the oxygen consumption of cultured cells with fluorescence detection [33]. Two versions of the jet ring cell with radial and axial probing light paths have been developed and used in FIA measurements for bioligand interaction assays [29, 42]. In the same way a leaky piston design can be employed, where a solid rod is inserted parallel or perpendicular (Figure 3.2B). Both types of cell, however, are difficult to use for particles with diameter less than 20 μm.

In another design with frit restriction (Figure 3.2C(i)) smaller particles can also be used. The flow of the suspension in this case is controlled by external microvalves. This is a simple solution without moving parts, which has been applied, for example, in nucleic acid extraction and purification [28]. In a slightly different version this concept was realized with the use of a two-position valve modified with a frit restriction (Figure 3.2C (ii) and (iii)) for SIA determination of radionuclides with radiometric detection [22]. In yet another design (Figure 3.2D) the flow of beads was stopped by inserting a beveled or chamfered rod into the flow cell. 180 ° rotation of the rod opens the side arm channel and releases the packed bed.

A different solution for retaining moveable beads in a flow system is shown in Figure 3.3, where the particles are immobilized by a plug furnished with a nozzle that does not allow the transfer of beads [43], while liquid can flow through and around the plug. One direction of flow retains the beads above the plug, while reversing the flow results in elution of the particles.

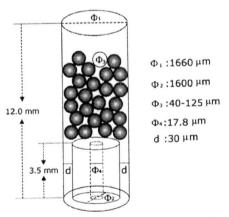

Φ_1 :1660 μm

Φ_2 :1600 μm

Φ_3 :40-125 μm

Φ_4:17.8 μm

d :30 μm

Figure 3.3 Schematic diagram of the concept of retaining microbeads in the flow system by the use of a piece of tubing insert with nozzle and outer diameter Φ_2, slightly smaller than the inner diameter Φ_1 of the tubing of the flow system [43].

Some modifications of these designs for the manipulation and immobilization of beads were used in more highly developed rotary valves for SIA measurements, which were additionally equipped with miniature fiber optic-based flow cuvettes, called by their designers lab-on-valves (LOV) [31–33].

3.4
Detection Methods Employed in Flow Systems with Particle Suspensions

The manipulation of suspensions in flow systems offers numerous possibilities for modification of conventional methods of detection in flow systems, and also some examples can be found of the development of new detection methods in flow measurements that cannot be used without moveable beads (reflectance spectroscopy, SERS). Especially interesting is the development of methods for direct detection on particles, as well as examples of renewable detectors with solid particles.

The most commonly used means of detection in all versions and generations of instrumentation for flow analysis is molecular absorption spectroscopy in the UV/Vis range. In the simplest mode in systems with particle suspensions the absorbance of the solution eluted from the bed is measured. This was used, for instance, for the detection of the product of the enzymatic reaction in the determination of organophosphorus pesticide based on the inhibition of AChE immobilized on magnetic particles [36]. Inhibition in this case is irreversible, hence renewal of the bed after each measurement is necessary. Similarly, absorptive measurement in the eluate was carried out in the systems where a SIA-LOV configuration was employed with renewal of the bed and hyphenation to HPLC. In one example on a renewable bed, low pressure affinity chromatography was carried out with Sepharose Protein A beads with UV detection [33]. The system was developed for the separation of immunoglobulin G from bovine serum albumin. In another example, where the FIA-LOV configuration was hyphenated to HPLC with UV detection, the renewing of the hydrophobic bed was used for preconcentration of analytes determined by HPLC [26].

A particular improvement of absorptive spectrophotometric detection is solid-phase absorptiometry. Invented in 1976 [44], it quickly found numerous applications [45]. Its main advantage is combining, in one step, absorptive detection and preconcentration of the absorbing compound on the solid sorbent. Additionally, such measurements favor elimination of matrix effects in complex natural samples. The sensitivity of such measurements can be improved by the use of appropriate geometry of the particle bed and optical system in the spectrophotometer [46]. At the end of the 1980s solid-phase absorptiometry found increasing application in flow measurements with cuvettes with packed sorbents, for example [47], including FIA systems [48–50] and multicomponent determinations, employing multi-variable analysis [51, 52]. At the end of the 1990s such measurements were applied in flow systems with particle suspensions. With appropriate design of the flow cell for retaining and releasing beads and passing the light beam (Figure 3.4), such measurements were developed, for instance for bioligand interaction assay [29, 42]

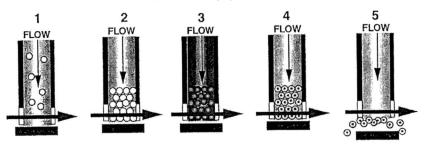

Figure 3.4 Illustration of the sequence of events in a jet ring cell in absorptiometric measurements in a FIA system with a suspension of microbeads [42]. (1) Introduction of beads, (2) capturing of beads, (3) perfusing beads with analyte solution, (4) perfusing beads with carrier solution, (5) discharge of beads. Horizontal arrow indicates a light beam for absorptiometric detection.

and monitoring of immobilization of proteins on solid particles [53]. In the latter case measurement was based on injections of beads periodically sampled from the microreactor where the protein immobilization process was carried out. During several years of research by Ruzicka's group, different designs of photometric cuvettes for retaining microbeads were developed. The radially probed jet-ring, illustrated in Figure 3.4, was replaced by an axially probed cell [29] that improved the repeatability of measurements and the detection limit. Further modifications of the optical flow cell were reported for optic-fiber-based detectors, built into a multi-position rotary switching valve (LOV) used in SIA measurements. A schematic diagram of such a simple absorptive detector, together with the whole measuring system, is shown in Figure 3.5. This system was used in a developed method of label dilution for studies of bioligand interactions [32]. In such a cell the diameter of the channel where the optic fiber is inserted is slightly larger than the diameter of the fiber in order to allow the flow of solution while microbeads from the suspension are retained. The third optic fiber mounted in the cell allows measurement of fluorescence. Such a flow cell was used in the SIA system for investigation of affinity capture-release connected in series with electrospray ionization mass spectrometry [30]. In another modification of an optic fiber flow cell incorporated in a rotary valve reflectance measurements are also possible [31].

In the design of measuring flow systems with microbead suspensions much attention is paid to photoluminescence detection. Instrumentally, photoluminescence measurements are not more difficult than absorptive measurements but they often give much better limits of detection. Several fluorimetric determinations were reported by Ruzicka *et al.* including the use of a jet ring cell where the bead chamber was fixed to the stage of an inverted epifluorescence microscope (Figure 3.2A) [13, 34, 54]. The microscope was connected to a luminator to provide an excitation beam. Such a system was used in the previously described detection of oxygen utilizing quenching of the phosphorescence of a Pt-porphyrin complex immobilized on microbeads [33]. Using beads with cells attached, the equilibrium and kinetic measurements of selected receptor antagonist were carried out, leading to quantification of both equilibrium and kinetic pharmacological values of the

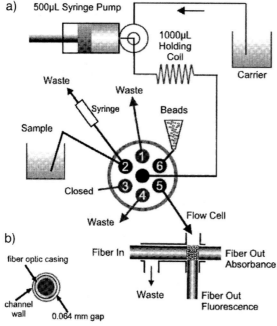

Figure 3.5 Schematic diagram of an SIA-LOV measuring system (a) and a flow cell for absorptiometric and fluorimetric detection (b) used in the label dilution method for studies of bioligand interactions [32].

examined antagonist [54]. Obviously, moveable microbeads function as a renewable fluorimetric detector. The chemiluminescence detection with magnetic beads was employed in the flow system shown in Figure 3.1. Retaining and releasing the beads is carried out by shifting the magnet up and down. From microscopic observations it was concluded that the magnetic beads are strictly controlled in the flow system, regardless of the flow rate used for pumping the solution after trapping the beads [37].

There have been several reports of SIA systems with bead suspensions employed for preconcentration of trace analytes with the use of ETAAS detection that did not require any modification of the construction of commercial detectors, but only selection of the mode of transport of eluate to the graphite furnace. Such determinations were reported for trace level chromium(VI) in natural waters [55] and nickel(II) in saline matrices [25]. In the case of Cr(VI) determinations in the SIA-LOV system it was shown that, by the use of renewable sorbent, it is possible to achieve a larger concentration efficiency than for sorbent packed into a column, but a smaller efficiency than with the use of a knotted PTFE reactor. In the determination of Ni (II) in environmental and biological samples the application of the elution of beads with preconcentrated analyte directly to the graphite furnace of an AAS detector was also reported [43]. In conclusion, from such an approach the authors indicate the need to lower the ashing/pyrolysis temperature and shorten the duration of this step.

The transportation of beads from the microcolumn to the furnace was effected by air-segmentation with a small volume of carrier solution.

In the SIA-LOV system with precipitate preconcentration with lanthanum hydroxide retained on immobilized C18 beads, cadmium(II) was determined by hydride generation atomic fluorescence spectrometry [56]. In this case the sorbent bed was renewed only after a set of determinations when a build up in pressure and drop in the collection efficiency was observed.

A particular spectroscopic detection developed for flow SIA determinations with beads suspension is surface-enhanced Raman spectroscopy (SERS) [23, 24]. This scattering method requires a stable solid surface, stabilized, for example, by appropriate electric polarization or use of noble metals such as gold or silver colloids as substrates. Such determinations can be carried out by the use of a suspension of microbeads and, in flow conditions, the formation on their surface of a SERS-active colloid attached to the bead matrix. Using various chemical procedures this modification can be made for both anion- and cation-exchange beads. In the case of the anion-exchange resin an aqueous solution of silver nitrate was reduced by sodium borohydride and the silver sols formed were adsorbed on microbeads [23]. In the case of the cation-exchange resin its beads were perfused in the flow-cell with silver nitrate and hydroxylamine solutions [24]. An example of a flow-cell for such detection is shown in Figure 3.6. The microbeads were retained by a piece of titanium foil inserted into the flow channel after the place where the laser beam hits the flow channel. After measurement the beads were pumped to the reaction chamber were they were reacted with nitric acid, rinsed with water and then used for the next determination.

In the recent analytical literature one can also find papers reporting flow systems with the use of bead suspensions for the design of electrochemical detectors. Three different concepts have been described for amperometric determinations in SIA systems [27]. A general scheme for the functioning of the jet ring amperometric detector is shown in Figure 3.7. For this purpose a planar concentric platinum electrode with screen-printed Pt layer surrounding a central Pt disk electrode was

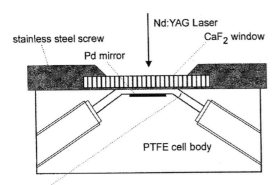

Figure 3.6 Schematic diagram of a flow cell for SERS detection in an SIA system operating with suspensions of microbeads [23].

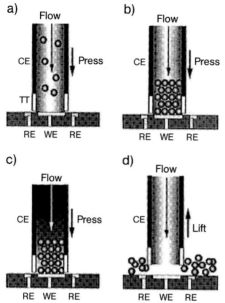

Figure 3.7 Schematic diagram of the operation of an amperometric jet ring detector for flow measurements with a suspension of beads [27]. (a) Introduction of beads, (b) accumulation of beads at the sensing surface, (c) perfusing beads with sample solution and monitoring of current, (d) discharge of beads. CE, counter electrode; TT, teflon tube; RE, reference electrode; WE, working electrode.

used. Microbeads pumped in as a suspension formed a packed-bed column. The outer ring Pt electrode served as a pseudoreference, while the stainless steel tube pressed by a spring electrode served as the auxiliary electrode. At the tip of the stainless steel tube a Teflon tube was installed to avoid a short circuit between the working and the counter electrode. The stainless steel tube was actuated via a computer-controlled solenoid. Three configurations of such an amperometric detector cell were examined using suspensions of different particles. In the first, using conducting glassy carbon microbeads 80–200 μm in diameter, a renewable working electrode was formed. Such packed-bed electrodes were widely examined earlier in electroanalysis [57], and used also as a conducting stationary phase in electromodulated liquid chromatography, for example Ref. [58], where polarization of microbeads was used for modification of the selectivity of retention. This may increase the sensitivity of amperometric response compared to a bare electrode without beads up to 12-fold. With the use of non-conducting particles with immobilized enzyme on the surface, an amperometric biosensing detector can be obtained for flow measurements with a renewable enzyme layer. In the most integrated system enzyme can be immobilized directly onto conducting microparticles and used as a renewable large surface of flow-through biosensor. In this case the measured signal was smaller than with the use of non-conducting microbeads, which can be attributed to less enzyme activity on conducting beads [27].

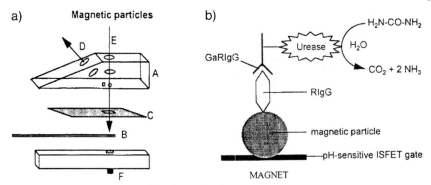

Figure 3.8 Schematic diagram of the flow cell with a potentiometric magnetoimmunosensor (a) and illustration of the concept of detection involving the use of magnetic beads with immobilized rabbit immunoglobulin G (RIgG) (b) [38]. A, methacrylate block; B, pH sensitive ISFET; C, silicone spacer; D, cell outlet; E, cell inlet; F, magnet.

The concept of accumulating and exchanging microbeads at the sensing surface of an electrode has also been employed with magnetic particles for potentiometric detection [38]. A schematic diagram of such a flow-through detector and a graphic illustration of its chemical functioning are shown in Figure 3.8. The sensing element of this detector at the bottom is a flat surface of ion-selective field-effect transistor (ISFET) sensitive to pH changes. Under the ISFET is located a magnet to retain modified magnetic beads at the sensing surface. On the surface of the magnetic beads rabbit immunoglobulin G (RIgG) was immobilized. In competitive immunoassays a solution containing goat anti-rabbit IgG (GaRIgG) urease conjugate and analyte RIgG was injected. Then injected substrate urea was hydrolyzed and the pH change was monitored with ISFET providing information about the amount of conjugate retained on the particles as a result of the competition of RIgG in solution and in the injected sample. After measurement of the potential the microparticles were released. Application of magnetic immunoparticles eliminates the need for chemical regeneration of the immunosensing surface.

In SIA systems operating with suspensions of beads ICP-MS [21] and radiometric [22] detections have been used. In both cases the use of suspensions did not require, and did not cause, any changes in the design of the detectors. With ICP-MS a direct injection high efficiency nebulizer was adopted from earlier work in the literature [59], and SIA measurements were made with a LOV rotary valve. In the case of the radiometric measurements a flow-through liquid scintillation detector was adopted with a 0.5 mL flow cell described in earlier FIA measurements [60].

3.5
Renewable Columns in Flow Systems

Because of the need for frequent switching of solutions, manipulation with small volumes and changing the direction of the flow, the majority of modern flow

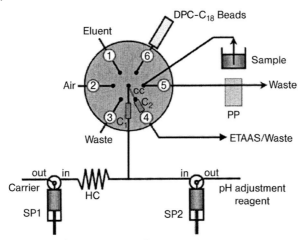

Figure 3.9 Schematic diagram of SIA-LOV system operating with a microbead suspension for the determination of Cr(VI) with ETAAS detection [55]. SP, syringe pumps; PP, peristaltic pump; HC, holding coil; DPC, 1,5-diphenylcarbazide immobilized on C18 beads.

measuring systems are designed with a sequential injection configuration. For this purpose multiposition rotary injection and switching valves are used (simple or more complex LOV, equipped with a miniaturized flow-through detector or/and a cavity for the sorbent), and a system of peristaltic and syringe pumps. A schematic diagram of such an SIA-LOV system with LOV, peristaltic pump and two syringe pumps, developed for ETAAS determinations of chromium(VI), is shown in Figure 3.9 [55]. Figure 3.10 shows a picture of a commercially available LOV valve with a cavity for microbeads, reproduced from a review article on SIA-LOV systems with renewable sorbent beds [61]. Two syringe pumps in the system in Figure 3.9 allow loading of the holding coil together with modification of the pH of the solution. The operation of such a system requires computer control. A procedure for single determination is composed of seven steps with numerous switching (15 times) of LOV and initializing two-directional operation of the injection valves. Transport of the eluate zone to the graphite furnace requires segmenting of the solution with air bubbles. The authors claim that the whole procedure of this mechanized determination for one sample takes only 4 min in spite of the complexicity of the procedure.

As described above, the application of particle suspension often provides improvement in the conditions of detection and is almost always employed for preconcentration of analyte by solid-phase extraction. The exchange of sorbent bed or performing the sorption each time on a new portion of sorbent has several important advantages. Early work on retaining and removal of a particle bed showed in measurements with fluorescence detection [13] a very quick rise of signal value and its return to the base-line – practically without any of the tailing typical for diffusion-convection processes in the liquid phase. In measurements with exchange of bed

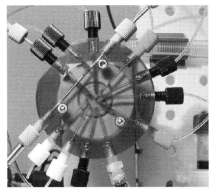

Figure 3.10 Picture of rotary switching valve incorporating cavity for retaining microbeads (LOV) [17].

and transport of suspension it is postulated that the microbeads should be perfectly spherical, uniform in size distribution and water-wettable [62]. However, this condition of hydrophilicity cannot be attained by all beads. In very early reported FIA determinations of trace level of metals with preconcentration by solid-phase extraction, the advantages of the application of hydrophobic sorbents with immobilized complexing ligands were demonstrated [63]. The selection of an appropriate ligand allows a very selective sorption, which can be used for speciation analysis, for example, at a given oxidation step of the determined analyte. An example of the application of this concept in an SIA system with particle suspension is the determination of Cr(VI) with preconcentration on a non-polar sorbent with immobilized 1,5-diphenylcarbazide. Application of sorbent with immobilized ligand does not improve significantly the kinetics of the whole preconcentration process, as usually the rate-determining process is the chelate-formation reaction, and not the sorption of chelate. A high selectivity of retention on beads modified with ligands also favors elimination of matrix effects.

For the preconcentration of trace analytes on a renewable bed with ETAAS detection direct introduction of a bead suspension with analyte to the graphite furnace can be utilized, instead of using eluate from the bed. Beads employed for this purpose usually consist primarily of organic material and can be pyrolyzed in a furnace [43]. It was shown, however, that this procedure also has some shortcomings [20]. Some amounts of products of pyrolysis can accumulate in the graphite tube, resulting in poor precision of the determination. The high temperature required to minimize the accumulation shortens the lifetime of the graphite tubes. Tolerance to some heavy metal ions may also be worse than in measurements in eluate from the bed. In determinations of Ni(II) with ETAAS detection it was demonstrated that, in comparison to measurements with injection of beads, for eluate injection the tolerance limits for Zn and Pb were improved up to 50-fold, and those for Cu and Mn up to 10-fold [20].

A similar clean-up effect resulting in reduction of spectroscopic and non-spectroscopic interferences was observed in the system with renewable sorbent bed and SIA measurements with ICP-MS detection [21]. A specific interference in this case can be isobaric. In the determination of ^{60}Ni such interferences can result, for example, from the formation of $^{44}Ca^{10}O$ and $^{36}Ar^{24}Mg$ ions. Although only a small part of the Ca and Mg was retained on the cationic resin used, it was necessary in a mechanized system to employ fractionated elution – first Ca and Mg with 1 : 80 diluted nitric acid, and then Ni and Bi with 6.25% nitric acid. It can be expected in this case, that application of a renewable bed in each measurement will improve the long term repeatability of the signal magnitude.

Several different factors have to be considered in biochemical applications of flow measurements in SIA systems with renewable sorbent beds. Generally, exchange of the bed is favorable when degradation of the bed may take place in multiple determinations, when elution of formed complexes is difficult, or when its application will improve the functional aspects of a particular procedure of determination. In the study of bioligand interactions the surface plasmon resonance (SPR) method is often used, which requires activation of the sensing surface for each determination. For comparison, in similar determinations carried out using SIA with a renewable column, for a large number of measurements a bulk of preactivated beads can be used, an obvious functional improvement [42].

In the separation of nucleic acids it was demonstrated, that the concept of SIA measurements with a renewable affinity microcolumn eliminates the need for routine operator intervention, disposables and consumables compared to other instruments that utilize microparticles [28]. This may lead to the design of convenient instruments, for example, for monitoring pathogenic microorganisms in food and the environment. This method was successfully employed for the development of a particular method of investigation of a bioligand interaction, namely label dilution [32]. The idea of this method is similar to the classical isotope dilution method. The analyzed sample is spiked with target molecule labeled, for instance, with chromophore. The spiked sample is then purified by interaction with selective bioligands immobilized on solid beads and interfering components are flushed away. The amount of unlabeled target compound, for example, protein, can be determined, based on the assumption that the labeled and non-labeled analogs of the target analyte react identically with immobilized bioligands. Renewal of the particle bed in such determinations, which can be an alternative for other formats of immunoassay for the determination of proteins of different binding affinities, is also helpful in obtaining satisfactory repeatability of determinations.

The SIA system with renewable column reported in the literature for affinity chromatography illustrates the possibility of a simple technical improvement of renewal of the stationary phase in low pressure chromatographic separations [33]. The same concept can be applied in other analytical systems for clean-up of samples of complex matrices prior to final detection.

3.6
Microfluidics with Processing of Particle Suspensions

As for many other devices used in various areas of modern life and for various purposes, if there is any real need, their miniaturization is unavoidable. Its scale and timing depend to a different degree on demand, as well as on progress in different areas of science and technology. In the case of chemical analytical instrumentation, the most important demand for miniaturization is in the transformation of large instruments used in specialized laboratories with highly trained personnel into devices for direct use by end-users. These demands are the main source of such trends in modern analytical instrumentation as miniaturization, mechanization and automation. It has to be admitted here that the last term in application to analytical instruments is still commonly misused, both in terms of general automation theory as well as in terms of the nomenclature recommendations formulated by IUPAC [64], but perhaps it might be tolerated as evidence of wishful thinking about the future of analytical instrumentation.

In a retrospective glance at half a century's history of laboratory flow analysis, that began with the pioneering work of Skeegs for clinical laboratories [65], the tendency towards miniaturization without doubt did not omit instrumentation for flow analysis. Nowadays, the most advanced examples of really miniaturized devices for flow analysis are microfluidic chips (see Chapter 6), which are systems mostly in the shape of thin plates with a network of capillary channels of micrometer diameters. Constructed and investigated since the beginning of the 1990s, they are nowadays available commercially, employed in commercial capillary electrophoresis instruments and find increasing use also in typical flow analysis in miniaturized format. Already in 1987 the theoretical limitations of the practical miniaturization of flow systems were discussed [66]. The intensive progress in the design and fabrication of microfluidics observed over the last two decades clearly indicates both the need for and technical possibilities of producing such devices [67, 68]. They are developed for applications such as portable diagnostic devices, DNA analysis, detection of microorganisms or biochemical warfare detection systems.

As the subject of this chapter is the application of suspensions of solid particles in flow analysis, only this aspect of the development of microfluidics will be discussed here. The number of published papers and technological achievements in this area has increased rapidly in recent years (see the review [69]). Technologically, design and fabrication of such devices are different to the analytical flow systems discussed above because of the scaling down, usually by at least one order of magnitude, in the dimensions of these systems.

As an example, one can find in the literature [70, 71] a microfluidic system for immunochemical determinations with the use of magnetic particles and amperometric detection, which is shown schematically in Figure 3.11, together with a diagram of all the steps of the determination. It is an integrated system for flow measurements with microvalves to produce the flow of solution, flow sensors, biofilters to achieve biosampling and biofiltering and an array of interdigitated microelectrodes for detection of the products of an enzymatic labeled reaction. The total volume of the fluidic chamber for biofiltration, reaction and detection is 750 nl.

The problem discussed first and developed in the application of suspensions of microparticles in microfluidics concerns the different concepts and solutions of transportation and retaining particles in such systems. Applications on the use of

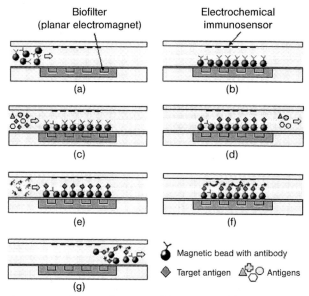

Figure 3.11 Schematic illustration of the concept of the immunoassay procedure [70]. (a) injection of magnetic beads, (b) separation and retention of beads, (c) flowing sample, (d) immobilization of target antigen, (e) flowing label antibody, (f) amperometric detection, (g) washing-out magnetic beads.

microfluidics with biomolecules immobilized on the walls of capillary channels are rare. More convenient is the use of suspensions of microparticles as they are an ideal reagent delivery system with a large surface area. Their advantage is also easier modification and characterization of the surface than capillary walls in microfluidics [72]. Similarly to the flow systems of commonly used size, an important problem to solve is transportation of the suspension and temporary immobilization of the particles. One of the typical difficulties in these systems is handling beads in etched channels. The movement and immobilization of microparticles is achieved by the design of appropriate architecture of the microfluidics or by the use of magnetic microparticles.

The simplest way to immobilize magnetic particles introduced by pressure-driven flow of the suspension is the application of a magnetic field. In the system developed for immunochemical determinations, shown in Figure 3.11, an embedded serpentine inductor structure with an electromagnet component was used to maximize magnetic forces on magnetic beads [70, 71]. A two-dimensional magnetic manipulation of the magnetic particles can be carried out with the use of a matrix of simple coils for the actuation [73]. A microfluidic with externally positioned magnets has been employed for the dynamic investigation of DNA hybridization [74]. A special way of manipulating magnetic beads for microfluidic mixing and assaying was developed by the use of a self-assembled structure of ferromagnetic beads that are retained within a microfluidic chip using a local alternating magnetic field [75]. The alternating field induces a rotational motion of the magnetic particles, which results in strong enhancement of the fluid perfusion through the bead structure. This strong particle–liquid interaction can be controlled by adjusting the magnetic field frequency and amplitude. This concept can be utilized for the design of microfluidics for bead-based assays.

Much attention in the design of microfluidics is focused on the transport of suspensions of non-magnetic particles and their immobilization for performing the determination. In the simplest case the microbeads are introduced into the microchannels by a syringe pump and retained there by the presence of weirs. This was used for instance for the hybridization of DNA to bead-immobilized probes using streptavidin-coated microbeads, conjugated with biotinylated single-strand DNA probes [76]. Trapping of non-magnetic particles is also possible by application of an ultrasonic transducer [77]. Standing ultrasonic waves are generated across a microfluidic channel by which particles are drawn to pressure minima in the acoustic field and trapped. Manipulation of the transport of microparticles can also be achieved by appropriate architecture of the microfluidics. Such structures should fulfill two purposes – to retain particles and, at the same time, allow transport of sample and reagents. Such designs have been developed, for example, for on-chip solid-phase extraction [78] or on-chip protein digestion prior to introduction to electrospray ionization mass spectrometry [79]. One can also find in the literature a description of surface patterning of a capillary wall by microcontact printing to alter the chemistry of the internal channel surface [80]. Such a modification enables the self-assembly of modified beads, for example, printing with biotinylated bovine serum albumin allows immobilization of straptavidin-coated beads. Submicrometer latex beads were immobilized with both a temperature-sensitive polymer and biotin [81]. Particles functionalized with the temperature-sensitive polymer and biotin are placed in a microchannel and when the temperature is raised the beads aggregate and adhere to the walls. After the reaction is finished and the temperature reduced to room temperature, the beads can be eluted from the channel for further analysis. The trapping and preconcentration of particles can also be carried out dielectrophoretically by the integration of microscopic electrodes into microfluidic devices, based on the generation of electrical cages attracting polarizable particles [82].

In the microfluidic format with bead processing the design of a special type of continuous biosensor with fluorescence detection has also been proposed [83]. Such a system allows one to detect biomarkers based on the concept of the "particle cross over" mechanism in a carefully designed microfluidic network. The device is based upon serial processing steps allowing binding of the analyte with cytometric beads activated with antibody to which antigen binds.

The few examples presented above from a wide literature indicate how many various concepts have been studied and realized in the competition for miniaturization of analytical instrumentation based on microfluidics. The application of particle suspensions in such systems has been widely examined, leading to better functioning of such devices and simplification of various analytical procedures (washing, operation with reagents, preparation of surface, etc.). The usually included step of preconcentration by solid-phase extraction facilitates the improvement of limits of detection. It seems that one can expect further improvement in the controlled movement of particles in microfluidics, which may result in better functional properties of these systems. Perhaps, further in the future, we can envisage individual addressing of microbeads in terms of a molecular recognition mechanism as suspension arrays, that can provide outstanding effectiveness of analytical determinations. One can also not forget the progress in the development of new sophisticated methods of detection for microfluidic technology, that additionally enhance their attractiveness as analytical instruments. Among expected applications are clinical diagnostics, physiological and biochemical investigations, and high-throughput drug screening.

3.7
Nanoparticles in Flow Systems

The large research and technological challenge observed in many branches of science and technology from the beginning of the 1990s has been the intensive development of nanotechnology. Self-organization of matter in artificially created conditions into objects of nanometer size may often be a source of new forms of elements or chemical compounds exhibiting unusual mechanical, optical or electrical properties. Interest in such materials has grown enormously together with the growth of numerous applications in various fields of science and technology.

Analytical chemistry, including flow analysis, has also made use of nanotechnology. Examples of nanostructures widely used for analytical purposes are self-organized monolayers on various supports used for the modification of electrodes in electro-analysis (also in flow measurements), in the flow method of surface plasmon resonance, and in piezoelectric detectors. Another example is artificial bilayer lipid membranes (BLM) with numerous applications in sensor technology.

Amongst widely investigated nanostructures are artificially synthesized nanofibers, nanotubes or nanoparticles. Besides many areas of science and technology, they also have numerous applications in biotechnology and biodiagnostics [84–86]. In analytical applications carbon nanotubes are especially frequently used, for example, for the design of gas detectors, the modification of electrodes in electroanalysis or as the sorption material in solid-phase extraction [87]. In the majority of these applications their success is attributed to their large surface area, electrocatalytic properties, strong sorption due to hydrophobic interactions and the ability to carefully tailor the physical properties. Additionally, numerous developed methods for chemical modification of surfaces provide the possibility of their wide application based on binding affinities to various biomolecules.

The advantages and various methods for the use of suspensions of microparticles in flow analysis and microfluidics have been discussed above. It is an open question, however, whether for these purposes a further decrease in the size of the particles down to nanosize might be advantageous. The chemical literature already provides data on numerous interesting applications of nanoparticles, which might be potentially useful in flow analysis with nanoparticle suspensions.

For analytical applications magnetic nanoparticles can be fabricated. Magnetic viral nanoparticles for the detection of viruses are composed of a superparamagnetic iron oxide core with a dextran coating onto which virus-surface-specific antibodies are attached via Protein G coupling [88]. After incubation with the corresponding virus, self-assembly occurs in solution causing changes in the measured water T_2 relaxation times. This allows detection of 50 viral particles in $100 \, \mu L$ of solution. Biomagnetic nanoparticles were obtained from strains which synthesize intracellular particles of magnetite (Fe_3O_4), aligned in chains and enveloped by a lipid membrane [89]. They are small in size (50–100 nm) and disperse very well due to the stable lipid membrane covering. They can be used for immobilization of enzymes and antibodies. They can be employed for sandwich chemiluminescence immunoassay of IgG [90]. One more example of the application of magnetic nanoparticles can be cited: particles were prepared from carbon nanotubes by deposition of iron nanoparticles inside the inner cavity of functionalized water-soluble multi-walled carbon nanotubes [91]. They were shown to be a good sorbent for several aromatic compounds, which can then be eluted from the sorbent with methanol. The possibility of separating the particles in a magnetic field makes them an attractive sorbent for analytical applications.

For the immunoassay of immunoglobulins the use of gold nanoshells has been reported [92]. These are layered, spherical nanoparticles, consisting of a dielectric silica core (96 nm diameter), surrounded by a thin metal (gold or silver) shell, and they possess a tunable surface plasmon resonant response. Their application in immunoassay was based, similarly to the method described above for latex particles, on aggregation of antibody/nanoshell conjugates in the presence of analyte with extinction spectra in the near-infrared. Successful detection of immunoglobulins was obtained in saline, serum, and whole blood.

Another type of nanoparticles developed for analytical purpose are the poly(decyl methacrylate)-based fluorescent particles PEBBLE (probes encapsulated by biologically localized embedding) of 150 nm–1 μm diameter range [93, 94]. After decyl methacrylate polymerization based on existing technologies of drug delivery, the spherical particles were modified with a highly selective ionophore and chromoionophore, which led to a fluorescent probe sensitive to potassium [93]. They were used successfully for the measurement of intracellular potassium activity, after delivery to the examined cells using a firing pressure method. The subsecond response time and full reversibility might also be well utilized in flow measurements, using a suspension of PEBBLE nanoprobes for various analytes, depending on the kind of ionochromophore used. Similarly, nanoparticles were also used in the measurement of dissolved oxygen in biosamples [94]. In this case platinum octaethylporphine ketone, the oxygen-sensitive dye, and octaethylporphyrin, the oxygen-insensitive dye, were incorporated into polymeric particles to make the sensors ratiometric. The developed probe was applied in oxygen measurements in human plasma. It showed very good stability and reversibility, very short response times and no perturbation by proteins.

At this point we should mention one additional aspect of nanostructures associated with further miniaturization of whole analytical flow measuring devices. This is a concept and the first experimental achievement in the design of nanofluidics. Attempts to construct such systems must involve consideration of the fact that movement of molecules of liquid inside channels of submicrometer diameter is associated with a change in the importance of physical and chemical interactions, compared to the micrometer scale or fractions of a millimeter. Forces interacting on the nanoscale involve not only interactions of the wall material with molecules of reagents, analyte or solvent, but also interactions between these molecules [95]. They may affect ionic equilibria in solution or kinetic phenomena. The electrostatic forces are determined by the formation of a double electric layer of thickness up to 10 nm and they may act in both directions. These forces may also induce some selectivity of transport of molecules in nanochannels. So far, as nanofluids, a nanoporous material employed in gel chromatography and a membrane with a network of nanochannels have been examined. Some designs based on such membranes have already been reported for analytical applications for various analytes with fluorescence detection [96, 97]. As another way of design in nano-fluidics the fabrication of GaN nanotubes integrated with a field transistor has been reported [98].

3.8
Conclusions

Operating with suspensions of solid particles may complicate to some degree the analytical flow instrumentation, but at the same time it allows one to gain evident advantages for analytical determinations, which are discussed in this chapter. The most important is easy to achieve renewal of the active surface of solid sorbents or the sensing surface of several different detectors. Another attractive advantage is the possibility, in some cases, to combine in one step of analytical procedure the act of detection and the preconcentration of trace analyte or sample clean-up by removal of complex and interfering matrices.

The degree of real utilization of such devices in routine chemical analysis depends essentially on the designers of analytical instrumentation and the ability of chemical analysts to employ various new materials and different chemical and biochemical interactions.

The future of these systems seems to be a further mechanization of their performance and their miniaturization by exploiting modern micromachining and nanotechnology. The miniaturized analyzers, integrating all necessary supporting instruments and devices, have a large potential for use in personal clinical diagnostics, as well as in personal monitoring devices, signaling environmental or any other chemical danger. Several steps towards the design of such devices have already been taken.

References

1 Thurman, E.M. and Mills, M.S. (1998) *Solid-Phase Extraction: Principles and Practice*, Wiley-Interscience, New York.

2 Bangs, L.B. (2001) Recent uses of microspheres in diagnostic tests and assays, Chapter 2, in *Novel Approaches in Biosensors and Rapid Diagnostic Assays* (eds Z. Liron, A. Bromberg and M. Fischer), Plenum, New York.

3 Lawruk, T.S., Lachman, C.E., Jourdan, S.W., Fleeker, J.R., Herzog, D.P. and Rubio, F.M. (1993) Determination of metolachlor in water and soil by a rapid magnetic particle-based ELISA. *Journal of Agricultural and Food Chemistry*, **41**, 1426–1431.

4 Rubio, F.M., Itak, J.A., Scutellero, A.M., Selisker, M.Y. and Herzog, D.P. (1991) Performance characteristics of a novel-magnetic-particle-based enzyme-linked immunosorbent assay for the quantitative analysis of atrazine and related triazines in water samples. *Food and Agricultural Immunology*, **3**, 113–118.

5 Nakamura, N., Hashimoto, K.K. and Matsunaga, T. (1991) Immunoassay method for the determination of immunoglobulin G using bacterial magnetic particles. *Analytical Chemistry*, **63**, 268–272.

6 Matsunaga, T., Kawasaki, M., Yu, X., Tsujimura, N. and Nakamura, N. (1996) Chemiluminescence enzyme immunoassay using bacterial magnetic particles. *Analytical Chemistry*, **68**, 3551–3554.

7 Sato, R., Takeyama, H., Tanaka, T. and Matsunaga, T. (2001) Development of high-performance and rapid immunoassay for model food allergen lysozyme using antibody-conjugated bacterial magnetic particles and a fully automated system. *Applied Biochemistry and Biotechnology*, **91–93**, 109–116.

8 Nolan, J.P., Lauer, S., Prossnitz, E.R. and Sklar, L.A. (1999) Flow cytometry: a versatile test for all phases of drug discovery. *Drug Discovery Today*, **4**, 171–180.

9 McBride, M.T., Gammon, S., Pitesky, M., O'Brien, T.W., Smith, T., Aldrich, J., Langlois, R.G., Colston, B. and Venkatewaran, K.S. (2003) Multiplexed liquid arrays for simultaneous detection of simulants of biological warfare agents. *Analytical Chemistry*, **75**, 1924–1930.

10 Cohen, E. and Stern, M. (1977) *Advances in Automated Analysis*, 1976 Technicon International Congress Mediad Press, Tarrytown, New York, p. 232.

11 Van Cleve, M., Ostrerova, N., Tietgen, K., Cao, W., Chang, C., Collins, M.L., Kolberg, J., Urdea, M. and Lohman, K. (1998) Direct quantification of HIV by flow cytometry using branched DNA signal amplification. *Molecular and Cellular Probes*, **12**, 243–247.

12 Pei, R., Lee, J., Chen, T., Rojo, S. and Terasaki, P.I. (1999) Flow cytometric detection of HLA antibodies using a spectrum of microbeads. *Human Immunology*, **60**, 1293–1302.

13 Ruzicka, J., Pollema, C.H. and Scudder, K.M. (1993) Jet ring cell: a tool for flow injection spectroscopy and microscopy on a renewable solid support. *Analytical Chemistry*, **65**, 3566–3570.

14 Ruzicka, J. (1995) Flow-injection renewable surface techniques. *Analytica Chimica Acta*, **308**, 14–19.

15 Sole, S., Mekoci, A. and Alegret, A. (2001) New materials for electrochemical sensing. III. Beads. *Trends in Analytical Chemistry*, **20**, 102–110.

16 Hartwell, S.K., Christian, G.D. and Grudpan, K. (2004) Bead injection with a simple flow injection system: an economical alternative for trace analysis. *Trends in Analytical Chemistry*, **23**, 619–623.

17 Ruzicka, J. (2006) Flow Injection Analysis, 3rd edn., Tutorial available from FIAlab Instruments, Inc. fialab@flowinjection.com.

18 Shapiro, H.M. (2003) *Practical Flow Cytometry*, 4th edn., Wiley-Liss, New York.

19 Yamada, M., Nakasima, M. and Seki, M. (2004) Pinched flow fractionation: continuous size separation of particles utilizing a laminar flow profile in a pinched microchannel. *Analytical Chemistry*, **76**, 5465–5471.

20 Wang, J. and Hansen, E.H. (2001) Coupling sequential injection on-line preconcentration by means of a renewable microcolumn with ion-exchange beads with detection by electrothermal atomic absorption spectrometry. *Analytica Chimica Acta*, **435**, 331–342.

21 Wang, J. and Hansen, E.H. (2001) Interfacing sequential injection on-line preconcentration using a renewable micro-column incorporated in a "lab-on-valve" system with direct injection nebulization inductively coupled plasma mass spectrometry. *Journal of Analytical Atomic Spectrometry*, **16**, 1349–1355.

22 Egorov, O., O'Hara, M.J. and Grate, J.W. (1999) Sequential injection renewable separation column instrument for automated sorbent extraction separations of radionuclides. *Analytical Chemistry*, **71**, 345–352.

23 Lendl, B., Ehmoser, H., Frank, J. and Schindler, R. (2000) Flow analysis-based surface-enhanced Raman spectroscopy employing exchangeable microbeads as SERS-active surfaces. *Applied Spectroscopy*, **54**, 1012–1018.

24 Canada, M.J.A., Medina, A.R., Frank, J. and Lendl, B. (2002) Bead injection for surface enhanced Raman spectroscopy: automated on-line monitoring of substrate generation and application in quantitative analysis. *Analyst*, **127**, 1365–1369.

25 Long, X.-B., Miro, M., Jensen, R. and Hansen, E.H. (2006) Highly selective micro-sequential injection lab-on-valve method for the determination of ultra-trace concentrations of nickel in saline matrices using detection by electrothermal atomic absorption spectrometry. *Analytical and Bioanalytical Chemistry*, **386**, 739–748.

26 Quintana, J.B., Miro, M., Estela, J.M. and Cerda, V. (2006) Automated on-line renewable solid-phase extraction-liquid chromatography exploiting multisyringe flow injection-bead injection lab-on-valve analysis. *Analytical Chemistry*, **78**, 2832–2840.

27 Mayer, M. and Ruzicka, J. (1996) Flow injection based renewable electrochemical sensor system. *Analytical Chemistry*, **68**, 3808–3814.

28 Chandler, D.P., Schuck, B.L., Brockman, F.J. and Brucken-Lea, C.J. (1999) Automated nucleic acid isolation and purification from soil extracts using renewable affinity microcolumns in a sequential injection system. *Talanta*, **49**, 969–983.

29 Ruzicka, J. (1998) Bioligand interaction assay by flow injection absorptiometry using a renewable biosensor system enhanced by spectral resolution. *Analyst*, **123**, 1617–1623.

30 Ogata, Y., Scampavia, L., Ruzicka, J., Scott, C.R., Gelb, M.H. and Turecek, F. (2002) Automated affinity capture-release of biotin-containing conjugates using a lab-on-valve apparatus coupled to UV/Visible and electrospray ionization mass spectrometry. *Analytical Chemistry*, **74**, 4702–4708.

31 Ruzicka, J. (2000) Lab-on-valve: universal microflow analyzer based on sequential and bead injection. *Analyst*, **125**, 1053–1060.

32 Carroll, A.D., Scampavia, L. and Ruzicka, J. (2002) Label dilution method: a novel tool for bioligand interaction studies using bead injection in the lab-on-valve format. *Analyst*, **127**, 1228–1232.

33 Erxleben, H. and Ruzicka, J. (2005) Sequential affinity chromatography miniaturized within a "lab-on-valve" system. *Analyst*, **130**, 469–471.

34 Lähdesmäki, I., Scampavia, L.D., Beeson, C. and Ruzicka, J. (1999) Detection of oxygen consumption of cultured adherent cells by bead injection spectroscopy. *Analytical Chemistry*, **71**, 5248–5252.

35 Kawaguchi, H. (2000) Functional polymer microspheres. *Progress in Polymer Science*, 25, 1171–1210.

36 Lui, J., Gunther, A. and Bilitewski, U. (1997) Detection of methamidophos in vegetables using a photometric flow injection system. *Environmental Monitoring and Assessment*, 44, 375–382.

37 Soh, N., Nishiyama, H., Asano, Y., Imato, T., Masadome, T. and Kurokawa, Y. (2004) Chemiluminescence sequential injection immunoassay for vitellogenin using magnetic microbeads. *Talanta*, 64, 1160–1168.

38 Santandreu, M., Sole, S., Fabregas, E. and Alegret, S. (1998) Development of electrochemical immunosensing systems with renewable surfaces. *Biosensors & Bioelectronics*, 13, 7–17.

39 Deng, T., Whitesides, G.M., Radhakrishnan, M., Zabow, G. and Prentiss, M. (2001) Manipulation of magnetic microbeads in suspension using micromagnetic systems fabricated with soft lithography. *Applied Physics Letters*, 78, 1775–1777.

40 Manz, A., Effenhauser, C.S., Burgraf, N., Harrison, D.J., Seiler, K. and Fluri, K. (1994) Electroosmotic pumping and electrophoretic separations for miniaturized chemical analysis system. *Journal of Micromechanics and Microengineering*, 4, 257–265.

41 Chandler, D.P., Brockman, F.J., Holman, D.A., Grate, J.W. and Bruckner-Lea, C.J. (2000) Renewable microcolumns for solid-phase nucleic acid separations and analysis from environmental samples. *Trends in Analytical Chemistry*, 19, 314–321.

42 Ruzicka, J. and Ivaska, A. (1997) Bioligand interaction assay by flow injection absorptiometry. *Analytical Chemistry*, 69, 5024–5030.

43 Wang, J. and Hansen, E.H. (2000) Coupling on-line preconcentration by ion-exchange with ETAAS. A novel flow injection approach based on the use of a renewable microcolumn as demonstrated for the determination of nickel in

environmental and biological samples. *Analytica Chimica Acta*, 424, 223–232.

44 Yoshimura, K., Waki, H. and Ohashi, S. (1976) Ion-exchange colorimetry – I. Microdetermination of chromium, iron, copper and cobalt in water. *Talanta*, 23, 449–454.

45 Yoshimura, K. and Waki, H. (1985) Ion-exchanger phase absorptiometry for trace analysis. *Talanta*, 32, 345–352.

46 Yoshimura, K. and Waki, H. (1987) Enhancement of sensitivity of ion-exchanger absorptiometry by using a thick ion-exchanger layer. *Talanta*, 34, 239–242.

47 Yoshimura, K. (1987) Implementation of ion-exchanged absorptiometric detection in flow analysis systems. *Analytical Chemistry*, 59, 2922–2924.

48 Lazaro, F., Luque de Castro, M.D. and Valcarcel, M. (1988) Integrated reaction/spectrophotometric detection in unsegmented flow analysis. *Analytica Chimica Acta*, 214, 217–227.

49 Matsuoka, S., Yoshimura, K. and Tateda, A. (1995) Application of ion-exchanger phase visible light absorption to flow analysis. Determination of vanadium in natural water and rock. *Analytica Chimica Acta*, 317, 207–213.

50 Teixeira, L.S.G. and Rocha, F.R.P. (2007) A green analytical procedure for sensitive and selective determination of iron n water samples by flow injection solid-phase spectrophotometry. *Talanta*, 71, 1507–1511.

51 Fernandez-Band, B., Linares, P., Luque de Castro, M.D. and Valcarcel, M. (1991) Flow-through sensor for the direct determination of pesticide mixtures without chromatographic separation. *Analytical Chemistry*, 63, 1672–1675.

52 Tennichi, Y., Matsuoka, S. and Yoshimura, K. (2000) Simultaneous determination of trace metals by solid phase absorptiometry: application to flow analysis of some rare earths. *Fresenius' Journal of Analytical Chemistry*, 368, 443–448.

53 Ruzicka, J., Carroll, A.D. and Lähdesmäki, I. (2006) Immobilization of proteins on

agarose beads, monitored in real time by bead injection spectroscopy. *Analyst*, **131**, 799–808.

54 Hodder, P.S., Beeson, C. and Ruzicka, J. (2000) Equilibrium and kinetic measurements of muscarinic receptor antagonism on living cells using bead injection spectroscopy. *Analytical Chemistry*, **72**, 3109–3115.

55 Long, X., Miro, M. and Hansen, E.H. (2005) Universal approach for selective trace metal determination via sequential injection-bead injection-lab-on-valve using renewable hydrophobic bead surface as reagent carriers. *Analytical Chemistry*, **77**, 6032–6040.

56 Wang, Y., Chen, M.L. and Wang, J.H. (2006) Sequential/bead injection lab-on-valve incorporating a renewable microcolumn for co-precipitate preconcentration of cadmium coupled to hydride generation atomic fluorescence spectrometry. *Journal of Analytical Atomic Spectrometry*, **21**, 535–538.

57 Sioda, R.E. and Keating, K.B. (1982) Flow electrolysis with extended-surface electrodes. Chapter, in *Electroanalytical Chemistry, A Series of Advances*, Vol. 12, Marcell Dekker, New York.

58 Harnish, J.A. and Porter, M.D. (2001) Electrochemically modulated liquid chromatography: electrochemical strategy for manipulating chromatographic retention. *Analyst*, **126**, 1841–1849.

59 McLean, J.A., Zhang, H. and Monaster, A. (1998) A direct injection high-efficiency nebulizer for inductively coupled plasma mass spectrometry. *Analytical Chemistry*, **70**, 1012–1020.

60 Grate, J., Strebin, R.S., Janata, J., Egorov, O. and Ruzicka, J. (1996) Automated analysis of radionulides in nuclear waste: Rapid determination of ^{90}Sr by sequential injection analysis. *Analytical Chemistry*, **68**, 333–340.

61 Wang, J. and Hansen, E.H. (2003) Sequential injection lab-on-valve: the third generation of flow injection analysis. *Trends in Analytical Chemistry*, **22**, 223–231.

62 Wang, J.H., Hansen, E.H. and Miro, M. (2003) Sequential injection-bead injection-lab-on-valve schemes for on-line solid-phase extraction and preconcentration of ultra-trace levels of heavy metals with determination by electrothermal atomic absorption and inductively coupled plasma mass spectrometry. *Analytica Chimica Acta*, **499**, 139–147.

63 Olbrych-Śleszyńska, E., Brajter, K., Matuszewski, W., Trojanowicz, M. and Frenzel, W. (1992) Modification of nonionic adsorbent with Eriochrome Blue-Black R for selective nickel(II) preconcentration in conventional and flow injection atomic-absorption spectrometry. *Talanta*, **39**, 779–787.

64 Kingston, H.M. and Kingston, M.L. (1994) Nomenclature in laboratory robotics and automation. *Pure and Applied Chemistry*, **66**, 609–630.

65 Skegges, L.T., Jr. (1957) An automated method for colorimetric analysis. *American Journal of Clinical Pathology*, **28**, 311–316.

66 Van der Linden, W.E. (1987) Miniaturisation in flow injection analysis. Practical limitations from theoretical point of view. *Trends in Analytical Chemistry*, **6**, 37–40.

67 Tabeling, P. (2006) *Introduction to Microfluidics*, Oxford University Press, USA.

68 Minteer, S.D. (2006) *Microfluidic Techniques: Reviews and Protocols*, Humana Press, Totowa, NJ.

69 Verpoorte, E. (2003) Beads and chips: new recipes for analysis. *Lab on a Chip*, **3**, 60N–68N.

70 Choi, J.W., Oh, K.W., Han, A., Okulan, N., Wijayawardhana, C.A., Lannes, C., Bhansali, S., Schlueter, K.T., Heineman, W.R., Halsall, H.B., Navin, J.H., Helmicki, A.J., Henderson, H.T. and Ahn, C.H. (2001) Development and characterization of microfluidic devices and systems for magnetic bead-based biochemical detection. *Biomedical Microdevices*, **3**, 191–200.

71 Choi, J.W., Oh, K.W., Thomas, J.U., Heineman, W.R., Halsall, H.B., Navin,

J.H., Helmicki, A.J., Henderson, H.T. and Ahn, C.H. (2002) An integrated microfluidic biochemical detection system for protein analysis with magnetic bead-based sampling capabilities. *Lab on a Chip*, **2**, 27–30.

72 Walsh, M.K., Wang, X. and Weimer, B.C. (2001) Optimizing the immobilization of single stranded DNA onto glass beads. *Journal of Biochemical and Biophysical Methods*, **47**, 221–231.

73 Lehmann, U., Vandevyver, C., Parashar, V.K. and Gijs, M.A.M. (2006) Droplet-based DNA purification in a magnetic lab-on-chip. *Angewandte Chemie-International Edition*, **45**, 3062–3067.

74 Fan, Z.H., Mangru, S., Granzow, R., Heaney, P., Ho, W., Dong, Q. and Kumar, R. (1999) Dynamic DNA hybridization on a chip using paramagnetic beads. *Analytical Chemistry*, **71**, 4851–4859.

75 Rida, A. and Gijs, M.A.M. (2004) Manipulation of self-assembled structures of magnetic beads for microfluidic mixing and assaying. *Analytical Chemistry*, **76**, 6239–6246.

76 Kim, J., Heo, J. and Crooks, R.M. (2006) Hybridization of DNA to bead-immobilized probes confined within a microfluidic channel. *Langmuir*, **22**, 10130–10134.

77 Lilliehorn, T., Simu, U., Nilsson, M., Almqvist, M., Stepinski, T., Laurell, T., Nilsson, J. and Johansson, S. (2005) Trapping of microparticles in the near field of an ultrasonic transducer. *Ultrasonics*, **43**, 293–303.

78 Oleschuk, R.D., Shultz-Lockyear, L.L., Ning, Y. and Harrison, D.J. (2000) Trapping of bead-based reagents within microfluidic systems: on-chip solid-phase extraction and electrochromatography. *Analytical Chemistry*, **72**, 585–590.

79 Wang, C., Oleschuk, R., Ouchen, F., Li, J., Thibault, P. and Harrison, D.J. (2000) Integration of immobilized trypsin bead beds for protein digestion within a microfluidic chip incorporating capillary electrophoresis separation and an electrospray mass spectrometry interface. *Rapid Communications in Mass Spectrometry: RCM*, **14**, 1377–1383.

80 Andersson, H., Jonsson, C., Moberg, C. and Stemme, G. (2002) Self-assembled and self-sorted array of chemically active beads for analytical and biochemical screening. *Talanta*, **56**, 301–308.

81 Malmstadt, N., Yager, P., Hoffman, A.S. and Stayton, P.S. (2003) A smart microfluidic affinity chromatography matrix composed of poly(N-isopropylacrylamide)-coated beads. *Analytical Chemistry*, **75**, 2943–2949.

82 Muller, T., Gradl, G., Howitz, S., Shirley, S., Schnelle, T. and Fuhr, G. (1999) A 3-D microelectrode system for handling and caging single cells and particles. *Biosensors & Bioelectronics*, **14**, 247–256.

83 Yang, S., Undar, A. and Zahn, J.D. (2007) Continuous cytometric bead processing within a microfluidic device for bead based sensing platform. *Lab on a Chip*, **7**, 588–595.

84 Baron, R., Willner, B. and Willner, I. (2007) Biomolecule-nanoparticle hybrids as functional units for nanobiotechnology. *Chemical Communications*, 323–332.

85 Rosi, N.L. and Mirkin, C.A. (2005) Nanostructures in biodiagnostics. *Chemical Reviews*, **105**, 1547–1562.

86 Niemeyer, C.M. (2001) Nanoparticles, proteins, and nucleic acids: Biotechnology meets materials science. *Angewandte Chemie-International Edition*, **40**, 4129–4158.

87 Trojanowicz, M. (2006) Analytical applications of carbon nanotubes. *Trends in Analytical Chemistry*, **25**, 480–489.

88 Perez, J.M., Simeone, F.J., Saeki, Y., Josephson, L. and Weissleder, R. (2003) Viral-induced self-assembly of magnetic nanoparticles allows the detection of vital particles in biological media. *Journal of the American Chemical Society*, **125**, 10192–10193.

89 Matsunaga, T. and Takeyama, H. (1998) Biomagnetic nanoparticle formation and

application. *Supramolecular Science*, **5**, 391–392.

90 Matsunaga, T., Kawasaki, M., Yu, X., Tsujimura, N. and Nakamura, N. (1996) Chemiluminescence enzyme immunoassay using bacterial magnetic particles. *Analytical Chemistry*, **68**, 3551–3554.

91 Jin, J., Li, R., Wang, H., Chen, H., Liang, K. and Ma, J. (2007) Magnetic Fe nanoparticle functionalized water-soluble multi-walled carbon nanotubes towards the preparation of sorbent for aromatic compounds removal. *Chemical Communications*, 386–388.

92 Hirsch, L.R., Jackson, J.B., Lee, A., Halas, N.J. and West, J.L. (2003) A whole blood immunoassay using gold nanoshells. *Analytical Chemistry*, **75**, 2377–2381.

93 Brasuel, M., Kopelman, R., Miller, T.J., Tjalkens, T. and Philbert, M.A. (2001) Fluorescent nanosensors for intracellular chemical analysis: decyl methylacrylate liquid polymer matrix and ion-exchange-based potassium PEBBLE sensors with real-time application to viable rat C6 glioma cells. *Analytical Chemistry*, **73**, 2221–2228.

94 Cao, Y., Koo, Y.E.L. and Kopelman, R. (2004) Poly(decyl methacrylate)-based fluorescent PEBBLE swarm nanosensors for measuring dissolved oxygen in biosamples. *Analyst*, **129**, 745–750.

95 Plecis, A., Schoch, R.B. and Renaud, P. (2005) Ionic transport phenomena in nanofluidics: experimental and theoretical study of the exclusion-enrichment effect on a chip. *Nano Letters*, **5**, 1147–1155.

96 Kuo, T.C., Kim, H.K., Cannon, D.M., Shannon, M.A., Sweeler, J.V. and Bohn, P.W. (2004) Nanocapillary arrays effect mixing and reaction in multilayer fluidic structures. *Angewandte Chemie-International Edition*, **43**, 1862–1865.

97 Chang, I.H., Tulock, J.J., Liu, J., Kim, W.S., Cannon, D.M., Lu, Y., Bohn, P.W., Sweedler, J.V. and Cropek, D.M. (2005) Miniaturized lead sensor based on lead-specific DNAzyme in a nanocapillary interconnected microfluidic device. *Environmental Science & Technology*, **39**, 3756–3761.

98 Goldberger, J., Fan, R. and Yang, P. (2006) Inorganic nanotubes: a novel platform for nanofluidics. *Accounts of Chemical Research*, **39**, 239–248.

4
Batch Injection Analysis

Christopher M.A. Brett

4.1
Introduction

Batch injection analysis (BIA), as its name suggests, is a hybrid between flow injection analysis (FIA) and batch, that is, discrete, analysis [1]. Its general aim, as for other flow analysis systems, is to furnish analytical data rapidly, and as close to real time as possible [2, 3]. Its particular objective is to increase the simplicity of the experiments undertaken, removing mechanically moving parts, and removing the constantly flowing carrier stream. For this reason the technique has also been referred to as "tubeless flow injection analysis" [4]. Samples are injected directly over a detector immersed in a suitable solution. The disadvantage is that, with one or two exceptions, chemical reactions cannot be carried out before the detector.

The beginnings of batch cells and injection over an inverted detector can be traced back to 1976 in a paper which sought to develop a simple methodology for analysing small, microliter volumes of solution, in this case by voltammetry, with injection of samples into a batch microcell specially developed for the purpose and allowing a throughput of 60 samples per hour [5]. The importance of the analysis of micro-volumes arose during the 1980s and developed significantly in the 1990s [6, 7]. Batch injection analysis, first described by Wang and Taha in 1991 [1] who coined the term, also concerns the analysis of microvolumes but the containing cell can be as large as desired without influencing the detection. Since then many papers on BIA have been published, nearly all of them using electrochemical voltammetric or potentiometric detectors. The applications using electrochemical detectors up to 2004 have been reviewed [8].

In this chapter, the theory and principles of the BIA technique will be described. Then, following a description of cell design and detection strategies, applications will be discussed with examples. A short comparison with FIA will be made and future perspectives will be indicated.

Advances in Flow Analysis. Edited by Marek Trojanowicz
Copyright © 2008 WILEY-VCH Verlag GmbH & Co. KGaA, Weinheim
ISBN: 978-3-527-31830-8

4.2
Theory of Batch Injection Analysis

BIA involves the injection of an aliquot of analyte, of microliter volume, directly over the centre of a stationary detector, usually a circular disk. During the injection and whilst the sample plug is flowing over the detector, the situation is equivalent to a continuous carrier solution into which the analyte sample has been introduced, except that the sample dispersion is zero. This means that the theory for the equivalent situation in continuous flow systems can be applied, with appropriate modifications, to the theoretical response.

The theory has been studied for electrochemical detectors. The hydrodynamics resulting from the imposed convection caused by the submerged impinging jet corresponds to wall-jet and wall-tube electrodes [9]. If the impinging jet is significantly smaller in diameter than the detector, this is designated as the wall-jet electrode and if the jet is the same size or larger than the electrode, thus guaranteeing uniform accessibility, it is designated as a wall-tube electrode. Particular practical advantages of both these flow-past hydrodynamic electrodes are that there is no reagent depletion while the sample plug passes the electrodes, and build-up of unwanted intermediates or products is avoided. With modern instrumentation it is also possible that kinetic and mechanistic, besides analytical, information may be obtained simultaneously in many experiments. It should be noted that in one report on electrochemical BIA, a rotating disk electrode, rather than a stationary electrode, was used to further increase the convection [10], but the extra benefits were not sufficient to warrant its continued development, particularly as the system involved mechanically moving parts.

The wall-jet electrode is an excellent detector for use in electroanalysis [11], owing to its high flow-rate dependence and low dead volume. Since typical impinging jet diameters in BIA are 0.3–0.5 mm and electrode diameters are 3.0–5.0 mm, the parameters for wall-jet hydrodynamic flow are fulfilled.

The theory of wall-jet electrodes will now be briefly reviewed. The fine jet of solution from the injection impinges perpendicularly on the center of the disk electrode and then spreads out radially over its surface, as shown in the calculated schematic streamlines of Figure 4.1. This figure demonstrates clearly that the wall in which the electrode is embedded is highly non-uniformly accessible and that only fresh solution from the impinging jet can reach the electrode surface – there is nothing from solution recirculation. This fact is crucial for the application of wall-jet disk and ring-disk electrodes [12, 13], and the cells must be designed to allow this hydrodynamic profile to develop correctly. There is, as mentioned in the introduction, no influence from the size of the electrochemical cell so long as it is large enough, the important parameter with respect to the response at the detector being the volume and volume flow rate of the jet of incoming solution.

The maximum response that can be obtained at the wall-jet electrode is the diffusion limited current, I_L.

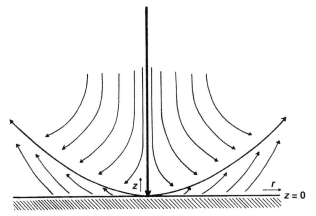

Figure 4.1 Schematic streamlines at a wall-jet electrode showing the impinging jet and the resultant radial convection. Solution from fluid recirculation within the cell cannot reach the electrode surface (from Ref. [12] with permission of Elsevier).

$$I_L = 1.38\, nFD^{2/3}v^{-5/12}a^{-1/2}V_f^{3/4}R^{3/4}c_\infty \tag{1}$$

in which n is the number of electrons transferred, F the Faraday constant, D the diffusion coefficient of the electroactive species, of concentration c_∞, a the diameter of the jet which impinges on the disk electrode of radius R at volume flow rate V_f, and v is the kinematic viscosity of the solution. The non-uniform accessibility – the diffusion layer thickness falls off with the (5/4) power of the radial distance – and high flow rate dependence can be extremely useful in the study of the kinetics of simple and complex electrode reactions. Analytical expressions to deduce the kinetics from the voltammetric profile for a simple irreversible reaction have been deduced [14, 15]. Computational procedures have also been employed [16] to deduce these parameters; the exploitation of the highly non-uniform accessibility for better discrimination between electrode reaction mechanisms than at uniformly accessible electrodes has been demonstrated [17]. Potential-step chronoamperometry [18, 19] and linear sweep voltammetry [20] have also been investigated and the limitations of using approximate theoretical responses examined.

In BIA experiments and during the injection, after a short initial period to reach steady-state, the hydrodynamics is wall-jet type and a time-independent current is registered. Thus, during the injection, once the steady state has been reached, wall-jet theory can be directly applied to deduce information concerning the magnitude of the current response [21]. In principle, kinetic and mechanistic information can also be obtained and the response analyzed according to the criteria described in Refs. [14–20]. The following section will discuss how cells should be designed to ensure that wall-jet hydrodynamics can be achieved and the various detection strategies which can be employed.

4.3
Experimental Aspects – Cell Design and Detection Strategies

A cell for a wall-jet continuous flow system is readily modified to discrete injection mode as shown in Figure 4.2, illustrated for electrochemical detection, where direct injection of the liquid sample from a micropipette tip replaces the continuous flow. Typical cell dimensions are given in the figure. In order for wall-jet hydrodynamics to be followed, the wall opposite the detector should be sufficiently far from the detector and the micropipette tip several millimeters distant from the detector; both of these criteria are to avoid interference in the hydrodynamics from solution recirculation within the cell. Generally the volume of electrolyte is around 35–40 cm^3, a size found to be convenient early on in cell design so as to permit easy portability of the batch injection system [21–23]. A sample of volume $\leq$100 μL – which normally corresponds to the maximum injectable volume from a micropipette – is injected directly from the micropipette, tip internal diameter ~0.5 mm, over the center of a macroelectrode in a configuration exactly the same as in a wall-jet system. Thus this is equivalent to a flow injection system with zero dispersion. The large volume of the cell electrolyte solution compared to the injected volume allows many injections, typically of 50 μL in analytical applications, to be performed without the necessity of removing solution or replenishing the electrolyte. As stated in the previous section and shown in Figure 4.1, there are no memory effects from the cell electrolyte solution that already contains the analyte species from previous injections.

A typical amperometric trace is shown in Figure 4.3. It shows the short initial period needed to reach steady-state, up to time t_1, after which the hydrodynamics is wall-jet type and a time-independent current is registered until the injection ends at time t_2. If the kinetics of the electrode reaction is not fast compared to mass transport, then a longer time will be needed to reach the maximum, plateau current. Programmable, motorised electronic micropipettes [24] enable the use of several injection

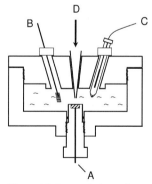

Figure 4.2 A typical electrochemical batch injection analysis cell with screw lid, made of Perspex; diameter ~12 cm, height 5 cm, internal volume ~40 cm^3. A: disk electrode contact; B: auxiliary electrode; C: reference electrode; D: injection from micropipette through micropipette tip.

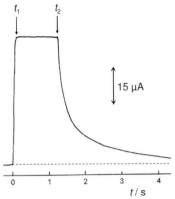

Figure 4.3 Amperometric transient corresponding to oxidation of 1 mM $K_4Fe(CN)_6$ in 0.4 M K_2SO_4 electrolyte, recorded at a fixed potential of $+0.6\,V$ vs. saturated calomel electrode on a Pt electrode (diameter 3.3 mm). Injection of 100 μL from electronic micropipette, jet diameter 0.47 mm, injection flow rate 75.3 μL s^{-1}. Injection begins at $t = 0$; maximum current is reached at $t = t_1$ and the injection finishes at $t = t_2$.

flow rates and the possibility of programming a series of consecutive injections, of equal or different volumes, up to the maximum total micropipette volume. Using this electronically-controlled injection procedure, volume accuracy and reproducibility are excellent.

Designs of cells have been described which do not place the tip of the micropipette so accurately above the detector and which use manual micropipettes. Whilst these approaches always give a response, it cannot be expected that the accuracy and precision will be as good as if the jet is properly centered at a known fixed distance and, additionally, if the rate of injection varies slightly (particularly so because the response is highly dependent on the volume flow rate – $V_f^{3/4}$).

Wall-jet hydrodynamics and the applicability of Equation (4.1) have been verified experimentally [21, 25]. The experimental confirmation of this primary wall-jet flow characteristic, that is, that no solution from recirculation can reach the electrode surface during the injection period, is extremely important. It means that the response to the injection of a sample does not depend on the solution composition close to the electrode before the injection, that is, there are no memory effects from previous injections or from the electrolyte in which the detector is immersed, see Figure 4.1.

An additional and very significant point is that the electrical resistance of a sample without added electrolyte does not make a significant contribution to the total cell resistance because it is only a thin layer of solution close to the working electrode. This has also been verified [25], and means that sample dilution is not necessary; besides the simplification this brings, it also avoids any change in the sample solution speciation due to adding electrolyte. This can be particularly important in diagnostic testing and full analysis of natural samples, which generally have complex matrices.

With this experimental arrangement, various types of detector – electrochemical and non-electrochemical – can be used to generate a profile similar to that shown in Figure 4.3. The constant response signal during the injection is monitored and is a direct indication of the amount of species present in the analyte sample.

It is also possible to alter the material and design of the detector. In Ref. [26], a BIA cell configuration was devised which uses thin film gold circular electrodes produced by radiofrequency sputtering on poly(vinyl chloride) films. Electrical contacts were added and the electrode held immersed in solution under the micropipette tip where the injection is performed. Excellent agreement with the results obtained with normal BIA cell designs was found using hydrogen peroxide measurements. The inexpensive electrode construction allows them to be used as disposable electrodes, if necessary.

The first applications of BIA were to amperometric [1], and potentiometric [27], detection. For electrochemical detection, in principle, all voltammetric techniques that can be used at wall-jet electrodes can be applied together with BIA, with the added possibility of obtaining information easily and without sample preparation or use of complex manifolds or sample dilution.

Apart from amperometric measurements at fixed potential, for example by applying a potential equal to that corresponding to the limiting current (Figure 4.3) [18], other experiments which can be carried out, as shown in Figure 4.4, can be summarized as [25]:

- Construction of a point-by-point pseudo-steady-state voltammetric curve by making consecutive injections during a slow linear potential scan (Figure 4.4a), followed by Tafel analysis or curve fitting.
- Cyclic voltammetry during or after sample injection to determine the kinetics of the electrode reaction and/or concentration determination (Figure 4.4b);
- Square wave voltammetric scans during the injection to give kinetic and analytical information (Figure 4.4c).

The detection limit with these simple types of experiment is 2×10^{-5} M for amperometric or pseudo-voltammetric experiments down to 5×10^{-6} M with square wave voltammetry. Fitting of the voltammetric curves can lead to the estimation of the relevant rate constants. After initial demonstration of the utility of these aspects in Refs. [21, 25], they have been less exploited than analytical applications. However, they could clearly be of particular use in situations where only a small amount of analyte is available.

An often-used way of increasing the sensitivity of voltammetric measurements and lowering detection limits is the use of pre-concentration techniques. These accumulate the analyte of interest over a period of time and so, in an electrochemical context, are essentially coulometric techniques. In electrochemical BIA this is equivalent to measuring the area under the amperometric transient curve. An advantage of this approach is that it can make use of the current response of the electrode reaction after injection has finished until the complete current transient decays to zero. Examples of chronoamperometric and chronocoulometric transients for the highest and lowest

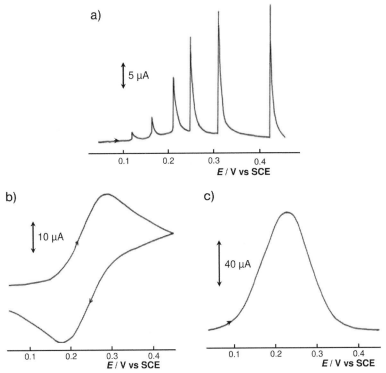

Figure 4.4 BIA voltammetry, illustrated by oxidation of 1 mM $K_4Fe(CN)_6$ in 0.4 M K_2SO_4 electrolyte at a Pt electrode (diameter 3.3 mm); injection flow rate 24.5 $\mu L\,s^{-1}$. Other cell parameters as in Figure 4.3. (a) Injections of 16 μL during a potential sweep at scan rate 10 mV s^{-1}; (b) background subtracted cyclic voltammograms recorded during injection of 100 μL; scan rate 2.0 V s^{-1}; (c) background-subtracted square wave (SW) voltammograms recorded during injection of 100 μL: SW increment 2 mV; frequency 100 Hz; SW amplitude 50 mV.

of the three different possible flow rates at one type of electronic micropipette are shown in Figure 4.5a. As generally happens in plug flow reactors, the efficiency of the accumulation by electrolysis is greatest at the lowest flow rate, where a very significant part of the charge comes after the end of the injection period. The reason for this is due to the lower linear velocity during injection over the detector: when injection has finished a greater fraction of the injected solution containing the electroactive species will not have been pushed radially far away from the electrode, and will be able to diffuse back to the detector surface. Again, because of the contribution to the total charge from the electrolysis is after injection has finished, there is no significant added benefit to injecting volumes greater than 50 μL. It has been shown that a total signal enhancement of a factor of 2.5 can be obtained by injecting 25 μL four times instead of injecting 100 μL once [28]. This coulometric approach has been applied with success in stripping voltammetry, as will be described below and can be seen in Figure 4.5b.

a)

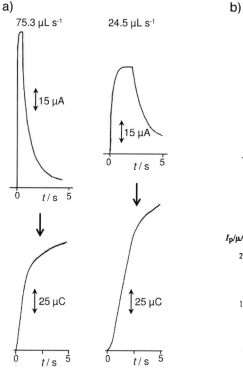

b)

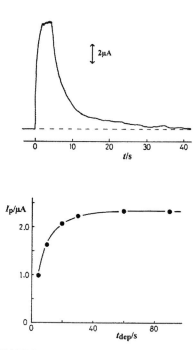

Figure 4.5 (a) BIA chronoamperometric transients and the corresponding charge curves for injection of 50 µL of 1.0 mM ascorbic acid in 0.2 M acetate buffer, pH 3.5, recorded at a fixed potential of +0.8 V vs. saturated calomel electrode at a Pt electrode at two different injection flow rates. Other parameters as in Figure 4.3. (b) BIA-SWASV of 50 nM Cd^{2+} in 0.1 M KNO$_3$ + 2 mM HNO$_3$ at a mercury thin film electrode on glassy carbon substrate ($d = 3$ mm). Injection of 100 µL at 24.5 µL s^{-1}; preconcentration at −1.0 V vs. SCE. Chronoamperometric transient and dependence of stripping peak current on deposition time (from Ref. [28] with permission of Elsevier).

4.4
Applications of Batch Injection Analysis

Various types of detector can be used in BIA, and have included calorimetric, and optical as well as the electrochemical voltammetric and potentiometric types mentioned above. A summary of applications up to 2003 is given in Ref. [8]: of the 33 tabulated applications, 29 are electrochemical (12 amperometric, 11 involve stripping voltammetry, 6 potentiometric), two involve calorimetry, and two are optical. It is notable that, since 2003, all the applications encountered in the literature have been electrochemical in nature. Some of the earlier applications will be referred to below when they describe new developments in BIA strategies and procedures; otherwise the emphasis will be on more recent papers published since 2003.

The emphasis on electrochemical detection can be explained partly through technical difficulties in using heat-based and optically based sensors and the advent of other strategies in injection systems or stationary systems, as well as the inherent simplicity of electrochemical BIA.

4.4.1
Stripping Voltammetry of Trace Metal Ions and Applications in Environmental Monitoring

As explained in Section 4.3, the basis of the stripping voltammetry (SV) technique is accumulation of the target species on an electrode surface which is carried out during the injection period and for a certain chosen extra period of time afterwards in order to maximize sensitivity and lower the detection limit as much as possible. Optimization of the parameters showed this to be usually of the order of 30 s for anodic stripping voltammetry (ASV) of trace metal ions co-deposited with a thin mercury film on a glassy carbon electrode substrate, see Figure 4.5b [28]. Square wave stripping in the determination step was shown to be effective and enables the whole experiment to be performed in a short time.

A recent, and rather unusual, example of BIA-ASV was carried out with a hanging mercury drop electrode, the sample being injected directly over the drop and with differential pulse voltammetry in the determination step, so as to enable fast mapping of lead from gunshot residues for forensic purposes [29].

In the analysis of natural samples, without pre-treatment or digestion, it was found that even with the short contact time between sample and electrode of 4 s there was still some fouling of the electrode surface. For this reason protection by polymer films was investigated [30], using a strategy developed previously for ultrasound-assisted anodic stripping voltammetry [31]. A small amount of Nafion cation exchange polymer dissolved in alcohols was dropped onto the glassy carbon electrode substrate, followed by dimethylformamide to aid in polymer curing; after solvent evaporation the assembly was heated at about 70 °C for a minute or so. Mercury was then electrodeposited *in situ* in the BIA cell through the Nafion film by injection of a solution containing mercury ions, forming tiny droplets of mercury, which are sufficiently close to each other to allow the observed behavior to be that of a mercury film whilst also permitting the polymer film to adhere to the electrode substrate, so that it did not lift. This strategy has three main benefits (i) the electrode assembly and mercury thin film need to be prepared only once per week – mercury ions do not need to be added to the sample solution – which means reduced possibilities of pollution by mercury ions; (ii) the chances of mercury being removed from the glassy carbon electrode substrate due to convection during injection do not exist and (iii) the assembly is sufficiently robust to be able to be transported in the dry state for measurements in the field.

Subsequent studies sought to improve the properties of the polymer film by use of different cation exchange polymers or incorporation of a higher density of sulfonate groups [32, 33]. In the former case, it was shown that dosing small amounts of poly(vinyl sulfonic acid) led to a modest improvement [32], but not sufficient to warrant its use for routine application. By contrast, a coating of poly(ester sulfonic acid), appropriately optimized, was shown to offer excellent discrimination against non-ionic surfactants, but had a similar performance to Nafion in the presence of other possible interferents [33]. Typical detection limits are of the order of 5 nM. The use of random arrays of carbon disk microelectrodes instead of a glassy carbon electrode was shown to improve the performance and resulted in a reduction in the detection limit by a factor of around 1.5 with significant increases in current sensitivity by a factor of 7 or more per unit electrode area [34]. These microelectrode arrays could be particularly applicable in extreme situations where the advantages conferred by microelectrodes are necessary.

BIA-ASV was successfully applied to the analysis of trace metal ions in various types of environmental sample [32, 35, 36] and also to the analysis of biotoxic heavy metal ions in nutrient solutions for ecotoxicological studies [37]. The latter showed very clearly the usefulness of having a simple methodology to estimate the fraction of uncomplexed or weakly complexed metal ions in such solutions, which is the fraction that is toxic to living organisms.

In a recent study, bismuth film electrodes on carbon paste substrates were coated with fibrinogen, a protein with properties similar to polyelectrolytes such as Nafion, to remove interference from surfactants. This allowed the voltammetric analysis of lead and cadmium by anodic stripping voltammetry in environmental samples down to the trace level [38].

Adsorptive stripping voltammetry in combination with BIA has also been evaluated for nickel and cobalt ions together using nioxime ligand as adsorptive accumulation agent, with detection limits of 5 nM and 2 nM, respectively [39] and for chromate using cupferron ligand, detection limit 32 nM [40].

More recent studies have focussed on the development of the stripping voltammetry technique with a view to including analyte sample pre-treatment before the electrochemical detection. Such an approach requires that the pre-treatment be done within the micropipette tip, that is, the tip must contain some substance with which components of the analyte sample interact physically or chemically. In Ref. [4], this strategy was used to remove the effect of unwanted contaminants in natural samples. The tip was packed with a bed of solid sorbent, chelating resin Chelex-100 resin, which adsorbed the metal ions on initial aspiration of the sample. The liquid, with contaminants, was then ejected, leaving the metal ions on the adsorbent. Aspiration of eluent led to release of the metal ions which were then injected onto the detector electrode in the BIA cell in the usual way. It was estimated that a 10-fold increase in pre-concentration was achieved. The results obtained were promising and further developments in this direction can be expected in the future.

4.4.2
Other BIA Environmental Monitoring Applications

Several environmental monitoring applications of BIA have recently been described which do not involve stripping voltammetry. Mercury (II) ions have been detected by inhibition of the invertase enzyme-catalyzed hydrolysis of sucrose into glucose and fructose down to the micromolar level, by measuring the glucose plus fructose oxidation electrocatalytic signal at a copper-modified glassy carbon electrode in alkaline solution [41].

Nitrite, an important indicator of water pollution, was measured indirectly by BIA after reaction with 2,3-diaminonaphthalene (DAN), recording the oxidation of the unreacted DAN, and permitting the analysis of up to 120 samples per hour [42].

Phosphate was determined in seawater, where contamination effects could appear, using fixed potential amperometric BIA with sample injection into acid electrolyte in the presence of molybdate by studying the reduction response [43].

A recent paper shows the first use of electrochemical BIA in the determination of pesticides – in this case paraquat [44]. One of the difficulties in the electrochemical determination of pesticides is their strong adsorption on electrode surfaces so that the combination of the small sample volumes in BIA and novel electrode materials or electrode coatings, which reduce the adsorption of the pesticide or its electrochemical reaction products, is of much interest for the future.

4.4.3
Analysis of Foodstuffs, Pharmaceuticals and Clinical Analysis

The measurement of components of foods has been the object of two recent studies. A BIA procedure for glucose was devised using the fact that glucose is oxidized electrocatalytically at electropolymerized tetraruthenated nickel-porphyrin-modified electrodes, with a submicromolar detection limit [45]. Chloride ion in meat products, as a means of probing the salt content, was monitored by a chloride ion-selective electrode with BIA after ultrasound-based sample extraction [46].

Pharmaceutical compounds have recently been extensively analyzed using electrochemical BIA with amperometric detection at fixed potential by Angnes *et al.* These have included acetaminophen at tetraruthenated porphyrin modified glassy carbon electrodes [47] and acetylsalicylic acid at copper electrodes in alkaline solution [48]. The same type of porphyrin-modified electrode was used to determine sodium metabisulfite in pharmaceutical formulations, which is normally added to prevent the oxidation of active ingredients [49] and to measure hydrogen peroxide in cosmetics and pharmaceutical product samples [50]. Salbutomol, a sympathomimetic drug [51], and isonazid, an antitubercular drug [52] were determined at glassy carbon electrodes in alkaline solution.

A BIA procedure was developed for the measurement of the antiparasitic drug niclosamide with a submicromolar detection limit [53], for field determinations in river streams and effluents.

Modification of the electrode surface of glassy carbon by redox polymers which interact with the injected sample can also be of interest for clinical applications. In Ref. [54], poly(methylene blue) was formed by potential cycling on the surface of a glassy carbon electrode in a solution containing the monomer. The modified electrode was then used to selectively determine the reduced forms of hemoglobin in whole blood by oxidation, without interference from other components of the blood, for example cytochrome-*c*. Analysis, by amperometry at fixed potential, can be

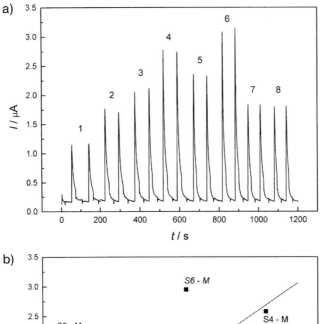

Figure 4.6 BIA of hemoglobin in whole blood by oxidation at poly(methylene blue) film-modified glassy carbon electrode, $E = 0.55$ V vs. SCE, in pH 8.2 phosphate buffer; for blood from patients 1 to 8 (M, male, F, female; patients 1–4 healthy, patients 5–8 potentially ill). (a) Chronoampero- metric transients for consecutive injections in BIA cell of 50 μL samples (diluted 1:5 in pH 8.2 phosphate buffer); (b) comparison between electrochemical and standard cyanidation method. (From Ref. [54] with permission of Elsevier).

done immediately after taking the blood sample and adding anti-coagulation agent, representing a significant time saving compared to current procedures. Figure 4.6 shows the transients obtained and a comparison between the electrochemical BIA method, which measures the amount of reduced hemoglobin, and the normally-used cyanidation method which takes ∼2 h and measures the total amount of hemoglobin. It can be seen that the potentially ill patients deviate from the line showing the correspondence between the two methods, suggesting that they have a different ratio between oxidized and reduced forms of hemoglobin. A full evaluation of the potential of BIA procedures, such as this one, in a clinical environment has yet to be carried out.

4.4.4
Capillary Batch Injection Analysis

In capillary batch injection analysis, a sample of volume of the order of 150 nL, but possibly as little as 20 nL, is introduced into a capillary, typically of internal diameter 100 μm, and then injected by means of a microliter syringe pump directly over an electrochemical detector immersed in electrolyte solution [55]. Amperometric, square wave voltammetric and potentiometric detection were evaluated; it was shown that, after optimization, the technique leads to analytical parameters as capillary FIA and can be used for measuring concentrations down to 10^{-5} M or lower [55, 56].

A recent development of this approach involved combining capillary batch injection analysis with capillary electrophoresis using short, 7 cm, internal diameter 25 μm, capillaries [57], see Figure 4.7. Micromolar detection limits were achieved and application to the determination of neurotransmitters [57] and peptides [58] was demonstrated.

4.4.5
Non-Electrochemical Methodologies and Detection Strategies

Thermal/calorimetric and optical sensors have been devised for batch injection analysis. The original description of batch injection thermal analysis involved the use of thermistors to measure the heat of chemical reactions which was sensed at the detector surface [59]. Calorimetric sensors were subsequently evaluated for the detection of heat changes in biomaterials for biosensors and could be applied in continuous monitoring [60]. This approach was also applied to the measurement of sucrose in sugar cane sap, where invertase enzyme was directly immobilized on

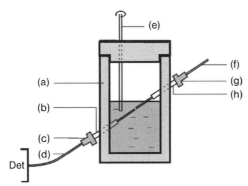

Figure 4.7 Schematic representation of the cell configuration for capillary batch injection–capillary electrophoresis–electrochemical detection (CBICE–ED): (a) injection cell; (b) stainless steel guiding tube for the separation capillary; (c) PTFE adaptor; (d) separation capillary; (e) PTFE cylinder ($d = 2$ mm); (f) injection capillary; (g) PTFE adaptor, and (h) guiding glass tube for the injection capillary. (From Ref. [57] with permission of Elsevier).

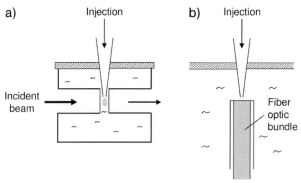

Figure 4.8 BIA with optical detection. (a) Adaptation of a spectrophotometer cavity for BIA optical detection with short optical path length ~3 mm, using disk-shaped optical windows in a large cell, with solution stirring; (b) fluorescence detection using fiber optic bundles: half of the fibers are source, half are collection; the solution is stirred.

the thermistor [61] – although good results were obtained the thermal BIA approaches appear not to have been exploited since then.

Spectrophotometric BIA has been described using a specially-designed cell placed in the cell cavity of a conventional spectrophotometer, filled with transparent liquid. The detector was the light beam so that injection was performed directly above the transmitted beam of light and the absorbance at a chosen wavelength was recorded over time [62] – see Figure 4.8a; calibration curves were successfully recorded. A more sophisticated approach which lends itself well to recent developments in optics is the use of optical fibers as source and detector, some of which supply exciting radiation and the others measure any fluorescence [63], see Figure 4.8b. On sample injection a fluorescent response is obtained; nevertheless, problems can occur owing to the small propagation distance through the solution and interactions of the excited molecules with other matrix components, affecting the response seen by the detector. As in the case of thermal BIA, apparently there has been no further development of these optical techniques.

4.5
Comparison of BIA with Flow Injection Techniques

Several essential points regarding BIA can be compared and discussed with respect to the equivalent FIA strategy. Both allow the use of small, microliter, volumes of sample and sometimes the signals generated by BIA can be larger than those generated by FIA, as was demonstrated early on in the development of BIA techniques.

Advantages of BIA can be summarized as the lack of necessity for a carrier stream with associated mechanical moving parts, which makes field measurements easier to perform, and the possibility of facile experimental procedures when no pre-treatment is needed. There is essentially zero dispersion of the sample plug and, thence, close to zero dead volume and no sample dilution before the sample reaches the detector, so that sensitivities can be higher. The disadvantages arise from the lack of easy-to-use detectors other than electrochemical ones – although this situation may change in the future, as well as the limitation of not being able to carry out derivitization reactions or other conversions "on-line" before the detector is reached. Nevertheless, there is no doubt that it represents a powerful tool for the situations in which it has been demonstrated.

As in FIA, since the contact time of the sample with the detector is only a maximum of several seconds, problems from poisoning of the electrode surface by components in the sample matrix are also less than in the equivalent wall-jet continuous flow system. The only evident limitation in this context is that the experiment must usually be carried out within the time period of the injection, whilst there is imposed convection.

4.6
Future Perspectives

Batch injection analysis is an interesting and potentially significant alternative to other injection analysis strategies and possesses many attractive features, such as not needing pumps and other mechanically moving parts. If a sample analyte can be measured directly without pre-treatment by injection over the detector then it is an almost ideal experimental strategy for rapid and accurate determinations and does not consume more than $100\,\mu L$ of sample solution. BIA is a robust and easily-performed analytical technique, both in the laboratory and in the field, particularly now that small portable, electronic instrumentation exists for a number of types of analytical experiment.

The potential for exploitation of this technique in the future is large given the ease of performing the experiments and that it lends itself to use by less-qualified personnel and in situations where expert experimentalists or sophisticated laboratory instrumentation are not available.

References

1 Wang, J. and Taha, Z. (1991) Batch injection analysis. *Analytical Chemistry*, **63**, 1053–1056.

2 Brett, C.M.A. (1999) Electroanalytical techniques for the future – the challenges of miniaturization and of real-time measurements. *Electroanalysis*, **11**, 1013–1016.

3 Brett, C.M.A. (2001) Electrochemical sensors for environmental monitoring: strategy and examples. *Pure and Applied Chemistry*, **73**, 1969–1977.

4 Trojanowicz, M., Koźmiński, P., Dias, H. and Brett, C.M.A. (2005) Batch injection stripping voltammetry (tube-less flow injection analysis) of trace metals with on-line sample pretreatment. *Talanta*, **68**, 394–400.

5 Karolczak, M., Dreiling, R., Adams, R.N., Felice, L.J. and Kissinger, P.T. (1976)

Electrochemical techniques for study of phenolic natural-products and drugs in microliter volumes. *Analytical Letters*, **9**, 783–793.

6 Tur'yan, Ya.I. (1997) Microcells for voltammetry and stripping voltammetry. *Talanta*, **44**, 1–13.

7 Brett, C.M.A. (1999) Electrode Reactions in Microvolumes, in *Comprehensive Chemical Kinetics*, Vol. 37 (eds R.G. Compton and G. Hancock), Elsevier, Amsterdam, Chapter 16.

8 Quintino, M.S.M. and Angnes, L. (2004) Batch injection analysis: an almost unexplored powerful tool. *Electroanalysis*, **16**, 513–523.

9 Brett, C.M.A. and Oliveira Brett, A.M. (1993) *Electrochemistry. Principles, Methods and Applications*, Oxford University Press, Oxford. Chapter 8.

10 Chen, L., Wang, J. and Angnes, L. (1991) Batch injection-analysis with the rotating-disk electrode. *Electroanalysis*, **3**, 773–776.

11 Brett, C.M.A. and Oliveira Brett, A.M. (1998) *Electroanalysis*, Oxford University Press, Oxford.

12 Albery, W.J. and Brett, C.M.A. (1983) The wall-jet ring-disc electrode. Part I. Theory. *Journal of Electroanalytical Chemistry*, **148**, 201–210.

13 Albery, W.J. and Brett, C.M.A. (1983) The wall-jet ring-disc electrode. Part II. Collection efficiency, titration curves and anodic stripping voltammetry. *Journal of Electroanalytical Chemistry*, **148**, 211–220.

14 Albery, W.J. (1985) The current distribution on a wall-jet electrode. *Journal of Electroanalytical Chemistry*, **191**, 1–13.

15 Aoki, K., Tokuda, K. and Matsuda, H. (1986) Theory of stationary current voltage curves of redox-electrode reactions in hydrodynamic voltammetry. 11. Wall jet electrodes. *Journal of Electroanalytical Chemistry*, **206**, 37–46.

16 Compton, R.G., Greaves, C.R. and Waller, A.M. (1990) A general computational method for mass-transport problems involving wall jet electrodes and its application to simple electron-transfer, ECE and Disp1 reactions. *Journal of Applied Electrochemistry*, **20**, 575–585.

17 Compton, R.G., Fisher, A.C. and Tyley, G.P. (1991) Nonuniform accessibility and the use of hydrodynamic electrodes for mechanistic studies – a comparison of wall-jet and rotating-disk electrodes. *Journal of Applied Electrochemistry*, **21**, 295–300.

18 Fisher, A.C., Compton, R.G., Brett, C.M.A. and Oliveira Brett, A.M.C.F. (1991) The wall-jet electrode: potential step chronoamperometry. *Journal of Electroanalytical Chemistry*, **318**, 53–59.

19 Brett, C.M.A., Oliveira Brett, A.M., Fisher, A.C. and Compton, R.G. (1992) Potential step chronoamperometry at the wall-jet electrode: experimental. *Journal of Electroanalytical Chemistry*, **334**, 57–64.

20 Compton, R.G., Fisher, A.C., Latham, M.H., Brett, C.M.A. and Oliveira Brett, A.M. (1992) Wall-jet electrode linear sweep voltammetry. *Journal of Physical Chemistry*, **96**, 8363–8367.

21 Brett, C.M.A., Oliveira Brett, A.M. and Mitoseriu, L.C. (1995) Amperometric batch injection analysis: theoretical aspects of current transients and comparison with wall-jet electrodes in continuous flow. *Electroanalysis*, **7**, 225.

22 Amine, A., Kauffmann, J.-M. and Palleschi, G. (1993) Investigation of the batch injection-analysis technique with amperometric biocatalytic electrodes using a modified small-volume cell. *Analytica Chimica Acta*, **273**, 213–218.

23 Wang, J. and Chen, L. (1994) Small-volume batch-injection analyser. *The Analyst*, **119**, 1345–1348.

24 Wang, J., Chen, L., Angnes, L. and Tian, B. (1992) Computerized pipettes with programmable dispension for batch injection-analysis. *Analytica Chimica Acta*, **267**, 171–177.

25 Brett, C.M.A., Oliveira Brett, A.M. and Mitoseriu, L.C. (1994) Amperometric and voltammetric detection in batch injection analysis. *Analytical Chemistry*, **66**, 3145–3150.

26 Oliveira Brett, A.M., Matysik, F.-M. and Vieira, M.T. (1997) Thin-film gold electrodes produced by magnetron sputtering. Voltammetric characteristics and application in batch injection analysis with amperometric detection. *Electroanalysis*, **9**, 209–212.

27 Wang, J. and Taha, Z. (1991) Batch injection analysis with potentiometric detection. *Analytica Chimica Acta*, **252**, 215–221.

28 Brett, C.M.A., Oliveira Brett, A.M. and Tugulea, L. (1996) Anodic stripping voltammetry of trace metals by batch injection analysis. *Analytica Chimica Acta*, **322**, 151–157.

29 De Donato, A. and Gutz, I.G.R. (2005) Fast mapping of gunshot residues by batch injection analysis with anodic stripping voltammetry of lead at the hanging mercury drop electrode. *Electroanalysis*, **17**, 105–112.

30 Brett, C.M.A., Oliveira Brett, A.M., Matysik, F.-M., Matysik, S. and Kumbhat, S. (1996) Nafion-coated mercury thin-film electrodes for batch injection analysis with anodic stripping voltammetry. *Talanta*, **43**, 2015–2022.

31 Matysik, F.-M., Matysik, S., Oliveira Brett, A.M. and Brett, C.M.A. (1997) Ultrasound-enhanced anodic stripping voltammetry using perfluorosulfonated ionomer-coated mercury thin-film electrodes. *Analytical Chemistry*, **69**, 1651–1656.

32 Brett, C.M.A., Fungaro, D.A., Morgado, J.M. and Gil, M.H. (1999) Novel polymer-modified electrodes for batch injection sensors and application to environmental analysis. *Journal of Electroanalytical Chemistry*, **468**, 26–33.

33 Brett, C.M.A. and Fungaro, D.A. (2000) Poly(ester sulfonic acid) coated mercury thin film electrodes: characterization and application in batch injection analysis stripping voltammetry of heavy metal ions. *Talanta*, **50**, 1223–1231.

34 Fungaro, D.A. and Brett, C.M.A. (1999) Microelectrode arrays: application in batch injection analysis. *Analytica Chimica Acta*, **385**, 257–264.

35 Brett, C.M.A. and Fungaro, D.A. (2000) Modified electrode voltammetric sensors for trace metals in environmental samples. *Journal of the Brazilian Chemical Society*, **11**, 298–303.

36 Fungaro, D.A. and Brett, C.M.A. (2000) Eletrodos modificados com polímeros perfluorados e sulfonados: aplicações em análises ambientais. *Quimica Nova*, **23**, 805–811.

37 Brett, C.M.A. and Morgado, J.M. (2000) Development of batch injection analysis for electrochemical measurements of trace metal ions in ecotoxicological test media. *Journal of Applied Toxicology*, **20**, 477–481.

38 Adraoui, I., El Rhazi, M. and Amine, A. (2007) Fibrinogen-coated bismuth film electrodes for voltammetric analysis of lead and cadmium using the batch injection analysis. *Analytical Letters*, **40**, 349–368.

39 Brett, C.M.A., Oliveira Brett, A.M. and Tugulea, L. (1996) Batch injection analysis with adsorptive stripping voltammetry for the determination of traces of nickel and cobalt. *Electroanalysis*, **8**, 639–642.

40 Brett, C.M.A., Filipe, O.M.S. and Neves, C.S. (2003) Determination of chromium (VI) by batch injection analysis and adsorptive stripping voltammetry. *Analytical Letters*, **36**, 955–969.

41 Mohammadi, H., El Rhazi, M., Amine, A., Oliveira Brett, A.M. and Brett, C.M.A. (2002) Determination of mercury (II) by invertase enzyme inhibition coupled with batch injection analysis. *The Analyst*, **127**, 1088–1093.

42 Idrissi, L., Amine, A., El Rhazi, M. and El Moursli Cherkaoui, F. (2005) Electrochemical detection of nitrite based on reaction with 2,3-diaminonaphthalene. *Analytical Letters*, **38**, 1943–1955.

43 Quintana, J.C., Idrissi, L., Palleschi, G., Albertano, P., Amine, A., El Rhazi, M. and Moscone, D. (2004) Investigation of amperometric detection of phosphate. Application in seawater and cyanobacterial biofilm samples. *Talanta*, **63**, 567–574.

44 Simões, F.R., Vaz, C.M.P. and Brett, C.M.A. (2007) Electroanalytical detection of the pesticide paraquat by batch injection analysis. *Analytical Letters*, **40**, 1800–1810.

45 Quintino, M.S.M., Winnischofer, H., Nakamura, M., Araki, K., Toma, H.E. and Angnes, L. (2005) Amperometric sensor for glucose based on electrochemically polymerized tetraruthenated nickel-porphyrin. *Analytica Chimica Acta*, **539**, 215–222.

46 Sucman, E. and Bednar, J. (2003) Determination of chlorides in meat products with ion-selective electrode using the batch injection technique. *Electroanalysis*, **15**, 866–871.

47 Quintino, M.S.M., Araki, K., Toma, H.E. and Angnes, L. (2002) Batch injection analysis utilizing modified electrodes with tetraruthenated porphyrin films for acetaminophen quantification. *Electroanalysis*, **14**, 1629–1634.

48 Quintino, M.S.M., Corbo, D., Bertotti, M. and Angnes, L. (2002) Amperometric determination of acetylsalicylic acid in drugs by batch injection analysis at a copper electrode in alkaline solutions. *Talanta*, **58**, 943–949.

49 Quintino, M.S.M., Araki, K., Toma, H.E. and Angnes, L. (2006) Amperometric quantification of sodium metabisulfite in pharmaceutical formulations utilizing tetraruthenated porphyrin film modified electrodes and batch injection analysis. *Talanta*, **68**, 1281–1286.

50 Quintino, M.S.M., Winnischofer, H., Araki, K., Toma, H.E. and Angnes, L. (2005) Cobalt oxide/tetraruthenated cobalt-porphyrin composite for hydrogen peroxide amperometric sensors. *The Analyst*, **130**, 221–226.

51 Quintino, M.S.M. and Angnes, L. (2004) Bia-amperometric quantification of salbutamol in pharmaceutical products. *Talanta*, **62**, 231–236.

52 Quintino, M.S.M. and Angnes, L. (2006) Fast BIA-amperometric determination of isoniazid in tablets. *Journal of Pharmaceutical and Biomedical Analysis*, **42**, 400–404.

53 Abreu, F.C., Goulart, M.O.F. and Oliveira Brett, A.M. (2002) Detection of the damage caused to DNA by niclosamide using an electrochemical DNA-biosensor. *Biosensors & Bioelectronics*, **17**, 913–919.

54 Brett, C.M.A., Inzelt, G. and Kertesz, V. (1999) Poly(methylene blue) modified electrode sensor for haemoglobin. *Analytica Chimica Acta*, **385**, 119–123.

55 Backofen, U., Hoffmann, W. and Matysik, F.-M. (1998) Capillary batch injection analysis: a novel approach for analyzing nanoliter samples. *Analytica Chimica Acta*, **362**, 213–220.

56 Backofen, U., Matysik, F.-M., Hoffmann, W. and Ache, H.-J. (1998) Capillary batch injection analysis and capillary flow injection analysis with electrochemical detection: a comparative study of both methods. *Fresenius Journal of Analytical Chemistry*, **362**, 189–193.

57 Matysik, F.-M. (2006) Capillary batch injection – a new approach for sample introduction into short-length capillary electrophoresis with electrochemical detection. *Electrochemistry Communications*, **8**, 1011–1015.

58 Psurek, A., Matysik, F.-M. and Scriba, G.K.E. (2006) Determination of enkephalin peptides by nonaqueous capillary electrophoresis with electrochemical detection. *Electrophoresis*, **27**, 1199–1208.

59 Wang, J. and Taha, Z. (1991) Batch injection analysis with thermistor sensing devices. *Analytical Letters*, **24**, 1389–1400.

60 Bataillard, P. (1993) Calorimetric sensing in bioanalytical chemistry – principles, applications and trends. *TRAC – Trends in Analytical Chemistry*, **12**, 387–394.

61 Thavarungkul, P., Suppapitnarm, P., Kanatharana, P. and Mattiasson, B. (1999) Batch injection analysis for the determination of sucrose in sugar cane juice using immobilized invertase and thermometric detection. *Biosensors & Bioelectronics*, **14**, 19–25.

62 Wang, J. and Angnes, L. (1993) Batch injection spectroscopy. *Analytical Letters*, **26**, 2329–2339.

63 Wang, J., Rayson, G.D. and Taha, Z. (1992) Batch injection-analysis using fiberoptic fluorometric detection. *Applied Spectroscopy*, **46**, 107–110.

5
Electroosmosis-Driven Flow Analysis

Petr Kubáň, Shaorong Liu, and Purnendu K. Dasgupta

5.1
Introduction

Flow analysis methods (SFA, FIA, SIA, etc.) are well established in analytical chemistry and consequently are widely applied in both routine analysis and research. They are used not only as stand-alone analytical techniques but also as powerful tools for liquid manipulation. There is a concurrent, sometimes conflicting, demand for miniaturizing the entire liquid handling system to reduce chemical consumption and decrease waste generation while simultaneously maintaining or even improving performance specifications. Low nanoliter volume injection valves and compatible detection cells or on-tube detectors are now commercially available and becoming more common in flow analysis practice. Often, however, these are not physically much smaller than their conventional size counterparts. Larger bore PTFE tubes, used in yesteryears for flow analysis systems, are being replaced by capillary conduits of fused silica and various polymers (with inner diameters in the tens to low hundreds of micrometers); alternatively, the flow analysis system is composed of microchannels on a glass or polymeric chip.

Pumping systems have also undergone considerable changes. In particular the new requirements include low flow rates that are highly stable over extended periods of time. Few of the new principles for pumping liquids in micro flow systems operate in a pulseless manner. When mixing noise is not a limiting factor then minimizing flow pulsations becomes the key to achieving the best detection sensitivity. Electroosmotic pumping provides one of the few means to achieve a stable, pulseless, low and adjustable flow in a miniature set-up that is particularly well suited for flow analysis. Here we will confine ourselves to such flow systems that use uninterrupted flow without gas or immiscible solvent segments. With rare exceptions [1], gas or organic solvent segmentation is not compatible with EOF pumping. Thus, we will focus mainly on the flow injection methods, such as FIA and SIA, in which electroosmotic pumping has been successfully applied.

The phenomenon of electroosmosis has been known for nearly two centuries [2] but it was not applied to FIA/SIA until the early 1990s [3]. This was only after capillary

Advances in Flow Analysis. Edited by Marek Trojanowicz
Copyright © 2008 WILEY-VCH Verlag GmbH & Co. KGaA, Weinheim
ISBN: 978-3-527-31830-8

electrophoresis emerged as a major technique with the seminal publication by Jorgensson and Lukacs in 1981 [4]. EOF pumping systems have evolved much over the years from the original publication by two of the present authors [3]. There is now a large number of contributions devoted to the development of EOF pumps, including those for high pressure applications, their characterization and, to a lesser extent, the application of EOF pumping for liquid propulsion in flow analysis systems. In the present contribution, the principles and applications of electroosmosis-driven flow analysis are described.

5.2
Pumping Systems

The creation of constant, reproducible flow over an extended period is not easy to achieve. Since both FIA and SIA rely on a precisely controlled dispersion and/or reaction in a flow system, temporal irregularities in pumping performance adversely affect the reproducibility of the analysis. The search for new, robust, stable pumping mechanisms is particularly important in miniaturized analysis systems. The available pumping mechanisms can be generally subdivided into two groups, (i) pressure-driven and (ii) electrokinetically driven, the latter being the main focus of this chapter.

5.2.1
Pressure Pumping Systems

The choice of a pumping system is typically governed by the scale of the flow system and required flow rate range. When operating at traditional FIA/SIA scales at relatively high flow rates (ml/min range), peristaltic or syringe pumps are most frequently used. Peristaltic pumps are especially useful: when operated at high flow rates and consequently high rotation speed of the rollers, pump pulsations and flow irregularities are minimized. Decreasing the size of the whole system and consequently also the flow velocities, however, means more stringent requirements on the flow characteristics of the pumping systems. The pump needs to reliably deliver reproducible flow rates in the $nl\,min^{-1}$ to $\mu l\,min^{-1}$ range. In such systems, peristaltic pumping is of little value. Syringe micropumps, piezoelectric pumps, or other types of micropumps may be used (see e.g., [5–7]). For systems with very low backpressure, fluid propulsion by gravity or pneumatic pressure and/or vacuum is also applicable and often preferred. As the hydraulic diameter decreases, the demands on the pumping system, especially its pressure capability, increase. Consider that the flow through a cylindrical tube can be expressed by the well known Hagen–Poiseuille equation:

$$\Delta p = \frac{8 Q_{(\Delta p)} \eta L}{\pi r^4} \tag{1}$$

where $Q_{(\Delta p)}$ is the flow rate $[ml\,s^{-1}]$, r is the tube radius [cm], Δp is the pressure difference $[g\,cm\,s^{-2}]$, η is the viscosity $[g\,cm^{-1}\,s^{-1}]$ and L is the tube length [cm].

In general, the residence time, t, in a flow system is an important factor, some minimum reaction time is usually needed whereas the reciprocal of t is typically directly related to the throughput rate. From geometric considerations, t is given by

$$t = \frac{\pi r^2 L}{Q_{(\Delta p)}} = \frac{8\eta L^2}{\Delta p r^2} \qquad (2)$$

To maintain a constant t using the same pressure drop Δp, the length to radius ratio L/r must be maintained the same as in macrosystems. If a constant L is retained, Δp must increase linearly with $1/r^2$. In practice, L is often reduced as r is reduced. In any case, although a variety of pressure-driven micropumps have been developed for liquid delivery in microfluidic platforms, there are none that would provide stable flow against very high pressures. High pressure piston pumps are available but are expensive, bulky and have a limited reservoir volume. The use of such pumps for microfluidic systems basically overwhelms the otherwise minute dimensions of the microsystem. Electroosmosis-based pumps can provide a viable solution to this problem; this is described in a subsequent section.

5.2.2
Electroosmosis-Driven Pumping Systems

5.2.2.1 Theoretical Considerations
To understand the nature of electroosmosis-driven flow, it is necessary to overview the fundamental principles of electroosmotic flow. A first prerequisite for EOF to occur is an electrically charged surface in contact with the fluid; typically this is the surface of a capillary conduit or a microchannel. Some polymeric materials (e.g., PDMS, PTFE, PE, etc.) and especially fused silica and glass exhibit a significant surface charge over a broad range of pH and are therefore suitable for EOF generation. Fused silica capillaries (FSCs) or microchannels fabricated on glass substrates are the most frequently used. FSCs are perhaps the most studied in the context of electroosmosis, and we will therefore use a FSC as a model to illustrate the basic principles of EOF.

The surface of an FSC consists of many silanol groups, $-Si-OH$, which behave as a weak acid with a pK_a of around 7.7 [8]. At all pH values above 2, there is perceptible negative charge on the capillary surface because of ionization of the silanol groups. The oppositely charged cations (counterions) from the solution are attracted to the negatively charged surface by strong electrostatic forces. An electrical double layer (EDL) is thus formed. The EDL is very thin (usually only a few nm thick), depending on the solution composition. While the counterions on the capillary surface (the so-called inner Helmholtz layer) are practically immobile, other counterions further towards the center of the capillary are in the so-called diffusion layer and are mobile. The potential at the layer of counterions next to the inner Helmholtz layer (the so-called outer Helmholtz layer) relative to the bulk solution is defined as the ζ (zeta) potential [V] of the surface. The counterion concentration in the diffusion layer and the electrokinetic behavior in the capillary depend on the ζ potential.

When an electric field, E, is imposed across the length of the FSC, the counterions in the diffusion layer are dragged towards the negatively charged electrode (cathode) and because of viscous coupling of the solvent molecules solvating such counterions with the bulk liquid, a net flow of the bulk liquid towards the cathode is induced in the capillary; this is called the electroosmotic flow. The electroosmotic flow velocity, v_{eof} $[m^2 V^{-1} s^{-1}]$, is expressed by the well known Smoluchowski equation:

$$v_{eof} = -\frac{\varepsilon_0 \varepsilon_r \zeta}{\eta} E \tag{3}$$

where ε_0 and ε_r are the relative permittivities of vacuum and the liquid and η is the viscosity. The EOF creates a bulk flow that carries all positively charged, uncharged and negatively charged species (provided their migration velocity is less than v_{eof}) in the liquid towards the cathode. Because the force creating EOF originates in the diffusion layer very close to the capillary surface, EOF displays a plug-like flow profile (Figure 5.1A). In contrast, in a pressure driven flow, the flow profile is parabolic (Figure 5.1B); the flow velocity is zero at the boundary layer and increases to twice the average velocity at the center of the tube. One consequence of plug flow is vastly diminished axial dispersion of an injected band relative to pressure driven flow; this has both positive and negative consequences.

It may be argued that plug flow or pulseless flow is not the most suitable for flow injection systems where axial dispersion in a single stream is sometimes used to achieve mixing of samples and reagents. However, this does not apply to the mixing of independent flow streams or when EOF merely serves as a pump to eventually induce a pressure driven flow in the system, In any case, EOF can be harnessed to provide pulse-free and high pressure flow.

5.2.2.2 Theoretical Considerations for an EOF-Induced Pumping System

In an EOF-driven system, the overall flow rate is the combination of a forward electroosmotic flow and a backward pressure-driven flow caused by the backpressure

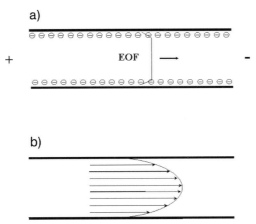

Figure 5.1 The profiles of electroosmosis-driven (A) and pressure-driven (B) flows.

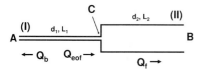

Figure 5.2 Flow distribution in a model system composed of two capillaries (I and II) connected in series. d-capillary inner diameter, L-capillary length, Q_{eof} electroosmotic flow induced in capillary I, Q_b, Q_f – backward and forward hydrodynamic flows in capillaries I and II.

represented by the "load," the network of capillary tubing or channels that form the rest of the flow system. Lazar and Karger [9] have modeled the flow characteristics of a simple system composed of two serially connected capillaries of different diameters. Such a system (Figure 5.2), is very illustrative for the understanding of general principles in EOF-driven pumps. In this model system, two capillaries of different diameters are connected in series, the narrow capillary tube having the dimensions d_1, L_1 and the large tube dimensions d_2, L_2. If a potential is applied only to the narrow capillary tube, for example, by connecting two electrodes of a voltage supply between the points A and C in the figure (the details on electrical isolation and practical arrangements will be deferred until later) an EOF-induced flow, Q_{eof} is generated. This EOF will create a pressure Δp at C that drives the solution in capillary II forward (Q_f) and the solution in capillary I backward (Q_b). Since EOF is the only input flow, $Q_{eof} = Q_f + Q_b$ if the liquids (the liquid in the two tubes need not be the same) are assumed to be incompressible. In other words, the EOF is split into a forward flow in the wide channel and a backward flow in the narrow channel. The efficiency of the pump, β, can be evaluated by the ratio of the forward flow Q_f (used to move the solutions in the fluidic system) to the total input flow Q_{eof}:

$$\beta = \frac{Q_f}{Q_{eof}} = \frac{\frac{L_1}{L_2}}{\frac{L_1}{L_2} + \left(\frac{d_1}{d_2}\right)^4} \tag{4}$$

The efficiency is dependent on the channel dimensions but not on the precise backpressure generated or the actual magnitude of the EOF in the system. This has some important practical implications shown in the following figure, where β is calculated as a function of capillary diameter and length ratios. In the extreme case, when the two capillaries (or channels) have the same diameter ($d_1/d_2 = 1$), β will change significantly even with a small change in the length ratio. The flow rate will thus be very sensitive to the backpressure in the microfluidic system. On the other hand, when the first channel is made sufficiently small, that is, 10 times smaller in diameter than the second channel ($d_1/d_2 = 0.1$), the flow will be practically independent of the length ratio (i.e., independent of the backpressure from the microfluidic system). It is imperative to have this in mind when constructing and designing an EOF-based pump.

Generally, single channel pumps depicted in Figure 5.3 are not effective, since there is a trade-off between the narrow channel diameter and the achievable flow rate.

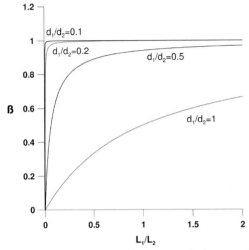

Figure 5.3 Pump efficiency (β) at various d_1/d_2 ratios calculated from Equation (4) for a single channel micropump depicted in Figure 5.2.

Too narrow a channel or capillary will provide sufficient pressure but the flow rates will be very low. For this reason, parallel pump channels are often used to boost the attainable flow rate. Consider a system comprised of N parallel narrow capillaries connected to a large bore tube as in Figure 5.4. In such a system, multiplexed narrow channels (in the calculations approximated as capillary tubes of diameter d_1) connected in parallel will provide the required flow rates while maintaining the pressure capabilities of the EOF-induced pump. The efficiency of a parallel channel EOF pump can be expressed as:

$$\beta = \frac{Q_f}{Q_{eof}} = \frac{\frac{L_1}{L_2}}{\frac{L_1}{L_2} + N\left(\frac{d_1}{d_2}\right)^4} \tag{5}$$

The larger the value of N, the larger the total EOF but β will become smaller. This effect is presented in Figure 5.5 under $d_1/d_2 = 0.1$. Thus an EOF pump that will

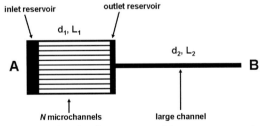

Figure 5.4 Flow distribution in a model system composed of multiple parallel channels connected in series to a large diameter channel. *d*-capillary inner diameter, *L*-capillary length, *n*-number of channels.

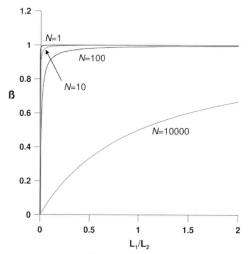

Figure 5.5 Pump efficiency (β) at $d_1/d_2 = 0.1$ calculated from Equation (5) for a parallel channel micropump depicted in Figure 5.4. Number of channels $N = 1$, 10, 100, and 10 000.

provide the desired flow rate can be constructed by connecting a number of narrow channels (capillaries) in parallel, however, the number of channels as well as other pump characteristics, such as the channel diameter ratio and the system back-pressure at the desired flow rate, must also be considered.

In this context, a packed bead column or monolithic column can be regarded as a system with a very large number of parallel microchannels. By closely packing the packing materials, such as μm-size silica particles, into a capillary column, a network of intertwining channels is created. These channels or micropores can be regarded as parallel with some approximations. The effective channel (pore) diameter is dependent on the packing density and particle size, but generally can be approximated as 1/5 of the mean particle diameter [10]. The length of the microchannels can be approximated by the so-called tortuosity of the porous media. If such a packed capillary EOF pump is connected to a micro flow injection system, an efficient, high pressure, high flow rate pump can be created. The types of pumps and pumping systems that utilize EOF are reviewed in the following section.

5.3
EOF Pumping Systems

5.3.1
EOF Pumping Without Electric Isolation

The simplest way of exploiting EOF for pumping in micro flow injection systems is to place a set of electrodes at the beginning and at the end of the system conduit. There is no isolation of any part of the system from the electric voltage and current applied.

This approach is commonly used in capillary electrophoresis (CE) or CE-like systems in which unbranched capillaries or channels with limited branches are used. Sample injections in such systems can be accomplished by simple hydrodynamic or various electrokinetic injection methods as practised in CE [11] or by using an electrically isolated rotary valve. In some cases, electric isolation of certain components such as the injection valve and detection system may be required for the system to function properly.

These systems have some limitations as voltage is present through the entire system. For example, the pump solution cannot be too conductive and the solution pH cannot be extreme since EOF generation depends on these parameters. High or low pH and/or high solution conductivity lead to excessive Joule heating in the capillary: bubble formation and disruption of EOF pumping may occur. Fluctuation of EOF is another limitation of such systems. Because the entire fluidic network is used to generate EOF and various solutions are passing through it, channel surfaces often cannot be maintained the same due to adsorption/desorption and/or chemical changes on the surfaces, resulting in changes in the EOF.

Additionally, since electrophoretic migrations always take place during analysis, the sample matrix components (other ions, solvent, water, etc.) may be electrophoretically separated from the analyte(s), producing system peaks. System peaks inherent to capillary electrophoresis arise, for example, from the differences in refractive index of the matrix and carrier electrolyte or simply from the absorbance change when UV–VIS detection is used. System peaks can be either positive or negative and their presence effectively reduces sample throughput because injection of the next sample can only be performed with sufficient time delay so as to not overlap with the system peak elicited by the previous sample.

Despite these limitations, such EOF pumping systems have been used for FIA/SIA applications. The use of EOF for pumping liquids in a flow injection analysis system was first demonstrated in 1992 [3]. The system was composed of a single 75 µm i.d. FSC and resembled a single-line FIA system that allows a reagent-sandwiched sample injection [12]. The sample plug containing the analyzed ion (Fe^{2+}) was sandwiched by two plugs of a colorimetric reagent (1,10-phenanthroline). A large number of samples were hydrodynamically injected (by elevating one capillary end) into a fused silica capillary and the injected samples were transported to the detector by EOF. During the passage through the capillary, the colored reaction product was formed and detected at 508 nm. Since each hydrodynamic injection of the sample and the reagents induced a laminar flow which caused additional dispersion to the samples already in the capillary, the samples loaded last exhibited less dispersion and had increasing peak heights. In this experiment, tetrabutylammonium perchlorate was used as the background electrolyte. The tetrabutylammonium cation has a small mobility compared to that of the large $Fe(o\text{-phen})_3^{2+}$ cation and hence helped to produce symmetric peaks without excessive tailing. Haswell et al. [13–15] have used EOF-driven micro flow injection analysis system in channels fabricated on a glass substrate. The system encompassed a microfluidic channel structure, 320 µm wide and 30 µm deep, of various geometries, that was used for sample, reagent and carrier electrolyte delivery purely by EOF. Electrodes were inserted in the respective reservoirs placed at the channel ends and the solutions were aspirated by EOF or a combination of EOF and electromigration. Since the chemistries usually involved working with solutions at low pH (acidic or neutral reagents) careful optimization of conditions to produce sufficient flow rates in the system for all the reagents and sample was necessary. The reaction products were detected by an LED-based absorbance detector coupled to one of the microchannels using a pair of optical fibers. However, there were limitations posed by the need for the solution composition to be compatible with EOF pumping and there was generally significant noise. Analyte separation/preconcentration can in theory be achieved in a more elegant fashion using traditional CE techniques. There may be more disadvantages than advantages in these sorts of systems and the real utility of non-isolated EOF-driven systems [12–16] is still to be proven.

Ramsey and Ramsey [17] have shown an interesting approach to achieve EOF-induced pressure pumping in a microfluidic system made on a glass slide. This system does not exactly fall into any of the pumping mechanisms previously discussed. A channel having a T-shape was formed on a glass chip, as shown in Figure 5.6. The electrodes were connected between the separation (sep) and ground (g) channels. In the system depicted in Figure 5.6a, the glass surface was not modified. Under these conditions, there is no EOF in the field free channel (ff). If, however, the electroosmotic flow in the side arm (g) is reduced, an excess pressure in the field free channel (ff) will be generated. One of the possible methods is to reduce the ζ potential of the side arm by modifying the surface with a polymer coating (e.g., linear polyacrylamide). This case is shown in Figure 5.6b. Ramsey and Ramsey demonstrated that there is indeed a pressure flow induced in the field free channel: they used it to supply fluid to an electrospray MS interface. In a subsequent paper Culbertson et al. [18] showed that such an electroosmotically induced hydraulic pumping can be used for differential ion transport, that is, depending on the amount

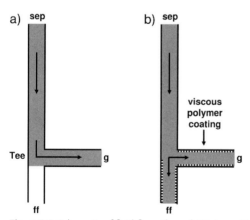

Figure 5.6 Schematic of fluid flows through T-intersection in which the electroosmotic velocity in the ground arm (g) is the same (A) or less than that found in the separation (sep) channel arm (B). Reprinted with permission from [20]. Copyright (2000) American Chemical Society.

of EOF reduction in the ground channel. The pump is shown to differentially transport ions down the two outlet channels. Under certain conditions the pump is able to differentially transport anions relative to cations down the field free channel. This may have some important implications, for instance in biochemical analysis problems that may require a continuous separation of cations from anions, such as continuous separation of proteins from DNA or various cell sorting schemes.

McKnight *et al.* [19] used a hybrid microchip device constructed from a glass slide bonded to a polymeric substrate (PDMS) with channel structure. A pair of closely spaced, interdigitated, electrodes was formed on the glass cover slide. The EOF was created between the electrodes and provided electric-free pumping of fluids in the sections of the channel outside the electrode pairs. The polymeric substrate was shown to effectively eliminate any electrolytically formed gas bubbles by gas permeation that allowed a rapid transport of gases from the microchannels during electroosmotic pumping. The pump was shown to tolerate very high field strengths (up to $4.5 \, kV \, cm^{-1}$).

5.3.2
Electrically Isolated EOF Pumps

Electric isolation of the EOF pump from the rest of the system achieves electric field free flow propulsion. In such systems, the EOF pump is electrically separated from the rest of the system through an interface made of a conductive material such as a polymeric membrane or porous glass frit. An ideal interface would have minimal electrical resistance, while exhibiting very high resistance to bulk fluid flow of liquid. If a pumping system is isolated from the rest of the flow system, there will be no influence of the electric field on the injected sample and the carrier. The liquid in the non-pump part of the system is transported by EOF-mediated hydrodynamic flow. The EOF pump fluid and the carrier/reagent fluids of analytical interest do not need to be the same. This allows for much greater flexibility in the selection of the carrier media: organic or hydro-organic solvents, concentrated electrolytes, and so on, can be used in the analytical part of the system. It is also possible to use conduits of different diameter and material from those used in the EOF pumps themselves.

The vast majority of the reported EOF-driven flow systems indeed use this principle, as it confers several advantages. An EOF pump electrically isolated from the rest of the system may be regarded as a "stand-alone EOF pump," a term now in common parlance. Indeed, most articles that deal with EOF pumping largely focus on the development and characterization of such pumps rather than their real applications. Practical applications of such EOF pumps still await further exploration and exploitation and may provide some exciting benefits to future miniaturized flow systems.

5.3.2.1   Open Tubular Electrically Isolated EOF Pumps
In open tubular EOF pumps one or more narrow bore capillaries or microchannels on a chip is hydraulically connected to the rest of the separation system (e.g., network of capillaries/channels, injection valves, detector). The EOF pumping channels are

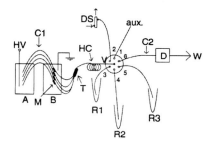

Figure 5.7 Schematic of an EOF-pumped SIA system. HV-high voltage power supply, A, B-pumping electrolyte solution containers, M-membrane joint, C1-pumping capillary, T-4 × 1 union, HC-holding coil, V1-four-way valve, S1, S2-syringes, V2-6 × 1 selector valve, R1, R2, R3-reagents, aux-unused auxiliary solution port typically dipped in the carrier electrolyte to prevent siphoning. Reprinted with permission from [21]. Copyright (1994) American Chemical Society.

electrically separated from the rest of the flow system. A basic requirement for this type of pump is that the EOF channel(s) have a minimum of two fluidic connections, one of which is responsible for the electrical contact and the other serves to provide the hydraulic connection with the rest of the system. The most commonly used geometries are T-shape interfaces or T-shape microfluidic channels on chips. As discussed in previous sections, a single channel EOF pump is often not practical because of the limitations of the pressure capabilities and maximum flow rates achievable. Multiple pumping capillaries/microchannels have been used to address this problem. As shown in Figure 5.7, Liu and Dasgupta [20, 21] used four FSCs in parallel to provide sufficiently large flow rates to a large bore flow injection channel. The terminal end of each of the pumping capillaries (75 μm i.d. FSC, (C1)) was inserted in a reservoir A filled with the pump electrolyte solution and a high voltage electrode. The other end of each capillary was joined at individual Nafion membrane joints M immersed in a second electrolyte reservoir B with a grounding electrode. Each joint M terminated in a second capillary that now constitutes part of the non-pump portion of the system. These four second capillaries are then joined together at a union T to form a single outlet. While the pump fluid is contained in reservoir A, it ends up pressure pumping the analytical stream fluid contained in the holding coil HC. The capacity of HC must be sufficient to accomplish one or more full analytical cycles. Normally during operation the ports in valve V1 are horizontally connected. When the cycle is finished, V1 is switched and the contents of HC and the contents of C1 are replenished from sources S1 and S2, respectively. Liu and Dasgupta [20] constructed two independent pump systems, using 2 mM $Na_2B_4O_7$ as the pump fluid. In one experiment this pump fluid itself was used as the carrier solution (HC not really needed) and an injector valve was located in this stream to introduce a chloride sample. This stream merged with a second EOF-pump driven system which pumped a solution of $Hg(SCN)_2$ – $Fe(NO_3)_3$ kept in HC by displacement.

It will be obvious to the astute reader that by merely changing the high voltage polarity, the flow direction of an EOF-pump will be reversible. It can perform the same aspirate or dispense function as a traditional syringe pump except without any mechanical inertia. It is therefore highly suitable for a SIA set-up where both the aspiration and dispense functions are needed. Indeed what is shown in Figure 5.7 is a SIA set-up where the multiport selector valve V2 is used to aspirate the sample and one or more reagents, mix them in the same conduit (if needed by reversing flow directions repeatedly) and then deliver the mixed bolus through detector D [21]. The authors then used a very similar SIA arrangement to aspirate a receiver fluid through a porous membrane tube around which a test gas (ammonia) was sampled [22]. The collected ammonia was mixed sequentially with other reagents by the same pump to form indophenol blue which was measured colorimetrically. Taking this yet one step further, these authors [23] used an EOF-driven pump in SIA configuration to form a small (6–18 µl) drop of dilute H_2SO_4 at the tip of a capillary and used it to sample ammonia from a test gas stream flowing past the droplet. The radial distribution of ammonia in the droplet was then probed by withdrawing 1 µl of the drop contents at a time and subjecting it to SIA-based colorimetric analysis for ammonia. These authors were able to perform 17 sequential analyses from an 18 µl droplet.

Interestingly, stand-alone EOF pumps can be used to augment or inhibit flows in other systems such as CE where EOF is intrinsic [24, 25], sometimes with beneficial results.

5.3.2.2 Microchannel-based Electrically Isolated EOF Pumps

The open tubular pump approach is very simple and is easily transferred onto microfluidic chip platforms, where a multitude of parallel channels of small cross-sectional area can be connected in parallel. The electrical isolation of the pump channels on a chip, however, is more challenging.

Takamura et al. [26] were among the first to develop a multiple channel EOF pump on a microchip. They used a conductive photopolymerized gel based on an acrylamide, N,N'-methylenebis(acrylamide), and 2,2'-dimethoxy-2-phenylacetophenone copolymer for electrical grounding. The primary goal of the device was in health care applications, for safety, low applied voltages were essential. Typically 10–40 V was used to generate EOF. The pump consisted of 6 or 15 pumping stages connected in series. Each pumping stage consisted of 10 narrow channels (5×20 µm) connected to a 50 µm wide 20 µm high channel. The pump was able to operate at moderate pressures of about 1 psi (at the low voltages used) and could be used in systems with low backpressures. In a subsequent publication [27], they modified the design of the pump to improve the pressure performance up to 5 psi backpressure and to attain flow rates up to 0.5 µL min^{-1}. The theoretical aspects of their low voltage cascade pump have been described [28]. Lazar and Karger [9] constructed a 100-parallel channel EOF pump able to operate at flow rates between 10 and 430 nl min^{-1} and at a pressure of up to 80 psi. Electrical isolation was achieved by a porous glass disk, placed in the buffer reservoir connected to the 100-channel network. Due to the large electrical resistance of the porous glass disk, the voltage drop across the disk was significant and the down stream fluid did not quite attain ground potential.

Liu *et al.* [29] described a truly grounded microchannel-based EOF pump in which they used a Nafion membrane for effective electrical grounding. Nafion exhibits low ohmic resistance and when electrolytic gases are generated at a low enough rate, it is apparently able to act as a permeative sink.

5.3.2.3 Electrically Isolated EOF Pumps Based on Packed Columns

Multiple parallel channels can be very efficiently formed by close packing of small particles, for example, µm-size silica particles in a capillary column, as is typically used in capillary chromatography. The interstitial spaces between the particles form a network of intertwined channels. Thus, if a fused silica capillary or a channel on a microfluidic chip is packed with nonporous, unmodified silica particles, the column behaves as a network of capillaries with the effective diameter of the pores being dependent on the packing density and particle size (it can generally be approximated to be 1/5 of the mean particle diameter [10]). The length of the microchannels is best expressed as the tortuosity of the porous media. Such a column can produce relatively large flows with the ability to withstand large backpressures. Paul *et al.* [10] first demonstrated such an EOF pump consisting of a short length of an FSC packed with C_{18} – functionalized or bare silica beads with diameters between 1.5 and 5 µm. The pump was able to deliver flow rates in the range of few nL min^{-1} to several µL min^{-1}, depending on the backpressure and operated against a backpressure in excess of 8000 psi, unprecedented for EOF-driven pumps. Porous borosilicate glass was used for isolation of the voltage imposed on the packed FSC EOF pump; practical applications of the pump were demonstrated [30].

Razunguzwa and Timperman [31] presented an interesting alternative to a silica pH-dependent EOF pump. Generally, the flow attained by a silica-based EOF-pump is dependent on the pH of the pump solution, because the extent of dissociation of silanol groups is pH dependent and consequently the EOF flow rate increases with pH until the silanol groups are fully ionized. Instead of silica, these authors relied on polymeric ion exchangers and their design also resulted in the remarkable attribute that the system was essentially self-grounding. In their design, shown in Figure 5.8, a T-shape channel was etched on a glass chip and 5 µm diameter polymeric anion exchanger and cation exchanger beads, respectively, were packed into the 500 µm wide short arms of the tee. One electrode each was connected at the termini of the T-legs packed with the ion exchanger beads, with a Teflon membrane serving to prevent bead escape as well as to allow access to pump inlet fluid. Both these exchangers are weak acid/base type. As such, at high pH only the cation exchanger leg was ionized and acted as a pump whereas at low pH only the anion exchanger was ionized and served as the pump. To prevent backflow through the non-pumping leg at either pH extreme, the packed legs were narrowed to a 50 µm width (which also effectively prevented bead escape). The voltage drop is effectively across the packed beds and thus the long leg of the T, 100 µm in width, is field-free and offers a far lower hydraulic resistance to flow than either of the packed legs. The liquid is thus pumped out through the long leg of the T. With the combination of the two different types of exchangers, the pump operated at acceptable flow rates over a much broader range of pH (2–12) compared to what is observed with

a)

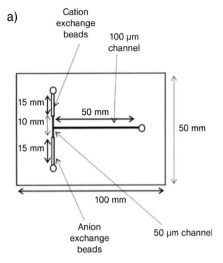

b)

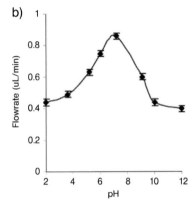

Figure 5.8 (a) Schematic diagram of improved version of the packed bed microchip EOF pump. The 50 μm constriction of the pumping channels reduces backflow through the non-pumping section at pH extremes by increasing the resistance to hydrodynamic flow into the non-pumping section. (b) The relationship between pH and flow rate at an applied voltage of 3 kV. The design shown in (a) was improved here by addition of a 1.5 mm section of 0.5 μm beads to further prevent leakage into the non-pumping channel. Reprinted with permission from [33]. Copyright (2004) American Chemical Society.

silica-based pumps. It is possible that had the authors used strong acid/base type ion exchangers, the performance across the pH range would have been even more impressive.

Zeng *et al.* [32] made a packed bed EOF pump by packing 3.5 μm nonporous silica particles into 500–700 μm bore capillaries. Retaining frits were made from sodium silicate solution by heating at 350 °C. The authors studied the pressure vs. flow rate characteristics. They achieved a maximum pumping pressure of 500 psi and attained flow rates between 4 and 18 μl min^{-1}. A subsequent study [33] was aimed at increasing the maximum attainable flow rate. The pump itself was of large bore: a 1 cm diameter acrylic chamber was packed with 1–3 μm diameter nonporous silica particles. The retaining frits were constructed by photopolymerization. Water was used as the pump fluid to minimize Joule heating and the pump was able to operate at very high flow rates of up to 1 ml min^{-1}. The attainable pressures, however, were much lower than those in the packed capillary EOF pump. The use of these pumps should therefore be suitable in traditional low pressure FIA/SIA systems. However, no practical flow system applications were demonstrated.

As in the case of open channel EOF pumps, the packed pumping channels or capillaries can be connected in parallel to provide higher pressures and increase flow rates. Chen et al. [34, 35] have connected in parallel 3 FSCs (530 μm i.d., 25 cm long) each packed with 2 μm porous silica beads and used the system for EOF pumping. In their latter version of the pump, a gas-release chamber was incorporated to release gas bubbles evolved from electrolysis. It could be operated at flow rates from a few nl min^{-1} to several μl min^{-1} with a maximum operating pressure of 5 MPa (about 725 psi). It was also shown that this EOF pump can be used to pump mixtures of organic and aqueous solutions without difficulties.

5.3.2.4 Electrically Isolated EOF Pumps Based on Porous Media and Monoliths

Microporous materials, such as porous glass frits or monolithic column materials can be used not only for electrical isolation but also to provide a network of parallel channels, as the packed bead columns. If the pump needs to generate large back-pressures, the pump packing material must also withstand it, thus placing requirements on the fragility and tensile strengths of the packing. Gan et al. [36] constructed an EOF pump based on a homemade 13 mm thick sintered borosilicate glass disk of relatively large diameter (3.5 cm). The pore size was between 2 and 5 μm. It resembles the large bore packed pump of Zeng et al. [33] previously discussed. The flow rates achieved were in the μl min^{-1} to ml min^{-1} range with moderate pressure capability (22 psi). The authors applied the pump for a traditional FIA application. One pump each was used to propel buffer and reagent solutions, respectively, at high volumetric flow rates (0.5–3.5 ml min^{-1}) in a two-line FIA system for the determination of Cr(VI) using the diphenylcarbazide colorimetric chemistry. Yao et al. [37] used a commercially available porous borosilicate glass disk with relatively narrow pore distribution around 1–2 μm to generate EOF. In their pump they also incorporated a gas recombination catalyst that served to recombine the evolved electrolysis gases (H_2, O_2) back to water. This significantly improved the long term pumping stability and alleviated pump failures from bubble formation and consequent interruption of electroosmosis.

Microporous monolithic columns prepared by various polymerization procedures provide an interesting alternative to porous glass materials. There are a large number of chemistries and procedures described in the literature that allow polymerization of materials with varying and controllable pore sizes. Recently Chen et al. [38] have used a silica-based monolith prepared by sol–gel polymerization inside a fused silica capillary for an EOF-pump bed. The pump was used for flow injection application. It could operate at pressures up to 58 psi and could also be used for delivering organic liquids or hydro-organic solvents.

5.4
EOF Injection Methods Utilized in FIA, SIA and Micro Total Analysis Systems (μ-TAS)

The injection of a sample in an EOF-driven flow analysis system can be classified into three general types, (i) valve injection, (ii) hydrodynamic/pressure injection and (iii) electrokinetic injection.

In valve injection, a rotary or slide valve containing a fixed volume sample loop is used. Low nanoliter volume injectors are commercially available with sample capacities ranging from 10 to 200 nL. When the EOF pumping system is electrically isolated from the rest of the system, injection by a small volume internal loop injector is the most straightforward option. Otherwise electrically nonconducting valves have to be used.

Hydrodynamic injection relies on immersion of one capillary end to the sample vial, elevating it to a specific height for a specific time interval. The sample volume in this type of injection method is not fixed, that is, it can be varied by varying the applied height/pressure/vacuum and the duration of the injection step. The sample is introduced into the fused silica capillary by gravity-induced flow. Since this flow is laminar in nature, with each consecutive injection it introduces additional zone broadening. This injection mode applies solely to a capillary-based system, because it can be accomplished by elevation of the fused silica capillary coated with flexible polyimide coating. Hydrodynamic injection, typical for capillary electrophoresis, was for instance used in the first publication on the topic [3]. In chip-based systems, pressure or vacuum can be applied and a small sample plug is introduced by either method.

In electrokinetic (EK) injection, samples are introduced into the capillary or a microchannel by a combination of EOF and (for charged analytes) electromigration. It can be used in any of the systems. The simplest EK injection approach is to apply high voltage over the capillary or channel section for a given period of time. The amount injected can be manipulated both by the choice of the applied voltage and the time. The second EK introduction type is based on defined volume. This type of sampling is frequently used in microfluidic structures and has been much studied in the context of microfluidic electrophoretic separations. In this mode, the sample volume is defined by the injection channel network geometry (typically a cross or a double-T). The sample is then introduced by the EOF into the flow system. Depending on the geometry and electrical field used, this mode can be further classified as floating or pinched. In the floating mode, the electric field is applied between the sample and the waste reservoirs: such an injection mode can, however, suffer from diffusional transport of the sample into other connecting parts of the system. In the pinched injection mode, a plurality of voltages are simultaneously applied to other parts of the system that help prevent diffusion and thus define an exact volume of the sample plug. In high efficiency electrophoretic separations, this mode is preferred because there is no diffusional broadening of the injected band.

Because an EOF-driven fluid delivery system is inherently bidirectional, it is possible to pick up a precise amount of a sample and then deliver it elsewhere. Recently, an EOF-driven nanopipettor was described by Byun *et al.* [39] – this device permits good precision even at low nl levels and it should be possible to configure this as a nanoinjector.

The many different ways EOF is exploited to achieve sample injection in μ-TAS systems are somewhat beyond the scope of this chapter, and the reader is referred to reviews by Haswell [40] and Roddy *et al.* [41].

5.5
Applications of EOF-Driven Pumping in Flow Analysis

Although the number of publications on EOF-mediated pumping and pump designs is relatively large, the number of real applications in flow analysis systems is rather limited. Most studies have focused on characterization of new pump designs rather then their application to real analytical chemistry problems.

Early on, Liu and Dasgupta [3] showed the flow injection analysis of Fe^{2+} based on its colored chelate formation with 1,10-phenanthroline; this was demonstrated in a single capillary system without electrical isolation. The calibration was linear between 5 and 200 mg l^{-1}. Compared to conventional FIA, this arrangement operates in a batch mode to obtain greater throughput; that is, multiple samples are hydrodynamically introduced, spaced by the reagent before EOF is used to drive the sample train to the detector. This results in variable dispersion of the sample plugs that are dependent on the position of the sample in the train. Peak heights for the same injected concentration thus varied along the train; however, peak areas were reproducible. Haswell *et al.* [13–15] have described micro flow analysis systems based on purely EOF-driven mobilization of reagents and samples. The determination of phosphate using the molybdenum blue reaction was studied. The same channel configuration was later used for the determination of nitrite using the Griess reaction. Greenway *et al.* have described a similar EOF-driven flow injection system for the chemiluminescence (CL) detection of codein with tris(2,2′-bipyridyl) ruthenium(II) [42] and the determination of cobalt based on its catalytic effect on oxidative CL reactions of luminol [43].

An electrically isolated EOF pump based on four parallel FSCs was used for sequential injection analysis of nitrite-nitrogen and ammonia-nitrogen [21]. The EOF pump was electrically isolated by a Nafion membrane. The schematic of the system is shown in Figure 5.7. The EOF pump electrolyte, 2 mM sodium tetraborate, was used as the carrier electrolyte in the SIA system. The sample and reagents were sequentially aspirated into a holding coil and then mixed and propelled into the detector. For the determination of ammonia, up to three reagents were aspirated after the sample. A careful optimization of reagent order and zone length was required; adaptation of batch methods is not a straightforward procedure in the optimization of SIA systems. In another study, these authors used an EOF-pumped SIA system with a gas sampling drop interface and demonstrated the determination of ammonia at ppb levels.

More than one EOF-based pump can be connected to form a multiple-line FIA system. Liu and Dasgupta [20] have utilized such a two-line system for the determination of chloride using the colorimetric reaction with mercuric thiocyanate and ferric nitrate solutions. The pumping electrolyte was 2 mM sodium tetraborate which pumped the chromogenic reagent held in the holding coil HC by displacement. The system could be run for 165 min before the chromogenic reagent became contaminated by the pump electrolyte and had to be replaced. By increasing the HC volume this time could be extended. Alternatively, an immiscible liquid plug can be put between the pump fluid and the reagent to prolong the period before replenishment is needed. In this paper, the authors also demonstrated electrostacking preconcentration of bromcresol green prior to flow injection analysis.

Pu and Liu [44] demonstrated capillary scale SIA on a chip using a microfabricated EOF pump. A serpentine reaction channel was fabricated on a borofloat glass wafer. They used an EOF pump with 32 parallel channels. The pump was connected to a selection valve that served to select sample, reagents, and so on. Fluorescence detection (λ_{ex} 470 nm, λ_{em} 520 nm) was used. The system was later exploited for the assay of an analyte that inhibits an enzymatic reaction. The hydrolysis of the substrate fluorescein di-(β-D-glucopyranoside) by β-galactosidase and its inhibition by DTPA was studied. The micro SIA approach utilized mixing of three separate zones of the enzyme, the inhibitor and the substrate. The stacked zone was subsequently aspirated into the microfluidic channel for detection. Figure 5.9 shows traces of the measured fluorescence intensity as a function of the DTPA concentration.

Gan *et al.* [36] used an EOF-pumped SIA system to determine Cr(VI) in waste water samples using the diphenylcarbazide reaction. They studied the effect of the carrier electrolyte composition on EOF velocity and found 0.35 mM NH_4OH to be highly suitable – it provided a flow rate of 3 ml min^{-1} at an applied voltage (V_{app}) of 500 V. The flow rate increased essentially linearly from $V_{app} = 100$ to 500 V. Two pumps were used in a SIA system connected to an 8-way selection valve. The sequence of injected zones was: carrier–reagent–sample–reagent–carrier. This zone sequence was propelled through the reaction coil and detected at 540 nm.

Zhao *et al.* [45] have reported the use of a home-made EOF pump (no details given) for SIA analysis of nitrite in drinking water by the Griess–Saltzman reaction.

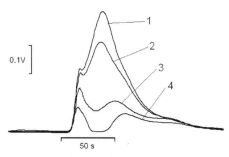

Figure 5.9 Peak shape of enzyme inhibition study. Signal obtained with 10 s aspiration time of enzyme (β-Gal), 10 s inhibitor (DTPA) of different concentration (0–3 mM), 10 s of substrate (FDG); (β-Gal: 0.1 mg ml^{-1} (with 83 µM MgCl$_2$); FDG: 91.3 µM; reaction buffer 10 mM TRIS at pH 7.3; operation voltage 4900 V. Trace 1:[DTPA] = 0 mM. 2:[DTPA] = 0.2 mM, 3: [DTPA] = 1.5 mM, 4:[DTPA] = 3.0 mM. Reprinted from [45] with permission from Elsevier.

The sample and reagent zones were aspirated into the conduit and the stacked zones were then delivered to the detector at 1.5 ml min^{-1} by polarity reversal of the voltage applied to the EOF pump. The response was linear from 10 to 800 μg l^{-1}; the LOD being 1 μg L^{-1}. Nitrite levels in drinking water are typically lower than this.

Although 15 years have passed since the first publication of an EOF-pumped flow analysis system, the use of EOF pumps in real analytical flow systems is still limited. As the drive to miniaturize continues, there are likely to be more applications.

5.6
Future Prospects

It has been shown by numerous researchers that electroosmosis can be efficiently used for fluid propulsion in traditional as well as micro flow analysis systems. Various designs of EOF pumps and their capabilities to perform pulseless pumping of liquids over a wide range of flow rates from a low nl min^{-1} to several ml min^{-1} have been demonstrated. Often there is a trade-off between the flow rates and pressures that can be achieved with such pumps. This may not be of concern, because typically a high flow rate FIA/SIA system has relatively low flow resistance, unless a packed bed reactor or a filtration system is used. EOF pumps happen to be ideally suited for microfluidic systems in which pulseless and stable flow is desired. The flow stability in an EOF-pumped system has been a problem in some applications, because solutions of different compositions go through the capillaries. Sometimes the flow changes because of changes in the pump surface due to absorption of high molecular weight analytes from the samples/sample matrix. This problem can be avoided if the pump fluid is separated from the necessary fluids in the analytical system. When properly designed, EOF pumps can be used to generate very high backpressures and thus be useful even as chromatographic pumps.

Acknowledgment

This work was supported by the Grant Agency of the Czech Republic Grant no. GA CR 203/07/0983 to PK, by NIH Grant RO1 GM078592-01 to SL and NSF Grant CHE-0518652 to PKD.

References

1 Zheng, H.J. and Dasgupta, P.K. (1994) Concentration and Optical measurement of Aqueous Analytes in an Organic Solvent Segmented Capillary under High Electric Field. *Analytical Chemistry*, **66**, 3997–4004.

2 Reuss, F.F. (1809) Sur un nouvel effect de l'èlectricite galvanique. *Memoires de la Societe Imperiale des naturalistes de Moskou*, **2**, 327–337.

3 Liu, S. and Dasgupta, P.K. (1992) Flow-injection analysis in the capillary

format using electroosmotic pumping. *Analytica Chimica Acta*, **268**, 1–6.

4 Jorgensson, J.W. and Lukacs, K.D. (1981) Zone electrophoresis in open-tubular glass capillaries. *Analytical Chemistry*, **53**, 1298–1302.

5 Ivaska, A. and Ruzicka, J. (1993) From flow injection to sequential injection: comparison of methodologies and selection of liquid drives. *Analyst*, **118**, 885–889.

6 Nguyen, N.T., Huang, X.Y. and Chuan, T.K. (2002) MEMS-micropumps: a review. *Journal of Fluids Engineering-Transactions of the ASME*, **124**, 384–392.

7 Laser, D.J. and Santiago, J.G. (2004) A review of micropumps. *Journal of Micromechanics and Microengineering*, **4**, R35–R64.

8 Hair, M.L. and Hertl, W. (1970) Acidity of surface hydroxyl groups. *The Journal of Physical Chemistry*, **74**, 91–95.

9 Lazar, I.M. and Karger, B.L. (2002) Multiple open-channel electroosmotic pumping system for microfluidic sample handling. *Analytical Chemistry*, **74**, 6259–6268.

10 Paul, P.H., Arnold, D.W. and Rakestraw, D.J. (1998) Electrokinetic generation of high pressures using porous microstructures, in *Micro Total Analysis Systems 1998* (eds D.J. Harrison and A. van den Berg), Kluwer, Boston MA, pp. 49–52.

11 Foret, F., Křivánková, L. and Boček, P. (1993) *Capillary Zone Electrophoresis*, Wiley-VCH, Weinheim, pp. 140–146.

12 Dasgupta, P.K. and Hwang, H. (1985) Application of a nested loop system for the flow injection analysis of trace aqueous peroxides. *Analytical Chemistry*, **57**, 1009–1012.

13 Daykin, R.N.C. and Haswell, S.J. (1995) Development of a micro flow injection manifold for the determination of orthophosphate. *Analytica Chimica Acta*, **313**, 155–159.

14 Doku, G.N. and Haswell, S.J. (1999) Further studies into the development of a micro-FIA (µFIA) system based on electroosmotic flow for the determination

of phosphate as orthophosphate. *Analytica Chimica Acta*, **382**, 1–13.

15 Greenway, G.M., Haswell, S.J. and Petsul, S.J. (1999) Characterisation of a micro-total analytical system for the determination of nitrite with spectrophotometric detection. *Analytica Chimica Acta*, **387**, 1–10.

16 Liu, S. and Dasgupta, P.K. (1993) Electroosmotically pumped capillary flow-injection analysis. *Analytica Chimica Acta*, **283**, 739–745.

17 Ramsey, R.S. and Ramsey, J.M. (1997) Generating electrospray from microchip devices using electroosmotic pumping. *Analytical Chemistry*, **69**, 1174–1179.

18 Culbertson, C.T., Ramsey, R.S. and Ramsey, J.M. (2000) Electroosmotically induced hydraulic pumping on microchips: Differential ion transport. *Analytical Chemistry*, **72**, 2285–2291.

19 McKnight, T.E., Culbertson, C.T., Jacobson, S.C. and Ramsey, J.M. (2001) Electroosmotically induced hydraulic pumping with integrated electrodes on microfluidic devices. *Analytical Chemistry*, **73**, 4045–4049.

20 Dasgupta, P.K. and Liu, S. (1994) Electroosmosis: A reliable fluid propulsion system for flow injection analysis. *Analytical Chemistry*, **66**, 1792–1798.

21 Liu, S. and Dasgupta, P.K. (1994) Sequential injection analysis in capillary format with electroosmotic pump. *Talanta*, **41**, 1903–1910.

22 Liu, S. and Dasgupta, P.K. (1995) Electroosmotically pumped capillary format sequential injection analysis with a membrane sampling interface for gaseous analytes. *Analytica Chimica Acta*, **308**, 281–285.

23 Liu, S. and Dasgupta, P.K. (1995) Liquid Droplet. A renewable gas sampling interface. *Analytical Chemistry*, **67**, 2042–2049.

24 Dasgupta, P.K. and Liu, S. (1994) Auxiliary electroosmotic pumping in capillary electrophoresis. *Analytical Chemistry*, **66**, 3060–3065.

25 Dasgupta, P.K. and Kar, S. (1999) Improving resolution in capillary zone electrophoresis through bulk flow control. *Microchemical Journal*, **62**, 128–137.

26 Takamura, Y., Onoda, H., Inokuchi, H., Adachi, A., Oki, A. and Horiike, Y. (2001) Low-voltage electroosmosis pump and its application to on-chip linear stepping pneumatic pressure source, in *Micro Total Analysis Systems 2001* (eds J.M. Ramsey and A. van den Berg), Kluwer, Boston MA, pp. 230–232.

27 Takamura, Y., Onoda, H., Inokuchi, H., Oki, A. and Horiike, Y. (2003) Low-voltage electroosmosis pump for stand-alone microfluidics devices. *Electrophoresis*, **24**, 185–192.

28 Brask, A., Goranovic, G. and Bruus, H. (2003) Theoretical analysis of the low-voltage cascade electro-osmotic pump. *Sensors and Actuators B-Chemical*, **92**, 127–132.

29 Liu, S., Pu, Q. and Lu, J.J. (2003) Electric field-decoupled electroosmotic pump for microfluidic devices. *Journal of Chromatography. A*, **1013**, 57–64.

30 Paul, P.H., Arnold, D.W., Neyer, D.W. and Smith, K.B. (2000) Electrokinetic pump application in micro-total analysis system mechanical actuation to HPLC, in *Micro Total Analysis Systems '2000* (eds A.W. van den Berg, P. Olthuis and P. Bergveld), Kluwer, Boston MA, pp. 583–590.

31 Razunguzwa, T.T. and Timperman, A.T. (2004) Fabrication and characterization of a fritless microfabricated electroosmotic pump with reduced pH dependence. *Analytical Chemistry*, **76**, 1336–1341.

32 Zeng, S., Chen, C.-H., Mikkelsen, J.C., Jr and Santiago, J.G. (2001) Fabrication and characterization of electroosmotic micropumps. *Sensors and Actuators B-Chemical*, **79**, 107–114.

33 Zeng, S., Chen, C.-H., Santiago, J.G., Chen, J.-R., Zare, R.N., Tripp, J.A., Svec, F. and Frechet, J.M.J. (2002) Electroosmotic flow pumps with polymer frits. *Sensors and Actuators B-Chemical*, **82**, 209–212.

34 Chen, L., Ma, J. and Guan, Y. (2003) An electroosmotic pump for packed capillary liquid chromatography. *Microchemical Journal*, **75**, 15–21.

35 Chen, L., Ma, J. and Guan, Y. (2004) Study of an electroosmotic pump for liquid delivery and its application in capillary column liquid chromatography. *Journal of Chromatography. A*, **1028**, 219–226.

36 Gan, W.-E., Yang, L., He, Y.-Z., Zeng, R.-H., Cervera, M.L. and de la Guardia, M. (2000) Mechanism of porous core electroosmotic pump flow injection system and its application to determination of chromium (VI) in waste-water. *Talanta*, **51**, 667–675.

37 Yao, S., Hertzog, D.E., Zeng, S., Mikkelsen, J.C., Jr and Santiago, J.G. (2003) Porous glass electroosmotic pumps: design and experiments. *Journal of Colloid and Interface Science*, **68**, 143–153.

38 Chen, Z., Wang, P. and Chang, H.-C. (2005) An electro-osmotic micro-pump based on monolithic silica for micro-flow analyses and electro-sprays. *Analytical and Bioanalytical Chemistry*, **382**, 817–824.

39 Byun, C.K., Wang, X., Pu, Q. and Liu, S. (2007) Electroosmosis-based nanopipettor. *Analytical Chemistry*, **79**, 3862–3866.

40 Haswell, S.J. (1997) Development and operating characteristics of micro flow injection analysis systems based on electroosmotic flow. A review. *Analyst*, **122**, 1R–10.

41 Roddy, E.S., Xu, H. and Ewing, A.G. (2004) Sample introduction techniques for microfabricated separation devices. *Electrophoresis*, **25**, 229–242.

42 Greenway, G.M., Nelstrop, L.J. and Port, S.N. (2000) Tris(2,2-bipyridyl)ruthenium (II) chemiluminescence in a microflow injection system for codeine determination. *Analytica Chimica Acta*, **405**, 43–50.

43 Nelstrop, L.J., Greenwood, P.A. and Greenway, G.M. (2001) An investigation of electroosmotic flow and pressure pumped luminol chemiluminescence detection for cobalt analysis in a miniaturised total

analytical system. *Laboratory on a Chip*, **1**, 138–142.

44 Pu, Q. and Liu, S. (2004) Microfabricated electroosmotic pump for capillary-based sequential injection analysis. *Analytica Chimica Acta*, **511**, 105–112.

45 Zhao, Y.-Q., He, Y.-Z., Gan, W.-E. and Yang, L. (2002) Determination of nitrite by sequential injection analysis of electrokinetic flow analysis system. *Talanta*, **56**, 619–625.

6
Flow Analysis in Microfluidic Devices

Manabu Tokeshi and Takehiko Kitamori

6.1
Introduction

The introduction of the concept of micro total analysis systems (μTAS) by Manz *et al.* [1] in 1990 triggered rapidly growing interest in the development of flow-based miniaturized analysis systems, also called Lab-on-a-Chip or microfluidic devices, in which all the stages of chemical analysis such as sample pre-preparation, chemical reactions, separation, purification, detection and analysis are performed in an integrated and automated fashion [2–5]. Flow-based analysis using microfluidic devices has attracted much attention for chemical, biological and medical applications. Microfluidic devices integrating all components required for analysis are ideal tools for microanalysis, and the development of such devices is highly desired.

On the other hand, a similar concept, called Lab-on-Valve (LOV), was proposed by Ruzicka in 2000, in which the sample/reagent injection ports, mixing coils, reaction columns and flow-through detector are integrated on a selector valve [6]. It may be said that this technique is the advanced method for conventional flow injection analysis. Although the LOV is also a miniaturized flow-based analysis system, it lies outside the scope of this chapter.

This chapter gives an overview of developments in flow-based analysis using microfluidic devices. The focus is on some applications in continuous flow chemical processing and flow injection analysis in microfluidic devices.

6.2
Continuous Flow Chemical Processing in Microfluidic Devices

6.2.1
Integration of Heavy Metal Analysis System

The integration of chemical processes on a chip by using continuous flow chemical processing (CFCP) in combination with micro-unit operations such as solvent

Advances in Flow Analysis. Edited by Marek Trojanowicz
Copyright © 2008 WILEY-VCH Verlag GmbH & Co. KGaA, Weinheim
ISBN: 978-3-527-31830-8

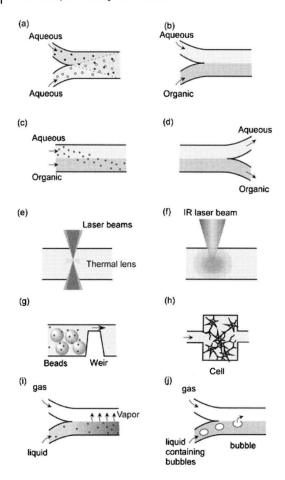

Figure 6.1 Series of micro-unit operations. (a) Confluence of two miscible solutions. Mixing by molecular diffusion. Reaction. (b) Confluence of two immiscible solutions. Formation of stable multi-phase laminar flow. (c) Solvent extraction. Phase-transfer reaction. (d) Phase separation. (e) Highly sensitive detection (using a thermal lens microscope). (f) Laser heating. (g) Solid phase extraction. Concentration and reaction on a surface of beads. (h) Cell culture. (i) Evaporation of solvent and concentration. (j) Bubble removal. (Figure 4 in *QSAR Comb. Sci.*, 24, 742, **2005**.)

extraction, phase separation, and so on (Figure 6.1) under continuous flow conditions is a strategy for the construction of real μTAS [7–13]. The integration of a Co(II) wet analysis is a good example of CFCP [7]. The microchip for this wet analysis (Figure 6.2) consists of two different areas: the reaction and extraction area and the decomposition and removal area (i.e., the washing area). In the former area, sample solutions containing Co(II) ions, a 2-nitroso-1-naphthol (NN) solution and *m*-xylene are introduced using a syringe pump at a constant flow rate through three inlets. These three liquids meet at the intersection point, and a parallel two-phase flow, having an organic/aqueous interface, forms in the microchannel. The chelating

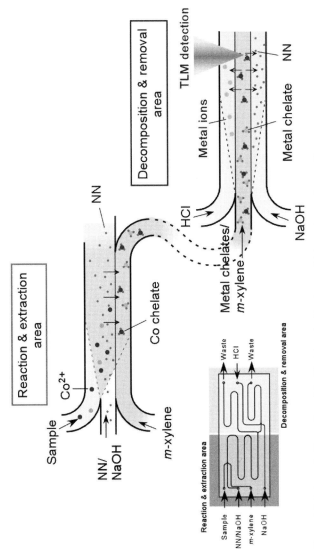

Figure 6.2 Schematic illustration of Co(II) determination by combining micro-unit operations. (Figure 5 in *Anal. Chem.,*, *74*, 1565, **2002**.)

reaction of Co(II) and NN and extraction of the resulting Co(II) chelate proceed as the reacting mixture flows along the microchannel. Since the NN reacts with the coexisting metal ions, such as Cu(II), Fe(II) and Pb(II), these coexisting metal chelates are also extracted into *m*-xylene. Therefore, a washing process is necessary after extraction for the decomposition and removal of coexisting metal chelates. The coexisting metal chelates decompose when they make contact with hydrochloric acid, and the metal ions pass into the HCl solution (backward extraction). The decomposed chelating reagent, NN, is dissolved in a sodium hydroxide solution. In contrast to the coexisting metal chelates, the Co chelate is stable in HCl and NaOH solutions and hence it remains. In the washing area, the *m*-xylene phase containing Co chelates and the coexisting metal chelates from the reaction and extraction area is interposed between the other two inlets at a constant flow rate. Then, the three-phase flow, HCl/*m*-xylene/NaOH, forms in the microchannel. The decomposition and removal of the coexisting metal chelates proceed along the microchannel in a similar manner as described above. Finally, the target chelates in *m*-xylene are detected downstream by a thermal lens microscope [8]. Cobalt in an admixture sample was successfully determined within 1 min [1]. Compared with a conventional method, this approach offers the advantages of simplicity and omission of time-consuming operations. In the conventional method, the acid and alkali solutions cannot be used simultaneously and alternative washing procedures must be repeated several times. The same effect can be obtained by using three-phase flow in the microchannel. In a subsequent paper, Kikutani *et al.* [12] expanded the concept of CFCP to integrate four parallel analyses of Co(II) and Fe(II) on a chip in a three-dimensional microchannel network.

6.2.2
Integration of Multi-Ion Sensing System

Since the liquid microspace provides a short diffusion distance and a large specific interfacial area as the liquid/liquid interface, novel attractive analytical features arise such as extremely fast ion sensing and ultra-small reagent solution volumes. In contrast to the slow response time of a standard ion-selective optode, in which the response time is basically governed by slow diffusion of ionic species in the viscous polymer membrane, that of the on-chip ion sensing system is clearly faster due to the short diffusion distance and the low viscosity of the organic solution. The ion-pair extraction scheme is an established methodology employed for ion-selective optodes which provides highly selective optical ion determination of various kinds of ions by using a single lipophilic pH indicator dye and highly selective neutral ionophores. However, exploitation of an ion-pair extraction reaction and chip technology provide attractive advantages which would not be achieved by conventional ion sensors. The advantages of the on-chip ion sensing system are:

1. Reaction time (response time) can be significantly reduced by the fast molecular transport in a microspace.

2. Since the required reagent solution volume is extremely small ($\sim$100 nL), fresh organic phase can be used in every measurement. Subsequently, response

degradation caused by the leaching of ion-sensing components, which is a typical problem of ion-selective optodes, does not occur. This merit is directly reflected in the excellent reproducibility of the response during continuous measurements and effective reduction in the amounts of expensive reagents consumed in one measurement.

3. Essentially, ion determination can be carried out by detection of the protonation/deprotonation process of a single lipophilic anionic dye in the organic phase. Therefore, no special color-changeable chelating reagents are required for measurement of another kind of analyte ion. It is only necessary to replace the neutral ionophore with different ion-selective ionophores. From the viewpoint of optical instrumentation, this merit is quite important; there is no need to change the excitation source to match the excitation wavelength of different chelating reagents.

4. Highly selective ionophores or carriers for various kinds of ions developed for application to ion-selective electrodes are commercially available and can be used without any chemical modification.

Figure 6.3 illustrates an experimental set-up and the ion-pair extraction scheme [13]. The organic solution containing a neutral ionophore and a lipophilic pH indicator dye and the aqueous solution containing sample ion (K^+ or Na^+) are

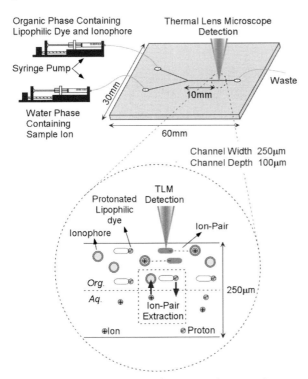

Figure 6.3 Schematic illustration of experimental set-up and ion-pair extraction model. (Figure 2 in *Anal. Chem.*, **73**, 1382, **2001**.)

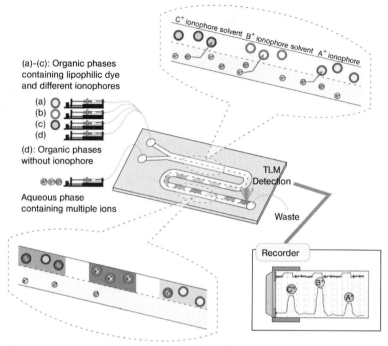

Figure 6.4 Concept of sequential ion-sensing system using a simple microchip. (Figure 1 in *Anal. Chem.*, *73*, 5551, **2001**.)

independently introduced into the microchannel to form an organic/aqueous interface. Then determination of the ion is carried out with a thermal lens microscope downstream in the organic phase under continuous flow conditions. The response time and minimum required reagent solution volume of the on-chip ion sensing system are about 8 s and 125 nL, respectively.

Furthermore, a sequential multi-ion sensing system using a single chip has been successfully realized by expanding the concept of on-chip ion sensing [14]. Figure 6.4 shows the basic concept of multi-ion sensing using a single chip. Different organic phases containing the same lipophilic pH indicator dyes but different ionophores are introduced sequentially, to avoid contamination, into the microchannel by on–off switching of syringe pumps. Aqueous sample solution containing different ions is introduced from the other inlet to form a parallel two-phase flow with the intermittently pumped organic phases. The selective ion-pair extraction reaction proceeds during flow; that is, different ions can be selectively extracted into different organic phases, depending on the selectivity of the neutral ionophores contained in the respective organic phases. Downstream in the flow, the ion-pair extraction reaction becomes equilibrated; thus, downstream detection of the color change of the organic phase allows sequential and selective multi-ion sensing in the single aqueous sample solution containing multiple ions.

Hisamoto *et al.* [14] used valinomycin and 2,6,13,16,19-pentaoxapentacyclo-[18.4.4.47,12.01.20.07,12]dotriacontane (DD16C5), which are known to exhibit high selectivity when used in conventional ion sensors, as highly selective potassium and sodium ionophores, respectively. Three types of aqueous sample solutions were analyzed with the system: buffer solutions containing 10^{-2} M K^+, 10^{-2} M Na^+, or both ions. When the aqueous phases containing a single type of ion were used, selective extractions occurred in each case; that is, potassium ions were extracted only for an organic phase segment containing valinomycin, and sodium ions were extracted only for that containing DD16C5. When the aqueous phase containing both ions was examined, both ions were independently extracted into different organic phases, depending on the nature of the ionophores in the respective organic phase. The minimum volume of single organic phase needed to obtain an equilibrium response without dilution by cross dispersion of two organic phases was about 500 nL in the system, indicating that the required amounts of expensive reagents in one measurement could be reduced to a few nanograms.

6.2.3
Integration of a Bioassay System

Goto *et al.* [15] demonstrated the integration of a bioassay system that realizes all processes required for a bioassay, that is, cell culture, chemical stimulation of cells, chemical and enzymatic reactions, and detection, on a chip using CFCP (Figure 6.5). By using a temperature control device, the system could maintain different temperatures for individual areas on the chip. Nitric oxide (NO) released from macrophage-like cells stimulated by lipopolysaccharide was successfully monitored by this system. Assay of NO was carried out using nitrate reductase and Griess reagents. The NO released from the cells to the medium was oxidized by dissolved oxygen, forming NO_2^- and NO_3^-. Then, for the NO assay in the microchannels, NO_3^- was reduced to NO_2^- by nitrate reductase. Next, all the NO_2^- was reacted with Griess reagents and

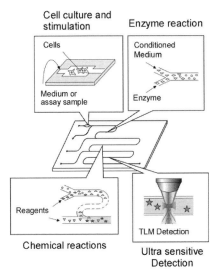

Figure 6.5 Concept of the microchip-based bioassay system.
(Figure 1 in *Anal. Chem.*, 77, 2125, **2005**.)

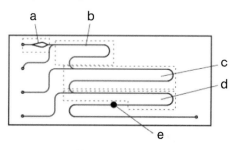

a) Cells $\xrightarrow{\text{LPS}}$ NO $\xrightarrow{O_2}$ NO$_2^-$ + NO$_3^-$

b) NO$_3^-$ $\xrightarrow{\text{Nitrate reductase}}$ NO$_2^-$

c) NO$_2^-$ + H$_2$NO$_2$S—⟨⟩—NH$_2$ $\xrightarrow{\text{H}^+}$ H$_2$NO$_2$S—⟨⟩—N≡N$^+$

d) H$_2$NO$_2$S—⟨⟩—N≡N$^+$ + ⟨NH, NH$_2$⟩ → H$_2$NO$_2$S—⟨⟩—N~N—⟨NH, NH$_2$⟩

e) Detection with a thermal lens microscope

Figure 6.6 Chemical processes carried out in the microchip for bioassay of macrophage-stimulating agent. (Figure 5 in *Anal. Chem.*, *77*, 2125, **2005**.)

the resulting colored product was detected by a thermal lens microscope (Figure 6.6). The total assay time was reduced from 24 h to 4 h, and the detection limit of NO was improved from 1×10^{-6} M to 7×10^{-8} M compared with conventional method. Moreover, the system could monitor a time course of the release, which is difficult to measure by conventional batch methods.

Schilling *et al.* [16] demonstrated the integration of the lysis of bacterial cells and protein extraction and detection on a chip under continuous flow conditions. Their chip had three inlets, two outlets and two straight (lysis and detection) microchannels (Figure 6.7). The width and depth of the microchannels were 1000 μm and 100 μm, respectively. A cell suspension (*E. coli*) and a chemical lytic agent enter through separate inlets into the lysis microchannel. These two fluid streams flow side by side in the microchannel with no mixing except by lateral diffusion. Because the cells are large, they do not diffuse any appreciable distance and therefore remain on the left half of the microchannel. In contrast with the cells, the lytic agent can quickly diffuse laterally from the right half of the lysis microchannel into the left half of the microchannel. The lytic agent permeabilizes the cell membranes and allows intracellular components (β-galactosidase) to exit the cells. These intracellular components are then free to diffuse in all directions and some diffuse into the right half of the lysis microchannel. Due to their relative size and diffusion coefficients, the smaller intracellular molecules travel further into the right half of the microchannel than do larger molecules. At the controlled outlet, the flow rate is such that all of the remaining cell fragments and very large molecules such as DNA flow out of the lysis

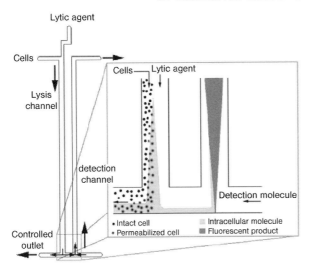

Figure 6.7 Schematic of microfluidic device for cell lysis and fractionation/detection of intracellular components. Pump rates are controlled at all inlets and one outlet. Lytic agent diffuses into the cell suspension, lysing the cells. Intracellular components then diffuse away from the cell stream and some are brought around the corner into the detection channel, where their presence can be detected by the production of a fluorescent species from a fluorogenic substrate. (Figure 1 in *Anal. Chem.*, **74**, 1798, **2002**.)

microchannel. All molecular species that have diffused into the right half of the lysis microchannel are carried around the bend into the detection microchannel, where they occupy the left portion of the microchannel. The large molecules are preferentially located near the centerline of the detection microchannel while smaller molecules have diffused away from the centerline. Entering the right half of the detection microchannel is a stream containing a detection molecule. As the two streams in the detection microchannel interdiffuse, the detection species allows the production of rusorufin (fluorescent molecule) through the reaction of resorufin β-D-galactopyranoside with β-galactosidase. The flow rates and the length of the two straight microchannels can be varied to control the average residence time in the chip, thereby allowing tuning of the system to any particular cell constituent.

6.2.4
Real-time Monitoring of a Chemical Reaction

If real-time monitoring of chemical reactions in a microchannel under continuous-flow conditions is realized, optimization of the reaction conditions (reaction time, concentration, pH, and so forth) and the acquisition of information on the reaction intermediates and the reaction kinetics is possible. Very recently, a real-time monitoring system for a chemical reaction was developed by using a unique microchip and a conventional NMR instrument [17]. Although many studies on microscale NMR, in which the microcoil was integrated on a chip, have been reported, the fabrication of these microscale NMRs is not easy and performance is not satisfactory. The big advantage of

a)

b)

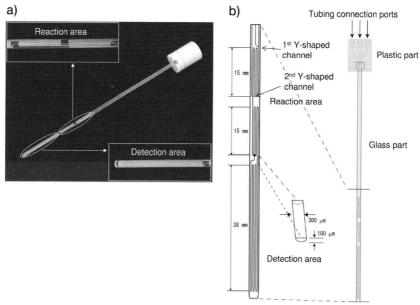

Figure 6.8 (a) Photograph and (b) schematic of the microchip for NMR. (Figure 1 in *Anal. Sci.*, *23*, 395–400, **2007**.)

the new system is being able to use standard 5 mm φ sample tubes without any modifications and a conventional NMR instrument. Figure 6.8 shows a photograph and a schematic illustration of the microchip. The width and depth of the microchannel were 300 and 100 μm, respectively. The length of the microchannel was 60 mm for reaction and 240 mm for detection. For the NMR measurement, the microchip was inserted into a standard sample tube, and was mounted into a sample holder and a rotor set in the NMR instrument. Deuterated solvents were introduced into a gap between the microchip and the standard sample tube, and used as an NMR lock (Figure 6.9). The

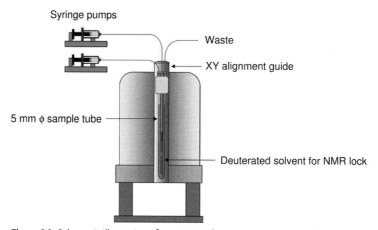

Figure 6.9 Schematic illustration of experimental set-up. (Figure 2 in *Anal. Sci.*, *23*, 395–400, **2007**.)

measurements were carried out without spinning of the microchip and the rotor set. Real-time monitoring of the products in the Wittig reaction and direct detection of intermediate in the Grignard reaction were successfully realized using the system.

6.3
Flow Injection Analysis in Microfluidic Devices

6.3.1
Flow Injection Analysis System: On-Chip PDMS Valve

In general, flow injection analysis (FIA) systems consist of pumps, an injection valve, a mixing column, and a detector [18]. The research and development of microchip-based FIA systems has not progressed much because the integration of all components required for the FIA system is difficult. A few studies have incorporated microchannels and mixing schemes coupled with conventional injection valves and pumps [19, 20]. While these were interesting approaches, several advantages of microchips, such as integration and low sample consumption, are lost. Veenstra *et al.* [21] have realized the integration of miniaturized peristaltic pumps on a chip for the determination of ammonium concentration. The integration of microvalves on a chip has been done by many researchers [22, 23]. Most of these microvalves were required for sophisticated chip designs and complicated fabrication processes. Unger *et al.* [24] have developed pneumatic valves fabricated with multilayer poly (dimethylsiloxane) (PDMS). These PDMS valves are based on two sets of microchannels (fluidic and control) that are bonded together, one on top and perpendicular to the other, with the two separated by a thin polymeric membrane in the microchannels which controls the flow of fluids. Increased pressure in the control channel results in the deformation of the PDMS membrane and closing of the fluidic channel. Simple valves and peristaltic pumps produced by a series of valves on a chip have been demonstrated using this technology, and several applications have been reported. Recently, Leach *et al.* [25] demonstrated the integration of peristaltic pumps, an injection valve, and a mixing column on a chip using the above-mentioned PDMS technology. Figure 6.10 shows schematic illustrations of the multilayer PDMS microfluidic FIA system and injection loop in the system. The low-volume injection loop, which is operated in a manner similar to a standard six-port, two-way valve commonly used in FIA, is a key component in this FIA system. The injection loop consists of two inputs and two outputs connected by multiple fluidic channels. Two sets of valves are used to control the state of the injection system, with one set of valves being closed at all times. In the load position (Figure 6.10(b) i and ii) one set of valves is closed, forcing the sample to flow through the injection volume while the carrier bypasses the loop and enters the mixing column. To inject a sample, the state of the valves is reversed (Figure 6.10(b) iii and iv), directing the carrier through the analyte-filled injection volume and into the mixing column. Because both the sample and carrier are constantly pumped into the system, the injection loop is refilled while the previous injection continues to move through the mixing region. The reproducibility of the repetitive injection of 10.6 nM fluorescein at an injection volume of 1.25 nL

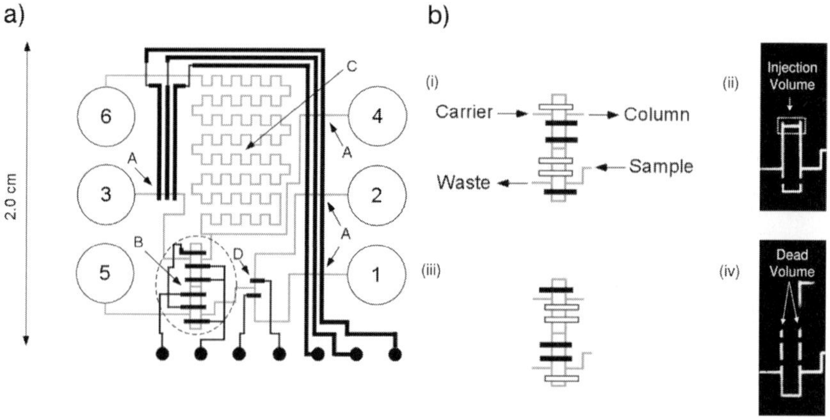

Figure 6.10 (a) Schematic of the multilayer PDMS microfluidic FIA system. Fluidic channels are shown in gray, and control channels (pump and valves) in black. Fluidic components include (A) peristaltic pumps, (B) injection valve, (C) mixing/reaction column, and (D) sample selector. The six fluid reservoirs contain (1, 2) two sample solutions, (3) carrier, (4) reactant, (5) sample waste, and (6) mixing column waste. (b) Operation of the microfluidic injection loop in the load (i) and inject (iii) positions. Fluidic channels are shown in gray, and microvalves are in black. A solid black bar signifies a closed valve; a dotted outline is an open valve. The valve system was imaged in the load (ii) and inject (iv) positions with 1 mM fluorescein dye. The dotted white rectangle in (ii) represents the injection volume, 1.25 nL. (Figures 1 and 2 in *Anal. Chem.*, *75*, 967, **2003**.)

(corresponding to 13.3 amol of fluorescein) is quite good. The peak height and area relative standard deviations were calculated to be 1.9 and 2.2%, respectively. Rapid chemical assay, hydrolysis of fluorescein diphosphate by alkaline phosphatase, was demonstrated using this FIA system.

6.3.2
Flow Injection Analysis System: Chip-based Slide Valve

Organic solvents are often used in FIA. Although the above-mentioned PDMS valves have good performance, the problem of chemical resistance of PDMS toward organic solvents exists. Kuwata and coworkers [26–28] have recently developed a novel glass chip integrated with a slide dispensing unit with zero dead volume. Figure 6.11 shows a photo and a schematic illustration of the microchip. The microchip consists of five parts. Parts A, C and E are fixed, while parts B and D can slide independently in parallel to the neighboring parts by using external actuators. Part B acts as volume-measuring channels, each of which is 10 nL. Since all contact surfaces of the sliding parts are coated with polytetra-fluoroethylene (PTFE), there is no leakage from the gaps between the parts. Therefore, this system can be applied to almost all kind of fluids including organic solvents, acids, alkalis and gases. The principle of liquid dispensing is shown in Figure 6.12. First, the sample solution is introduced into the upper channel

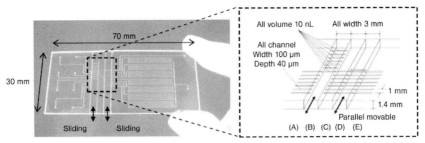

Figure 6.11 Schematic illustration of microfluidic sliding valve device. (Figure 1 in *Proc. Micro Total Analysis Systems 2005*, Transducer Research Foundation, San Diego, Vol.1, 602, 2005.)

(Figure 6.12(a)). Second, part D is slid down to stop all flows (Figure 6.12(b)). Third, part B is slid down step by step to cut off the sample solution at volumes of 10 nL (Figure 6.12(c),(d)). Finally, part D is slid up to inject all the sample solutions which are stored in the microchannel of part B into the analysis channels located

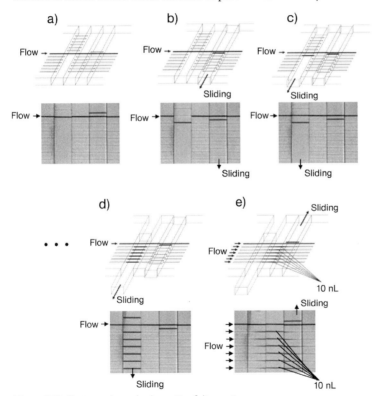

Figure 6.12 Photographs and schematic of dispensing procedures. (a) Sample introduction. (b) Flow stop. (c) Sample introduction into second channel. (d) Sample introduction into final channel. (e) Sample injection into analysis channel. (Figure 2 in *Proc. Micro Total Analysis Systems 2005*, Transducer Research Foundation, San Diego, Vol.1, 602, 2005.)

on the right-hand side of the valve unit (Figure 6.12(e)). The volume of the measuring channels can be easily reduced by changing the channel volume (e.g., width, depth, length). Actually, the injection volume of 1 nL is not so difficult to achieve (W: 100 μm, D: 10 μm, L: 1 mm), as there is no serious effect from pressure drop. The system showed good reproducibility for repetitive injection of 10 nL dye solution. Peak height and area relative standard deviations were calculated to be 0.94 and 0.98%, respectively. Using such systems, the determination of NO_2^- [19] and Cr(IV) [20] was successfully demonstrated.

6.3.3
Flow Injection Analysis System: Immunoassay

Immunoassays are sensitive analytical methods that harness the unique properties of antibodies. They were one of the most productive technological contributions to medicine and fundamental life science research in the twentieth century, and have provided us with a sophisticated biochemical technique which can be applied to investigate and determine minute concentrations of complex molecules such as proteins. Since the principles of immunoassay were first expounded by Yalow and Berson [29] in 1959, there has been an exponential growth in applications to pharmaceutical, forensic, veterinary and food sciences, and medical diagnosis. FIA-based immunoassays were developed in the early 1980s [12] and they have also been reviewed [30]. Recently, FIA-based immunoassays using microfluidic devices have been developed by several research groups. Yakovleva *et al.* [31] have demonstrated microfluidic immunoassay based on the affinity of proteins A and G which are immobilized on silicon microchannel surfaces, as shown in Figure 6.13. The analysis of atrazine, one of the most widely used herbicides, in orange juice samples was carried out using this system with a limit of detection of 0.006 μg L^{-1}. The total assay time was 10 min including sample injection, three substrate injections and regeneration. Kakuta *et al.* [32] have also reported the development of a microchip-based semi-automated heterogeneous immunoassay system. Figure 6.14 shows a schematic illustration of this system which consists of a pump, two injectors, a detector and a microchip. 34 μm spherical beads pre-coated with capture antibodies were packed into a microchannel. Brain natriuretic peptide (BNP), which is known as a heart failure biomarker, was successfully determined at concentrations of 0.1–100 pg mL^{-1} using this system. The assay was carried out within 15 min including sample injection, substrate injection and regeneration. Ohashi *et al.* [33] have developed a fully automated microchip-based ELISA (enzyme-linked immunosorbent assay) system for clinical diagnosis of allergy (IgE). A photo of the prototype system and the assay protocol are shown in Figures 6.15 and 6.16, respectively. The total assay time for 4 samples was 20 min. The sample (patient serum) injection volume was 5 μL. The prototype was subjected to a clinical trial, and showed good correlation ($R = 0.891$) with the conventional method for 82 allergy patients. The system can be applied for various immunoassays by changing the antigen or antibody on the beads in the microchannel.

a)

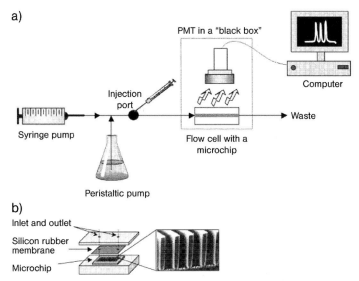

b)

Figure 6.13 (a) Scheme of the microfluidic immunosensor manifold: a syringe pump and a peristaltic pump were used for carrier buffer and regeneration solution at flow rate of 40 and 50 μL min^{-1}, respectively. The sample, containing the enzyme tracer, atrazine standard and antibody, pre-mixed off-line, was injected through a six-port injection valve, and the chemiluminescent signal was detected by a PMT place above the flow cell, containing the microfluidic immunosensor. (b) Detailed view of the plexiglass microchip flow cell and a magnified image of the microchip channel network. (Figure 2 in *Biosens. Bioelectron.*, *19*, 21, **2003**.)

6.4
Perspectives

Flow-based analysis using microfluidic devices has expanded rapidly over the last few years, although the devices are still in their developmental stage. However, much

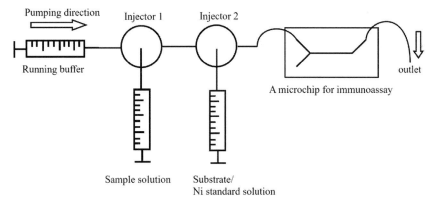

Figure 6.14 Schematic of fluidic system for the sequential injection method. (Figure 2 in *Meas. Sci. Technol.*, *17*, 3189, **2006**.)

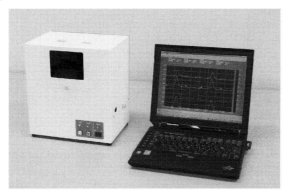

Figure 6.15 Photograph of a prototype fully automated microfluidic ELISA system.

noteworthy progress has been reported recently with respect to bioassays such as cell-based assay and immunoassay, DNA diagnosis, and large-scale integration. As technology and knowledge about integrating analysis functions are accumulated rapidly, it will not be too long before these devices are put into practice in various fields of analysis. In the very near future, the authors believe that microfluidic devices will become high performance portable instruments for on-site analysis and point-of-care testing, and fully automated unmanned analyzers for use in extreme environments (deep sea, space, etc.).

Moreover, flow-based analysis using nanofluidic devices has lately become of major interest as the fabrication techniques advance [34]. Of course, although liquid handling in nanofluidic devices is not so easy, this issue is scientifically interesting and will bring us large benefits.

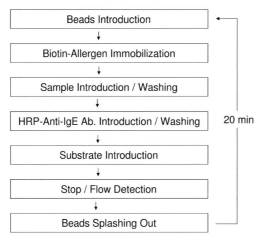

Figure 6.16 Diagram of assay procedures for microfluidic ELISA system.

References

1 Manz, A., Graber, N. and Widmer, H.M. (1990) Miniaturized total chemical analysis systems. A novel concept for chemical sensing. *Sensors and Actuators B-Chemical*, **1**, 244–248.

2 Reyes, D.R., Iossifidis, D., Auroux, P.-A. and Manz, A. (2002) Micro total analysis systems. 1. Introduction, theory, and technology. *Analytical Chemistry*, **74**, 2623–2636.

3 Auroux, P.-A., Iossifidis, D., Reyes, D.R. and Manz, A. (2002) Micro total analysis systems. 2. Analytical standard operations and applications. *Analytical Chemistry*, **74**, 2637–2652.

4 Vilkner, T., Janasek, D. and Manz, A. (2004) Micro total analysis systems. Recent developments. *Analytical Chemistry*, **76**, 3373–3386.

5 Dittrich, P.S., Tachikawa, K. and Manz, A. (2006) Micro total analysis systems. Latest advancements and trends. *Analytical Chemistry*, **78**, 3887–3907.

6 Ruzicka, J. (2000) *Analyst*, **125**, 1053–1060.

7 Tokeshi, M., Minagawa, T., Uchiyama, K., Hibara, A., Sato, K., Hisamoto, H. and Kitamori, T. (2002) Continuous-flow chemical processing on a microchip by combining microunit operations and a multiphase flow network. *Analytical Chemistry*, **74**, 1565–1571.

8 Kitamori, T., Tokeshi, M., Hibara, A. and Sato, K. (2004) Thermal lens microscopy and microchip chemistry. *Analytical Chemistry*, **76**, 52A–60A.

9 Kikutani, Y., Ueno, M., Hisamoto, H., Tokeshi, M. and Kitamori, T. (2005) Continuous-flow chemical processing in three-dimensional microchannel network for on-chip integration of multiple reaction in a combinatorial mode. *QSAR & Combinatorial Science*, **24**, 742–757.

10 Sato, K., Hibara, A., Tokeshi, M., Hisamoto, H. and Kitamori, T. (2003) Integration of chemical and biochemical analysis systems into a glass microchip. *Analytical Sciences*, **19**, 15–22.

11 Tokeshi, M., Kikutani, Y., Hibara, A., Sato, K., Hisamoto, H. and Kitamori, T. (2003) Chemical process on microchips for analysis, synthesis and bioassay. *Electrophoresis*, **23**, 3583–3594.

12 Kikutani, Y., Hisamoto, H., Tokeshi, M. and Kitamori, T. (2004) Micro wet analysis system using multi-phase laminar flows in three-dimensional microchannel network. *Laboratory on a Chip*, **4**, 328–332.

13 Hisamoto, H., Horiuchi, T., Tokeshi, M., Hibara, A. and Kitamori, T. (2001) On-chip integration of neutral ionophore-based ion pair extraction reaction. *Analytical Chemistry*, **73**, 1382–1386.

14 Hisamoto, H., Horiuchi, T., Uchiyama, K., Tokeshi, M., Hibara, A. and Kitamori, T. (2001) On-chip integration of sequential ion-sensing system based on intermittent reagent pumping and formation of two-layer flow. *Analytical Chemistry*, **73**, 5551–5556.

15 Goto, M., Sato, K., Murakami, A., Tokeshi, M. and Kitamori, T. (2005) Development of a microchip-based bioassay system using cultured cells. *Analytical Chemistry*, **77**, 2125–2131.

16 Schilling, E.A., Kamholz, A.E. and Yager, P. (2002) *Analytical Chemistry*, **74**, 1798–1804.

17 Takahashi, Y., Nakakoshi, M., Sakurai, S., Akiyama, Y., Suematsu, H., Utsumi, H. and Kitamori, T. (2007) Development of an NMR interface microchip (MICCS) for direct detection of reaction products and intermediates of micro-syntheses using a MICCS-NMR. *Analytical Sciences*, **23**, 395–400.

18 Trojanowicz, M. (2000) *Flow Injection Analysis*, World Scientific, Singapore.

19 Yokovleva, J., Davidsson, R., Lobanova, A., Bengtsson, M., Eremin, S., Laurell, T. and Emnéus, J. (2002) Microfluidic enzyme immunoassay using silicon microchip with immobilized antibodies and chemiluminescence detection. *Analytical Chemistry*, **74**, 2994–3004.

20 Kerby, M. and Chien, R.-L. (2001) A fluorogenic assay using pressure-driven flow on a microchip. *Electrophoresis*, **22**, 3916–3923.

21 Tiggelaar, R.M., Veenstra, T.T., Sanders, R.G.P., Berenschot, E., Gardeniers, H., Elwenspoek, M., Prak, A., Mateman, R., Wissink, J.M. and van den Berg, A. (2003) Analysis systems for the detection of ammonia based on micromachined components modular hybrid versus monolithic integrated approach. *Sensors and Actuators B-Chemical*, **92**, 25–36.

22 Madou, M.J. (2002) *Fundamentals of Microfabrication: the Science of Miniaturization*, 2nd edn, CRC Press, Boca Raton.

23 Nguyen, N.-T. and Wereley, S.T. (2006) *Fundamentals and Applications of Microfluidics*, 2nd edn, Artech House, Boston, Chapter 6.

24 Unger, M.A., Chou, H.P., Thorsen, T., Scherer, A. and Quake, S.R. (2000) Monolithic microfabricated valves and pumps by multilayer soft lithography. *Science*, **288**, 113–116.

25 Leach, A.M., Wheeler, A.R. and Zare, R.N. (2003) Flow injection analysis in a microfluidic format. *Analytical Chemistry*, **75**, 967–972.

26 Kuwata, M., Kawakami, T., Morishima, K., Murakami, Y., Sudo, H., Yoshida, Y. and Kitamori, T. (2004) Sliding Micro Valve Injection Device for Quantitative Nano Liter Volume, *Proceedings of Micro Total Analysis System 2004*, Vol. 2 Kluwer Academic Publishers, Dordrecht, pp. 342–344.

27 Kuwata, M., Sakamoto, K., Murakami, Y., Morishima, K., Sudo, H., Kitaoka, M. and Kitamori, T. (2005) Sliding Quantitative Nanoliter Dispensing Device for Multiple Analysis, *Proceedings of Micro Total Analysis Systems 2005*, Transducer Research Foundation, San Diego, Vol. 1, pp. 602–604.

28 Kuwata, K. and Kitamori, T. (2006) Sliding Micro Valve Device for Nano Liter Handling in Microchip, *Proceedings of Micro Total Analysis Systems 2006*, CHEMINAS, Tokyo, Vol. 2, pp. 1130–1132.

29 Yalow, R.S. and Berson, S.A. (1959) Assay of plasma insulin in human subjects by immunological methods. *Nature*, **184**, 1648–1649.

30 Fintschenko, Y. and Wilson, G.S. (1998) Flow injection immunoassay: a review. *Mikrochimica Acta*, **129**, 7–18.

31 Yakovleva, J., Davidsson, R., Bengtsson, M., Laurell, T. and Emnéus, J. (2003) Microfluidic enzyme immunosensors with immobilised protein A and G using chemiluminescence detection. *Biosensors & Bioelectronics*, **19**, 21–34.

32 Kakuta, M., Takahashi, H., Kazuno, S., Murayama, K., Ueno, T. and Tokeshi, M. (2006) Development of the microchip-based repeatable immunoassay system for clinical diagnosis. *Measurement Science & Technology*, **17**, 3189–3194.

33 Ohashi, T., Matsuoka, Y., Mawatari, K., Kitaoka, M., Enomoto, T. and Kitamori, T. (2006) Automated Micro-ELISA System for Allergy Checker: a Prototype and Clinical Test, *Proceedings of Micro Total Analysis Systems 2006*, CHEMINAS, Tokyo, Vol. 1, pp. 858–860.

34 Tamaki, E., Hibara, A., Kim, H.-B., Tokeshi, M. and Kitamori, T. (2006) Pressure-driven flow control system for nanofluidic chemical process. *Journal of Chromatography. A*, **1137**, 256–262.

7
The Concept of Multi-commutation in Flow Analysis

Mário A. Feres, Elias A.G. Zagatto, João L.M. Santos, and José L.F.C. Lima

7.1
Introduction

Flow systems are excellent tools for solution handling [1] and hence are especially suitable for wet chemical analysis. A relevant aspect is that system versatility can be enhanced by resorting to specially designed valves, injectors, commuters, pumps or other active devices or, in other words, by exploiting commutation.

Every flow system is characterized by at least one time window, the time interval available for development of given events [2]. As precise timing is concerned, partial, and reproducible development of some physicochemical processes becomes feasible and reaction kinetics can be better exploited. Other typical features of flow analysis are the low consumption of sample and reagents, the feasibility of gradient exploitation, the low susceptibility to sample contamination, and so on [3]. These features, as well as the ever expanding application range, are driving forces towards the overall acceptance of flow analysis [2, 4, 5].

The number of time windows per channel may undergo a pronounced increase by exploiting commutation, thus expanding the potentialities of flow analysis. It should be recalled that the sampling arm of the early segmented flow analyzers selected either the sample or the carrier solution to be aspirated towards the analytical path, and can be considered as a commuting device. The sample and carrier streams were commuted between each other, and this solution interchange can be regarded as the cornerstone of commutation in flow analysis [6].

During the development of segmented-flow analysis the commutation concept underwent little expansion. In fact, the chemical processing was always unaltered from sample to sample, active devices were not generally present in the manifold, external timing[1] for defining commutation times such as, for instance, the STOP period in procedures involving sample stopping, were not generally exploited. No feedback mechanisms were applied.

---

1) External timing refers to selection of a pre-defined available time interval for development of a given physico-chemical process by an external component, usually the computer.

Advances in Flow Analysis. Edited by Marek Trojanowicz
Copyright © 2008 WILEY-VCH Verlag GmbH & Co. KGaA, Weinheim
ISBN: 978-3-527-31830-8

With the advent of unsegmented flow analysis, commutation underwent a remarkable expansion because the analytical path – including the flowing sample – behaves as an uncompressible liquid column, allowing different processes, such as stream redirecting, stream splitting, flow reversal, sample stopping, zone sampling, as well as addition/removal of manifold components, to be easily and efficiently accomplished. Consequently, several approaches relying on merging zones, zone sampling, stopped-flow and others have been proposed [3], thus remarkably expanding the application range of flow analysis. A comprehensive review of commutation in flow injection analysis was published in 1986 [1].

Operation of the simple commuters – usually sliding bars or rotary valves – involved two resting positions, each corresponding to a different manifold state [6]. These states are evident in Figure 7.1 that shows the use of a simple commuter for sample insertion: the load and inject situations correspond to the different states.

Two or more commuting actions can be performed with a single commuter; therefore it is possible to exploit several commuting sections [7] linked to its moving element. This potential is here regarded as *solidary commutation*. Figure 7.2 shows the implementation of the zone sampling approach relying on a single commuter. Although two commuting sections are involved, the number of possible manifold states is still two. However, flow systems comprising a single commuter lack versatility, even if solidary commutation is concerned.

With several discretely operated devices in the manifold, several states can be attained [6], thus enhancing the system versatility and improving the figures of merit of the analytical procedure. Implementation of zone sampling is a good example to illustrate this aspect. With solidary mechanical commutation (Figure 7.2) only one sample aliquot per injection is re-sampled, whereas a number of aliquots per injection can be selected by utilizing discretely operated commuters [8].

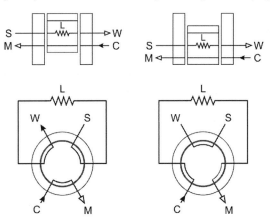

Figure 7.1 Sample introduction with a sliding bar commuter (A) or a six-port rotary valve (B). (a) and (b) refer to load and insert positions. S = sample; L = sampling loop; C = carrier/wash stream; M = towards manifold; W = towards waste; filled arrows = sites where pumping is applied; empty arrows = flow directions. S is aspirated to fill L, which precisely defines the inserted volume. The sample excess is discarded. Switching of the commuter (or valve) introduces the selected sample aliquot into C.

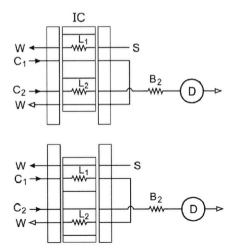

Figure 7.2 Flow diagram of a system exploiting zone sampling and utilizing a sliding bar commuter. S = sample; C_1, C_2 = carrier/wash streams; B_1, B_2 = reactors; L_1, L_2 = sampling loops; D = detector; IC = injector-commuter; upper and lower portions = different states of the flow system; filled arrows = sites where pumping is applied; empty arrows = flow directions. S is inserted into C_1, undergoes dispersion inside B_1 and flows through L_2. After a pre-selected time interval, the commuter is switched, and the selected fraction of the dispersed sample inside L_2 is inserted into C_2 for further handling and monitoring.

During the 1990s, multi-commutation underwent a pronounced expansion, as a consequence of the increased degree of instrument automation, the need for enhanced system versatility and the diversity of analytical applications. The landmark article on multi-commutation [9] was essential in relation to the consolidation of the concept.

Multi-commutation is compatible with the different modalities of flow analysis (segmented-flow, flow injection, sequential injection) and flowing streams (unsegmented, segmented, pulsed, tandem). It is amenable to external timing [10] and to concentration-oriented feedback mechanisms [11]. The processes inherent to the specific analytical procedure can be implemented in-line by modifying the time course related to the computer-operated discrete devices. High versatility is then attained [6]. Here, it is interesting to mention the use of both solenoid pumps and valves to enhance system performance [12].

7.2
Concepts

In a broader context, commutation is defined in the Websters' Dictionary of the English Language [13] in relation to four circumstances:

"(i) *a passing from one state to another.*" This definition holds also in the context of flow analysis, as at least two states are involved. As illustration, addition of an intermittent stream [14] can be selected (Figure 7.3). Without addition of this stream,

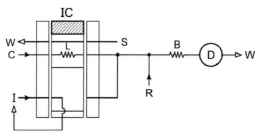

Figure 7.3 Flow diagram of a system exploiting an intermittent stream and utilizing a sliding bar commuter. S = sample; L = sampling loop; C, R, I = carrier/wash, reagent, intermittent streams; IC = injector-commuter; B = reactor; D = detector; filled arrows = sites where pumping is applied; empty arrows = flow directions; dashed area = alternative IC position.

the total flow rate is lower and relatively slow chemical reactions can be better exploited. After attainment of the analytical signal, the injector-commuter is switched in order to direct the intermittent stream towards the main channel, thus improving system washing and enhancing sampling rate.

Intermittent streams are also useful for reagent addition [4]. The reagent stream flows only when the sample zone is passing through the confluence site where it is added. The strategy is analogous to *merging zones* [15], and leads to a pronounced reduction of the reagent consumption. Exploitation of intermittent streams may also rely on the pump operation [16].

Sample stopping can be similarly implemented (Figure 7.4). The approach permits long residence times to be efficiently attained without increasing sample dispersion. The sensitivity of analytical procedures relying on relatively slow chemical reactions is then improved. Sample stopping was initially exploited to attain the successive measurements involved in kinetic analysis [17]. It can be alternatively implemented through external timing: the pump is stopped during a pre-selected time interval and thereafter switched again.

"(ii) *the act of giving one thing for another* and (iii) *the act of substituting.*" These definitions are also valid in the context of flow analysis with commutation, but the

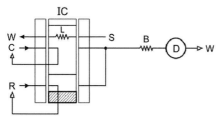

Figure 7.4 Flow diagram of a system exploiting sample stopping and utilizing a sliding bar injector-commuter. S = sample; L = sampling loop; C, R = carrier/wash, reagent streams; IC = injector-commuter; B = reactor; D = detector; full arrows = sites where pumping is applied; empty arrows = flow directions; dashed area = alternative IC position. After a pre-set time interval has elapsed after the S introduction, C is re-directed in order to recycle, thus stopping the sample zone either inside the analytical path or inside the detector.

meanings (ii) and (iii) are equivalent to each other. The situation involving addition/removal of manifold components can be considered for illustrative purposes, and the spectrophotometric determination of nitrate and nitrite relying on a single commuter can be highlighted in this context [18]. The sample aliquot was injected twice, after every commuter switching. One manifold state corresponded to insertion of a Cd/Cu mini-column into the manifold, for quantitative reduction of nitrate to nitrite, whereas the other state corresponded to insertion of a transmission line. The analytical signals reflected then the nitrate plus nitrite concentration, or the nitrite concentration.

"(iv) *in electricity: a change of direction of a current by a commutator.*" The definition holds in the context of flow analysis providing that the flowing stream is taken into account instead of the electric current. Flow reversal is a good example, as it constitutes itself as a useful strategy in relation to multiple peak recording [19], kinetic assays [20], and so forth. With successive fast flow reversals, a situation analogous to sample shaking is approached; this aspect is important for improving liquid–liquid extraction [21].

One can then perceive that the above-mentioned definitions are closely related to flow systems exploiting commutation, holding also in the context of multi-commutation. The related flow system involves n discretely operated devices (<n for solidary commutation) that allow the establishment of up to 2^n manifold states [9]. These states provide the different conditions required for carrying out the steps inherent to the specific analytical application, and can be changed in real-time through the control software (external timing) [22] or, eventually, according to a concentration-oriented feedback mechanism [11].

Multi-commutation is not a modality of flow systems. The authors consider it as an attribute to a given flow system. When this multi-commutation principle is more evident, the flow system can be referred to as a *multi-commuted flow system*.

In general, a high number of active devices in the manifold has been regarded as indicative of a multi-commuted flow system. This is not a *sine qua non* condition [6], as it is possible to design a multi-commuted system with just one valve [23]. On the other hand, there are analytical flow systems with several commuters where the aspect of multi-commutation is not highlighted [24].

7.3
The Discretely Operated Devices

The valves, pumps, timers and other devices often used in multi-commuted flow systems can be operated in an active or passive manner, depending on whether concentration-oriented feedback mechanisms are exploited or not.

7.3.1
Passively Operated Devices

These devices are needed in order to provide proper conditions for sample handling. Their operations are defined prior to sample introduction into the analytical path.

Consequently, conditions for sample handling do not vary from one sample to another. Even so, several time-dependent analytical steps can be efficiently carried out, in view of the system time window [2].

There are situations where external timing is exploited in order to, for instance, attain increased sample incubation time. Sample procedures requiring sample stopping [17] can be selected for illustrative purposes: the pre-defined time interval for development of a given physico–chemical process is set by an external component, usually the computer, which defines the duration of the STOP period.

Another possibility is to provide via a keyboard the information needed for handling samples in a diverse and specific way, as in flow systems with random reagent access [25].

7.3.2
Actively Operated Devices

The potentialities of the flow analyzer are expanded by real-time modification of the conditions for sample handling according to a prior measurement. A typical example is the individual sample conditioning [26]: relevant information (acidity, ion strength, presence of a potential interfering species) is roughly gathered and the result is taken as the basis for real-time sample conditioning through reagent additions, pH adjustment, sample dilution, and so forth. Analogously, the need for multiple STOP periods [27] or cascade dilutions [28] can be real-time confirmed.

Self-optimization of the flow system can also be performed by taking advantage of automatically operated devices [29].

7.4
System Design

Multi-commuted flow systems can be designed in two different modes [30], herein referred to as propelling or aspirating modes, depending on the selected strategy for fluid delivery.

In the propelling mode (Figure 7.5), the solutions are pumped through directing valves towards either the analytical path or recycling [31]. All sampling strategies already used in flow analysis can be implemented. In fact, with S and R simultaneous inlet, a merging zones configuration is attained; with sequential S/R inlet, sandwich techniques [32] can be implemented; with inlet of large C, S, and C plugs, a sampling strategy similar to loop-based injection is attained; with successive S/R inlets, a tandem stream is established; with simultaneous S, R, C inlet, a situation providing high and controlled dilution is attained, and so forth.

In flow systems designed in the propelling mode, the hydrodynamic pressure is higher than atmospheric, and the expression *positive pressure* has often been used in this context. As a rule, a different propelling channel corresponds to each delivered solution. This fluid delivery mode is compatible with both confluent and straight flow systems.

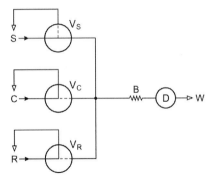

Figure 7.5 A multi-commuted flow system designed in the propelling mode. S, C, R = sample, carrier/wash, reagent streams; V_S, V_C, V_R = three-way solenoid valves; x = confluence site; B = reactor; D = detector; filled arrows = sites where pumping is applied; empty arrows = flow directions; traced line = alternative position of the redirecting valves. S, C, and/or R are pumped through the corresponding valves which direct them either towards the analytical path or towards recycling.

With a peristaltic pump, several solutions can be simultaneously pumped through different pumping tubes. When syringes or solenoid pumps are used, a given device corresponds to each delivered solution. Another possibility is to exploit solidary syringes as fluid propellers, and the innovation is the heart of the *multi-syringe flow analyzer* [33].

In the aspirating mode (Figure 7.6), different commuted devices are used in order to select one of the involved solutions to be aspirated towards the main channel each time. Hydrodynamic pressure is lower than atmospheric, and the expression *negative pressure* has often been used in this context. As a consequence, the flow system is more susceptible to eventual inlet of air through the junctions or to delivery of formed gaseous species as micro-bubbles. The flow system is usually designed in the straight configuration, thus requiring only one pump application site [34]. One pumping

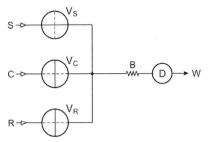

Figure 7.6 A multi-commuted flow system designed in the aspirating mode. S, C, R = sample, carrier/wash, reagent streams; V_S, V_C, V_R = two-way solenoid valves; x = confluence site; B = reactor; D = detector; filled arrow = site where pumping is applied; empty arrow = flow direction; traced line = alternative position of the redirecting valves. S, C and/or R are aspirated through the corresponding valves towards the analytical path.

tube of a peristaltic pump or a single syringe suffices. Consequently the system is characterized by enhanced simplicity and ease of control.

7.5
Tandem Streams

An important characteristic of multi-commutation is the establishment of tandem streams. This unique stream initially comprises several neighboring plugs of different solutions, which undergo fast mixing while being transported through the analytical path [35]. As these plugs are usually very small, almost total overlap between them is efficiently attained.

With tandem streams, the flow system can be designed in the single-line configuration without the drawbacks inherent to it, such as limitations in sample volume, poor sample/reagent interactions, double-peaking, and so on.

For a given volume of the sample zone (sample plus carrier/wash or sample plus reagent/wash solutions), mixing can be improved by lessening the plug initial volumes and increasing the number of plugs accordingly. Insertion of n pairs of plugs results in $2n - 1$ liquid interfaces that mix rapidly, even in short lengths of straight tubing [36].

Tandem stream was conceived in relation to the spectrophotometric determinations of relevant parameters in natural waters [37]. Several plugs of different reagent solutions were selected and sequentially introduced in tandem with the sample plugs, allowing the flow system to rely on the random-access reagent selection. Tandem streams were also exploited for controlling sample dilution prior to ICP-OES or ICP-MS [38]. Several sample plugs were introduced in tandem with plugs of a diluting solution, and the resulting binary string was directed towards the instrument nebuliser. Good mixing conditions were attained in spite of the undulations noted on the recorded tracing, leading to precise results. The approach is referred to as *tandem injection*. Another possibility for implementing a tandem stream is to merge different streams towards a fast switching three-way valve. A binary string is established, and the strategy is referred to as *binary sampling* [9]. It can also be implemented by resorting from a reciprocating pump [39]. Other expressions such as *multi-insertion principle* [40] and *time-division multiplex technique* [41] have also been used in this context.

A tandem stream can also be established inside the sampling loop of a flow injection system [42] or inside the holding coil of a sequential injection system [43]. The interaction of the sample with the reagent starts already during the sampling step; therefore the mean available time for chemical reaction development is increased without affecting sampling rate.

A similar approach involving insertion of sample/reagent plugs at a high frequency through nozzles was proposed for in-line potentiometric titration of calcium in waters [44]. Instant turbulence was attained, leading to improved mixing conditions and low axial dispersion.

Tandem flows are also useful for accomplishing true titrations, as the number and length of carrier/wash, titrant, and titrand plugs are easily modifiable. The end point was found by real-time modification of the number of plugs in accordance with a concentration-oriented feed-back mechanism and an extrapolative mathematical algorithm. This approach was named *binary search* [45].

7.6
Processes Involving Multi-Commutation

Different processes inherent to a given analytical procedure are more efficiently accomplished by taking advantage of multi-commutation, as discussed further.

7.6.1
Sample Introduction

Sample introduction into its carrier stream usually relies on time-based or loop-based injection [1]. A critical comparison between these strategies in a sequential injection system is given elsewhere [46].

Regarding time-based injection, the sample is aspirated or pumped (depending on whether a propelling or aspirating mode flow analyzer is concerned) at a constant flow rate during a pre-selected time interval. Hence, the introduced sample aliquot is proportional to the aspirating flow rate and sampling time. The strategy is very attractive for process monitoring, as the sampling tube remains in the investigated medium, and carryover effects tend to be less relevant. For serial assays however, this sampling strategy may present some limitations, as the sampling tube is moved from one sample to another. Carryover effects, as well as eventual inlet of air, may manifest themselves. The drawback can be circumvented by including an additional valve (V_W, Figure 7.7A) which directs the remainder of the sampled stream towards waste, without passing through the analytical path [9]. Alternatively, a re-directing valve can be placed in the analytical path [10] as shown in Figure 7.7B. The main stream is stopped during sample directing towards waste, and this aspect is beneficial in relation to detection techniques requiring higher measurement time intervals, such as fluorescence measurements.

Loop-based injection relies on insertion of an external loop into the analytical path. Although the feasibility of this strategy was demonstrated in relation to flow systems using stream redirecting valves as the only discretely operated devices [47], it has been seldom exploited. In view of the favorable characteristics of loop-based injection, use of an ordinary injector (rotary valve or sliding bar commuter) for sample insertion and other active devices to enhance system versatility is strongly recommended.

Hydrodynamic injection [48] is a strategy analogous to loop-based injection (Figure 7.8); nevertheless the sample volume is defined in a manifold region without movable parts. Large or small volumes (as low as 0.6 ul [49]) can be inserted with good precision. The approach can be efficiently accomplished by resorting from stream directing valves [46].

Other strategies for accomplishing the sampling step exploiting, for instance, gas diffusion or dialysis, can also be accomplished by multi-commutation. To date, however, this potential seems not to have been exploited.

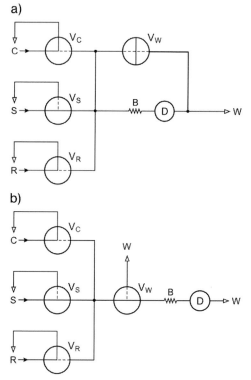

Figure 7.7 Carryover minimization in flow systems exploiting time-based injection by means of detector by-passing (A) or sample flowing directly towards waste recycling (B). Figure refers to the propelling mode. S, C, R = sample, carrier/wash, reagent streams; V_S, V_C, V_R, V_W = three-way solenoid valves; B = reactor; D = detector; filled arrows = sites where pumping is applied; empty arrows = flow directions. S is directed towards B whereas C and R are recycled. With fast valve switching, a tandem stream is established. Alternatively, R and S can be simultaneously inserted. For details, see text.

Multi-commuted flow systems are also able to carry out multiple injections, allowing different approaches such as merging zones [50] and zone sampling [51] to be efficiently accomplished. Successive injections of different aliquots taken from a dispersing sample can also be performed (Figure 7.9). The approach has been exploited to, for example, widen the concentration range of an analytical procedure [52], perform sequential determinations involving random reagent selection [37] or different masking agents [53], to perform speciation [54].

7.6.2
Sample Dispersion

In flow analysis, the sample zone undergoes a continuous dispersion process, and controlled dispersion is a key issue for system design. The inserted volume is the parameter of utmost relevance in this context as higher dispersions are associated

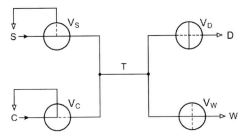

Figure 7.8 Hydrodynamic injection. S, C = sample, carrier/wash streams; D, W = towards detection, waste; V_S, V_C, V_D, V_W = solenoid valves; T = portion of the H-shaped injection module which defines the sample volume; filled arrows = sites where pumping is applied; empty arrows = flow directions. S is aspirated in order to fill T, and its excess is discarded. Meanwhile, the in- and outlets related to the carrier/wash stream remain closed. When the valves are switched, the sample selected volume is pushed forwards by the carrier/wash stream towards detection.

with lower volumes. For very low inserted volumes, the analytical signal, usually associated with the most concentrated portion of the flowing sample, tends to be linearly proportional to the sample inserted volume [3, 4]. Achievement of the analytical curve by using a single standard solution then becomes feasible, and the approach is better implemented by taking advantage of multi-commutation [55]. The high versatility of multi-commuted flow systems permits the utilization of

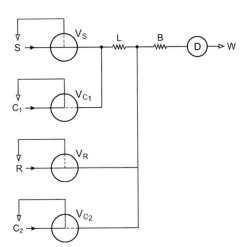

Figure 7.9 Successive insertions of aliquots taken from a dispersing solution. The figure refers to the propelling mode. S, R, C_1, C_2 = sample, reagent, carrier/wash streams; V_S, V_{C1}, V_R, V_{C2} = three-way solenoid valves; L = trapping line; B = reactor; D = detector; filled arrows = sites where pumping is applied; empty arrows = flow directions. For details, see text. Switching V_S and V_{C1} directs S towards L where it is stopped. Different sample aliquots are then selected and delivered towards B by slicing the main stopped zone. To this end, V_{C1} is switched in synchrony with V_R. Thereafter, V_{C2} is switched in order to transport the aliquot of the sample zone towards detection. Multiple injections can then be attained from only one sample aliquot.

several sample inserted volumes in the same manifold. Different analytical curves are then attained and this is a useful means of widening the dynamical concentration range [56]. It should be emphasized that two or more analytical results relying on different analytical curves can be taken into account in order to attain an additional accuracy assessment [57].

Dispersion is also influenced by the length of the analytical path, which can be efficiently modified by resorting to commuting facilities. Different lengths – thus different sample handling times – can be efficiently attained [58], and this is important for instance when expansion of the analytical dynamical range is required.

Re-sampling at the sample zone is also relevant for attaining high degrees of sample dispersion, and implementation of zone sampling and zone slicing in multi-commuted flow systems is straightforward (Figure 7.2).

7.6.3
Solution Additions to the Sample

Most of the analytical procedures require additions of diluents, reagents, and buffers to the handled sample. In straight flow systems, axial sample dispersion is the main driving force for accomplishing these additions, whereas in confluent flow systems, additions of confluent streams constitutes another possibility. Multi-commutation plays an important role in this context, especially with respect to implementation of random access reagent selection for sequential determinations [59].

In analogy with sample introduction, other solutions can also be introduced as a plug. The established zones coalesce inside the analytical path. Exploitation of this strategy has led to several achievements such as a pronounced reduction in reagent consumption, baseline stabilization, and others. Multi-commuted flow systems have been designed to this end, especially with regard to time-based introduction and/or tandem flow exploitation. Figure 7.10 exemplifies a simple multi-commuted flow system for sample/reagent introduction relying on two convergent carrier streams.

Implementation of the standard addition method to overcome matrix interference is also efficiently implemented [60]. The different aliquots to be added to the sample contain the chemical species under determination. This holds also for flow titrations [61] and for analytical procedures requiring spiking [62] or sample conditioning [26]. In this latter situation, several three-way directional valves permit conditioning of every sample in an individual manner.

7.6.4
Sample Incubation

Implementation of analytical procedures involving relatively slow physical-chemical processes may require a relatively long handling time in order to get a higher degree of completion of the involved processes. This aspect becomes more critical when sensitivity is a limiting factor.

To increase the mean sample residence time in the analytical path, the sample zone can be stopped at a specific region of the manifold (Figure 7.7A). In a

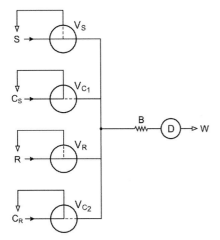

Figure 7.10 A multi-commuted system exploiting merging zones. The figure refers to the propelling mode. S, R, C_1, C_2 = sample, reagent, related carrier/wash streams; V_S, V_{C1}, V_R, V_{C2} = three-way solenoid valves; B = reactor; D = detector; filled arrows = sites where pumping is applied; empty arrows = flow directions. For details, see text. S and R are pushed in order to reach B simultaneously. After a pre-selected time interval, all valves are simultaneously switched allowing both C_1 and C_2 to push the merged zones towards detection and waste. With successive fast switching, a tandem stream with sample plus reagent or C_1 plus C_2 plugs is established (tandem injections).

multi-commuted flow system, different STOP periods can be set and this innovation may expand the application range of a given analytical procedure.

Another possibility to attain long sample handling times without excessive dispersion is to take advantage of *zone trapping* [63]. The sample zone is removed from the original path and trapped inside a parallel reactor during a pre-selected time interval. The approach was conceived in relation to the sliding bar injector-commuter, and is more efficiently performed by resorting to discretely operated redirecting valves [64].

7.6.5
Analyte Separation/Concentration

A noteworthy consequence of the ability of multi-commuted flow system in managing solutions is the easy implementation of in-line analyte concentration/separation steps.

Regarding solid-phase extraction, the sample, conditioning, and eluting solutions can be efficiently directed towards the mini-column, allowing advanced flow systems to be designed. Liquid–liquid extraction can also be efficiently implemented, as the organic and aqueous streams are also efficiently managed. Whenever necessary, stream stopping for proper phase separation is easily attained. Exploitation of tandem streams is worthwhile, as the multiple interfaces between plugs and the multiple interactions with the tubing wall play a relevant role in the extraction process.

The potentialities of electrolytic dissolution, gas diffusion, dialysis, precipitation, and other processes are also expanded in relation to multi-commutation. A deeper discussion of this theme is outside the scope of this chapter.

7.7
Applications

7.7.1
Selection Criteria

The analytical procedures listed in this section were selected by taking into account the following aspects. The flow system should clearly exploit multi-commutation, although it is not imperative that the author(s) emphasize(s) this term in the article. The article, preferably written in English, should be easily accessible, and indexed in important data bases such as the Institute for Science Information. The method should present superior figures of merit; it should be properly validated, although special situations where spiking was needed are acceptable. No more than two analytical procedures are selected per analyte per sample matrix. All concentration units are presented as provided by the author(s).

Among the different application areas considered in this text, those more related to environmental, agronomical, metallurgical, and pharmaceutical applications were selected. Pharmaceutical analysis is dealt with in another chapter, therefore the related applications are not highlighted. The applications do not reflect the whole available methodology, as their number is very high and tends to increase. The authors apologize for the omission of any relevant contribution.

7.7.2
Selected Applications

Enhanced multi-commuted flow systems have been designed for determinations of metals in ores, often involving electrolytic dissolution and detection by atomic spectrometric techniques or inductively-coupled plasma mass spectrometry. To this end, multi-commutation has allowed the implementation of, for example, analytical curves based on a single multi-element standard solution [65], external calibration and sequential addition of electrolytic solution at different manifold sites [66], in-line programmable isotope dilution [67] and expansion of the concentration dynamical range through valve triggering [68]. Regarding UV–Vis spectrophotometry, applications exploiting prior assay for improving flow titrations [69], two flow systems sharing the injection unit [70] and use of a single reagent and different masking agents [53] were also proposed for alloys and ore analysis.

Analytical applications related to analyses of foods and similar have often been proposed, mainly with regard to food quality control. Table 7.1 highlights the most relevant applications, including some specific examples such as milk, yogurt, wine, beer, sweeteners, and so forth.

Applications related to water analysis are summarized in Table 7.2. As all samples are obviously water, the kind of assayed sample is specified in the second column. Recently, analyses of gasoline and detergents using Fourier transform infrared spectrophotometry [71, 72], and of lubricating oils using UV-Vis spectrophotometry [73] have also been proposed: no prior sample treatment was required.

Plant (including silage) and soil analysis are presented together (Table 7.3) in view of their relevance in agro-industrial laboratories. Multi-commuted flow systems proposed for biological fluids are presented in Table 7.4.

7.8
Conclusions

The number of applications of multi-commuted flow-based analytical procedures is increasing and the rate of increase has become more pronounced in recent years [74]. A literature survey of recently published innovations reveals an increased number of articles reporting novel flow systems with a high degree of automation.

This aspect is probably due to the high versatility inherent in the multi-commuted flow systems as well as the easy implementation of concentration-oriented feed-back mechanisms [26]. This latter aspect has been more evident in relation to expert flow systems [11], especially those designed for titrations [45].

As the flowing streams are often added or removed from the manifold, transient variations in baseline may occur. Therefore special attention should be paid during the monitoring step, as undesirable concentration gradients may impair detection, especially in relation to potentiometric or spectrophotometric (Schlieren effect) analytical procedures.

The multi-commuted concept has been recognized as relevant in the field of flow analysis, as can be verified by checking the referenced quotations in international data bases. This recognition will certainly increase by incorporating the concept in commercially available flow analyzers.

7.9
Trends

There are presently several modalities of flow analysis, for example, segmented flow, flow injection, sequential injection, multi-syringe flow injection, batch injection, mono-segmented flow, flow-batch, multi-pumping flow, bead injection, among others. All of them have been assigned acronyms, and this aspect is strongly discouraged by IUPAC [75]. The existence of so many modalities was recently the theme for a jocose foreword [76] in a prestigious journal devoted to flow analysis.

In view of the high number of modalities and acronyms, and considering the existence of several features common to the different modalities, one expects that most of the modality names will vanish.

Table 7.1 Multi-commuted flow systems for food analysis.

Analyte	Sample	Detection technique	Detection limit or range	Sampling rate/h^{-1}	Remarks	Ref.
acetic acid	vinegar, soft drinks	pot.	0.4–$9.0\,mol\,l^{-1}$	(a)	monosegmented flow titration; algorithm for end point search real-time implemented	[77]
Ag, Au, Te, U	several	ICP-MS	$0.82,\ 0.64,\ 2.24,\ 0.05\,pg\,ml^{-1}$	21 (18 for Au)	in-line analyte concentration; applicable also to water analysis	[78]
Al	fruits	UV–Vis	$0.1\,mg\,l^{-1}$	variable	dynamical range expanded by split zone	[79]
ascorbic acid	fruit juices, soft drinks	UV–Vis	0.6–$6.0\,mmol\,l^{-1}$	5–30	no prior sample treatment	[80]
azoxystrobin	grape, must, wine	fluor.	$0.021\,mg\,kg^{-1}$,18, $8\,ug\,l^{-1}$	28	solid-phase detection; sample stopping at the detector	[81]
Ba, Cu, Pb, Zn	honey	ICP-MS	0.2–$1.24,\ 0.49$–$1.23,\ 0.61$–$2.28,$ 0.5–1.51%	30	isotope dilution	[82]
Bi	milk shake	HG-AFS	$1.67\,ng\,g^{-1}$	72	in-line neutralization of waste effluent	[83]
bitertanol	banana	fluor.	$0.014\,mg\,kg^{-1}$	(b)	screening analysis	[84]
Cd, Ni, Pb	various	ICP-OES	$1,\ 2\,ng\,ml^{-1}$	90	in-line ion-exchange for analyte concentration	[34]
chloride	milk, wine	pot.	10–$1200\,mg\,l^{-1}$	(b)	argentimetric monosegmented flow titration	[61]
cyclamate	sweetener	UV–Vis	$30\,umol\,l^{-1}$	60	solenoid pumps as fluid propellers	[85]
diphenylamine	apple, pear	fluor.	$0.06\,mg\,kg^{-1}$	(b)	analyte concentration on a CV18 silica gel minicolumn	[86]
ethanol	wine	CL	2.5–$25\%\,v/v$	23	laboratory-made luminometer	[87]

ethanol	red wine	UV–vis	$0.05\,mol\,l^{-1}$	50	reagent-less procedure; exploitation surface tension	[88]
glycerol	wine	UV–Vis	$0.006\,g\,l^{-1}$	33	enzyme immobilized onto amino-propyl glass beads	[89]
Hg	milk	CV-AFS	$0.011\,ng\,g^{-1}$	70	ultrasound-assisted sample preparation	[90]
Hg	fish	CV-AFS	$7\,mg\,kg^{-1}$	(b)	multi-commutation relying on six-way rotary valves	[91]
lactic acid	yoghurt	CL	$10–125\,mg\,l^{-1}$	55	enzyme immobilized onto porous silica beads	[92]
lactate	sugar-cane juice	UV–Vis	$5.0–100.0\,mg\,l^{-1}$	36	exploitation of tandem stream	[93]
Pb	several	FAAS	$3.7\,ug\,l^{-1}$	(b)	in-line analyte concentration	[94]
tartaric acid	wine	UV–Vis	$0.50–10.0\,g\,l^{-1}$	28	applicable also to red wine	[95]
Te	milk	HG-AFS	$0.20\,ng\,l^{-1}$	85	Environmentally friendly procedure; exploitation of tandem flow	[96]
Te	milk	HG-AFS	$0.57\,ng\,g^{-1}$	24	free Te(IV)/total Te speciation	[97]
thiabendazole	citrus fruits	fluor.	$0.09\,mg\,kg^{-1}$	(b)	sample stopping at the spectrofluorimeter	[98]
total acidity	fruit juices, soft drinks	pot.	$1–100\,mmol\,l^{-1}$	22	monosegmented wide-range flow titration	[99]

pot. = potentiometry; ICP-MS = inductively-coupled plasma mass spectrometry; UV–Vis = UV–Vis spectrophotometry; fluor. = fluorimetry or fluorescence; HG-AFS = hydride generation atomic fluorescence spectrometry; CL = chemiluminescence; CV-AFS = cold vapor atomic fluorescence spectrometry; FAAS = flame atomic absorption spectrometry. (a) Not pertinent; (b) Not informed.

Table 7.2 Multi-commuted flow systems for water analysis.

Analyte	Water	Detection technique	Detection limit or range	Sampling rate/h⁻¹	Remarks	Ref.
Ag, Au, Te, U	sea, river	ICP-MS	0.82, 0.64, 2.24, 0.05 pg ml⁻¹	21 (18 for Au)	in-line analyte concentration; applicable also to food analysis	[78]
Al	drinking	fluor.	0.5 ug l⁻¹	up to 154	micellar enhanced luminescence	[100]
Al	plant nutrition solution, natural	fluor.	0.04 mg l⁻¹	60	exploitation of monosegmented flow	[101]
aldicarb	mineral	CL	0.069 ug l⁻¹	17	sample stopping at the detector	[102]
anionic surfactants	filtered wastewater	UV–Vis	0.034 mg l⁻¹	60	ion-pair formation with cetyl pyridine ion	[103]
anionic surfactants	lake	UV–Vis	10 ng ml⁻¹	2	in-line analyte concentration (factor of 65)	[104]
azulan	irrigation, tap	CL	40 ug l⁻¹	30	in-line photodegradation; sample stopping at the detector	[105]
cationic surfactants	natural	UV–Vis	0.035 mg l⁻¹	72	no prior sample treatment	[106]
cationic surfactants	natural	UV–Vis	8 exp(−8) mol l⁻¹	(b)	tandem flow for solvent extraction; use of the film on the tubing inner wall	[107]
chloride	river	UV–Vis	0.50–10.0 mg l⁻¹	25	accuracy assessment relying on two methods; optional in-line spiking	[62]
chlorine	drinking water, bleach tablets	UV–Vis	0.05 ug ml⁻¹	38	in-line gas diffusion	[108]
chlorsulfuron	mineral	CL	0.06 mg l⁻¹	25	in-line photodegradation	[109]

Analyte	Matrix	Technique	LOD		Notes	Ref.
Cr, Co	river, coastal, harbour, wastewater	CL	$0.2\ \mu g\,l^{-1}$	(b)	loop-based injection; sample stopping	[110]
Cu	sea	ET-AAS	$5\ ng\,l^{-1}$	(b)	valves operation in synchrony with the auto sampler	[111]
Cu, Cd, Pb, Bi, Se(IV)	sea	ICP-MS	$5, 0.2, 0.3, 0.06, 5\ ng\,l^{-1}$	22	in-line analyte concentration	[112]
Fe	natural	UV–Vis	$0.1–35\ mg\,l^{-1}$	(b)	speciation; analyte concentration onto chelating disks; expert system	[54]
Fe	tap, seawater	UV–Vis	$8.4\ ng\,ml^{-1}$	22	two different concentration ranges, in-line speciation or concentration, no manifold reconfiguration	[113]
Fe(II), Fe(III)	river	UV–Vis	$0.4\ mg\,l^{-1}$ (total Fe)	60	Ni determination in steels with the same manifold	[42]
free chlorine	wastewater	UV–Vis	$0.05\ \mu g\,ml^{-1}$	38	in-line gas diffusion; applicable also to industrial formulations	[108]
fuberidazoli, o-phenylphenol	river, well	fluor.	$0.18–6.1\ ng\,ml^{-1}$	12	optotrode (resin mini-column inside the flow cell)	[114]
heavy metals	river	ICP-MS	(a)	21	in-line analyte concentration; applicable also to analysis of biological tissues	[115]
Hg	sea	ICP-MS	$5\ ng\,l^{-1}$	21	in-line analyte concentration	[116]
Hg	natural	CV-AFS	$1.3\ ng\,l^{-1}$	63	argon inlet modified through commutation.	[117]
hydroquinone	surface waters	CL	$0.1–15.0\ mg\,l^{-1}$	103	pioneer CL application in relation to tandem flow	[118]
methyl parathion	natural	UV–Vis	50 ppt	(b)	flow ELISA with superior characteristics relative to plate ELISA	[119]

(Continued)

Table 7.2 (*Continued*)

Analyte	Water	Detection technique	Detection limit or range	Sampling rate/h^{-1}	Remarks	Ref.
Na, K, Ca, Mg	parenteral and hemodialysis solutions	FAAS/FOES	500–3500, 50–150, 30–120, 20–40 mg l^{-1}	70, 75, 70, 58	high and variable in-line sample dilution	[120]
NH$_4$, P-PO$_4$	natural	UV–Vis	1.0, 1.0 ug l^{-1}	40	simultaneous in-line concentration	[121]
nitrate, nitrite	lake, fountain	UV–Vis	1.9 (exp-7) mol l^{-1}	15	tandem stream for sampling and in-line photoreduction	[40]
nitrite	natural	UV–Vis	25 or 8 ug l^{-1}	108 or 44	comparison of two analytical methods	[122]
NO$_2$, NO$_3$, Cl, P-PO$_4$	natural	UV–Vis	6, 40, 400, 30 ug l^{-1}	50	optional in-line analyte concentration or nitrate reduction; sequential determinations	[123]
NO$_3$, NO$_2$, NH$_4$	river	UV–vis	5, 15, 25 ug l^{-1}	60	gravity as fluid propeller	[124]
phenols	natural	CL	5 ng ml^{-1}	12 or 60	optional addition of a resin minicolumn	[125]
phenols	natural	UV–Vis	1.0 ug l^{-1}	90	long optical path for improving sensitivity	[126]
phenols	wastewater, spring, tap, rain	UV–Vis	13 ng ml^{-1}	65	greener procedure	[127]
P-PO$_4$	mineral, ground, tap	CL	4 ug l^{-1}	11	flow-through solid-phase sensor	[128]
propanil and related herbicides	natural	CL	8 ug l^{-1}	20	tandem stream exploitation	[129]
Se	natural, drinking	HG-AFS	50 ng l^{-1}	40	use of a tungsten filament	[130]
sulfate	natural	turb.	0.1–2.0 mg l^{-1}	50	ion-line analyte concentration	[131]

sulfide	environmental, wastewater	UV–Vis	$0.15, 0.09\,\mathrm{mg\,l^{-1}}$	80	two concentration dynamical ranges	[132]
Ssulfide	sea, ground, wastewater	UV–Vis	$1.3\,\mathrm{ug\,l^{-1}}$	(b)	in-line gas diffusion	[133]
Warfarin	drinking	fluor.	$50–64\,000\,\mathrm{ng\,l^{-1}}$	12	in-line analyte concentration	[134]

ICP-MS = inductively-coupled plasma mass spectrometry; fluor. = fluorimetry or fluorescence; CL = chemiluminescence; UV–Vis = UV–Vis spectrophotometry; ET-AAS = electrothermal vaporization atomic absorption spectrometry; CV-AFS = cold vapor atomic fluorescence spectrometry; FAAS = flame atomic absorption spectrometry; FOES = flame optical emission spectrometry; HG-AFS = hydride generation atomic fluorescence spectrometry; turb. = turbidimetry. (a) Not pertinent; (b) Not informed.

Table 7.3 Multi-commuted flow systems for plant and soil analysis. Concentration data refer to plant digests or soil extracts.

Analyte	Sample	Detection technique	Detection limit or range	Sampling rate/h^{-1}	Remarks	Ref.
Al, Fe	plant	UV–Vis	1.0–15.0, 2.0–12.0 mg l^{-1}	60	in-line adjustment of a critical parameter (pH)	[135]
Bo	plant	UV–Vis	0.25–6.00 mg l^{-1}	35	simultaneous zone trapping inside three different analytical paths	[58]
carbohydrates, reducing sugars	forage materials	UV–Vis	0.2–0.8% w/v	32	heating of a tandem stream	[136]
Cd	plant	UV–Vis	0.23 mg l^{-1}	20	minimization of interferences by in-line electrolytic deposition.	[137]
Cd, Pb, Ni	plant	ICP-OES	1, 4, 2 ng ml^{-1}	90	simultaneous operation of three ion-exchange mini-column	[34]
Cu	plant	UV–Vis	50–400 ug l^{-1}	30	in-line solvent extraction	[138]
Cu	plant	FAAS	1 ng ml^{-1}	48	multi-purpose flow system	[139]
Cu, Zn	plant	UV–Vis	0.05, 0.04 mg l^{-1}	45	computer-assisted splitting of the flowing sample	[140]
exchangeable K	soil	pot.	6–390 mg l^{-1}	50	in-line extraction	[141]
Fe, B	soil	UV–Vis	0.50–10.0, 0.20–4.0 mg l^{-1}	34, 15	hydrodynamic injection	[142]
Hg	agro-industrial products	CV-AAS	0.8 ng l^{-1}	25	solidary multicommutation implemented by a sliding bar commuter	[143]
lactate	silage materials	UV–Vis	10.0–100.0 mg l^{-1}	16	cucurbita pepo as a natural source of the enzyme	[144]
Mn	plant	UV–Vis	1.2 mg l^{-1}	50	exploitation of monosegmented flow	[145]
NH$_4$, P-PO$_4$	plant	UV–Vis	25.0–125.0, 2.5–12.5 mg l^{-1}	80	random access reagent selection	[146]
nitrate, nitrite	soil, fertilizer,	UV–Vis	1.9 (exp-7) mol l^{-1}	15	tandem stream for sampling and in-line photoreduction	[40]

Pb	plant	UV-Vis	12 ug l^{-1}	15	in-line solvent extraction; gaseous washing stream	[147]
P-PO$_4$	plant	UV-Vis	24 ug l^{-1}	38	monosegmented flow, different extracting fractions	[148]
P-PO$_4$	soil, sediments	UV-Vis	0.02 mg l^{-1}	(b)	serial extraction	[149]
SO$_4$	plant	turb.	10–150 mg l^{-1} SO$_4$	100	possibility of expanding dynamical range	[52]
Total acidity	silage materials	UV-Vis	0.001–0.1 mol l^{-1}	typically 16	flow titration of colored solutions	[150]

UV–Vis = UV–Vis spectrophotometry; ICP-OES = inductively-coupled plasma optical emission spectrometry; FAAS = flame atomic absorption spectrometry; pot. = potentiometry, CV-AAS = cold vapor atomic absorption spectrometry; turb. = turbidimetry. (a) Not pertinent; (b) Not informed.

Table 7.4 Multi-commuted flow systems for biological fluids.

Analyte	Sample	Detection Technique	Detection limit or range	Sampling rate/h^{-1}	Remarks	Ref.
3-hydroxybutyrate	animal serum, plasma	UV–Vis	2 mg l^{-1}	600	no prior sample treatment	[151]
3-hydroxybutyrate, glucose, cholesterol	animal whole blood	UV–Vis	1.5, 14, 4 mg l^{-1}	55, 40, 40	polyvalent flow system; reagents from commercial kits	[152]
acetaminophen	blood serum	volt.	1.7 exp(−5) mol l^{-1}	24	use of a modified tubular electrode	[153]
albumin, total protein	animal plasma	UV–Vis	Up to 15 g l^{-1}	45	portable systems with a LED photometer	[154]
cholesterol	animal serum	CL	3.7 mg l^{-1}	40	two enzyme mini-columns	[155]
creatinine	urine	UV–Vis	0.50–2.00 g l^{-1}	24	exploitation of zone splitting	[156]
Cu	urine	UV–Vis	3 ug l^{-1}	14	sol-gel optotrode sensor	[157]
Cu	urine, serum	FAAS	0.035 mg l^{-1}	24	two flow systems sharing the spectrometer	[158]
furosemide, triamterene	urine, serum	fluor.	15, 0.1 ng ml^{-1}	(c)	in-line analyte separation in a Sephadex mini-column	[159]
glucose	blood serum	UV–Vis	(b)	22	multicommutation achieved by special rotary valve; enzyme commercial kit	[20]
heavy metals	Urine, liver, muscles	ICP-MS	(a)	21	in-line analyte concentration; applicable also to water analysis	[115]

flufenamic acid	human serum, urine	fluor.	1.12 exp(−9) mol l^{-1}	38	applicable also to pharmaceutical formulations	[160]
karbutilate	human urine	CL	10 ug l^{-1}	17	in-line photo-degradation	[161]
naproxen, salicylic acid	urine, serum	fluor.	0.3, 1.3 ng ml^{-1}	8	sample stopping at the detection unit for native fluorescence measurements	[10]
turic acid	urine	amp.	9.90 exp(−4) mol l^{-1}	(b)	large (2500-fold) sample dilution	[162]

UV–Vis = UV–Vis spectrophotometry; volt. = voltammetry; CL = chemiluminescence; FAAS = flame atomic absorption spectrometry; Fluor. = fluorimetry or fluorescence; ICP-MS = inductively-coupled plasma mass spectrometry; amp. = amperometry. (a) Not pertinent; (b) Not informed.

Multi-commutation does not refer to a modality, but to an attribute to the already existing modalities. In fact, a literature survey reveals expressions such as multi-commuted flow injection systems, multi-commuted segmented flow systems, and so on.

In the near future, therefore, one expects a pronounced reduction in the number of modalities. Just a few will be maintained, perhaps only two: segmented and unsegmented flow analysis. For a proper characterization of a given flow system, the attribute *multi-commutation* will be included in order to specify the system versatility, degree of automation, system analytical potential and manifold complexity.

References

1 Krug, F.J., Bergamin Filho, H. and Zagatto, E.A.G. (1986) Commutation in flow injection analysis. *Analytica Chimica Acta*, **179**, 103–118.

2 Zagatto, E.A.G. and Worsfold, P.J. (2005) Flow Analysis: Overview, in *Encyclopedia of Analytical Science* 2nd edn, 3, (eds P.J. Worsfold, A. Townshend and C.F. Poole), Elsevier, Oxford, pp. 24–31.

3 Trojanowicz, M. (2000) *Flow Injection Analysis: Instrumentation and Applications*, World Scientific, London.

4 Ruzicka, J. and Hansen, E.H. (1988) *Flow Injection Analysis*, 2nd edn, Wiley Interscience, New York.

5 Smith, J.P. and Hinson-Smith, V. (2002) Flow injection analysis: Quietly pushing ahead. *Analytical Chemistry*, **74**, 385A–388.

6 Rocha, F.R.P., Reis, B.F., Zagatto, E.A.G., Lima, J.L.F.C., Lapa, R.A.S. and Santos, J.L.M. (2002) Multicommutation in flow analysis: concepts, applications and trends. *Analytica Chimica Acta*, **468**, 119–131.

7 Reis, B.F., Zagatto, E.A.G., Jacintho, A.O., Krug, F.J. and Bergamin Filho, H. (1980) Merging zones in flow injection analysis. Part 4. Simultaneous spectrophotometric determination of total nitrogen and phosphorus in plant material. *Analytica Chimica Acta*, **119**, 305–311.

8 Rocha, F.R.P., Martelli, P.B., Frizzarin, R.M. and Reis, B.F. (1998) Automatic multicommutation flow system for wide range spectrophotometric calcium determination. *Analytica Chimica Acta*, **366**, 45–53.

9 Reis, B.F., Gine, M.F., Zagatto, E.A.G., Lima, J.L.F.C. and Lapa, R.A.S. (1994) Multicommutation in flow analysis. Part 1. Binary sampling: concepts, instrumentation and spectrophotometric determination of iron in plant digests. *Analytica Chimica Acta*, **293**, 129–138.

10 Garcia-Reyes, J.F., Ortega-Barrales, P. and Molina-Diaz, A. (2007) Multi-commuted fluorometric multiparameter sensor for simultaneous determination of naproxen and salicylic acid in biological fluids. *Analytical Sciences*, **23**, 423–428.

11 Grassi, V., Dias, A.C.B. and Zagatto, E.A.G. (2004) Flow systems exploiting in line prior assays. *Talanta*, **64**, 1114–1118.

12 Rocha, F.R.P., Infante, C.M.C. and Melchert, W.R. (2006) A multi-purpose flow system based on multi-commutation. *Spectroscopy Letters*, **39**, 651–668.

13 McKechnie, J.L. (ed.) (1983) *Websters' New Twentieth Century Dictionary of the English Language*, 2nd edn, Prentice Hall, New York.

14 Zagatto, E.A.G., Jacintho, A.O., Mortatti, J. and Bergamin Filho, H. (1980) An improved flow injection determination of nitrite in waters by using intermittent flows. *Analytica Chimica Acta*, **120**, 399–403.

15 Bergamin Filho, H., Zagatto, E.A.G., Krug, F.J. and Reis, B.F. (1978) Merging zones in flow injection analysis. Part 1. Double proportional injector and reagent consumption. *Analytica Chimica Acta*, **101**, 17–23.

16 Ruzicka, J. and Hansen, E.H. (1980) Flow-injection analysis: principles, applications and trends. *Analytica Chimica Acta*, **114**, 19–44.

17 Ruzicka, J. and Hansen, E.H. (1979) Stopped flow and merging zones – a new approach to enzymatic assay by flow injection analysis. *Analytica Chimica Acta*, **106**, 207–224.

18 Gine, M.F., Bergamin Filho, H., Zagatto, E.A.G. and Reis, B.F. (1980) Simultaneous determination of nitrate and nitrite by flow injection analysis. *Analytica Chimica Acta*, **114**, 191–197.

19 Valcarcel, M., Luque de Castro, M.D., Lazaro, F. and Rios, A. (1989) Multiple peak recordings in flow injection analysis. *Analytica Chimica Acta*, **216**, 275–288.

20 Toei, J. (1988) Determination of glucose in clinical samples by flow reversal flow injection analysis. *Analyst*, **113**, 475–478.

21 Schindler, D.R., Rios, A., Valcarcel, M. and Grasserbauer, M. (1997) Simple and rapid screening of total aromatic hydrocarbons in polluted water samples by the flow reversal liquid-liquid extraction technique. *International Journal of Environmental Analytical Chemistry*, **66**, 285–297.

22 Lapa, R.A.S., Lima, J.L.F.C., Reis, B.F. and Santos, J.L.M. (1998) Continuous sample recirculation in an opened-loop multicommutated flow system. *Analytica Chimica Acta*, **377**, 103–110.

23 Almeida, C.M.N.V., Lapa, R.A.S., Lima, J.L.F.C., Zagatto, E.A.G. and Araujo, M.C.U. (2000) An automatic titrator based on a multicommutated unsegmented flow system. Its application to acid-base titration. *Analytica Chimica Acta*, **407**, 213–223.

24 Stewart, K.K., Brown, J.F. and Golden, B.M. (1980) A microprocessor control system for automated multiple flow injection analysis. *Analytica Chimica Acta*, **114**, 119–127.

25 Bennaoui, N., Periou, C., Harault, C. and Lemoel, G. (1993) Evaluation of the Technicon DAX 48, a multiparametric biochemical analyzer. *Annales de Biologie Clinique*, **51**, 713–720.

26 Carneiro, J.M.T., Dias, A.C.B., Zagatto, E.A.G. and Honorato, R.S. (2002) Spectrophotometric catalytic determination of Fe(III) in estuarine waters using a flow-batch system. *Analytica Chimica Acta*, **455**, 327–333.

27 Lapa, R.A.S., Lima, J.L.F.C. and Santos, J.L.M. (2000) Dual-stopped-flow spectrophotometric determination of amiloride hydrochloride in a multicommutated flow system. *Analytica Chimica Acta*, **407**, 225–231.

28 Whitman, D.A. and Christian, G.D. (1989) Cascade system for rapid on-line dilutions in flow injection analysis. *Talanta*, **36**, 205–211.

29 Rius, A., Callao, M.P. and Rius, F.X. (1995) Self-configuration of sequential injection analytical systems. *Analytica Chimica Acta*, **316**, 27–37.

30 Lavorante, A.F., Feres, M.A. and Reis, B.F. (2006) Multi-commutation in flow analysis: A versatile tool for the development of the automatic analytical procedure focused on the reduction of reagent consumption. *Spectroscopy Letters*, **39**, 631–650.

31 Martelli, P.B., Reis, B.F., Kronka, E.A.M., Bergamin Filho, H., Korn, M., Zagatto, E.A.G., Lima, J.L.F.C. and Araujo, A.N. (1995) Multi-commutation in flow

analysis. Part 2. Binary sampling for spectrophotometric determination of nickel, iron and chromium in steel alloys. *Analytica Chimica Acta*, **308**, 397–405.

32 Alonso, J. and Bartroli, J. (1987) M. del Valle, M. Escalada, R. Barber, Sandwich techniques in flow injection analysis. Part 1. Continuous recalibration techniques for process control. *Analytica Chimica Acta*, **199**, 191–196.

33 Cerda, V. and Pons, C. (2006) Multicommutated flow techniques for developing analytical methods. *Trends in Analytical Chemistry*, **25**, 236–242.

34 Miranda, C.E.S., Reis, B.F., Baccan, N., Packer, A.P. and Gine, M.F. (2002) Automated flow analysis system based on multicommutation for Cd, Ni and Pb on-line pre-concentration in a cationic exchange resin with determination by inductively coupled plasma atomic emission spectrometry. *Analytica Chimica Acta*, **453**, 301–310.

35 Vicente, S., Borges, E.P., Reis, B.F. and Zagatto, E.A.G. (2001) Exploitation of tandem streams for carry-over compensation in flow analysis. I. Turbidimetric determination of potassium in fertilizers. *Analytica Chimica Acta*, **438**, 3–9.

36 Dias, A.C.B., Santos, J.L.M., Lima, J.L.F.C., Quintella, C.M., Lima, A.M.V. and Zagatto, E.A.G. (2007) A critical comparison of analytical flow systems exploiting streamlined and pulsed flows. *Analytical and Bioanalytical Chemistry*, **388**, 1303–1310.

37 Malcome-Lawes, D.J. and Pasquini, C. (1988) A novel approach to non-segmented flow analysis. Part 2. A prototype high-performance analyser. *Journal of Automatic Chemistry*, **10**, 192–197.

38 Israel, Y., Lasztity, A. and Barnes, R.M. (1989) On-line dilution, steady-state concentrations for inductively coupled plasma atomic emission and mass

spectrometry achieved by tandem injection and merging-stream flow injection. *Analyst*, **114**, 1259–1265.

39 Korenaga, T., Zhou, X., Moriwake, T., Muraki, H., Naito, T. and Sanuki, S. (1994) Computer-controlled micropump suitable for precise microliter delivery and complete in-line mixing. *Analytical Chemistry*, **66**, 73–78.

40 Calatayud, J.M., Mateo, J.V.G. and David, V. (1998) Multi-insertion of small controlled volumes of solutions in a flow assembly for determination of nitrate (photoreduction) and nitrite with proflavin sulfate. *Analyst*, **123**, 429–434.

41 Wang, X.D., Cardwell, T.J., Cattrall, R.W., Dyson, R.P. and Jenkins, G.E. (1998) Time-division multiplex technique for producing concentration profiles in flow analysis. *Analytica Chimica Acta*, **368**, 105–111.

42 Martelli, P.B., Reis, B.F., Korn, M. and Rufini, I.A. (1997) The use of ion exchange resin for reagent immobilization and concentration in flow systems. Determination of nickel in steel alloys and iron speciation in waters. *Journal of the Brazilian Chemical Society*, **8**, 479–485.

43 Fernandes, R.N., Sales, M.G.F., Reis, B.F., Zagatto, E.A.G., Araújo, A.N. and Montenegro, M.C.B.S.M. (2001) Multi-task flow system for potentiometric analysis: its application to the determination of vitamin B-6 in pharmaceuticals. *Journal of Pharmaceutical and Biomedical Analysis*, **25**, 713–720.

44 Wang, X.D., Cardwell, T.J., Cattrall, R.W. and Jenkins, G.E. (1998) Pulsed flow chemistry. A new approach to the generation of concentration profiles in flow analysis. *Analytical Communications*, **35**, 97–101.

45 Korn, M., Gouveia, L.F.B.P., Oliveira, E. and Reis, B.F. (1995) Binary search in flow titration employing photometric end-point detection. *Analytica Chimica Acta*, **313**, 177–184.

46 Segundo, M.A., Oliveira, H.M., Lima, J.L.F.C., Almeida, M.I.G.S. and Rangel, A.O.S.S. (2005) Sample introduction in multi-syringe flow injection systems: comparison between time-based and volume-based strategies. *Analytica Chimica Acta*, **537**, 207–214.

47 Pasquini, C. and Faria, L.C. (1991) Operator-free flow injection analyser. *Journal of Automatic Chemistry*, **13**, 143–146.

48 Ruzicka, J. and Hansen, E.H. (1983) Recent developments in flow injection analysis – gradient techniques and hydrodynamic injection. *Analytica Chimica Acta*, **145**, 1–15.

49 Zagatto, E.A.G., Bahia Fo, O., Gine, M.F. and Bergamin Filho, H. (1986) A simple procedure for hydrodynamic injection in flow injection analysis applied to the atomic absorption spectrometry of chromium in steels. *Analytica Chimica Acta*, **181**, 265–270.

50 Carneiro, J.M.T., Zagatto, E.A.G., Santos, J.L.M. and Lima, J.L.F.C. (2002) Spectrophotometric determination of phytic acid in plant extracts using a multi-pumping flow system. *Analytica Chimica Acta*, **474**, 161–166.

51 Lapa, R.A.S., Lima, J.L.F.C. and Santos, J.L.M. (2000) Fluorimetric determination of isoniazid by oxidation with cerium(IV) in a multicommutated flow system. *Analytica Chimica Acta*, **419**, 17–23.

52 Vieira, J.A., Reis, B.F., Kronka, E.A.M., Paim, A.P.S. and Gine, M.F. (1998) Multicommutation in flow analysis. Part 6. Binary sampling for wide concentration range turbidimetric determination of sulphate in plant digests. *Analytica Chimica Acta*, **366**, 251–255.

53 Rocha, F.R.P., Reis, B.F. and Rohwedder, J.J.R. (2001) Flow-injection spectrophotometric multideterminationof metallic ions with a single reagent exploiting multicommutation and multidetection. *Fresenius' Journal of Analytical Chemistry*, **370**, 22–27.

54 Pons, C., Forteza, R. and Cerda, V. (2004) Expert multi-syringe flow injection system for the determination and speciation analysis of iron using chelating disks in water samples. *Analytica Chimica Acta*, **524**, 79–88.

55 Oliveira, F.S. and Korn, M. (2003) Employment of a single standard solution for analytical curves in flow injection analysis system coupled to solid phase spectrophotometry. *Quimica Nova*, **26**, 470–474.

56 Leal, L.O., Elsholz, O., Forteza, R. and Cerda, V. (2006) Determination of mercury by multisyringe flow injection system with cold-vapor atomic absorption spectrometry. *Analytica Chimica Acta*, **573**, 399–405.

57 Martinelli, M., Bergamin Filho, H., Arruda, M.A.Z. and Zagatto, E.A.G. (1989) A new approach for wide-range flow injection spectrophotometry: determination of cobalt in livestock mineral supplements. *Quimica Analitica*, **8**, 153–161.

58 Tumang, C.D., Luca, G.C., Fernandes, R.N., Reis, B.F. and Krug, J.F. (1998) Multicommutation in flow analysis exploiting a multizone trapping approach: spectrophotometric determination of boron in plants. *Analytica Chimica Acta*, **374**, 53–59.

59 Feres, M.A. and Reis, B.F. (2005) A downsized flow set up based on multicommutation for the sequential photometric determination of iron(II)/iron(III) and nitrite/nitrate in surface water. *Talanta*, **68**, 422–428.

60 Ventura-Gayete, J.F., Armenta, S., Garrigues, S., Morales-Rubio, A. and de la Guardia, M. (2006) Multicommutation-NIR determination of Hexythiazox in pesticide formulations. *Talanta*, **68**, 1700–1706.

61 Vieira, J.A., Raimundo, I.M., Reis, B.F., Montenegro, M.C.B.S.M. and Araujo, A.N. (2003) Monosegmented flow potentiometric titration for the determination of chloride in milk and

wine. *Journal of the Brazilian Chemical Society*, **14**, 259–264.

62 Oliveira, C.C., Sartini, R.P., Zagatto, E.A.G. and Lima, J.L.F.C. (1997) Flow analysis with accuracy assessment. *Analytica Chimica Acta*, **350**, 31–36.

63 Krug, F.J., Reis, B.F., Gine, M.F., Zagatto, E.A.G., Ferreira, J.R. and Jacintho, A.O. (1983) Zone trapping in flow injection analysis Spectrophotometric determination of low levels of ammonium ion in natural water. *Analytica Chimica Acta*, **151**, 39–48.

64 Fernandes, R.N. and Reis, B.F. (2002) Flow system exploiting multi-commutation to increase sample residence time for improved sensitivity. Simultaneous determination of ammonium and ortho-phosphate in natural water. *Talanta*, **58**, 729–737.

65 Giacomozzi, C.A., Queiroz, R.R.U., Souza, I.G. and Gomes No, J.A. (1999) High current-density anodic electro-dissolution in flow injection systems for the determination of aluminium and zinc in non-ferroalloys by flame atomic absorption spectrometry. *Journal of Automated Methods & Management in Chemistry*, **21**, 17–22.

66 Gervasio, A.P.G., Luca, G.C., Menegario, A.A., Reis, B.F. and Bergamin Filho, H. (2000) On-line electrolytic dissolution of alloys in flow injection analysis. Determination of iron, tungsten, molybdenum, vanadium and chromium in tool steels by inductively coupled plasma atomic emission. *Analytica Chimica Acta*, **405**, 213–219.

67 Packer, A.P., Gervasio, A.P.G., Miranda, C.E.S., Reis, B.F., Menegario, A.A. and Gine, M.F. (2003) On-line electrolytic dissolution for lead determination in high-purity copper by isotope dilution inductively coupled plasma mass spectrometry. *Analytica Chimica Acta*, **485**, 145–153.

68 Silva, J.B.B., Giacomelli, M.B.O., Souza, I.G. and Curtius, A.J. (1998) Automated determination of tin and nickel in brass by on-line anodic electro-dissolution and electrothermal atomic absorption spectrometry. *Talanta*, **47**, 1191–1198.

69 Honorato, R.S., Zagatto, E.A.G., Lima, R.A.C. and Araujo, M.C.U. (2000) Prior assay as an approach to flow titrations. Spectrophotometric determination of iron in alloys and ores. *Analytica Chimica Acta*, **416**, 231–237.

70 Fernandes, R.N., Reis, B.F. and Campos, L.F.P. (2003) Automatic flow system for simultaneous determination of iron and chromium in steel alloys employing photometers based on LED's as radiation source. *Journal of Automated Methods & Management in Chemistry*, **25**, 1–5.

71 Rodenas-Torralba, E., Ventura-Gayete, J., Morales-Rubio, A., Garrigues, S. and de la Guardia, M. (2004) Multicommutation Fourier transform infrared determination of benzene in gasoline. *Analytica Chimica Acta*, **512**, 215–221.

72 Ventura-Gayete, J.F., Reis, B.F., Garrigues, S., Morales-Rubio, A. and de la Guardia, M. (2004) Multicommutation ATR-FTIR: determination of sodium alpha-olefin sulfonate in detergent formulations. *Microchemical Journal*, **78**, 47–54.

73 Reis, B.E., Knochen, M., Pignalosa, G., Cabrera, N. and Giglio, J. (2004) A multicommuted flow system for the determination of copper, chromium, iron and lead in lubricating oils with detection by flame AAS. *Talanta*, **64**, 1220–1225.

74 Rodenas-Torralba, E., Morales-Rubio, A. and de la Guardia, M. (2006) Scientometric picture of the evolution of the literature of automation in spectroscopy and its current state. *Spectroscopy Letters*, **39**, 513–532.

75 van der Linden, W.E. (1994) Classification and definition of analytical methods based on flowing media (IUPAC Recommendations 1994). *Pure and Applied Chemistry*, **66**, 2493–2500.

76 Polasek, M. (2006) Playful reflections on the use of abbreviations and acronyms in

analytical flow methods. *Journal of Flow Injection Analysis*, **23**, 81.

77 Martelli, P.B., Reis, B.F., Korn, M. and Lima, J.L.F.C. (1999) Automatic potentiometric titration in monosegmented flow system exploiting binary search. *Analytica Chimica Acta*, **387**, 165–173.

78 Dressler, V.L., Pozebon, D. and Curtius, A.J. (2001) Determination of Ag, Te, Au and U in waters and in biological samples by FI-ICP-MS following on-line preconcentration. *Analytica Chimica Acta*, **438**, 235–244.

79 Toth, I.V., Rangel, A.O.S.S., Santos, J.L.M. and Lima, J.L.F.C. (2004) Determination of aluminum(III) in crystallized fruit samples using a multicommutated flow system. *Journal of Agricultural and Food Chemistry*, **52**, 2450–2454.

80 Paim, A.P.S. and Reis, B.F. (2000) An automatic spectrophotometric titration procedure for ascorbic acid determination in fruit juices and soft drinks based on volumetric fraction variation. *Analytical Sciences*, **16**, 487–491.

81 Flores, F.L., Diaz, A.M. and Cordova, M.L.F. (2007) Determination of azoxystrobin residues in grapes, musts and wines with a multicommuted flow-through optosensor implemented with photochemically induced fluorescence. *Analytica Chimica Acta*, **585**, 185–191.

82 Packer, A.P. and Gine, M.F. (2001) Analysis of undigested honey samples by isotopic dilution inductively coupled plasma mass spectrometry with direct injection nebulization (ID-ICP-MS). *Spectrochimica Acta Part B-Atomic Spectroscopy*, **56**, 69–75.

83 Ventura-Gayete, J.F., Rodenas-Torralba, E., Morales-Rubio, A., Garrigues, S. and de la Guardia, M. (2004) A multi-commutated flow system for determination of bismuth in milk shakes by hydride generation atomic fluorescence spectrometry incorporating on-line neutralization of waste effluent. *Journal of AOAC International*, **87**, 1252–1259.

84 Llorent-Martinez, E.J., Garcia-Reyes, J.F., Ortega-Barrales, P. and Molina-Diaz, A. (2007) Multicommuted fluorescence based optosensor for the screening of bitertanol residues in banana samples. *Food Chemistry*, **102**, 676–682.

85 Rocha, F.R.P., Rodenas-Torralba, E., Morales-Rubio, A. and de la Guardia, M. (2005) A clean method for flow injection spectrophotometric determination of cyclamate in table sweeteners. *Analytica Chimica Acta*, **547**, 204–208.

86 Garcia-Reyes, J.F., Ortega-Barrales, P. and Molina-Diaz, A. (2005) Rapid determination of diphenylamine residues in apples and pears with a single multicommuted fluorometric optosensor. *Journal of Agricultural and Food Chemistry*, **53**, 9874–9878.

87 Fernandes, E.N. and Reis, B.F. (2004) Automatic flow procedure for the determination of ethanol in wine exploiting multicommutation and enzymatic reaction with detection by chemiluminescence. *Journal of AOAC International*, **87**, 920–926.

88 Borges, S.S., Frizzarin, R.M. and Reis, B.F. (2006) An automatic flow injection analysis procedure for photometric determination of ethanol in red wine without using a chromogenic reagent. *Analytical and Bioanalytical Chemistry*, **385**, 197–202.

89 Fernandes, E.N., Moura, M.N.C., Lima, J.L.F.C. and Reis, B.F. (2004) Automatic flow procedure for the determination of glycerol in wine using enzymatic reaction and spectrophotometry. *Microchemical Journal*, **77**, 107–112.

90 Cava, P., Rodenas-Torralba, E., Morales-Rubio, A., Cervera, M.L. and de la Guardia, M. (2004) Cold vapour atomic fluorescence determination of mercury in milk by slurry sampling using multicommutation. *Analytica Chimica Acta*, **506**, 145–153.

91 Cava, P., Dominguez-Vidal, A., Cervera, M.L., Pastor, A. and de la Guardia, M. (2004) On-line speciation of mercury in

fish by cold vapor atomic fluorescence through ultrasound-assisted extraction. *Journal of Analytical Atomic Spectrometry*, **19**, 1386–1390.

92 Martelli, P.B., Reis, B.F., Araujo, A.N. and Montenegro, M.C.B.S.M. (2001) A flow system with a conventional spectrophotometer for the chemiluminescent determination of lactic acid in yoghurt. *Talanta*, **54**, 879–885.

93 Kronka, E.A.M., Paim, A.P.S., Tumang, C.A., Latanze, R. and Reis, B.F. (2005) Multicommutated flow system for spectrophotometric L(+) lactate determination in alcoholic fermented sugar cane juice using enzymatic reaction. *Journal of the Brazilian Chemical Society*, **16**, 46–49.

94 Ferreira, S.L.C., Lemos, V.A., Santelli, R.E., Ganzarolli, E. and Curtius, A.J. (2001) An automatic on-line flow system for the pre-concentration and determination of lead by flame atomic absorption spectrometry. *Microchemical Journal*, **68**, 41–46.

95 Fernandes, E.N. and Reis, B.F. (2006) Automatic spectrophotometric procedure for the determination of tartaric acid in wine employing multicommutation flow analysis process. *Analytica Chimica Acta*, **557**, 380–386.

96 Rodenas-Torralba, E., Cava, P., Morales-Rubio, A., Cervera, M.L. and de la Guardia, M. (2004) Multicommutation as an environmentally friendly analytical tool in the hydride generation atomic fluorescence determination of tellurium in milk. *Analytical and Bioanalytical Chemistry*, **379**, 83–89.

97 Rodenas-Torralba, E., Morales-Rubio, A. and de la Guardia, M. (2005) Multicommutation hydride generation atomic fluorescence determination of inorganic tellurium species in milk. *Food Chemistry*, **91**, 181–189.

98 Garcia-Reyes, J.F., Llorent-Martinez, E.J., Ortega-Barrales, P. and Molina-Diaz, A. (2006) Determination of thiabendazole residues in citrus fruits using a

multicommuted fluorescence-based optosensor. *Analytica Chimica Acta*, **557**, 95–100.

99 Borges, E.P., Martelli, P.B. and Reis, B.F. (2000) Automatic stepwise potentiometric titration in a monosegmented flow system. *Mikrochimica Acta*, **135**, 179–184.

100 Armas, G., Miro, M., Cladera, A., Estela, J.M. and Cerda, V. (2002) Time-based multisyringe flow injection system for the spectrofluorimetric determination of aluminium. *Analytica Chimica Acta*, **455**, 149–157.

101 Paim, A.P.S., Reis, B.F. and Vitorello, V.A. (2004) Automatic fluorimetric procedure for the determination of aluminium in plant nutrient solution and natural water employing a multicommutated flow system. *Mikrochimica Acta*, **146**, 291–296.

102 Palomeque, M., Bautista, J.A.G., Icardo, M.C., Mateo, J.V.G. and Calatayud, J.M. (2004) Photochemical-chemiluminometric determination of aldicarb in a fully automated multicommutation based flow-assembly. *Analytica Chimica Acta*, **512**, 149–156.

103 Lavorante, A.F., Morales-Rubio, A., de la Guardia, M. and Ries, B.F. (2005) Micro-pumping flow system for spectrophotometric determination of anionic surfactants in water. *Analytical and Bioanalytical Chemistry*, **381**, 1305–1309.

104 Hu, Y.Y., He, Y.Z., Qian, L.L. and Wang, L. (2005) On-line ion pair solid-phase extraction of electrokinetic multicommutation for determination of trace anion surfactants in pond water. *Analytica Chimica Acta*, **536**, 251–257.

105 Chivulescu, A., Catala-Icardo, M., Mateo, J.V.G. and Calatayud, J.M. (2004) New flow-multicommutation method for the photo-chemiluminometric determination of the carbamate pesticide asulam. *Analytica Chimica Acta*, **519**, 113–120.

106 Lavorante, A.F., Pires, C.K., Morales-Rubio, A., de la Guardia, M. and Ries, B.F. (2006) A spectrophotometric flow procedure for the determination of cationic surfactants in natural waters using a solenoid micro-pump for fluid propulsion. *International Journal of Environmental Analytical Chemistry*, **86**, 723–732.

107 Lindgren, C.C. and Dasgupta, P.K. (1992) Flow injection and solvent extraction with intelligent segment separation. Determination of quaternary ammonium ions by ion-pairing. *Talanta*, **39**, 101–111.

108 Icardo, M.C., Mateo, J.V.G. and Calatayud, J.M. (2001) Selective chlorine determination by gas diffusion in a tandem flow assembly and spectro-photometric detection with o-dianisidine. *Analytica Chimica Acta*, **443**, 153–163.

109 Mervartova, K., Calatayud, J.M. and Icardo, M.C. (2005) A fully automated assembly using solenoid valves for the photodegradation and chemiluminometric determination of the herbicide chlorsulfuron. *Analytical Letters*, **38**, 179–194.

110 Tortajada-Genaro, L.A., Campins-Falco, P. and Bosch-Reig, F. (2003) Analyser of chromium and/or cobalt. *Analytica Chimica Acta*, **488**, 243–254.

111 Queiroz, Z.F., Rocha, F.R.P., Knapp, G. and Krug, F.J. (2002) Flow system with in-line separation/preconcentration coupled to graphite furnace atomic absorption spectrometry with W-Rh permanent modified for copper determination in seawater. *Analytica Chimica Acta*, **463**, 275–282.

112 Pozebon, D., Dressler, V.L. and Curtius, A.J. (1998) Determination of copper, cadmium, lead, bismuth and selenium (IV) in sea-water by electrothermal vaporization inductively coupled plasma mass spectrometry after on-line separation. *Journal of Analytical Atomic Spectrometry*, **13**, 363–369.

113 Pons, C., Miro, M., Becerra, E., Estela, J.M. and Cerda, V. (2004) An intelligent flow analyser for the in-line concentration, speciation and monitoring of metals at trace levels. *Talanta*, **62**, 887–895.

114 Llorent-Martinez, E.J., Ortega-Barrales, P. and Molina-Diaz, A. (2006) Multi-commutated flow-through multi-optosensing: A tool for environmental analysis. *Spectroscopy Letters*, **39**, 619–629.

115 Dressler, V.L., Pozebon, D. and Curtius, A.J. (1998) Determination of heavy metals by inductively coupled plasma mass spectrometry after on-lie separation and preconcentration. *Spectrochimica Acta Part B-Atomic Spectroscopy*, **53**, 1527–1539.

116 Seibert, E.L., Dressler, V.L., Pozebon, D. and Curtius, A.J. (2001) Determination of Hg in seawater by inductively coupled plasma mass spectrometry after on-line pre-concentration. *Spectrochimica Acta Part B-Atomic Spectroscopy*, **56**, 1963–1971.

117 Reis, B.F., Rodenas-Torralba, E., Sancenon-Buleo, J., Morales-Rubio, A. and de la Guardia, M. (2003) Multicommutation cold vapour atomic fluorescence determination of Hg in water. *Talanta*, **60**, 809–819.

118 Corominas, B.G.T., Icardo, A.C., Zamora, L.L., Mateo, J.V.G. and Calatayud, J.M. (2004) A tandem-flow assembly for the chemiluminometric determination of hydroquinone. *Talanta*, **64**, 618–625.

119 Kumar, M.A., Chouhan, R.S., Thakur, M.S., Rani, B.E.A., Mattiasson, B. and Karanth, N.G. (2006) Automated flow enzyme-linked immunosorbent assay (ELISA) system for analysis of methyl parathion. *Analytica Chimica Acta*, **560**, 30–34.

120 Piston, M., Dol, I. and Knochen, M. (2006) Multiparametric flow system for the automated determination of sodium, potassium, calcium, and magnesium in large-volume parenteral solutions and concentrated hemodialysis solutions. *Journal of Automated Methods & Management in Chemistry*, **47627**, 1–6.

121 Rocha, F.R.P., Martelli, P.B. and Reis, B.F. (2004) Simultaneous in-line concentration for spectrophotometric determination of cations and anions. *Journal of the Brazilian Chemical Society*, **15**, 38–42.

122 Melchert, W.R., Infante, C.M.C. and Rocha, F.R.P. (2007) Development and critical comparison of greener flow procedures for nitrite determination in natural waters. *Microchemical Journal*, **85**, 209–213.

123 Rocha, F.R.P., Martelli, P.B. and Reis, B.F. (2001) An improved flow system for spectrophotometric determination of anions exploiting multicommutation and multidetection. *Analytica Chimica Acta*, **438**, 11–19.

124 Rocha, F.R.P. and Reis, B.F. (2000) A flow system exploiting multicommutation for speciation of inorganic nitrogen in waters. *Analytica Chimica Acta*, **409**, 227–235.

125 Michalowski, J., Halaburda, P. and Kojlo, A. (2000) Determination of phenols in natural waters with a flow-analysis method and chemiluminescence detection. *Analytical Letters*, **33**, 1373–1386.

126 Lupetti, K.O., Rocha, F.R.P. and Fatibello, O. (2004) An improved flow system for phenols determination exploiting multicommutation and long pathlength spectrophotometry. *Talanta*, **62**, 463–467.

127 Rodenas-Torralba, E., Morales-Rubio, A. and de la Guardia, M. (2005) Determination of phenols in waters using micro-pumped multicommutation and spectrophotometric detection: an automated alternative to the standard procedure. *Analytical and Bioanalytical Chemistry*, **383**, 138–144.

128 Morais, I.P.A., Miro, M., Manera, M., Estela, J.M., Cerda, V., Souto, M.R.S. and Rangel, A.O.S.S. (2004) Flow-through solid-phase based optical sensor for the multisyringe flow injection trace determination of orthophosphate in waters with chemiluminescence

detection. *Analytica Chimica Acta*, **506**, 17–24.

129 Albert-Garcia, J.R., Icardo, M.C. and Calatayud, J.M. (2006) Analytical strategy photodegradation/chemiluminescence/continuous-flow multicommutation methodology for the determination of the herbicide propanil. *Talanta*, **69**, 608–614.

130 Barbosa, F., Souza, S.S. and Krug, F.J. (2002) In-situ trapping of selenium hydride in rhodium-coated tungsten coil electrothermal atomic absorption spectrometry. *Journal of Analytical Atomic Spectrometry*, **17**, 382–388.

131 Santos Fa, M.M., Reis, B.F., Krug, F.J., Collins, C.H. and Baccan, N. (1993) Sulfate preconcentration by anion-exchange resin in flow injection and its turbidimetric determination in water. *Talanta*, **40**, 1529–1534.

132 Ferrer, L., Armas, G., Miro, M., Estela, J.M. and Cerda, V. (2004) A multisyringe flow injection method for the automated determination of sulfide in waters using a miniaturised optical fiber spectrophotometer. *Talanta*, **64**, 1119–1126.

133 Ferrer, L., Armas, G., Miro, M., Estela, J.M. and Cerda, V. (2005) Interfacing in-line gas-diffusion separation with optrode sorptive preconcentration exploiting multisyringe flow injection analysis. *Talanta*, **68**, 343–350.

134 Armas, G., Miro, M., Estela, J.M. and Cerda, V. (2002) Multisyringe flow injection spectrophotometric determination of warfarin at trace levels with on-line solid-phase preconcentration. *Analytica Chimica Acta*, **467**, 13–23.

135 Kronka, E.A.M. and Reis, B.F. (1998) Spectrophotometric determination of iron and aluminium in plant digests employing binary sampling in flow analysis. *Quimica Analitica*, **17**, 15–20.

136 Tumang, C.A., Tomazzini, M.C. and Reis, B.F. (2003) Automatic procedure exploiting multicommutation in flow analysis for simultaneous spectrophotometric determination of nonstructural carbohydrates and reducing sugar in forage materials. *Analytical Sciences*, **19**, 1683–1686.

137 Gomes No, J.A., Oliveira, A.P., Freshi, G.P.G., Dakuzaku, C.S. and Moraes, M. (2000) Minimization of lead and copper interferences on spectrophotometric determination of cadmium using electrolytic deposition and ion-exchange in multi-commutation flow system. *Talanta*, **53**, 497–503.

138 Blanco, T., Maniasso, N., Gine, M.F. and Jacintho, A.O. (1998) Liquid–liquid extraction in flow injection analysis using open-phase separator for the spec-trophotometric determination of copper in plant digests. *Analyst*, **123**, 191–193.

139 Miranda, C.E.S., Olivares, S., Reis, B.F. and Luzardo, F.M. (2000) On-line preconcentration employing a tannin resin for copper determination in plant material and food stuff by atomic absorption spectrometry. *Journal of the Brazilian Chemical Society*, **11**, 44–49.

140 Oliveira, C.C., Sartini, R.P., Reis, B.F. and Zagatto, E.A.G. (1996) Multicommutation in flow analysis. Part 4. Computer-assisted splitting for spectrophotometric determination of copper and zinc in plants. *Analytica Chimica Acta*, **332**, 173–178.

141 Almeida, M.I.G.S., Segundo, M.A., Lima, J.L.F.C. and Rangel, A.O.S.S. (2006) Potentiometric multi-syringe flow injection system for determination of exchangeable potassium in soils with in-line extraction. *Microchemical Journal*, **83**, 75–80.

142 Gomes, D.M.C., Segundo, M.A., Lima, J.L.F.C. and Rangel, A.O.S.S. (2005) Spectrophotometric determination of iron and boron in soil extracts using a multi-syringe flow injection system. *Talanta*, **66**, 703–711.

143 Gomes Neto, J.A., Zara, L.F., Rocha, J.C., Santos, A., Dakuzaku, C.S. and Nobrega, J.A. (2000) Determination of mercury in agroindustrial samples by flow injection cold vapor atomic absorption spectrometry using ion exchange and reductive elution. *Talanta*, **51**, 587–594.

144 Tumang, C.A., Borges, E.P. and Reis, B.F. (2001) Multicommutation flow system for spectrophotometric L(+)lactate determination in silage material using an enzymatic reaction. *Analytica Chimica Acta*, **438**, 59–65.

145 Smiderle, M., Reis, B.F. and Rocha, F.R.P. (1999) Monosegmented flow system exploiting multicommutation applied to spectrophotometric determination of manganese in soybean digests. *Analytica Chimica Acta*, **386**, 129–135.

146 Kronka, E.A.M., Reis, B.F., Korn, M. and Bergamin Filho, H. (1996) Multicommutation in flow analysis. Part 5. Binary sampling for sequential spectrophotometric determination of ammonium and phosphate in plant digests. *Analytica Chimica Acta*, **334**, 287–293.

147 Comitre, A.L.D. and Reis, B.F. (2005) Automatic flow procedure based on multicommutation exploiting liquid-liquid extraction for spectrophotometric lead determination in plant material. *Talanta*, **65**, 846–852.

148 Maruchi, A.K. and Rocha, F.R.P. (2006) An improved procedure for phosphorous fractionation in plant materials exploiting sample preparation and monosegmented flow analysis. *Microchemical Journal*, **82**, 207–213.

149 Buanuam, J., Miro, M., Hansen, E.H., Shiowatana, J., Estela, J.M. and Cerda, V. (2007) A multisyringe flow-through sequential extraction system for on-line monitoring of orthophosphate in soils and sediments. *Talanta*, **71**, 1710–1719.

150 Tumang, C.A., Paim, A.P.S. and Reis, B.F. (2002) Automatic flow system titration based on multicommutation for spectrophotometric determination of total acidity in silage extracts. *Journal of AOAC International*, **85**, 328–332.

151 Pires, C.K., Martelli, P.B., Reis, B.F., Lima, J.L.F.C. and Saraiva, M.L.M.F.S. (2003) An automatic flow procedure for

the determination of 3-hydroxybutyrate in animal serum and plasma. *Journal of Agricultural and Food Chemistry*, **51**, 2457–2460.

152 Pires, C.K. and Reis, B.F. (2005) Enzyme immobilization using a commercial kit: Determination of metabolic parameters in animal blood employing a multicommutation flow system. *Quimica Nova*, **28**, 414–420.

153 Silva, M.L.S., Garcia, M.B.Q., Lima, J.L.F.C. and Barrado, E. (2006) Modified tubular electrode in a multi-commuted flow system. Determination of acetaminophen in blood serum and pharmaceutical formulations. *Analytica Chimica Acta*, **573**, 383–390.

154 Luca, G.C. and Reis, B.F. (2004) Simultaneous photometric determination of albumin and total protein in animal blood plasma employing a multicommutated flow system to carried out on line dilution and reagents solutions handling. *Spectrochimica Acta Part A-Molecular and Biomolecular Spectroscopy*, **60**, 579–583.

155 Pires, C.K., Reis, B.F., Galhardo, C.X. and Martelli, P.B. (2003) A multicommuted flow procedure for the determination of cholesterol in animal blood serum by chemiluminescence. *Analytical Letters*, **36**, 3011–3024.

156 Araujo, A.N., Lima, J.L.F.C., Reis, B.F. and Zagatto, E.A.G. (1995) Multicommutation in flow analysis. Part 3. Spectrophotometric determination of creatinine in urine exploiting a novel zone sampling approach. *Analytica Chimica Acta*, **310**, 447–452.

157 Jeronimo, P.C.A., Araujo, A.N., Montenegro, M.C.B.S.M., Pasquini, C.

and Raimundo, I.M. (2004) Direct determination of copper in urine using a sol-gel optical sensor coupled to a multicommutated flow system. *Analytical and Bioanalytical Chemistry*, **380**, 108–114.

158 Lopes, C.M.P.V., Almeida, A.A., Santos, J.L.M. and Lima, J.L.F.C. (2006) Automatic flow system for the sequential determination of copper in serum and urine by flame atomic absorption spectrometry. *Analytica Chimica Acta*, **555**, 370–376.

159 Llorent-Martinez, E., Ortega-Barrales, P. and Molina-Diaz, A. (2005) Multi-commuted flow-through fluorescence optosensor for determination of furosemide and triamterene. *Analytical and Bioanalytical Chemistry*, **383**, 797–803.

160 Lopez-Flores, J., Cordova, M.L.F.D. and Molina-Diaz, A. (2007) Multicommutated flow-through optosensors implemented with photochemically induced fluorescence. Determination of flufenamic acid. *Analytical Biochemistry*, **361**, 280–286.

161 Amorim, C.M.P.G., Albert-Garcia, J.R., Montenegro, M.C.B.M.S., Araujo, A.N. and Calatayud, J.M. (2007) Photo-induced chemiluminometric determination of karbutilate in a continuous-flow multicommutation assembly. *Journal of Pharmaceutical and Biomedical Analysis*, **43**, 421–427.

162 Silva, M.L.S., Garcia, M.B.Q., Lima, J.L.F.C., Santos, J.L.M. and Barrado, E. (2005) Multicommutated flow system with amperometric detection. Determination of uric acid in urine. *Electroanalysis*, **17**, 2156–2162.

8
Advanced Calibration Methods in Flow Injection Analysis

Paweł Kościelniak

8.1
Introduction

In spite of broad improvements in methods and instrumentation, calibration still remains a key element in analytical chemistry. The calibration strategy undertaken affects the precision and accuracy of the analytical results. Furthermore, the calibration procedure carried out influences the whole analytical procedure in terms of labor, time and reagent consumption. Hence the search for more reliable and effective calibration approaches is one of the most important analytical tasks.

Analytical calibration can be regarded as the reconstruction of the real dependence (calibration dependence) of the analytical signal on the analyte(s) concentration, and transformation of measurement data obtained into the concentration of the analyte(s) in the sample analyzed. In an empirical version of calibration the calibration dependence is reconstructed with the use of standards (of well-known analyte(s) concentrations) measured under the same instrumental conditions as the sample. In analytical practice the samples are usually analyzed with respect to a single component; in such a case (being the main subject of consideration here) the calibration dependence is reconstructed in the form of a two-dimensional graph, as presented in Figure 8.1.

The empirical calibration can be performed according to many characteristic procedures, which are different from each other in terms of treatment (i. e., preparation and measurement) of the standards and samples and, consequently, in the way of interpreting measurement data and calculating analytical results. However, according to the classification recommended recently [1, 2] all these procedures can be divided into several categories, the most often used being the interpolative and extrapolative ones.

In the common calibration method (usually called "the calibration curve method" or "the standard addition method", SSM) the standard solutions and the sample are prepared and measured separately from each other. As a consequence, the calibration graph can be constructed in any required concentration range and the analytical

Advances in Flow Analysis. Edited by Marek Trojanowicz
Copyright © 2008 WILEY-VCH Verlag GmbH & Co. KGaA, Weinheim
ISBN: 978-3-527-31830-8

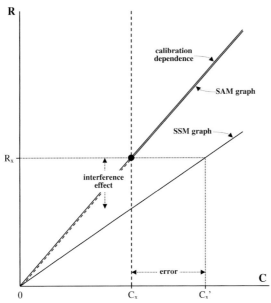

Figure 8.1 Comparison of the SSM and SAM calibration methods: if the multiplicative interference effect occurs, the calibration dependence can be exactly reconstructed by the SAM calibration graph but not by the SSM graph; consequently, a signal measured for a sample (R_x) leads to accurate determination of an analyte by the extrapolative method (C_x) but not by the interpolative method (C'_x).

result can be calculated interpolatively (as seen in Figure 8.1). The other calibration strategy (named "the standard addition method", SAM) comprises the addition of the standards into the same portions of the sample and measurement of the analytical signal for the total amount of the analyte in each portion. By doing so, the calibration dependence is reconstructed over the original concentration of the analyte in the sample and the analytical result has to be estimated extrapolatively (see Figure 8.1).

The essential difference between interpolative (SSM) and extrapolative (SAM) calibration strategies is related to the interference effect. If the SSM method is used, the interference effect can be eliminated only by matching the sample composition in the standards or by dosing the sample with special chemical reagents. If using the SAM method, the analyte can be accompanied by interferents present in a sample and, as a result, the interference effect can be compensated without any chemical treatment. The great advantage of the SAM method is that it can be used for overcoming interferences when the composition of a sample is very complex or even completely unknown. In such cases the SSM method usually fails. On the other hand, when the calibration dependence is non-linear the application of the SAM method is very limited and the interpolative strategy is then recommended.

Flow analysis is characterized by specific features that appear to be very useful for calibration purposes [3–11]. It offers the automation of chemical processing such as sampling, addition of reagents, dilution, preconcentration, and separation. In flow injection analysis (FIA) the analytical signal is produced as a transient peak consisting of signals corresponding to different local analyte concentration in the dispersed sample zone and for this reason it is a richer source of analytical information than the steady-state signal typical for a conventional measurement mode. Furthermore, the dynamic response of FIA allows the selectivity of the detection to be improved by a kinetic discrimination of analytes. Consequently, flow calibration has the potential for much general improvement in comparison with batch calibration.

Despite the advantages mentioned, both calibration methods, SSM and SAM, are carried out in FIA in a traditional manner, imitating the batch calibration procedures. These comprise the preparation of a set of standard solutions, separately from the sample or added to the sample, introduction of them one after the other into the flow injection system and measurement of the analytical signals for each of them individually. In this way typical single multi-point calibration graphs are constructed and the signal measured for a sample serves for calculation of the analytical result in either an interpolative or extrapolative way. Such an approach remains laborious and does not offer any analytical benefits over the conventional batch calibration.

In this chapter it is shown how the SSM and SAM calibration methods can be accomplished, exploiting the specific facilities of the flow injection technique more effectively and ingeniously.

8.2
Advanced Calibration Procedures

Based on the above mentioned definition of empirical calibration it can be stated that the entire calibration procedure, independently of the particular calibration method used, consists of three main stages: (i) preparation of the calibration solutions (preparation stage), (ii) construction of the calibration graph (reconstruction stage), and (iii) calculation of the analytical results (transformation stage). The FIA offers specific ways for realization of each of these stages.

8.2.1
Preparation Stage

The characteristic feature of flow systems is the possibility of separating a solution zone from the greater volume of this solution (usually from the solution stream introduced continuously into the system) and injection of this zone into the carrier stream. If the injector used is able to produce the zones accurately, precisely and reproducibly, it can be used for calibration purposes. The "calibration" injector commonly used in FIA is the rotary valve with a loop of well-defined volume installed. However, some other devices incorporated in the flow systems, such as solenoid valves or different kinds of pumps (e.g., peristaltic, syringe, piston, solenoid) can play the same role. Recently, some quite novel and ingenious instrumental tools have been designed and recommended for this aim. An example is the system allowing the

Table 8.1 Different methods of preparation of the standard solutions in FIA.

Case	Number of standard solutions introduced to a FIA system	Number of standard streams flowing to the injector	Number of zones injected	Parameter differentiating analytical signals
A	several	several	several	analyte concentration
B	one	several	several	analyte concentration
C	one	one	several	volume of initial zone
D	one	one	several	dispersion
E	one	one	one	local analyte concentration

calibration solutions to be taken from the horizontally positioned conical microvials with the aid of a capillary probe scanned sequentially through the slots on the vials and using gravity [12], electroosmotic [13], or capillary evaporation [14] forces. In another interesting approach the use of a microprocessor-controlled electronic pipette is proposed [15].

As the examples found in the literature show, the calibration solutions can be prepared in the form of separate zones in various manners. In Table 8.1 are collected the methods used for preparation of the standard solutions for the SSM calibration method. Mode A corresponds to the traditional manner, imitating the batch procedure. Much more interesting and advanced are those approaches (B–E) allowing a single standard solution to be exploited for realization of the entire calibration procedure.

Instead of preparing a set of standard solutions of different analyte concentrations outside a FIA system, it is possible to generate them from a single standard inside a system and then to inject them one after the other in the form of identical zones (Table 8.1, mode B). For this purpose dedicated modules are usually used, allowing the standard solution to be gradually and controllably diluted. For instance, in the system with the closed-loop installed, the standard solution is circulated in the loop with the aid of a pump and progressively diluted with a diluent successively dosed to this solution with known volume [16]. Another possibility for achieving a similar effect is to use the fully rotary valve (FRV), which is an eight-channel valve capable not only of 45° (as commonly applied) but also 360° rotation in relation to the stator [17]. For calibration purposes the standard and diluent solutions, both propelled into the FRV by separate tubes, are controllably mixed with each other when the FRV is rotated gradually through the particular eight positions. It has been demonstrated that as many as eight standards with different concentrations can be prepared by this means [18].

The FIA calibration can also be based on the standard zones differentiated in terms of volume (mode C) by controllably changing the parameters of the injection tools mentioned above (e.g., time and/or speed of pumping, distance between two solenoid valves, volumes of the injection loops, etc.). In some cases the injectors are purposely modified, such as the peristaltic pump driven by a stepper motor that is actuated by pulses generated by a microcomputer [19] or the time-based injector that allows preparation of zones of strictly defined volumes by a fixed air pressure [20]. As revealed, the FRV can be useful for the same aim if it is equipped with four loops of different volumes and fully rotated through all eight positions [17, 21]. It is also suggested to inject a standard solution sequentially from several loops [22] or from two loops separated from each other by an intermediate loop [23].

A unique feature of flow injection analysis is the possibility of controllable changing of the dispersion of a zone flowing toward the detector. For calibration purposes the standard zone is occasionally modified in terms of dispersion (mode D) by changing appropriate instrumental parameters of the FIA system used. For example, the standard zone can be directed to three tubes of different lengths that are merged before the detector: the three sub-zones produced by this means are then typified by different dispersions because of both the different routes to the detector and the different residence times in the tubes [24]. The same effect is achieved in the manifold, allowing a standard solution to be propelled by tubes of different lengths installed between two six-way stream switching valves [25]. If the FRV is equipped not only with four loops of different volumes but also with an intermediate loop and it is fully rotated clockwise and inversely, as many as eight different sub-zones can be produced due to their different volumes and different ways of transporting them from the FRV to the detector [17, 21]. In the FIA system, working on the basis of the zone-sampling technique, the standard solution injected initially is then re-injected after several defined time values [26]. In one more FIA system it is proposed to merge the zone of the standard solution inserted into a carrier stream with another carrier stream; in this case dispersion is controlled by the flow rates of the merging streams [27].

In some exceptional approaches it is suggested to achieve full calibration information from a single zone of the standard solution only (mode E). The basis for this is the concept of the gradient technique, that is, the zone dispersed in the carrier stream is treated as consisting of a set of solutions of different local analyte concentrations. In such cases the zone is usually required to be produced with the use of a single-line FIA system under low dispersion conditions [28] or with a mixing chamber installed [29, 30].

In principle, the aforementioned modes of preparing the standard solutions are related to calibration by the SSM method. If it is decided to use this method, the sample solution is usually recommended to be prepared in the same manner and in the same instrumental conditions as is the standard solution. In the case of calibration by the SAM method an additional and specific problem arises consisting in the proper preparation of a sample with the standard solutions added. The SAM method is capable of compensating for the interference effect most effectively if the standard solutions are added to the sample in such a way that the sample components (including interferents) are kept at the same concentration.

It would appear that the simplest way to solve this problem could be to prepare the calibration solution(s) (sample with standard(s)) outside the FIA system and then to treat it (them) according to one of the modes described above. Certainly, such an approach is acceptable if several calibration solutions of different analyte concentrations (including zero concentration) are prepared initially and introduced to an FIA system individually (mode A) or a single zone is required to be produced from a single calibration solution (mode E). However, if it is intended to prepare a set of calibration solutions from a single solution (modes B–D), one should be especially careful to do this in such a way as to fulfill the mentioned prerequisite of the SAM method.

The proper strategy related to the preparation stage of the SAM calibration is to prepare the standard solutions according to one of above modes (similarly as for the SSM calibration) in equal volumes and then merge them with the sample solution inside the FIA system. For instance, if several standard solutions of different concentrations prepared initially are injected one after the other by a valve but the sample is injected by another valve, each standard zone has a chance to meet the sample zone and to be mixed with it totally. Such an approach (mode A) was successfully applied to the determination of calcium, magnesium and potassium in plant materials by flame emission spectrometry [31]. The system working in accordance with mode B was used for the determination of iron in mineral water by UV–VIS spectrophotometry: the standards were generated by the FRV (preceded by the diluent stream) and then merged with a continuous stream of the sample before the injection valve [32].

In another approach the standard, sample and diluent solutions are proposed to be pumped independently with defined flow rates by three peristaltic pumps and then merged, mixed and injected; the analyte concentration in the standard stream added to the sample is changed by controllably changing the flow rates of both the standard and diluent stream [27]. The in-line preparation of the SAM calibration solutions can also be accomplished with the aid of a multi-syringe FIA system [33]. The zone-sampling technique can be used in such a way that the standard zone is re-injected after a well-defined time (mode D) and merged with a sample flowing in the form of either a zone [33] or the continuous stream [34]. Finally, an example of applying the gradient technique (mode E) is the proposal to insert a zone of the sample between the diluent and standard solutions, both flowing continuously [36].

In some cases in the literature the calibration solutions are proposed to be prepared in such a way that the prerequisite of the SAM method is not fulfilled. For example, in accordance with the procedure called "the interpolative standard addition method" [37] the standard solutions of different analyte concentrations, prepared outside the FIA system, are injected one after the other into the sample stream flowing continuously to the detector. The weak point of this approach is simply that the standards injected cause dilution of the sample, hence the concentrations of interferents in the sample alone and in the sample dosed by each of the standards are different. It was proved for the example of the determination of magnesium in the presence of aluminum by flame AAS that the analytical results obtained in such a way can be seriously inaccurate [38].

8.2.2
Reconstruction Stage

As mentioned above, the calibration dependence is, as a rule, reconstructed in the form of a calibration graph representing the relationship between the analytical response and the analyte concentration. In FIA the analytical signal can be measured in different ways. In addition, the analyte concentration of the calibration solutions prepared as the injected zones is not always well defined. Therefore, the reconstruction stage of the FIA calibration procedure is commonly not so clear as in the case of conventional batch calibration.

Independently of the preparation mode (listed in Table 8.1) a calibration zone produces a peak, which can be measured by means of either height or area. The exception is when the zones are injected from a single calibration solution and differentiated in terms of dispersion (mode D): the peaks produced are then of equal (or at least similar) areas, they can be measured by means of height only. Furthermore, it has been proved and exploited for calibration purposes [39, 40] that, in a flow system working under high dispersion conditions, the width of the injection peak can be measured and this can be connected mathematically to the analyte concentration. If a single-zone calibration is performed based on the gradient technique (mode E), a peak is analyzed in its entirety (or along its downward slope at least), that is, each individual point of a peak is considered as a source of measurement information. In some exceptional cases, that is, if the zone penetration technique is used, two or several (partly overlapped) peaks are produced and it is recommended to measure these not only at the maximum but also at the minimum points [22–24].

A separate problem of FIA calibration is how to attribute the analytical signal measured to the concentrations of an analyte present in the zones injected. The problem is that a zone is dispersed after injection and when reaching the detector it is of different local analyte concentrations which are in fact unknown. If the zones are produced from solutions of well-defined concentrations (modes A and B), one can assume that the signals correspond to the analyte concentrations in the calibration solutions flowing continuously to the injector. However, it is clearly not possible to do so if a single calibration solution is used and the analytical signals are measured either for a set of zones of different volumes or dispersion factors (modes C and D) or for different parts of a single zone (mode E). In such cases different approaches are used to solve this problem. They can be divided into three categories: mathematical, experimental and interpretative.

It has been proved, for instance, that if the signals are measured for a series of well-known volumes of a standard solution injected by means of the controlled stepper-motor driven peristaltic pump and injected by partially filling a sample loop, the concentration of an analyte in each standard can be mathematically expressed by the concentration of the solution introduced into the FIA system [41]. The same relationship can be strictly defined when using a peristaltic pump equipped with small bore tubes and turning at a low speed [42, 43].

The experimental approach to finding the local analyte concentration consists usually in calculation of the dispersion coefficient, D. This is found as the ratio of the steady state signal generated by the solution propelled continuously through a FIA system and the transient signal measured for the peak (e.g., maximum point or a point corresponding to a defined delay time) produced under the same instrumental conditions by the zone injected from this solution. Applying this approach it is believed that the unknown local analyte concentration is D-times less that the analyte concentration in the solution introduced to the FIA system. In the literature many examples of using the D-value for calibration purposes can be found, including calibration by both the SSM [19, 25, 33, 43, 44] and SAM [22, 35, 36, 45, 46] methods.

The analyte concentrations in the differently dispersed zones can also be found through the calibration graph constructed initially by conventional means (i.e., according to mode A). This is done in the proposed FIA calibration procedure encompassing the change in dispersion of the standard solution by means of propelling it by a permeation tube with various flow rates [47]. The same approach has been applied to the calibration performed on the basis of the gradient technique [29, 30].

If a single calibration solution is used for generation of different transient signals measured in well-controlled conditions defined by different values of a parameter (P), the measurement data can be interpreted in such a way that both mathematical and experimental efforts can be omitted. Namely, the series of signals obtained can be attributed to a single known concentration of an analyte present in the solution introduced to the system and the calibration graph consisting of a set of two-point lines can be constructed. This strategy is presented in Figure 8.2. In the literature it is shown how to apply this approach when the zones are differentiated in terms of both volume (mode C) [20, 22, 48] and dispersion (mode D). In the latter case both the injector–detector distance [24] and the time between two injections [26] are proposed to be chosen as parameter P. The FRV is recommended to be used for realization of

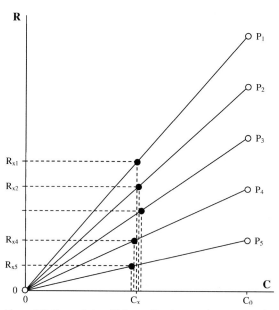

Figure 8.2 Two-point multi-line calibration graph constructed with the aid of a single standard solution (of concentration C_0) injected in conditions defined by different parameters (P_1–P_5); signals obtained for a sample (R_{x1}–R_{x5}) allow the analytical result (C_x) to be estimated by several values.

this strategy for calibration by both the SSM [17, 21] and SAM [17] method. It has also been proved that if a series of peaks corresponds to the zones produced from solutions of different concentrations, the multi-line calibration graph can be constructed from the signals measured at a given defined time (as parameter *P*) after injection of each of these zones [49].

8.2.3
Transformation Stage

As shown above, FIA offers quite new opportunities compared to batch analysis, both in terms of handling calibration solutions and in processing analytical responses. Specifically, in accordance with some FIA calibration procedures, neither a typical calibration graph is constructed (as in Figure 8.2) nor is a standard usually added to a sample (as in the case of "the interpolative standard addition method"). This may be a reason for some confusion when calculating the analytical results.

As mentioned, there are two general methods of calculating the analyte concentration in a sample: interpolative and extrapolative. Independently of the detailed procedures of the preparative and reconstruction stages the rule for which calculation method is applied in a given case and the consequences related to the interference effect is the following:

- If the standard solutions and the sample are prepared and measured separately from each other (the SSM method), the analytical result is calculated interpolatively but the interference effects cannot be eliminated (without additional efforts).

- If the standard solutions are merged with the sample in such a manner that the sample components are kept in the same concentration (the SAM method), the analytical result is calculated extrapolatively and the interference effects of multiplicative character can be compensated,

- If the standard solutions are merged with the sample in a manner different from above, the analytical result can be calculated interpolatively but the possibility of eliminating the interference effects is very limited.

Since the solutions can be merged in FIA in various ways, special attention should be paid as to whether a given calibration procedure can overcome interferences to the extent expected.

In most of the SSM calibration procedures met in the FIA literature the analytical result is obtained by measuring a single signal (neglecting repetitions) for a sample. However, in such cases, when a set of standard zones of different known analyte concentrations is produced by means of controllable gradual dilution of a single standard (Table 8.1, mode B), a set of sample zones is prepared in the same manner from the sample introduced to the FIA system and a series of different signals is measured [21]. If relating each signal to the calibration graph, several estimations of the analyte concentration in the sample are obtained, as shown in Figure 8.3. After their "equalization" by taking the dilution factors into account they can serve to evaluate the analytical precision but their average value can be considered as the analytical result.

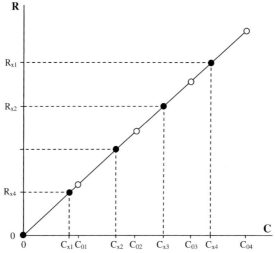

Figure 8.3 Calibration case when both the sample and standard solutions are controllably in-line diluted from the single solutions; the calibration graph is constructed with the aid of standards of different analyte concentrations $(C_{01}-C_{04})$ and the signals obtained for the sample $(R_{x1}-R_{x4})$ allow the analytical result to be estimated by several values $(C_{x1}-C_{x4})$.

If the measurement data obtained for the standard zones prepared in different conditions are interpreted in the form of multi-line two-point calibration graphs, the sample prepared and measured in the same conditions provides the signals, which can be individually related to each of the lines, as shown in Figure 8.2. In this way the analyte concentration in the sample is then also estimated by several values, which are the basis (if assumed to be randomly different) for evaluation of the analytical result in terms of precision. It is worth noticing that such a manner of transformation of measurement information can also be applied in the case of the SAM calibration [17].

Special attention should be paid at the transformation step of the SAM procedures, if the sample is proposed to be merged with the standard solutions inside a FIA system [32]. One should remember in such cases, that not only is the sample diluted with the standard added, but also each standard is diluted with the sample. Therefore, both dilution factors should be well-defined and taken into consideration during interpolative calculation of the analytical result.

8.2.4
Critical View

All the procedures considered above have great advantages over the conventional batch calibration: they can be significantly improved at the preparation stage and realized fully automatically under the control of dedicated computer software (even if some of them are not originally proposed in such a form). For both reasons they are generally more efficient in terms of time- and work consumption (Table 8.2). If the

Table 8.2 Comparison between conventional batch calibration (CBC), conventional FIA calibration (A) and advanced calibration procedures realized by different methods (B÷E).

Case	Speed	Work consumption	Precision	Accuracy
CBC	+	++++	++++	++++
A	++	+++	+++	+++
B	++++	+	++	++
C	++++	+	++	++
D	+++	++	++	++
E	++++	+	+	+

analytical signals are produced with the use of a single standard solution (Table 8.1, modes B–E), the whole calibration procedure can be performed in a few minutes (including preparation of this solution) instead of tens of minutes (i.e., when several standards are prepared) or even more (i.e., when the SAM calibration needs to be performed).

The price which is in general paid for making the calibration procedure faster and more operationally convenient is less precision of the analytical results. If a single standard solution is used to generate a set of different analytical signals, a FIA system has to be more instrumentally sophisticated than the conventional one working with a set of different standard solutions. Each modification of a simple flow system in terms of either an additional instrumental module (pump, injector, tubes, etc.) or specific mode of operation (injection from partially filling sample loop, sequential injection synchronized in time etc.) is a potential source of additional random errors influencing the analytical precision. Special attention should be paid to the repeatability of analytical signals when using the gradient technique as, in this case, the analytical result is calculated taking into account each individual point of a peak.

The more sophisticated the FIA calibration system and the more complex the operation of the system, the greater the risk of systematic errors and, consequently, greater inaccuracy. The good examples are the systems involving successive dilution of the calibration solution: if a dilution module is constructed with even slight inaccuracy (e.g., the recirculation loop is longer then expected), the successive standards are prepared with this error, which gradually accumulates. Some additional experimental procedures required in the calibration procedure (e.g., leading to evaluation of the dispersion coefficient) can also influence analytical accuracy.

However, the main general sources of systematic errors in calibration are nonlinearity of the calibration dependence and the interference effect. If the calibration dependence is well-known to be non-linear or if its linearity depends strongly on instrumental and chemical conditions, such calibration procedures can be applied only when using calibration graphs based on at least several experimental points. For the same reason the procedures using two-point multi-line graphs (i.e., when different signals are measured in conditions defined by different values of an instrumental parameter) should be avoided in such cases. If the interference effect is expected to occur in the analytical system assayed one should remember that any calibration interpolative approach (SSM) cannot be recommended (without using some additional chemical efforts) irrespective of how the procedure is realized in the preparation, reconstruction and interpretation and transformation stages. It is only possible to overcome the interferences when a procedure of the extrapolative method (SAM) is used providing that it is realized properly at the preparation stage (i.e., as mentioned above).

8.3
Advanced Calibration Concepts

The aforementioned FIA calibration procedures have been developed for the basic application of analytical calibration: to determine a single analyte precisely, accurately and more efficiently in terms of labor, time and reagent consumption than by the use of batch calibration procedures. However, FIA calibration can be improved not only procedurally but also more essentially allowing, for instance, the interference effect to be detected and examined as well as the analytical results to be controlled and verified in terms of accuracy. FIA also allows the calibration procedures to be ingeniously based on reactions occurring with analyte participation and also adapted to multi-component analysis. Some examples of how such concepts are accomplished are shown below.

8.3.1
Integrated Calibration

The simplest method of detecting and evaluating the interferences is to compare the calibration graphs constructed for the SSM and SAM calibration. If both graphs are presented in the same coordinate system, that is, as shown in Figure 8.1, the difference between their slopes can be a measure of the multiplicative interference effect occurring in an analytical system assayed. However, in spite of its simplicity such a procedure is surprisingly not commonly used in batch analysis.

In FIA the simultaneous use of the SSM and SAM methods, that is, integrating them in a single calibration procedure, can be easily accomplished [1, 50, 51]. Specifically, the integrated calibration method can be carried out with the use of the FIA system working on the basis of the merging zones technique [50, 51]. In this case the standard and sample solutions are simultaneously injected from two loops, then both zones are directed with different flow rates to two tubes of different lengths and finally merged with each other in such a way as to be partly overlapped. As a result a complex zone is produced, and exposed to measurement, that is composed of three sections containing standard, standard with sample and sample. In the second step of the calibration procedure both solutions are injected again but they are propelled with the reversed flow rates, hence they are merged in a different ratio than previously. Consequently, the three-section zone is produced again but with different analyte concentrations in each section. Using such a procedure it is possible from the data

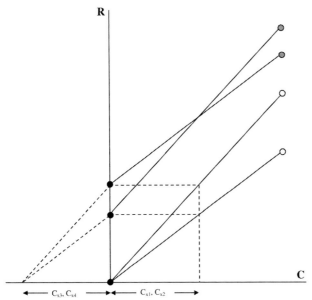

Figure 8.4 Calibration graph constructed in the integrated calibration method: measurement data produced in two different instrumental conditions by the standard solution (white circles), the sample with standard solution (gray circles) and the sample solution (black circles) allow the analytical result to be estimated by several values (C_{x1}–C_{x4}) obtained by the interpolative (C_{x1}, C_{x2}) and extrapolative (C_{x3}, C_{x4}) methods.

collected to construct as many as four calibration lines and to evaluate the analyte concentration in the sample by four independent estimations obtained in both interpolative and extrapolative ways, as shown in Figure 8.4.

In the approach described the interference effect can be easily detected through comparison of the calibration lines in pairs in terms of their mutual positions. The interference effect can be detected more quantitatively by comparison of the estimations of the analytical result. If all four values are statistically equal to each other, they can be considered as free of errors caused by the interference effect. This is because the estimations are obtained according to different calibration strategies (i.e., by the SSM and SAM methods) and in different dilution conditions (i.e., for differently diluted calibration solutions). Consequently, if the concentrations obtained interpolatively are different from those obtained extrapolatively, the interference effect can be surely expected. A scheme to continue the calibration in such a situation has also been suggested [51].

It is worth noticing that the above conclusions on the interference effect can be related to the accuracy of the analyte determination. If no other sources of systematic analytical errors are suspected, the analytical result estimated by four independent equal values obtained by different calibration methods can be considered as accurate with especially great probability. Thus, the integrated calibration method can serve not only for detection and evaluation of interference effects but also for verifying and controlling the analytical results in terms of accuracy. These advantages have been proved experimentally for the spectrophotometric determination of iron in drugs and the determination of magnesium in the presence of titanium by flame AAS [51].

8.3.2
Gradient Dilution Calibration

As already mentioned, the flow injection peak can be considered as a collection of individual points corresponding to local analyte concentrations in the standard (sample) zone injected and exposed to measurement at a given moment. If the falling part of a peak is analyzed from the maximum to zero points, the information gained is related in fact to the standard (sample) successively diluted from the maximum to zero concentration of the analyte in the zone. It can be believed, in addition, that each pair of transient signals measured for the standard and then for the sample (both treated in the same conditions) at a given defined time after injection represents both solutions as diluted to the same extent. This is the basis of the gradient dilution calibration.

In the basic form of this approach both calibration solutions are injected one after the other in the low dispersion FIA system and the two peaks recorded are synchronised with each other with respect to their base-widths [28, 52–54]. The values of each two adequate transient signals and of the analyte concentration in the standard solution injected are then taken for interpolative estimation of the analyte concentration in the sample. A series of estimation values are presented versus dilution factor, they are then fitted by a function and extrapolated to the value corresponding to the infinite dilution of the sample. This value is assumed as the measure of the analytical result.

Such a calibration approach is exceptional for a few reasons. It is extremely simple and informative, as a single standard zone is a source of information on the analyte concentrations in different parts of the sample zone, that is, in the sample progressively diluted. Furthermore, when the sample contains an interferent and the interference effect is successively diminished with the sample dilution, the estimated values of the analyte concentration in the sample – if examined as described above – are closer and closer to the true analyte concentration in the sample.

In some cases when the interferences are not able to be eliminated in such a way (e.g., when determining calcium in the presence of aluminum by flame AAS), it is recommended that both calibration solutions are injected into the carrier stream of a buffering reagent, which can be chosen in terms of kind and concentration through the estimated analyte concentration vs. the dilution factor relationship currently defined. Thus, as shown experimentally [54], the method is capable not only of determining the analyte accurately but also of doing so in the presence of interferents and examining the interference effects.

A factor reducing the accuracy is the adjustment of the sample and standard peaks to each other. Both peaks should theoretically have the same width at their bases but some lack of conformity can occur in practice. Moreover, even if the two peaks are compatible with each other, deciding which criteria should be taken into account to adjust them accurately enough is always debatable.

The gradient dilution calibration method can also be performed in the extrapolative version [55]. In this case the procedure is the same as above except that the sample and the sample with standard are the calibration solutions. It has been proved for the example of the determination of calcium in the presence of various interferents by flame AAS that, in this version, one can have greater confidence in terms of accuracy, as the interference effect can be reduced by the dilution of the sample and the presence of an analyte with the interferents in both calibration solutions.

8.3.3
Reaction-based Calibration

The most popular analytical technique in which the analytical result is found on the basis of a reaction occurring between an analyte and a reagent added to the analytical system assayed is titration. According to the classification of the calibration strategies titration is considered a calibration method of indicative (in contrast to interpolative and extrapolative) character [1]. The reason is that the titration procedure encompasses, as a common empirical calibration procedure, the reconstruction of the calibration dependence (in the form of a titration curve) with the use of the standard solution (titrant) added to the sample (or vice versa). The characteristic feature of titration is only that the standard reacts with an analyte in the sample and the analytical result is calculated from a single measurement point corresponding to the equivalence point of this reaction.

FIA titration can be accomplished in various modes, imitating the batch titration procedure. In particular, the sample can be mixed with the titrant solution in different ways, including discrete titrant addition to a continuous sample stream [56] or to a single sample aliquot [57] and volumetric fraction manipulation [58–61]. The titration curves showing signal vs. volume of titrant [56–58], or signal vs. time [59–61] relationships are constructed.

A more popular mode of FIA titration is that based on the gradient technique. It was the first application of the FIA gradient concept [62] and has since been frequently used in analytical practice. According to this concept, a sample is, as a rule, injected into the titrant stream, hence reproducible concentration gradients are formed between the analyte present in the injected sample zone and the titrant. If the analyte concentration is higher than the titrant concentration then, in both parts of the dispersed zone, an element of fluid exists, within which the analyte is entirely titrated by titrant. These two points of equivalence form a pair of the same dispersion and the time interval between them is proportional to the log of the analyte concentration [62].

In gradient titration knowledge of the precise volume and concentration of the titrant in unnecessary [63]. On the other hand, the analytical response cannot be related directly to the concentration of an analyte in the sample but is measured as a distance between two points (corresponding to the equivalence points) at a fixed peak height. In order to transform such measurement data the common calibration technique is applied with the use of standard solutions and a calibration graph. Regarding both above aspects the approach is considered as pseudo-titration [64]. From the formal point of view it can be found as an example of two calibration strategies – indicative and interpolative – linked in a single procedure [2].

Another FIA technique based on the reaction occurring in the system examined is the stopped-flow technique. It is mostly used to obtain chemical kinetics information, such as order of reaction and rate constant, but its application to calibration is increasing in popularity. In this approach the sample zone is merged and mixed with a reagent and then it is desirable to stop the flow in the detector cell. By doing so the reaction can be observed as a change in the peak profile. If no reaction occurs the signal is essentially a flat line parallel to the existing baseline. However, if the analyte is still produced due to the reaction, the observed signal increases. The basis of calibration is that this increase is strictly related to the analyte concentration. The common flow-stop calibration procedure encompasses the use of a series of standard solutions, construction of the calibration graph and calculation of the analytical result interpolatively.

An important feature of the reaction-based calibration approaches is that they enable accurate determination of analyte in spite of interference effects. The only condition is to use a reagent which is not able to react with the interferents in a given analytical system.

8.3.4
Multi-Component FIA Calibration

Flow injection analysis has been proved to be a highly appealing technique for the determination of two or more analytes in a sample within a single analytical procedure, that is, for performing multi-component analysis. Although the multi-component calibration approaches are sometimes conceptually specific, they are accomplished with the use of FIA systems designed in the same way and working on the basis of the same techniques as those for a single-component calibration.

In general, there are three main ways to realize multi-component analysis within FIA:

- Install two or more separate detectors and measure the signals individually for each of the analytes in conditions characteristic for them.
- Use a single detector, to chemically transform the analytes to a form adequate for the detector and to measure the signals sequentially for each of the analytes in conditions characteristic for this form.
- Use a single detector and measure the signals in different instrumental conditions (i.e., perform analysis on the basis of multivariate measurement data).

In first two cases each analyte is determined according to an individual calibration method, that is, single-component calibration procedures are applied. The third approach is more complex and advanced and needs to be accomplished specifically at all three stages (preparation, reconstruction and transformation) of the calibration procedure.

In the case of multi-component multivariate calibration the standard solutions are usually prepared in such a way that each contains all examined analytes at different concentration levels. They can be composed in accordance with a selected experimental plan [65] allowing measurement data to be effectively interpreted at the transformation stage.

Multivariate responses are generated from the standard and sample solutions by taking a time array over the peak under defined instrumental conditions (e.g., at a certain wavelength) or by changing the instrumental parameters (e.g., recording the whole spectrum) at a pre-selected time. Both measurement domains are also taken into consideration simultaneously [66, 67]. In any case the general idea is to enable one to distinguish the analytes as much as possible on the basis of their measurement data. Therefore, the peak profile is not considered as the dilution gradient but rather its discrimination power is increased by generation of the chemical gradient (usually pH gradient) or kinetically dependent process. For this purpose both the zone penetration technique [68] and the stopped-flow technique [69, 70] are often used.

Multivariate measurement information is used for reconstruction of multidimensional calibration dependence. In order to transform these data to analytical information, chemometric techniques are applied. Among these, the partial least-square method (PLS) is most widely used [71, 72], but various other chemometric approaches (such as classical least-square regression, principal component regression, artificial neural networks, factor analysis) can also be useful.

The majority of multivariate calibration procedures proposed in the literature are accomplished in such a manner that the analytical results are obtained (independently of the chemometric approach used) in the interpolative manner. Using the extrapolative version of multivariate calibration, named the generalized standard addition method (GSAM), it is suggested to prepare the multi-component calibration solutions by means of the merging-zones technique [73] and the zone-sampling technique [74]. It is also shown how to produce measurement data informative enough for GSAM calibration by preparing initially only one standard solution per analyte [75].

The great advantage of multivariate calibration approaches is their capability of improving determinations in terms of selectivity. The point is that if the standard solutions are prepared according to the rule mentioned above and the analytes examined are discriminated at the measurement stage, their mutual interferences of not only multiplicative but also additive character (e.g., spectral interferences) can be eliminated. This is so even when the interpolative calibration approach is applied. Furthermore, if it is decided to use the GSAM approach for calibration, the interference effects caused by the matrix components can also be overcome.

8.4
Trends and Perspectives

It can easily be stated (e.g., on the basis of reports presented at the Tenth International Conference "Flow Analysis," Porto, 2006), that, in general, there are two directions of

FIA progress, both related to the instrumental aspects: the design of increasingly complex and sophisticated FIA systems and the miniaturization of these systems. Both tendencies are strictly correlated with defined calibration problems.

Present day advances in both the technology of new materials and automation offer interesting ways to develop FIA systems of instrumental complexity, such as multi-commutated, multi-syringe, multi-pumping and multi-detector flow injection systems. The high degree of automation, low reagent consumption, high sample throughput and enhanced selectivity achieved by these techniques should be highlighted. Moreover, FIA manifolds are more and more frequently coupled with analytical instruments, which play the role not only of detectors (as e.g., AAS or UV/VIS spectrometers) but of the devices preparing the samples for detection. Such hyphenated techniques as FIA-ICP-MS, FIA-CE, FIA-HPLC, FIA-GC-MS are becoming increasingly popular in applications where multi-component analyses (including speciation) as well as analyses of complex matrices with low detection limits and high specificity are expected.

However, it is important to realize that, as mentioned above, the more sophisticated and complex the analytical system, the greater the number of sources of both systematic and random analytical errors. Therefore, it is very important in such cases to use especially robust calibration approaches so that the analytical results can be highly reproducible, repeatable and accurate. The calibration methods should also be relatively fast and low cost in order to allow frequent recalibration.

Recently, much attention has been paid to the miniaturization of FIA systems, either in the form of micro-total analysis systems (μTAS) or as miniaturized modules (e.g., "lab-on-valve") of the FIA systems. The miniaturized devices enable the performance of analyses on the nanoliter scale. All liquids, including standard and sample solutions, are transported to the detector (sensor) with very low flow rate in nanoliter amounts. In such cases the problem is both to prepare the standard solutions in-line and to dose them into the system with sufficiently high precision and accuracy. As the miniaturized systems are usually dedicated to biochemical analyses, the calibration approaches are expected to be helpful in eliminating matrix effects.

There are, of course, many more new ideas and concepts making progress in FIA. Furthermore, when reviewing the FIA literature many interesting and really ingenious calibration approaches are recommended to be exploited in analysis. However, it may be clearly observed that the methodological and instrumental FIA development is not, unfortunately, consistent with that occurring in the calibration domain. In spite of efforts to make the calibration process simpler, faster and analytically improved, the fact is that FIA calibration is mostly carried out in the traditional manner, that is, preparing standard solutions individually outside the FIA system, using conventional several-point one-line calibration graphs and calculating the analytical result in the interpolatively. Certainly, the main reason is simply psychological: if a common procedure is generally so popular then people have greater confidence in it even if another procedure seems to be better in some respects. However, another reason is quite real: apparently, none of the novel calibration approaches are sufficiently advanced to be competitive with the conventional approach.

The fact is that alternative FIA calibration methods proposed in the literature are usually typified by specific drawbacks [10, 11], which are perhaps not so noticeable at the stage of laboratory development but certainly create some limitations in analytical practice. The result is that they arouse distrust and suspicion. When authors pay more attention to ingenious FIAs rather than considering the practical aspects, the impression is that FIA calibration designs are developed as a kind of toy rather than as a tool which could be effectively and reliably introduced to routine analysis. Therefore, the development of new calibration ideas with improved analytical performance and greater confidence in terms of routine applicability is still a great analytical challenge.

Evidently, there are two fundamental conditions for the novel FIA calibration approaches to have a chance of gaining popularity among analysts: first, they should be conceptually and procedurally similar as far as possible to traditional calibration methods, and secondly, they should offer some additional analytical advantages over the traditional methods, that is, they should detect, examine and eliminate interferences as well as verifying the accuracy of the analytical results. The necessary research direction is to construct dedicated flow calibration manifolds in the form of prototypes. They should be easily operated and typified by a high degree of automation and high sample throughput. They should also be versatile in terms of the different calibration methods that can be accomplished and the various analytical instruments that can be coupled as detectors. Such an ambitious calibration task is not able to be realized by individual small research teams, but it should be the subject of collaboration for consortia consisting of scientists, analysts performing routine analysis and representatives of the analytical industry. This is the only way to achieve real progress in the field of FIA calibration.

References

1 Kościelniak, P. (2001) Univariate Calibration Techniques in Flow Injection Analysis. *Analytica Chimica Acta*, **438**, 323–333.

2 Kościelniak, P. (2003) New Horizons and Challenges in Environmental Analysis and Monitoring, in *Centre of Excellence in Environmental Analysis and Monitoring*, Gdańsk, Chapter 8.

3 Růžička, J. and Hansen, E.H. (1988) *Flow Injection Analysis*, 2nd edn, John Wiley, New York, Chapter 2.

4 Fang, Z. (1989) in *Flow Injection Atomic Spectroscopy* (ed. J.L. Burguera), Marcel Dekker, New York, Chapter 5.

5 De la Guardia, M. (1999) in *Flow Analysis with Atomic Spectrometric Detectors* (ed. A. Sanz-Mendel), Elsevier, Amsterdam, Chapter 4.

6 Tyson, J.F. (1988) Flow Injection Calibration Techniques. *Fresenius' Journal of Analytical Chemistry*, **329**, 663–667.

7 Tyson, J.F. (1991) Flow Injection Atomic Spectrometry. *Spectrochimica Acta Reviews*, **14**, 169–233.

8 Trojanowicz, M. and Olbrych-Śleszyńska, E. (1992) Flow Injection Sample Processing in Atomic Absorption Spectrometry. *Chemia Analityczna (Warsaw)*, **37**, 111–138.

9 Fang, Z. (1995) *Flow Injection Atomic Absorption Spectrometry*, Wiley, Chichester, Chapter 10.

10 Kościelniak, P. and Kozak, J. (2004) Review and Classification of the Univariate Interpolative Calibration Procedures in Flow Analysis. *Critical Reviews in Analytical Chemistry*, **34**, 25–37.

11 Kościelniak, P. and Kozak, J. (2006) Review of Univariate Standard Addition Calibration Procedures in Flow Analysis. *Critical Reviews in Analytical Chemistry*, **36**, 27–40.

12 Du, W., Fang, Q., He, Q. and Fang, Z. (2005) High-Throughput Nanoliter Sample Introduction Microfluidic Chip-Based Flow Injection Analysis System with Gravity-Driven Flows. *Analytical Chemistry*, **77**, 1330–1337.

13 He, Q., Fang, Q., Du, W., Huang, Y. and Fang, Z.L. (2005) An Automated Electrokinetic Continuous Sample Introduction System for Microfluidic Chip-based Capillary Electrophoresis. *Analyst*, **130**, 1052–1058.

14 Guan, Y., Xu, Z., Dai, J. and Fang, Z. (2006) The Use of a Micropump Based on Capillary and Evaporation Effects in a Microfluidic Flow Injection Chemiluminescence System. *Talanta*, **68**, 1384–1389.

15 Daniel, D. and Gutz, I.G.R. (2003) Quick Production of Gold Electrode Sets or Arrays and of Microfluidic Flow Cells Based on Heat Transfer of Laser Printed Toner Masks onto Compact Discs. *Electrochemistry Communications*, **5**, 782–786.

16 Agudo, M., Rios, A. and Valcarcel, M. (1992) Automatic Calibration and Dilution in Unsegmented Flow Systems. *Analytica Chimica Acta*, **264**, 265–273.

17 Kościelniak, P., Janiszewska, J. and Fang, Z. (1996) Flow-injection Calibration Procedures with the Use of Fully Rotary Valve. *Chemia Analityczna (Warsaw)*, **41**, 85–93.

18 Kościelniak, P., Herman, M. and Janiszewska, J. (1999) Flow Calibration System with the Use of Fully Rotary Directive Valve. *Laboratory Robotics and Automation*, **11**, 111–119.

19 Sherwood, R.A., Rocks, B.F. and Riley, C. (1985) Controlled-dispersion Flow Analysis with Atomic Absorption Detection for the Determination of Clinically Relevant Elements. *Analyst*, **110**, 493–496.

20 Burguera, J.L., Burguera, M., Rivas, C., De la Guardia, M. and Salvador, A. (1990) Simple Variable-volume Injector for Flame Injection Systems. *Analytica Chimica Acta*, **234**, 253–257.

21 Kościelniak, P. and Janiszewska, J. (1997) Design of the Fully Rotary Valve for Calibration Purposes in Flow Analysis. *Laboratory Robotics and Automation*, **9**, 47–54.

22 Zagatto, E.A.G., Giné, M.F., Fernández, E.A.N., Reis, B.F. and Krug, F.J. (1985) Sequential Injections in Flow Systems as an Alternative to Gradient Exploitation. *Analytica Chimica Acta*, **173**, 289–297.

23 Fang, Z., Sperling, M. and Welz, B. (1992) Comparison of Three Propulsion Systems for Application in Flow Injection Zone Penetration Dilution and Sorbent Extraction Preconcentration for Flame Atomic Absorption Spectrometry. *Analytica Chimica Acta*, **269**, 9–19.

24 Tyson, J.F. and Bysouth, S.R. (1988) Network Flow Injection Manifolds for Sample Dilution and Calibration in Flame Atomic Absorption Spectrometry. *Journal of Analytical Atomic Spectrometry*, **3**, 211–215.

25 Tyson, J.F., Adeeyinwo, C.E., Appleton, J.M.H., Bysouth, S.R., Idris, A.B. and Sarkissian, L.L. (1985) Flow Injection Techniques of Method Development for Flame Atomic Absorption Spectrometry. *Analyst*, **110**, 487–492.

26 Garrido, J.M.P.J., Lapa, R.A.S., Lima, J.L.F.C., Delerue-Matos, C. and Santos, J.L.M. (1996) FIA Automatic Dilution System for the Determination of Metallic Cations in Waters by Atomic Absorption and Flame Emission Spectrometry. *Journal of Automatic Chemistry*, **18**, 17–21.

27 Novic, M., Berregi, I., Rios, A. and Valcarcel, M. (1991) A New Sample-injection/Sample-dilution System for the Flow Injection Analytical Technique. *Analytica Chimica Acta*, **381**, 287–295.

28 Sperling, M., Fang, Z. and Welz, B. (1991) Expansion of Dynamic Working Range and Correction for Interferences in Flame Atomic Absorption Spectrometry Using Flow Injection Gradient Ratio Calibration with a Single Standard. *Analytical Chemistry*, **63**, 151–158.

29 De la Guardia, M., Carbonell, V., Morales, A. and Salvador, A. (1989) Direct Determination of Calcium, Magnesium, Sodium and Potassium in Water by Flow Injection Flame Atomic Spectroscopy Using a Dilution Chamber. *Fresenius' Journal of Analytical Chemistry*, **335**, 975–979.

30 Carbonell, V., Sanz, A., Salvador, A. and De la Guardia, M. (1991) Flow Injection Flame Atomic Spectrometric Determination of Aluminium, Iron, Calcium, Magnesium, Sodium and Potassium in Ceramic Material by On-line Dilution in a Stirred Chamber. *Journal of Analytical Atomic Spectrometry*, **6**, 233–238.

31 Zagatto, E.A.G., Krug, F.J., Bergamin, F.H., Jorgensen, S.S. and Reis, B.F. (1979) Merging Zones in Flow Injection Analysis. Part 2. Determination of Calcium, Magnesium and Potassium in Plant Material by Continuous Flow Injection Atomic Absorption and Flame Emission Spectrometry. *Analytica Chimica Acta*, **104**, 279–284.

32 Kościelniak, P. and Herman, M. (1999) Simple Flow System for Calibration by the Standard Addition Method. *Chemia Analityczna (Warsaw)*, **44**, 773–784.

33 Segundo, M.A. and Magalhães, L.M. (2006) Multisyringe Flow Injection Analysis: State-of-the-Art and Perspectives. *Analytical Sciences*, **22**, 3–8.

34 Giné, M.F., Reis, B.F., Zagatto, E.A.G., Krug, F.J. and Jacintho, A.O. (1983) A Simple Procedure for Standard Additions in Flow Injection Analysis Spectrophotometric Determination of Nitrate in Plant Extracts. *Analytica Chimica Acta*, **155**, 131–138.

35 Reis, B.F., Giné, M.F., Krug, F.J. and Filho, H.B. (1992) Multipurpose Flow Injection System. Part 1. Programmable Dilutions and Standard Additions for Plant Digests Analysis by Inductively Coupled Plasma Atomic Emission Spectrometry. *Journal of Analytical Atomic Spectrometry*, **7**, 865–868.

36 Fang, Z., Harris, J.M., Růžička, J. and Hansen, E.H. (1985) Simultaneous Flame Photometric Determination of Lithium, Sodium, Potassium, and Calcium by Flow Injection Analysis with Gradient Scanning Standard Addition. *Analytical Chemistry*, **57**, 1457–1461.

37 Tyson, J.F. (1981) Low Cost Continuous Flow Analysis. *Analytical Proceedings*, **18**, 542–545.

38 Kościelniak, P. (1996) Critical Look at the Interpolative Standard Addition Method. *Analusis*, **24**, 24–27.

39 Tyson, J.F. (1984) Extended Calibration of Flame Atomic Absorption Instruments by a Flow Injection Peak Width Method. *Analyst*, **109**, 319–321.

40 Bysouth, S.R. and Tyson, J.F. (1986) A Microcomputer-Based Peak-Width Method of Extended Calibration for Flow Injection Atomic Absorption Spectrometry. *Analytica Chimica Acta*, **179**, 481–486.

41 Fang, Z., Xu, S. and Bai, X. (1996) A New Flow Injection Single Standard Calibration Method for Flame Atomic Absorption Spectrometry Based on Dilution by Microsample Dispersion. *Analytica Chimica Acta*, **326**, 49–55.

42 López-García, I., Viñas, P., Campillo, N. and Hernández-Córdoba, M. (1996) Extending the Dynamic Range of Flame Atomic Absorption Spectrometry: a Comparison of Procedures for the Determination of Several Elements in Milk and Mineral Waters using On-line

Dilution. *Fresenius' Journal of Analytical Chemistry*, **355**, 57–64.

43 López-García, I., Viñas, P. and Hernández-Córdoba, M. (1994) Flow Injection Dilution System for the Analysis of Highly Concentrated Samples Using Flame Atomic Absorption Spectrometry. *Journal of Analytical Atomic Spectrometry*, **9**, 1167–1172.

44 Reis, B.F., Jacintho, A.O., Morlatti, J., Krug, F.J., Zagatto, E.A.G., Bergamin, E.H. and Pessenda, L.C.R. (1981) Zone-sampling Processes in Flow Injection Analysis. *Analytica Chimica Acta*, **123**, 221–228.

45 Beauchemin, D. (1995) On-line Standard Addition Method with ICPMS Using Flow Injection. *Analytical Chemistry*, **67**, 1553–1557.

46 Lavilla, I., Perez-Cid, B. and Bendicho, C. (1999) Use of Flow-injection Sample-to-standard Addition Methods for Quantification of Metals Leached by Selective Chemical Extraction from Sewage Sludge. *Analytica Chimica Acta*, **381**, 297–305.

47 Chalk, S.J., Tyson, J.F. and Olson, D.C. (1993) Permeation Tubes for Calibration in Flow Injection Analysis. *Analyst*, **118**, 1227–1231.

48 De la Guardia, M., Morales-Rubio, A., Carbonell, V., Salvador, A., Burguera, J.L. and Burguera, M. (1993) Flow Injection Flame Atomic Spectrometric Determination of Iron, Calcium, Magnesium, Sodium and Potassium in Ceramic Materials by Using a Variable-volume Injector. *Fresenius' Journal of Analytical Chemistry*, **345**, 579–584.

49 Olsen, S., Růžička, J. and Hansen, E.H. (1982) Gradient Techniques in Flow Injection Analysis. Stopped-flow Measurement of the Activity of Lactate Dehydrogenase with Electronic Dilution. *Analytica Chimica Acta*, **136**, 101–112.

50 Kościelniak, P., Kozak, J., Herman, M., Wieczorek, M. and Fudalik, A. (2004) Complementary Dilution Method – a New

Version of Calibration by the Integrated Strategy. *Analytical Letters*, **37**, 1233–1253.

51 Kościelniak, P., Wieczorek, M., Kozak, J. and Herman, M. (2007) Versatile Flow Injection Manifold for Analytical Calibration. *Analytica Chimica Acta*, **600**, 6–13.

52 Fan, S. and Fang, Z. (1990) Compensation of Calibration Graph Curvature and Interference in Flow Injection Spectrophotometry Using Gradient Ratio Calibration. *Analytica Chimica Acta*, **241**, 15–22.

53 Kościelniak, P., Sperling, M. and Welz, B. (1996) High Efficient Calibration Procedure for Flow Injection Flame Atomic Absorption Spectrometry. *Chemia Analityczna (Warsaw)*, **41**, 587–600.

54 Kościelniak, P. (1998) Calibration Procedure for Flow Injection Flame Atomic Absorption Spectrometry with Interferents as Spectrochemical Buffers. *Analytica Chimica Acta*, **367**, 101–110.

55 Kościelniak, P. and Kozak, J. (2002) Simple Flow Injection Titration Method Based on Variation of the Sample Volume. *Analytica Chimica Acta*, **460**, 235–245.

56 Bartroli, J. and Alerm, L. (1992) Automated Continuous-flow Titration. *Analytica Chimica Acta*, **269**, 29–34.

57 Vidal de Aquino, E., Rohwedder, J.J.R. and Pasquini, C. (2001) Monosegmented Flow Titrator. *Analytica Chimica Acta*, **438**, 67–74.

58 Paim, A.P.S. and Reis, B.F. (2000) An Automatic Spectrophotometric Titration Procedure for Ascorbic Acid Determination in Fruit Juices and Soft Drinks Based on Volumetric Fraction Variation. *Analytical Sciences*, **16**, 487–491.

59 Almeida, C.M.N.V., Lapa, R.A.S., Lima, J.F.L.C., Zagatto, E.A.G. and Araújo, M.C.U. (2000) An Automatic Titrator Based on a Multicommutated Unsegmented Flow System: Its Application to Acid–base Titrations. *Analytica Chimica Acta*, **407**, 213–223.

60 Assali, M., Rajmundo, I.M., Jr and Facchin, I. (2002) Simultaneous Multiple

Injection to Perform Titration and Standard Addition in Monosegmented Flow Analysis. *Journal of Automated Methods & Management in Chemistry*, **23**, 83–89.

61 Almeida, C.M.N.V., Lapa, R.A.S. and Lima, J.F.L.C. (2001) Automatic Flow Titrator Based on a Multicommutated Unsegmented Flow System for Alkalinity Monitoring in Wastewaters. *Analytica Chimica Acta*, **438**, 291–298.

62 Růžička, J., Hansen, E.H. and Mosbaek, H. (1977) Flow Injection Analysis: Part IX. A New Approach to Continuous Flow Titrations. *Analytica Chimica Acta*, **92**, 235–249.

63 Toei, J. (1988) A New Flow Injection Titration Analysis. *Fresenius' Journal of Analytical Chemistry*, **330**, 484–488.

64 Pardue, H.L. and Fields, B. (1981) Kinetic Treatment of Unsegmented Flow Systems: Part 1. Subjective and Semiquantitative Evaluations of Flow-injection Systems with Gradient Chamber. *Analytica Chimica Acta*, **124**, 39–63.

65 Kościelniak, P. and Herman, M. (2000) Flow System for Calibration and Interference Examination in Multicomponent Analysis. *Laboratory Robotics and Automation*, **12**, 228–235.

66 Beck, H.P. and Wiegand, C. (1995) Development and Optimization of a Multichannel FIA-cell Allowing the Simultaneous Determination with a Multiwavelength Photometric Device Based on Light Emitting Diodes. *Fresenius' Journal of Analytical Chemistry*, **351**, 701–707.

67 Polster, J., Prestel, G., Wollenweber, M., Kraus, G. and Gauglitz, G. (1995) Simultaneous Determination of Penicillin and Ampicillin by Spectral Fibre-optical Enzyme Optodes and Multivariate Data Analysis Based on Transient Signals Obtained by Flow Injection Analysis. *Talanta*, **42**, 2065–2072.

68 Whitman, D.A., Seasholtz, M.B., Christian, G.D., Růžička, J. and Kowalski, B.R. (1991) Double-injection Flow Injection Analysis Using Multivariate Calibration for Multicomponent Analysis. *Analytical Chemistry*, **63**, 775–781.

69 Muñoz De la Peña, A., Acedo-Valenzuela, M.I., Espinosa-Mansilla, A. and Sánchez-Maqueda, R. (2002) Stopped-flow Fluorimetric Determination of Amoxycillin and Clavulanic Acid by Partial Least-squares Multivariate Calibration. *Talanta*, **56**, 635–642.

70 Pistonesi, M., Centurión, M.E., Fernández-Band, B.S., Damiani, P.C. and Olivieri, A.C. (2004) Simultaneous Determination of Levodopa and Benserazide by Stopped-flow Injection Analysis and Three-way Multivariate Calibration of Kinetic-spectrophotometric Data. *Journal of Pharmaceutical and Biomedical Analysis*, **36**, 541–547.

71 Azubel, M., Fernández, F.M., Tudino, M.B. and Troccoli, O.E. (1999) Novel Application and Comparison of Multivariate Calibration for the Simultaneous Determination of Cu, Zn and Mn at Trace Levels Using Flow Injection Diode Array Spectrophotometry. *Analytica Chimica Acta*, **398**, 93–102.

72 Hernández, O., Jiménez, F., Jiménez, A.I., Arias, J.J. and Havel, J. (1996) Multicomponent Flow Injection Based Analysis With Diode Array Detection and Partial Least Squares Multivariate Calibration Evaluation. Rapid Determination of Ca(II) and Mg(II) in Waters and Dialysis Liquids. *Analytica Chimica Acta*, **320**, 177–183.

73 Zagatto, E.A.G., Jacintho, A.O., Krug, F.J., Reis, B.F., Bruns, R.E. and Araújo, M.C.U. (1983) Flow Injection Systems with Inductively-coupled Argon Plasma Atomic Emission Spectrometry: Part 2. The Generalized Standard Addition Method. *Analytica Chimica Acta*, **145**, 169–178.

74 Giné, M.F., Krug, F.J., Bergamin, H., Reis, B.F., Zagatto, E.A.G. and Bruns, R.E. (1988) Flow Injection Calibration of Inductively Coupled Plasma Atomic Emission Spectrometry Using the Generalized Standard Additions Method.

Journal of Analytical Atomic Spectrometry, **3**, 673–678.

75 Silva, E.C., Martins, V.L., Araújo, A.F. and Araújo, M.C.U. (1999) Implementation of a Generalized Standard Addition Method in a Flow Injection System Using Merging-Zones and Gradient Exploitation. *Analytical Sciences*, **15**, 1235–1240.

9
Multicomponent Flow Injection Analysis

Javier Saurina

9.1
Introduction

Since its introduction in the 1970s, flow injection analysis (FIA) has continuously gained popularity for a wide variety of applications, ranging from research activities to routine analysis [1, 2]. In parallel to the development of conventional FIA methodology, derivative branches, such as sequential injection analysis, bead injection analysis, microfluidic devices and multi-pumping, multisyringe and multi-commuted flow systems have been proposed for tackling diverse analytical issues and new challenges [3–5]. Also, in the last few years, the combination of FIA and related flow systems with separation techniques offers new possibilities to enhance the analytical potential of hyphenation. Pioneering studies proposed continuous flow or FIA-HPLC combinations, but nowadays there is great interest in the development of novel interfaces, especially with capillary electrophoresis [6, 7]. These topics are discussed in specific chapters of this book.

Among the new trends in flow methods, multicomponent analysis has received increasing attention [8, 9]. This chapter illustrates the application of FI and related flow techniques to multicomponent analysis. Although most of the FI methods described in the literature are focused on the quantification of one analyte, the potential of FIA for multicomponent analysis cannot be underestimated. But what is the significance of multicomponent FIA in the context of modern analytical chemistry? Typically, chromatography and electrophoresis are considered a good choice when facing multicomponent determinations in a wide variety of analytical problems, owing to of their excellent separation efficiency. However, it is clear that in certain cases the separation methods are somewhat tedious and time-consuming, so simpler and faster methods might be preferred. It is in this context that FIA and related flow techniques may have advantages. It should be pointed out that the principles for multicomponent determinations, relying on separation methods, are completely different to those occurring in flow injection methods. In chromatography and electrophoresis, typically, components of the sample are first separated

Advances in Flow Analysis. Edited by Marek Trojanowicz
Copyright © 2008 WILEY-VCH Verlag GmbH & Co. KGaA, Weinheim
ISBN: 978-3-527-31830-8

and the corresponding peak signals are used as selective responses for quantification. Conversely, separation is hardly ever exploited in multicomponent flow determinations in which the required selectivity is accomplished from a wide variety of mechanisms ranging from specific chemicals to mathematical approaches.

The excellent analytical features of flow methods regarding precision, sample throughput, low sample and reagent consumption, automation, simplification and miniaturization can be extrapolated to multicomponent analysis. Moreover, the possibility of using unstable reagents or analyzing unstable compounds deserves attention. Complementary aspects such as the on-line implementation of a wide variety of chemical operations (e.g., solid-phase extraction, liquid–liquid extraction, dialysis, gas-diffusion, derivatization, etc.) may be exploited for multicomponent analysis [10, 11]. As a result, many papers are published on this topic every year in the fields of environmental, pharmaceutical, clinical, food and biochemical analysis.

The first studies on multicomponent FIA were carried out in the 1970s. In 1976 Steward and Ruzicka published a spectrophotometric method for the simultaneous determination of nitrogen and phosphorus in acid digests using a dual-channel manifold [12]. In the 1980s, various methods for the analysis of binary mixtures of metal ions and drugs in aqueous samples and pharmaceuticals were developed [13–16]. In these examples, the presence of interferences and matrix effects was not dramatic, and the selectivity of each compound of the mixture was easily accomplished through, for instance, kinetic discrimination or enzymatic reactions. The topic was first reviewed by Kuban at the beginning of the 1990s [17]. Since then, significant advances have been made with an extension of applications towards the determination of more than two components in a wide variety of samples. In these new applications, additional problems of multiple and unknown interferences and strong matrix effects have arisen and, in parallel, strategies for solving these shortcomings have been proposed [9, 18].

The suitability of FIA for multicomponent analysis depends on several experimental circumstances such as the number of components, the presence of interferences and the complexity of the sample matrix. First, achieving selectivity for an excessively large number of analytes may lead to a practical limit in the development of FI multicomponent applications. The effort entailed in optimizing the method may not be compensated by the advantages of flow systems. Most of the analytical methods proposed in the literature are for the analysis of two components while the number of publications involving more than four components is low (see Figure 9.1).

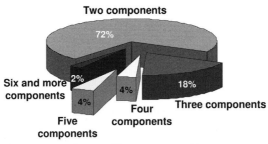

Figure 9.1 Distribution of flow injection multicomponent determinations according to the number of analytes quantified.

The presence of multiple sources of interferences may result in another practical hindrance of FI multicomponent determinations. Furthermore, complementary factors, such as the number of samples, cost per analysis, detection limits, analysis time, availability and simplicity of the equipment and so on, should be borne in mind in the final choice of method.

9.2
Principal Strategies for Multicomponent Analysis

Obtaining selectivity for all analytes is a primary objective in the development of FI multicomponent methods. The main experimental mechanisms that have been proposed for gaining selectivity rely on the use of specific reagents, multi-way or fast-scanning detectors and multi-channel manifolds. Beyond these strategies, the application of chemometrics to data analysis is complementary (sometimes a last resort) in those cases where selectivity has not been accomplished experimentally. Mechanisms to enhance the performance of the method are frequently combined. For instance, the use of specific reagents is often connected with the design of multichannel manifolds for developing each reaction separately. Also, multi-way or fast-scanning detectors are frequently combined with chemometrics since the huge amounts of data generated with these instruments may be better analyzed with mathematical algorithms.

Most of the multicomponent FIA and SIA determinations have been adapted from previous batch and flow methods established for the analysis of single components. Modifications of the experimental conditions are often required to achieve the proper selectivity for each component. The adaptation and re-optimization of conditions for FI multicomponent analysis may become difficult as multiple experimental factors influence the process. As a result, the optimization of the experimental conditions is a fundamental aspect for achieving selectivity. In the following section this topic is discussed at length.

9.2.1
Optimization of FI Multicomponent Methods

The steps towards optimization of the experimental conditions of FIA, SIA and related methodologies are schematized in Figure 9.2. For a more exhaustive review of optimization methods, see Refs. [19, 20].

The starting point is the definition of optimization criteria. Although in practice this aspect is often simplified, the election of a proper objective or response to be maximized deserves thorough attention. Currently, the optimization of flow methods consists of searching for the highest instrumental response (e.g., maximum peak height) in order to enhance the sensitivity of detection. However, many other complementary issues may be of interest. For instance, sampling frequency, peak shapes, detection limits, precision and so on, should also be kept in mind throughout the optimization process. In the case of FI multicomponent methods, the selectivity for each analyte is the main goal, as this property determines the accuracy of the

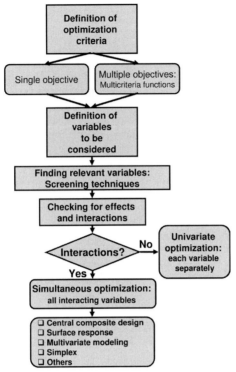

Figure 9.2 Flow-chart of optimization process in FI multicomponent determinations.

analysis. Hence, apart from the complementary objectives cited above, searching for selectivity is a priority in the optimization.

The second point for consideration is the definition of potential experimental variables that may influence the objective response. In the case of FI multicomponent methods, significant factors include the configuration of the manifold (i.e., scheme of channels and connections) as well as experimental variables such as flow rates, reactor dimensions, injection volume and chemical conditions.

9.2.1.1 Types of Objective Functions in Optimization

As pointed out, the optimization of FI multicomponent methods relies on the maximization of various complementary objectives. In this section, strategies for dealing simultaneously with these objectives and reaching a suitable compromise for all of them are discussed.

When two or more responses or objectives are considered to be relevant in a given process, the simplest optimization approach consists of an independent evaluation of each of these aspects. In many cases, the information is processed separately and, in a further stage, a suitable compromise satisfying the predefined objectives is sought. However, the way of reaching this compromise is often arbitrary and the process may even incorporate some prejudices. For instance, peak height and analysis time are current objectives in the optimization of FIA methods. However, these objectives are sometimes contradictory, as conditions leading to high peaks may correspond to poor sample throughput and vice versa. Hence, the final selection of the optimum experimental conditions combining maximum peak height with minimum analysis time is a decision analysts make on the basis of their own experience. This issue becomes much more complex when more objectives have to be taken intoaccount and more experimental factors have to be examined.

An excellent way to deal with multiple objectives while minimizing the arbitrariness of the previous approach is based on multi-criteria decision making. It consists of the definition of a response function that measures the overall suitability or quality of the experimental results. Hence, single objectives such as sensitivity, analysis time, and so on are combined in an objective response function that provides overall optimal conditions.

Multicriteria response functions have been implemented as mathematical expressions involving, for instance, the sum of the weighted contributions of each objective response according to a generic equation $\mu = \sum w_i \times r_i$, where μ is the overall objective response, r_i represents an individual response and w_i the weighting coefficient of the said response. In this case, certain arbitrariness still persists as the w_i values depend on the criterion of the analyst.

The construction of objective functions can also be based on multiplying the individual objective responses. This is the case of Derringer desirability functions which have been widely used for multi-criteria decision making. The overall result is a product of desirability values calculated from the experimental values. The individual desirabilities are then combined using the geometric mean, which gives the overall desirability D as:

$$D = (d_1 \times d_2 \times d_3 \ldots d_k)^{1/k}$$

where d_j is the desirability of each particular objective and k represents the number of objectives considered. The desirability values range between 1 and 0, with 1 indicating the best results and 0 the unacceptable ones. Note that if one of the individual objectives is completely unacceptable, the value of the Derringer function is zero, independently of the rest of the objectives. In contrast, the maximum of D corresponds to the optimum experimental conditions reached from the combined objectives.

9.2.1.2 Univariate versus Multivariate Optimization

The most typical optimization procedure consists of a univariate approach in which each given variable (e.g., pH, flow-rate, temperature, etc.) is studied separately within a desired range while the remaining experimental conditions are kept constant. Unfortunately, this strategy may be costly and time-consuming and, more importantly, univariate optimization is wrong when interactions appear. Indeed, in the case of interacting variables, conditions finally found in the univariate way may be far away from the optimum ones since the effect of any given variable depends on the magnitude of the other(s).

Chemometric methods for experimental design are highly efficient in ascertaining which variables are significant as well as in detecting possible interactions between them from a reduced number of experiments.

The number of variables involved in the optimization of an FI multicomponent manifold is expected to be high, so a preliminary screening is recommended. Screening methods like fractional or Plackett–Burman designs provide simple models with information about variables having a significant influence on the response. Only a few experiments are required to gain relevant information. Factors found to be irrelevant can be excluded from the optimization and, thus, the study can be simplified. The experimental runs are usually performed in random order to ensure that uncontrolled factors do not bias the results.

Screening designs are usually a prelude to further optimization experiments in which the main effects and interactions are evaluated more deeply. At this stage, full factorial designs are usually used for a more exhaustive study of the statistical significance of effects and interactions. The number of experiments to be performed is L^f, where L is the number of levels and f the number of variables to be evaluated. Investigating more than five factors/variables with the full factorial design can be tedious and time-consuming and alternative fractional factorial design might be preferred.

According to the scheme shown in Figure 9.2, the presence or absence of interactions conditions the strategy to be followed. Hence, those variables which are independent of the rest can be studied in a univariate way. In contrast, the simultaneous optimization of interacting variables requires the use of such methods as response surface, central composite and related designs. The construction of surface responses involving two variables may be based on a grid structure of experiments in which various levels are defined for each variable. The resulting data are then used to fit the surface response. When dealing with three or more variables, three-dimensional and higher order structures can be modeled. Obviously, representing more than two dimensions is difficult, so slices of pairs of variables can be plotted at different levels. Central composite designs are also widely used for the optimization of FIA conditions because they offer the possibility of evaluating the curvature of the data and fitting the experimental points to response surfaces with a minimum number of runs. Analogous cubic designs can be adopted for three variables. The multivariate optimization with the simplex algorithm seeks the optimum step-by-step on the basis of searching in the direction of maximum improvement of the response. Nowadays, however, simplex has fallen into disuse due to the problems with local maxima and the impossibility of planning the experiments in advance.

Various examples of the application of mathematical and statistical tools to the optimization step are given in Refs. [21, 22].

9.2.1.3 A Practical Case

A practical example of the application of experimental design and multi-criteria functions to the optimization of an FI method for the simultaneous determination of aniline and cyclohexylamine impurities in cyclamate products is now described.

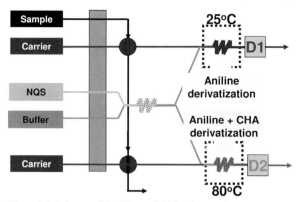

Figure 9.3 Scheme of the FI manifold for the simultaneous determination of aniline and cyclohexylamine in sweeteners. Carrier = water, NQS = 1,2-naphthoquinone-4-sulfonate reagent, Buffer = tetraborate solution, D1 = detector 1 for monitoring the selective reaction of aniline, D2 = detector 2 for monitoring the nonselective derivatization of aniline + cyclohexylamine. Flow rate, 0.5 ml min^{-1} each channel.

The method is based on the derivatization of amines with 1,2-naphthoquinone-4-sulfonate under selective and nonselective conditions using an FIA scheme, as depicted in Figure 9.3. Selectivity can be achieved by modifying experimental conditions such as pH or temperature. For instance, only aniline reacts at $20\,°C$, whilst higher temperatures ($80\,°C$) lead to a nonselective derivatization of the two analytes. Analogous discrimination of aniline can be accomplished via pH if the reaction is developed in acid media.

The example provides an overview of the most significant optimization issues according to the general guidelines shown in Figure 9.2, and can be extrapolated to a wide variety of multicomponent flow methods. The key point in this optimization is the achievement of a selective derivatization of aniline without any interference from CHA. However, apart from the thorough study of experimental conditions preserving the selectivity, at the same time the sensitivity (peak height) and sample throughput need to be maximized. According to these objectives, a tentative desirability function could be $D = (d_1 \times d_2 \times d_3)^{1/3}$, with d_1, d_2 and d_3 being the individual desirabilities associated with selectivity, peak height and sampling frequency. The next step consists in the definition of reasonable limits of the optimal ($d_i = 1$) and unacceptable ($d_i = 0$) results for these three objectives according to the user criteria. An example of graphic representation of d_1, d_2 and d_3 functions is shown in Figure 9.4.

The overall D function can be used for the optimization of all variables influencing the signal of aniline, A_{ANI}, including, for instance, injection volume, pH, reagent concentration, buffer concentration, flow-rate, reactor dimensions, temperature, and so on. Here, the simultaneous optimization of flow-rate and reactor dimensions is shown as an example of interacting variables. Indeed, these variables are strongly interrelated as they determine the reaction time between analyte and labeling agent.

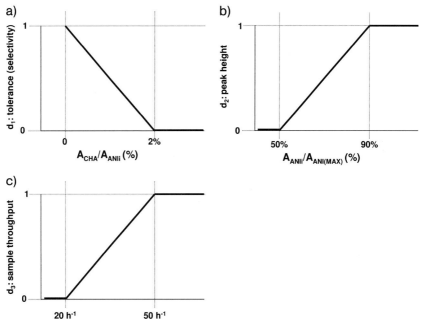

Figure 9.4 Desirability functions for (a) selectivity, (b) peak height and (c) sample throughput.

Table 9.1 shows the data generated according to a factorial design as well as desirability values calculated as in Figure 9.4. The results of the individual desirability surfaces as well as the overall objective function are given in Figure 9.5. The maximum of this function, marked with an arrow, corresponds to those experimental conditions with high RC1 lengths and high flow-rates. In the example, a total flow-rate of 1.4 mL min^{-1} and a RC1 length of 200 cm are finally selected as the optimum according to the multi-criteria approach defined on the basis of selectivity, sensitivity and sampling frequency.

9.2.2
Reagents

The huge arsenal of selective and sensitive reagents available commercially for labeling purposes can be used for FI multicomponent analysis. In this section, the main contributions of the reagents to multicomponent FI methods are discussed.

Classical labeling compounds that have been used for decades in other batch and flow methods as well as in chromatographic and electrophoretic derivatizations can be readapted to multicomponent analysis. Typically, these reagents are used in solutions which are continuously pumped through the FI manifolds or injected as discrete segments in SIA modes. In general, each analyte reaction is developed separately to facilitate its detection without interference, using manifolds such as those shown in Figures 9.6–9.8. The possibility of combining reagents in a universal labeling mixture for the simultaneous derivatization of all analytes is highly

Table 9.1 Simultaneous optimization of reactor coil length and flow rate as interacting variables according to grid experimental design and multi-criteria decision making with a Derringer desirability function.

Variables			Experimental responses		Desirabilities			
Reactor length (cm)	Flow rate (ml min⁻¹)	Selectivity A_{CHA}/A_{ANI} (%)	A_{ANI} (A.U.)	Sampling frequency (samples h⁻¹)	Selectivity d_1	A_{ANI} d_2	Sampling frequency d_3	Overall desirability
100	0.7	0.5	0.0744	33	0.61	0.44	0.75	0.59
	1.0	0.1	0.0664	39	0.41	0.62	0.95	0.63
	1.4	0.0	0.0592	54	0.23	1	1	0.62
	1.75	0.0	0.0631	55	0.33	1	1	0.69
200	0.7	0.7	0.0878	32	0.94	0.39	0.65	0.63
	1.0	0.4	0.0735	36	0.59	0.54	0.8	0.64
	1.4	0.2	0.0787	48	0.72	0.92	0.9	0.84
	1.75	0.0	0.0750	54	0.62	1	1	0.86
300	0.7	0.6	0.0926	29	1	0.31	0.7	0.61
	1.0	0.4	0.0969	36	1	0.54	0.8	0.76
	1.4	0.2	0.0851	43	0.88	0.77	0.9	0.85
	1.75	0.1	0.0810	46	0.78	0.87	0.95	0.86

Overall desirability $D = (d_1 \times d_2 \times d_3)^{1/3}$, with d_1, d_2 and d_3 being the individual desirabilities of selectivity, aniline peak height and sampling frequency, respectively, calculated as schematized in Figure 9.4. A_{ANI}, aniline peak height at 5×10^{-5} M; A_{CHA}, cyclohexylamine peak height 5×10^{-4} M.

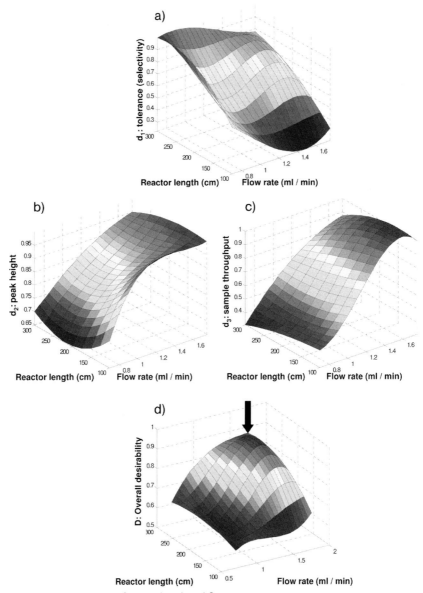

Figure 9.5 Optimization of reactor length and flow-rate conditions in the determination of aniline according to the FI scheme of Figure 9.3. (a) Selectivity desirability, (b) peak height desirability, (c) sample throughput desirability, (d) overall desirability.

attractive. Thus, the labeling reactions of all compounds occur together in a common reactor and the corresponding products can be further monitored. However, despite being fast and straightforward, mixtures of labeling agents have not been widely applied to the analysis of real samples. The loss of selectivity during the detection step

Step 1

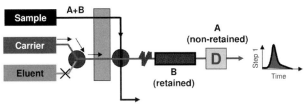

Step 2

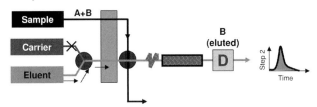

Figure 9.6 Flow manifold for a two-component (A + B) sequential determination based on retention/elution processes. Step 1: selective retention of B, A is detected; Step 2: elution of B, B is detected.

due to the lack of selective instrumental responses and, especially, the incompatibility between reagents or reaction conditions are among the main shortcomings of reagent mixtures.

Immobilization results in an attractive way of using reagents often applied to multicomponent determinations [23–25]. A wide range of reagents can be packed in micro-column reactors, linked on the surface of open tubular reactors or directly attached to detection devices [26]. The principal advantages of immobilized formats in the field of multideterminations arise in the reusable nature as well as in the possibility of being easily integrated in flow manifolds. Note that even linked reactions occurring in successive steps can be implemented with reactors in series. Typical cases of immobilized reagents in multicomponent flow methods include packed reactors of inorganic fillers (e.g., reducing agents), polymeric absorbent materials, ion-exchangers and bioreactors [27]. For example, the simultaneous determination of nitrite and nitrate in waters and other samples can be carried out in a flow manifold incorporating a reducing cadmium column to transform nitrate into nitrite [28, 29].

Biochemical reactions are increasingly used for multicomponent purposes as they may take advantage of the specificity of a given reagent towards the corresponding target analyte. Immobilized biomacromolecules such as proteins and nucleic acids open up new possibilities of developing methods based on enzymatic and other recognition processes [30, 31]. A representative example illustrates a fully automated FIA set-up with five parallel reactors for on-line monitoring of glucose, lactate, glutamine, ammonia and xantine in animal cell cultures [32]. Apart from enzymatic processes, immunoassay-based FI multicomponent determinations are emerging. For instance, a method for the determination of prolactin and glycosylated prolactin

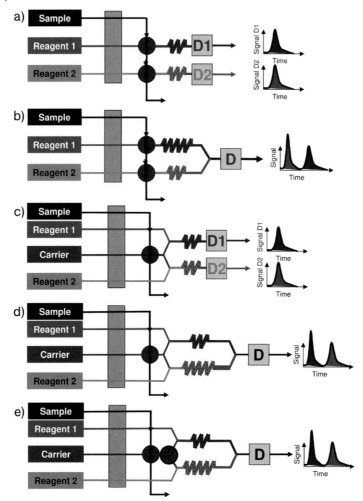

Figure 9.7 Flow manifolds for the simultaneous determination of analytes based on parallel configurations. (a) multi-injection system; (b) multi-injection + joining system; (c) splitting system; (d) splitting-delay-joining system; (e) splitting through selection valve.

in cell cultures using specific antibodies immobilized on flow-through cartridges has been described [33]. In another case, an amperometric immunosensing SIA method has been proposed for the simultaneous analysis of L-thyroxine, D-thyroxine and L-triiodothyronine [34].

Beyond the apparent success of highly selective reagents, the specificity of the (bio) recognition reaction has to be accompanied by the specificity of the instrumental response. We have to be aware of the limitations associated with the further detection stage in which interferences may arise. For instance, H_2O_2 specifically generated in a

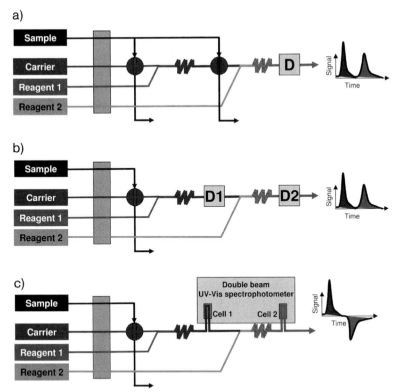

Figure 9.8 Flow manifolds for the simultaneous determination of analytes based on serial configurations. (a) multi-injection system; (b) multi-detection system; (c) double-beam spectrophotometric system.

given enzymatic reaction may undergo interference by other electroactive components present in the sample during the electrochemical detection. Analogous problems have been encountered in many other detection examples. In consequence, the selectivity has to be preserved throughout the whole method, from the reaction process to the detection step, by a suitable design of the flow manifold in combination with appropriate detection systems.

9.2.3
Manifolds for Multicomponent Determinations

The manifold in flow methods is the key element that enables the development of the processes and reactions in the desired way. Often, flow methods have been adapted from batch methods previously established through a suitable conversion of experimental batch conditions into FIA and SIA systems. The design of the manifold and its operation mode are fundamental aspects in multicomponent analysis. The possibilities in the construction of the manifold are almost unlimited and, in consequence, the analytical potential and versatility of flow methods are outstanding.

Possibilities for the simultaneous determination of compounds have been classified into representative manifold configurations which are characterized by sequential, parallel and serial schemes (see Figures 9.6–9.8). A wider description of the different cases is treated in what follows.

9.2.3.1 Sequential Configurations

The term "sequential determination" refers to the assay of components one by one in successive runs or steps. Hence, after processing one of the components, the system is modified to accommodate the proper experimental conditions for the quantification of another. In some cases, multicomponent sequential analysis relies on the construction of a general manifold that is adapted to the successive determination of various compounds. The selectivity of each analysis can be accomplished by changing reagents or experimental conditions such as pH or temperature [35]. Complementarily, analytes can be discriminated kinetically by using stopped-flow approaches. In this case, one stage may involve measurements under dynamic conditions while the other consists of a kinetic development of a given reaction. For instance, aniline and cyclohexylamine have been determined simultaneously using this strategy [35]. In more detail, a selective derivatization of aniline can be achieved in the continuous-flow mode while the kinetic development of the cyclohexylamine reaction occurs during the stopped-flow step. When full kinetic differentiation of components is difficult to achieve, chemometric treatments may be required [36–38].

Other interesting applications of sequential methods are briefly mentioned as follows. The analysis of thiamine and ascorbic acid in commercial vitamin complexes has been carried out using Hg(II) and quinoxaline as selective reagents [39]. The inhibitory effect of cysteine on the oxidation of thiamine by Hg(II) has been exploited for the simultaneous quantification of cysteine and cystine [40]. Another example describes the quantification of fluoride and monofluorophosphate in toothpaste by a sequential method in which, in a first run, free F^- is directly quantified; in a second run, monofluorophosphate is hydrolyzed to release F^- which is determined spectrophotometrically [41]. In another case, the sequential determination of Zn and Co mixtures in clinical, food and pharmaceutical samples has been based on the formation of ion aggregates with malachite green [42].

The incorporation of additional elements, such as packed reactors and selection valves, offers more possibilities of implementing sequential multideterminations. Several methods have been based on the differential retention of the compounds in a column reactor filled with appropriate materials such as octadecylsilica or ion exchangers through controlled absorption/elution steps (see scheme in Figure 9.6). In the case of two-component mixtures, typically, one of the components is retained in the system while the other reaches the flow-cell. Hence, the first step consists of the quantification of the nonretained species without interference from the other component. In a second step, the analyte retained in the column is eluted with a suitable solvent and then measured in the flow-cell [43]. This strategy can be extended to more complex mixtures if selective retention/elution conditions can be found for each analyte [44–46].

Similar systems based on the differential elution of analytes and further detection of the corresponding peaks can be adapted to perform simple chromatographic separations using home-made columns working at low pressure. A different issue is the coupling of flow techniques and commercial HPLC instruments in which the main task of flow systems consists of the sample processing while the chromatography is used for the separation of the components.

9.2.3.2 Parallel Configurations

Parallel configurations comprise a wide variety of schemes developed for FI multicomponent analysis. Double-injection and multi-port valves as well as multicommutation systems [47] can be used to perform simultaneous injection of two (or more) sample aliquots into independent parallel flow lines, as shown in Figure 9.7a. Each sample bolus flows through a given section of the assembly devoted to the development of a specific reaction. Note that additional channels for delivering reagents, reaction coils and other components and pieces may be needed to complete the manifold. Further, each line is equipped with suitable flow-cells and detectors. This kind of set-up has been used in multiple applications, such as the simultaneous analysis of fluoride and phosphate in toothpaste [48]. In the example, a double-injection valve has been combined with a double line manifold furnished with potentiometric and spectrophotometric detectors.

As shown in Figure 9.7b, when various analytes are detectable with a common technique, reunifying channels prior to the detection simplify the manifold. Flow-splitting pieces are used to divide and distribute the sample bolus in various lines. This option may provide an attractive alternative to multi-injection systems. Once the sample is distributed, each flowing portion is used for the determination of a given species. Appropriate reagents can be delivered to develop the corresponding reactions. Subsequently, each flow-line can be coupled to an appropriate detection system, as schematized in Figure 9.7c. The main disadvantage of this system is the need for various detectors to carry out the multicomponent analysis which is, in practice, complex and costly. Again, if reactions are monitored with the same technique, joining channels facilitates the detection as only one instrument is needed (see Figure 9.7d). This kind of manifold operates in combination with delay systems which prevent the simultaneous confluence of all sample segments in the detector. Hence, dimensions and residence times of each parallel line are adjusted to avoid overlap between the different sample segments inside the flow cell. As a result, successive peaks corresponding to each channel/analyte are obtained from a single sample injection.

Examples of applications of parallel assemblies based on splitting/delay coils/joining systems have been described for the binary determination of sulfur dioxide and ascorbic acid in beverages [49], glucose and lactate in neuronal monitoring [50], nitrate and nitrite in food samples [28, 29], Fe(II) and total iron in human hair [51], malic and lactic acids in juices [52] choline and acetylcholine in biological fluids [53], alkaline ions in blood serum [54] and Ca, Mg, K and Na in milk [55].

A weakness of splitting systems arises in the poor reproducibility of the sample division into the different parallel channels. This limitation may influence the precision and accuracy of the method. The robustness of sample division can be improved with the aid of a selection valve which alternately sends the sample bolus to one or another channel in a configuration as depicted in Figure 9.7e. This system has been used for the simultaneous determination of ammonium and nitrate [56].

9.2.3.3 Serial Configurations

The combination of various reactors and detectors in series can be used in multianalyte determinations. The basic schemes shown in Figure 9.8 can be complemented by the inclusion of auxiliary channels and elements for developing the desired reactions and processes. Also, when the analytes are detected with the same instrument, the assembly of various injection valves in different locations of a common flow line may avoid the need for multiple detectors. Examples of this type of manifold have been used in the simultaneous determination of phosphate and total phosphate in nucleoside hydrolysates [57], creatinine and creatine in serum [58], hypoxanthine, inosine and inosine 5'-monophosphate in fish [59], Ca and Mg in hemodialysis preparations [60], salicylic and acetylsalicylic acids in pharmaceuticals [61].

A particular case of serial systems for two-component analysis is based on the use of the double-beam spectrophotometer. This instrument can be adapted to work analogously to two independent photometric devices. Two flow cells are placed in the sample and reference holders (see Figure 9.8c) to monitor the corresponding analytical reactions. Such a configuration has been proposed for the determination of ascorbic acid and cysteine in pharmaceutical formulations [62]. Typical registers consist of a double peak (positive + negative) produced in the passage of the sample through each flow cell. Aniline and cyclohexylamine have also been determined spectrophotometrically using a similar approach [21, 22]. The first part of the set-up performs selective derivatization of aniline with subsequent on-line detection. The outlet of the first flow-cell is then coupled to the second part of the manifold to monitor aniline and cyclohexylamine simultaneously.

9.2.4
Detectors

Part II of this book comprises several chapters describing state-of-the-art detection, with a discussion of advances and further trends. Recently, there has been an irruption of massive techniques with high analytical performance (e.g., mass spectrometry, inductively-coupled plasma atomic emission spectroscopy, etc.) for enhancing the quality of the characterization and quantification of analytes. However, due to the simplicity, availability and low cost of UV–vis spectrophotometric and electroanalytical instruments, these traditional techniques continue to dominate the panorama of the detection in flow analysis. Indeed, one of the most notable features of FIA and related flow modes is that they facilitate simple and cheap methods. A chart summarizing the use of the principal detection techniques in this

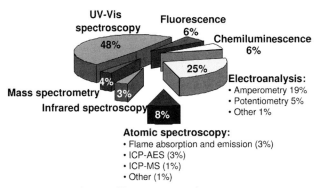

Figure 9.9 Distribution of flow injection multicomponent determinations according to the detection technique.

type of application is shown in Figure 9.9. Additionally, the location of the detector in the manifold is not trivial and multi-site set-ups as well as relocation of the detector may result in significant advantages over traditional systems [63]. In this section, a brief reference to the principal aspects of detection focusing on multicomponent analysis is given.

In the case of UV–vis spectrophotometry, colorimeters and conventional spectrophotometers operate at a prefixed wavelength to monitor the desired reaction. This means that they are used as single detection devices often integrated in multichannel or serial manifolds giving typical FIA-grams. Fast scanning spectrophotometers such as diode array and charge-coupled devices are especially suitable for multicomponent purposes as they may provide additional information through the spectral domain. This option has been widely exploited in combination with chemometric techniques for data analysis. Novel trends in UV–vis detection include the development of optosensors displaying high specificity towards the corresponding analytes. The sensitive component can be attached on optical fibers or incorporated in the sensing zone of a conventional flow cell [43, 44].

Considerations regarding fluorimetric detection are similar to those concerning molecular absorption spectroscopy. Conventional instruments allow one to register the FI peaks at fixed excitation and emission wavelengths. In contrast, fast-scanning fluorimeters may provide spectral data over time, which is very valuable as an additional source of information for multicomponent determinations. Either excitation or emission spectra can be recorded throughout the entire FI peak although, in general, emission data seem to be preferred. The apparent advantages of multi-way fluorimeters contrast with their poorer detection limits in comparison with the conventional counterparts. Hence, mono-channel monitoring modes combined with traditional highly sensitive fluorogenic reagents are currently applied to fluorimetric FI multicomponent analysis [39, 40]. Methods relying on fluorimetric monitoring of enzymatic cofactors such as NADH have also been proposed [32, 52].

As shown in Figure 9.9, the contribution of chemiluminescence to FI multi-determinations is also significant. Most of the applications are based on the reaction of luminol with H_2O_2 produced from enzymatic degradation of the substrate [64]. The luminescence emitted is measured over time, giving a typical FI peak of intensity versus time. For instance, chemiluminometric methods have been developed for the analysis of glutamate, lysine and urate [65] and choline and acetylcholine [66].

Electrochemical detection in FIA is mainly exploited through amperometric, potentiometric and voltammetric techniques [67, 68]. Apart from the wide variety of commercial electrodes, home-made devices can be easily prepared in analytical laboratories. In the amperometric mode, current intensities from the oxidation or reduction of electroactive compounds at a selected working potential are monitored versus time, providing typical FI intensity peaks. Most of the applications involve enzymatic reactions, after which the H_2O_2 is measured in various ways. Recent developments for improving the performance of the electrochemical detection include multi(bio)sensor arrays for obtaining multiparametric measurement and the integration of various enzymatic systems to develop linked reactions [69, 70]. For instance, a triple electrode has been constructed for the simultaneous *in vivo* monitoring of glucose, lactate and pyruvate in rat brain [71]. A microsensor array has also been proposed for the assay of glucose, lactate, glutamate and hypoxanthine in serum [72]. Potentiometric detection with ion-selective electrodes offers great possibilities for the determination of inorganic cation and anion species such as NH_4^+, F^-, Ca^{2+} at moderate concentrations [48, 73]. In addition, NH_3 or NH_4^+ ISEs can be readapted to prepare potentiometric biosensors of interest in biological processes and food analysis [74]. Finally, voltammetric detection yields richer data, as full voltammograms can be recorded over time to provide three-dimensional peaks consisting of current intensity values as a function of potential and time.

Flame atomic absorption and emission techniques have mainly been applied to the FI multidetermination of metals in water and environmental samples. Metal-containing organic compounds of significance in pharmaceutical, biological and clinical samples can also be quantified [75]. The main functions of flow systems in these methods are: sample introduction, addition of auxiliary reagents, and making the flow rate compatible with the flame. Additional tasks such as analyte pre-concentration and matrix removal, including solid-phase, solvent and membrane extraction, hydride generation and so on, can be implemented. Then, the emerging flow stream is directly coupled to nebulizer-flame devices of both absorption and emission techniques. Note that, in this case, one instrument or one run is needed per component to be analyzed. Other instrumental techniques including inductively coupled plasma-atomic emission spectrometry (ICP-AES) and inductively coupled plasma-mass spectrometry (ICP-MS) [76–78] have recently been introduced for multidetection in FIA. The excellent analytical performance of these techniques has permitted, for instance, the simultaneous determination of 18 elements in water samples [79]. However, the costs of the instruments themselves as well as the costs per analysis are sometimes prohibitive, so simpler and cheaper alternatives are usually preferred.

Other detectors that can be used in FIA are mass spectrometers (MS), and infrared (IR) and nuclear magnetic resonance (NMR) instruments. In many cases, the interfaces enabling these hyphenations have been adapted from those developed previously for HPLC. Coupling FIA and MS is facilitated with commercial electrospray and atmospheric pressure chemical ionization sources [80]. In the 1990s, FIA-MS methods were first proposed for a quantification of trace components in environmental analysis [81, 82]. Since then, other applications to pharmaceutical, food and clinical fields have been reported. For example, proanthocyanidins in grape products have been determined by FIA-MS [83]. Monitoring of the biological treatment of waste water by FIA-MS/MS has also been proposed [84]. The coupling of flow systems to IR and NMR results in a much more complex issue and off-line procedures based on fraction collectors and robotic devices may be required [85–87].

9.2.5
Chemometric Techniques for Data Analysis

Flow-injection and related flow techniques are especially effective for generating huge amounts of data. A typical FIA system equipped with a fast scanning spectrometer is able to record full spectra at regular steps over the entire FIA-gram. Hence, the three-dimensional data obtained result in a rich source of information to be interpreted or analyzed with chemometric techniques. To reinforce this aspect, the introduction of computers has been fundamental in controlling the elements of the set-up as well as facilitating data acquisition, storage and analysis. As a result, the analytical potential of flow systems is greatly enhanced in combination with mathematical tools for extracting information [18, 88].

This section reviews the application of chemometrics to FI multicomponent determinations, with special emphasis on the types of data that can be obtained and the corresponding calibration methods.

9.2.5.1 Data Sets from Flow Systems
Flow data can be classified according to their dimensions or domains of measurement [89]. As shown in Figure 9.10, the complexity of the data increases from zero-way to multi-way systems, but the information contained and the analytical possibilities also increase.

Zero-Way Data Typically, flow injection and related methods use peak heights or peak areas, that is, scalar values, as analytical data for quantitative purposes. Despite the simplicity of the so-called zero-way data, the information contained is quite limited. Apart from strict requirements of selectivity, studies for checking the presence of outlier samples or detecting interferences are not compatible. The simultaneous determination of various components from zero-way data is also not possible.

One-Way Data One-way or first-order data consist of a vector array such as a peak profile at a given wavelength or a spectrum. In contrast to zero-way data, they constitute a rich source of variance that can be exploited for more efficient

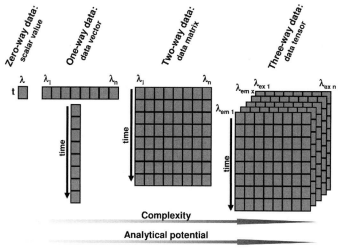

Figure 9.10 Types of data in FI and related flow modes.

characterization of samples and quantification of analytes. Some advantages derived from such data affect multicomponent determination in the presence of unknown interferences. Additionally, one-way data are compatible with sample classification and outlier detection.

Two-Way Data Two-way data (also called second-order data) combine instrumental signals registered over both spectral and time domains. The spectroscopic response is thus arranged in a data matrix or table of values in which each column corresponds to a wavelength and each row corresponds to a time point. Two-way data offer excellent possibilities in multicomponent determinations in the presence of interferences as well as in studies of peak purity and mixture resolution.

Three-Way Data Progressing in the complexity of the structure of data, three-way data, sometimes called data tensors, involve three domains of measurement. This is the case, for instance, for excitation-emission fluorescence spectra or MS-MS registered over time. Applications based on such types of data are still under research.

Data Augmentation Apart from the typical instrumental responses commented on above, the combination of data sets from different sources contributes to the generation of highly structured data arrangements. This combination, usually referred to as augmentation, results in a valuable way to enrich the information content as well as enhancing the analytical possibilities [90]. As schematized in Figure 9.11, data augmentation involving n-way data sets leads to $(n + 1)$-way arrangements. For instance, joining data vectors results in data matrices, joining data matrices provides two-way data sets and so on.

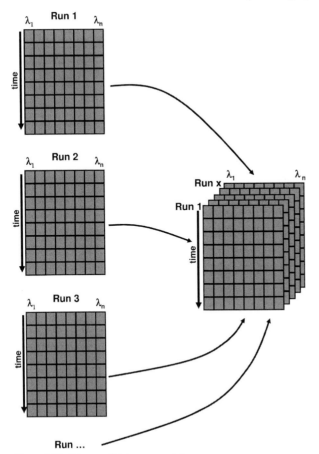

Figure 9.11 Scheme of data augmentation.

9.2.5.2 Calibration Methods Applied to Flow Data

The determination of the concentration of one or several analytes by flow methods is based on the establishment of appropriate calibration models. Hence, in parallel to the different types of data described above, mathematical procedures have been developed to extract information and quantify the components of interest. Obviously, the complexity of calibration methods increases with the order or dimension of the data.

The use of multivariate calibration methods opens up excellent opportunities in the field of multicomponent FI determinations. However, there is no reason to use complex methods if simpler options are valid. Hence, the use of univariate calibration will always be preferred to the multivariate counterparts if the circumstances (in particular the selectivity) allow this possibility.

A summary of the principal features of the calibration methods is given in Table 9.2. In the following sections, a brief description of calibration methods, with the most relevant pros and cons, is also provided.

Table 9.2 Principal characteristics of calibration methods.

	Zero-order calibration methods	First-order calibration methods	Second-order calibration methods
Features	use of scalar data limited information	use of data vectors information from one domain of measurement (spectral or time domain)	use of data matrices information from two domains of measurement (spectral and time domains)
	few standards are needed	large sets of standards needed	few standards needed
	use of pure standards	use of standards of similar nature to the test samples	use of synthetic standards
	high simplicity of the algorithms	moderately complex algorithms	complex algorithms
	full selectivity is required	full selectivity not required	full selectivity not required
	simultaneous determination of analytes not possible	simultaneous determination of analytes is possible	simultaneous determination of analytes is possible
Examples	linear regression	principal component regression (PCR)	parallel factor analysis (PARAFAC)
	nonlinear regression	partial least square regression (PLS)	direct trilinear decomposition
	standard addition method	artificial neural networks (ANN)	Tucker models
			multivariate curve resolution based on alternating least squares (MCR-ALS)

Univariate Calibration (Zero-Order Calibration)   Univariate or zero-order calibration in FIA and related flow modes is analogous to any other analytical technique. Typically, the calibration model is established by linear regression using peak heights as analytical data. In certain cases, nonlinear calibration or standard addition methods may be required. As mentioned above, the key point of univariate calibration is the selectivity for the analyte(s) of interest so that interferences have to be removed or masked. In FI multicomponent determinations the desired selectivity can be gained using a wide variety of strategies, as illustrated in this chapter.

In particular cases, interferences can be avoided by a simple transformation of the instrumental response. For instance, derivative spectra in combination with the zero-crossing method have occasionally been used to achieve selectivity. For a given mixture

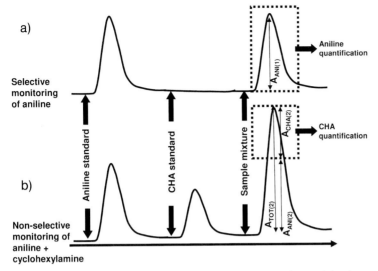

Figure 9.12 Differential analysis for the determination of aniline and cyclohexylamine using the flow manifold of Figure 9.3. FIA-grams obtained in the injection of aniline and cyclohexylamine standard and a sample mixture under selective (a) and nonselective conditions (b). $A_{ANI(1)}$ = aniline peak height in system (a); $A_{TOT(2)}$ = total peak height in system (b); $A_{ANI(2)}$ and $A_{CHA(2)}$ = aniline and cyclohexylamine contributions in system (b). $A_{CHA(2)}$ is calculated as $A_{TOT(2)} - A_{ANI(2)}$, where $A_{ANI(2)} = A_{ANI(1)} \times$ Sensitivity$_{ANI(2)}$/Sensitivity$_{ANI(1)}$.

of A + B, the derivative spectrum of component A is selective at the wavelength at which the derivative spectrum of component B is zero, that is, the so-called zero-crossing point. However, the performance of this approach is practically limited to two- and three-component mixtures in simple pharmaceutical products [91].

Differential analysis is another strategy for enhancing the selectivity based on the isolation of the contributions of the components of the system in a nonselective gross response through mathematical signal subtraction. Differential analyses can be implemented either through the spectral domain by working at appropriate wavelengths or using FIA peaks. The procedure is illustrated with an example of the quantification of aniline and cyclohexylamine in mixture solutions according to the flow manifold depicted in Figure 9.3 [21, 22]. As shown in Figure 9.12, the concentration of aniline in the sample can be directly quantified from the FIA-gram (a) as its response is selective under appropriate pH or temperature conditions. In contrast, the derivatization of cyclohexylamine cannot be achieved independently from aniline. Thus, the cyclohexylamine concentration has to be determined from its associated contribution as in FIA-gram (b) after subtracting the signal corresponding to aniline.

Multivariate Calibration Multivariate calibration comprises both first-order and second-order methods. For decades, first-order methods have been applied to multicomponents to overcome the lack of selectivity in a wide variety of situations in which univariate calibration fails. More recently, the proliferation of instruments providing second-order data has contributed to the development of a new type of calibration methods. In these cases, instead of a physical or chemical separation of analytes, chemometric methods contribute to a mathematical discrimination relying on the features of spectral and peak profiles.

Unfortunately, the performance of chemometric methods is limited, since it depends on the possibility of discriminating between the experimental signal of a given analyte and the rest of the contributions. In practice, this means the contributions of all components should be reasonably different. In order to evaluate numerically the degree of differentiation of profiles, correlation coefficients between pairs of spectra or FI peaks can be calculated. Correlation values close to 1 correspond to highly similar components with poor possibility of successful resolution. In contrast, low correlations are consistent with dissimilar profiles and, thus, the corresponding compounds are expected to be distinguishable mathematically.

Differences in the spectra of components depend on the characteristics and functional groups of the molecules. In contrast, differences in the peak profiles are experimentally accomplished with a gradient of reagent that provokes chemical changes along the peak [92]. This gradient contributes to discrimination of the analytes as a function of their chemical behavior. The most typical example consists of a pH gradient obtained from the mixing and controlled dispersion of acidic and basic solutions using flow schemes such as those shown in Figure 9.13. As a result, the acid–base features of the analytes or the variations of reactivity as a function of pH can be exploited to create differences in the shapes of the peak profiles [93].

Since the 1980s, principal component regression (PCR) and partial least squares regression (PLS) are the first-order methods most extensively used in FI multicomponent analysis. PCR and PLS have essentially been developed for linear modeling of relationships between responses and concentrations [94, 95]. In the calibration step, the algorithms decompose the experimental matrix of responses into relevant factors related to sample and variable features. The resulting scores and loading matrices contain information that can be interpreted to extract sample patterns and correlations. Subsequently, in the prediction step, the model is applied to quantify the desired analytes in unknown test samples. Apart from PCR and PLS, nonlinear algorithms have sometimes been used, especially for dealing with certain electroanalytical responses displaying severe nonlinearities. Among the nonlinear algorithms, artificial neural networks (ANNs) have been applied to empirical data modeling [96]. The almost unlimited capacity of modeling all types of data has been greatly appreciated by scientists. However, the actual performance of ANNs may have been overestimated as they operate as black boxes and overfitting may appear. For these reasons, a careful validation of ANN models is always required to ensure their reliability.

Another fundamental issue of first-order multivariate calibration is the design and preparation of the set of standards. As a common requirement, unknown samples and standards must have the same characteristics and matrix composition. This is because all sources of variability present in the samples, including analyte concentration range and sample matrix complexity, should be covered by the model. In practice, standards are indeed other samples analyzed beforehand by an independent method. For this reason, the calibration is often a costly experimental step. Moreover,

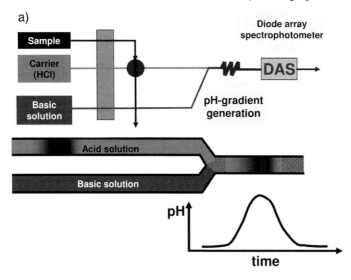

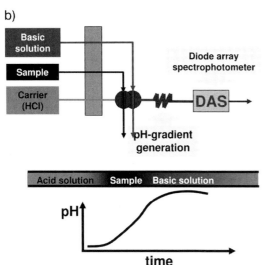

Figure 9.13 Flow manifolds for the generation of a (reagent) pH-gradient. (a) Double pH-gradient (acid–base–acid), (b) monotonic pH-gradient (acid to base).

in comparison with univariate calibration, the number of standards required for building reliable models is relatively high (from 20 to 500 standards are usually used).

Representative examples involving first-order calibration in multicomponent flow analysis include the determination of vitamins in pharmaceutical preparations [97], drug mixtures (caffeine + dimenhydrinate + acetaminophen and paracetamol caffeine + acetylsalicylic acid) [98, 99], penicillin and ampicillin [100], cysteine and methionine [101], metal mixtures in blood serum and dialysis liquids [102, 103].

Several chemometric methods such as parallel factor analysis (PARAFAC) [104] and multivariate curve resolution based on alternating least squares (MCR-ALS) [105] are addressed in quantifying the analytes through a second-order calibration procedure. These algorithms are more complex than those required for the analysis of scalars and data vectors. The mathematical complexity is compensated in terms of superior analytical possibilities as they take advantage of differences in both spectral and time domains of measurement.

As summarized in Table 9.2, in general, second-order calibration methods use only a few calibration standards to build the model. Furthermore, unless there are chemical matrix effects, pure standard solutions are often used. When dealing with strong matrix effects, a generalization of the standard addition method to second-order calibration can be applied [106].

The application of second-order calibration to FI multicomponent analysis has been described for the quantification of mixtures of nucleic acid components [107], quinonic compounds [108, 109], amoxicillins [110], benzoic and sorbic acids [111] and antiretroviral drugs [90]. Other references deal with the determination of chromium species in tanning samples using SIA and MCR-ALS [112–114]. For a wider review, see Refs. [115, 116].

9.2.5.3 A Practical Example of Second-Order Calibration

This section illustrates a practical example of application of second-order calibration with MCR-ALS to the determination of drug mixtures of therapeutic interest. In particular, the characterization and quantification of two antiretroviral drugs (zidovudine and didanosine) in plasma is examined. The main goal of the study is the establishment of a feasible analytical method for monitoring plasmatic levels of drugs in a rapid and simple way as an alternative to chromatographic procedures. The method is based on a pH gradient FI system equipped with a diode array spectrophotometer which takes advantage of the spectral features of the zidovudine and didanosine as well as their acid–base characteristics.

In more detail, the set-up consists of a two-channel manifold in which 250 μL of sample are injected into a HCl carrier. The carrier channel joins the basic channel and solutions are mixed in a 200 cm × 0.5 mm i.d. PTFE mixing coil (RC) coupled on-line to the detection flow-cell. The corresponding acid–base reactions between carrier and basic components as well as the sample dispersion are responsible for the generation of the pH-gradient. The set-up is similar to that schematized in Figure 9.13a. Accordingly, the pH gradient created shows a maximum acidity in the boundaries of the sample bolus. Then, the pH decreases progressively until reaching a basic medium in the center of the peak. The spectroscopic data generated in each run are arranged in a data matrix in which each row represents a time along the FI peak and each column a wavelength. Hence, the elements of the matrix consist of absorbance values as a function of pH (through the pH domain) and wavelength (spectral domain).

The experimental data (individual or augmented matrices) are further analyzed with MCR-ALS to recover the spectral and peak profiles of species, as shown in Figure 9.14. The algorithm is implemented as a MATLAB program available at

a)

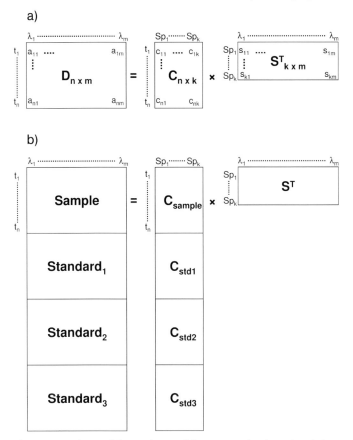

b)

Figure 9.14 Scheme of the resolution of the spectroscopic data matrices into the peak profiles over time C and the spectra S^T of species. (a) Resolution of an individual data matrix D. (b) Resolution of an augmented data matrix [$Sample$; $Standard_1$; $Standard_2$; $Standard_3$, ...]. Sp_i = chemical species; a_{ij} = absorbance value at time i and wavelength j; c_{ik} = arbitrary concentration of species k at time i; s_{kj} = absorptivity of species k at wavelength j. C_{sample}, C_{std1}, C_{std2}, C_{std3}, ... contain the peak profiles of chemical species recovered in the sample, standard 1, standard 2, standard 3, ..., respectively.

www.ub.es/gesq/mcr/mcr.htm. The principal steps of MCR-ALS are schematized in Figure 9.15 (see Refs. [90, 105] for a more exhaustive description).

When dealing with complex samples, the sensitivity of the calibration curve may be affected by matrix effects. In these circumstances, the application of the standard addition method is recommended. As shown in the example of Figure 9.16, in this approach the test sample is analyzed simultaneously with the data sets of standard additions of each analyte as a source of information (see Ref. [117] for details). The concentration profiles of both acid zidovudine and didanosine species show characteristic double peaks, consistent with the existence of two acid zones in the peak front and tail. The profiles of basic zidovudine and didanosine species display single peaks

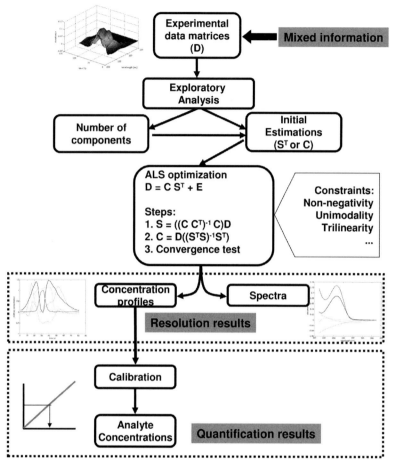

Figure 9.15 Flow chart of the multivariate curve resolution
method based on alternating least squares.

in accordance with the extension of the basic range in the center of the FIA peak. After
resolution, the concentration of analyte in the sample is calculated by comparing the
peak areas with those obtained for appropriate standards, as schematized in
Figure 9.16. Results obtained in the analysis of these drugs in plasma demonstrate
the performance of the method, giving overall prediction errors of about 5%.

9.3
Trends and Perspectives

The evolution of multicomponent FI analysis is closely connected with general
advances in the flow methodology. The potential of flow systems as a way of
facilitating sample pretreatment is in marked expansion for achieving selective

a)

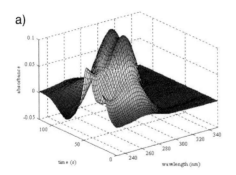

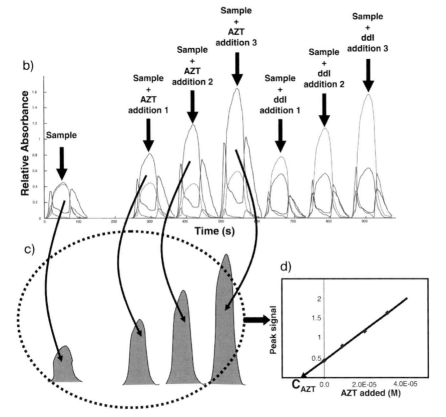

Figure 9.16 Quantification of zidovudine and didanosine in plasma samples using the extension of the standard addition method to second-order calibration. (a) Example of the three-dimensional FI peak obtained from the injection of a plasma sample into the pH-gradient FI manifold. (b) Concentration profiles recovered from the resolution of the augmented data matrix [S; SA_{zid1}; SA_{zid2}; SA_{zid3}; SA_{did1}; SA_{did2}; SA_{did3}] by MCR-ALS. (c) Example of the extraction of the peak areas of the basic species of zidovudine as analytical information. (d) Calculation of the concentration of zidovudine in the test sample according to the standard addition method.

conditions. Some attractive ideas such as relocating the detection systems and lab-in-a-valve for carrying out the reactions and sample treatments may enhance the possibilities of multicomponent methods. Also, the use of reagents such as biomacromolecules fixed on solid supports is gaining popularity in multianalyte determinations as a way of saving chemicals and decreasing dispersion.

Regarding detection, the wide variety of electrochemical and optical (bio)sensors available is leading to a proliferation of commercial or home-made multisensing devices implemented in flow systems. In this field, the use of sensor arrays and electronic tongues may facilitate enormously the development of new multicomponent applications, especially for characterizing complex properties. A different concept detection focused on multideterminations relies on the hyphenation of massive instruments to the flow systems. Although this possibility has scarcely been explored, the performance of coupled systems seems to be excellent. As a result, the number of FI methods involving MS, IR and NMR hyphenations is expected to increase in the near future.

Multivariate calibration is one of the most solid and reasonable options for performing FI multicomponent determinations. In particular, PLS offers great possibilities for simplifying the preliminary steps of the methods and minimizing the sample treatment. The current incidence of second-order calibration in multicomponent analysis is still a matter under intensive research. Efforts are needed to bring second-order methods closer to their potential users, as the higher mathematical complexity may discourage their use. In any case, the analytical performance of these methods is exceptional so that, in the next few years, expansion and generalization will certainly be forthcoming.

References

1 Ruzicka, J. and Hansen, E.H. (1988) *Flow Injection Analysis*, 2nd edn, John Wiley and Sons, New York.

2 Trojanowicz, M. (1999) *Flow Injection Analysis: Instrumentation and Applications*, World Scientific, River Edge, New Jersey.

3 Hansen, E.H. and Miró, M. (2007) How flow injection analysis (FIA) over the past 25 years has changed our way of performing chemical analyses. *Trends in Analytical Chemistry*, **26**, 18–26.

4 Wang, J. and Hansen, E.H. (2003) Sequential injection lab-on-valve: the third generation of flow injection analysis. *Trends in Analytical Chemistry*, **22**, 225–231.

5 Hartwell, S.K., Christian, G.D. and Grudpan, K. (2004) Bead injection with a simple flow injection system: an economical alternative for trace analysis. *Trends in Analytical Chemistry*, **23**, 619–623.

6 Hanrahan, G., Dahdouh, F., Clarke, K. and Gomez, F.A. (2005) Flow injection-capillary electrophoresis (FI-CE): Recent advances and applications. *Current Anal. Chem.*, **1**, 321–328.

7 Saurina, J. (2002) Hyphenation in capillary electrophoresis: from sample pretreatment to data analysis. *LC GC Europe*, **15**, 734.

8 Hlabangana, L., Hernandez-Cassou, S. and Saurina, J. (2006) Multicomponent determination of drugs using flow injection analysis. *Current Pharmaceutical Analysis*, **2**, 127–140.

9 van Staden, J.F. and Stefan, R.I. (2004) Chemical speciation by sequential injection analysis: an overview. *Talanta*, **64**, 1109–1113.

10 Economou, A. (2005) Sequential-injection analysis (SIA): a useful tool for on-line sample-handling and pre-treatment. *Trends in Analytical Chemistry*, **24**, 416–425.

11 Pyrzynska, K. and Trojanowicz, M. (1999) Functionalized cellulose sorbents for preconcentration of trace metals in environmental analysis. *Critical Reviews in Analytical Chemistry*, **29**, 313–321.

12 Stewart, J.W.B. and Ruzicka, J. (1976) Flow injection analysis. 5. Simultaneous determination of nitrogen and phosphorus in acid digests of plant material with a single spectrophotometer. *Analytica Chimica Acta*, **82**, 137–144.

13 Betteridge, D. and Fields, B. (1983) 2 point kinetic simultaneous determination of cobalt(II) and nickel(II) in aqueous-solution using flow injection analysis (FIA). *Fresenius' Zeitschrift für Analytische Chemie*, **314**, 386–390.

14 Gutierrez, M.C., Gómez Hens, A. and Pérez Bendito, D. (1987) Individual and simultaneous stopped-flow fluorometric-determination of perphenazine and chlorpromazine. *Analytical Letters*, **20**, 1847–1865.

15 Saitoh, K., Hasebe, T., Teshima, N., Kurihara, M. and Kawashima, T. (1998) Simultaneous flow injection determination of iron(II) and total iron by micelle enhanced luminol chemiluminescence. *Analytica Chimica Acta*, **376**, 247–254.

16 Matsumoto, K., Asada, W. and Murai, R. (1998) Simultaneous biosensing of inosine monophosphate and glutamate by use of immobilized enzyme reactors. *Analytica Chimica Acta*, **358**, 127–136.

17 Kuban, V. (1992) Simultaneous determination of several components by flow injection analysis. *Critical Reviews in Analytical Chemistry*, **23**, 15–53.

18 Saurina, J. and Hernández-Cassou, S. (2001) Quantitative determinations in conventional flow injection analysis based on different chemometric calibration statregies: a review. *Analytica Chimica Acta*, **438**, 335–352.

19 Massart, D.L., Vandeginste, B.G.M., Buydens, L.M.C., de Jong, S., Lewi, P.J. and Smeyers-Verbeke, J. (1997) *Handbook of Chemometrics and Qualimetrics Part A*, Elsevier, Amsterdam.

20 Sentellas, S. and Saurina, J. (2003) Chemometrics in capillary electrophoresis. Part A: Methods for optimisation. *Journal of Separation Science*, **26**, 875–885.

21 Hlabangana, L., Saurina, J. and Hernández-Cassou, S. (2005) Flow-injection differential spectrophotometric pH selectivity system for the determination of cyclamate contaminants. *Microchimica Acta*, **150**, 115–123.

22 Saurina, J., Hlabangana, L., García-Milla, D. and Hernández-Cassou, S. (2004) Flow-injection determination of amine contaminants in cyclamate samples based on temperature for controlling selectivity. *Analyst*, **129**, 468–474.

23 Miró, M. and Hansen, E.H. (2006) Solid reactors in sequential injection analysis: recent trends in the environmental field. *Trends in Analytical Chemistry*, **25**, 267–281.

24 Kandimalla, V.B., Tripathi, V.S. and Ju, H.X. (2006) Immobilization of biomolecules in sol-gels: biological and analytical applications. *Critical Reviews in Analytical Chemistry*, **36**, 73–106.

25 Miró, M. and Frenzel, W. (2004) Flow-through sorptive preconcentration with direct optosensing at solid surfaces for trace-ion analysis. *Trends in Analytical Chemistry*, **23**, 11–20.

26 Prieto-Simon, B., Campas, M., Andreescu, A. and Marti, J.L. (2006) Trends in flow-based biosensing systems for pesticide assessment. *Sensors*, **6**, 1161–1186.

27 Luque de Castro, M.D. (1992) Solid-phase reactors in flow injection analysis. *Trends in Analytical Chemistry*, **11**, 149–155.

28 Ensafi, A.A. and Kazemzadeh, A. (1999) Simultaneous determination of nitrite and nitrate in various samples using flow injection with spectrophotometric detection. *Analytica Chimica Acta*, **382**, 15–21.

29 Monser, L., Sadok, S., Greenway, G.M., Shah, I. and Uglow, R.F. (2002) A simple simultaneous flow injection method based on phosphomolybdenum chemistry for nitrate and nitrite determinations in water and fish samples. *Talanta*, **57**, 511–518.

30 Pividori, M.I. and Alegret, S. (2005) Electrochemical genosensing based on rigid carbon composites. A review. *Analytical Letters*, **38**, 2541–2565.

31 Spohn, U., Preuschoff, F., Blankenstein, G., Janasek, D., Kula, M.R. and Hacker, A. (1995) Chemiluminometric enzyme sensors for flow injection analysis. *Analytica Chimica Acta*, **303**, 109–120.

32 Fu, Z., Liu, H. and Ju, H.X. (2006) Flow-through multianalyte chemiluminescent immunosensing system with designed substrate zone-resolved technique for sequential detection of tumor markers. *Analytical Chemistry*, **78**, 6999–7005.

33 Reinecke, M. and Stephanopoulos, G. (2000) Flow injection analysis for simultaneous quantification of prolactin concentration and glycosylation macroheterogeneity in cell culture samples. *Cytotechnology*, **23**, 237–242.

34 Stefan, R.I., van Staden, J.F. and Aboul-Enein, H.Y. (2004) Simultaneous determination of L-thyroxine (L-T-4), D-thyroxine (D-T-4), and L-triiodothyronine (L-T-3) using a sensors/sequential injection analysis system. *Talanta*, **64**, 151–155.

35 Saurina, J. and Hernández-Cassou, S. (1999) Flow-injection and stopped-flow completely continuous flow spectrophotometric determinations of

aniline and cyclohexylamine. *Analytica Chimica Acta*, **396**, 151–159.

36 Fortes, P.R., Meneses, S.R.P. and Zagatto, E.A.G. (2006) A novel flow-based strategy for implementing differential kinetic analysis. *Analytica Chimica Acta*, **572**, 316–320.

37 Amigo, J.M., Coello, J. and Maspoch, S. (2005) Three-way partial least-squares regression for the simultaneous kinetic-enzymatic determination of xanthine and hypoxanthine in human urine. *Analytical and Bioanalytical Chemistry*, **382**, 1380–1388.

38 Magni, D.M., Olivieri, A.C. and Bonivardi, A.L. (2005) Artificial neural networks study of the catalytic reduction of resazur: stopped-flow injection kinetic-spectrophotometric determination of Cu(II) and Ni(II). *Analytica Chimica Acta*, **528**, 275–284.

39 Pérez-Ruiz, T., Martínez-Lozano, C., Sanz, A. and Guillén, A. (2004) Successive determination of thiamine and ascorbic acid in pharmaceuticals by flow injection analysis. *Journal of Pharmaceutical and Biomedical Analysis*, **34**, 551–557.

40 Pérez- Ruiz, T., Martínez-Lozano, C., Tomás, V. and Lambertos, G. (1991) Flow-injection successive determination of cysteine and cystine in pharmaceutical preparations. *Talanta*, **38**, 1235–1239.

41 Themelis, D.G. and Tzanavaras, P.D. (2001) Simultaneous spectrophotometric determination of fluoride and monofluorophosphate ions in toothpastes using a reversed flow injection manifold. *Analytica Chimica Acta*, **429**, 111–116.

42 Aggarwal, S.G. and Patel, K.S. (1998) Flow injection analysis of Zn and Co in beverages, biological, environmental and pharmaceutical samples. *Fresenius' Journal of Analytical Chemistry*, **362**, 571–576.

43 Ortega-Barrales, P., Ruiz-Medina, A., Fernández de Córdova, M.L. and Molina Díaz, A. (2002) A flow-through solid-phase spectroscopic sensing device implemented with FIA solution measurements in the same flow cell: determination of binary mixtures of thiamine with ascorbic acid or acetylsalicylic acid. *Analytical and Bioanalytical Chemistry*, **373**, 227–232.

44 Fernández de Córdova, M.L., Ortega-Barrales, P., Rodríguez Torné, G. and Molina Díaz, A. (2003) A flow injection sensor for simultaneous determination of sulfamethoxazole and trimethoprim by using Sephadex SP C-25 for continuous on-line separation and solid phase UV transduction. *Journal of Pharmaceutical and Biomedical Analysis*, **31**, 669–677.

45 García-Jiménez, J.F., Valencia, M.C. and Capitán-Vallvey, L.F. (2006) Improved multianalyte determination of the intense sweeteners aspartame and acesulfame-K with a solid sensing zone implemented in an FIA scheme. *Analytical Letters*, **39**, 1333–1347.

46 Capitán-Vallvey, L.F., Valencia, M.C., Nicolás, E.A. and García-Jiménez, J.F. (2006) Resolution of an intense sweetener mixture by use of a flow injection sensor with on-line solid-phase extraction. *Analytical and Bioanalytical Chemistry*, **385**, 385–391.

47 Rocha, F.R.P., Reis, B.F., Zagatto, E.A.G., Lima, J.L.F.C., Lapa, R.A.S. and Santos, J.L.M. (2002) Multicommutation in flow analysis: concepts, applications and trends. *Analytica Chimica Acta*, **468**, 119–131.

48 Tzanavaras, P.D. and Themelis, D.G. (2002) Simultaneous flow injection determination of fluoride, monofluorophosphate and orthophosphate ions using alkaline phosphatase immobilized on a cellulose nitrate membrane and an open-circulation approach. *Analytica Chimica Acta*, **467**, 83–89.

49 Cardwell, T.J. and Christophersen, M.J. (2000) Determination of sulfur dioxide and ascorbic acid in beverages using a dual channel flow injection electrochemical detection system. *Analytica Chimica Acta*, **416**, 105–110.

50 Jones, D.A., Parkin, M.C., Langemann, H., Landolt, H., Hopwood, S.E., Strong, A.J. and Boutelle, M.G. (2002) On-line monitoring in neurointensive care – enzyme-based electrochemical assay for simultaneous, continuous monitoring of glucose and lactate from critical care patients. *Journal of Electroanalytical Chemistry*, **538**, 243–252.

51 Saitoh, K., Hasebe, T., Teshima, N., Kurihara, M. and Kawashima, T. (1998) Simultaneous flow injection determination of iron(II) and total iron by micelle enhanced luminol chemiluminescence. *Analytica Chimica Acta*, **376**, 247–254.

52 Mataix, E. and Luque de Castro, M.D. (2001) Determination of L-(−)-malic acid and L-(+)-lactic acid in wine by a flow injection-dialysis-enzymic derivatisation approach. *Analytica Chimica Acta*, **428**, 7–14.

53 Hasebe, T., Nagao, J. and Kawashima, T. (1997) Simultaneous flow injection determination of acetylcholine and choline based on luminol chemiluminescence in a micellar system with on-line dialysis. *Analytical Sciences*, **13**, 93–98.

54 Doku, G.N. and Gadzekpo, P.Y. (1996) Simultaneous determination of lithium, sodium and potassium in blood serum by flame photometric flow injection analysis. *Talanta*, **43**, 735–739.

55 Lima, J.L.F.C., Matos, C.D. and Vaz, M.C.V.F. (1996) Determination of Ca, Mg, Na, and K in milk by AAS and flame emission spectroscopy using high dilution FIA manifolds based on stream splitting or a dialysis unit. *Atomic Spectroscopy*, **17**, 196–200.

56 Haghighi, B. and Kurd, S.F. (2004) Sequential flow injection analysis of ammonium and nitrate using gas phase molecular absorption spectrometry. *Talanta*, **64**, 688–694.

57 Yao, T., Takashima, K. and Nanjyo, Y. (2003) Simultaneous determination of orthophosphate and total phosphates

(inorganic phosphates plus purine nucleotides) using a bioamperometric flow injection system made up by a 16-way switching valve. *Talanta*, **60**, 845–851.

58 Yao, T. and Kotegawa, K. (2002) Simultaneous flow injection assay of creatinine and creatine in serum by the combined use of a 16-way switching valve, some specific enzyme reactors and a highly selective hydrogen peroxide electrode. *Analytica Chimica Acta*, **462**, 283–291.

59 Park, I.S. and Kim, N. (1999) Simultaneous determination of hypoxanthine, inosine and inosine 5'-monophosphate with serially connected three enzyme reactors. *Analytica Chimica Acta*, **394**, 201–210.

60 Domínguez Vidal, A., Ortega Barrales, P. and Molina Díaz, A. (2003) Simultaneous Determination of Paracetamol, Caffeine and Propyphenazone in Pharmaceuticals by Means of a Single Flow-Through UV Multiparameter Sensor. *Microchimica Acta*, **141**, 157–163.

61 Catarino, R.I.L., Garcia, M.B.Q., Lapa, R.A.S. and Lima, J.L.F.C. (2002) Sequential determination of salicylic and acetylsalicylic acids by amperometric multisite detection flow injection analysis. *Journal of AOAC International*, **85**, 1253–1259.

62 Teshima, N., Nobuta, T. and Sakai, T. (2001) Simultaneous flow injection determination of ascorbic acid and cysteine using double flow cell. *Analytica Chimica Acta*, **438**, 21–29.

63 Grassi, V., Zagatto, E.A.G. and Lima, J.L.F.C. (2005) Flow-injection systems with multi-site detection. *Trends in Analytical Chemistry*, **24**, 880–886.

64 Qin, W. (2002) Flow injection chemiluminescence-based chemical sensors. *Analytical Letters*, **35**, 2207–2220.

65 Kiba, N., Miwa, T., Tachibana, M., Tani, K. and Koizumi, H. (2002) Chemiluminometric sensor for simultaneous determination of L-glutamate and L-lysine with immobilized

oxidases in a flow injection system. *Analytical Chemistry*, **74**, 1269–1274.

66 Kiba, N., Ito, S., Tachibana, M., Tani, K. and Koizumi, H. (2003) Simultaneous determination of choline and acetylcholine based on a trienzyme chemiluminometric biosensor in a single line flow injection system. *Analytical Sciences*, **19**, 1647–1651.

67 Pérez-Olmo, R., Olmo, J.C., Zarate, N., Araujo, A.N. and Montenegro, M.C.B.S.M. (2005) Sequential injection analysis using electrochemical detection: A review. *Analytica Chimica Acta*, **554**, 1–16.

68 Trojanowicz, M., Szewczynska, M. and Wcislo, M. (2003) Electroanalytical flow measurements-Recent advances. *Electroanalysis*, **15**, 347–365.

69 Curey, T.E., Goodey, A., Tsao, A., Lavigne, J., Sohn, Y., McDevitt, J.T., Anslyn, E.V., Neikirk, D. and Shear, J.B. (2001) Characterization of multicomponent monosaccharide solutions using an enzyme-based sensor array. *Analytical Biochemistry*, **293**, 178–184.

70 Maestre, E., Katakis, I., Narváez, A. and Domínguez, E. (2005) A multianalyte flow electrochemical cell: application to the simultaneous determination of carbohydrates based on bioelectrocatalytic detection. *Biosensors & Bioelectronics*, **21**, 774–781.

71 Yao, T., Yano, T. and Nishino, H. (2004) Simultaneous *in vivo* monitoring of glucose, L-lactate, and pyruvate concentrations in rat brain by a flow injection biosensor system with an on-line microdialysis sampling. *Analytica Chimica Acta*, **510**, 53–59.

72 Silber, A., Bisenberger, M., Bräuchle, C. and Hampp, N. (1996) Thick-film multichannel biosensors for simultaneous amperometric and potentiometric measurements. *Sensors and Actuators B*, **30**, 127–132.

73 Lapa, R.A.S., Lima, J.L.F.C., Vela, M.H. and Barrado, E. (1997) Sequential determination of calcium and magnesium cations in haemodialysis solutions by FIA. *Analytical Sciences*, **13**, 409–414.

74 Campmajó, C., Cairó, J.J., Sanfeliu, A., Martínez, E., Alegret, S. and Gòdia, F. (1994) Determination of ammonium and l-glutamine in hybridoma cell-cultures by sequential flow injection analysis. *Cytotechnology*, **14**, 177–182.

75 Sanz-Medel, A. (1999) *Flow Analysis with Atomic Spectrometric Detectors*, Elsevier, Amsterdam.

76 Wang, Y., Chen, M.L. and Wang, J.H. (2007) New developments in flow injection/sequential injection on-line separation and preconcentration coupled with electrothermal atomic absorption spectrometry for trace metal analysis. *Applied Spectroscopy Reviews*, **42**, 103–118.

77 Wang, J.H. and Hansen, E.H. (2005) Trends and perspectives of flow injection/ sequential injection on-line sample-pretreatment schemes coupled to ETAAS. *Trends in Analytical Chemistry*, **24**, 1–8.

78 Wang, J. and Hansen, E.H. (2003) On-line sample-pre-treatment schemes for trace-level determinations of metals by coupling flow injection or sequential injection with ICP-MS. *Trends in Analytical Chemistry*, **22**, 836–846.

79 Vassileva, E. and Furuta, N. (2001) Application of high-surface-area ZrO_2 in preconcentration and determination of 18 elements by on-line flow injection with inductively coupled plasma atomic emission spectrometry. *Fresenius' Journal of Analytical Chemistry*, **370**, 52–59.

80 Halvorsen, T.G., Pedersen-Bjergaard, S., Reubsaet, J.L.E. and Rasmussen, K.E. (2001) Liquid-phase microextraction combined with flow injection tandem mass spectrometry-Rapid screening of amphetamines from biological matrices. *Journal of Separation Science*, **24**, 615–622.

81 Schroder, H.F. (1995) Polar organic-compounds in the river elbe-development of optimized concentration methods using substance-specific detection

techniques. *Fresenius' Journal of Analytical Chemistry*, **353**, 93–97.

82 Geerdink, R.B., Berg, P.J., Kienhuis, P.G.M., Niessen, W.M.A. and Brinkman, U.A.T. (1996) Flow-injection analysis thermospray tandem mass spectrometry of triazine herbicides and some of their degradation products in surface water. *International Journal of Environmental Analytical Chemistry*, **64**, 265–278.

83 Wu, Q.L., Wang, M.F. and Simon, J.E. (2005) *Rapid Communications in Mass Spectrometry*, **19**, 2062.

84 Schroder, H.F. (1999) Substance-specific detection and pursuit of non-eliminable compounds during biological treatment of waste water from the pharmaceutical industry. *Waste Management*, **19**, 111–123.

85 Armenta, S., Garrigues, S. and de la Guardia, M. (2007) Recent developments in flow-analysis vibrational spectroscopy. *Trends in Analytical Chemistry*, **26**, 775–787.

86 Gallignani, M. and Brunetto, M.D. (2004) Infrared detection in flow analysis-developments and trends. *Talanta*, **64**, 1127–1146.

87 Quintas, G., Armenta, S., Morales-Noe, A., Garrigues, S. and de la Guardia, M. (2003) Simultaneous determination of Folpet and Metalaxyl in pesticide formulations by flow injection Fourier transform infrared spectrometry. *Analytica Chimica Acta*, **480**, 11–21.

88 Karlberg, B. and Torgrip, R. (2003) Increasing the scope and power of flow injection analysis through chemometric approaches. *Analytica Chimica Acta*, **500**, 299–306.

89 Booksh, K.S. and Kowalski, B.R. (1994) Theory of analytical-chemistry. *Analytical Chemistry*, **66**, 782A–791A.

90 Checa, A., Oliver, R., Saurina, J. and Hernández-Cassou, S. (2006) Flow-injection spectrophotometric determination of reverse transcriptase inhibitors used for acquired immuno deficiency syndrome (AIDS) treatment. Focus on strategies for dealing with the background components. *Analytica Chimica Acta*, **572**, 155–164.

91 Tomsu, D., Catalá Icardo, M. and Martínez Calatayud, J. (2004) Automated simultaneous triple dissolution profiles of two drugs, sulphamethoxazole-trimethoprim and hydrochlorothiazide-captopril in solid oral dosage forms by a multicommutation flow-assembly and derivative spectrophotometry. *Journal of Pharmaceutical and Biomedical Analysis*, **36**, 549–557.

92 Saurina, J. (2000) Analytical application of pH-gradients in flow injection analysis and related techniques. *Reviews in Analytical Chemistry*, **19**, 157–178.

93 Checa, A., González-Soto, V., Hernández-Cassou, S. and Saurina, J. (2005) Fast determination of pK(a) values of reverse transcriptase inhibitor drugs for AIDS treatment by using pH-gradient flow injection analysis and multivariate curve resolution. *Analytica Chimica Acta*, **554**, 177–183.

94 Vandeginste, B.G.M., Massart, D.L., Buydens, L.M.C., de Jong, S., Lewi, P.J. and Smeyers-Verbeke, J. (1998) *Handbook of Chemometrics and Qualimetrics: Part B*, Elsevier, Amsterdam.

95 Martens, M. and Naes, T. (1989) *Multivariate Calibration*, John Wiley & Sons, New York.

96 Zupan, J. and Gasteiger, J. (1993) *Neural Networks for Chemists: an Introduction*, VCH, Weinhein.

97 Li, W., Chen, J., Xiang, B. and An, D. (2000) Simultaneous on-line dissolution monitoring of multicomponent solid preparations containing vitamins B1 and B2. *Analytica Chimica Acta*, **408**, 39–47.

98 Ruiz Medina, A., Fernández de Córdova, M.L. and Molina-Díaz, A. (1999) Simultaneous determination of paracetamol, caffeine and acetylsalicylic acid by means of a FI ultraviolet PLS multioptosensing device. *Journal of Pharmaceutical and Biomedical Analysis*, **21**, 983–992.

99 Ayora Cañada, M.J., Pascual Reguera, M.I., Molina Díaz, A. and Capitán Vallvey, L.F. (1999) Solid-phase UV spectroscopic multisensor for the simultaneous determination of caffeine, dimenhydrinate and acetaminophen by using partial least squares multicalibration. *Talanta*, **49**, 691–701.

100 Polster, J., Prestel, G., Wollenweber, M., Kraus, G. and Gauglitz, G. (1995) Simultaneous determination of penicillin and ampicillin by spectral fibre-optical enzyme optodes and multivariate data analysis based on transient signals obtained by flow injection analysis. *Talanta*, **42**, 2065–2072.

101 Blasco, F., Medina-Hernández, M.J. and Sagrado, S. (1997) Use of pH gradients in continuous-flow systems and multivariate regression techniques applied to the determination of methionine and cysteine in pharmaceuticals. *Analytica Chimica Acta*, **348**, 151–159.

102 Hernández, O., Jiménez, A.I., Jiménez, F. and Arias, J.J. (1995) Evaluation of multicomponent flow injection analysis data by use of a partial least-squares calibration method. *Analytica Chimica Acta*, **310**, 53–61.

103 Hernández, O., Jiménez, F., Jiménez, A.I. and Arias, J.J. (1996) Multicomponent analysis by flow injection using a partial least-squares model. Determination of copper and zinc in serum and metal alloys. *Analyst*, **121**, 169–172.

104 Bro, R. (2006) Review on multiway analysis in chemistry-2000–2005. *Critical Reviews in Analytical Chemistry*, **36**, 279–293.

105 de Juan, A., Casassas, E. and Tauler, R. (2000) in *Encyclopedia of Analytical Chemistry: Instrumentation and Applications* (ed. R.A. Meyers), Wiley, Chichester. 9800–9837.

106 Saurina, J. and Tauler, R. (2000) Strategies for solving matrix effects in the analysis of triphenyltin in sea-water samples by three-way multivariate curve resolution. *Analyst*, **125**, 2038–2043.

107 Saurina, J., Hernández-Cassou, S., Tauler, R. and Izquierdo-Ridorsa, A. (1999) Continuous-flow and flow injection pH gradients for spectrophotometric determinations of mixtures of nucleic acid components. *Analytical Chemistry*, **71**, 2215–2220.

108 Norgaard, L. and Ridder, C. (1994) Rank annihilation factor-analysis applied to flow injection analysis with photodiode-array detection. *Chemometrics and Intelligent Laboratory Systems*, **23**, 107–114.

109 Smilde, A.K., Tauler, R., Saurina, J. and Bro, R. (1999) Calibration methods for complex second-order data. *Analytica Chimica Acta*, **398**, 237–251.

110 Pasamontes, A. and Callao, M.P. (2003) Determination of amoxicillin in pharmaceuticals using sequential injection analysis (SIA) – evaluation of the presence of interferents using multivariate curve resolution. *Analytica Chimica Acta*, **485**, 195–204.

111 Marsili, N.R., Lista, A., Fernández Band, B.S., Goicoechea, H.C. and Olivieri, A.C. (2004) New method for the determination of benzoic and sorbic acids in commercial orange juices based on second-order spectrophotometric data generated by a pH gradient flow injection technique. *Journal of Agricultural and Food Chemistry*, **52**, 2479–2484.

112 Gómez, V. and Callao, M.P. (2005) Use of multivariate curve resolution for determination of chromium in tanning samples using sequential injection analysis. *Analytical and Bioanalytical Chemistry*, **382**, 328–334.

113 Gómez, V., Larrechi, M.S. and Callao, M.P. (2006) Chromium speciation using sequential injection analysis and multivariate curve resolution. *Analytica Chimica Acta*, **571**, 129–135.

114 Gómez, V., Cuadros, R., Ruisánchez, I. and Callao, M.P. (2007) Matrix effect in second-order data-determination of dyes in a tanning process using vegetable tanning agents. *Analytica Chimica Acta*, **600**, 233–239.

115 Pasamontes, A. and Callao, M.P. (2006) Sequential injection analysis linked to multivariate curve resolution with alternating least squares. *Trends in Analytical Chemistry*, **25**, 77–85.

116 Gómez, V. and Callao, M.P. (2007) Multicomponent analysis using flow systems. *Trends in Analytical Chemistry*, **26**, 767–774.

117 Checa, A., Oliver, R., Hernández-Cassou, S. and Saurina, J. (2007) Flow-injection determination of zidovudine in plasma samples using multivariate curve resolution. *Analytica Chimica Acta*, **592**, 173–180.

10
Flow Processing Devices Coupled to Discrete Sample Introduction Instruments

M. Valcárcel, S. Cárdenas, B.M. Simonet, and R. Lucena

10.1
Introduction: The Problem of Sample Treatment

In recent years, analytical instrumentation has evolved exponentially. As a result a wide gamut of instruments, capable of performing highly accurate determinations, based on a variety of physico/chemical principles, currently exists. Such instruments include molecular and atomic absorption and emission spectrophotometers, vibrational and mass spectrometers, and conductometers, to name a few. Advances in instrumentation have run in parallel with substantial improvements in separation science, particularly as regards liquid chromatography and capillary electrophoresis. Also, the growing trend to miniaturization has facilitated the development of analytical equipment enabling accurate, expeditious separation of a variety of components by using reduced amounts of sample and reagents.

Despite the previous advances, work in routine analytical laboratories continues to be slowed down by the preliminary operations of the analytical process. Pretreating complex samples is often labour-intensive and time-consuming. Therefore, automating or integrating such operations can help simplify routine analyses. In this context, flow systems provide an elegant, effective way of automating/mechanizing sample treatment. This chapter is concerned with the way sample processing can be improved by using a flow system connected to a discrete sample introduction system.

10.2
Roles of Flow Processing Devices

Simplification is another clear trend in analytical chemistry, where it has so far been aimed at automating/mechanizing sample preparation for (bio)chemical measurement. As stated in other chapters, flow systems are especially relevant to sample treatment and method calibration. On-line combinations of flow sample pretreatment units and discrete sample introduction devices are highly attractive in this context as they allow the whole analytical sequence to be developed in a single

Advances in Flow Analysis. Edited by Marek Trojanowicz
Copyright © 2008 WILEY-VCH Verlag GmbH & Co. KGaA, Weinheim
ISBN: 978-3-527-31830-8

instrumental assembly while avoiding the typical problems associated with conventional sample treatment approaches.

Simplification also influences the quality of the information delivered by analytical laboratories. Flow analysis systems can help consolidate it as a trend. In fact, the intrinsic advantages of flow systems, which typically include low costs, a high flexibility and throughput, and the ability to easily minimize errors, make them especially suitable for developing fast-response analytical systems. A flow analyser can provide a total index for a family of compounds that can be used for sample classification and qualification in accordance with preset threshold values imposed either by clients or by legislation [1]. To this end, the flow system can be on-line coupled to a detector (*vanguard configuration*) or a high-performance instrument (*vanguard/rearguard configuration*), the flow processing unit being used to treat samples [2]. Vanguard/rearguard configurations are the more flexible inasmuch as the sample treatment unit and the high performance instrument can share the same detector or use a different one, which facilitates altering the system if required to perform additional or alternative confirmation measurements. The potential of flow systems in this emerging area of analytical chemistry has been examined [3].

10.3
Ways of Coupling Flow Processing Devices to Discrete Sample Introduction Instruments

Depending on the degree of human participation and the particular hardware used, samples can be treated for analysis by using batch procedures, integrated devices or coupled devices. In the last case, devices can be coupled at-line, on-line or in-line.

At-line coupling involves the use of a robotic interface to link the flow system to the measuring instrument. Usually, the robotic interface operates by placing a fraction of treated or conditioned sample into an autosampler vial or inserting the sample directly into the injector of the analytical instrument.

Unlike at-line coupling, on-line coupling involves direct contact between a flowing stream coming from the flow system, a valve (or an appropriate interface) and a part of the instrument that is usually the injection region. The flow system and the instrument are usually linked via a flow element such as a valve, T-connector or split-flow interface.

Finally, in-line coupling involves complete, close integration of the flow system and instrument. This is accomplished by incorporating a critical element of the flow system (e.g., an extraction unit) into the instrument.

Manufacturers have favored at-line coupling in commercial instruments. One typical example is the device used to couple solid-phase microextraction in fibers to liquid or gas chromatographs. Because the commercial instrument is of the closed type, the analyst has no access to specific parts, which restricts its potential for linking to other devices. This is probably why flow systems and instruments are most often coupled by inserting a fraction of sample into the injector of the measuring instrument.

Within the discrete sample introduction instruments, two specific techniques can be highlighted, namely: electrothermal atomic absorption and inductively coupled plasma (ICP). Although they have traditionally been regarded as discrete sample introduction techniques, recent advances have enabled their direct coupling to flow systems. For more detailed information about both hyphenations, readers are referred to Chapter 18.

Therefore, this chapter is devoted only to the coupling of flow processing devices to instruments based on chromatographic and electrophoretic separations.

10.4
Coupling Flow Processing Devices to Gas Chromatographs

The main difficulty in coupling flow systems to gas chromatographs arises from the different aggregation state of the fluids involved and the relatively low sample volume to be inserted (a few microliters) in order to maintain chromatographic resolution [4]. This problem can be avoided by using the large volume injection (LVI) technique, which affords injections of hundreds of microliters with a negligible effect on resolution. This section deals with the on-line coupling of flow configurations of the low- and high-pressure type to gas chromatographs. Most existing configurations have evolved from previous off-line arrangements which can be deemed at-line systems when an autosampler is used to inject samples.

10.4.1
Interfaces and Types of Coupling

Properly coupling a flow processing device to a gas chromatograph entails considering the following: (i) the volatility and thermal stability of the target analytes; (ii) the compatibility of the medium to be used with liquid samples; (iii) the extract volume, which should be as low as possible in order to minimize analyte dilution and avoid unduly decreasing the sensitivity as a result; and (iv) the transfer of the analytes from the pretreatment module to the gas chromatograph, which should be quantitative. A number of interfaces have been used for this purpose, the three most popular are based on on-column injection, solvent vaporization and multiport valves [4, 5].

In the on-column injection, a continuous high-pressure flow system based mainly on solid-phase extraction is coupled to a gas chromatograph via a six-port injection valve controlled by the GC software; the valve transfers the analyte, in an appropriate medium, via a metal or fused-silica capillary typically 20–30 cm long × 0.1 mm i.d. that is inserted into the septum of the on-column injector. Because this type of system is usually employed with the LVI technique to inject large volumes of solvent (typically 100 µl), the on-column injector must be connected to a retention gap (viz. a 3–5 m long × 0.53 mm i.d. deactivated fused silica capillary) followed by a 1–2 × 0.25 mm i.d. retaining pre-column and, finally, the analytical column. Usually, the pre-column and the analytical column contain the same stationary phase. A solvent vapor exit (SVE) kit is also installed between the two columns in order to remove most injected solvent and minimize losses of the most volatile compounds. Alternatively, a programmable temperature vaporizer (PTV) can be used instead of the on-column injector, the SVE being kept even though it is unnecessary as excess organic solvent can be removed in the split valve. When a multiport injection valve is used, the organic phase that leaves the flow processing device, which contains the highest analyte concentrations, is used to fill the loop of a high-pressure injection valve (typically 5 μl in volume). A desiccating column is placed in front of the valve in order to prevent water traces from reaching the chromatographic column. The carrier gas of the chromatograph is also used to transfer the analytes from the loop to the injector via a PTFE or stainless steel tube furnished with an injection needle that is inserted into the septum of the split/splitless injector. Depending on the particular instrument and injector configuration, the carrier gas should be split (between the injection valve and the chromatograph) or not in order to maintain peak resolution. In some cases, additional heating of the interface is required. The following section describes the use of non-chromatographic separation techniques for sample treatment in flow systems and discusses the most salient advantages and disadvantages of the previous interfaces.

10.4.2
Analytical Uses

The main purpose of coupling a flow processing device on-line to a gas chromatograph is usually to facilitate the development of a fully mechanized sample pretreatment process. Obviously, a continuous non-chromatographic separation technique will most often be required for this purpose. Solid-phase extraction (SPE) is the most common choice in this context. This is hardly surprising as SPE is compatible with gas chromatography by virtue of the analytes, which are isolated on a packed sorbent material, usually being eluted with a few microliters of an organic solvent that can be directly transferred to the instrument for separation prior to detection. Water traces can be removed from the sorbent by drying prior to elution. This on-line trace enrichment operation can be performed by using various flow configurations. The earliest approaches used a low-pressure continuous flow system where the sorbent was packed in a laboratory-made miniaturized column accommodated in the loop of an injection valve, the sample and reagents (conditioning solvents, eluent and derivatizing reagents –if needed–) being directly aspirated into the manifold. A wide variety of applications have been developed with this basic configuration including the determination of pollutants, vitamins, fatty acids and drugs in samples of environmental, clinical, toxicological and agrifood interest [6–14]. Table 10.1 lists selected examples.

The flow manifold can also be constructed by using a combination of high-pressure injection valves and pumps in order to deliver the sample and organic solvents required to condition the columns and a syringe pump for the eluent. The sorbent column is hand-packed (usually with a polymer) and placed between the high-pressure injection valves for the sample and eluent. Typical determinations with these systems include pesticides [15] and endocrine disruptors [16] in waters.

Table 10.1 Selected examples of low pressure flow processing devices coupled to gas chromatography.

Samples	Analytes	Separation technique	Refs.
edible oils	sterols	SPE	[6]
pharmaceutical preparations	vitamins D_2 and D_3	SPE	[7]
waters	chlorophenols, phenols, N-methylcarbamates	SPE	[8]
			[9]
			[10]
biological fluids	drugs	SPE	[11]
dairy products	fatty acids	SPE/LLE/derivatization	[12]
biological fluids	triglicerides and fatty acids	SPE/LLE/derivatization	[13]
human serum	cholesterol (total and HDL)	immobilized enzymatic reactor	[14]
waters	pesticides	SPE	[15]
waters	endocrine disrupters	SPE	[16]
waters	s-triazines	SPE (Prospekt using inmunoaffinity sorbents)	[17]
waters	pesticides		[18, 19]
air samples	organophosphate esters	dynamic microwave-assisted extraction coupled to SPE	[20]
air samples	organophosphate esters	dynamic sonication-assisted extraction coupled to SPE	[21]
Soils and sludges	Me_2Hg, Et_2Hg and MeHgCl	pervaporation	[22]
food samples	acetaldehyde and acetone	pervaporation	[23]
Orange juice	Flavour and off-flavour	pervaporation	[24]
water	phenols	LLE	[25]
fats	fatty acids	LLE	[26]
dairy products	carbamic pesticides	LLE	[27]
edible oils and fats	cholesterol and tocopherols	LLE/derivatization	[28]
waters	organochlorine pesticides	MMLLE	[29]
red wine	organochlorine pesticides	MMLLE	[30]
soils	PAHs	hot water extraction-hollow fiber liquid extraction	[31]
grapes	pesticides	hot water extraction-hollow fiber liquid extraction	[32]

The extraction selectivity can be increased by using an immuno-sorbent material such as that employed in the determination of s-triazines in aqueous samples [17] with the Prospekt (Programmable on-line solid-phase extraction technique) system, commercially available from Spark Holland for sample preparation, a Midas Auto-sampler, also from Spark Holland, and an HPLC pump for analyte elution. A PTV interface was used in combination with the LVI technique for the determination of pesticides in water [18, 19]. Dynamic microwave-assisted extraction [20] and dynamic sonication-assisted solvent extraction [21] have been coupled on-line prior to SPE-LVI-GC in order to determine organophosphate esters in air samples.

Analytical pervaporation has proved a reliable alternative to headspace techniques for the direct analysis of volatile fractions when coupled on-line to gas chromatography [22]. Usually, the pervaporation module consists of a lower unit for sample introduction and an upper part insulated by a membrane permeable to the target analytes. A spacer is usually needed to facilitate formation of the headspace. For coupling, the upper chamber is placed in the loop of an injection valve in order to allow the acceptor gas (viz. the chromatographic carrier gas) to pass through and drive the analytes to the injector. Table 10.1 shows selected applications of these on-line configurations [22–24].

Liquid/liquid extraction (LLE) can also be implemented in a continuous-flow manifold. However, it cannot be a first choice as it provides poorer enrichment factors owing to the need to maintain the organic-to-aqueous phase ratio as close to unity as possible in order to ensure efficient phase separation, and also because of its lack of robustness, the weakest element in the system being the phase separator (a T-piece, membrane or sandwich connector). In any case, on-line coupled LLE has been used with or without derivatization to determine phenols [25], fatty acids [26], pesticides [27] and cholesterol [28] in waters, fats and dairy products (see Table 10.1).

Emerging non-chromatographic separation techniques have also been on-line coupled to gas chromatography with a view to circumventing the shortcomings of some well-established extraction techniques. Such is the case with microporous membrane liquid/liquid extraction (MMLLE), which has some advantages over LLE including the formation of no emulsions, the use of smaller sample volumes and the obtainment of clean extracts. Also, selectivity can be increased by choosing an appropriate membrane material and its pore size. Reported applications use planar flat-sheet polypropylene membranes sandwiched between two blocks about 10–100 µl in volume. The extraction unit and the GC instrument are interfaced via a multiport valve accommodating the sample loop. The membrane should be replaced every 50 to100 uses, depending on the particular application, in order to minimize potential adsorption problems. Configurations typically use LVI and SVE. These systems have been employed to determine organochlorine pesticides in waters [29] and red wines [30]. Hollow-fiber membranes, which provide increased extraction surfaces – and improved extraction efficiency as a result – have been used for the extraction of polycyclic aromatic hydrocarbons and pesticides from soils and grapes, respectively, following on-line pressurized hot water extraction [31, 32]. The water phase, containing the analytes, was driven to the donor side of the membrane and the analytes were extracted to the acceptor solution, the concentrated extract then being on-line transferred to the GC instrument via an on-column interface.

10.4.3
Critical Discussion

Flow processing devices for coupling to gas chromatographs use a low- or high-pressure mode in combination with split/splitless or large volume injection.

In general, they afford the introduction of untreated samples, which simplifies the analytical process and results in improved productivity-related analytical properties. The most common non-chromatographic techniques can be easily coupled to GC via an appropriate interface; alternatively, a tandem flow system-interface can be used for the on-line coupling of a previous auxiliary energy-based extraction system.

10.5
Coupling Flow Processing Devices to Liquid Chromatographs

The combined use of flow processing devices and liquid chromatographs has proved a powerful tool for solving a number of problems in analytical science. The nature of the two partners facilitates their coupling, which produces a synergistic effect that enhances the analytical properties of the ensuing methods. The similarities between the two partners are self-apparent from their associated equipment [33].

Flow processing devices and liquid chromatographs can be coupled in two main types of configurations, namely:

1. *Pre-column arrangements*, where the flow system is used mainly to improve sensitivity and selectivity by preconcentration or clean-up, but can also be employed for other purposes such as saving reagents or introducing problematic or hazardous samples.

2. *Post-column arrangements*, which are intended to facilitate detection of the target analytes by derivatization in the flow system following separation on the chromatographic column.

This section focuses on pre-column arrangements, where the flow device is used to introduce treated sample aliquots into the chromatographic system. The high-pressure injection valve is the central element of the combined system as it acts as the interface between the low-pressure (flow) and high-pressure (chromatographic) lines. The autoinjector of a liquid chromatograph constitutes the simplest example of pre-column arrangements (in the on-line mode); an automatic syringe similar to those employed in sequential injection analysis is used to fill the sample loop of the injection valve.

10.5.1
Interfaces and Types of Coupling

In practice, coupled flow processing/liquid chromatography systems depart markedly from the previously described example of the liquid chromatograph autoinjector. In fact, the flow processing device is used not only to introduce samples into the chromatographic system, but also for sample pretreatment. The pretreatment, which is mainly intended to adapt the sample to the chromatographic separation conditions, involves one or more operations including a non-chromatographic procedure. Treated sample aliquots are then driven to the high-pressure injection valve, which is actuated to introduce the plug into the column (see Figure 10.1). Reproducible injection

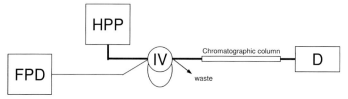

Figure 10.1 General flow device-liquid chromatographic coupling. FPD: flow processing device, HPP: high pressure pump, IV: high pressure injection valve, D: detector.

into the chromatograph relies on accurate timing and synchronized operation of the whole system.

A wide variety of sample treatment protocols on-line coupled to the chromatographic system can be used to increase the sensitivity and selectivity of determinations. Using a solid-phase extraction (SPE) protocol for analyte preconcentration and clean-up allows a different type of coupling to be implemented (see Figure 10.2). A pre-column packed with sorbent material is accommodated in the loop of the high-pressure injection valve to retain the target analytes. The flow system allows the initial steps of the SPE protocol (viz. conditioning, equilibration, sample loading and sorbent clean-up) to be developed in a continuous manner. Elution is carried out by switching the valve to its injection position. As noted in Section 10.4.2, fully automated SPE-based configurations such as the Prospekt [34] or Aspec [35] can also be on-line coupled to a liquid chromatograph by using a high-pressure injection valve as the interface between the two.

Other coupled systems use two or three injection valves. Thus, a configuration with two valves was used to implement sample screening and confirmation methods in the same manifold; this operational mode has been referred to as the dual use of flow configurations in liquid chromatography [36, 37]. To this end, an SPE column is placed in the first valve and the chromatographic column in the second, the two being on-line connected, as shown in the general scheme of Figure 10.3. The second valve is essential with a view to switching between the screening (load position) and confirmation mode (inject position). Essentially, the FI system is first employed to screen all samples for the target analytes or analyte families and then those with total concentrations close to or above the preset cut-off level are individually analysed by liquid chromatography. In the confirmation step, the FI system can be used either to preconcentrate samples or as a post-column derivatization system.

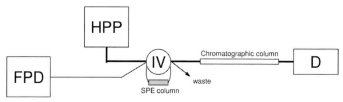

Figure 10.2 Continuous solid-phase extraction coupled to liquid chromatography. FPD: flow processing device, HPP: high pressure pump, IV: high pressure injection valve, D: detector.

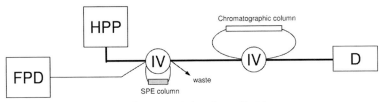

Figure 10.3 Flow processing device coupled on-line to liquid chromatographic involving both a screening and confirmation method. FPD: flow processing device, HPP: high pressure pump, IV: high pressure injection valve, D: detector.

Before analytical injection techniques were developed, continuous segmented flow configurations were used in a pre-column combination with liquid chromatography. This allowed samples to be processed in a mechanized manner. Such a combination can thus be considered a precursor of current flow approaches based on sample injection. Flow injection (FI) techniques evolved from a straightforward operational basis, used inexpensive hardware, were easy and convenient to operate, and highly flexible, and provided a high sample throughput and cost-effective performance [38]. Sequential injection (SI) techniques evolved from FI and shared some of their basic features. Both FI and SI manifolds have been massively used to design flow processing devices coupled on-line to liquid chromatographs. Both techniques can also be used in combination to assemble complex manifolds for specific applications.

The recently developed technique sequential injection chromatography (SIC) [39] testifies to the highly symbiotic nature of flow manifolds and liquid chromatographs. The SIC technique (see Figure 10.4) uses an SI manifold to introduce samples and deliver the mobile phase, in addition to a monolithic chromatographic column to separate analytes at a low pressure. This combination has been used to determine naphazoline nitrate and methylparaben [40], and lidocaine and prilocaine [41], in pharmaceuticals.

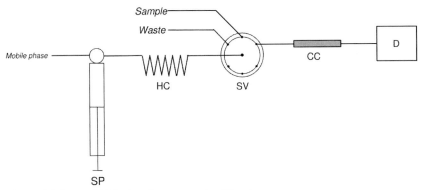

Figure 10.4 Sequential injection chromatography. SP: syringe pump, HC: holding coil, SV: selection valve, CC: chromatographic column, D: detector.

10.5.2
Analytical Uses

The flexibility of FI and SI techniques has been exploited for the efficient pre-treatment of samples for chromatographic separation, implementing several non-chromatographic techniques.

Membrane-based separation techniques such as dialysis and gas diffusion have been used for the determination of anions [42], and ammonia and methylamines [43] in natural waters, respectively. A typical sandwich diffusion cell was placed in the low-pressure line and the acceptor stream was driven to the loop of the high-pressure injection valve. Automated sequential trace enrichment of dialysate (ASTED) systems, which perform dialysis and preconcentration, have been successfully used to determine tetracycline in eggs [44], and ioxidanol in human, rat and monkey plasma [45]. On-line membrane extraction alone [46] and in combination with pervaporation [47] has been used to preconcentrate analytes prior to HPLC processing.

In recent years, microdialysis coupled on-line to HPLC has proved effective for the analysis of some species in complex matrices with operational ease, expeditious isolation of components from the sample matrix, potential enrichment and the use of little or no organic solvent [48]. The determination of sulfonamides in milk [49] and that of organic acids in milk fermentation products [50] constitute two salient examples.

The SPE/HPLC couple is a consolidated choice which has been applied for the determination of pesticides in water [51], lanthanides in rocks [52], and caffeine and selected anilines and phenol compounds in aquatic systems [53]. This operational approach is in continuous development, particularly as regards sorbent materials and protocols. Some authors have used molecular imprinted polymers (MIPs) [54] and carbon nanotubes [55] as sorbents for selective analyte retention. Similarly, new protocols such as renewable solid-phase extraction are being developed and combined with HPLC [56].

10.5.3
Critical Discussion

The synergetic hyphenation between flow processing devices and liquid chromatography has been extensively used in analytical sciences, since it is useful and easy to implement taking into account the nature of the partners. Flow processing devices provide an evident selectivity and sensitivity improvement to the developed methods. Several non-chromatographic techniques have been mechanized and coupled to liquid chromatography, proving the evident versatility of the described arrangement.

10.6
Coupling Flow Processing Devices to Capillary Electrophoresis Equipment

Capillary electrophoresis (CE) is a highly efficient, flexible separation technique which has become a serious competitor for other separation techniques including chromatographies [57, 58]. However, sample requirements are more stringent in CE than they are in other separation techniques. In fact, appropriate sample treatment is a critical step towards obtaining accurate, reproducible results; this entails avoiding clogging of the capillary and adsorption of macromolecules on its walls, among other cautions.

Electroosmotic flow is one of the driving forces in CE. This phenomenon arises from the presence of surface charges on capillary walls. The result is a net flow of buffer solution in the direction of the negative electrode. Electroosmotic flow is quite robust and occurs at a rate of $0.5–4\,nl\,s^{-1}$ depending on the buffer pH [59]. One other important factor is the small inner diameter of the capillary, which affords the use of very low volumes of samples and reagents, and also effective miniaturization [58, 60, 61].

The ability to couple flow processing devices to capillary electrophoresis equipment is limited by the following factors, all of which warrant careful consideration:

1. Compatibility of hydrodynamic flow in the processing device with electroosmotic flow in the capillary.
2. Compatibility of the high flow-rates typically used in processing devices with the low rates of electroosmotic flow in a CE system.
3. Compatibility of the sample plug coming from the processing device with the small sample volume to be introduced in the CE capillary.
4. Decoupling of the high voltages and currents applied to the electrophoretic separation system and the flow processing device.

The following sections describe selected coupled systems and their interfaces, and also some of their more salient uses.

10.6.1
Interfaces and Types of Coupling

Flow processing devices have been coupled in-, at- and on-line to CE equipment [62]. Overall, on-line coupled systems are the most commonplace. Interested readers can find a review of sample treatment devices used in combination with commercially available CE equipment elsewhere [63].

Coupling a flow processing device at-line with CE equipment entails using a robotic arm interface to place the sample coming from the former in an empty vial in the latter [4, 58, 64]. This combination is subject to several restrictions, the most significant of which are the accessibility of the CE autosampler and the minimum volume that can be delivered to the vial. It also requires the use of an electronic interface and appropriate control software in order to synchronize the movements of the robotic interface and autosampler. As can be seen in Figure 10.5, the needles of the programmable robotic arm can be positioned in two ways with respect to the auto sampler vials. While the flow processing device is working, the needles are down and the sample is prepared and transferred to the CE vial. When the flow processing device stops, the needles are lifted and CE analysis is started by moving the previously filled sample vial to the position where the capillary end and electrode are located.

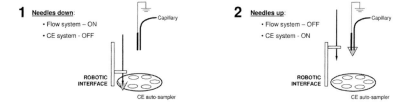

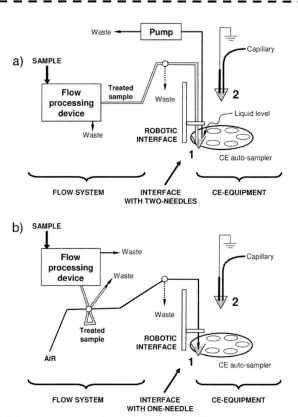

Figure 10.5 Schematic depiction of the at-line coupling of flow processing devices to capillary electrophoresis equipment. Configuration of robotic interface in (1) sample-collection mode and (2) sample-injection mode for electrophoretic analysis. Description of flow system manifold for (A) robotic interfaces with two needles and (B) robotic interfaces with one needle.

As shown in Figure 10.5, the programmable arm can be fitted with one or two needles. In the latter case, the needles can be identical or different in length, the longer one being used to fill the autosampler vials and the shorter one to drain them in order to maintain a preset liquid level. When a single needle is used, a constant, preset volume is delivered to each vial (e.g., by using air as carrier).

Alternatively to a programmable robotic arm, processing devices and CE equipment can be coupled at-line via the replenishment system used to empty vials and fill them with fresh buffer in some commercial instruments [65]. This entails disconnecting the replenishment needles from the Teflon tube coming from the replenishment bottles and replacing it with one coming from the flow processing device [63, 65].

The most salient advantage of at-line coupled systems is that the CE equipment is run in its normal operation mode. This allows samples to be introduced hydrodynamically or electrokinetically into the capillary. It also facilitates other CE operations such as capillary conditioning.

Coupling a flow processing device on-line to CE involves inserting the capillary end in a continuous stream of the former. Because the flow processing device and the electrophoretic system operate at a different flow-rate, the two require a split-flow interface for coupling. The split-flow interface was originally developed simultaneously in a vertical configuration by Fang's group [66–68] and in a horizontal configuration by Karlberg's group [69–71]. Figure 10.6 compares the two types of interface, which possess a low dead volume and are electrically grounded. Samples and electrolyte solutions are introduced into the electrophoretic capillary by the effect of electroosmotic flow and electrophoretic mobility. Hydrodynamic flow should be avoided by ensuring that the liquid level at the interface (viz. the capillary inlet) coincides with that in the capillary outlet.

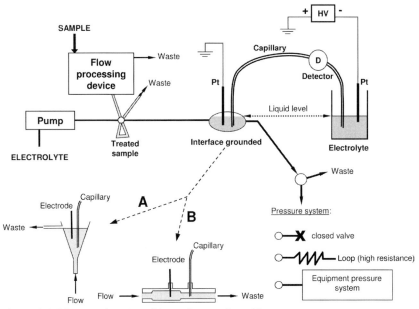

Figure 10.6 Schematic depiction of the on-line coupling of flow processing devices to capillary electrophoresis equipment. Detail of split flow interface (A) in vertical design and (B) in horizontal design. Description of systems to perform hydrodynamic injection mode for electrophoretic analysis.

A split-flow interface can be readily constructed from dielectric materials such as Teflon, methacrylate or Plexiglas. Alternatively, a piece of Tygon tubing can be used to insert the electrophoretic capillary and attach it with glue [72].

The flow processing device is made compatible with the electrophoretic system by grounding the interface. This avoids voltage differences between the interface and flow system. Electrophoretic separation is accomplished by applying a voltage difference to the capillary end, as shown in Figure 10.6. This configuration is highly recommended when an optical detector is used with the electrophoretic system, but scarcely useful with other types of detectors (e.g., mass spectrometers). According to Valcárcel's group, the split-flow interface must be subjected to a voltage difference in order to obtain one in the electrospray needle of the electrospray ionization interface of a mass spectrometer [73]. Figure 10.7 shows two configurations including a mass spectrometer with a grounded electrospray needle and electrospray chamber, respectively. The current arising from the voltage difference between the grounded flow processing device and the split-flow interface – which is connected to a voltage source – must be interrupted in order to avoid arc discharges in valves and pumps in the flow system. Alternatively, the current can be interrupted by using a plug of a substance with a high electrical resistance, such as pure water or air [73]. It is also advisable to insert a safety line consisting of a grounded electrode immediately in front of the pumps [63, 73].

In principle, the use of on-line coupled systems linked via a split-flow interface is limited to the electrokinetic introduction of samples. However, this injection mode

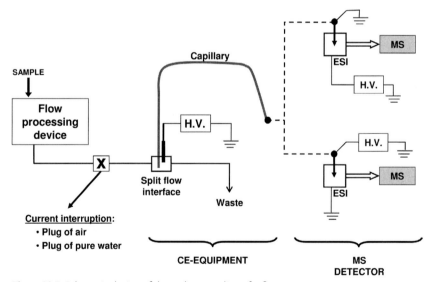

Figure 10.7 Schematic design of the on-line coupling of a flow processing device to capillary electrophoresis-mass spectrometry equipment by using a split flow interface. This configuration requires voltage isolation by using a plug of air or a large plug or pure water.

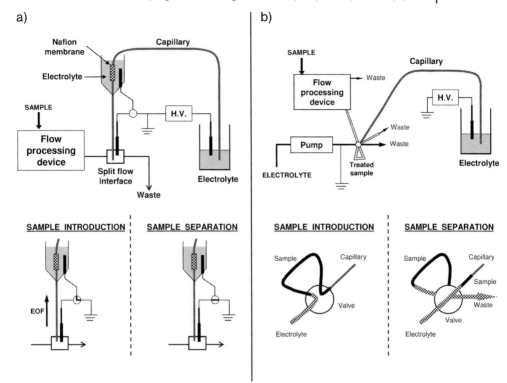

Figure 10.8 Alternatives to perform automatic hydrodynamic sample introduction for electrophoretic analysis in on-line flow system-capillary electrophoresis combinations. (A) use of Nafion membranes and (B) use of injection valve.

provides biassed results in some cases. This has promoted the use of hydrodynamic injection instead. Thus, Pu *et al.* [74] used a split-flow cell affording hydrodynamic injection of samples into the CE capillary by electroosmotic flow. The interface included a Nafion joint to connect the CE capillary to the tube of the flow system (see Figure 10.8A). An electronic time relay system was used to control switching of the high voltage. Kuban *et al.* [71] inserted a valve at the end of a horizontal split-flow interface. Switching the valve off caused the sample to be forced into the capillary. The greatest limitations of this approach arise from the need to control the time during which the valve is on and off, and also the pressure generated in the system. Santos *et al.* [73] proposed controlling the pressure by using a valve loop capable of withstanding higher pressures. With commercial CE instruments, one can also connect the pressure system of the equipment [63, 73] (see Figure 10.6). It should be noted that commercial CE equipment affords injection at a low pressure (0.5 psi) with a high precision (0.01 psi).

The split-flow interface can be used for the on-line coupling of both flow injection and sequential injections systems. Based on a common approach, split-flow interfaces, microsequential injection systems or integrated Lab-on-a-valve systems [75, 76] have been coupled to CE. Sequential injection systems have also been on-line coupled via a microvalve allowing the insertion of a constant volume of sample into the capillary [77, 78]. However, the loop volume is high relative to those typically used in CE work and must be reduced by switching the valve or flushing the sample from the interface in order to facilitate electrokinetic insertion of a portion of sample (see Figure 10.8B).

A modified version of the split-flow interface was recently used to accommodate a typical solid-phase microextraction (SPME) fiber precisely at the capillary end [79]. Although samples were treated on the fiber and the flow system was only used to couple SPME and CE, they can also be processed at the interface connecting the fiber to the flow system [80].

In-line coupling a flow processing device to CE involves inserting the capillary end (inlet region) into the treated sample or vice versa. This can be done by using hollow fibers to perform liquid-phase microextraction (LPME). As shown in Figure 10.9, the electrophoretic capillary can be inserted into the lumen of a hollow fiber or the extraction unit into the capillary [81]. In the latter case, the fiber can be fitted to the capillary by heating [63]

Flow systems have also been coupled to microchip electrophoretic systems [82], using a capillary as the interface between a sequential injection system and the microchip; for this purpose, one end of the capillary was attached to the microchip via a Teflon fitting. The principal shortcoming of this combination is the presence of residual hydrodynamic flow in the microchip separation channel [82, 83].

10.6.2
Analytical Uses

The use of flow processing devices constitutes one the most reliable ways of improving performance in analytical methods through automation, miniaturization

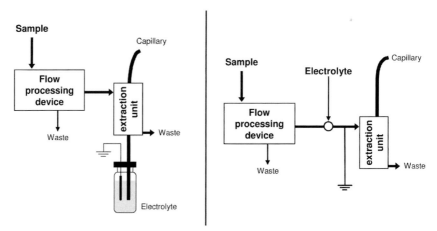

Figure 10.9 Schematic depiction of the in-line coupling of flow processing devices to capillary electrophoresis equipment.

and simplification of the preliminary operations of the analytical process – which are critical in CE work. Coupling flow processing devices to CE equipment facilitates three types of tasks, namely: automatic calibration [64, 84, 85], screening of samples [86–88], and sample preconcentration and clean-up [89–91].

Throughput in this context is limited by electrophoretic time. In fact, the high applied voltage must be interrupted and the capillary rinsed before each new run. Kuban *et al.* [92], however, have demonstrated that consecutive runs are feasible. The ability to perform multiple injections without rinsing the capillary after each run has been examined by Roche *et al.* [93] and Kaljurand *et al.* [94].

Table 10.2 shows selected examples of flow processing devices coupled to CE equipment. As can be seen, coupled equipment includes extraction, filtering, dialysis, membrane, gas extraction and gas diffusion units; hollow fibers and even auxiliary devices. Rather than an exhaustive compilation, the table provides selected entries illustrating the potential of this methodology for the analysis of various types of real samples with complex matrices. The principal conclusion is that choosing an appropriate flow processing device can facilitate the treatment and analysis of virtually any type of sample (biomedical, pharmaceutical, environmental, food). Readers interested in a more detailed description of the types of devices coupled to CE so far are referred to Ref. [62].

One of the most critical factors in selecting a flow processing device is compatibility of the chemical composition of the conditioned sample with the electrophoretic analysis to be performed. A number of sample treatment techniques (e.g., SPE, LLE) require the use of organic solvents such as methanol, direct insertion of which into an electrophoretic buffer can result in current interruptions. This problem can be avoided by changing the buffer pH, adding a modifier (e.g., a surfactant to the organic solvent) or supplying the buffer with a small amount of solvent. In this way, at-line coupled systems afford slight modification of treated or conditioned samples by collection into vials containing a modifier. Alternative approaches involve modifying samples within the flow system. Other influential factors to be considered include the sample volume and dilution by the effect of diffusion phenomena occurring in the flow system.

Coupling flow processing devices to microelectrophoretic chips or electrophoretic systems is of special interest with a view to developing *in vivo* determinations; this is facilitated by the ability of electrophoretic systems to use small volumes of sample and provide rapid analyses. For example, a microdialysis needle on-line coupled to CE enabled the *in vivo* monitoring of the evolution of a drug [102, 103]. In fact, this approach allows the characteristic pharmacokinetic response of a simple animal to the drug to be recorded. One other useful advantage is the ability to analyse special samples such as brain tissue [103].

10.6.3
Critical Discussion

In general, the use of flow processing devices improves the overall efficiency, selectivity and sensitivity of CE methods. In addition, it facilitates special determinations

Table 10.2 Selected example of coupling flow processing device to capillary electrophoresis.

Samples	Analytes	Flow processing device	Coupling	Type CE	LOD	RSD (%)	Refs.
milk samples	sulfonamides	solid-phase extraction	at-line	commercial	0.6 µg/L	7.1	[88]
food sample	myo-inositol phosphates	anion exchange resin	at-line	commercial	11 µmol/L	7.9	[91]
human urine	chlorophenols	solid-phase extraction	at-line	commercial	0.08 µg/L	7.2	[95]
human plasma	pseudoephedrine	solid-phase extraction	on-line	home-made	12 µg/L	1.2	[96]
human plasma	bambuterol	supported liquid membrane	in-line	home-made	<4 nM	5	[81]
biological samples	acidic drugs	dialysis unit	on-line	home-made	0.05 µg/mL	10	[97]
tissue samples	aspartate isomers	dialysis unit	on-line	home-made	10 nM	5	[98]
fish samples	trimethylamine	gas extraction	at-line	commercial	0.5 mg/L	4	[99]
serum samples	quinolones	solid-phase extraction	on-line	commercial	0.5 µg/mL	3.2	[100]
blood	glucose	microdialysis	in-line	home-made	<0.5 mM	1.1	[101]

such as *in vivo* pharmacokinetic tests. No doubt, miniaturization of flow processing devices will help further expand the potential of this combination.

10.7
Future Prospects

As noted earlier, simplification is a clear-cut trend in today's analytical chemistry. Simplification can no doubt bring substantial advantages to routine analytical laboratories, where throughput is compromised mainly by the need to pretreat samples. Sample treatments are often labor-intensive, time-consuming and a source of errors. Therefore, the ability to treat samples and transfer them directly to a measuring instrument is highly desirable. The large number of scientific publications describing flow systems capable of treating complex samples testifies to the usefulness of flow techniques with a view to performing this preliminary operation of the analytical process. New efforts can be expected to be made in the near future towards overcoming some limitations (e.g., those encountered in the direct analysis of solid samples). One other major shortcoming of automated/mechanized systems is the need for frequent replacement of membranes, extraction columns and various other elements.

Probably, the use of new materials and nanomaterials will help develop more efficient and useful interfaces compatible with micro- or nano-instrumental configurations. Coupled systems are also bound to become increasingly compatible with routine analytical work provided the current gap between R&D lines and clients are effectively bridged. Basic research on this topic has already been conducted and widely documented. It is thus time to develop practical applications with a view to facilitating its implementation in routine laboratories. The parties engaged in this transfer process should obviously include the developers of analytical tools.

References

1 Valcárcel, M., Cárdenas, S. and Gallego, M. (1999) Sample screening systems in analytical chemistry. *Trends in Analytical Chemistry*, **18**, 685–694.

2 Valcárcel, M. and Cárdenas, S. (2005) Vanguard-rearguard analytical strategies. *Trends in Analytical Chemistry*, **24**, 67–74.

3 Valcárcel, M., Cárdenas, S. and Gallego, M. (2002) Continuous flow systems for rapid sample screening. *Trends in Analytical Chemistry*, **21**, 251–258.

4 Valcárcel, M., Gallego, M. and Ríos, A. (1998) Coupling continuous flow systems to instruments based on discrete sample introduction. *Fresenius' Journal of Analytical Chemistry*, **362**, 58–66.

5 Hyötyläinen, T. and Riekkola, M. (2004) Approaches for on-line coupling of extraction and chromatography. *Analytical and Bioanalytical Chemistry*, **378**, 1962–1981.

6 Ballesteros, E., Gallego, M. and Valcárcel, M. (1995) Simultaneous determination of sterols in edible oils by use of a continuous separation module coupled to a gas-chromatograph. *Analytica Chimica Acta*, **308**, 253–260.

7 Ballesteros, E., Gallego, M. and Valcárcel, M. (1995) Gas-chromatographic flow method for preconcentration and determination of vitamins D2 and D3 in pharmaceutical preparations. *Chromatographia*, **40**, 425–431.

8 Crespín, M.A., Ballesteros, E., Gallego, M. and Valcárcel, M. (1996) Automatic preconcentration of chlorophenols and gas chromatographic determination with electron capture detection. *Chromatographia*, **43**, 633–639.

9 Crespín, M.A., Ballesteros, E., Gallego, M. and Valcárcel, M. (1997) Trace enrichment of phenols by on-line solid-phase extraction and gas chromatographic determination. *Journal of Chromatography A*, **757**, 165–172.

10 Ballesteros, E., Gallego, M. and Valcárcel, M. (1996) On line preconcentration and gas chromatographic determination of N-methylcarbamates and their degradation products in aqueous samples. *Environmental Science & Technology*, **30**, 2071–2077.

11 Cárdenas, S., Gallego, M. and Valcárcel, M. (1997) An automated preparation device for the determination of drugs in biological fluids coupled on-line to a gas chromatograph mass spectrometer. *Rapid Communications in Mass Spectrometry: RCM*, **11**, 973–980.

12 Ballesteros, E., Cárdenas, S., Gallego, M. and Valcárcel, M. (1994) Determination of free fatty-acids in dairy-products by direct coupling of a continuous preconcentration ion-exchange-derivatization module to a gas-chromatograph. *Analytical Chemistry*, **66**, 628–634.

13 Cárdenas, S., Ballesteros, E., Gallego, M. and Valcárcel, M. (1994) Sequential determination of triglycerides and free fatty-acids in biological-fluids by use of a continuous pretreatment module coupled to a gas-chromatograph. *Analytical Biochemistry*, **222**, 332–341.

14 Cárdenas, S., Ballesteros, E., Gallego, M. and Valcárcel, M. (1995) Automatic gas-chromatographic determination of the high-density-lipoprotein cholesterol and total cholesterol in serum. *Journal of Chromatography B*, **672**, 7–16.

15 Pocurrull, E., Aguilar, C., Borrull, F. and Marcé, R.M. (1998) On-line coupling of solid-phase extraction to gas chromatography with mass spectrometric detection to determine pesticides in water. *Journal of Chromatography A*, **818**, 85–93.

16 Brossa, L., Marcé, R.M., Borrull, F. and Pocurrull, E. (2002) Application of on-line solid-phase extraction–gas chromatography–mass spectrometry to the determination of endocrine disruptors in water simples. *Journal of Chromatography A*, **963**, 287–294.

17 Dallüge, J., Hankemeier, T., Vreuls, R.J.J. and Brinkman, U.A.T. (1999) On-line coupling of immunoaffinity-based solid-phase extraction and gas chromatography for the determination of s-triazines in aqueous samples. *Journal of Chromatography A*, **830**, 377–386.

18 Sasano, R., Hamada, T., Kurano, M. and Furuno, M. (2000) On-line coupling of solid-phase extraction to gas chromatography with fast solvent vaporization and concentration in an open injector liner: analysis of pesticides in aqueous samples. *Journal of Chromatography A*, **896**, 41–49.

19 Brondi, S.H.G., Spoljaric, F.C. and Lanas, F.M. (2005) Ultratraces analysis of organochlorine pesticides in drinking water by solid-phase extraction coupled with large volume injection/gas chromatography/mass spectrometry. *Journal of Separation Science*, **28**, 2243–2246.

20 Ericsson, M. and Colmsjö, A. (2003) Dynamic microwave-assisted extraction coupled on-line with solid-phase extraction and large-volume injection gas chromatography: Determination of organophosphate esters in air simples. *Analytical Chemistry*, **75**, 1713–1719.

21 Sánchez, C., Ericsson, M., Carlsson, H. and Colmsjö, A. (2003) Determination of organophosphate esters in air samples by dynamic sonication-assisted solvent extraction coupled on-line with large-volume injection gas chromatography utilizing a programmed-temperature vaporizer. *Journal of Chromatography A*, **993**, 103–110.

22 Bryce, D.W., Izquierdo, A. and Luque de Castro, M.D. (1997) Pervaporation as an alternative to headspace. *Analytical Chemistry*, **69**, 844–847.

23 Priego-López, E. and Luque de Castro, M.D. (2002) Pervaporation–gas chromatography coupling for slurry samples: Determination of acetaldehyde and acetone in food. *Journal of Chromatography A*, **976**, 399–407.

24 Gómez-Ariza, J.L., García-Barrera, T. and Lorenzo, F. (2004) Determination of flavour and off-flavour compounds in orange juice by on-line coupling of a pervaporation unit to gas chromatography–mass spectrometry. *Journal of Chromatography A*, **1047**, 313–317.

25 Ballesteros, E., Gallego, M. and Valcárcel, M. (1990) On-line coupling of a gas-chromatograph to a continuous liquid–liquid extractor. *Analytical Chemistry*, **62**, 1587–1591.

26 Ballesteros, E., Gallego, M. and Valcárcel, M. (1993) Automatic method for online preparation of fatty-acid methyl-esters from olive oil and other types of oil prior to their gas-chromatographic determination. *Analytica Chimica Acta*, **282**, 581–588.

27 Ballesteros, E., Gallego, M. and Valcárcel, M. (1993) Automatic gas-chromatographic determination of n-methylcarbamates in milk with electron-capture detection. *Analytical Chemistry*, **65**, 1773–1778.

28 Ballesteros, E., Gallego, M. and Valcárcel, M. (1996) Gas chromatographic determination of cholesterol and tocopherols in edible oils and fats with automatic removal of interfering triglycerides. *Journal of Chromatography A*, **719**, 221–227.

29 Lüthje, K., Hyötyläinen, T. and Riekkola, M. (2004) On-line coupling of microporous membrane liquid–liquid extraction and gas chromatography in the analysis of organic pollutants in water. *Analytical and Bioanalytical Chemistry*, **378**, 1991–1998.

30 Hyötyläinen, T., Lüthje, K., Rautiainen-Rämä, M. and Riekkola, M. (2004) Determination of pesticides in red wines with on-line coupled microporous membrane liquid–liquid extraction-gas chromatography. *Journal of Chromatography A*, **1056**, 267–271.

31 Kuosmanen, K., Hyötyläinen, T., Hartonen, K. and Riekkola, M. (2003) Analysis of polycyclic aromatic hydrocarbons in soil and sediment with on-line coupled pressurised hot water extraction, hollow fibre microporous membrane liquid–liquid extraction and gas chromatography. *Analyst*, **128**, 434–439.

32 Lüthje, K., Hyötyläinen, T., Rautiainen-Rämä, M. and Riekkola, M. (2005) Pressurised hot water extraction–microporous membrane liquid–liquid extraction coupled on-line with gas chromatography–mass spectrometry in the analysis of pesticides in grapes. *Analyst*, **130**, 52–58.

33 Luque, M.D. and Valcarcel, M. (1992) New approaches to coupling flow-injection analysis and high-performance liquid chromatography. *Journal of Chromatography A*, **600**, 183–188.

34 Barret, Y.C., Akinanya, B., Chang, S.Y. and Vesterqvist, O. (2005) Automated on-line SPE LC-MS/MS method to quantitate 6beta-hydroxycortisol and cortisol in human urine: Use of the 6beta-hydroxycortisol to cortisol ratio as an indicator of CYP3A4 activity. *Journal of Chromatography B*, **821**, 159–165.

35 Halme, K., Lindfors, E. and Peltronen, K. (2007) A confirmatory analysis of malachite green residues in rainbow trout with liquid chromatography–electrospray

tandem mass spectrometry. *Journal of Chromatography B*, **845**, 74–79.

36 Criado, A., Cárdenas, S., Gallego, M. and Valcárcel, M. (2004) Direct automatic screening of soils for polycyclic aromatic hydrocarbons based on microwave-assisted extraction/fluorescence detection and on-line liquid chromatographic confirmation. *Journal of Chromatography A*, **1050**, 111–118.

37 Criado, A., Cárdenas, S., Gallego, M. and Valcárcel, M. (2002) Biological fluid screening and confirmation of bile acids by use of an integrated flow injection-LC-evaporative light-scattering system. *Chromatographia*, **55**, 49–54.

38 Grudpan, K. (2004) Some recent developments on cost-effective flow-based analysis. *Talanta*, **64**, 1084–1090.

39 Satínský, D., Huclová, J., Solich, P. and Karlíek, R. (2003) Reversed-phase porous silica rods, an alternative approach to high-performance liquid chromatographic separation using the sequential injection chromatography technique. *Journal of Chromatography A*, **1015**, 239–244.

40 Chocholous, P., Satinsky, D. and Solich, P. (2006) Fast simultaneous spectrophotometric determination of naphazoline nitrate and methylparaben by sequential injection chromatography. *Talanta*, **70**, 408–413.

41 Klimundová, J., Šatinský, D., Sklenářová, H. and Solich, P. (2006) Automation of simultaneous release tests of two substances by sequential injection chromatography coupled with Franz cell. *Talanta*, **69**, 730–735.

42 Grudpan, K., Jakmuneea, J. and Sooksamitib, P. (1999) Flow injection dialysis for the determination of anions using ion chromatography. *Talanta*, **49**, 215–223.

43 Gibb, S.W., Fauzi, R., Mantoura, C. and Liss, P.S. (1995) Analysis of ammonia and methylamines in natural waters by flow injection gas diffusion coupled to ion chromatography. *Analytica Chimica Acta*, **316**, 291–304.

44 Zurhelle, G., Muller-Seitz, E. and Petz, M. (2000) Automated residue analysis of tetracyclines and their metabolites in whole egg, egg white, egg yolk and hen's plasma utilizing a modified ASTED system. *Journal of Chromatography B*, **739**, 191–203.

45 Jacobsen, F.B. (2000) On-line dialysis and quantitative high-performance liquid chromatography analysis of iodixanol in human, rat and monkey plasma. *Journal of Chromatography B*, **749**, 135–142.

46 Wang, X. and Mitra, S. (2005) Development of a total analytical system by interfacing membrane extraction, pervaporation and high-performance liquid chromatography. *Journal of Chromatography A*, **1068**, 237–242.

47 Guo, X. and Mitra, S. (2000) On-line membrane extraction liquid chromatography for monitoring semi-volatile organics in aqueous matrices. *Journal of Chromatography A*, **904**, 189–196.

48 Jen, J.F. and Liu, T.C. (2006) Determination of phthalate esters from food-contacted materials by on-line microdialysis and liquid chromatography. *Journal of Chromatography A*, **1130**, 28–33.

49 Yang, T.C.C., Yang, I.L. and Liao, L.J. (2004) Determination of sulfonamide residues in milk by on-line microdialysis and HPLC. *Journal of Liquid Chromatography & Related Technologies*, **27**, 501–510.

50 Wei, M.C., Chang, C.T. and Jen, J.F. (2001) Determination of organic acids in fermentation products of milk with high performance liquid chromatography/on-lined micro-dialysis. *Chromatographia*, **54**, 601–605.

51 Koal, T., Asperger, A., Efer, J. and Engewald, W. (2003) Simultaneous determination of a wide spectrum of pesticides in water by means of fast on-line SPE-HPLC-MS-MS - a novel approach. *Chromatographia*, **57**, S93–S101.

52 Buchmeiser, M.R., Seeber, G. and Tessadri, R. (2000) Quantification of lanthanides in rocks using succinic acid-derivatized sorbents for on-line SPE-RP-ion-pair HPLC. *Analytical Chemistry*, **72**, 2595–2602.

53 Papadopoulou-Mourkidou, E., Patsias, J., Papadakis, E. and Koukourikou, A. (2001) Use of an automated on-line SPE-HPLC method to monitor caffeine and selected aniline and phenol compounds in aquatic systems of Macedonia-Thrace, Greece. *Fresenius' Journal of Analytical Chemistry*, **371**, 491–496.

54 Koeber, R., Fleischer, C., Lanza, F., Boos, K.S., Sellergren, B. and Barcelo, D. (2001) Evaluation of a multidimensional Solid-Phase extraction Platform for highly selective on-line cleanup and high-throughput LC-MS analysis of triazines in river water samples using molecularly imprinted polymers. *Analytical Chemistry*, **73**, 2437–2444.

55 Fang, G.Z., He, J.X. and Wang, S. (2006) Multiwalled carbon nanotubes as sorbent for on-line coupling of solid-phase extraction to high-performance liquid chromatography for simultaneous determination of 10 sulfonamides in eggs and pork. *Journal of Chromatography A*, **1127**, 12–17.

56 Quintana, J.B., Miró, M., Estela, J.M. and Cerdá, V. (2006) Automated on-line renewable solid-phase extraction-liquid chromatography exploiting multisyringe flow injection-bead injection lab-on-valve analysis. *Analytical Chemistry*, **78**, 2832–2840.

57 Marina, M.L., Ríos, A. and Valcárcel, M. (2005) *Analysis and Detection by Capillary Electrophoresis*, Comprehensive Analytical Chemistry, Elsevier.

58 Simonet, B.M., Ríos, A. and Valcárcel, M. (2003) Enhancing sensitivity in capillary electrophoresis. *Trends in Analytical Chemistry*, **22**, 605–614.

59 Marina, M.L. and Torre, M. (1994) Capillary electrophoresis. *Talanta*, **41**, 1411–1433.

60 Marina, M.L., Ríos, A. and Valcárcel, M. (2005) Chapter 1: Fundamentals of capillary electrophoresis, *Comprehensive Analytical Chemistry, Analysis and Detection by Capillary Electrophoresis*. Elsevier, **45**, pp. 1–30.

61 Ríos, A., Escarpa, A., González, M.C. and Crevillén, G. (2006) Challenges of analytcial microsystems. *Trends in Analytical Chemistry*, **25**, 467–479.

62 Simonet, B.M., Ríos, A. and Valcárcel, M. (2005) Chapter 4: Coupling continuous flow systems to capillary electrophoresis, *Comprehensive Analytical Chemistry, Analysis and detection by capillary electrophoresis*. Elsevier, **45**, pp. 173–223.

63 Santos, B., Simonet, B.M., Ríos, A. and Valcárcel, M. (2006) Automatic sample preparation in commercial capillary-electrophoresis equipment. *Trends in Analytical Chemistry*, **25**, 968–976.

64 Arce, L., Hinsmann, P., Novic, M., Ríos, A. and Valcárcel, M. (2000) Automatic calibration in capillary electrophoresis. *Electrophoresis*, **21**, 556–562.

65 Santos, B., Simonet, B.M., Ríos, A. and Valcárcel, M. (2004) Direct automatic determination of biogenic amines in wine by flow injection-capillary electrophoresis-mass spectrometry. *Electrophoresis*, **25**, 3427–3433.

66 Fang, Z.L., Chen, H.W., Fang, Q. and Pu, Q.S. (2000) Developments in flow injection-capillary electrophoresis systems. *Analytical Sciences*, **16**, 197–203.

67 Fang, Z.L., Liu, Z.S. and Shen, Q. (1997) Combination of flow injection with capillary electrophoresis. Part I. The basic system. *Analytica Chimica Acta*, **346**, 135–143.

68 Fang, Q., Wang, F.R., Wang, S.L., Liu, S.S., Xu, S.K. and Fang, Z.L. (1999) Sequential injection sample introduction microfluidic-chip based capillary

electrophoresis system. *Analytica Chimica Acta*, **390**, 27–37.

69 Kuban, P. and Karlberg, B. (1998) Interfacing of flow injection pre-treatment system with capillary electrophoresis. *Trends in Analytical Chemistry*, **17**, 34–41.

70 Kuban, P., Engström, A., Olsson, J.C., Thorsén, G., Tryzell, R. and Karlberg, B. (1997) New interface for coupling flow injection and capillary electrophoresis. *Analytica Chimica Acta*, **337**, 117–124.

71 Kuban, P., Pirmohammadi, R. and Karlberg, B. (1999) Flow injection analysis capillary electrophoresis system with hydrodynamic injection. *Analytica Chimica Acta*, **378**, 55–62.

72 Fan, L., Cheng, Y., Li, Y., Chen, H., Chen, X. and Hu, Z. (2005) Head-column field amplified sample stacking in a capillary electrophoresis flow injection system. *Electrophoresis*, **26**, 4345–4354.

73 Santos, B., Simonet, B.M., Lendl, B., Ríos, A. and Valcárcel, M. (2006) Alternatives for coupling sequential injection systems to commercial capillary electrophoresis mass spectrometry equipment. *Journal of Chromatography A*, **1127**, 278–285.

74 Pu, Q.S. and Fang, Z.K. (1999) Combination of flow injection with capillary electrophoresis. Part 6. A bias-free sample introduction system based on electroosmotic flow traction. *Analytica Chimica Acta*, **398**, 65–74.

75 Wu, C.H., Scampavia, L. and Ruzicka, J. (2002) Microsequential injection: anion separations using "Lab-on-valve" coupled with capillary electrophoresis. *Analyst*, **127**, 898–905.

76 Wu, C.H., Scampavia, L. and Ruzicka, J. (2003) Microsequential injection: automated insulin derivatization and separation using a lab-on-valve capillary electrophoresis system. *Analyst*, **128**, 1123–1130.

77 Wuersig, A., Kuban, P., Khaloo, S.S. and Hauser, P.C. (2006) Rapid electrophoretic separations in short capillaries using contactless conductivity detection and a

sequential injection analysis manifold for hydrodynamic sample loading. *Analyst*, **131**, 944–949.

78 Zacharis, C.K., Tempels, F.W.A., Theodoridis, G.A., Voulgaropulos, A.N., Underberg, W.J.M., Somsen, G.W. and de Jong, G.J. (2006) Combination of flow injection analysis and capillary electrophoresis – Laser induced fluorescence via a valve interface for on-line derivatization and analysis os amino acids and peptides. *Journal of Chromatography A*, **1132**, 297–303.

79 Santos, B., Simonet, B.M., Ríos, A. and Valcárcel, M. (2007) On-line coupling of solid phase microextraction to commercial CE-MS equipment. *Electrophoresis*, **28**, 1312–1318.

80 Portillo, M., Prohibas, N., Salvadó, V. and Simonet, B.M. (2006) Vial position in the determination of chlorophenols in water by solid phase microextraction. *Journal of Chromatography A*, **1103**, 29–34.

81 Palmarsdottir, S., Thordarson, E., Edholm, L.E., Jonson, J.A. and Mathiasson, L. (1997) Miniaturized supported liquid membrane device for selective on-line enrichment of basic drugs in plasma combined with capillary zone electrophoresis. *Analytical Chemistry*, **69**, 1732–1737.

82 Chen, Y., Lu, W., Chen, X. and Hu, Z. (2007) Combination of flow injection with electrophoresis using capillaries and chips. *Electrophoresis*, **28**, 33–44.

83 Li, C.C., Lee, G.B. and Chen, S.H. (2002) Automation for continuous analysis on microchip electrophoresis using a flow through sampling. *Electrophoresis*, **23**, 3550–3557.

84 Tyson, J.F. (1988) Flow injection calibration techniques, Fresenius. *Analytical Chemistry*, **329**, 663–667.

85 Kuban, P., Tennberg, K., Tryzell, R. and Karlberg, B. (1998) Calibration principles for flow injection analysis capillary electrophoresis systems with electrokinetic injection. *Journal of Chromatography A*, **808**, 219–227.

86 Manganiello, L., Arce, L., Ríos, A. and Valcárcel, M. (2002) Piezoelectric screening coupled on line to capillary electrophoresis for detection and speciation of mercury. *Journal of Separation Science*, **25**, 319–327.

87 Peña, R., Alcaraz, M.C., Arce, L., Ríos, A. and Valcárcel, M. (2002) Screening of aflotoxins in feed samples using a flow system coupled to capillary electrophoresis. *Journal of Chromatography A*, **967**, 303–314.

88 Santos, B., Lista, A., Simonet, B.M., Ríos, A. and Valcárcel, M. (2005) Screening and analytical confirmation of sulfonamide residues in milk by capillary electrophoresis-mass spectrometry. *Electrophoresis*, **26**, 1567–1575.

89 Suárez, B., Simonet, B.M., Cárdenas, S. and Valcárcel, M. (2007) Determination of non-antiinflamatory drugs in urine by combining an immobilized carboxylated carbon nanotubes minicolumn for solid-phase extraction with capillary electrophoresis mass spectrometry. *Journal of Chromatography A*, **1159**, 203–207.

90 Nozal, L., Arce, L., Simonet, B.M., Ríos, A. and Valcárcel, M. (2004) Rapid determination of trace levels of tetracyclines in surface water using a continuous flow manifold coupled to a capillary electrophoresis system. *Analytica Chimica Acta*, **517**, 89–94.

91 Simonet, B.M., Ríos, A., Grases, F. and Valcárcel, M. (2003) Determination of myo-inositol phosphates in food samples by flow injection capillary zone electrophoresis. *Electrophoresis*, **24**, 2092–2098.

92 Kuban, P., Engström, A., Olsson, J.C., Thorsen, G., Tryzell, R. and Karlberg, B. (1997) New interface for coupling flow injection and capillary electrophoresis. *Analytica Chimica Acta*, **337**, 117–124.

93 Roche, M.E., Oda, R.P., Machacek, D., Lawson, G.M. and Landers, J.P. (1997)

Enhanced throughput with capillary electrophoresis via continuous sequential sample injection. *Analytical Chemistry*, **69**, 99–104.

94 Kaljurand, M., Ebber, M. and Somer, T. (1995) An automatic sampling device for capillary zone electrophoresis. *Journal of High Resolution Chromatography*, **18**, 263–265.

95 Mardones, C., Ríos, A. and Valcárcel, M. (1999) Determination of chorophenols in human urine based on the integration of on-line automated clean-up and preconcentration unit with micellar electrokinetic chromatography. *Electrophoresis*, **20**, 2922–2929.

96 Chen, H.W. and Fang, Z.L. (1999) Combination of flow injection with capillary electrophoresis. Part 5. Automated preconcentration and determination of pseudoephedrine in human plasma. *Analytica Chimica Acta*, **394**, 13–22.

97 Veraart, J.R., Groot, M.C.E., Gooijer, C., Lingeman, H., Vethorts, N.H. and Brinkman, U.A.Th. (1999) On-line dialysis-SPE-CE of acidic drugs in biological samples. *Analyst*, **124**, 115–118.

98 Thompson, J.E., Vickrory, T.W. and Kennedy, R.T. (1999) Rapid determination of aspartate enantiomers in tissue samples by microdialysis coupled on-line with capillary electrophoresis. *Analytical Chemistry*, **71**, 2379–2384.

99 Lista, A.G., Arce, L., Ríos, A. and Valcárcel, M. (2001) Analysis of soil samples by capillary electrophoresis using a gas extraction sampling device in a flow system. *Analytica Chimica Acta*, **438**, 315–322.

100 Priego Capote, F. and Luque de Castro, M.D. (2007) On-line preparation of microsamples prior to CE. *Electrophoresis*, **28**, 1214–1220.

101 Chen, H., Yu, Y., Xia, Z., Tang, S., Mu, X. and Long, S. (2006) The fabrication and evaluation of inline coupling of microdialysis with capillary electrophoresis and its application in the determination of blood glucose. *Electrophoresis*, **27**, 4182–4187.

102 Wang, L., Zhang, Z. and Yang, W. (2005) Pharmakocinetic study of trimebutine maleate in rabbit blood using in vivo microdialysis coupled to capillary electrophoresis. *Journal of*

Pharmaceutical and Biomedical Analysis, **39**, 399–403.

103 Ciriacks, C.M. and Bowser, M.T. (2004) Monitoring D-serine dynamics in the rabit brain using on-line microdialysis-capillary electrophoresis. *Analytical Chemistry*, **76**, 6582–6587.

11
On-line Sample Processing Methods in Flow Analysis

Manuel Miró and Elo Harald Hansen

11.1
Introduction

The low concentration levels of target measurands and the complexity of the matrices of the real samples commonly encountered in any field of analytical chemistry, encompassing environmental, biological, industrial and biotechnological applications including food analyses, process monitoring and quality control testing, frequently impede the direct determination of core parameters even by exploiting modern analytical instrumentation. This is a consequence of the influence of the concomitant matrix components on the analytical signal, and the fact that the concentration of the target species is often below the dynamic linear range of the detection device. Thus, there is a need for the development of rugged and effective sample pretreatment procedures, prior to the quantification step, aimed at removing interfering matrix constituents and at the same time improving measurand detectability by preconcentration. When performed in a manual fashion, these preliminary operations are labor-intensive and time consuming, difficult to control systematically, and a source of major bias and accidental errors (e.g., sample contamination) that might have a decisive impact on the accuracy and precision of the analytical results. Therefore, they are viewed as the major bottleneck of the overall analytical process as regards the reliability of the analytical data [1].

The development of on-line sample pretreatment procedures exploiting the various methods of flow injection, that is, flow injection analysis (FIA), sequential injection analysis (SIA), multicommutation-based approaches and hyphenated techniques, has opened new avenues regarding automation and miniaturization of sample handling with the added advantage of saving sample and reagent consumption and reducing waste generation [2–5].

This chapter presents and discusses with selected examples the progress of the state-of-the-art within the last decade in implementing flow injection systems for on-line matrix separation and/or preconcentration (and possibly dilution) as interfaced to a variety of detection instruments. These schemes comprise the use of solvent

Advances in Flow Analysis. Edited by Marek Trojanowicz
Copyright © 2008 WILEY-VCH Verlag GmbH & Co. KGaA, Weinheim
ISBN: 978-3-527-31830-8

extraction, including salient variants such as micelle-mediated extraction, flow-batch extraction and wetting-film extraction; solid-phase extractions embracing sorbent optosensing and bead injection, on-wall molecular sorption and precipitate/(co) precipitate retention using knotted reactors; hydride/vapor generation; membrane-based separations involving (micro)dialysis, gas-diffusion, pervaporation and supported-liquid membranes; and digestion protocols, such as microwave-, ultrasound- and UV-irradiation assisted extractions. It should be noted that several of the methods herein reported hold equally well for the earlier methodology of air-segmented flow analysis though the FIA/SIA counterparts offer unique facilities. Special attention is also paid to the potential of flow-based approaches for accommodation of sample processing steps, regardless of the aggregate state of the sample medium. To this end, new trends for on-line handling of solid samples, particle and colloidal suspensions as well as air samples are thoroughly described. The chapter concludes with a discussion on further downscaling of these preliminary operations by tailoring solid-phase microextraction and solvent microextraction protocols into the flow system, thus rendering the so-called green chemistry approaches where the use of hazardous chemicals and/or organic solvents is eliminated or kept to a minimum.

11.2
On-line Sample Pretreatment Protocols for Aqueous and Air Samples

11.2.1
On-line Dilution

Sample dilution is the simplest unit operation in flow systems. It is inherent to the concept of controlled dispersion and reproducible timing in flow analysis. The idea behind it is to perform single standard calibration using gradient dilution [6], to minimize matrix interfering effects, or to bring the measurand concentration within the dynamic linear range of the detection device. Moderate dilution factors might be readily accomplished via microfluidic manipulation within the flow network using flow-splitting, zone sampling or merging zone procedures, as demonstrated with the atomic absorption/emission determination of metal elements in saline matrices [7]. Another elegant alternative is to exploit flow-reversal approaches in SIA aimed at dispersing the sample segment into the carrier stream to different extents [8, 9]. The degree of dilution may be tailored to the requirement of the assays by selecting the number of forward and backward movements and the lengths of the sample displacements [10]. However, sample dilutions higher than 20-fold are difficult to achieve.

The use of a small mixing chamber with an inner volume of a few milliliters and clustered to a peripheral port of the selection valve of an SIA manifold might be regarded as the most reproducible and reliable way of performing on-line controlled sample dilution within a wide concentration range [10, 11]. Dilution factors of a few hundreds might be attained in a fully automatic fashion by appropriate selection of that part of the diluted zone to be entrapped within the holding coil [11]. Particularly interesting possibilities for dilution of polar species arise from using single or serial arrangements of flow-through parallel-plate dialyzers furnished with regenerated cellulose (Cuprophan) membranes [12].

11.2.2
Derivatization Reactions

Since the early development of flow analysis, many analytical applications have been based on its role as a mechanized system for homogeneous chemical derivatization. FIA and related approaches thereof have proven to be ideal vehicles for the conversion of non-detectable species into detectable ones through kinetically controlled chemical reactions. Both the highly reproducible mixing of streams and controllable timing enable novel applications never thought of before. The monitoring of unstable intermediate reaction products clearly illustrates the powerful capabilities of flow analysis. For example, the classical spectrophotometric determination of cyanide is based on the halogenation of the target species with chloramine-T, whereafter it reacts with a mixture of pyrazolone or barbituric acid to form a violet polymethine dye. However, determination of the metastable red-colored intermediate product allows a significant sensitivity improvement since the color development is more intense.

The use of solid-phase reactors for heterogeneous chemical derivatization should, however, be considered as one of the most rapidly developing and challenging areas in flow analysis research [13]. The implementation of packed-bed microcolumns in the flow network for on-line reduction or oxidation of targeted compounds has proven to be an excellent avenue for speciation studies [2]. A worthwhile virtue of FIA, as opposed to batch procedures, is the likelihood of using unstable reagents in solution via entrapment in suitable matrices, thus obtaining highly stable reagent sources, or in-line generation of the active species at a solid-phase redox reactor [14]. Displacement reactions based on the detection of the delivered species following passage of the sample through a packed column containing an insoluble salt have attracted the interest of a plethora of researchers [15, 16] owing to the feasibility of reagent reutilization, precipitate collection in the reactor as well as on-line hyphenation with various detection instruments such as spectrophotometers or atomic absorption/emission spectrometers.

The coupling of catalytic heterogeneous reactions using enzyme packed-bed or open-tubular reactors with numerous detection techniques has been presented as an appealing approach for determination of substrates of enzymatic reactions and the activity of enzymes [17]. The most relevant benefits gained by the exploitation of flow analysis are the strict repeatability of time-dependent operations and the limited consumption of the often costly biocomponents via immobilization procedures, advantage being taken of the increased enzyme stability after physical or chemical binding to appropriate supports [18]. An increased rigidity of the structure of the immobilized enzyme results in preservation of the native configuration of the protein and makes unfold less probable. The advantage of immobilization of the enzyme in flow-through column reactors via Schiff base formation in comparison to integrated biosensors is the ability to entrap a much larger amount of catalyst. The ensuing high activity facilitates an extensive and rapid conversion of substrate into the detectable product, thus yielding lower detection limits [19].

11.2.3
Solvent Extraction

Liquid–liquid extraction was the first sample processing approach for isolation and preconcentration of target species adapted to automatic flow systems. The foremost asset was to overcome the main shortcomings that had arisen from their batch counterparts, namely, loss of measurand by manipulation, handling of large volumes of hazardous reagents, low sampling frequency, contamination of the laboratory atmosphere by organic vapors, and generation of large amounts of residual solvents. The essential steps of a batch liquid–liquid extraction procedure, that is, introduction of defined volumes of the aqueous and organic phases, transfer of derivatized species by bringing the immiscible solutions into intimate contact, and physical separation of the two phases, are conventionally carried out in continuous flow systems using a phase segmentor, an extraction coil and a phase separator, as shown schematically in Figure 11.1A.

Despite the initial progress in the separation and preconcentration of transition metals from a variety of matrices [20], along with the establishment of official standard flow-based methods for phenol index [21] and anionic surfactant [22] determinations, its further development has not enjoyed the interest it deserves. In fact, the repeatability, sensitivity and accuracy of the methods are drastically limited by the existence of the above-mentioned components of the flow-through extractor. The phase separator should actually be viewed as the Achilles' Heel of the flow analyzer due to fouling and carry-over effects and low efficiency in phase separation [23, 24]. Although the overall performance of solvent extraction in flow-through systems can be improved by exploiting either novel separation designs for quantitative recovery of both the low-density and high-density phases [25, 26] or back-extraction schemes for enhanced selectivity and versatility in the coupling to detection devices [27], recent trends are directed towards the development of novel concepts and strategies for improving the ruggedness of phase separation with no need for the classical extractor components.

Figure 11.1 Sketches of typical flow-systems for automatic liquid–liquid extraction: (a) Conventional flow injection set-up furnished with a segmentor and phase separator, (b) supported liquid membrane-based scheme, (c) flow injection assembly for cloud-point extraction, (d) sequential injection system for solvent extraction and phase separation under steady-state regime. The insets in (a) illustrate the merging-tube and coaxial-type segmentors as well as the T-tube gravitational and sandwich-type phase separators. S: sample, PP: peristaltic pump, C: carrier, R: reagent, W: water, IV: injection valve, EC: extraction coil, Seg: segmentor, Sep: phase separator, DB: displacement bottle, Org: organic phase, Aq: aqueous phase, SP: segmented phase, D: detector; SLM: supported-liquid membrane, TB: thermostatic bath, FC: filtration column, HC: holding coil, S: syringe pump, MV: multiposition valve. (Adapted from Ref. [39] by permission of Bentham Science. Publishers.)

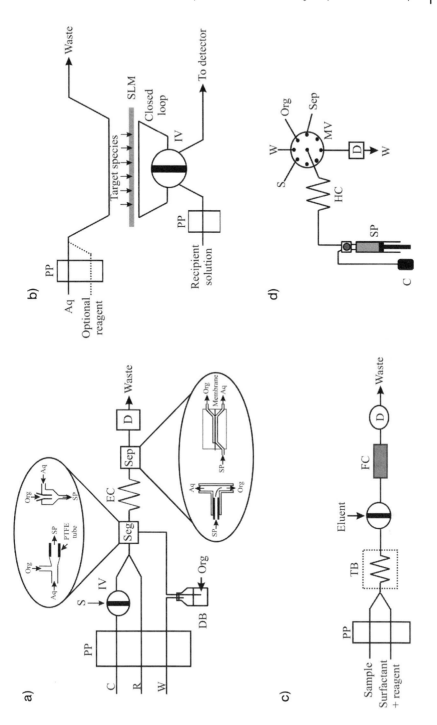

The utilization of a microporous, hydrophobic material to hold the organic extractant by capillary forces (hence called liquid membrane) offers an appealing alternative to segmented flow injection systems for the construction of efficient and automatic sample pretreatment set-ups [28]. The most accepted methodologies are the so-called supported liquid membrane extraction (SLME) [29, 30], or hollow-fiber assisted liquid–liquid microextraction [31] based on the extraction of targeted species from the donor stream into another aqueous, but frequently temporally stagnant, phase through the immobilized organic phase. For SLME, closed-loop type arrangements for the acceptor volume are frequently exploited, as depicted in Figure 11.1B. It should be emphasized that the applicability of SLME is not solely restricted to the preconcentration of ionizable organic compounds, because permanently charged compounds (e.g., metal ions) can be equally well determined by carrier mediated extraction. Readers are referred to the review articles [28, 32] for an exhaustive description of the fundamentals of on-line liquid membrane extraction and application for clinical and environmental research.

Single-phase liquid–liquid continuous extraction is an optimal approach for expediting pre-extraction derivatization reactions, as the sample is processed in a homogeneous medium. In this context, in-line micelle-mediated extraction techniques [33], such as flow injection cloud-point extraction (CPE), have recently attracted considerable attention for the isolation and enrichment of hydrophobic organic species or metal ions, which are either in their native form or in non-charged covalent chelates or ion-pairs [34, 35]. In CPE, following the generation of derivatives of the target compounds, the increase in temperature causes the single phase system to be broken down into two distinct phases; one of them, the so-called surfactant-enriched phase, entraps the target species from the bulk medium into organized entities by hydrophobic or electrostatic interactions [36]. This mixture is commonly delivered to a microcolumn packed with a suitable filtering material, namely, glass wool, cotton or nylon fibers, for retention of the large-size measurand-entrapped surfactant aggregates, as illustrated in Figure 11.1C. However, it should be borne in mind that the adaptation of CPE to FIA spectrophotometric/spectrofluorimetric assemblies might be hindered by the vapor bubbles generated within the flow network when raising the temperature. This shortcoming can be overcome by using salting-out agents for inducing on-line phase separations [37]. Other organized assemblies encompassing vesicles, admicelles and microemulsions have also been utilized in flow systems for tuning the sensitivity and selectivity of analytical methods by concentration or solubilization of measurands, respectively [36].

A noteworthy advance in terms of enhanced sensitivity and avoidance of carryover between samples has been achieved with the advent of the novel SIA-wetting film methodology [38]. This is based on coating the inner walls of a polytetrafluoroethylene tubular reactor with a thin layer of organic phase, as facilitated by the hydrophobic interactions between the solvent and the reactor material, which will cause a delay of the organic phase with respect to that of the aqueous solutions. The enhanced aqueous/organic phase volume ratio as compared with classical FIA solvent extraction renders high concentration factors. This film is also responsible for high extraction efficiencies, avoiding the axial dispersion of the measurand that occurs in segmented FIA extraction. Wetting-film extraction has proven suitable for the preconcentration and speciation of metal traces in hyphenation with atomic or molecular spectrometric detection, for the simultaneous determination of various phenolic isomers and for monitoring of radionuclides at low activity levels [39]. The most serious shortcoming of this approach is the pseudo-stationary nature of the organic phase [40], which requires careful optimization of the hydrodynamic variables for each specific analytical application. Moreover, the capacity of the film is rather limited.

The so-called extraction chromatography (EC) has also exploited the virtues of SIA for sample handling. It is utilized for the isolation of target radioisotopes from inactive matrix ingredients and interfering radioactive fission products by using inert polymeric columns impregnated with selective chelating agents or macrocyclic ionophores. Thus, Grate and coworkers [41, 42] have highlighted the potential of SIA-EC as a front-end to ICP-MS for actinide isotopic measurements and to non-selective liquid scintillation spectrometers for on-line radiometric detection, as demonstrated via separation, identification and/or quantification of ^{90}Sr, ^{99}Tc, and actinide isotopes, viz., Am, Cm, Pu, Th, Np and U, in nuclear waste waters. Not least, enhanced chromatographic resolution of actinide species may be accomplished by application of multiple elution schemes entailing the on-line modification of the chemical composition of the mobile phase [42].

The implementation of microextraction chambers in a flow manifold furnished with a set of commutators and optical sensors constitutes the basis of a fully automatic solvent extraction approach supplementary to continuous FIA extraction [43]. The resulting microscale liquid-phase analyzers operating under discontinuous-flow mode encompass a suite of advantages, including minimization of solvent consumption, gravitational phase separation under stagnant in lieu of dynamic regime, and reliable transportation of the enriched organic phase to the flow-through detector [44]. Identical assets are to be gained by clustering a conventional separation module to one of the peripheral ports of the multiposition valve in SIA systems [24] (see Figure 11.1D).

The analytical capabilities of other innovative extraction modules and flow-through arrangements for on-line solvent extraction, involving iterative flow reversal extraction, on-tube detection, chromatomembrane extraction and extractive optrodes are described in detail in a recent fundamental review article [39]. The influence of external energy sources, such as ultrasounds, on mass transfer in LLE systems has also been explored, and a significant effect on those systems involving chemical derivatization has been reported [45].

11.2.4
Sorbent Extraction

On-line solid-phase extraction (SPE), also called sorbent extraction, is the predominant sample processing method that has been growing rapidly as a consequence of its straightforward operation and high separation and preconcentration capabilities. Most often it is employed by using packed-bed or disk-phase-based microcolumns,

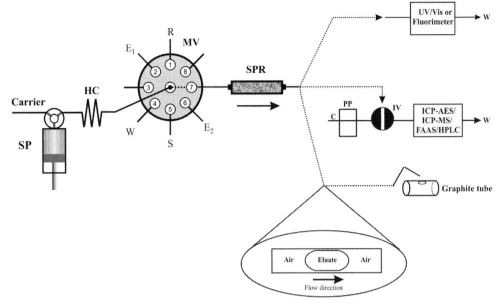

Figure 11.2 On-line hyphenation of sequential injection systems containing a packed bed column with various optional detection devices. The inset illustrates the principle of air-sandwiched elution for electrothermal atomic absorption detection. SP: syringe pump, HC: holding coil, MV: multiposition valve, SPR: solid-phase reactor, S: sample, R: reagent, E: eluent, PP: peristaltic pump, C: carrier, IV: injection valve, W: waste. ICP-AES: inductively coupled plasma-atomic emission spectrometry, ICP-MS: inductively coupled plasma-mass spectrometry, FAAS: flame atomic absorption spectrometry, HPLC: high performance liquid chromatography. (Reproduced from Ref. [54] by permission of Elsevier Science Publishers.)

which are filled with appropriate sorptive materials and placed within the flow network prior to the detection device (see Figure 11.2 for a schematic illustration of an SIA system furnished with a packed column for SPE purposes). The ultimate goal is to improve the sensitivity of the analytical procedure and/or overcome the inherent low tolerance of the detection system to sample constituents by preconcentrating the measurand and/or removing interfering components from complex biological, industrial or environmental matrices. Whenever trace analysis is pursued, matrix separation occurs concomitantly with measurand enrichment.

Sorbent extraction plays an outstanding role in combination with atomic spectrometric detection, as has been demonstrated in several recent review articles [2, 25, 46, 47]. The discontinuous operation of SIA makes this approach well suited for hyphenation of SPE with ETAAS in a column-in-tip or traditional column-in-manifold mode as a consequence of the discrete, non-continuous nature of the detector [47, 48] (see Figure 11.2). It should be stressed that inorganic metal speciation analyses are compatible with FIA/SIA-SPE by employing selective sorption or chemical elution protocols as demonstrated with the determination of, for example, Cr(III)/Cr(VI), As(III)/As(V), Fe(II)/Fe(III), Se(IV)/Se(VI) and Sb(III)/Sb(V) [49–52].

A plethora of solid-phase reactors with different designs, mainly uniformly-bored or conical micro-columns, have been successfully implemented in flow-based assemblies. The temporary retention of low level concentrations of individual metal ions, nutrients and charged chelates by electrostatic interactions on ion-exchange microcolumns or chelating reactors (generally containing iminodiacetic acid or 8-hydroxyquinoline moieties) has been utilized in flow-through SPE methods applied to a variety of analytical fields [53, 54]. Derivatized non-polar chelates (mainly dithiocarbamates or dithiophosphates) or hydrophobic species, for example, organic pollutants, dyes and drugs, have been preconcentrated on reversed-phase materials (e.g., octadecyl-chemically modified silicagel, polytetrafluoroethylene (PTFE) beads or turnings or co-polymeric sorbents with monomers of variable polarity) by partitioning, hydrophobic, or π–π interactions. The target compounds are either directly retained on the active phase after appropriate sorbent conditioning or in-line derivatized into a suitable chemical form to be sorbed onto the reactive surfaces. The latter alternative gives rise to selectivity enhancement for trace metal assays via intelligent selection of the derivatization agent, as demonstrated by the uptake of metal species onto reversed-phase materials or modified copolymers [2, 47].

Highly selective, tailor-made sorbent materials based on molecular or ionic recognition properties are currently being synthesized, yielding the so-called molecular or ionic imprinted polymers [55, 56]. However, the actual applicability in flow systems has been rather limited owing to bleeding of the unleached template with time. Special SPE supports possessing restricted access properties, whereby proteins and macromolecular matrix components are size excluded while low-molecular weight measurands are enriched, have been developed and implemented in flow systems for direct injection and automatic analysis of untreated biological samples [57].

The SPE extraction mode involving sorption/elution has commonly been utilized in hyphenation with various optical detection techniques. Optical sensing at solid surfaces, involving direct measurement of light attenuation of the sorbent phase following the uptake of the target compounds conveniently derivatized, is an excellent alternative to conventional procedures with eluate detection [58, 59]. The partial loss of the preconcentration capabilities gained during the sorption step as a result of dilution in the eluate phase is thus circumvented. The current state-of-the-art of the flow-through solid-phase optosensing concept has been reviewed recently [60], where potentials and limitations of the various optode/sensor configurations for practical applications have been thoroughly described.

Solid-phase extraction of soluble metallo-organic complexes has also proved feasible without the aid of a particle packed column using three-dimensionally disorientated (knotted) reactors made of naked or precoated PTFE tubing as a sorptive medium [61–63] (see Figure 11.3). The main benefits of such systems are the relatively low flow resistance in comparison with packed columns, the high concentration efficiencies as a consequence of the large flow rates allowable, and their virtually unlimited lifetime. However, the lower retention efficiencies achieved (typically <40%) force the injection of large sample volumes for ultratrace level determinations.

Metal ion

Reagent

Sample

Matrix

Loading of target species

Waste

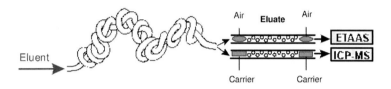

Eluent

Air   Eluate   Air

ETAAS

ICP-MS

Carrier   Carrier

Elution

Figure 11.3 Schematic illustration of the analytical procedure for isolation and preconcentration of metal traces by on-line (co) precipitation. The precipitate formed by reaction of the target species with a derivatization reagent is entrapped in a knotted reactor. In order to proceed with the analytical measurement by atomic spectrometry, the precipitate is then dissolved in a metered volume of eluent. (Reproduced from Ref. [3] by permission of Elsevier Science Publishers.)

Conventionally, the sorbent column is regarded as an integral component of the flow network, being used repeatedly for the sample loading and elution sequences, and being replaced or repacked only after long-term operation. However, the repeated use of sorbent reactors in flow systems may give rise to several problems associated with the build-up of backpressure, measurand carryover or irreversible modification of the surface properties of the sorptive entities as a consequence of contamination, deactivation or even loss of active sites. A superb alternative to overcome these drawbacks is the so-called bead-injection (BI) (or bead renewable) approach [64, 65] where the contents of the SPE column are withdrawn on-line and replaced for each analytical run. Further information on the various configurations designed for BI, that is, the jet-ring cell and the lab-on-a-valve scheme, and relevant applications of this approach, typically linked to SIA, can be found in Chapter 3.

11.2.5
Precipitation/Co-Precipitation

Precipitation or coprecipitation is yet another commonly used approach for isolation and preconcentration of trace elements. Current flow systems capitalize on the filter-less collection of the (co)precipitate onto an open tubular knotted reactor [62], which is a consequence of the centrifugal force created in the flowing stream that carries the particles towards the tubing walls (see Figure 11.3). The sample is initially mixed in-line with a reagent giving rise to the formation of a water-insoluble species, that is entrapped within the reactor. Provided that the (co)precipitate and the tube material are compatible, that is, either both hydrophobic or both hydrophilic, the target elements will be collected at the inner walls in a thin layer, thereby not giving rise to the creation of flow resistance [46, 47]. After washing of the retained particles, the (co)precipitate is eluted by a minute, well-defined volume of eluent, and transferred discretely to the detection device by sandwiching it with air-segments or liquid zones. However, it should be borne in mind that the high level concentrations of (co)precipitation agent utilized may be incompatible with the detection technique used as a result of the generation of spectroscopic and/or non-spectroscopic interfering effects. Thus, (co)precipitation schemes have been traditionally coupled to hydride generation approaches for the determination of on-line evolved volatile compounds which are effectively isolated from interfering matrix constituents [66, 67] (see next section).

11.2.6
Gas–Liquid Separation

The use of flow manifolds furnished with gas–liquid separators for evolving and isolation of vapor species from a liquid reaction medium into a stream of noble gas with further transport to a continuously operating detector has attracted considerable interest and enjoyed increasing popularity in recent years when hyphenated to inductively coupled plasma-atomic emission spectrometry (ICP-AES) and atomic fluorescence spectrometry (AFS) [68, 69]. The most widely used approach for on-line generation of the volatile compounds employs a suitable chemical reducing agent, such as sodium tetrahydroborate, however, on-line electrochemical hydride generation (HG) involving the application of an electrolysis current has been recently described [70, 71]. Both schemes can completely eliminate concomitant matrix ingredients, thus being most favorable for analyzing biological samples where serious spectroscopic and non-spectroscopic interferences are frequently encountered.

Vapor generation techniques for hydride-forming elements (namely, As, Se, Sb, Bi, Te, Sn and Pb) and volatile species of metals (namely, Hg, Au, Zn, Cd, Ag, and Ni) have profited greatly from flow-based approaches in terms of a dramatic decrease in consumption of sample and costly chemicals as compared to the batch counterparts, and enhanced tolerance to interfering transition metals, for example, Ni(II), Fe(III), Co(II) and Ni(II), by application of kinetic discrimination schemes [2]. It should be noted that the aforementioned metal ions might become reduced to colloidal free metals which, in batch protocols, have been shown to act as superb catalysts for degrading the hydrides.

As to the determination of metalloid species, flow injection assemblies have been presented as suitable vehicles to implement speciation protocols for both inorganic and organic species via the careful control of the experimental variables, such as pH and tetrahydroborate concentration for the formation of the volatile species [72–74], the sequential volatilization of hydride species following cryogenic trapping [72, 75], the hyphenation of flow injection with liquid chromatographic separations [76, 77] or the operational control of focused microwave ovens or ultrasonic probes [78, 79]. Successive *in situ* trapping of hydride species onto a Pd-, W-, Zr- or Ir-precoated graphite tube of an electrothermal atomic absorption spectrometer [80] and mercury enrichment as a gold amalgam prior to detection [81] have led to the improvement of detection limits by more than one decade. Although the in-atomizer sequestration might appear attractive, it is, from an operational point of view, preferable to effect all the fluidic manipulations in a single cycle exploiting either on-line sorbent extraction or (co)precipitation protocols. Not only because the sample processing protocol might be optimized regardless of the vapor generation scheme used, but also because it yields much higher enrichment factors in considerably shorter sampling times [2].

11.2.7
Membrane-based Separation

The implementation of permselective barriers into flow networks has opened up new possibilities for the automation of sample processing steps. Amongst the various membrane separation techniques including filtration, ultrafiltration, reverse osmosis, pervaporation, membrane extraction, (micro)dialysis and gas-diffusion (also termed isothermal distillation), the latter two methodologies have found the largest number of applications in flow analysis. Sandwich-type units embodying flat sheet membranes are the preferred configuration for flow-through analyzers. Concentric or linear-type arrangements with hollow-fiber membranes have also been employed for the construction of dynamic headspace devices and microdialysis probes [82]. Unconventional membraneless units housing an internal headspace zone have also been devised by proper design of the donor and acceptor channels [83]. The hyphenation of flow-through membrane-based modules to modern analytical instrumentation is shown schematically in Figure 11.4, and discussed in more detail in the following.

11.2.7.1 Gas Diffusion
This term was introduced in flow-analysis to describe a method whereby endogenous or generated gaseous species in a donor stream, are transported, via a microporous or homogenous hydrophobic membrane (e.g., polytetrafluoroethylene or silicone, respectively), into a receiver liquid solution. Since only relatively few compounds are sufficiently volatile at room temperature, gas diffusion is associated with a high degree of selectivity enhancement. Ionic compounds are totally excluded and colored and cloudy samples can generally be analyzed, even with spectrophotometric methods, without any additional pretreatment. Common procedures involving distillative separation or sample digestion (e.g., Kjeldahl nitrogen, free and total cyanide) [84, 85] or the evolution of hydrides or volatile species (e.g., AsH_3, Hg) [86, 87] can be readily tailored to flow systems for the interference-free analysis of highly

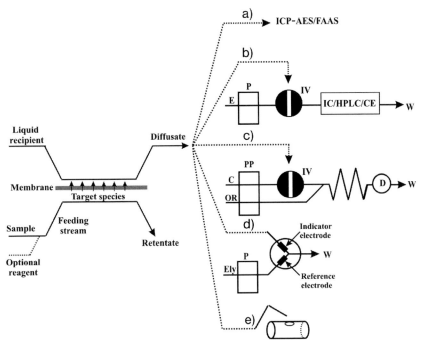

Figure 11.4 Coupling of membrane-based sample processing techniques with analytical instrumentation exploiting flow analysis approaches. On-line hyphenation to (a) continuously-operating atomic spectrometers, (b) column separation systems, (c) secondary flow injection manifolds (with optional derivatization reactions), (d) miniaturized potentiometric devices or micro-total analysis systems and (e) electrothermal atomic absorption spectrometers. ICP-AES: inductively coupled plasma-atomic emission spectrometry, FAAS: flame atomic absorption spectrometry, IV: injection pump, P: pump, CE: capillary electrophoresis, IC: ion-chromatography, HPLC: high performance liquid chromatography, El: eluent, C: carrier, OR: optional reagent, PP: peristaltic pump, Dt: detector, Ely: electrolyte, W: waste. (Adapted from Ref. [82] by permission of Elsevier Science Publishers.)

particle-loaded samples, such as wastewaters, industrial effluents, soil slurries and plant materials.

The intrinsic high selectivity of flow-based gas diffusion analyzers facilitates the utilization of non-discriminating detectors or non-selective analytical procedures, where conductivity changes using conventional or contactless detectors [88] or the color change of acid–base indicators [82, 89] have been employed for quantitative measurements. Regarding the preconcentration capabilities, gas-diffusion procedures may take full advantage of SIA or multicommutated approaches by resorting to stopped-flow or flow-reversal schemes [90–92].

The analytical applicability of this membrane-based separation technique is not restricted to on-line processing of liquid samples, but can be used also for automatic gas sampling. In the so-called diffusion scrubbers or permeation denuders [93, 94], soluble atmospheric trace gases passing through a gas-permeable polymeric membrane are trapped within the recipient aqueous stream and can be further processed by chemical derivatization [95]. Alternatively, the scrubber solution may feed the injection system of an ion- or gas-chromatograph allowing multicomponent determinations of entrapped protolytic gases (e.g., ammonia, sulfur dioxide, nitric acid, nitrous acid and hydrochloric acid) [93, 96]. Using appropriate configurations and barrier dimensions both fast response times and near-quantitative collection of the measurand can be attained [96]. A striking feature of diffusion scrubbers as compared with conventional filtration, impinger or sorbent sampling is the fact that potential interactions between the gas and the concomitantly present particulate matter during sampling are circumvented.

11.2.7.2 Dialysis

The use of passive dialysis has a long tradition in flowing systems, and a huge number of papers have appeared since its first application in air-segmented continuous flow analysis. It is aimed at separating low-molecular weight compounds from interfering macromolecules, humic substances, colloidal matter and suspended particles. Dialysis efficiency is related to the cell configuration and its geometrical dimensions, the properties of the hydrophilic membrane (namely, kind of material, porosity and thickness), and the flow rates and chemical composition of the donor and receiver streams [12, 97]. Depending on the intended application, flow dialysis can be tuned not only to obtain a high degree of sample dilution but also to give almost complete recovery of the measurand by temporarily halting the receiver solution, or implementation of either flow reversal procedures or sequestration reactions in the acceptor phase in order to maintain the concentration gradient [98, 99]. Particularly interesting possibilities arise from the fact that multidetection schemes can be implemented into the flow manifold, permitting simultaneous detection in a single sample plug (e.g., the determination of total and dialysable metal ions) by coupling the donor and acceptor streams to appropriate flow-through detectors, or by parallel or serial arrangement of two or more separation units [12, 100].

Donnan or active dialysis involving transfer of species through permanently charged surfaces has been less frequently employed in flow systems. However, for ionic compounds it offers preconcentration capabilities and improved selectivity due to electrostatic interactions between the sample components and the ionogenic moieties of the membrane. Typical applications encompass trace metal enrichment followed by atomic spectrometric detection [101, 102].

Microdialysis is a specialized application of dialysis used traditionally in neurochemistry and pharmacokinetic studies for dynamic monitoring of extracellular chemical events in living tissues. Flow systems have been utilized for hyphenation of the miniaturized sampling unit to chromatographic and electrophoretic separations [103, 104]. However, the microdialysis concept can be expanded beyond its current applications, as recent reports on the use of microdialysers for automatic microsampling of transition metals, organic acids and saccharides in environmentally relevant matrices have indicated [105].

11.2.7.3 Pervaporation

The combination of continuous evaporation and gas-diffusion, that is, pervaporation, has been established as a powerful membrane-based separation technique for on-line processing of liquid samples containing high particle load and solid substrates, as demonstrated extensively by Luque de Castro and coworkers [106, 107]. To generate volatile species in the closed system (e.g., hydrogen sulfide, ammonia, carbon dioxide, hydrogen cyanide, sulfur dioxide and metalloid hydrides), a releasing chemical agent merges in-line with the sample stream feeding the module [108, 109]. Alternatively, thermal volatilization and microwave-assisted digestion have been used as releasing procedures for the determination/speciation analysis of inorganic and organic species in foodstuffs and environmentally relevant solid substrates [110, 111] The sample does not contact the hydrophobic membrane, but the volatile reaction product evaporates into the headspace of the assembled device before diffusion through the membrane into the acceptor solution, wherein it reacts with proper derivatization agents prior to detection, as in gas-diffusion procedures.

11.2.8
Digestion Protocols

Flow-based approaches should be regarded as ideal vehicles for on-line sample digestion procedures as has been proven via their coupling to powerful analytical techniques such as HG-AAS, ICP-AES, ICP-MS or to hybrid techniques such as HPLC-HG-ICP-MS or HPLC-HG-AAS/AFS for speciation studies of metal and metalloid species [76, 77]. The idea behind these procedures is to mineralize complex environmental and biological samples (e.g., whole blood, urine, sewage sludges, soils, seafood, beverages) with minimum operation attention in order to decompose organic matrices that might interfere with the analytical measurement, or to transform the measurand into a chemical form that can be monitored by an element specific detection technique, for example, HG-AAS/AFS/ICP-MS [112]. They might also be utilized for improving the yield of liquid-phase derivatization reactions with slow kinetics [113].

Three sample processing procedures involving the application of external energy sources are currently being extensively used in an on-line fashion, that is, UV-photooxidation [114, 115], microwave-assisted decomposition [78, 112, 116] and ultrasound-assisted sample preparation [117, 118]. The first two approaches, in combination with oxidizing agents, have proven crucial in the environmental field for monitoring purposes of, for instance, dissolved organic phosphorus, total dissolved phosphorus, chemical oxygen demand, bound cyanide, total nitrogen and inorganic and organic mercury [5]. Recent trends are focused on expanding the analytical applicability of optical techniques, such as spectrofluorimetry and chemiluminescence, for determination of pesticides or drugs which do not show intrinsic luminescence, by direct on-line photolysis with UV light in order to induce fluorescence or chemiluminescence [119–121].

However, it should be borne in mind that the implementation of in-line sample digestion protocols is not as straightforward as anticipated due to the limitation of short reaction times, evolution of gas bubbles during sample processing and the high percentage of non-absorbed radiation. These shortcomings have been alleviated to some extent by careful optimization of the chemical variables and reactor configuration as well as by tailoring a debubbler at the outlet of the flow-through digestion device.

11.3
On-line Processing of Solid Samples: Leaching/Extraction Methods

In recent years, much effort has been focused on the design and characterization of sample processing units coupled to flowing systems aiming to enable direct introduction of solid samples (e.g., foodstuffs and environmental substrates) in a mechanized fashion [122]. In this respect, various sample pretreatment techniques including electrolytic dissolution [123], on-line dialysis [97, 98], in-line filtration or ultrafiltration [124] and pervaporation (see above) have been successfully implemented in continuous flow systems. The interfacing of flow injection/continuous flow systems with alternative treatments such as conductive heating [125], microwave- [126, 127] or ultrasound-assisted extraction [118, 128], supercritical carbon dioxide extraction [129] or subcritical water extraction [130, 131] has proven useful for monitoring organic pollutants (namely, polycyclic aromatic hydrocarbons, phenolic species, chlorinated biphenyls, pesticides and endocrine disruptors) as well as metal and metalloid traces in soils, seeds, plant materials and foodstuffs. These methods usually provide similar efficiencies to those endorsed by the regulatory authorities (e.g., US-Environmental Protection Agency), but with dramatic reduction in both the extraction time and sample handling.

It should be stressed that there is current interest in adapting the classical end-over-end sequential extraction schemes and *in vitro* gastrointestinal digestion protocols for metal, metalloid and nutrient fractionation in solid substrates to an on-line dynamic process. It is aimed at shortening the time required to execute the batch methods and fostering a more accurate quantification of bioavailable pools while getting relevant knowledge on the extraction kinetics and the efficiency of the extracting agent [132, 133]. Besides chemical fractionation, cross-flow ultrafiltration protocols have been studied and optimized over the last decade for size fractionation of particles, macromolecules and colloids [134, 135].

In the pharmaceutical area, special attention has been devoted to the development of flow systems for monitoring dissolution profiles of tablets [136] and/or the *in vitro* testing of the release and permeation of topical semisolid preparations through a membrane mimicking the skin via the use of the so-called Franz diffusion cell [137].

With respect to the sampling modalities, the implementation of microcolumn reactors packed with solid materials [124–132], and the coupling of the slurry sampling strategy with AAS [138] or liquid-phase detectors via on-line dialysis separation [98], are especially worth mentioning. The capabilities of flow systems for direct processing and analysis of solid materials when hyphenated to appropriate analytical instrumentation are addressed in comprehensive review articles [122, 133].

11.4
Trends and Perspectives

The promise of revolutionizing sample processing through automation of micro-fluidic handling, miniaturization of the flow network and improvement of the ruggedness of both the analytical methods and the equipment used has inspired the development and progress of the various types of flow injection analysis. The pursued aim is to speed the preliminary operations within the analytical process while operating at the micro-, or submicro-level in order to reduce the usage of large amounts of chemicals and organic solvents, thus yielding the so-called "green analytical methods". Relevant analytical applications of flow-based systems furnished with sample pretreatment/enrichment modules are compiled in Table 11.1.

In the field of microminiaturization of flow systems, the Lab-on-a-valve (LOV) concept has opened up new possibilities. The rigid and compact SIA-LOV coupling behaves as a portable and versatile laboratory that enhances the repeatability of any desired sample processing sequence at the μL-scale level. Most importantly, LOV is currently emerging as a downscaled tool to overcome the dilemma of Lab-on-a-Chip microsystems to admit real-life samples [139]. This is nurtured via its intrinsic flexibility for accommodation of sample pretreatment schemes, for example, sorbent extraction via integrated microcolumn reactors [65, 140], which are imperative for monitoring trace level concentrations of target species in complex environmental and biological samples.

The use of various external energy sources within the flow manifold, such as ultraviolet, ultrasound and microwave radiation has facilitated the on-line handling of samples, irrespective of the complexity of the matrix and aggregation state prior to further quantification of target species by a downstream detector. In this context, on-line electro-stacking based on the different migration of ionic species within an electrical field has recently been proposed as an alternative means for preconcentration and speciation analysis [141].

Current efforts are being directed to the launching of novel microextraction techniques for sample processing, such as solid-phase microextraction (SPME) and liquid–liquid microextraction (LLME), including microdrop extraction and hollow-fiber assisted extraction, which have been utilized mostly for off-line sample treatment prior to chromatographic separations. The coupling of microextraction techniques with flow analysis and chromatographic separations is attracting increasing attention in order to develop entirely automatic systems, as demonstrated by the recent works dealing with on-line single-drop LLME [142] and in-tube SPME [143] methods as hyphenated to HPLC or GC.

Table 11.1 Relevant applications of flow analysis as an on-line sample pretreatment tool for the isolation and enrichment of target species in aqueous solutions.

Separation/ preconcentration technique	Detection principle	Typical measurands	Features	Relevant references
solvent extraction/cloud-point extraction	spectrophotometry, ETAAS, FAAS, ICP-AES	metal ions, surfactants, phenolic derivatives, aliphatic and aromatic hydrocarbons, pesticides, vitamins	(I) Formation of neutral chelates or ion-paired complexes with metal traces	3, 4, 5, 20, 23, 33, 34, 35, 36, 39, 142
			(II) The larger the phase volume ratio the larger the enrichment factors	
			(III) Use of sequential injection systems for phase separation under static regime	
			(IV) Potential miniaturization by exploitation of flow-through single-drop microextraction methods	
sorbent extraction	mainly atomic spectrometry but also photometry, spectrofluorimetry and chemiluminescence	nutrients, inorganic species, metal ions, radionuclides, drugs, vitamins	(I) Frequently used to improve selectivity of AAS, ICP-MS, ICP-AES and luminescence methods with concomitant analyte enrichment	2, 3, 4, 5, 46, 47, 48, 53, 54, 57, 59, 60, 61, 64
			(II) Employing solid-phase optosensing methodologies for sensitivity improvement	
			(III) Used for speciation analysis of metal species (e.g., Fe, Al, Cr, Cd)	
			(IV) Better than liquid–liquid extraction because of the higher separation efficiencies and enrichment factors	
			(V) Application of column renewable schemes using bead injection approaches	
			(VI) Coupled to column separation systems, usually HPLC, for multiparametric determination of traces of pollutants	

Table 11.1 (*Continued*)

Separation/ preconcentration technique	Detection principle	Typical measurands	Features	Relevant references
			(VII) Potential miniaturization by exploitation of in-tube solid-phase microextraction	
vapor generation (cold vapor and hydride generation)	AAS, ICP-AES, ICP-MS and AFS	metalloids (As, Se, Te, Sb, Bi), Pb, and Sn with hydride generation	(I) Construction of dedicated configurations for gas–liquid separations	3, 5, 67, 68, 71, 75, 78, 79, 80
		Hg with cold vapor techniques	(II) Implementation of various speciation strategies in flowing systems for metalloids and Hg (inorganic/organic species)	
		Cd, Zn, Au, Ag and Ni with vapour generation	(III) FI is preferred in HG-AFS aiming to feed the flame continuously with tetrahydroborate, which is used as a hydrogen source	
			(IV) Preconcentration of Hg via gold-trap collection	
gas-diffusion/ pervaporation	spectrophotometry (also in the gas phase), conductivity, potentiometry, amperometry and chemiluminescence	ammonium, carbonate, sulfide, cyanide and sulfite, As(III), Hg, low-molecular weight alcohols, amines, and phenolic compounds	(I) Mostly employed for selectivity enhancement (II) Adaptable to non-discriminating detectors (III) Useful for particle-containing aqueous solutions (IV) Analyte enrichment via stopped-flow approaches in the recipient stream	4, 82, 83, 84, 86, 87, 88, 90, 92, 107, 108, 110
dialysis (passive and Donnan)	spectrophotometry, spectrofluorimetry, FAAS, ICP-MS, ICP-AES, potentiometry, flame photometry and amperometry	alkaline and alkaline earth metals, heavy metals, anionic species (e.g., chloride, phosphate and sulfate), glucose, lactate, glutamate, glutamine	(I) Measurand separated from cells and high molecular weight matrix compounds (e.g., proteins)	5, 12, 82, 96, 97, 98, 101, 102, 103, 105

(*Continued*)

Table 11.1 *(Continued)*

Separation/ preconcentration technique	Detection principle	Typical measurands	Features	Relevant references
			(II) Adaptable to slurry sampling schemes	
			(III) Useful for enhancing the selectivity, stability and lifetime of ion selective electrodes	
			(IV) On-line analyte dilution via passive dialysis	
			(V) On-line preconcentration of transition metals via Donnan dialysis	
			(VI) Applicable to simultaneous multi-elemental determinations	
digestion procedures	spectrophotometry, amperometry, spectrofluorometry, chemiluminescence, HG-AAS, ICP-AES, ICP-MS, FAAS, ETAAS, and AFS	N-total, Hg-total, CN-total, phosphorus species, chemical oxygen demand, metal traces in organic matrices, pesticides, and pharmaceutical compounds	(I) Mainly UV-, ultrasound and microwave assisted digestion	5, 75, 76, 77, 112, 113, 114, 115, 116, 118, 119
			(II) Careful optimization of the experimental conditions to obtain high yields under dynamic regime	
			(III) Suitable for speciation studies	
			(IV) Photochemical degradation is used to enhance the sensitivity of luminescence methods	

Acronyms: FAAS: flame atomic absorption spectrometry, ETAAS: electrothermal atomic absorption spectrometry, AFS: atomic fluorescence spectrometry, ICP-MS: inductively coupled plasma-mass spectrometry, ICP-AES: inductively coupled plasma-atomic emission spectrometry, HG: hydride generation.

List of Abbreviations

FIA : flow injection analysis
SIA : sequential injection analysis

LOV : Lab-on-a-valve
BI : bead-injection
AAS : atomic absorption spectrometry
ETAAS : electrothermal atomic absorption spectrometry
AFS : atomic fluorescence spectrometry
HG : hydride generation
ICP-AES : inductively coupled plasma-atomic emission spectrometry
ICP-MS : inductively coupled plasma-mass spectrometry
HPLC : high-performance liquid chromatography
GC : gas chromatography
SLME : supported liquid membrane extraction
LLE : liquid–liquid extraction
SPE : solid-phase extraction
SPME : solid phase microextraction
LLME : liquid–liquid microextraction
CPE : cloud point extraction
EC : extraction chromatography
PTFE : polytetrafluoroethylene

References

1 Valcárcel, M. (2000) *Principles of Analytical Chemistry*, Springer-Verlag, Heildeberg.

2 Hansen, E.H. and Miró, M. (2007) How flow injection analysis (FIA) over the past 25 years has changed our way of performing chemical analyses. *Trends in Analytical Chemistry*, **26**, 18–26.

3 Hansen, E.H. and Wang, J.-W. (2002) Implementation of suitable flow injection/sequential-sample separation/ preconcentration schemes for determination of trace metal concentrations using detection by electrothermal atomic absorption spectrometry and inductively coupled plasma mass spectrometry. *Analytica Chimica Acta*, **467**, 3–12.

4 Economou, A. (2005) Sequential-injection analysis (SIA): a useful tool for on-line sample-handling and pre-treatment. *Trends in Analytical Chemistry*, **24**, 416–425.

5 Miró, M. and Frenzel, W. (2004) What Flow Injection has to Offer in the Environmental Analytical Field. *Microchimica Acta*, **148**, 1–20.

6 Ruzicka, J. and Hansen, E.H. (1988) *Flow Injection Analysis*, 2nd edn, Wiley-Interscience, New York, Chapter 2, pp. 44–52.

7 Holliday, A.E. and Beauchemin, D. (2003) Preliminary investigation of direct sea-water analysis by inductively coupled plasma mass spectrometry using a mixed-gas plasma, flow injection and external calibration. *Journal of Analytical Atomic Spectrometry*, **18**, 1109–1112.

8 Silva, M.S.P. and Masini, J.C. (2002) Exploiting monosegmented flow analysis to perform in-line standard additions using a single stock standard solution in spectrophotometric sequential injection procedures. *Analytica Chimica Acta*, **466**, 345–352.

9 Economou, A., Panoutsou, P. and Themelis, D.G. (2006) Enzymatic chemiluminescent assay of glucose by sequential-injection analysis with soluble

enzyme and on-line sample dilution. *Analytica Chimica Acta*, **572**, 140–147.

10 Mas-Torres, F., Estela, J.M., Miró, M., Cladera, A. and Cerdà, V. (2004) Sequential injection spectrophotometric determination of orthophosphate in beverages, wastewaters and urine samples by electrogeneration of molybdenum blue using tubular flow-through electrodes. *Analytica Chimica Acta*, **510**, 61–68.

11 Albertús, F., Horstkotte, B., Cladera, A. and Cerdà, V. (1999) A robust multisyringe system for process flow analysis. Part I. On-line dilution and single point titration of protolytes. *Analyst*, **124**, 1373–1381.

12 Van Staden, J.F. (1995) Membrane separation in flow injection systems. 1. Dialysis. *Fresenius' Journal of Analytical Chemistry*, **352**, 271–302.

13 Valcárcel, M. and Luque de Castro, M.D. (1995) Use of Solid Reagents in Flow Injection Analysis, in *Automation in the Laboratory* (ed. W.J. Hurst), VCH, Weinheim, Chapter 3, pp. 74–77.

14 Yang, M., Xu, Y. and Wang, J.-H. (2006) Lab-on-valve system integrating a chemiluminescent entity and in situ generation of nascent bromine as oxidant for chemiluminescent determination of tetracycline. *Analytical Chemistry*, **78**, 5900–5905.

15 van Staden, J.F. and Kluever, L.G. (1998) Determination of sulphide in effluent streams using a solid-phase lead(II) chromate reactor incorporated into a flow injection system. *Analytica Chimica Acta*, **369**, 157–161.

16 López-Gómez, A.V. and Martínez-Calatayud, J. (1998) Determination of cyanide by a flow injection analysis-atomic absorption spectrometric method. *Analyst*, **123**, 2103–2107.

17 Trojanowicz, M. (2000) Enzymatic Methods of Detection, in *Flow Injection Analysis-Instrumentation, and Applications*, World Scientific, Singapore, Chapter 4, pp. 165–177.

18 Manera, M., Miró, M., Estela, J.M. and Cerdà, V. (2004) A multisyringe flow injection system with immobilized glucose oxidase based on homogeneous chemiluminescence detection. *Analytica Chimica Acta*, **508**, 23–30.

19 Hansen, E.H. (1994) Flow injection analysis: a complementary or alternative concept to biosensors. *Talanta*, **41**, 939–948.

20 Trojanowicz, M. (2000) Solvent Extraction, in *Flow Injection Analysis, Instrumentation and Applications*, World Scientific, Singapore, Chapter 6, pp. 223–236.

21 EN/ISO 14402 (1999) Water quality – Determination of phenol index by flow analysis and photometric detection.

22 EN/ISO 16264 (2000) Water quality – determination of anionic surfactants by flow analysis and photometric detection.

23 Fang, Z.-L. (1993) *Flow Injection Separation and Preconcentration*, VCH, Weinheim, Chapter 3, pp. 47–83.

24 Marshall, G., Wolcott, D. and Olson, D. (2003) Zone fluidics in flow analysis: potentialities and applications. *Analytica Chimica Acta*, **499**, 29–41.

25 Wang, J.-H. and Hansen, E.H. (2002) Development of an automated sequential injection on-line solvent extraction-back extraction procedure as demonstrated for the determination of cadmium with detection by electrothermal atomic absorption spectrometry. *Analytica Chimica Acta*, **456**, 283–292.

26 Anthemidis, A.N., Zachariadis, G.A. and Stratis, J.A. (2003) Development of an on-line solvent extraction system for electrothermal atomic absorption spectrometry utilizing a new gravitational phase separator. Determination of cadmium in natural waters and urine samples. *Journal of Analytical Atomic Spectrometry*, **18**, 1400–1403.

27 Wang, J.-H. and Hansen, E.H. (2003) On-line sample-pre-treatment schemes for trace-level determinations of metals by coupling flow injection or sequential injection with ICP-MS. *Trends in Analytical Chemistry*, **22**, 836–846.

28 Jönsson, J.Å. and Mathiasson, L. (2003) Sample preparation perspectives – Membrane extraction for sample preparation. *LC-GC Europe*, **16**, 683–690.

29 Kocherginsky, N.M., Yang, Q. and Seelam, L. (2007) Recent advances in supported liquid membrane technology. *Separation and Purification Technology*, **53**, 171–177.

30 Jönsson, J.Å. and Mathiasson, L. (1999) Liquid membrane extraction in analytical sample preparation – I. Principles. *Trends in Analytical Chemistry*, **18**, 318–324.

31 Rasmussen, K.E. and Pedersen-Bjergaard, S. (2004) Developments in hollow fiber-based, liquid-phase microextraction. *Trends in Analytical Chemistry*, **23**, 1–10.

32 Jönsson, J.Å. and Mathiasson, L. (1999) Liquid membrane extraction in analytical sample preparation – II. Applications. *Trends in Analytical Chemistry*, **18**, 325–334.

33 Paleologos, E.K., Giokas, D.L. and Karayannis, M.I. (2005) Micelle-mediated separation and cloud-point extraction. *Trends in Analytical Chemistry*, **24**, 426–436.

34 Bezerra, M.A., Arruda, M.A.Z. and Ferreira, S.L.C. (2005) Cloud point extraction as a procedure of separation and pre-concentration for metal determination using spectroanalytical techniques: a review. *Applied Spectroscopy Reviews*, **40**, 269–299.

35 Silva, M.F., Cerutti, E.S. and Martinez, L.D. (2006) Coupling cloud point extraction to instrumental detection systems for metal analysis. *Microchimica Acta*, **155**, 349–364.

36 Burguera, J.L. and Burguera, M. (2004) Analytical applications of organized assemblies for on-line spectrometric determinations: present and future. *Talanta*, **64**, 1099–1108.

37 Garrido, M., Di Nezio, M.S., Lista, A.G., Palomeque, M. and Fernández-Band, B.S. (2004) Cloud-point extraction/preconcentration on-line flow injection method for mercury determination. *Analytica Chimica Acta*, **502**, 173–177.

38 Luo, Y.-Y., Al-Othman, R., Ruzicka, J. and Christian, G.D. (1996) Solvent extraction-sequential injection without segmentation and phase separation based on the wetting film formed on a teflon tube. *The Analyst*, **121**, 601–606.

39 Miró, M., Estela, J.M. and Cerdà, V. (2005) Recent advances in on-line solvent extraction exploiting flow injection/sequential injection analysis. *Current Analytical Chemistry*, **1**, 329–343.

40 Miró, M., Cladera, A., Estela, J.M. and Cerdà, V. (2001) Dual wetting-film multi-syringe flow injection analysis extraction Application to the simultaneous determination of nitrophenols. *Analytica Chimica Acta*, **438**, 103–116.

41 Egorov, O.B., O'Hara, M.J., Farmer, O.T., III and Grate, J.W. (2001) Extraction chromatographic separations and analysis of actinides using sequential injection techniques with on-line inductively coupled plasma mass spectrometry (ICP MS) detection. *Analyst*, **126**, 1594–1601.

42 Grate, J.W., Egorov, O.B. and Fiskum, S.K. (1999) Automated extraction chromatographic separations of actinides using separation-optimized sequential injection techniques. *Analyst*, **124**, 1143–1150.

43 Ródenas-Torralba, E., Reis, B.F., Morales-Rubio, A. and de la Guardia, M. (2005) An environmentally friendly multicommutated alternative to the reference method for anionic surfactant determination in water. *Talanta*, **66**, 591–599.

44 Comitre, A.L.D. and Reis, B.F. (2005) Automatic flow procedure based on multicommutation exploiting liquid–liquid extraction for

spectrophotometric lead determination in plant material. *Talanta*, **65**, 846–852.

45 Luque de Castro, M.D. and Priego-Capote, F. (2007) Ultrasound assistance to liquid–liquid extraction: a debatable analytical tool. *Analytica Chimica Acta*, **583**, 2–9.

46 Benkhedda, K., Goenaga-Infante, H., Adams, F.C. and Ivanova, E. (2002) Inductively coupled plasma mass spectrometry for trace analysis using flow injection on-line preconcentration and time-of-flight mass analyzer. *Trends in Analytical Chemistry*, **21**, 332–342.

47 Burguera, M. and Burguera, J.L. (2007) On-line electrothermal atomic absorption spectrometry configurations: recent developments and trends. *Spectrochimica Acta Part B-Atomic Spectroscopy*, **62**, 884–896.

48 Wang, J.-H. and Hansen, E.H. (2005) Trends and perspectives of flow injection/ sequential injection on-line sample-pretreatment schemes coupled to ETAAS. *Trends in Analytical Chemistry*, **24**, 1–8.

49 Jitmanee, K., Oshima, M. and Motomizu, S. (2005) Speciation of arsenic(III) and arsenic(V) by inductively coupled plasma-atomic emission spectrometry coupled with preconcentration system. *Talanta*, **66**, 529–533.

50 Marqués, M.J., Morales-Rubio, A., Salvador, A. and de la Guardia, M. (2001) Chromium speciation using activated alumina microcolumns and sequential injection analysis-flame atomic absorption spectrometry. *Talanta*, **53**, 1229–1239.

51 Long, X.-B., Miró, M. and Hansen, E.H. (2005) An automatic micro-sequential injection bead injection Lab-on-Valve (μSI-BI-LOV) assembly for speciation analysis of ultra trace levels of Cr(III) and Cr(VI) incorporating on-line chemical reduction and employing detection by electrothermal atomic absorption spectrometry (ETAAS). *Journal of Analytical Atomic Spectrometry*, **20**, 1203–1211.

52 Bosch-Ojeda, C., Sánchez-Rojas, F. and Cano-Pavón, J.M. (2005) Use of 1,5-bis(di-2-pyridyl)methylene thiocarbohydrazide immobilized on silica gel for automated preconcentration and selective determination of antimony(III) by flow injection electrothermal atomic absorption spectrometry. *Analytical and Bioanalytical Chemistry*, **382**, 513–518.

53 Trojanowicz, M. (2000) Solid-phase Extraction, in *Flow Injection Analysis: Instrumentation and Applications*, World Scientific, Singapore, Chapter 6, pp. 236–247.

54 Miró, M. and Hansen, E.H. (2006) Solid reactors in sequential injection analysis: recent trends in the environmental field. *Trends in Analytical Chemistry*, **25**, 267–281.

55 Xiong, Y., Zhou, H.-J., Zhang, Z.-J., He, D.-Y. and He, C. (2006) Molecularly imprinted on-line solid-phase extraction combined with flow injection chemiluminescence for the determination of tetracycline. *Analyst*, **131**, 829–834.

56 Bravo, J.C., Fernández, P. and Durand, J.S. (2005) Flow injection fluorimetric determination of β-estradiol using a molecularly imprinted polymer. *Analyst*, **130**, 1404–1409.

57 Huclova, J., Satinsky, D., Maia, T., Karlicek, R., Solich, P. and Araujo, A.N. (1087) Sequential injection extraction based on restricted access material for determination of furosemide in serum. *Journal of Chromatography. A*, **2005**, 245–251.

58 Bosch-Ojeda, C. and Sánchez-Rojas, F. (2006) Recent development in optical chemical sensors coupling with flow injection analysis. *Sensors*, **6**, 1245–1307.

59 Tzanavaras, P.D. and Themelis, D.G. (2007) Review of recent applications of flow injection spectrophotometry to pharmaceutical analysis. *Analytica Chimica Acta*, **588**, 1–9.

60 Miró, M. and Frenzel, W. (2004) Flow-through sorptive preconcentration with direct optosensing at solid surfaces for trace-ion analysis. *Trends in Analytical Chemistry*, **23**, 11–20.

61 Yan, X.-P. and Jiang, Y. (2001) Flow injection on-line preconcentration and separation coupled with atomic (mass) spectrometry for trace element (speciation) analysis based on sorption of organo-metallic complexes in a knotted reactor. *Trends in Analytical Chemistry*, **20**, 552–562.

62 Cerutti, S., Martinez, L.D. and Wuilloud, R.G. (2005) Knotted reactors and their role in flow injection on-line preconcentration systems coupled to atomic spectrometry-based detectors. *Applied Spectroscopy Reviews*, **40**, 71–101.

63 Dimitrova-Koleva, B., Benkhedda, K., Ivanova, E. and Adams, F. (2007) Determination of trace elements in natural waters by inductively coupled plasma time of flight mass spectrometry after flow injection preconcentration in a knotted reactor. *Talanta*, **71**, 44–50.

64 Kradtap-Hartwell, S., Grudpan, K. and Christian, G.D. (2004) Bead injection with a simple flow injection system: an economical alternative for trace analysis. *Trends in Analytical Chemistry*, **23**, 619–623.

65 Wang, J.-H., Hansen, E.H. and Miró, M. (2003) Sequential injection-bead injection-lab-on-valve schemes for on-line solid-phase extraction and preconcentration of ultra-trace levels of heavy metals with determination by electrothermal atomic absorption spectrometry and inductively coupled plasma mass spectrometry. *Analytica Chimica Acta*, **499**, 139–147.

66 Wu, H., Hong, J., Yan, S., Shi, Y.-Q. and Bi, S.-P. (2007) On-line organoselenium interference removal for inorganic selenium species by flow injection coprecipitation preconcentration coupled with hydride generation atomic fluorescence spectrometry. *Talanta*, **71**, 1762–1768.

67 Wang, Y., Chen, M.-L. and Wang, J.-H. (2006) Sequential/bead injection lab-on-valve incorporating a renewable microcolumn for co-precipitate preconcentration of cadmium coupled to hydride generation atomic fluorescence spectrometry. *Journal of Analytical Atomic Spectrometry*, **21**, 535–538.

68 Hernandez, P.C., Tyson, J.F., Uden, P.C. and Yates, D. (2007) Determination of selenium by flow injection hydride generation inductively coupled plasma optical emission spectrometry. *Journal of Analytical Atomic Spectrometry*, **22**, 298–304.

69 Burguera, J.L. and Burguera, M. (2002) On-line flow injection-atomic spectroscopic configurations: road to practical environmental analysis. *Quimica Analitica*, **20**, 255–273.

70 Bings, N.H., Stefanka, Z. and Rodríguez-Mallada, S. (2003) Flow injection electrochemical hydride generation inductively coupled plasma time-of-flight mass spectrometry for the simultaneous determination of hydride forming elements and its application to the analysis of fresh water samples. *Analytica Chimica Acta*, **479**, 203–214.

71 Zhang, W.-B., Gan, W., Shao, L.-J. and Lin, X.-Q. (2006) Flow-Injection online reduction atomic fluorescence spectrometry determination of Se(IV) and Se(VI) with electrochemical hydride generation. *Spectroscopy Letters*, **39**, 533–545.

72 Burguera, J.L., Burguera, M., Rivas, C. and Carrero, P. (1998) On-line cryogenic trapping with microwave heating for the determination and speciation of arsenic by flow injection/hydride generation/atomic absorption spectrometry. *Talanta*, **45**, 531–542.

73 Sigrist, M.E. and Beldomenico, H.R. (2004) Determination of inorganic arsenic species by flow injection hydride generation atomic absorption spectrometry with variable sodium tetrahydroborate concentrations.

Spectrochimica Acta Part B-Atomic Spectroscopy, **59**, 1041–1045.

74 Coelho, N.M.M., Cósmen da Silva, A. and Moraes da Silva, C. (2002) Determination of As(III) and total inorganic arsenic by flow injection hydride generation atomic absorption spectrometry. *Analytica Chimica Acta*, **460**, 227–233.

75 Hsiung, T.-M. and Wang, J.-M. (2004) Cryogenic trapping with a packed cold finger trap for the determination and speciation of arsenic by flow injection/hydride generation/atomic absorption spectrometry. *Journal of Analytical Atomic Spectrometry*, **19**, 923–928.

76 Tsalev, D.L. (1999) Hyphenated vapour generation atomic absorption spectrometric techniques. *Journal of Analytical Atomic Spectrometry*, **14**, 147–162.

77 Simon, S., Lobos, G., Pannier, F., De Gregori, I., Pinochet, H. and Potin-Gautier, M. (2004) Speciation analysis of organoarsenical compounds in biological matrices by coupling ion chromatography to atomic fluorescence spectrometry with on-line photooxidation and hydride generation. *Analytica Chimica Acta*, **521**, 99–108.

78 Gallignani, M., Valero, M., Brunetto, M.R., Burguera, J.L., Burguera, M. and Petit de Peña, Y. (2000) Sequential determination of Se(IV) and Se(VI) by flow injection-hydride generation-atomic absorption spectrometry with HCl/HBr microwave aided pre-reduction of Se(VI) to Se(IV). *Talanta*, **52**, 1015–1024.

79 Fernandez, C., Conceiçao, A.C.L., Rial-Otero, R., Vaz, C. and Capelo, J.L. (2006) Sequential flow injection analysis system on-line coupled to high intensity focused ultrasound: green methodology for trace analysis applications as demonstrated for the determination of inorganic and total mercury in waters and urine by CVAAS. *Analytical Chemistry*, **78**, 2494–2499.

80 Burguera, J.L. and Burguera, M. (2001) Volatile species generation in flow injection for the on-line determination of species in environmental samples by electrothermal atomic absorption spectrometry. *Journal of Flow Injection Analysis*, **18**, 5–12.

81 Kan, M., Willie, S.N., Scriver, C. and Sturgeon, R.E. (2006) Determination of total mercury in biological samples using flow injection CVAAS following tissue solubilization in formic acid. *Talanta*, **68**, 1259–1263.

82 Miró, M. and Frenzel, W. (2004) Automated membrane-based sampling and sample preparation exploiting flow injection analysis. *Trends in Analytical Chemistry*, **23**, 624–636.

83 Choengchan, N., Mantim, T., Wilairat, P., Dasgupta, P.K., Motomizu, S. and Nacapricha, D. (2006) A membraneless gas diffusion unit: design and its application to determination of ethanol in liquors by spectrophotometric flow injection. *Analytica Chimica Acta*, **579**, 33–37.

84 Miralles, E., Compañó, R., Granados, M. and Prat, M.D. (1999) Photodissociation/gas-diffusion separation and fluorimetric detection for the analysis of total and labile cyanide in a flow system. *Fresenius' Journal of Analytical Chemistry*, **365**, 516–520.

85 Lima, J.L.F.C., Montenegro, M.C.B.S.M. and Pinto, A.P.M.M.O. (1999) Determination of total nitrogen in food by flow injection analysis (FIA) with a potentiometric differential detection system. *Fresenius' Journal of Analytical Chemistry*, **364**, 353–357.

86 Amini, N. and Kolev, S.D. (2007) Gas-diffusion flow injection determination of Hg(II) with chemiluminescence detection. *Analytica Chimica Acta*, **582**, 103–108.

87 Lomonte, C., Currell, M., Morrison, R.J.S., McKelvie, I.D. and Kolev, S.D. (2007) Sensitive and ultra-fast determination of arsenic(III) by gas-diffusion flow injection analysis with chemiluminescence detection. *Analytica Chimica Acta*, **583**, 72–77.

88 Hohercakova, Z. and Opekar, F. (2005) A contactless conductivity detection cell for flow injection analysis: determination of total inorganic carbon. *Analytica Chimica Acta*, **551**, 132–136.

89 Moskvin, L.N. and Nikitina, T.G. (2004) Membrane Methods of Substance Separation in Analytical Chemistry. *Journal of Analytical Chemistry*, **59**, 2–16.

90 de Armas, G., Ferrer, L., Miró, M., Estela, J.M. and Cerdà, V. (2004) In-line membrane separation method for sulfide monitoring in wastewaters exploiting multisyringe flow injection analysis. *Analytica Chimica Acta*, **524**, 89–96.

91 Catalá-Icardo, M., García-Mateo, J.V. and Martínez-Calatayud, J. (2001) Selective chlorine determination by gas diffusion in a tandem flow assembly and spectrophotometric detection with o-dianisidine. *Analytica Chimica Acta*, **443**, 153–163.

92 Mesquita, R.B.R. and Rangel, A.O.S.S. (2005) Gas diffusion sequential injection system for the spectrophotometric determination of free chlorine with o-dianisidine. *Talanta*, **68**, 268–273.

93 Toda, K. (2004) Trends in atmospheric trace gas measurement instruments with membrane-based gas diffusion scrubbers. *Analytical Sciences*, **20**, 19–27.

94 Dasgupta, P.K. (2002) Automated Diffusion-based Collection and Measurement of Atmospheric Trace Gases, in *Sampling and Sample Preparation Techniques for Field and Laboratory* (ed. J. Pawliszyn), Wilson and Wilson's Comprehensive Analytical Chemistry Series, Vol. 37, Elsevier, The Netherlands, pp. 97–160.

95 Amornthammarong, N., Jakmunee, J., Li, J.-Z. and Dasgupta, P.K. (2006) Hybrid fluorometric flow analyzer for ammonia. *Analytical Chemistry*, **78**, 1890–1896.

96 Takeuchi, M., Li, J.-Z., Morris, K.J. and Dasgupta, P.K. (2004) Membrane-based parallel plate denuder for the collection and removal of soluble atmospheric gases. *Analytical Chemistry*, **76**, 1204–1210.

97 Miró, M. and Frenzel, W. (2003) A novel flow-through microdialysis separation unit with integrated differential potentiometric detection for the determination of chloride in soil samples. *Analyst*, **128**, 1291–1297.

98 van Staden, J.F. and Tlowana, S.I. (2002) On-line separation, simultaneous dilution and spectrophotometric determination of zinc in fertilisers with a sequential injection system and xylenol orange as complexing agent. *Talanta*, **58**, 1115–1122.

99 Araújo, A.N., Lima, J.L.F.C., Rangel, A.O.S.S. and Segundo, M.A. (2000) Sequential injection system for the spectrophotometric determination of reducing sugars in wine. *Talanta*, **52**, 59–66.

100 da Silva, J.E., Pimentel, M.F., da Silva, V.L., Montenegro, M.C.B.S.M. and Araújo, A.N. (2004) Simultaneous determination of pH, chloride and nickel in electroplating baths using sequential injection analysis. *Analytica Chimica Acta*, **506**, 197–202.

101 Antonia, A. and Allen, L.B. (2001) Extraction and analysis of lead in sweeteners by flow injection Donnan dialysis with flame atomic absorption spectroscopy. *Journal of Agricultural and Food Chemistry*, **49**, 4615–4618.

102 Pyrzynska, K. (2006) Preconcentration and recovery of metal ions by Donnan dialysis. *Microchimica Acta*, **153**, 117–126.

103 Ruiz-Jiménez, J. and Luque de Castro, M.D. (2006) Coupling microdialysis to capillary electrophoresis. *Trends in Analytical Chemistry*, **25**, 563–571.

104 Haskins, W.E., Watson, C.J., Cellar, N.A., Powell, D.H. and Kennedy, R.T. (2004) Discovery and neurochemical screening of peptides in brain extracellular fluid by chemical analysis of in vivo microdialysis samples. *Analytical Chemistry*, **76**, 5523–5533.

105 Miró, M. and Frenzel, W. (2005) The potential of microdialysis as an automatic sample-processing technique for environmental research. *Trends in Analytical Chemistry*, **24**, 324–333.

106 Caballo-López, A. and Luque de Castro, M.D. (2007) Determination of cadmium in leaves by ultrasound-assisted extraction prior to hydride generation, pervaporation and atomic absorption detection. *Talanta*, **71**, 2074–2079.

107 Ruiz-Jiménez, J. and Luque de Castro, M.D. (2006) Pervaporation as interface between solid samples and capillary electrophoresis. *Journal of Chromatography. A*, **1110**, 245–253.

108 Rupasinghe, T., Cardwell, T.J., Cattrall, R.W., Potter, I.D. and Kolev, S.D. (2004) Determination of arsenic by pervaporation-flow injection hydride generation and permanganate spectrophotometric detection. *Analytica Chimica Acta*, **510**, 225–230.

109 Satienperakul, S., Sheikheldin, S.Y., Cardwell, T.J., Cattrall, R.W., Luque de Castro, M.D., McKelvie, I.D. and Kolev, S.D. (2003) Pervaporation-flow injection analysis of phenol after on-line derivatisation to phenyl acetate. *Analytica Chimica Acta*, **485**, 37–42.

110 Fernandez-Rivas, C., Muñoz-Olivas, R. and Camara, C. (2001) Coupling pervaporation to AAS for inorganic and organic mercury determination. A new approach to speciation of Hg in environmental samples. *Fresenius' Journal of Analytical Chemistry*, **371**, 1124–1129.

111 Luque de Castro, M.D. and Papaefstathiou, I. (1998) Pervaporation – A useful tool for speciation analysis. *Spectrochimica Acta Part B-Atomic Spectroscopy*, **53**, 311–319.

112 Burguera, M. and Burguera, J.L. (1998) Microwave-assisted sample decomposition in flow analysis. *Analytica Chimica Acta*, **366**, 63–80.

113 Luque de Castro, M.D. and Priego-Capote, F. (2006) Ultrasound-assisted preparation of liquid samples. *Talanta*, **72**, 321–334.

114 Dan, D.-Z., Sandford, R.C. and Worsfold, P.J. (2005) Determination of chemical oxygen demand in fresh waters using flow injection with on-line UV-photocatalytic oxidation and spectrophotometric detection. *Analyst*, **130**, 227–232.

115 Tue-Ngeun, O., Ellis, P., McKelvie, I.D., Worsfold, P.J., Jakmunee, J. and Grudpan, K. (2005) Determination of dissolved reactive phosphorus (DRP) and dissolved organic phosphorus (DOP) in natural waters by the use of rapid sequenced reagent injection flow analysis. *Talanta*, **66**, 453–460.

116 Almeida, M.I.G.S., Segundo, M.A., Lima, J.L.F.C. and Rangel, A.O.S.S. (2004) Multi-syringe flow injection system with in-line microwave digestion for the determination of phosphorus. *Talanta*, **64**, 1283–1289.

117 Priego-Capote, F. and Luque de Castro, M.D. (2007) Ultrasound in analytical chemistry. *Analytical and Bioanalytical Chemistry*, **387**, 249–257.

118 Priego-Capote, F. and Luque de Castro, M.D. (2007) Ultrasound-assisted digestion: a useful alternative in sample preparation. *Journal of Biochemical and Biophysical Methods*, **70**, 299–310.

119 López-Flores, J., Fernández de Córdova, M.L. and Molina-Diaz, A. (2005) Implementation of flow-through solid phase spectroscopic transduction with photochemically induced fluorescence: determination of thiamine. *Analytica Chimica Acta*, **535**, 161–168.

120 Gómez-Taylor, B., Palomeque, M., García-Mateo, J.V. and Martínez-Calatayud, J. (2006) Photoinduced chemiluminescence of pharmaceuticals. *Journal of Pharmaceutical and Biomedical Analysis*, **41**, 347–357.

121 Pérez-Ruiz, T., Martínez-Lozano, C. and García, M.D. (2007) Determination of propoxur in environmental samples by automated solid-phase extraction followed by flow injection analysis with tris(2,2'-bipyridyl)ruthenium(II) chemiluminescence detection. *Analytica Chimica Acta*, **584**, 275–280.

122 Zhi, Z.-L., Ríos, A. and Valcárcel, M. (1996) Direct processing and analysis of solid and other complex samples with automatic flow injection systems [Review]. *Critical Reviews in Analytical Chemistry*, **26**, 239–260.

123 Packer, A.P., Gervasio, A.P.G., Miranda, C.E.S., Reis, B.F., Menegario, A.A. and Gine, M.F. (2003) On-line electrolytic dissolution for lead determination in high-purity copper by isotope dilution inductively coupled plasma mass spectrometry. *Analytica Chimica Acta*, **485**, 145–153.

124 Caballo-López, A. and Luque de Castro, M.D. (2003) Continuous ultrasound-assisted extraction coupled to on line filtration-solid-phase extraction-column liquid chromatography-post column derivatisation-fluorescence detection for the determination of N-methylcarbamates in soil and food. *Journal of Chromatography. A*, **998**, 51–59.

125 Sweileh, J.A. (2007) On-line flow injection solid sample introduction, leaching and potentiometric determination of fluoride in phosphate rock. *Analytica Chimica Acta*, **581**, 168–173.

126 Morales-Muñoz, S., Luque-García, J.L. and Luque de Castro, M.D. (2004) A continuous approach for the determination of Cr(VI) in sediment and soil based on the coupling of microwave-assisted water extraction, preconcentration, derivatization and photometric detection. *Analytica Chimica Acta*, **515**, 343–348.

127 Silva, M., Kyser, K. and Beauchemin, D. (2007) Enhanced flow injection leaching of rocks by focused microwave heating with in-line monitoring of released elements by inductively coupled plasma mass spectrometry. *Analytica Chimica Acta*, **584**, 447–454.

128 Yebra-Biurrun, M.C., Moreno-Cid, A. and Cancela-Pérez, S. (2005) Fast on-line ultrasound-assisted extraction coupled to a flow injection-atomic absorption spectrometric system for zinc determination in meat samples. *Talanta*, **66**, 691–695.

129 Tena, M.T., Luque de Castro, M.D. and Valcárcel, M. (1996) Screening of polycyclic aromatic hydrocarbons in soil by on-line fiber-optic-interfaced supercritical fluid extraction spectrofluorometry. *Analytical Chemistry*, **68**, 2386–2391.

130 Priego-López, E. and Luque de Castro, M.D. (2004) Superheated water extraction of linear alkylbenzene sulfonates from sediments with on-line preconcentration/derivatization/detection. *Analytica Chimica Acta*, **511**, 249–254.

131 Morales-Muñoz, S., Luque-García, J.L. and Luque de Castro, M.D. (2006) Pure and modified water assisted by auxiliary energies: an environmental friendly extractant for sample preparation. *Analytica Chimica Acta*, **557**, 278–286.

132 Jimoh, M., Frenzel, W. and Müller, V. (2005) Microanalytical flow-through method for assessment of the bioavailability of toxic metals in environmental samples. *Analytical and Bioanalytical Chemistry*, **381**, 438–444.

133 Miró, M., Chomchoei, R., Hansen, E.H. and Frenzel, W. (2005) Dynamic flow-through approaches for metal fractionation assays in environmentally relevant solid samples. *Trends in Analytical Chemistry*, **24**, 759–771.

134 Liu, R.-X., Lead, J.R. and Baker, A. (2007) Fluorescence characterization of cross flow ultrafiltration derived freshwater colloidal and dissolved organic matter. *Chemosphere*, **68**, 1304–1311.

135 Hassellöv, M., Buesseler, K.O., Pike, S.M. and Dai, M.-H. (2007) Application of cross-flow ultrafiltration for the determination of colloidal abundances in suboxic ferrous-rich ground waters. *The Science of the Total Environment*, **372**, 636–644.

136 Tomsu, D., Catalá-Icardo, M., Martínez-Calatayud, J., (2004) Automated simultaneous triple dissolution profiles of two drugs, sulphamethoxazoletrimethoprirm and hydrochlorothiazide-captopril in solid oral dosage forms by a multicommutation flow assembly and derivative spectrophotometry. *Journal of Pharmaceutical and Biomedical Analysis*, **36**, 549–557.

137 Motz, S.A., Klimundová, J., Schaefer, U.F., Balbach, S., Eichinger, T., Solich, P. and Lehr, C.-M. (2007) Automated measurement of permeation and dissolution of propranolol HCl tablets using sequential injection analysis. *Analytica Chimica Acta*, **581**, 174–180.

138 Rio-Segade, S. and Tyson, J.F. (2007) Determination of methylmercury and inorganic mercury in water samples by slurry sampling cold vapor atomic absorption spectrometry in a flow injection system after preconcentration on silica C18 modified. *Talanta*, **71**, 1696–1702.

139 Miró, M. and Hansen, E.H. (2007) Miniaturization of environmental chemical assays in flowing systems: the Lab-on-a-Valve approach vis-a-vis Lab-on-a-Chip microfluidic devices. *Analytica Chimica Acta*, **600**, 46–57.

140 Hansen, E.H., Miró, M., Long, X.-B. and Petersen, R. (2006) Recent developments in automated determinations of trace level concentrations of elements and on-line fractionations schemes exploiting the micro-sequential injection – lab-on-valve approach. *Analytical Letters*, **39**, 1243–1259.

141 Coelho, L.M., Coelho, N.M.M., Arruda, M.A.Z. and de la Guardia, M. (2007) On-line bi-directional electrostacking for As speciation/preconcentration using electrothermal atomic absorption spectrometry. *Talanta*, **71**, 353–358.

142 Liu, Y., Hashi, Y. and Lin, J.-M. (2007) Continuous-flow microextraction and gas chromatographic-mass spectrometric determination of polycyclic aromatic hydrocarbon compounds in water. *Analytica Chimica Acta*, **585**, 294–299.

143 Kataoka, H. (2002) Automated sample preparation using in-tube solid-phase microextraction and its application – a review. *Analytical and Bioanalytical Chemistry*, **373**, 31–45.

12
Flow Analysis and the Internet – Databases, Instrumentation, and Resources

Stuart J. Chalk

12.1
Introduction

It is interesting to realize that the evolution of flow analysis technology has occurred over the same basic time frame as the development of the Internet. Just after Leonard Skeggs [1] published the first article on segmented continuous flow analysis in 1957 [2], J. C. R. Licklider developed the "Galactic Network" concept [3] which was the template for ARPANET (Advanced Research Projects Agency Network) the great-grandfather of today's internet. Of course the Internet has grown significantly more than flow analysis in the time between 1957 and 2007, and that is the reason for the inclusion of this chapter.

The internet is now *the* medium for communication on a global scale and for flow analysis it has been a boon for the dissemination of the knowledge, practice, and research in the area. However, if you do a search on "Flow injection analysis" (performed 6/1/07) on Google [4] you get 424 000 hits! There's too much information! Thus, this chapter is focused on highlighting a select few internet resources that the reader can use for learning about flow analysis technology, identifying applications of flow analysis to real world samples, and identify vendors of instruments/components for addition to your laboratory.

This chapter will also introduce some of the new "Web 2.0" [5] concepts and technologies that are starting to revolutionize the ways in which we use/reuse/access the Internet. While there is a risk in discussing new technologies before they have been proven (and therefore become standard), it is important to convey to the reader the direction the internet is going so that the researchers' area of flow analysis can continue to thrive as a global community.

12.2
Databases

The growth of research in the area of flow analysis has been astounding. Since 1975 it has been estimated that there are some 21 000 research articles on all aspects of

Advances in Flow Analysis. Edited by Marek Trojanowicz
Copyright © 2008 WILEY-VCH Verlag GmbH & Co. KGaA, Weinheim
ISBN: 978-3-527-31830-8

the general area of flow analysis. A consequence of this (akin to the wealth of information on the internet) is how difficult and time consuming it is to find all pertinent literature on a specific research topic. The growth of the internet and the advent of journal publishers digitizing research articles in portable document format (PDF) has both helped and hindered this problem.

The accessibility of research articles (both current and legacy) through publisher websites has become standard and consequently it is easy for anyone in the world with internet access to download (legally of course) a paper they need. However, the flip side of instant full-text access is that finding the right article can be very difficult because searches currently are not context sensitive (this will change in the future – see the last section). Searching for papers on iron analysis by sequential injection analysis will likely contain a number of "hits" that are inappropriate as iron may show up as a reagent, an interferent, or as part of the word "environment." We won't even mention legacy papers that have been scanned and converted to PDF where "iron" has been misidentified as "lion".

Thus, important tools on the web for accurate searching of the research literature are databases focused on specialized areas, like flow analysis. Currently, there are three of note for flow analysis, as described below.

12.2.1
Dr. Elo Hansen's Flow Bibliography
http://www.flowinjection.com/database1_interface/fia_database/results_page.asp

One of the two original inventors of flow injection analysis, Dr. Elo Hansen, has made available his extensive bibliography through the FIAlab instruments website for the last eight years. Updated for 2006 and with 16 500 citations (6/1/07), this comprehensive database can be searched for keywords in the title of the research article and/or by the author name. Results obtained are shown in chronological order (as far back as 1974) and contain author, title, language, and citation information.

12.2.2
Google Scholar Beta
http://scholar.google.com/

A relative newcomer, but one that has a massive database of information, is Google Scholar. Announced in 2005, Google Scholar specifically applies the Google search technology to research article data and metadata. Although this is not a database specific to flow analysis technology the comprehensive nature of the database and the ease of searching make it an extremely useful tool for finding literature on any research topic, let alone flow analysis. Table 12.1 below shows the hits for flow analysis terms found in Google Scholar as of 6/1/07.

In addition to the regular search screen, the advanced search screen (Figure 12.1) allows results to be limited by author, journal, date, and subject area. Additionally, the results contain links to citing articles and you can find those by clicking the "cited by" link.

Table 12.1 Searches in Google Scholar (6/1/7).

Search term	Hits
+"flow injection analysis"	17 600
+"sequential injection analysis"	957
+"continuous flow analysis"	1790
+"segmented flow analysis"	290
+"lab on valve"	150

12.2.3
The Flow Analysis Database[1] (http://www.fia.unf.edu/)

The Flow Analysis Database is a website that went online in 1997 with roughly 3000 citations. Since that time it has evolved into a more comprehensive website by inclusion of a larger legacy archive of citations as well as current ones. This database covers all areas of flow analysis including articles that describe research on post-column derivatization in high performance liquid chromatography (HPLC) (flow analysis after separation).

As the database is specifically targeted at flow analysis research literature it has been designed to serve the need of its user population. The front page (Figure 12.2) contains a number of features for both browsing and searching the database of over 17 000 articles.

The basic search allows users to find citations by author, journal title, analyte, sample matrix, detection technique, keyword (flow analysis specific), language, and publication year. Additionally, on the left-hand side, users can select a browsable index for each of these fields, which shows them terms that are used in the database to identify citations. More sophisticated searching (Figure 12.3) can be done on the advanced search page using boolean (AND/OR) logic.

Search results are presented in a condensed view when retrieved for search and the user can select one or more of these results to view in more detail. Figure 12.4 shows the detailed results page.

In addition to the obvious fields of information about the citation there is also a link to the paper online (if one exists) making it easy to access the research. Optionally users may export the results to a number of formats including EndNote/Procite, PDF and comma separated variable (CSV) so that the information can be imported into other applications.

12.3
Journals

As of 2006, articles on flow analysis had appeared in over 1000 peer reviewed journals throughout the world. The majority of the articles are found in analytical chemistry

1) Disclaimer: This website has been developed by
   the author of this chapter.

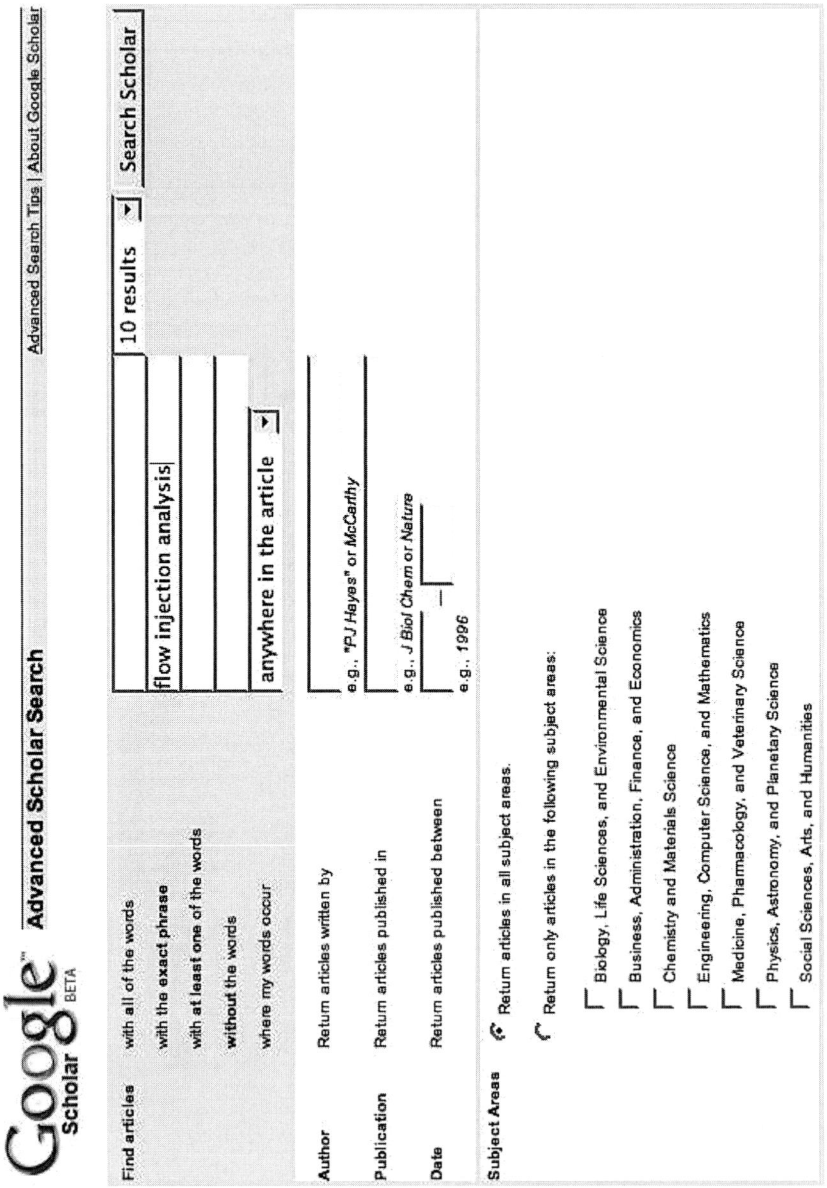

Figure 12.1 Google scholar advanced search page.

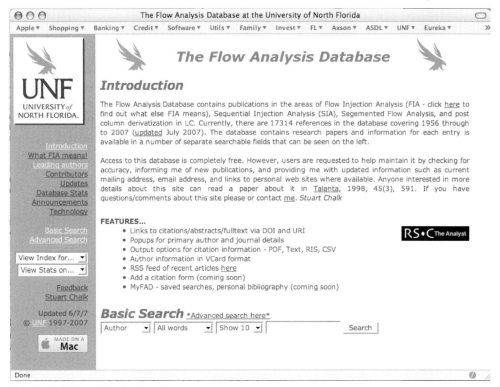

Figure 12.2 Flow analysis database homepage.

 The Flow Analysis Database

Advanced Search

Use this to perform a more complicated boolean search. For each field enter a single term

- OR as an operator takes prescedence over AND (i.e. 'Chalk OR Tyson AND Iron' is searched as '(Chalk OR Tyson) AND Iron')
- multiple words in the same field wil be searched as a phrase (e.g. ' flow injection ' would not match 'flow through the injection port')
- use >, <, >=, <= to search for years greater than or less than a specific year (year field only)
- use -- to indicate a date range (year field only)
- for specific authors names use the following syntax - Chalk, S.J.

| FIND | | ☐ Exact | in Author ▼ |

in Author
in Primary Author
in Journal
in Title
in Analyte
in Matrix
in Technique
in Keyword
in Language
in FADID**
in DOI
in Year

| AND ▼ | | ☐ Exact |
| AND ▼ | | ☐ Exact |

Show 10

**The FADID is generated as follows JournalCode+Year+V+Volume(four digits)+P+Pages(five

Figure 12.3 Flow analysis database advanced search page.

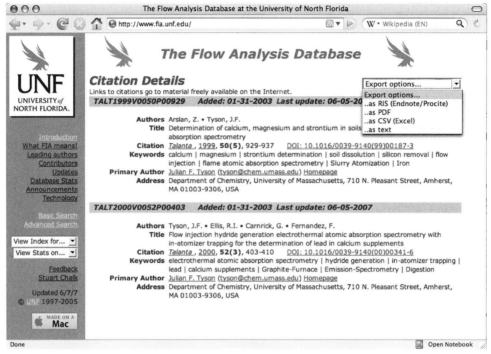

Figure 12.4 Flow analysis database detailed results page.

journals but quite a number are found in application specific journals like *Clinical Chemistry* (see below), the *Journal of Pharmaceutical and Biomedical Analysis* [6], the *Journal of Agricultural and Food Chemistry* [7], and the *Journal of Biological Chemistry* [8]. Below is a list of the most important journals for publication of flow analysis research.

12.3.1
Analytica Chimica Acta (Elsevier)
http://www.sciencedirect.com/science/journal/00032670

Analytica Chimica Acta (ACA) was the first journal to publish an article on flow injection analysis back in 1975 [9] and in fact published the entire series of ten articles by Ruzicka and Hansen. As a consequence of these articles appearing in ACA the flow analysis community adopted it as the primary journal of publications in the area. Today ACA publishes significantly more papers on flow analysis than any other journal and this is likely to continue considering the makeup of the current editorial board.

12.3.2
Talanta (Elsevier)
http://www.sciencedirect.com/science/journal/00399140

Talanta has become an internationally recognized journal for high quality publications in analytical chemistry, a significant proportion of which are flow analysis (starting in 1976), thanks in part to the editorial guidance of Chief Editor Gary Christian.

12.3.3
Analytical Chemistry (American Chemical Society – ACS)
http://pubs.acs.org/acs/journals/ancham/

Analytical Chemistry from the ACS consistently ranks as the top analytical chemistry journal in the world. Although its main focus is far toward instrumentation design and development there are many important flow analysis papers within its pages, especially in the field of microfluidics.

12.3.4
The Analyst (Royal Society of Chemistry – RSC)
http://www.rsc.org/publishing/journals/an/

The premier analytical journal from the Royal Society of Chemistry, The Analyst continues the tradition of its roots as a forum for publishing fundamental research as well as educational reviews of important concepts in analytical science.

12.3.5
Journal of Flow Injection Analysis (Japanese Association of Flow Injection Analysis)
http://aitech.ac.jp/~jafia/english/jfia/

The *Journal of Flow Injection Analysis* (JFIA) is the only journal dedicated to the area of flow injection analysis and is published by the Japanese Association of Flow Injection Analysis. This leading edge journal has been a catalyst for research in and application of flow analysis in Japan since its introduction in 1984. It also publishes a bibliography of citations in flow injection analysis in every (semi-annual) issue.

12.3.6
Journal of Automatic Methods and Management in Chemistry (Hindawi)
http://www.hindawi.com/journals/jammc/

Originally published as the Journal of Automatic Chemistry, the *Journal of Automatic Methods and Management in Chemistry* (JAMMC) also published a large number of articles on SFA and other automated chemical analysis methods. Today, JAMMC is an open access journal with all issues of the journal on-line and available free.

12.3.7
Analytical and Bioanalytical Chemistry (Springer)
http://www.springerlink.com/content/1618-2642/

Previously titled *"Fresenius' Journal of Analytical Chemistry,"* Analytical and Bioanalytical Chemistry contains a large number of interesting technical articles on the application of flow analysis. The journal is now owned by a consortium of chemical societies and this makes for a very well balance analytical journal.

12.3.8
Analytical Sciences (Japanese Society for Analytical Chemistry)
http://www.jsac.or.jp/cgi-bin/analsci/toc/

A young (first published in 1985) analytical journal, Analytical Sciences is rapidly growing into a major international analytical chemistry journal. Recently, Analytical Sciences passed both Bunseki Kagaku and The Journal of Flow Injection Analysis (see below) as the publication with the most papers in flow analysis. Major factors that have helped it grow are the availability of its complete archive (open access) and it being an English language journal.

12.3.9
Analytical Letters (Taylor and Francis)
http://www.informaworld.com/0003-2719

Originally part of Marcel Dekker, *Analytical Letters* publishes research in all areas of analytical chemistry, specializing in rapid publication of short communication articles.

12.3.10
Journal of Chromatography A (Elsevier)
http://www.sciencedirect.com/science/journal/00219673

Even though the *Journal of Chromatography A* is primarily a separations journal, there are a number of articles on the use of post-column reaction chemistry (derivatization) with liquid chromatography that are relevant to flow analysis, in addition to hybrid flow analysis-liquid chromatography and flow analysis-capillary electrophoresis papers.

12.3.11
Electroanalysis (Wiley)
http://www3.interscience.wiley.com/journal/26571/home

Under the editorial leadership of Joe Wang, *Electroanalysis* publishes research in all areas of electrochemical analysis. Given that almost 20% of the publications in flow analysis use electrochemical detection, it is not surprising that *Electroanalysis* is a major journal in the area.

12.3.12
Journal of Analytical Atomic Spectrometry (RSC)
http://www.rsc.org/publishing/journals/ja/

The impact of flow analysis on the area of atomic spectrometry cannot be understated. Automation of the introduction of samples into flames, plasmas, and discharges has helped the evolution of these techniques significantly. The *Journal of Analytical Atomic Spectrometry* (JAAS) publishes a significant proportion of papers on interfacing flow analysis with these detection techniques.

12.3.13
Fenxi Huaxue (Wanfang Data)
http://www.analchem.com

Fenxi Huaxue (*Analytical Chemistry*) is the most prominent analytical journal in China today and flow analysis papers are a large part of what the journal publishes. Unfortunately, as of this writing only the last eight years of the journal are online (paid access). Note: This website and the papers available are written in Chinese.

12.3.14
Bunseki Kagaku (Japanese Society for Analytical Chemistry)
http://www.jsac.or.jp/bunka/bunsekikagaku_e.html

This Japanese language journal "*Bunseki Kagaku*" ("*Japan Analyst*") was the first to publish papers on flow analysis in Japan. As of this writing full-text articles are available back to 1975, however some of the volumes are on JSTAGE (see below) and it is likely the full archive of papers in *Bunseki Kagaku* will be available soon.

12.3.15
Clinical Chemistry (American Association for Clinical Chemistry)
http://www.clinchem.org

Clinical Chemistry published many of the original articles on automated flow analysis using segmented continuous flow analysis (SFA), developed by Skeggs, starting in the mid 1960s. *Clinical Chemistry* is now available on-line at the link above and the archive of all the issues of the journal is free access (as of July 2007).

It is also important to mention there there are a number of journal archive sites that contain vast amounts of research in all areas of science (and beyond). Table 12.2 gives a list of the most important ones for flow analysis.

Table 12.2 Journal archive sites.

Publisher	Site
Elsevier	http://www.sciencedirect.com
Wiley	http://www.interscience.wiley.com
Taylor and Francis	http://www.informaworld.com/smpp/subjecthome?db=jour
Springer	http://www.springerlink.com
Blackwell	http://www.blackwell-synergy.com/
American Chemical Society	http://pubs.acs.org/archives/
Royal Society of Chemistry	http://www.rsc.org/Publishing/Journals/DigitalArchive/Index.asp
J-STAGE (Japan)	http://www.jstage.jst.go.jp/
SciElo	http://www.scielo.org/
Wanfang Data (China)	http://www.wanfangdata.com

12.4
Instrumentation

Flow analyzers have been around since the late 1960s (AutoAnalyzer) based on the work by Leonard Skeggs. Today, instrumentation for all types of flow-based analysis is available throughout the world. Table 12.3 shows the major vendors of equipment and supplies and the technologies that support the flow analysis instrumentation market.

Table 12.3 Vendors of flow analysis instrumentation.

Vendor (website)	Technology[a]
Astoria Pacific (http://www.astoria-pacific.com)	SFA, DA
Alliance (http://www.alliance-instruments.com/)	SFA
FIALab (http://www.flowinjection.com)	FIA, SIA, LOV, PA
Foss (http://www.foss.dk)	FIA
Global FIA (http://www.globalfia.com)	FIA, SIA, ZF, DA, PA, Custom
Lachat (http://www.lachatinstruments.com)	FIA
OI Analytical (http://www.oico.com)	FIA, SIA, SFA, DA
Seal Analytical (http://www.seal-analytical.com)	SFA, DA, PA
Skalar (http://www.skalar.com)	SFA, PA
Bran + Luebbe (http://www.bran-luebbe.de)	PA
Ogawa (http://www.ogawajapan.com)	FIA

[a]FIA – Flow injection analysis, SIA – sequential injection analysis, LOV – Lab-on-valve, SFA – segmented flow analysis (CFA), DA – discrete analyzer, PA – process analyzer.

12.5
Standard Methods

Over the course of the last 30 years flow analysis methodologies have been promulgated as standard methods around the world. The following tables summarize the methods published by the major standards organizations around the world (this list is not comprehensive of all standards agencies).

12.5.1
International Standards Organization (http://www.iso.org/)

Milk and milk products	
14673-2:2004	Determination of nitrate and nitrite contents – Part 2: Method using segmented flow analysis (Routine method) http://www.iso.org/iso/en/CatalogueDetailPage.Catalogue Detail?CSNUMBER=38703

Milk and milk products

14673-3:2004	Determination of nitrate and nitrite contents – Part 3: Method using cadmium reduction and flow injection analysis with in-line dialysis (Routine method) http://www.iso.org/iso/en/CatalogueDetailPage.Catalogue Detail?CSNUMBER=38704

Soil quality

17380:2004	Determination of total cyanide and easily released cyanide – Continuous-flow analysis method http://www.iso.org/iso/en/CatalogueDetailPage.Catalogue Detail?CSNUMBER=33033

Tobacco

15517:2003	Determination of nitrate content – Continuous-flow analysis method http://www.iso.org/iso/en/CatalogueDetailPage.Catalogue Detail?CSNUMBER=27377
15152:2003	Determination of the content of total alkaloids as nicotine – Continuous-flow analysis method http://www.iso.org/iso/en/CatalogueDetailPage.Catalogue Detail?CSNUMBER=26506
15153:2003	Determination of the content of reducing substances – Continuous-flow analysis method http://www.iso.org/iso/en/CatalogueDetailPage.Catalogue Detail?CSNUMBER=26505
15154:2003	Determination of the content of reducing carbohydrates – Continuous-flow analysis method http://www.iso.org/iso/en/CatalogueDetailPage.Catalogue Detail?CSNUMBER=26507

Water quality

11732:2005	Determination of ammonium nitrogen – Method by flow analysis (CFA and FIA) and spectrometric detection http://www.iso.org/iso/en/CatalogueDetailPage.Catalogue Detail?CSNUMBER=38924
13395:1996	Determination of nitrite nitrogen and nitrate nitrogen and the sum of both by flow analysis (CFA and FIA) and spectrometric detection http://www.iso.org/iso/en/CatalogueDetailPage.Catalogue Detail?CSNUMBER=21870
14402:1999	Determination of phenol index by flow analysis (FIA and CFA) http://www.iso.org/iso/en/CatalogueDetailPage.Catalogue Detail?CSNUMBER=23708
14403:2002	Determination of total cyanide and free cyanide by continuous flow analysis http://www.iso.org/iso/en/CatalogueDetailPage.Catalogue Detail?CSNUMBER=23709

Milk and milk products

15681-1:2003	Determination of orthophosphate and total phosphorus contents by flow analysis (FIA and CFA) – Part 1: Method by flow injection analysis (FIA) http://www.iso.org/iso/en/CatalogueDetailPage.Catalogue Detail?CSNUMBER=35050
15681-2:2003	Determination of orthophosphate and total phosphorus contents by flow analysis (FIA and CFA) – Part 2: Method by continuous flow analysis (CFA) http://www.iso.org/iso/en/CatalogueDetailPage.Catalogue Detail?CSNUMBER=35051
15682:2000	Determination of chloride by flow analysis (CFA and FIA) and photometric or potentiometric detection http://www.iso.org/iso/en/CatalogueDetailPage.Catalogue Detail?CSNUMBER=27984
16264:2002	Determination of soluble silicates by flow analysis (FIA and CFA) and photometric detection http://www.iso.org/iso/en/CatalogueDetailPage.Catalogue Detail?CSNUMBER=30224
22743:2006	Determination of sulfates – Method by continuous flow analysis (CFA) http://www.iso.org/iso/en/CatalogueDetailPage.Catalogue Detail?CSNUMBER=38339
23913:2006	Determination of chromium(VI) – Method using flow analysis (FIA and CFA) and spectrometric detection http://www.iso.org/iso/en/CatalogueDetailPage.Catalogue Detail?CSNUMBER=37017

12.5.2
US Environmental Protection Agency (http://www.epa.gov/)

130.1	Total Hardness by Spectrophotometer http://infotrek.er.usgs.gov/pls/apex/f?p=119:38:3112691435248230::::P38_METHOD_ID:5211
245.2	Mercury by CVAA (Automated) http://infotrek.er.usgs.gov/pls/apex/f?p=119:38:3785554329050224::::P38_METHOD_ID:4822
310.2	Alkalinity by Autoanalyzer http://infotrek.er.usgs.gov/pls/apex/f?p=119:38:3112691435248230::::P38_METHOD_ID:5231

325.1   Chloride by Automated Colorimetry
http://infotrek.er.usgs.gov/pls/apex/f?p=119:38:1941274247707098::::
P38_METHOD_ID:5771

325.2   Chloride by Automated Colorimetry
http://infotrek.er.usgs.gov/pls/apex/f?p=119:38:1941274247707098::::
P38_METHOD_ID:5765

335.3   Cyanide by Automated Colorimetry
http://infotrek.er.usgs.gov/pls/apex/f?p=119:38:1941274247707098::::
P38_METHOD_ID:5404

335.4   Cyanide, total, by Colorimeter
http://infotrek.er.usgs.gov/pls/apex/f?p=119:38:3112691435248230::::
P38_METHOD_ID:5759

340.3   Fluoride by Colorimetry
http://infotrek.er.usgs.gov/pls/apex/f?p=119:38:3112691435248230::::
P38_METHOD_ID:5775

350.1   Ammonia by Automated Colorimetry
http://infotrek.er.usgs.gov/pls/apex/f?p=119:38:3112691435248230::::
P38_METHOD_ID:5405

351.1   TKN by Automated Colorimetry
http://infotrek.er.usgs.gov/pls/apex/f?p=119:38:3112691435248230::::
P38_METHOD_ID:4872

351.2   TKN by Semi-Automated Block Digestion and Colorimetry
http://infotrek.er.usgs.gov/pls/apex/f?p=119:38:3112691435248230::::
P38_METHOD_ID:4702

353.1   Nitrate-Nitrite by Colorimetry
http://infotrek.er.usgs.gov/pls/apex/f?p=119:38:3112691435248230::::
P38_METHOD_ID:5250

353.2   Nitrate-Nitrite Nitrogen by Colorimetry
http://infotrek.er.usgs.gov/pls/apex/f?p=119:38:3112691435248230::::
P38_METHOD_ID:4873

353.4   Nitrate and Nitrite in Estuarine and Coastal Waters by Automated
Colorimetry
http://infotrek.er.usgs.gov/pls/apex/f?p=119:38:3112691435248230::::
P38_METHOD_ID:7225

365.1   Phosphorus (all forms) by Semi-Automated Colorimetry
http://infotrek.er.usgs.gov/pls/apex/f?p=119:38:3785554329050224::::
P38_METHOD_ID:5313

365.4   Phosphorus by Automated Colorimetry
http://infotrek.er.usgs.gov/pls/apex/f?p=119:38:3785554329050224::::
P38_METHOD_ID:4823

365.5   Orthophosphate in Estuarine and Coastal Waters by Colorimetry
http://infotrek.er.usgs.gov/pls/apex/f?p=119:38:3112691435248230::::
P38_METHOD_ID:7231

375.2   Sulfate by Colorimetry
http://infotrek.er.usgs.gov/pls/apex/f?p=119:38:3112691435248230::::
P38_METHOD_ID:4824

420.2   Phenolics by Automated Colorimetry
http://infotrek.er.usgs.gov/pls/apex/f?p=119:38:3112691435248230::::
P38_METHOD_ID:5265

OIA-1677   Available Cyanide by Flow Injection with Ligand Exchange
http://infotrek.er.usgs.gov/pls/apex/f?p=119:38:1941274247707098::::
P38_METHOD_ID:4836

12.5.3
American Society for Testing and Materials (http://www.astm.org/)

D3867-04   Standard Test Methods for Nitrite-Nitrate in Water
http://www.astm.org/DATABASE.CART/REDLINE_PAGES/D3867.
htm

D7237-06   Standard Test Method for Aquatic Free Cyanide with Flow Injection
Analysis (FIA) Utilizing Gas Diffusion Separation and Amperometric
Detection
http://www.astm.org/DATABASE.CART/REDLINE_PAGES/D7237.
htm

C1310-01   Standard Test Method for Determining Radionuclides in Soils by
(2007)   Inductively Coupled Plasma-Mass Spectrometry Using Flow Injection
Preconcentration
http://www.astm.org/DATABASE.CART/REDLINE_PAGES/C1310.
htm

D6888-04   Standard Test Method for Available Cyanide with Ligand Displacement and Flow Injection Analysis (FIA) Utilizing Gas Diffusion
Separation and Amperometric Detection
http://www.astm.org/DATABASE.CART/REDLINE_PAGES/D6888.
htm

12.5.4
APHA/AWWA/WEF Standard Methods (http://www.standardmethods.org/)

4500-Br D   Flow Injection Analysis (Bromide)
http://www.standardmethods.org/Store/ProductView.cfm?
ProductID=178

4500-Cl E   Chloride by Automated Ferricyanide Method

4500-Cl G   Mercury Thiocyanate Flow Injection Analysis
http://www.standardmethods.org/store/ProductView.cfm?
ProductID=182

4500-CN N	Total Cyanide after Distillation, by Flow Injection Analysis
4500-CN O	Total Cyanide and Weak Acid Dissociable Cyanide by Flow Injection Analysis http://www.standardmethods.org/store/ProductView.cfm?ProductID=180
4500-F G	Ion-Selective Electrode Flow Injection Analysis http://www.standardmethods.org/store/ProductView.cfm?ProductID=184
4500-NH3 G	Ammonia by Automated Phenate
4500-NH3 H	Flow Injection Analysis (Ammonia) http://www.standardmethods.org/store/ProductView.cfm?ProductID=191
4500-NO3 F	Nitrate by Automated Cadmium Reduction Method
4500-NO3 G	Nitrate by Automated Hydrazine Reduction
4500-NO3 I	Cadmium Reduction Flow Injection Method http://www.standardmethods.org/store/ProductView.cfm?ProductID=193
4500-N B	In-Line UV/Persulfate Digestion and Oxidation with Flow Injection Analysis http://www.standardmethods.org/Store/ProductView.cfm?ProductID=189
4500-N_{org} D	Block Digestion and Flow Injection Analysis http://www.standardmethods.org/store/ProductView.cfm?ProductID=194
4500-P F	Phosphorus by Automated Ascorbic Acid Reduction
4500-P G	Flow Injection Analysis for Orthophosphate
4500-P H	Manual Digestion and Flow Injection Analysis for Total Phosphorus
4500-P I	In-line UV/Persulfate Digestion and Flow Injection Analysis for Total Phosphorus http://www.standardmethods.org/Store/ProductView.cfm?ProductID=197
4500-SiO2 E	Silica, Automated Method for Molybdate-Reactive Silica
4500-SiO2 F	Flow Injection Analysis for Molybdate-Reactive Silicate http://www.standardmethods.org/store/ProductView.cfm?ProductID=199
4500-SO42 G	Methylthymol Blue Flow Injection Analysis http://www.standardmethods.org/store/ProductView.cfm?ProductID=202
4500-S2 I	Distillation, Methylene Blue Flow Injection Analysis Method http://www.standardmethods.org/store/ProductView.cfm?ProductID=200

12.5.5
US Geological Survey Standard Methods (http://www.usgs.gov/)

I-2302	Cyanide, whole water recoverable; colorimetric http://infotrek.er.usgs.gov/pls/apex/f?p=119:38:3112691435248230:::: P38_METHOD_ID:5660
I-2371	Iodide, dissolved in water by colorimetry http://infotrek.er.usgs.gov/pls/apex/f?p=119:38:3112691435248230:::: P38_METHOD_ID:5661
I-2525	Nitrogen, ammonia, dissolved, low ionic-strength, colorimetric, ASF http://infotrek.er.usgs.gov/pls/apex/f?p=119:38:3112691435248230:::: P38_METHOD_ID:5480
I-2542	Nitrogen, nitrite, dissolved, colorimetric, ASF, low ionic-strength http://infotrek.er.usgs.gov/pls/apex/f?p=119:38:3112691435248230:::: P38_METHOD_ID:5481
I-2545	Nitrogen, nitrite plus nitrate, dissolved, colorimetric, ASF http://infotrek.er.usgs.gov/pls/apex/f?p=119:38:3112691435248230:::: P38_METHOD_ID:5482
I-2602	Phosphorus, orthophosphate plus hydrolyzable, dissolved, colorimetric, ASF http://infotrek.er.usgs.gov/pls/apex/f?p=119:38:3112691435248230:::: P38_METHOD_ID:5483
I-2606	Phosphorus, orthophosphate plus hydrolyzable, dissolved, colorimetric, ASF http://infotrek.er.usgs.gov/pls/apex/f?p=119:38:3112691435248230:::: P38_METHOD_ID:5484
I-2607	Phosphorus, dissolved, low ionic-strength, colorimetric, ASF http://infotrek.er.usgs.gov/pls/apex/f?p=119:38:3112691435248230:::: P38_METHOD_ID:5485
I-4302	Cyanide, whole water recoverable; colorimetric http://infotrek.er.usgs.gov/pls/apex/f?p=119:38:3112691435248230:::: P38_METHOD_ID:5671
I-4602	Phosphorus, orthophosphate plus hydrolyzable, total, colorimetric, ASF http://infotrek.er.usgs.gov/pls/apex/f?p=119:38:3112691435248230:::: P38_METHOD_ID:5502
I-4607	Phosphorus, total, colorimetric, ASF http://infotrek.er.usgs.gov/pls/apex/f?p=119:38:3112691435248230:::: P38_METHOD_ID:5503
I-6302	Cyanide, recoverable-from-bottom-material, dry wt, colorimetric http://infotrek.er.usgs.gov/pls/apex/f?p=119:38:3112691435248230:::: P38_METHOD_ID:5690
I-6545	Nitrogen, nitrite plus nitrate, total-in-bottom-material, dry weight, col- orimetric, ASF http://infotrek.er.usgs.gov/pls/apex/f?p=119:38:3112691435248230:::: P38_METHOD_ID:5511

12.6
Other Useful Websites

12.6.1
Tutorials

- http://www.lachatinstruments.com/products/qcfia/fiaprimer.asp "Primer on FIA" – A brief but useful discussion of the points of FIA.
- http://www.flowinjection.com/freeCD.html "Flow injection Analysis" – Third edition of an in-depth and comprehensive CD based tutorial on flow methods.
- http://www.globalfia.com/tutorial.html FIA/SIA Tutorial – An excellent online tutorial on the fundamentals and applications of FIA/SIA/ZF.
- http://www.sci.monash.edu.au/wsc/fia/ Intro to FIA – A useful overview of the principles and practice of FIA.

12.6.2
Books (Chronological Order)

- "Continous Flow Analysis (Clinical and Biochemical Analysis Series: Vol 3)" W.B. Furman 1976 ISBN 9780824763206 http://www.amazon.com/dp/0824763203.
- "Flow Injection Analysis" J. Ruzicka and E. Hansen 1st. Edition 1981 ISBN 9780471081920 http://www.amazon.com/dp/0471081922.
- "Flow-injection Analysis: Principles and Applications" M. Valcarcel, M.D. Luque de Castro, and A. Losada 1987 ISBN 9780133219364 http://www.amazon.com/dp/0133219364.
- "Flow Injection Analysis" J. Ruzicka and E. Hansen 2nd. Edition 1988 ISBN 9780471813559 http://www.amazon.com/dp/0471813559.
- "Flow Injection Atomic Spectrometry" J. L. Burguera 1989 ISBN 9780824780593 http://www.amazon.com/dp/0824780590.
- "Flow Injection Analysis: A Practical Guide" B. Karlberg and G. Pacey 1989 ISBN 9780444880143 http://www.amazon.com/dp/0444880143.
- "Flow Injection Analysis (FIA) Based on Enzymes or Antibodies" R. L. Schmid 1991 ISBN 9783527282494 http://www.amazon.com/dp/3527282491.
- "Flow Injection Separation and Preconcentration" Z. L. Fang 1993 ISBN 9781560811473 http://www.amazon.com/dp/1560811471.
- "Flow Injection Atomic Absorption Spectrometry" Z. L. Fang 1995 ISBN 9780471953319 http://www.amazon.com/dp/0471953318.
- "Flow Injection Analysis of Pharmaceuticals" J. M. Calatayud 1996 ISBN 9780748404452 http://www.amazon.com/dp/0748404457.
- "Flow Injection Analysis: Instrumentation and Application" M. Trojanowicz 2000 ISBN 9789810227104 http://www.amazon.com/dp/9810227108.

12.6.3
Webpages of Prominent Researchers (Alphabetical)

• http://www.uv.es/~martinej/personal/index1.html	Dr. Jose Martinez Calatayud
• http://www.uib.es/depart/dqu/dquiweb/grupo_e.html	Dr. Victor Cerda
• http://depts.washington.edu/chem/people/faculty/christian.html	Dr. Gary Christian
• http://www3.uta.edu/faculty/dasgupta/index.htm	Dr. Sandy Dasgupta
• http://www.css.zju.edu.cn/~imas/faculty/fzl.html	Dr. Zhaolun Fang
• http://www.aua.gr/georgiou/index.htm	Dr. Constantinos Georgiou
• http://www.uv.es/solinqui/	Dr. Miguel De La Guardia
• http://www.anchem.su.se/staffcard.asp?ID=16	Dr. Bo Karlberg
• http://www.chemistry.unimelb.edu.au/staff/spas/research/index.html	Dr. Spas Kolev
• http://www.chem.monash.edu.au/staff/mckelvie/research.html	Dr. Ian McKelvie
• http://chem1.chem.okayama-u.ac.jp/staff_e/motomizu.html	Dr. Shoji Motomizu
• http://depts.washington.edu/chem/people/faculty/ruzicka.html	Dr. Jarda Ruzicka
• http://www.chem.uw.edu.pl/labs/lfac_head.htm	Dr. Marek Trojanowicz
• http://www.hull.ac.uk/chemistry/academic_staff.php?id=at	Dr. Alan Townshend
• http://www.chem.umass.edu/Faculty/tyson.htm	Dr. Julian Tyson
• http://www.chemistry.nmsu.edu/~research/sensors/srg/srg.html	Dr. Joseph Wang
• http://www.plymouth.ac.uk/staff/pworsfold	Dr. Paul Worsfold
• http://www.uco.es/grupos/FQM-215/index.htm	Dr. Miguel Valcarcel
• http://www.up.ac.za/academic/chem/koos.htm	Dr. Koos Van Staden

(NOTE: No web page was found for Dr. Elo Hansen)

12.6.4
Other

- http://aitech.ac.jp/~jafia/ – Japanese Association of Flow injection Analysis.
- http://www.uv.es/~martinej/Flow-Analysis/ – Flow Analysis Forum.
- http://www.iupac.org/publications/pac/2002/7404/7404x0585.html Elias A. G. Zagatto, Jacobus F. van Staden, Nelson Maniasso1, Raluca I. Stefan, and Graham D. Marshall "Information essential for characterizing a flow-based analytical system (IUPAC Technical Report)" *Pure Appl. Chem.* 2002 74(4) p. 585.
- http://www.iupac.org/publications/pac/2004/7606/7606x1119.html K. Tóth, K. Stulík, W. Kutner, Z. Fehér, and E. Lindner "Electrochemical detection in liquid flow analytical techniques: Characterization and classification (IUPAC Technical Report)" *Pure Appl. Chem.* 2004 76(6) p. 1119.

12.7
Future Directions

Looking into the internet "Crystal Ball" what does the future hold for flow analysis research? Many good things. Technologies that are being developed for general enhancement of the internet experience will help scientific research across all disciplines. Here is a look at what's coming.

12.7.1
The Semantic Web

The Semantic Web [10] is the next revolution of the internet. In essence the Semantic Web means that machines (computers) will be able to understand information and make sense of it just like humans can. An example of this is that to a human 1-904-620-1938 is obviously a telephone number, but all a computer sees is that it is a string of digits and dashes. The computer can understand that the string is a telephone number if we add some metadata (data about data), for example <telephone>1-904-620-1938</telephone>. The addition of the metadata has added a context around the information and thus made it more useful.

The digital revolution in publishing has changed the way we "mine" the research literature. With the availability of full-text searching we no longer need to rely on the abstract of a paper for a glimpse of its content. However, finding the word "phosphate" in a PDF does not tell us whether it is an analyte (orthophosphate), a buffer (sodium hydrogenphosphate), or a sample (phosphate rock). Again we need context, and so the Semantic Web, as abstract a concept as it appears, will impact digital publishing by allowing researchers to search by context as well as compound, matrix and so on.

Controlled vocabularies, thesauri, and ontologies are important parts of the Semantic Web that we, the practitioners of flow analysis research, need to develop so we may take advantage of this revolution. Controlled vocabularies are exactly what the name implies, standard lists of terms that are used for specific topics in a discipline area. Publishers and authors have been working on this for years in the guise of keywords, yet they are not of course controlled! To make controlled vocabularies work in a practical sense you need thesauri in addition. As much as chemists like to use the term absorbance (Beer's law), biologists insist on using optical density. So, an intelligent computer system must be able to translate the non-standard term (optical density) into a controlled vocabulary term (absorbance) for searching, that is, using a thesaurus. Finally, the relationship of controlled vocabulary terms to each other and how all the terms fit together can be defined in an ontology. If we as a global community can organize and developed each of these pieces then we can significantly advanced our science.

12.7.2
Extensible Markup Language (XML)

The extensible markup language (XML) [11] is a web technology born out of the need to contextualize information. It is an elegant approach to identification of information in a platform neutral "American standard code for information interchange" format (ASCII) and flexible enough for any application you can dream up. So, here are a couple of interesting examples that could easily be added to research articles on flow analysis and would enhance searching for and comparing research papers.

- *Analytical Figures of Merit* – A concise summary of the important analytical parameters in the development of a new method.

```
<codedisplay><![CDATA[
    <calibration analyte="iron" xunit="ppm"
    yunit="absorbance">
        <slope>0.219</slope>
        <intercept>0.0004</intercept>
        <rsquared>0.999903</rsquared>
        <detectionlimit unit="ppb">5.03</detectionlimit>
        <limitoflinearity>4.0</limitoflinearity>
    </calibration>
    <analysis replicates="5">
        <analyte>iron</analyte>
        <matrix>Seawater</matrix>
        <dilution>10 mL to 100 mL</dilution>
        <result unit="ppm">1.74</result>
        <resulterror unit="ppm">0.0074</resulterror>
    <analysis>
]]></codedisplay>
```

- *Manifold Flow Diagram* – A textual representation of the parameters of a manifold.

```
<manifold analyte="cyanide">
      <line id="1" type="input" name="carrier">
             <flowrate unit="mL/min">1.0</flowrate>
             <solution>Borate buffer ph 10.8</solution>
      </line>
      <line id="2" type="input" name="reagent">
             <flowrate unit="mL/min">0.1</flowrate>
             <solution>Copper(II)/EDTA/Phenolphthalin
             </solution>
      </line>
      <line id="3" type="output" name="mixingcoil">
             <coil>
             <length unit="cm">100</length>
             <diameter unit="mm">0.1</diameter>
</coil>
      </line>
      <confluencePoint>
             <input id="1"/>
             <input id="2"/>
             <output id="3"/>
             <geometry>Tee</geometry>
      <confluencePoint>
      <detector>
             <input id="3"/>
             <type>spectrophotometer</type>
             <wavelength unit="nm">512</pathlength>
             <pathlength unit="cm">1.0</pathlength>
      </detector>
</manifold>
```

In conclusion, the Internet has a lot to offer for the flow analysis community. Not only are we able to exchange ideas more easily, keep up with the literature, and collaborate across continents, there are still many opportunities for building a well integrated and cyber-aware community. What will the picture be in another 50 years?

References

1 Skeggs, L.T. (2000) Persistence... and Prayer: From the Artificial Kidney to the AutoAnalyzer. *Clinical Chemistry*, **46**, 1425–1436. http://www.clinchem.org/cgi/content/full/46/9/1425.

2 Skeggs, L.T. Jr. (1957) An automatic method for colorimetric analysis. *American Journal of Clinical Pathology*, **28**, 311–322.

3 The History of Web Hosting – A Galactic Network, Accessed July 1, 2007,

http://www.thehistoryof.net/history-of-web-hosting.html.

4 Google Search Engine, Accessed July 1, 2007, http://www.google.com.

5 Web 2.0, Accessed July 1, 2007, http://en.wikipedia.org/wiki/Web_2.

6 Journal of Pharmaceutical and Biomedical Analysis, Accessed July 1, 2007, http://www.sciencedirect.com/science/journal/07317085.

7 Journal of Agricultural and Food Chemistry, Accessed July 1, 2007, http://pubs.acs.org/journals/jafcau/index.html.

8 Journal of Biological Chemistry, Accessed July 1, 2007, http://www.jbc.org.

9 Ruzicka, J. and Hansen, E. (1975) Flow injection analyses. Part I. A new concept of fast continuous flow analysis. *Analytica Chimica Acta*, **78**, 145–147 http://dx.doi.org/10.1016/S0003-2670(01)84761-9.

10 W3C Semantic Web Activity, Accessed July 1, 2007, http://www.w3.org/2001/sw/.

11 Extensible Markup Language (XML), Accessed July 1, 2007, http://www.w3.org/xml/.

II

Advances in Detection Methods in Flow Analysis

Advances in Flow Analysis. Edited by Marek Trojanowicz
Copyright © 2008 WILEY-VCH Verlag GmbH & Co. KGaA, Weinheim
ISBN: 978-3-527-31830-8

13
Luminescence Detection in Flow Analysis

Antonio Molina-Díaz and Juan Francisco García-Reyes

13.1
Introduction

Luminescence phenomena (fluorescence, phosphorescence and chemiluminescence) are the basis of widely used analytical detection techniques. Molecular
fluorescence (FL) is a process in which a luminophore absorbs a suitable-energy
photon to raise an electron from the ground state to a higher energy vacant orbital
(excited state), followed by the electron returning back to the original ground state
energy level, then emitting a quantum of light. The electron spin remains unchanged
throughout the entire process. Therefore, the molecule is always in either its
ground or excited singlet states (a magnetic field does not split the energy levels
of these electronic configurations). The process is fast with luminescence lifetimes of
the order of nanoseconds.

Sometimes, a much slower radiative deactivation of a molecule can occur. This
involves the excited electron reversing its spin while it is transitioned to a lower-
energy state (called a triplet state, because the energy level is split into three levels in a
magnetic field). When the electron comes back to the ground state level its spin must
be reversed again. The time scale of the process, called phosphorescence (PH), is
much slower than that of an FL process, with lifetime ranging from some tens of
microseconds up to some milliseconds.

Chemiluminescence (CL) is defined [1] as electromagnetic radiation produced
when a chemical reaction yields an electronically excited intermediate or product,
which either luminesces (direct CL) or donates its energy to another molecule
responsible for the emission (indirect or sensitized CL). When this emission
originates from living organisms or from chemical systems derived from them, it
is called bioluminescence. Basically, the process is identical to that for photoluminescence (fluorescence or phosphorescence) except that no excitation source is
needed; thus, CL detectors do not suffer perturbations caused by the background.
High sensitivity is common to practically all of the luminescence detection techniques and it is not unusual to get very low detection limits, even in sub-picomolar

Advances in Flow Analysis. Edited by Marek Trojanowicz
Copyright © 2008 WILEY-VCH Verlag GmbH & Co. KGaA, Weinheim
ISBN: 978-3-527-31830-8

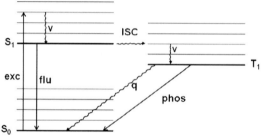

Figure 13.1 Schematic representation of photophysical processes relevant to directly excited RTP. S_0, singlet ground state; S_1, lowest excited singlet state; T_1, lowest excited triplet state; exc, excitation; flu, fluorescence; v, vibrational relaxation, ISC, intersystem crossing; q, biomolecular quenching; phos, phosphorescence. (Adapted from [77] with permission from Elsevier, © 2003.)

levels. A scheme showing the different luminescence phenomena (Jablonski diagram) is shown in Figure 13.1.

Molecular luminescence phenomena have been extensively implemented as detection systems in continuous flow analysis, including continuous segmented flow methods [2]. In fact, fluorescence detection was already present in many early FIA applications [3]. Luminescence detection is compatible with FIA [4] and with those more recently developed flow injection techniques such as sequential injection analysis (SIA) [5], multisyringe flow injection analysis (MSFIA) [6], multicommutation flow systems [7] and, multi-pumping flow injection analysis (MPFIA) [8]. Figure 13.2 shows a scheme of luminescence detection techniques in flow analysis. In this chapter, the use of molecular luminescence detection techniques in flow analysis systems is described and the most significant advances in the last few years are presented and discussed.

13.2
Luminescence Detection in Continuous Flow Systems

13.2.1
Fluorescence

13.2.1.1 Introduction
Fluorescence spectrometry has become established as a routine technique in many specialized applications. It is a relatively simple and rapid analytical technique, particularly adequate for quantitation of aromatic, or highly unsaturated, organic molecules present in trace concentrations, especially in biological and environmental samples. The most salient analytical features of fluorescence detection are its high sensitivity, accompanied by high/moderate selectivity in many cases. In contrast, the main drawback of fluorimetric techniques is the small number of substances that exhibit native fluorescence. However, a few straightforward procedures/reactions

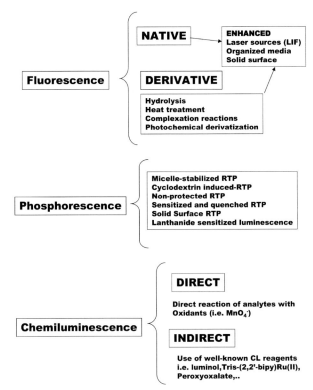

Figure 13.2 Luminescence detection in flow injection systems.

have enormously expanded the range of available fluorophores, enabling the application of this technique to a wide variety of organic and inorganic compounds via chemical labeling and derivatization procedures.

13.2.1.2 Fluorescence Detection in Flow Methods
The most immediate classification of fluorimetric determinations is based on whether the substance concerned is fluorimetrically active or must be activated by chemical derivatization.

Flow Methods Based on Native Fluorescence   The direct methods based on the fluorescence detection of analytes are essentially applied to those molecules exhibiting an intrinsic fluorescence, strong enough to be analytically useful. For substances which exhibit this property, the FIA assembly required is straightforward: it typically consists of a carrier line and an injection valve. The fluorescence behavior of species containing acid or basic groups is pH dependent, so that the carrier channel is used to provide a suitable pH and medium. Some applications of native fluorescence flow injection methods are shown in Table 13.1 [9–24].

Enhanced Native Fluorescence. The three main analytical principles used to improve the capabilities of native fluorescence methods are: (1) use of laser sources; (2) micellar or organized media, and (3) solid supports.

Table 13.1 Selected flow methods based on native fluorescence measurements [9–24].

Analyte	Description	Limit of Detection	Ref.
fuberidazole and thiabendazole	native FL using a SIA assembly and resolution of mixtures by multiple linear regression (MLR) calibration model	LOD: 0.04 (FBZ) and 0.08 μg L^{-1} (TBZ)	[9]
benfuresate	native FL (278/316 nm)	0.001–0.5 mg L^{-1}	[10]
3-indolyl acetic acid	native FL (280/364 nm)	LDR: 0.005–0.6 mg L^{-1}	[11]
asulam	native FL (258/342 nm)	LDR: 0.005–15 mg L^{-1}	[12]
bumetanide	native FL in strong alkaline medium (314/370 nm)	0.05–10.0 μg mL^{-1}	[13]
scopoletine and umbelliferone	native FL (350/418 nm) at different pH values	LOD < 1 μM	[14]
ergotamine tartrate	SIA assembly, native FL (236/390 nm)	LOD: 10 ng mL^{-1}	[15]
salicylic acid	SIA assembly, native FL (299/405 nm)	LDR: 1–100 mg L^{-1}	[16]
prazosin	native FL (244–389 nm) using a SIA assembly	LOD: 7 ng mL^{-1}	[17]
PDT (sunscreen agent)	on-line SPE followed by native FL detection (330/454 nm)	LOD: 10 ng mL^{-1}	[18]
PBS (sunscreen agent)	native FL (301–681 nm) using a SIA assembly; on-line purification using SAX minicolumns	LOD: 12 ng mL^{-1}	[19]
17β-estradiol (E2)	native FL; on-line SPE using a molecularly imprinted polymer for matrix removal (281/305 nm)	LOD: 1.12 ng mL^{-1}	[20]
Enhanced native FL flow methods			
carbendazim	micellar enhanced native FL detection using CTAB	LOD: 13 nM	[21]
naproxen	native FL of naproxen enhanced by complexation with β-cyclodextrin, using a SIA assembly	0.19 μM	[22]
warfarin	native FL of warfarin enhanced by formation of inclusion complexes with β-cyclodextrin, using a SIA assembly	LOD: 0.02 μg mL^{-1}	[23]
PAHs (mixtures of)	native FL, laser 335 nm, and multivariate calibration (SIMCA and PLS-1)	—	[24]

PDT: disodium phenyldibenzimidazoletetrasulfonate; PBS: 2-phenylbenzimidazole-5 sulfonic acid; CTAB: cetyltrimethylammonium bromide; SPE: solid-phase extraction; FL: fluorescence.

1. *Laser induced fluorescence* (LIF) detection in flow systems. Lasers sources significantly expand the potential of fluorescence, because of the instrumental sensitivity
is directly proportional to the intensity of the exciting radiation. Some applications
using a laser beam for excitation allow the detection of ultratrace amounts
of substances including rhodamine 6 G [25], riboflavin [26], vitamins and
tryptophan [27]) and porphyphyrins [28]. Another application for the determination of ethanol, based on FIA-LIF was developed by Imasaka *et al.* [29]. Amador-
Hernández *et al.* developed a method for the multi-analyte determination of
polycyclic aromatic hydrocarbons (PAH) using FIA-LIF and multivariate calibration without prior chromatographic separation. The FIA assembly used is shown
in Figure 13.3.

2. *Fluorescence in organized media.* The fluorescence of some organic species is
strongly affected by the physicochemical environment. This has fostered the use
of organized media. Surfactants and cyclodextrins provide the fluorophore with a
microenvironment that enhances the fluorescence of many solutes [30], lowering
the detection limits of the assay and, at the same time, avoids undesirable
phenomena related to the decrease in the fluorescence emission intensity.

   Micellar media are prepared by adding appropriate amounts of surfactants
to water or water–organic mixtures. Surfactants are compounds that possess
amphiphilic character. These molecules have two parts; one, the head, is polar and
hydrophilic, while the other, the tail, is hydrophobic. On the other hand,
cyclodextrins (CDs) are water-soluble, cyclic oligosaccharides formed by the
connection of individual glucopyranose units through α-1,4-glycosidic oxygen

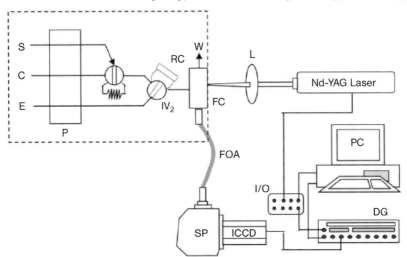

Figure 13.3 Experimental set-up for the
screening of PAHs using flow injection analysis
with laser induced fluorescence detection. Laser
arrangement; L, lens; FC, flow cell; FOA, fiber
optic assembly; SP, spectrograph; ICCD,
intensified charge coupled device; I/O, multi IO
box; DG, delay generator; PC, personal
computer; S, sample solution; C, carrier solution;
E, eluting solution; P, peristaltic pump; IV,
injection valve; RC, retention column; W, waste.
(Reproduced from [24] with permission from
Elsevier, © 2001.)

bridges. Depending on their size, these CDs can form inclusion complexes with certain organic molecules, producing changes in their photophysical properties.

The use of organized media in flow systems can improve the sensitivity of luminescence detection by enhancing the luminescence signal from the analyte. Selected applications of flow methods with fluorescence detection in the presence of organized media are outlined in Table 13.1. It should be noted that organized media are not only applied to native fluorescence methods but also to other derivatization methods such as the complexation reaction or photochemically induced fluorescence.

3. *Solid Phase Fluorescence.* This technique involves measuring the luminescence emitted by species adsorbed or bound to a suitable support. The main features of this strategy and selected applications are described in detail in Section 13.2.4 of this chapter.

Flow Methods with Derivatization of the Analyte   Chemical derivatization is based on the transformation of non- or weakly fluorescent compounds into strongly fluorescent species by means of a chemical reagent or physical treatments (i.e., UV irradiation or heating). Thus, a wide variety of reactions have been used including the formation of fluorescent chelates, displacement of fluorescent ligands, formation of ion-pairs, redox processes, enzyme catalysis, hydrolysis and various typical organic synthesis reactions (e.g., condensation-cyclization, labeling, antigen–antibody reactions). Some of the main derivatization strategies are:

1. *Hydrolysis reactions.* Hydrolysis is one of the most simple treatment methods to convert non-fluorescent compounds to fluorescent ones. It is usually accomplished in a strongly alkaline aqueous medium (NaOH) and, in some cases, at high temperature (50–100 °C), resulting in the formation of fluorescent anions. For instance, acetylsalicylic acid (aspirin) has weak native fluorescence while salicylate ion, fluoresces strongly at ~400 nm upon excitation at 310 nm and can be used to determine aspirin and salicylates in different samples. Hydrolysis has also been successfully applied to the determination of naptalam in natural waters, with detection limits in the low nanogram range [31]. The flow assembly used is shown in Figure 13.4.

2. *Photochemically induced fluorescence (PIF).* The concept of PIF is based on the conversion upon UV irradiation of non-fluorescent analytes into strongly fluorescent photoproducts. PIF is more recent than chemical derivatization. The main photoreactions are photocyclization, photooxidation, photoisomerization, photolysis and photoreduction. Typical examples of this approach are the fluorimetric determination of the anabolic agent diethylstilbesterol in urine and the fluorimetric determination of different pesticides.

Some parameters affect the conversion of nonfluorescent compounds into fluorescent ones, for example, UV irradiation time and the nature of the solvent used. For instance, pesticides such as fenvalerate, diflubenzuron and deltamethrin are efficiently converted into fluorescent products (high signal with short irradiation time) in protic solvents, while for fenitrithion and chlorpyrifos polar aprotic solvents are the choice.

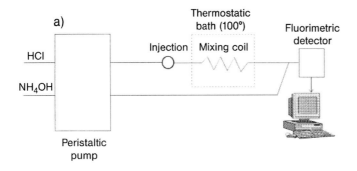

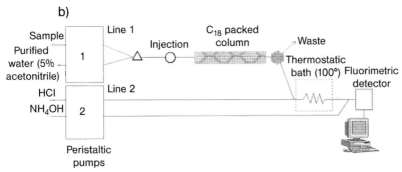

Figure 13.4 Manifolds used for the determination of naptalam based on its hydrolysis to 1-naphthylamine: (a) normal FIA configuration and (b) configuration with on-line preconcentration by solid-phase extraction. Line 1 refers to effluents pumped by the first pump and Line 2 refers to effluents pumped by the second pump. (Reproduced from [31] with permission from Elsevier, © 1999.)

The main advantages of photochemical derivarization are as follows [32]: (i) there is no need to use chemicals, (ii) fast reaction rates, (iii) inexpensive equipment required, (iv) suitable for flow analysis methods and (v) the structure of the fluorescent compound need not necessarily be known provided that reproducible fluorescence signals are obtained.

3. *Complexation.* Complexation involves the combination of a metal ion with a weakly fluorescent or a nonfluorescent compound containing chelate-forming electron donating functional groups to form highly fluorescent metal chelates. Complexation reactions have been extensively applied to the fluorimetric determination of both metal ions and organic compounds.

Table 13.2 shows some selected applications of recent flow injection fluorescence-based methods using different derivatization strategies which have been applied to pharmaceutical, biomedical, food and environmental analysis [33–76].

Table 13.2 Selected fluorimetric flow methods with derivatization of the analyte [33–76].

Analyte	Description	Limit of Detection	Ref.
Oxidation/Reduction reactions			
thioridazine	on-line oxidation of thioridazine using a lead dioxide solid-phase reactor (349/429 nm)	LOD: 5.5 ng mL^{-1}	[33]
ascorbic acid and cysteine	on-line reduction of Tl(III) with cysteine and/or ascorbic acid in acidic media (227/419 nm)	LDR: 0.015–2 µg mL^{-1} LOD: 0.8 µM (ascorbic acid)	[34]
thiamine	oxidation with potassium hexa-cyanoferrate(III) immobilized on an anionic exchange resin (Thiocrome)	0.7 µM (cysteine) LDR: 0.1–4 mg L^{-1}	[35]
penicillamine and thiopronin	oxidation with Tl(III) in hydro-chloric acid medium (227/419 nm)	LOD < 1 µM	[36]
1,4-benzodiazepines	fluorimetric detection after hydrolysis with sulfuric acid in ethanolic or methanolic medium at room temperature	LOD 0.005–0.01 mg mL^{-1}	[37]
thiamine and as-corbic acid	oxidation with Hg(II) to yield fluorescent derivates (thiocrome (thiamine) and quinoxaline derivate (ascorbic acid) (356/440 nm)	LDR: 2–100 mg L^{-1} (B$_1$); 5–100 mg L^{-1} (ascorbic acid)	[38]
fluphenazine	Ce(IV) arsenite as a strong oxidizing solid-bed reactor	LDR: 0.05–100 mg L^{-1}	[39]
adrenaline	poly vinyl chloride coil with iodine	5–25 mg L^{-1}	[40]
iproanizid and isoanizid	oxidation with hydrogen peroxide	LOD: 0.008 mg L^{-1} (iproanizid); 0.005 mg L^{-1} (isoanizid)	[41]
carbamazepine	use of a solid-phase reactor con-taining lead dioxide for on-line oxidation of carbamazepine into a strongly fluorescent compound in an acid medium	LOD: 57 µM	[42]
Hydrolysis			
indomethacin	alkaline hydrolysis in micellar medium, using a SIA system	16 nM	[43]
naptalam	hydrolisis of naptalam in acidic medium and measure-ment of the product in alkaline medium; use of a C$_{18}$ column on-line to preconcen-trate the analyte	LOD: 3 µM	[31]

Table 13.2 (*Continued*)

Analyte	Description	Limit of Detection	Ref.
PIF			
phylloquinone	on-line reduction of phylloquinone in dodecylsulfate micelles after irradiation with UV light	LOD: 0.05 μg mL^{-1} LDR: 0.09–45 μg mL^{-1}	[44]
phenylurea herbicides (isoproturon, neburon, linuron and diuron)	on-line photoconversion under UV irradiation of these herbicides into strongly fluorescent photoproducts in buffered aqueous solutions in the presence of surfactant (SDS and CTAC)	LOD: 0.33–0.93 mg L^{-1}	[45]
chlorophenoxyacid herbicides	UV irradiation (PIF) in the presence of CTAC	LOD: 73.2 and 33.5 ng mL^{-1}	[46]
imidacloprid	photochemical conversion of imidacloprid into the fluorophore 1-(6-chloro-3-pyridylmethyl)-2-hydroxyimino)-3,4-didehydroimidalozolidene) (334/377 nm)	LOD: 0.3 ng mL^{-1}	[47]
sulfonylurea herbicides	micellar enhanced photochemically induced fluorescence using CTAC and SDS	LOD: 0.1–1 ng mL^{-1}	[48]
fluometuron	on-line photoreaction at basic pH medium, using a multicommuted flow assembly (247/325 nm)	LDR: 0.01–4 mg L^{-1}	[49]
methamidophos	decomposition of methamidophos in the presence of peroxydisulfate upon irradiation with UV light; the phosphate generated is used to form phosphomolybdic acid which oxidizes thiamine to thiocrome	LOD: 1.7 ng mL^{-1}	[50]
malathion	decomposition of malathion upon irradiation with UV light; the phosphate generated is used to form phosphomolybdic acid which oxidizes thiamine to thiocrome	LOD: 0.02–2 mg L^{-1}	[51]
diazepam	on-line photoreaction (328/382 nm) in basic media containing Cu(II)	LDR: 0.5–50 μg mL^{-1}	[52]
ascorbic acid	photooxidation of ascorbic acid sensitized by Thionine Blue to yield Leucothionine Blue	LDR: 0.8–50 μM	[53]

(*Continued*)

Table 13.2 (*Continued*)

Analyte	Description	Limit of Detection	Ref.
tianeptine	photochemically induced fluorescence in acidic water alcohol mixtures using a FIA setup	LOD: $15 \, ng \, mL^{-1}$	[54]
chloroquine	photochemical derivatization in an alkaline medium and fluorescence using a pulsed Nd:YAG laser at 355 nm	LOD: $8 \, \mu g \, L^{-1}$	[55]
reserpine	photochemical decomposition of reserpine to form a highly fluorescent derivate in the presence of acetone	LOD: $0.45 \, ng \, mL^{-1}$	[56]
arsanilic acid	on-line decomposition of arsanilic acid in the presence of peroxydisulfate on irradiation with UV light	LOD: $10 \, ng \, mL^{-1}$	[57]
promethazine	on-line photoderivatization	LDR: $0.05–20 \, mg \, L^{-1}$	[58]
aminophenotiazines	Photochemical induced fluorescence detection (ethopropazine, trifluoperazine, levomepropazine and thioproperazine)	LOD: $60–90 \, ng \, mL^{-1}$	[59]
Menadione	on-line photo-reduction of menadione in SDS micelles upon irradiation with UV light (340/410 nm)	LOD: $0.18 \, ng \, mL^{-1}$	[60]

Complexation reactions

Analyte	Description	Limit of Detection	Ref.
Al	complex with lumogallion in the presence of surfactant Brij-35. On-line preconcentration using a column with 8-hydroxyquinoline	LOD: 0.15 nM	[61]
B	complexation of boric acid with chromotropic acid (313/360 nm)	LOD: $3 \, \mu g \, L^{-1}$	[62]
Al	complex formation between Al (III) and 8-HQSA	LOQ: $2.8 \, \mu g \, L^{-1}$	[63]
Al	complexation with salicyl-aldehide carbohydrazone in Triton X-100	LOD: $2.2 \, ng \, mL^{-1}$	[64]
Al	metal complex with 8-HQSA; on-line SPE with XAD-4.	LOD: $0.2 \, ng \, mL^{-1}$	[65]
Mg(II)	metal complex with 8-HQSA in the presence of CTAC and HTAC	LOD: $12 \, ng \, mL^{-1}$	[66]
Zn	complexation with salicyl-aldehide carbohydrazone in Triton X-100	LOD: $5 \, ng \, mL^{-1}$	[67]
Zn	complexation of Zn(II) with 8-(benzenesulfonamido) quinoline in SDS	LOD: $0.2 \, ng \, mL^{-1}$	[68]

Table 13.2 (Continued)

Analyte	Description	Limit of Detection	Ref.
V(V)	cloud point extraction of V/8-HQ complex	LOD: 0.007 ng mL^{-1}	[69]
ammonium	reaction with OPA and sulfite with ammonia	LOD: 7 nM	[70]
cyanide	interaction between Cu-calcein complex and Cyanide	LOQ: 0.4 mM	[71]
aminocaproic acid	reaction with OPA and N-acetyl-cysteine to yield a fluorescent product (350/450 nm), using a SIA assembly	LOD: 0.25 µM	[72]
clenbuterol	reaction with OPA and mercaptoethanol	LOD: 0.06 mg L^{-1}	[73]
chloroxine	complex formation between chloroxine and Al(III) in a micellar medium	LOD: 5 nM	[74]
sulfamethoxazole	fluorescent reaction with OPA and β-mercaptoethanol	LOD: 7 ng mL^{-1}	[75]
ranitidine	fluorescent reaction of the drug with sodium hypochlorite followed by reaction with OPA and β-mercaptoethanol	LOD: 13 ng mL^{-1}	[76]

Abbreviations: CTAC: cetyltrimethylammonium chloride; SDS: sodium dodecyl sulfate; PIF: photochemically induced fluorescence; β- CD: β- cyclodextrin; FL: fluorescence; HTAC: hexadecyltrimethylammonium chloride; SPE: solid-phase extraction; OPA: o-phthaldialdehyde; 8-HQ: 8-hydroxyquinoline; 8-HQSA: 8-hydroxy-quinoline 5- sulfonic acid.

13.2.2
Phosphorescence

13.2.2.1 Introduction

In the past decades, room temperature phosphorescence (RTP) in the liquid state [77] has evolved into a sensitive and versatile tool in analytical chemistry. RTP covers a wide range of strategies which can be implemented in flowing streams such as: micelle-stabilized-RTP (MS-RTP), cyclodextrin-induced RTP (CD-RTP), non-protected RTP (NP-RTP) and solid-surface RTP (SS-RTP). Although these techniques are usually applied in batch or stand alone format, they can also be implemented in flowing streams, likewise SS-RTP optosensing.

Phosphorescence signals in the liquid state are usually too poor to be used for analytical purposes, particularly when using flowing streams, considering the intrinsic drawbacks associated with RTP methodologies and the presence of oxygen. Therefore, the literature and methodologies available are scarce when compared with other luminescence techniques such as fluorescence and chemiluminescence. From amongst all the available RTP methodologies, the more widely used in flowing streams is SS-RTP. Over 80% of the applications described are based on the use of solid supports to develop flow-through room temperature phosphorescence optosensors [78]. The main features and applications of SS-RTP detection in flowing streams will be described in detail in Section 13.2.4 of this chapter.

13.2.2.2 Room Temperature Phosphorescence in Ordered Media

Cline-Love *et al.* did systematic research on the development of RTP in liquid aqueous solution, exploring the use of sodium dodecyl sulfate (SDS) and showed that there are three fundamental requirements for observing MS-RTP for most luminophores [79]: (i) securing the presence of micellar aggregates (using a surfactant concentration above its critical micellar concentration (CMC)); (ii) the presence of a heavy atom or a heavy species; and (iii) ensuring oxygen scavenging.

Although this approach is compatible with flowing streams, the use of MS-RTP as a detection scheme in FIA or related continuous flow systems is scarce [80]. Sanz-Medel and coworkers used MS-RTP for the determination of aluminum in a continuous flow assembly [81]. The method was successfully applied to determine Al(III) in dialysis fluids.

As for MS-RTP, CD-RTP was also described as a detection strategy in liquid chromatography. However, unless the problems of low efficiency with CD mobile phases in chromatography are resolved, CD-RTP will not be a competitive alternative to the MS-RTP detection [82]. Although over 100 papers on RTP methods using cyclodextrins have been described, none of these uses a flow injection assembly [83].

13.2.2.3 Non-Protected Room Temperature Phosphorescence

A further development towards the practical application of homogeneous RTP is one in which the presence of a protecting agent in the medium is not a necessary condition for RTP emission. In 1997, Li *et al.* reported a phosphorimetric method to determine dansyl chloride in an aqueous system by adding thallium nitrate as heavy-atom perturber and sodium sulfite as deoxygenator [84]. This new concept of room temperature phosphorescence in liquid solution is called non-protected RTP [85] or heavy atom induced room-temperature phosphorescence (HAI-RTP) [86]. This approach is an important technical advance, since numerous organic molecules such as plant growth regulators (α- and β-naphthoxyacetic acids), pharmaceuticals (naphazoline, naproxen) and pesticides (thiabendazole, carbaryl), PAHs and other compounds (tryptamine, tryptophan, indole-3-butyric acid) can be determined through a very simple technique [87].

Cañabate-Díaz *et al.* [88] developed a FIA-HAI-RTP method to determine naphazoline in pharmaceuticals using 1 M KI and 10 mM Na_2SO_3 obtaining an LOD of 1.6 ng ml^{-1}. The method was found to be more sensitive than the analogous FIA-RTP methods for the same compound based on FIA-MS-RTP and a RTP flow-through optosensor using Amberlite XAD-7 as solid support. Fernández-González *et al.* [89] developed a FIA-HAI-RTP method to determine nafcillin in milk with an LOD of 0.4 µM. The development of flow-through optosensors using HAI-RTP has been reported and is described in Section 13.2.4.

13.2.3
Chemiluminescence

13.2.3.1 Introduction

Since CL involves both a luminescence process and a chemical reaction, the observed intensity depends upon the rate of the chemical reaction, the number of species excited, and their light emission efficiency. These reactions exhibit analytical utility because the emission intensities are a function of the concentrations of the chemical species involved. In this sense, CL observed signals are dependent upon the reaction kinetics so that reproducible times and appropriate mixing can be provided by using flow injection systems.

Hundreds of applications have been described in the last 25 years on the development of flow injection methods and applications based on CL detection [90–92]. This section will cover the main chemiluminescent reactions and discuss those characteristics that are important for analytical chemistry.

Flow-cell design in CL-flow systems is a key aspect. The cell location and distance to the detector must be highly reproducible. The best-suited flow-cell design most frequently used in FIA is a spiral geometry placed as close to the photomultiplier tube (PMT) as possible. The main requirement is that the maximum light intensity should be emitted while the analyte–reagent mixture is in front of the detector. This requires rapid mixing (particularly with fast reactions) and a cell of an appropriate volume for the emission peak to be measured. The analytical signal increases with increasing spiral length; the minimum acceptable length is dictated by the reaction rate. The distance (and thus volume) between the confluence point of the sample and reagent streams and the detection cell needs to be optimized for the kinetics of the reaction used. The flow cell must be transparent to the wavelength of the CL emission and inert to the chemical reaction or solvent system. Glass, quartz, and Teflon tubing are commonly employed.

Figure 13.5 shows a basic FIA assembly for CL determinations. The sample is inserted into a carrier at an appropriate pH. The reagent, which is circulated along a different line, is merged with the sample-carrier stream as close to the mixing chamber as possible; once in the chamber, the reaction mixture is monitored by means of the phototube [93].

Sample–Reagent Reaction A CL reaction in an FIA system takes place as usual, that is, on merging two streams containing the sample and reagent, respectively. Mixing should be done near the cell for obvious hydrodynamic reasons. CL measurements can be made within a few seconds after mixing, which is particularly useful for monitoring rapid kinetics. Heterogeneous systems involving solid-phase reactors are one of the commonest strategies in continuous-flow CL applications. The CL intensity depends on the kinetics of the reaction involved and the way in which the sample is brought into contact with the reagent.

Immobilized Reagents Immobilization techniques [94] are usually based on the use of reactors (with immobilized reagents including enzymes or CL reagent using,

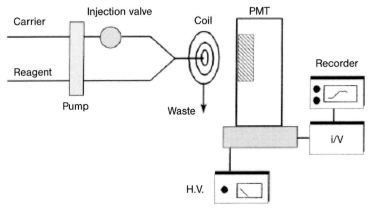

Figure 13.5 Schematic diagram of flow injection chemiluminometer (PMT: photomultiplier, *I/V*: current-to-voltage converter, H.V.: high voltage). (Reproduced from. [93] with permission from Elsevier, © 2000.)

typically, non-polar sorbent or ion-exchanger to retain the reagents), which can be positioned before the CL reaction takes place. When using enzyme reactors, the lack of selectivity that may occur when a given CL reagent yields emission for a variety of compounds is circumvented. The analyte is the substrate of the enzymatic reaction and one of the products will sensitively participate in the CL reaction. As an example, Figure 13.6, shows a typical manifold for glucose determination based on the CL emission produced when hydrogen peroxide formed by immobilized glucose oxidase reacts with luminol in the presence of potassium hexacyanoferrate(III). Using the luminol reaction, glucose can be detected by immobilization of pyranose oxidase and hydrogen peroxide or sulfite by immobilization of luminol and Co(II) [92].

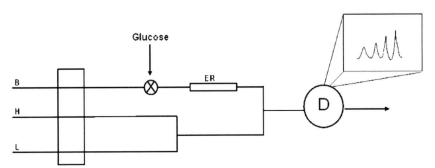

Figure 13.6 Schematic diagram for a FIA manifold for the analysis of glucose incorporating an enzymatic reactor with detection based on the luminol reaction. B, H and L: buffer, hexacyanoferrate (III) and luminol solutions, ER: enzymatic reactor; D: detector. (Adapted from [92] with permission from EDP Sciences, © 2000.)

13.2.3.2 CL Flow Methods

Amongst the range of chemistries used in FI-CL systems it is clear that the luminol reaction is the most popular, but there are also a significant number of papers that utilize the oxidizing power of permanganate and cerium(IV). The analytical applications of liquid-phase CL can be divided into those procedures that involve well-known CL reactions and those that involve light emission from the reaction oxidant–analyte, which is the substrate in the CL reaction and corresponds to the determinations identified as "direct CL determinations" [95].

Reagent-Based CL Determinations In the first modality, the analyte interacts with the CL reaction, usually as a reagent, a catalyst, a quencher, and even an enhancer. There is a limited number of such types of systems and they are responsible for the majority of analytical applications: these include luminol, peroxyoxalate, tris-(2,2'-bipyridyl) ruthenium(II), acridinium esters, cerium(IV), and so on.

Luminol. Luminol is one of the most commonly used liquid-phase chemiluminescence reagents [90–92]. Oxidants such as permanganate, hypochlorite, or iodine can be used, but hydrogen peroxide is the most common. A catalyst is often required with this chemistry and these include transition-metal ions (Ti(IV), V(II), Cr(III), Mn(II), Fe(II), Fe(III), Co(II), Ni(II), Cu(II)), hexacyanoferrate(III), hemin and heme proteins (hemoglobin, peroxidases, catalase, and cytochromes). The optimum reaction pH varies between 8 and 11. This chemistry can be used to sensitively determine the oxidant, catalyst, or species derivatized with luminol or related compounds.

Peroxyoxalate. The most commonly used reagents are bis-(2,4,6-trichlorophenyl) oxalate (TCPO) and bis-(2,4-dinitrophenyl)oxalate (DNPO). Fluorescent compounds (including anthracene, perylene, aminoanthracenes and aminopyrenes) and suitably derivatized analytes (such as amines, steroids with dansyl chloride; thiols with N-[4-(6-dimethylamino-2-bezofuranyl)-phenylmaleimide and catecholamines with fluorescamine) can be sensitively detected. The main drawback is that organic solvents should be used. This CL reagent is more popular for HPLC-CL than for FIA-CL determinations [96].

Tris(2,2'-bipyridyl)ruthenium(II). Although a large variety of compounds can reduce tris(2,2'-bipyridyl)ruthenium(III), only certain species (e.g., aliphatic amines, amino acids, NADH, some alkaloids, aminoglycoside or tetracycline antibiotics, and the oxalate ion) will produce the characteristic orange luminescence with this reagent. Small differences in chemical structure can have a dramatic effect on CL intensity [97].

Cerium(IV)-Based CL systems. This CL reagent can be regarded as a direct CL system or a reagent-based CL system. The cerium(IV)-based CL reaction in an acidic medium has been applied for the determination of biochemical species in dosage forms. A few pharmaceuticals can reduce the Ce(IV) directly and produce luminescence. Many FI-CL methods have been established for such species as naproxen [98] or acetaminophen [99]. However, most determinations involving Ce(IV) as the oxidant are of the indirect type, which are based on the enhancement effect of some analytes on the CL intensity of the Ce(IV)–sulfite system, Ce(IV)–Ru(bpy)$_3^{2+}$ system or Ce(IV)–Tween 20 system, and so on. This type of process, called sensitization, is used to determine reducing compounds, such as ofloxacin [100], mefenamic acid [101] or salicylic acid [102]. In particular, light emission resulting from the chemical reaction of Ce(IV) with some mercapto-containing compounds can be enhanced by certain fluorometric reagents such as quinine [103], rhodamine B or 6G [104] or by lanthanide ions [105].

Direct (Oxidation) CL Methods   These CL methods are based on the direct CL light emission involving the reaction between an oxidant (sometimes a reductant) and the analyte [106]. A dedicated preliminary experimental study based on testing CL reactions between the analyte and a wide range of oxidants in different media should be performed prior to the characterization of CL reactions of analytical interest. Strong oxidants (e.g., MnO_4^-, ClO^-, Ce(IV), H_2O_2, IO_4^-, Br_2 and N-bromosuccinimide) and reductants, are used under different chemical conditions in order to produce CL emissions from different analytes. In recent years, the use of FI methodologies with direct CL detection based on direct oxidations in an acidic medium presents a great potential for performing different analyses in food, drinks, pharmaceuticals and body fluids [107].

Permanganate-Based CL Systems in FIA. The use of acidic potassium permanganate as a CL reagent has been employed under a variety of conditions to generate CL from a wide range of analytes. $KMnO_4$ FI-CL methods are mainly based either on the direct oxidation of the analyte by $KMnO_4$ [108], or on the enhancement of CL intensity of the $KMnO_4$–sulfite system by the analyte [109], using quinine, formaldehyde, glyoxal, formic acid and some surfactants as a sensitizer. Acidic potassium permanganate can sensitively detect molecules containing phenolic and/or amine moieties and therefore it has considerable potential for the determination of a wide range of important analytes [110].

The advantages of CL-FIA analysis include low detection limits and a wide linear range; both can generally be achieved with simple, robust, and relatively inexpensive commercial instrumentation. Table 13.3 shows selected examples of different CL detection reactions developed for FIA-CL methods, applied to different samples including pharmaceutical, biomedical, environmental and food analysis [111–163].

Other FI-CL Systems   *Electrogenerated Chemiluminescence (ECL).* In recent years, CL flow systems in acidic media with electrogenerated unstable oxidation reagents have received much attention and some analytical applications have appeared in the literature [164, 165]. In ECL, one or more of the reagents is generated *in situ* in an electrolytic process. ECL shares many analytical advantages with CL mainly because no expensive excitation optics or sophisticated instrumentation are needed. Some of the FIA-ECL methods developed have been described for the analyses of substances of interest such as pharmaceuticals [166–168], lactate, glucose, cholesterol and choline [169], aminoacids [170, 171], metal ions [172, 173], and so on.

Photoinduced chemiluminescence is a novel strategy applied to pesticides and pharmaceuticals based on an on-line photodegradation step prior to direct CL reaction using, typically, potassium permanganate in sulfuric acid medium as oxidant [174, 175]. This strategy has been proposed for the determination of

Table 13.3 Selected flow methods based on direct CL or reagent-based CL detection [111–163].

Analyte	Sample matrix	CL Reaction/Reagents	Limit of detection	Ref.
Direct CL detection				
chloroquine	urine	direct CL of chloroquine in hydrogen peroxide-nitrite-sulfuric acid medium.	0.086 µM	[111]
cefradroxil hydrochloride	pharmaceuticals and biological fluids	potassium permanganate-sulfuric acid CL with quinine as sensitizer	0.05 µg mL^{-1}	[112]
chloramphenicol	pharmaceuticals	on-line photodegradation of chloramphenicol and result-ing phot-fragments are detected by direct CL using potassium permanganate in sulfuric acid medium as oxidant	30 ng mL^{-1}	[113]
ergonovine maleate	pharmaceuticals	direct oxidation with hexa-cyanoferrate(III) in sodium hydroxide, enhanced by a cationic surfactant (hexade-cylpyridinium chloride)	0.07 µg L^{-1}	[114]
2-ethyl-hexyl-4-(N, N-dimethylami-no)-benzoate	sunscreen formulations	direct CL of analyte oxidation by potassium permanganate in sulfuric acid	25 ng mL^{-1}	[115]
iproanizid	pharmaceuticals and urine	oxidation of the drug by cerium(IV) in sulfuric acid medium at room tempera-ture in presence of sulfite		[116]
naltrexone	pharmaceuticals	acidified permanganate CL	2.5 ng mL^{-1}	[117]
phenolic compounds	wines	oxidation and direct CL detection with acidic potassium permanganate	0.4–700 nM	[118]
propanolol	pharmaceuticals	direct CL by oxidation with potassium permanganate in a sulfuric acid medium	0.87 mg L^{-1}	[119]
salicylamide	human urine and pharmaceuticals	acidified potassium perman-ganate CL	30 ng mL^{-1}	[120]
tyrosine	—	potassium permanganate (pH 6.75) and acidic potassi-um permanganate in ortho-phosphoric acid (pH 2)	10–50 nM	[121]
Reagent-Based CL detection				
Pharmaceutical and Biomedical Applications				
bilirubin	aqueous	N-bromosuccinimide or sodium hypochlorite CL	1.75 µg mL^{-1}	[122]
captopril	pharmaceuticals	cerium(IV)-sulfuric acid CL	0.2 µM	[123]

(*Continued*)

Table 13.3 (*Continued*)

Analyte	Sample matrix	CL Reaction/Reagents	Limit of detection	Ref.
catecholamines	plasma	imidazole-catalyzed decomposition by catecholamines producing hydrogen peroxide which is detected using peroxyoxalate CL	—	[124]
choline and acetylcholine	rat brain tissue	use of two reactors containing acetylcholinesterase and choline oxidase immobilized on glass beads; detection using Co(II)-luminol CL	500–600 fmol	[125]
clomipramine	pharmaceuticals	sensitizing effect on the chemiluminescent oxidation of sulfite by Ce(IV)	2.5 mg L^{-1}	[126]
DNA	—	enhancement of the CL of Rhodamine B-cerium(IV) complex by DNA in acidic media	8.3 fg mL^{-1}	[127]
D-amino acids	human plasma	immobilized enzyme column reactor with peroxyoxalate CL detection	0.4–30 pmol (10 μL injected)	[128]
dopamine	biochemical samples	imidazole-peroxyoxalate CL	10 nmol (20 μL injected)	[129]
glucose	rabbit fluid and blood	microdialysis followed by reaction in an immobilized glucose oxidase reactor to produce hydrogen peroxide, detected with luminol-hexacyanoferrate CL	10 μM	[130]
glycerol-3-phosphate	aqueous	glycerol-3-phosphate oxidase immobilized on controlled pore glass; detection with luminol-Co(II) CL	0.5 μM	[131]
glutathione	—	enhancing effect on luminol hydrogen peroxide CL system	68 nM	[132]
isoanizid	pharmaceuticals	inhibition of the luminol-hydrogen peroxide-potassium hexacyanoferrate(III) reaction	5 mg L^{-1}	[133]
metopropol tatrate	pharmaceuticals and human urine	sensitizer of the CL of Ce(IV)-sulfite CL system in acidic medium	4.7 nM	[134]
naproxen	pharmaceuticals	Ce(IV)-sulfuric acid CL		[98]
oxymetazoline	pharmaceuticals	inhibition of the CL from the oxidation of luminol in alkaline medium		[135]

Table 13.3 (*Continued*)

Analyte	Sample matrix	CL Reaction/Reagents	Limit of detection	Ref.
phenotiazines	pharmaceuticals and biological fluids	cerium(IV)-acid CL with Rhodamine B as a sensitizer	0.01–$0.1\,\mu g\,mL^{-1}$	[136]
proline	wine	Ru(bipy)$_3^{3+}$ in sulfuric acid	10 nM	[137]
proteins	bovine serum albumin, human serum albumin, γ-globulin and egg albumin	1,10-phenanthroline-hydrogen peroxide-Cu(II) cetyltrimethylammonium bromide CL	$0.02\,\mu g\,mL^{-1}$	[138]
pyridoxine hydrochloride	tablets	luminol-hydrogen peroxide CL	$6\,\mu g\,mL^{-1}$	[139]
rutin (vitamin P)	pharmaceuticals	enhancing effect on the luminol-ferricyanide system	$30\,ng\,mL^{-1}$	[140]
tannic acid	pharmaceuticals	inhibition of the CL of luminol-H_2O_2-Mn tetrasulfonatophthalocyanine system by tannic acid	0.8 nM	[141]
terbutaline	pharmaceuticals, plasma and urine	enhancement by terbutaline sulfate of the chemiluminescence emission of the luminol-permanganate system under alkaline conditions	—	[142]
tetracyclines	pharmaceutical preparations	cerium(IV)-sulfuric acid CL with quinine as sensitizer	0.25–25 nmol	[143]
thiamine	pharmaceuticals and urine	luminol and KIO$_4$ immobilized on an anion-exchange column; controlled reagent release technology in FIA system	1 p.m.	[144]
triclosan	toothpaste samples	phototransformation of triclosan to a light-emitting precursor in the presence of fluorescein in an alkaline medium, and CL reaction with N-bromosuccinimide	50 nM	[145]
tryptophan	tissue	cerium(IV)-sulfuric acid CL	$0.1\,\mu g\,mL^{-1}$	[146]
uric acid	human urine and serum	luminol and periodate immobilized on anion-exchange resin packed in a column, used as reagents; inhibition of this CL system by uric acid	$1.8\,ng\,mL^{-1}$	[147]
vitamin B$_{12}$	pharmaceuticals and human serum	enhancing effect of Co(II) on the CL reaction between luminol and dissolved oxygen in a flow injection system	$50\,pg\,L^{-1}$	[148]

(*Continued*)

Table 13.3 (*Continued*)

Analyte	Sample matrix	CL Reaction/Reagents	Limit of detection	Ref.
Environmental applications				
Cu (complexed)	sea water	1,10-phenanthroline-hydrogen peroxide CL	0.1 nM	[149]
Fe and Mn (dissolved)	underground water	luminol-potassium periodate CL	3 and 5 pg mL^{-1}	[150]
Fe(II) and Fe(III)	natural waters	Fe(III) reduced to Fe(II) with Cu-coated Zn; luminol immobilized on an anion-exhange resin; eluted with sodium hydroxide for CL detection	0.4 ng L^{-1}	[151]
hydrogen peroxide	rain water	immobilized Co(II) and luminol are eluted by hydrolysis for CL detection	12 nM	[152]
nitrite	natural waters	nitrite reacts with hydrogen peroxide to form peroxynitrile, which produces CL with luminol	1 nM	[153]
phenol	water samples	enhancement effect of phenol on the luminol-hexacianoferrate CL system; on-line SPE to increase selectivity	4.7 ng L^{-1}	[154]
phosphate	natural waters	hydrogen peroxide produced from the reaction of immobilized pyruvate oxidase with phosphate is detected using luminol-horseradish peroxidase CL	74 nM	[155]
V(V)	Geochemical and hair samples	Luminol and hexacyanoferrate(II), are both immobilized on an anion-exchange resin column, and are eluted with phosphoric acid to produce CL	5.4 ng/ml	[156]
V(IV)	Water samples	Catalytic effect of vanadium (IV) on the oxidation of purpurogallin by periodate to produce light emission	0.05 ng mL^{-1}	[157]
Food and Beverage applications				
chlorpyrifos	fruits and waters	suppression effect on the luminol-periodate CL reaction, using immobilized reagents on an anion exchange column	0.18 ng mL^{-1}	[158]

Table 13.3 (Continued)

Analyte	Sample matrix	CL Reaction/Reagents	Limit of detection	Ref.
L-malate	wines	malate dehydrogenase/ reduced nicotinamide adenine dinucleotide oxidase co-immobilized on polymer beads to produce hydrogen peroxide for detection using luminol-hexacyano-ferrate (III) CL	0.08 μM	[159]
sulfite	beers and wines	auto-oxidation sensitized by Rhodamine 6G (immobilized on cation exchange resin) enhanced with Tween 80	0.03 mg L^{-1}	[160]
Sudan I	hot chilli sauce	enhancing effect of CL from luminol-H_2O_2 system	3 pg mL^{-1}	[161]
tannic acid	hop pellet samples	inhibition of luminol-hydrogen peroxide-Cu(II) CL	9 nM	[162]
tetracyclines	fish	enhancing effect of tetracycline on the reaction between Ce(IV) and Rhodamine B; use of on-line SPE with a MIP-based column	1 ng mL^{-1}	[163]

aldicarb [176] and asulam [177], with photodegradation of the pesticide followed by an oxidation reaction into a fully automated multicommutation based flow assembly.

13.2.3.3 Bioluminescence

Bioluminescence (BL) is an enzyme-catalyzed process found in biological systems, in which a catalytic protein increases the efficiency of a luminescent reaction [178]. There is a wide variety of BL organisms, from bacteria and fungi to molluscs, fishes and insects. The most extensively studied BL organism is the firefly (*photinus pyralis*) because fireflies are easy to collect and their light organs are replete with the chemical constituents that make up the luminescent reaction. The use of BL systems provides two main advantages: the sensitivity is often increased because of higher quantum yields, and the selectivity is increased because of the enzymes involved [179].

The sensitivity and versatility of bioluminescent reactions has led to a wide range of applications, including clinical assays for enzymes and substrates (glucose, trygli-cerides, glycerol, etc.), steroid analysis, aquatic toxicity testing, screening bacteria in raw milk, and so on. Three bioluminescent reactions, firefly luciferase, marine bacterial luciferase, and the aequorin reaction account for more than 90% of the applications. A flow injection system for the analysis of low levels of ATP based on the reaction with luciferase-luciferin was described by Hansen [180]. A dual injection valve was used, which permitted the simultaneous injection of sample and enzyme solutions. A LOD of 0.1 nM was obtained. Moreover, Blum *et al.* [181] described a fiber optic sensor based on enzyme-catalyzed light-emitting reactions integrated in a flow injection analysis system.

13.2.4
Solid-phase Luminescence Based Detection in Flowing Streams

The coupling of FIA and solid-surface luminescence (SSL) [182] overcomes the procedural SPS limitations and shows remarkable advantages over discrete manual measurements: (i) the sample treatment process (reagent mixing, removal of interferences, pH adjustments, etc.) can all be carried out on-line; (ii) improved sensor lifetime, as in flow analysis the sensing material can be exposed to the sample for only a short time and maintained in a friendlier matrix between measurements; (iii) automation, which in turn, offers a series of advantages such as high sample throughput (short response times), high precision and accuracy, saving of reagents and solid sensing support, and so on. In recent years, the coupling FIA-SSL has become a research area of great interest and luminescence flow-through sensors have been described for a range of (inorganic and organic) analytes and applications in relevant fields such as agricultural, environmental, biomedical and pharmaceutical analysis have been described.

This coupling uses an ordinary instrumentation equipped with a commercially available cell in which the sensing material is placed and requires a very simple manifold. The signal is developed as the analyte is retained on the solid microbeads placed in the flow cell. After measurement, the active solid surface has to be regenerated in order to keep it available for the next determination.

13.2.4.1 Solid-phase Fluorescence Detection
Most solid surface fluorescence detection systems are based on measurements of the native fluorescence of the analyte [183]. One of the first systems described was based on this principle and allowed the selective determination of micro-amounts of europium, terbium, dysprosium or samarium by using the weak-acid cation-exchange gel, CM-Sephadex, as sensing support in the luminescence detection area. A series of solid-phase native fluorescence detection systems applied to the determination of pharmaceutical active principles have been described [183]. Other application fields include environmental and food analysis as well as pH measurements and oxygen sensing. Table 13.4 shows a selection of the most relevant systems of this type.

A reaction product can also be the species detected in systems based on native fluorescence signals. Thus, the native fluorescence of the product of the reaction between glycerol and NAD^+ catalyzed by glycerol dehydrogenase is the basis of a solid-phase fluorimetric detection system for the indirect determination of glycerol in wine [184]. The coupling of SIA and solid-phase fluorescence detection has recently been described. Native and derivative solid phase fluorescence measurements have been used in this coupling [185]. The implementation of native solid-phase

Table 13.4 Selected representative examples of analytical flow injection methodologies based on solid-phase luminescence detection [203–226].

Analyte	Sample matrix	Description	Limit of Detection	Ref.
Fluorescence-Based optosensors				
diphenhydramine	pharmaceutical preparations	native FL detection; solid support: Sephadex G-15	$0.1–1.2 \, mg \, L^{-1}$	[203]
quinine and quinidine	pharmaceuticals, water, shampoo, soft drinks	native FL detection; solid support: Sephadex SPC-25	quinine: $0.4–20 \, \mu g \, L^{-1}$ quinidine: $0.9–20 \, \mu g \, L^{-1}$	[204]
PAHs	tap and mineral water	native FL detection; solid support: Amberlite XAD-4	—	[205]
α- and β-naphthol	water	native FL detection; second derivative synchronous fluorescence + PLS; solid support: Sephadex QAE-A25	—	[206]
imidacloprid	peppers and environmental water	PIF detection. Solid support: C_{18} silica gel beads	LOD: $1.8 \, \mu g \, L^{-1}$	[192]
thiamine	pharmaceutical preparations	bead injection system: renewable surface sensor; FL detection based on thiochrome complex (Fe$(CN)_6^{3-}$/OH^- (385/ 433 nm); solid support: C18-PS-DP	LDR: $0.06–8 \, mg \, L^{-1}$	[202]
flufenamic acid	serum, urine and pharmaceuticals	PIF detection (258/ 442 nm); multicommuted flow assembly; solid support: C_{18} silica gel beads	LOD: 0.55 nM	[207]
RTP-based optosensors				
tetracyclines	urine	lanthanide-sensitized Luminescence. Eu(III)-TC lomplex, (390/ 622 nm); solid support: Amberlite XAD-2	LOD: $0.25–0.4 \, \mu g \, L^{-1}$	[208]
nafcillin (β-lactamic antibiotic)	milk-based products	RTP detection; 283/ 505 nm solid support: nafcillin imprinted sol–gel or imprinted ormosils (organically modified silanes)	LOD: 5.8 μM (aqueous milk) 33 μM (skimmed milk)	[209–211]

(Continued)

Table 13.4 (*Continued*)

Analyte	Sample matrix	Description	Limit of Detection	Ref.
naphazoline	eye lotions	optosensor based on HAI-RTP (290/520 nm) using Amberlite XAD-7 as sensing support. [KI] 1.6 M; 15 mM sodium sulfite	LOD: 9.4 ng ml^{-1}	[88], [212]
fluoranthene	river water samples	RTP optosensing using a molecularly imprinted polymer based on iodinated monomers	LOD: 35 ng L^{-1}	[213]
			LDR: 0–100 µg L^{-1}	
benzo[a]pyrene	natural water	RTP optosensing using a molecularly imprinted polymer based on iodinated monomers	LOD: 10 ng L^{-1}	[214]
			LDR: 0–100 µg/L	
benzo[a]pyrene	water	optosensor based on HAI-RTP (390/690 nm) using Amberlite XAD-7 as sensing support. [KI] 4 M; 20 mM sodium sulfite	LOD: 12 ng ml^{-1}	[215]
2-naphthoxyacetic acid	water	optosensor based on HAI-RTP (276/516 nm) using Amberlite XAD-7 as sensing support. [Tl(I)] = 175 mM; 10 mM sodium sulfite	LOD: 4.9 ng ml^{-1}	[216]
1-naphthyl-phthlamic acid and metabolite	water	optosensor based on HAI-RTP (276/516 nm) using Amberlite XAD-7 as sensing support. [Tl(I)] = 0.2 M; 15 mM sodium sulfite	LOD: 8.1–11.2 ng ml^{-1}	[217]
naphthalene acetic acid	natural water and apples	on-line retention of 1-NAA on the packed resin and measurements of the heavy atom induced (HAI)- RTP emission (290/490 nm) from this native luminescent compound are taken	LDR: 0–500 µg L^{-1}	[218]
			LOD: 1.2 µg L^{-1}	

Table 13.4 (*Continued*)

Analyte	Sample matrix	Description	Limit of Detection	Ref.
Chemiluminescence-based optosensors				
epinephrine	serum	CL reaction between luminol and potassium ferrocyanide potassium: enhancing effect of epinephrine retained on a solid support	LDR: 5–100 nM	[219]
		solid support: MIP polymer	LOD: 3 nM	
norfloxacin	urine samples	norfloxacin imprinted polymer as recognition element; Ce(IV)–sodium sulfite norfloxacin CL reaction	LDR: 0.1–10 µM	[220]
			LOD: 30 nM	
isoniazid	human urine	isoniazid MIP is synthesized through thermal radical copolymerization of metharylic acid (MAA) and ethylene glycol dimethacrylate (EGDMA) in the presence of isoniazid template molecules. Solid support: Isoniazid MIP CL detection using MIP recognition	LDR: 2–200 nM	[221]
salbutamol	urine sample	CL reaction between luminol and ferricyanide potassium: enhancing effect of salbutamol. MIP as recognition element	LDR: 0.05–10 µM	[222]
			LOD: 16 ng mL^{-1}	
hidralazine	human urine	CL reaction between luminol periodote: enhancing effect of hidralazine. MIP as recognition element	LOD: 0.6 nM	[223]
			LDR: 2–800 nM	
terbutaline	human serum	CL detection based on luminol CL reaction with ferrocyanide, using a MIP as recognition material for terbutaline; microflow sensor on a chip. 12 µL injected	LDR: 8–100 ppb	[224]
			LOD: 4 ppb	

Table 13.4 (*Continued*)

Analyte	Sample matrix	Description	Limit of Detection	Ref.
orthophosphate	mineral, ground, tap and pond water	CL detection based on luminol reaction	LDR: 5–50 µg L^{-1} P;	[225]
		solid support: *N*-vinyl-pyrrolidone/divinylben-zene sorbent (Oasis HLB)	LOQ: 4 µg L^{-1} P	
salicylic acid	pharmaceutical preparations	CL from direct oxidation of salicylic acid with permanganate on a solid-support (Sephadex QAE-A25)	LDR: 1–30 mg L^{-1}	[226]
			LOD: 0.30 mg L^{-1}	

fluorescence detection in multicommutation systems has been evaluated by developing both single and multi-detection measurements [186, 187].

Solid-phase fluorescence detection systems based on chemical derivatization of the analyte require a more complex configuration involving either an additional channel for the derivative reagent merging with the injected sample or a bifurcation of the carrier stream with a simultaneous double injection (sample and reagent) [188].

However, a simplification can be achieved if the fluorogenic reagent (usually a chelating agent) can be permanently immobilized on the support. In this case, the regeneration of the sensing support after each measurement is performed by dissociating the fluorescent complex and eluting only the analyte [189]. Recently, a SIA solid-phase fluorimetric detection system based on the use of a derivative reagent has been described for the determination of paracetamol [190]. The reaction product between the analyte and sodium nitrite in an acidic medium is inserted, after alkalinization, into the system.

Solid-phase detection has also been implemented with PIF detection. This implementation PIF-SPS-FIA has been described (including multicommutation approaches) for the determination of drugs and pesticides [191, 192].

There are only a few solid-phase fluorescence multi-detection flow systems (able to detect several species in a sample). Practically all these systems are based on the measurements of native fluorescence signals. Up to three analytes in a sample could be detected [193]. On the other hand, a few multicommutated fluorimetric solid phase multidetection systems have been proposed [194].

The solid phase is always viewed as a permanent component in a solid surface fluorescence detection system. After measurement, the solid support beads have to be regenerated so that the sensing surface remains ready for the next sample, thus making the solid phase reusable. Nevertheless, when the species of interest is strongly retained (i.e., multicharged species on ion exchange beads) the regeneration of the sensing surface becomes extraordinarily difficult to achieve. Then, highly concentrated saline or acid solutions have to be used [195, 196] reducing the life-time of the sensing device, reproducibility and baseline stability [197]. Surface deactivation always occurs after a number of measurements. These drawbacks can be avoided by using the so-called bead injection spectroscopy (BIS) or flow injection-renewable solid surface fluorescence methodology [198], envisaged as the third generation of FI microanalytical techniques [199]. This BIS-FIA fluorimetric methodology has been described for the determination of berillium and aluminum [200] and vanadium (V) [201]. A BIS-SIA fluorimetric system has been described for the prior conversion of thiamine to thiochrome [202].

13.2.4.2 RTP-Based Optosensing

Different SS-RTP methodologies have been introduced recently for the continuous and *in situ* monitoring of different analytes, based on flow injection systems, to determine analytes such as oxygen, metals ions, lanthanides, organic compounds, and enzyme substates like glucose.

The sensing of molecular oxygen based on luminescence quenching is regarded as one of the most typical and widespread optosensing applications. In all luminescence quenching detection schemes, a strong and stable luminescence background signal is generated, which is dynamically quenched in the presence of molecular oxygen or other quenchers. The experimental manifolds used for the detection of molecular oxygen [227, 228] based on quenched RTP can also be used for the indirect detection of enzyme substrates like glucose [229, 230] and cholesterol [231]. In the case of glucose, RTP was obtained after the consumption of oxygen in the enzymatic reaction of glucose to form gluconic acid and hydrogen peroxide in the presence of the enzyme glucose oxidase, which implies that the obtained RTP is an indirect measurement of glucose concentration.

Methods for the detection of several metal ions based on the SS-RTP obtained from a complex of the metal with ferron have been described [232]. These phosphorescent complexes were retained on an anion exchange resin. Regeneration of the resin is performed by injecting a small volume of a strongly acidic (HCl) solution.

Native phosphorescent properties based on SS-RTP using an oxygen scavenger and a heavy atom perturber were applied to develop different optosensors to determine pesticides and PAHs in different samples. Recently, the use of molecularly imprinted materials, widely used in separation techniques, has emerged as a new tool to develop novel sensing materials with higher selectivity. This recognition process has been recently applied with RTP-based transduction in flowing streams [211, 213, 214].

The use of luminescence from lanthanide–ligand complexes enables detection of the rare earth metals as well as the ligands. Sensitized lanthanide luminescence was applied for the determination of Eu(III) [233], Gd(III) [234] and Tb(III) [235]. In these examples, a complex formed between the lanthanide and acetylacetone was transiently immobilized on a resin, Chelex 100. Immobilization of the complex resulted in increased luminescence intensities and lifetimes. Using the same strategy, analytes (used as ligand) such as tetracycline antibiotics [236–238] and anthracyclines [239] were detected using the luminescence of their complexes with Eu(III).

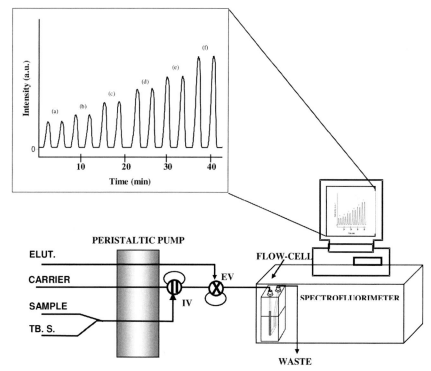

Figure 13.7 Flow-injection assembly used for the terbium-sensitized luminescence based optosensor for the determination of norfloxacin in pharmaceuticals and biological fluids. IV: injection valve; EV: eluting valve; ELUT: eluting solution (0.08 M EDTA); T.B.S.: Tb(III) (4 mM); CARRIER (0.1 M acetate buffer, pH 5.6); SAMPLE (NFX in 0.1 M acetate buffer pH 5.6). Diagram showing the peaks, duplicated, obtained for (a) the blank; (b) 10 ng ml^{-1}, (c) 30 ng ml^{-1}, (d) 50 ng ml^{-1}, (e) 70 ng ml^{-1} and (f) 100 ng ml^{-1} of NFX. Reproduced from [238] with permission from Elsevier, © 2005.

Figure 13.7 shows the manifold used in an optosensor for the determination of norfloxacin in pharmaceuticals and biological fluids based on terbium-sensitized luminescence.

13.2.4.3 Solid-phase Chemiluminescence Detection

The immobilization of reagents in flow injection systems with CL detection has been described with an increasing number of analytical applications [240, 241]. In many of these cases analytes are determined by the CL reaction with the dissolved reagents which are released from the immobilized substrate by appropriate eluents. In such cases, no solid support is used in the flow cell, the detection being performed in homogeneous solution, so these systems cannot really be considered to be flow-through CL sensors [94].

CL Sensors Based on Immobilization of CL Reagents Reagent immobilization onto proper substrates plays an important role in the development of CL sensors.

Ion-exchange resins are widely used to immobilize the CL reagents. The immobilization of luminol and other CL reagents on these types of support have been extensively employed for developing a series of CL sensors for both inorganic and organic analytes. Usually, the resin with the immobilized reagent(s) is packed into a piece of glass tube which serves as the flow cell and placed in front of the detection window of a photomultiplier [242, 243]. A very interesting way of increasing the lifetime of these CL sensors is to immobilize CL reagents that can be reused: Ru (bpy)$_3^{2+}$ immobilized on Dowex-50W ion exchange resin could be used for at least six months [244]. Lin *et al.* [245] studied the effect of several kinds of anion exchange resins used for the immobilization of anionic reagents on the CL response in the determination of H_2O_2 based on its CL reaction with periodate: Amberlite IRA series, Amberlyst A series and Muromac series resins were compared. Resins with gel structure or with macroreticular structure showed different CL emission, the strongest CL signal corresponding to a gel acrylic type strong anionic exchanger base resin (Amberlite IRA 458).

Several flow-through sensors have been developed based on the sensitizing effect of the analyte on the CL oxidation of sulfite in H_2SO_4 medium by a solid oxidant: PbO_2 [246], $NaBiO_3$ [247] or MnO_2 [248] have been used as oxidants absorbed on a sponge (choloroprene) and packed into a glass column used as the flow cell. CL flow sensors based on the use of luminol immobilized in a suitable polymeric sorbent have also been used for gas analysis [249, 250]. Several types of nanosized particles have recently been proposed as catalysts for developing cataluminescence gas sensors [251, 252].

CL Sensors Based on Immobilization of Enzymes Enzymes act as highly selective catalysts, hence their (co)immobilization is one of the most interesting ways to develop chemical and biochemical CL sensors. (Co)immobilization of enzymes on tresylated poly(vinyl alcohol) beads packed into a transparent PTFE tube (used as a spiral flow cell) has been described for several CL flow sensors: glucose [253], branched chain amino acids (L-leucine, L-valine and L-isoleucine) [254], L-glutamate in serum [255], histamine determination in fish meat extracts [256].

Recently, the use of magnetic microbeads immobilized with an antibody has been proposed for CL determination of anionic surfactants [257]. The magnetic microbeads are trapped in the flow cell by using a neodymium magnet, the CL reaction taking place on them. After monitoring the signal, the magnetic beads are discarded by shifting the magnet downwards. The use of sol–gel to immobilize enzymes has become a familiar process for developing CL sensors. One of the advantages of sol–gel is its suitability for optical sensors due to its optical transparency and chemical stability. As an example, this advantage has been demonstrated by the use of sol–gel immobilized hemoglobin as a catalyst [258].

CL Sensors Based on Immobilization of the Analyte on Molecular Imprinted Polymers (MIPs) MIPs have been used as sensing materials to design CL flow sensors. The imprinted cavities of a defined shape and functional groups in the molecularly imprinted polymer are expected to be developed not only for their molecule recognition function but also as a special CL reaction medium. These sensing materials have been applied to the determination of several analytes including 1,10-phenanthroline [259], clenbuterol [260], epinephrine [219], norfloxacin [220] and isoniazid [221].

Other flow analysis methodologies such as multicommutation and multisyringe flow injection analysis have been introduced in the development of CL sensing systems. Thus, the first multicommuted CL sensor was described for salicylic acid, based on the use of four three-way solenoid valves controlled by a microcomputer by means of appropriate software [226]. Similarly, the multisyringe flow injection analysis concept combined with multicommutation was exploited to develop a flow-through solid-phase-based CL sensor for orthophosphate trace determination in environmental water samples [225].

13.3
Recent Trends and Perspectives

Steps are being taken in order to increase both the selectivity and the sensitivity of flow methods. The main areas that are being explored are miniaturization aiming at lowering absolute LODs and reducing the volume consumption per analysis, together with the development of new materials such as molecularly imprinted polymers used as specific recognition systems in order to design selective recognition systems which can be applied to sense target species in complex matrices, or the use of quantum dots with unique optical features for sensing purposes. Luminescence is an attractive and cost-effective detection tool, fully complementary with these advanced strategies.

13.3.1
Miniaturization

Miniaturization is generally recognized as one of the most important trends in the development of analytical instrumentation, and the ultimate purpose is the integration of the entire analytical process on a micro-scale. In recent years, research and development in miniaturized total analysis systems (μTAS) on silica and glass microchips has grown dramatically in chemical reaction and analytical chemistry due to its inherent advantages of being portable, requiring low reagent consumption and reduction of analysis time [261].

Although microfluidic systems have been designed mainly for separation devices, such as capillary electrophoresis and microflow injection analysis (μFIA) systems, one form of microfluidic system has been developed dramatically in recent years for its low reagent consumption, simple operation and suitability for miniaturized systems. In this sense, CL is a promising method of detection in micro-total analytical systems. No light source is required in CL measurements, hence the instrumentation for the CL method is much simpler and it can be easily integrated onto the chip for detection.

13.3.2
Molecularly Imprinted Materials

Another strategy to enhance selectivity in flow injection methods with luminescence detection is the use new materials and sorbents specific for a target species. Recently, the use of MIPs (organic polymers and sol–gels), widely used in separation techniques, has emerged as a new tool to develop novel sensing materials with higher selectivity. This recognition process can be implemented in flow-through optical sensors with luminescence detection. In this sense, imprinted materials provide highly selective recognition and sensitive determinations due to the analyte pre-concentration measured inside the specific cavities of the polymeric structure. Moreover, this recognition mode provides the rigidity needed for RTP. The use of imprinted materials as solid supports with a selective luminescence detection mode such as RTP is an attractive combination for the design of highly sensitive, selective optical sensors [78].

Although imprinted sol–gel matrices may offer a series of advantages (e.g., the possibility of using an aggressive washing step, such as combustion, to better remove the imprint molecule), such materials have not yet achieved the success of imprinted organic polymers in chemical applications. The general use of imprinted sol–gels as active phases (supports) for optical sensing is still in the early stages and more research is necessary to explore their potential.

13.3.3
Quantum Dots

Quantum dots (QDs) are nanostructured materials with unique attractive optoelectronics characteristics especially suited to analytical applications in the (bio)chemical field [262]. QDs typically exhibit higher fluorescence quantum yields than conventional organic fluorophores, allowing greater analytical sensitivity. The popularity of QDs as photoluminescent probes for optical sensing is increasing, as research efforts are devoted to expanding the unique properties of this new class of luminophores. Most of the work on QD applications so far has been restricted to solution-sensing assays. A step forward towards the development of more sensitive and selective optosensing schemes could be based on the immobilization of these QDs in appropriate solid supports to fabricate "active" solid phases for working in flowing solutions [263]. Optosensing technologies can combine the advantages of QDs with flow analysis techniques. QDs can be integrated into appropriate solid supports, a process that has only just begun, in order to develop reliable "active" phases and optosensors able to provide useful flow-through optical sensing or fiber-optic-based sensing applications.

As an example of these recent trends, state-of-the-art CL µFIA systems on a chip have recently been described by Zhang and coworkers [264]. Figure 13.8 shows the scheme of the flow assembly used. The technology of laser ablation was used to fabricate the microchannels on a polymethyl methacrylate (PMMA) chip. A micro-column for specific molecular recognition, including molecularly imprinted

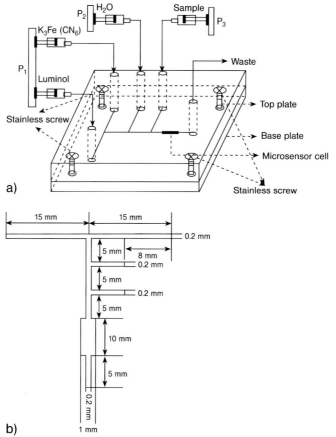

Figure 13.8 a) Schematic diagram of microflow sensor on chip for determination of terbutaline (P1, P2, P3: Ts2–60 syringe pump). b) Schematic diagram showing dimensions of the microchannel. Reproduced from [224] with permission from Elsevier, © 2006.

polymer, enzyme and bacteria, was incorporated into the flow assembly in order to enhance selectivity. These μFIA systems have been applied to clinical analysis, detection of food safety, *in vivo* and real time determination of drugs and pharmacokinetics studies [224, 265].

As an example of the sensing schemes used in these CL microflow sensors, the sensing mechanism principle used for the CL determination of terbutaline in human serum is shown in Figure 13.9 [221]. It is based on a MIP as the recognition element. The isoniazid MIP was synthesized through thermal radical copolymerization of methacrylic acid (MAA) and ethylene glycol dimethacrylate (EGDMA) in the presence of isoniazid template molecules. Isoniazid could be selectively adsorbed by the MIPs and the adsorbed isoniazid was sensed by its great enhancing effect on the weak CL reaction between luminol and periodate, which were mixed in the flow cell.

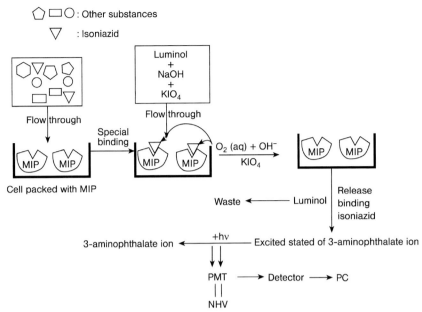

○ ▢ ○ : Other substances

▽ : Isoniazid

Figure 13.9 Proposed sensing mechanism for the flow injection CL sensor based on an isoniazid MIP. Reproduced from [221] with permission from Elsevier, © 2007.

The sensor was reversible and reusable. It brings a remarkable improvement in both sensitivity and selectivity for CL analysis. These specific sensing schemes will assist the development of new more selective and sensitive flow injection techniques in the near future, with the possibility of manufacturing portable instruments associated with luminescence transduction, either using a CL system with no light source needed or FL or RTP schemes using fiber optics and light emitting diodes.

References

1 Hurtubise, R.J. (1988) in *Molecular Luminescence Spectroscopy: Methods and Applications, Part 2* (ed. S.J. Shulman), J. Wiley, New York, Chapter 1.

2 Valcárcel, M. and Luque de Castro, M.D. (1988) *Automatic Methods of Analysis*, Elsevier, Amsterdam.

3 Kina, K., Shiraishi, K. and Ishibashi, N. (1978) Ultramicro solvent extraction and fluorimetry based on the flow injection method. *Talanta*, **25**, 295–297.

4 Ruzicka, J. and Hansen, E.H. (1975) Flow injection analyses: Part I. A new concept of fast continuous flow analysis. *Analytica Chimica Acta*, **78**, 145–157.

5 Ruzicka, J. and Marshall, G. (1990) Sequential injection: a new concept for chemical sensors, process analysis and laboratory assays. *Analytica Chimica Acta*, **237**, 329–343.

6 Cerdà, V., Estela, J.M., Forteza, R., Cladera, A., Becerra, E., Altimira, P. and

Sitjar, P. (1999) Flow techniques in water analysis. *Talanta*, **50**, 695–705.

7 Reis, B.F., Gine, M.F., Zagatto, E.A.G., Lima, J.L.F.C. and Lapa, R.A. (1994) Multicommutation in flow analysis. Part 1. Binary sampling: concepts, instrumentation and, spectrophotometric determination of iron in plant digests. *Analytica Chimica Acta*, **293**, 129–138.

8 Lapa, R.A.S., Lima, J.L.F.C., Reis, B.F., Santos, J.L.M. and Zagatto, E.A.G. (2002) Multicommutation in flow analysis: concepts, applications and trends. *Analytica Chimica Acta*, **466**, 125–132.

9 De Armas, G., Becerra, E., Cladera, A., Estela, J.M. and Cerdá, V. (2001) Sequential injection analysis for the determination of fuberidazole and thiabendazole by variable-angle scanning fluorescence spectrometry. *Analytica Chimica Acta*, **427**, 83–92.

10 Albert-García, J.R. and Martínez-Calatayud, J. (2006) FIA-fluorimetric determination of the herbricide Benfuresate. *Journal of Flow Injection Analysis*, **23**, 19–24.

11 Martinez-Calatayud, J., Goncalves Ascencao, J. and Albert-García, J.R. (2006) FIA-fluorimetric determination of the Pesticide 3-indolyl acetic Acid. *Journal of Fluorescence*, **16**, 61–67.

12 Subova, I., Assandas, K.A., Catala-Icardo, M. and Martínez-Calatayud, J. (2006) Fluorescence determination of the pesticide asulam by flow injection Analysis. *Analytical Sciences*, **22**, 21–24.

13 Solich, P., Polydorou, C.K., Koupparis, M.A. and Efstathiou, C.E. (2001) Automated flow injection fluorimetric determination and dissolution studies of bumetanide in pharmaceuticals. *Analytica Chimica Acta*, **438**, 131–136.

14 Solich, P., Polasek, M. and Karlicek, R. (1995) Sequential flow injection spectrofluorimetric determination of coumarins using a double-injection single-line system. *Analytica Chimica Acta*, **308**, 293–298.

15 Legnerova, Z., Sklenarova, H. and Solich, P. (2002) Determination of rhodamine 123 by sequential injection technique for pharmacokinetic studies in the rat placenta. *Talanta*, **58**, 1151–1155.

16 Klimundova, J., Mervartova, K., Sklenarova, H. and Solich, P. (2006) Automated sequential injection fluorimetric set-up for multiple release testing of topical formulation. *Analytica Chimica Acta*, **573–574**, 366–370.

17 Legnerova, Z., Huclova, J., Thun, R. and Solich, P. (2004) Sensitive fluorimetric method based on sequential injection analysis technique used for dissolution studies and quality control of prazosin hydrochloride in tablets. *Journal of Pharmaceutical and Biomedical Analysis*, **34**, 115–121.

18 Balaguer, A., Chisvert, A. and Salvador, A. (2006) Sequential-injection determination of traces of disodium phenyl dibenzimidazole tetrasulphonate in urine from users of sunscreens by on-line solid-phase extraction coupled with a fluorimetric detector. *Journal of Pharmaceutical and Biomedical Analysis*, **40**, 922–927.

19 Vidal, M.T., Chisvert, A. and Salvador, A. (2003) Sensitive sequential-injection system for the determination of 2-phenylbenzimidazole-5-sulphonic acid in human urine samples using on-line solid-phase extraction coupled with fluorimetric detection. *Talanta*, **59**, 591–599.

20 Bravo, J.C., Fernández, P. and Durand, J.S. (2005) Flow injection fluorimetric determination of β-estradiol using a molecularly imprinted polymer. *Analyst*, **130**, 1404–1409.

21 Sacenon, J.F. and De la Guardia, M. (1994) Micellar enhanced fluorimetric determination of carbendazim in natural waters. *Analytica Chimica Acta*, **287**, 49–57.

22 Zisiou, E.-P., Pinto, P.C.A.G., Saravia, M.L.M.F.S., Siquet, C. and Lima, J.L.F.C. (2005) Sensitive sequential injection determination of naproxen based on interaction with β-cyclodextrin. *Talanta*, **68**, 226–230.

23 Tang, L.X. and Rowell, F.J. (1998) Rapid determination of warfarin by sequential injection analysis with cyclodextrin-enhanced fluorescence detection. *Analytical Letters*, **31**, 891–901.

24 Amador-Hernández, J., Fernández-Romero, J.M. and Luque de Castro, M.D. (2001) Flow injection screening and semiquantitative determination of polycyclic aromatic hydrocarbons in water by laser induced spectrofluorimetry — chemometrics. *Analytica Chimica Acta*, **448**, 61–69.

25 Bradley, A.B. and Zare, R.N. (1976) Laser fluorimetry. Sub-part-per-trillion detection of solutes. *Journal of the American Chemical Society*, **98**, 620–621.

26 Richardson, J.H., Tallin, B.W., Jonson, D.C. and Hrubesh, L.W. (1976) Sub-part-per-trillion detection of riboflavin by laser-induced fluorescence. *Analytica Chimica Acta*, **86**, 263–267.

27 Richardson, J.H. (1977) Sensitive assay of biochemicals by laser-induced molecular fluorescence. *Analytical Biochemistry*, **83**, 754–762.

28 Huie, C.W., Airen, J.H. and Williams, W.R. (1991) Rapid screening of porphyrins using flow injection analysis and visible laser fluorimetry. *Analytica Chimica Acta*, **254**, 189–196.

29 Imasaka, T., Higashijima, T. and Ishibashi, N. (1991) Dehydrogenase and ethanol assay based on visible semiconductor laser spectrometry. *Analytica Chimica Acta*, **251**, 191–195.

30 Santana-Rodríguez, J.J., Halko, R., Batancort-Rodríguez, J.R. and Aaron, J.J. (2006) Environmental analysis based on luminescence in organized supramolecular systems. *Analytical and Bioanalytical Chemistry*, **385**, 525–545.

31 Galeano-Díaz, T., Acedo-Valenzuela, M.I. and Salinas, F. (1999) Determination of the pesticide Naptalam, at the ppb level, by FIA with fluorimetric detection and

on-line preconcentration by solid-phase extraction on C_{18} modified silica. *Analytica Chimica Acta*, **384**, 185–191.

32 Coly, A. and Aaron, J.J. (1998) Fluorimetric analysis of pesticides: methods, recent developments and applications. *Talanta*, **46**, 815–843.

33 Zhang, Z.-Q., Ma, J., Lei, Y. and Lu, Y.-M. (2007) Flow-injection on-line oxidizing fluorimetry and solid-phase extraction for determination of thioridazine hydrochloride in human plasma. *Talanta*, **71**, 2056–2061.

34 Rezaei, B., Ensafi, A.A. and Nouroozi, S. (2005) Flow-injection determination of ascorbic acid and cysteine simultaneously with spectrofluorometric detection. *Analytical Sciences*, **21**, 1067–1071.

35 Martínez-Calatayud, J., Gómez-Benito, C. and Gaspar-Gimenez, D. (1990) FIA—fluorimetric determination of thiamine. *Journal of Pharmaceutical and Biomedical Analysis*, **8**, 667–670.

36 Pérez-Ruiz, T., Martínez-Lozano, C., Tomás, V. and Sidrach de Cardona, C. (1996) Flow-injection fluorimetric determination of penicillamine and tiopronin in pharmaceutical preparations. *Journal of Pharmaceutical and Biomedical Analysis*, **15**, 33–38.

37 Dolejsova, J., Solich, P., Polydorou, C.K., Koupparis, M.A. and Efstathiou, C.E. (1999) Flow-injection fluorimetric determination of 1,4-benzodiazepines in pharmaceutical formulations after acid hydrolysis. *Journal of Pharmaceutical and Biomedical Analysis*, **20**, 357–362.

38 Perez-Ruiz, T., Martinez-Lozano, C., Sanz, A. and Guillen, A. (2004) Successive determination of thiamine and ascorbic acid in pharmaceuticals by flow injection analysis. *Journal of Pharmaceutical and Biomedical Analysis*, **34**, 551–557.

39 Laredo Ortiz, S., Gómez-Bendito, C. and Martínez-Calatayud, J. (1993) Determination of fluphenazine hydrochloride in a flow assembly incorporating cerium(IV) arsenite as a

solid-bed sector. *Analytica Chimica Acta*, **276**, 281–286.

40 Kojlo, A. and Martínez-Calatayud, J. (1995) Spectrofluorimetric flow injection determination of adrenaline with an iodine solid-phase reactor. *Analytica Chimica Acta*, **308**, 334–338.

41 García-Bautista, J.A., García-Mateo, J.V. and Martínez-Calatayud, J. (1998) Spectrofluorimetric determination of iproniazid and isoniazid in a FIA system provided with a solid-phase reactor. *Analytical Letters*, **31**, 1209–1218.

42 Zhang, Z.-Q., Liang, G.-X., Ma, J., Lei, Y. and Lu, Y.-M. (2006) A sensitive flow injection fluorimetry for the determination of carbamazepine in human plasma. *Analytical Letters*, **39**, 2417–2428.

43 Pinto, P.C.A.G., Saraiva, M.L.M.F.S., Santos, J.L.M. and Lima, J.L.F.C. (2005) A pulsed sequential injection analysis flow system for the fluorimetric determination of indomethacin in pharmaceutical preparations. *Analytica Chimica Acta*, **539**, 173–179.

44 Pérez-Ruiz, T., Martínez-Lozano, C., Marín, J. and García, M.D. (2006) Automatic determination of phylloquinone in vegetables and fruits using on-line photochemical reduction and fluorescence detection via solid-phase extraction and flow injection. *Analytical and Bioanalytical Chemistry*, **384**, 280–285.

45 Irace-Guigand, S., Leverend, E., Seye, M.D.G. and Aaron, J.-J. (2005) A new on-line micellar-enhanced photochemically-induced fluorescence method for determination of phenylurea herbicide residues in water. *Luminescence*, **20**, 138–142.

46 García-Campaña, A.M., Aaron, J.-J. and Bosque-Sendra, J.M. (2001) Micellar-enhanced photochemically induced fluorescence detection of chlorophenoxyacid herbicides. Flow injection analysis of mecoprop and 2,4-dichlorophenoxyacetic acid. *Talanta*, **55**, 531–539.

47 Vílchez, J.L., Valencia, M.C., Navalón, A., Molinero-Morales, B. and Capitán-Vallvey, L.F. (2001) Flow injection analysis of the insecticide imidacloprid in water samples with photochemically induced fluorescence detection. *Analytica Chimica Acta*, **439**, 299–305.

48 Coly, A. and Aaron, J.-J. (1999) Sensitive and rapid flow injection analysis of sulfonylurea herbicides in water with micellar-enhanced photochemically induced fluorescence detection. *Analytica Chimica Acta*, **392**, 255–264.

49 Cydzik, I., Albert-García, J.R. and Martinez-Calatayud, J. (2007) Photo-induced fluorescence of fluometuron in a continuous-flow multicommutation assembly. *Journal of Fluorescence*, **17**, 29–36.

50 Perez-Ruiz, T., Martínez-Lozano, C., Tomas, V. and Martin, J. (2001) Flow injection determination of methamidophos using on-line photo-oxidation and fluorimetric detection. *Talanta*, **54**, 989–995.

51 Perez-Ruiz, T., Martínez-Lozano, C., Tomas, V. and Martin, J. (2002) Flow injection spectrofluorimetric determination of malathion in environmental samples using on-line photooxidation. *Analytical Letters*, **35**, 1239–1250.

52 Segarra Guerrero, R., Gómez Benito, C. and Martínez Calatayud, J. (1993) On-line photoreaction and fluorimetric determination of diazepam. *Journal of Pharmaceutical and Biomedical Analysis*, **11**, 1357–1360.

53 Perez-Ruiz, T., Martínez-Lozano, C., Tomas, V. and Sidrach, C. (1997) Flow injection fluorimetric determination of ascorbic acid based on its photooxidation by thionine Blue. *Analyst*, **122**, 115–118.

54 Nair, M.B., Aaron, J.J., Prognon, P. and Mahuzier, G. (1998) Photochemically induced fluorimetric detection of tianeptine and some of its metabolites.

Application to pharmaceutical preparation. *Analyst*, **123**, 2267–2270.

55 Amador-Hernández, J., Fernández-Moreno, J.M. and Luque de Castro, M.D. (2001) Continuous determination of chloroquine in plasma by laser-induced photochemical reaction and fluorescence. *Fresenius' Journal of Analytical Chemistry*, **369**, 438–441.

56 Chen, H. and He, Q. (2000) Flow injection spectrofluorimetric determination of reserpine in tablets by on-line acetone sensitized photochemical reaction. *Talanta*, **53**, 463–469.

57 Perez-Ruiz, T., Martínez-Lozano, C., Tomas, V. and Martin, J. (2002) Fluorimetric determination of arsanilic acid by flow injection analysis using on-line photo-oxidation. *Analytical and Bioanalytical Chemistry*, **372**, 387–390.

58 Mellado-Romero, A., Gómez-Bendito, C. and Martínez-Calatayud, J. (1992) Photochemical derivatization and fluorimetric determination of promethazine in a FIA Assembly. *Analytical Letters*, **25**, 1289–1308.

59 Laasis, B. and Aaron, J.-J. (1997) Flow-injection fluorimetric analysis of several aminophenothiazines based on photooxidation. *Analusis*, **25**, 183–188.

60 Pérez-Ruiz, T., Martínez-Lozano, C., Tomás, V. and Martín, J. (2004) Flow-injection fluorimetric determination of menadione using on-line photo-reduction in micellar media. *Analytica Chimica Acta*, **514**, 259–264.

61 Resing, J.A. and Measures, C.I. (1994) Fluorometric determination of Al in seawater by flow injection analysis with in-line preconcentration. *Analytical Chemistry*, **66**, 4105–4111.

62 Economou, A., Themelis, D.G., Bikou, H., Tzanavaras, P.D. and Rigas, P.G. (2004) Determination of boron in water and pharmaceuticals by sequential-injection analysis and fluorimetric detection. *Analytica Chimica Acta*, **510**, 219–224.

63 Brach-Papa, C., Coulomb, B., Boudenne, J.-L., Cerdá, V. and Theraulaz, F. (2002) Spectrofluorimetric determination of aluminum in drinking waters by sequential injection analysis. *Analytica Chimica Acta*, **457**, 311–318.

64 Sanchez-Rojas, F., Cristofol Alcaraz, E. and Cano Pavon, J.M. (1994) Determination of aluminium in water by flow injection with fluorimetric detection by using salicylaldehyde carbohydrazone as reagent in a micellar medium. *Analyst*, **119**, 1221–1223.

65 Brach-Papa, C., Coulomb, B., Branger, C., Margaillan, A., Theraulaz, F., Van Loot, P. and Boudenne, J.-L. (2004) Fluorimetric determination of aluminium in water by sequential injection through column extraction. *Analytical and Bioanalytical Chemistry*, **378**, 1652–1658.

66 De Armas, G., Cladera, A., Becerra, E., Estela, J.M. and Cerdá, V. (2000) Fluorimetric sequential injection determination of magnesium using 8-hydroxiquinoline-5-sulfonic acid in a micellar medium. *Talanta*, **52**, 77–82.

67 Gañan-Gutierrez, N., Sánchez-Rojas, F. and Cano-Pavón, J.M. (1991) Determination of zinc by flow injection with fluorimetric detection in a micellar medium. *Fresenius' Journal of Analytical Chemistry*, **355**, 88–91.

68 Compañó, R., Ferrer, R., Guitera, J. and Prat, M.D. (1996) Flow injection method for the fluorimetric determination of Zn with 8-(benzenesulphonamido) quinoline. *Mikrochimica Acta*, **124**, 73–79.

69 Paleologos, E.K., Vlessidis, A.G., Karayannis, M.I. and Veltsistas, P.G. (2001) Nonaqueous catalytic fluorimetric trace determination of vanadium based on the pyronine B-hydrogen peroxide reaction and flow injection after cloud point extraction. *Analytical Chemistry*, **73**, 4428–4433.

70 Watson, R.J., Butler, E.C.V., Clementson, L.A. and Berry, K.M. (2005) Flow-injection analysis with fluorescence detection for the determination of trace levels of

ammonium in seawater. *Journal of Environmental Monitoring*, **7**, 37–42.

71 Recalde, D.L., Andres-García, E. and Díaz-Garcia, M.E. (2000) Fluorimetric flow injection and flow-through sensing systems for cyanide control in waste water. *Analyst*, **125**, 2100–2105.

72 Pinto, P.C.A.G., Saraiva, M.L.F.F.S., Santos, J.L.M. and Lima, J.L.F.C. (2006) Fluorimetric determination of aminocaproic acid in pharmaceutical formulations using a sequential injection analysis system. *Talanta*, **68**, 857–862.

73 López-Error, C., Viñas, P., Cerdan, F.J. and Hernández-Córdoba, M. (2000) Determination of clenbuterol in pharmaceutical preparations by reaction with o-phthalaldehyde using a flow injection fluorimetric procedure. *Talanta*, **53**, 47–53.

74 Pérez-Ruiz, T., Martínez-Lozano, C., Tomas, V. and Carpena, J. (1996) Fluorimetric determination of chloroxine using manual and flow injection methods. *Journal of Pharmaceutical and Biomedical Analysis*, **14**, 1505–1511.

75 López-Erroz, C., Viñas, P. and Hernández-Córdoba, M. (1994) Flow-injection fluorimetric analysis of sulfamethoxazole in pharmaceutical preparations and biological fluids. *Talanta*, **41**, 2159–2164.

76 López-Erroz, C., Viñas, P., Campillo, N. and Hernández-Córdoba, M. (1996) Flow injection–fluorimetric method for the determination of ranitidine in pharmaceutical preparations using o-phthalaldehyde. *Analyst*, **121**, 1043–1046.

77 Kuijt, J., Ariese, F., Brinkman, U.A.Th. and Gooijer, C. (2003) Room temperature phosphorescence in the liquid state as a tool in analytical chemistry. *Analytica Chimica Acta*, **488**, 135–171.

78 Sánchez-Barragán, I., Costa-Fernández, J.M., Valledor, M., Campo, J.C. and Sanz Medel, A. (2006) Room-temperature phosphorescence (RTP) for optical sensing. *Trends in Analytical Chemistry*, **25**, 958–967.

79 Cline-Love, L.J., Skrilec, M. and Habarta, J.G. (1980) Analysis of micelle-stabilized room temperature phosphorescence in solution. *Analytical Chemistry*, **52**, 754–759.

80 Díaz-García, M.E. and Sanz Medel, A. (1986) Facile chemical deoxygenation of micellar solutions for room temperature phosphorescence. *Analytical Chemistry*, **58**, 1436–1439.

81 Liu, Y.M., Fernández de la Campa, M.R., Díaz-García, M.E. and Sanz Medel, A. (1990) Phosphorescence detection in flowing systems: selective determination of aluminium by flow injection liquid room-temperature phosphorimetry. *Analytica Chimica Acta*, **234**, 233–238.

82 Weinberger, R., Yarmchuk, P. and Cline Love, L.J. (1982) Liquid chromatographic phosphorescence detection with micellar chromatography and postcolumn reaction modes. *Analytical Chemistry*, **54**, 1552–1558.

83 Muñoz de la Peña, A., Mahedero, M.C. and Bautista-Sánchez, A. (2000) Room temperature phosphorescence in cyclodextrins. *Analytical Applications, Analusis*, **28**, 670–678.

84 Li, L., Chen, Y., Zhao, Y. and Tong, A. (1997) Room-temperature phosphorescence of dansyl chloride solution in the absence of protective medium and its medium effect. *Analytica Chimica Acta*, **341**, 241–249.

85 Li, L., Zhao, Y., Yu, Y. and Tong, A. (1998) Non-protected fluid room temperature phosphorescence of several naphthalene derivatives. *Talanta*, **46**, 1147–1154.

86 Segura-Carretero, A., Cruces-Blanco, C., Cañabate-Díaz, B. and Fernández-Gutierrez, A. (1998) An innovative way of obtaining room-temperature phosphorescence signals in solution. *Analytica Chimica Acta*, **361**, 217–222.

87 Segura-Carretero, A., Salinas-Castillo, A. and Fernández-Gutierrez, A. (2005) A review of heavy-atom-induced room-temperature phosphorescence a straightforward phosphorimetric method. *Critical Reviews in Analytical Chemistry*, **35**, 3–14.

88 Cañabate-Díaz, B., Casado-Terrones, S., Segura-Carretero, A., Costa-Fernández, J.M. and Fernández-Gutierrez, A. (2004) Comparison of three phosphorescent methodologies in solution for the analysis of naphazoline in pharmaceutical preparations. *Analytical and Bioanalytical Chemistry*, **379**, 30–34.

89 Fernández-González, A., Badía, R. and Díaz-García, M.E. (2003) Sensitive flow injection system for nafcillin determination based on non-protected room temperature phosphorescence. *Analytica Chimica Acta*, **498**, 69–77.

90 Fletcher, P., Andrew, K.N., Calokerinos, A.C., Forbes, S. and Worsfold, P.J. (2001) Analytical applications of flow injection with chemiluminescence detection-a review. *Luminescence*, **16**, 1–23.

91 Bowie, A.R., Sanders, M.G. and Worsfold, P.J. (1996) Analytical applications of liquid phase chemiluminescence reactions – a review. *Journal of Bioluminescence and Chemiluminescence*, **11**, 61–90.

92 García-Campaña, A.M. and Baeyens, W.R.G. (2000) Principles and recent analytical applications of chemiluminescence. *Analusis*, **28**, 686–698.

93 Palilis, L.A. and Calokerinos, A.C. (2000) Analytical applications of chemiluminogenic reactions. *Analytica Chimica Acta*, **413**, 175–186.

94 Qin, W. (2002) Flow injection chemiluminescence-based chemical sensors. *Analytical Letters*, **35**, 2207–2220.

95 Lahuerta-Zamora, L., Fuster-Mestre, Y., Duart, M.J., Antón-Fos, G.M., García-Domenech, R., Gálvez Álvarez, J. and Martínez-Calatayud, J. (2001) prediction of the chemiluminescent behavior of pharmaceuticals and pesticides. *Analytical Chemistry*, **73**, 4301–4306.

96 Tsunoda, M. and Imai, K. (2005) Analytical applications of peroxyoxalate chemiluminescence. *Analytica Chimica Acta*, **541**, 13–23.

97 Gerardi, R.D., Barnett, N.W. and Lewis, S.W. (1999) Analytical applications of tris (2,2′-bipyridyl)ruthenium(III) as a chemiluminescent reagent. *Analytica Chimica Acta*, **378**, 1–41.

98 Campiglio, A. (1998) Determination of naproxen with chemiluminescence detection. *Analyst*, **123**, 1571–1574.

99 Koukli, L.L., Calokerinos, A.C. and Hadjiioannou, T.P. (1989) Continuous-flow chemiluminescence determination of acetaminophen by reduction of cerium(IV). *Analyst*, **114**, 711–714.

100 Rao, Y., Tong, Y., Zhang, X.R., Luo, G.A. and Baeyens, W.R.G. (2000) Determination of ofloxacin using a chemiluminescence flow injection method. *Analytica Chimica Acta*, **416**, 227–230.

101 Aly, F.A., Al-Tamimi, S.A. and Alwarthan, A.A. (2000) Determination of flufenamic acid and mefenamic acid in pharmaceutical preparations and biological fluids using flow injection analysis with tris(2,2′-bipyridyl) ruthenium(II) chemiluminescence detection. *Analytica Chimica Acta*, **416**, 87–96.

102 Cui, H., Li, S.F., Li, F., Sun, Y.G. and Lin, X.Q. (2002) A novel chemiluminescent method for the determination of salicylic acid in bactericidal solutions. *Analytical and Bioanalytical Chemistry*, **372**, 601–604.

103 Capitan-Vallvey, L.F., Valencia-Miron, M.C. and Acosta, R.A. (2000) Chemiluminescence determination of sodium 2-mercaptoethane sulfonate by flow injection analysis using cerium(IV) sensitized by quinine. *Talanta*, **51**, 1155–1161.

104 Zhang, Z.D., Baeyens, W.R.G., Zhang, X.R. and Weken, G.V.D. (1996) Chemiluminescence flow injection analysis of captopril applying a sensitized rhodamine 6G method. *Journal of Pharmaceutical and Biomedical Analysis*, **14**, 939–945.

105 Wang, X., Zhao, H.C., Nie, L.H., Jin, L.P. and Zhang, Z.L. (2001) Europium sensitized chemiluminescense determination of rufloxacin. *Analytica Chimica Acta*, **445**, 169–175.

106 Hindson, B.J. and Barnett, N.W. (2001) Analytical applications of acidic potassium permanganate as a chemiluminescence reagent. *Analytica Chimica Acta*, **445**, 1–19.

107 Chen, J. and Fang, Y. (2007) Flow injection technique for biochemical analysis with chemiluminescence detection in acidic media. *Sensors*, **7**, 448–458.

108 Deftereos, N.T., Grekas, N. and Calokerinos, A.C. (2000) Flow injection chemiluminometric determination of albumin. *Analytica Chimica Acta*, **403**, 137–143.

109 Li, B.X., Zhang, Z.J. and Zhao, L.X. (2002) Flow-injection chemiluminescence detection for studying protein binding for drug with ultrafiltration sampling. *Analytica Chimica Acta*, **468**, 65–70.

110 Barnett, N.W., Hindson, B.J. and Lewis, S.W. (1998) Determination of 5-hydroxytryptamine (serotonin) and related indoles by flow injection analysis with acidic potassium permanganate chemiluminescence detection. *Analytica Chimica Acta*, **362**, 131–139.

111 Liang, Y.-D., Song, J.-F., Yang, Z.-F. and Guo, W. (2004) Flow-injection chemiluminescence determination of chloroquine using peroxynitrous acid as oxidant. *Talanta*, **62**, 757–763.

112 Aly, F.A., Alarfaffj, N.A. and Alwarthan, A.A. (1998) Permanganate-based chemiluminescence analysis of cefadroxil monohydrate in pharmaceutical samples and biological fluids using flow injection. *Talanta*, **48**, 471–478.

113 Catalá-Icardo, M., Misiewicz, M., Ciucu, A., García-Mateo, J.V. and Martínez-Calatayud, J. (2003) FI-on-line photochemical reaction for direct chemiluminescence determination of photodegradated chloramphenicol. *Talanta*, **60**, 405–414.

114 Fuster-Mestre, Y., Fernández-Band, B., Lahuerta-Zamora, L. and Martínez-Calatayud, J. (1999) Flow injection analysis–direct chemiluminescence determination of ergonovine maleate enhanced by hexadecylpyridinium chloride. *Analyst*, **124**, 413–416.

115 Townshend, A., Wheatley, R.A., Chisvert, A. and Salvador, A. (2002) Flow injection-chemiluminescence determination of octyl dimethyl PABA in sunscreen formulations. *Analytica Chimica Acta*, **462**, 209–215.

116 Sanfeliu Alonso, M.C., Lahuerta Zamora, L. and Martínez-Calatayud, J. (2001) Flow-injection with chemiluminescence detection for the determination of iproniazid. *Analytica Chimica Acta*, **437**, 225–231.

117 Campiglio, A. (1998) Chemiluminescence determination of naltrexone based on potassium permanganate oxidation. *Analyst*, **123**, 1053–1056.

118 Costin, J.W., Barnett, N.W., Lewis, S.W. and McGillivery, D.J. (2003) Monitoring the total phenolic/antioxidant levels in wine using flow injection analysis with acidic potassium permanganate chemiluminescence detection. *Analytica Chimica Acta*, **499**, 47–56.

119 Townshend, A., Murillo-Pulgarín, J.A. and Alañon-Pardo, M.T. (2003) Flow injection-chemiluminescence determination of propranolol in pharmaceutical preparations. *Analytica Chimica Acta*, **488**, 81–88.

120 Fuster-Mestre, Y., Lahuerta-Zamora, L. and Martínez-Calatayud, J. (1999) Direct flow injection chemiluminescence determination of salicylamide. *Analytica Chimica Acta*, **394**, 159–163.

121 Costin, J.W., Francis, P.S. and Lewis, S.W. (2003) Selective determination of amino acids using flow injection analysis coupled with chemiluminescence detection. *Analytica Chimica Acta*, **480**, 67–77.

122 Palilis, L.P., Calokerinos, A.C. and Grekas, N. (1996) Chemiluminescence arising from the oxidation of bilirubin in aqueous media. *Analytica Chimica Acta*, **333**, 267–275.

123 Zhang, Z.D., Baeyens, W.R.G., Zhang, X.R. and VanDerWeken, G. (1996) Chemiluminescence flow injection analysis of captopril applying a sensitized rhodamine 6G method. *Journal of Pharmaceutical and Biomedical Analysis*, **14**, 939–945.

124 Nozaki, O., Iwaeda, T., Moriyama, H. and Kato, Y. (1999) Chemiluminescent detection of catecholamines by generation of hydrogen peroxide with imidazole. *Luminescence*, **14**, 123–127.

125 Fan, W.Z. and Zhang, Z.J. (1996) Determination of acetylcholine and choline in rat brain tissue by FIA with immobilized enzymes and chemiluminescence detection. *Microchemical Journal*, **53**, 290–295.

126 Marques, K.L., Santos, J.L.M. and Lima, J.L.F.C. (2004) Multicommutated flow system for the chemiluminometric determination of clomipramine in pharmaceutical preparations. *Analytica Chimica Acta*, **518**, 31–36.

127 Ma, Y., Zhou, M., Jin, X., Zhang, Z., Teng, X. and Chen, H. (2004) Flow-injection chemiluminescence assay for ultra-trace determination of DNA using rhodamine B–Ce(IV)-DNA ternary system in sulfuric acid media. *Analytica Chimica Acta*, **501**, 25–30.

128 Wada, M., Kuroda, N., Akiyama, S. and Nakashima, K. (1997) A sensititve and rapid FIA with an immobilized enzyme column reactor and peroxyoxalate chemiluminescence detection for the determination of total D-amino acids in human plasma. *Analytical Sciences*, **13**, 945–950.

129 Nozaki, O., Iwaeda, T. and Kato, Y. (1996) Amines for detection of dopamine by generation of hydrogen peroxide and peroxyoxalate chemiluminescence.

Journal of Bioluminescence and Chemiluminescence, **11**, 309–313.

130 Fang, Q., Shi, X.T., Sun, Y.Q. and Fang, Z.L. (1997) A flow injection microdialysis sampling chemiluminescence system for in vivo on-line monitoring of glucose in intravenous and subcutaneous tissue fluid microdialysates. *Analytical Chemistry*, **69**, 3570–3577.

131 Yaqoob, M., Nabi, A. and Masoom, Y.M. (1997) Flow-injection chemiluminescent determination of glycerol-3-phosphate and glycerophosphorylcholine using immobilized enzymes. *Journal of Bioluminescence and Chemiluminescence*, **12**, 1–5.

132 Wang, L., Li, Y., Zhao, D. and Zhu, C. (2003) A Novel Enhancing Flow-Injection Chemiluminescence Method for the Determination of Glutathione Using the Reaction of Luminol with Hydrogen Peroxide. *Microchimica Acta*, **141**, 41–45.

133 Alapont, A.G., Jiménez, E.A., Zamora, L.L. and Martinez-Calatayud, J. (1998) Inhibition of the system luminol-H_2O_2-Fe$(CN)_6^{3-}$ chemiluminescence by the Mn(II) indirect determination of isoniazid in a pharmaceutical formulation. *Journal of Bioluminescence and Chemiluminescence*, **13**, 131–137.

134 Liu, H., Ren, J., Hao, Y., Ding, H., He, P. and Fang, Y. (2006) Determination of metoprolol tartrate in tablets and human urine using flow injection chemiluminescence method. *Journal of Pharmaceutical and Biomedical Analysis*, **42**, 384–388.

135 García-Campaña, A.M., Bosque-Sendra, J.M., Bueno-Vargas, M.P., Baeyens, W.R.G. and Zhang, X. (2004) Flow injection analysis of oxymetazoline hydrochloride with inhibited chemiluminescent detection. *Analytica Chimica Acta*, **516**, 245–249.

136 Aly, F.A., Alarfaj, N.A. and Alwarthan, A.A. (1998) Flow-injection chemiluminometric determination of some phenothiazines in dosage forms and biological fluids. *Analytica Chimica Acta*, **358**, 255–262.

137 Costin, J.W., Barnett, N.W. and Lewis, S.W. (2004) Determination of proline in wine using flow injection analysis with tris(2,2'-bipyridyl)ruthenium(II) chemiluminescence detection. *Talanta*, **64**, 894–898.

138 Ping, L.Z., An, L.K. and Yang, T.S. (1998) Microdetermination of proteins with the 1,10-phenanthroline –H_2O_2–cetyltrimethylammonium bromide–Cu(II) chemiluminescence system. *Microchemical Journal*, **60**, 217–223.

139 Alwarthan, A.A. and Aly, F.A. (1998) Chemiluminescent determination of pyridoxine hydrochloride in pharmaceutical samples using flow injection. *Talanta*, **45**, 1131–1138.

140 Du, J., Li, Y. and Lu, J. (2001) Flow Injection Chemiluminescence Determination of Rutin based on its enhancing effect on the luminol-ferricyanide/ferrocyanide system. *Analytical Letters*, **34**, 1741–1748.

141 Li, S., Chen, H., Wei, X., Lu, X. and Zhang, L. (2006) Determination of tannic acid by flow injection analysis with inhibited chemiluminescence detection. *Microchimica Acta*, **155**, 427–430.

142 Wang, Z., Zhang, Z., Fu, Z. and Zhang, X. (2004) Sensitive flow injection chemiluminescence determination of terbutaline sulfate based on enhancement of the luminol–permanganate reaction. *Analytical and Bioanalytical Chemistry*, **378**, 834–840.

143 Zhang, X.R., Baeyens, W.R.G., Vandenborre, A., VanDerWeken, G., Calokerinos, A.C. and Schulman, S.G. (1995) Chemiluminescence determination of tetracyclines based on their reaction with hydrogen peroxide catalysed by the copper ion. *Analyst*, **120**, 463–466.

144 Song, Z. and Hou, S. (2002) Determination of picomole amounts of thiamine through flow injection analysis based on the suppression of luminol–KIO_4 chemiluminescence system. *Journal of Pharmaceutical and Biomedical Analysis*, **28**, 683–691.

145 Song, S., Song, Q.J. and Chen, Z. (2007) On-line phototransformation–flow injection chemiluminescence determination of triclosan. *Analytical and Bioanalytical Chemistry*, **387**, 2917–2922.

146 Alwarthan, A.A. (1995) Chemiluminescent determination of tryptophan in a flow injection system. *Analytica Chimica Acta*, **317**, 233–237.

147 Song, Z. and Hou, S. (2002) Chemiluminescence assay for uric acid in human serum and urine using flow injection with immobilized reagents technology. *Analytical and Bioanalytical Chemistry*, **372**, 327–332.

148 Song, Z. and Hou, S. (2003) Sub-picogram determination of Vitamin B_{12} in pharmaceuticals and human serum using flow injection with chemiluminescence detection. *Analytica Chimica Acta*, **488**, 71–79.

149 Zamzow, H., Coale, K.H., Johnson, K.S. and Sakamoto, C.M. (1998) Determination of copper complexation in seawater using flow injection analysis with chemiluminescence detection. *Analytica Chimica Acta*, **377**, 133–144.

150 Zhou, Y.X. and Zhu, G.Y. (1997) Rapid automated in-situ monitoring of total dissolved iron and total dissolved manganese in underground water by reverse-flow injection with chemiluminescence detection during the process of water treatment. *Talanta*, **44**, 2041–2049.

151 Qin, W., Zhang, Z.J. and Wang, F.C. (1998) Chemiluminescence flow system for the determination of Fe(II) and Fe(III) in water. *Fresenius' Journal of Analytical Chemistry*, **360**, 130–132.

152 Qin, W., Zhang, Z.J., Li, B.X. and Liu, S.N. (1998) Chemiluminescence flow-sensing system for hydrogen peroxide with immobilized reagents. *Analytica Chimica Acta*, **372**, 357–363.

153 Mikuska, P., Vecera, Z. and Zdrahal, Z. (1995) Flow-injection chemiluminescence determination of ultra low concentrations of nitrite in water. *Analytica Chimica Acta*, **316**, 261–268.

154 Qi, H., Lv, J. and Li, B. (2007) Determination of phenol at ng l^{-1} level by flow injection chemiluminescence combined with on-line solid-phase extraction. *Spectrochimica Acta Part A*, **66**, 874–878.

155 Ikebuko, K., Wakamura, H., Karube, I., Kubo, I., Inagawa, M., Sugawara, T., Arikawa, Y., Suzuki, M. and Takeuchi, T. (1996) Phosphate sensing system using pyruvate oxidase and chemiluminescence detection. *Biosensors & Bioelectronics*, **11**, 959–965.

156 Qin, W., Zhang, Z.J. and Zhang, C.J. (1997) Chemiluminescence flow system for vanadium(v) With immobilized reagents. *Analyst*, **122**, 685–688.

157 Nakano, S., Sakamoto, K., Takenobu, A. and Kawashima, T. (2002) Flow-injection chemiluminescent determination of vanadium(IV) and total vanadium by means of catalysis on the periodate oxidation of purpurogallin. *Talanta*, **58**, 1263–1270.

158 Song, Z., Hou, S. and Zhang, N. (2002) A new green analytical procedure for monitoring sub-nanogram amounts of chlorpyrifos on fruits using flow injection chemiluminescence with immobilized reagents. *Journal of Agricultural and Food Chemistry*, **50**, 4468–4474.

159 Kiba, N., Inagaki, J. and Furusawa, M. (1995) Chemiluminometric flow injection method for determination of free L-malate in wine with co-immobilized malate dehydrogenase/NADH oxidase. *Talanta*, **42**, 1751–1755.

160 Huang, Y.M., Zhang, C., Zhang, X.R. and Zhang, Z.J. (1999) Chemiluminescence of sulfite based on auto-oxidation sensitized by rhodamine 6G. *Analytica Chimica Acta*, **391**, 95–100.

161 Liu, Y., Song, Z., Dong, F. and Zhang, L. (2007) Flow injection chemiluminescence determination of Sudan I in hot chilli sauce. *Journal of Agricultural and Food Chemistry*, **55**, 614–617.

162 Cui, H., Li, Q., Meng, R., Zhao, H.Z. and He, C.X. (1998) Flow injection analysis of tannic acid with inhibited chemiluminescent detection. *Analytica Chimica Acta*, **362**, 151–155.

163 Xiong, Y., Zhou, H., Zhang, Z., He, D. and He, C. (2006) Molecularly imprinted on-line solid-phase extraction combined with flow injection chemiluminescence for the determination of tetracycline. *Analyst*, **131**, 829–834.

164 Zheng, X.W. and Zhang, Z.J. (2002) Flow-injection chemiluminescence detecting sulfite with *in situ* electrogenerated Mn^{3+} as the oxidant. *Sensors and Actuators B*, **84**, 142–147.

165 Zhang, C., Huang, J., Zhang, Z. and Aizawa, M. (1998) Flow injection chemiluminescence determination of catecholamines with electrogenerated hypochlorite. *Analytica Chimica Acta*, **374**, 105–110.

166 Zhu, L., Li, Y. and Zhu, G. (2002) Flow injection determination of dopamine based on inhibited electrochemiluminescence of luminol. *Analytical Letters*, **35**, 2527–2537.

167 Tomita, I.N. and Bulhoes, L.O.S. (2001) Electrogenerated chemiluminescence determination of cefadroxil antibiotic. *Analytica Chimica Acta*, **442**, 201–206.

168 Zheng, X., Guo, Z. and Zhang, Z. (2001) Flow-injection electrogenerated chemiluminescence determination of epinephrine using luminol. *Analytica Chimica Acta*, **441**, 81–86.

169 Marquette, C.A., Leca, B.D. and Blum, L.J. (2001) Electrogenerated chemiluminescence of luminol for oxidase-based fibre-optic biosensors. *Luminescence*, **16**, 159–165.

170 Zhu, L., Li, Y. and Zhu, G. (2003) Electrochemiluminescent determination of L-cysteine with a flow injection analysis system. *Analytical Sciences*, **19**, 575–578.

171 Waseem, A., Yaqoob, M., Nabi, A. and Greenway, G.M. (2007) Determination of thyroxine using tris(2,2'-bipyridyl) ruthenium(iii)-NADH enhanced electrochemiluminescence detection. *Analytical Letters*, **40**, 1071–1083.

172 Whitchurch, C. and Andrews, A. (2000) Development and characterization of a novel electrochemiluminescent reaction involving cadmium. *Analyst*, **125**, 2065–2070.

173 Lv, J., Zhang, Z. and Luo, L. (2003) An On-line Galvanic Cell-generated Electrochemiluminescence and Flow Injection Determination of Calcium in Milk and Vegetables. *Analytical Sciences*, **19**, 883–886.

174 Gómez-Taylor, B., Palomeque, M., García-Mateo, J.V. and Martínez-Calatayud, J. (2006) Photoinduced chemiluminescence of pharmaceuticals. *Journal of Pharmaceutical and Biomedical Analysis*, **41**, 347–357.

175 Sahuquillo-Ricart, I., Antón-Fos, G.M., Duart, M.J., García-Mateo, J.V., Lahuerta-Zamora, L. and Martínez-Calatayud, J. (2007) Theoretical prediction of the photoinduced chemiluminescence of pesticides. *Talanta*, **72**, 378–386.

176 Palomeque, M., García-Bautista, J.A., Catalá-Icardó, M., García-Mateo, J.V. and Martínez-Calatayud, J. (2004) Photochemical-chemiluminometric determination of aldicarb in a fully automated multicommutation based flow-assembly. *Analytica Chimica Acta*, **512**, 149–156.

177 Chivulescu, A., Catalá-Icardó, M., García-Mateo, J.V. and Martínez-Calatayud, J. (2004) New flow-multicommutation method for the photo-chemiluminometric determination of the carbamate pesticide asulam. *Analytica Chimica Acta*, **519**, 113–120.

178 Roda, A., Pasini, P., Guardigli, M., Baraldini, M., Musiani, M. and Mirasoli, M. (2000) Bio- and chemiluminescence in bioanalysis. *Fresenius' Journal of Analytical Chemistry*, **366**, 752–759.

179 Roda, A., Guardigli, M., Michelini, E., Mirasoli, M. and Pasini, P. (2003) Analytical Bioluminescence and Chemiluminescence. *Analytical Chemistry*, **75**, 462A–470A.

180 Gamborg, G. and Hansen, E.H. (1994) Flow-injection bioluminescent determination of ATP based on the use of the luciferin-luciferase system. *Analytica Chimica Acta*, **285**, 321–328.

181 Blum, L.J., Gautier, S.M. and Coulet, P.R. (1993) Design of bioluminescence-based fiber optic sensors for flow injection analysis. *Journal of Biotechnology*, **31**, 357–368.

182 Yoshimura, K. (1987) Implementation of ion-exchanged absorptiometric detection in flow analysis systems. *Analytical Chemistry*, **59**, 2922–2924.

183 Molina-Díaz, A. Ruiz-Medina, A. Fernández-de-Córdova, M.L. (2002) The potential of flow-through optosensors in pharmaceutical analysis. *Journal of Pharmaceutical and Biomedical Analysis*, **29**, 399–419.

184 Cañizares, P. and Luque de Castro, M.D. (1995) Flow-through spectrofluorimetric sensor for the determination of glycerol in wine. *Analyst*, **120**, 2837–2840.

185 Llorent-Martínez, E.J., Satinsky, D. and Solich, P. (2007) Fluorescence optosensing implemented with sequential injection analysis: a novel strategy for the determination of labetalol. *Analytical and Bioanalytical Chemistry*, **387**, 2065–2069.

186 García Reyes, J.F., Ortega Barrales, P. and Molina Diaz, A. (2005) Rapid determination of diphenylamine residues in apples and pears with a single multicommuted fluorometric optosensor. *Journal of Agricultural and Food Chemistry*, **53**, 9874–9878.

187 García Reyes, J.F., Ortega Barrales, P. and Molina Díaz, A. (2007) Multicommuted fluorometric multiparameter sensor for

simultaneous determination of naproxen and salicylic acid in biological fluids. *Analytical Sciences*, **23**, 423–427.

188 Chen, D., Luque de Castro, M.D. and Valcárcel, M. (1990) Flow-through sensor for the fluorimetric determination of cyanide. *Talanta*, **37**, 1049–1055.

189 de la Torre, M., Fernández Gámez, F., Lázaro, F., Luque de Castro, M.D. and Valcárcel, M. (1991) Spectrofluorimetric flow-through sensor for the determination of beryllium in alloys. *Analyst*, **116**, 81–83.

190 Llorent-Martínez, E.J., Šatínský, D., Solich, P., Ortega-Barrales, P. and Molina-Díaz, A. (2007) Fluorimetric SIA optosensing in pharmaceutical analysis: determination of paracetamol. *Journal of Pharmaceutical and Biomedical Analysis*, **45**, 318–321.

191 López-Flores, J., Fernández de Córdova, M.L. and Molina-Díaz, A. (2005) Implementation of flow-through solid phase spectroscopic transduction with photochemically induced fluorescence: determination of thiamine. *Analytica Chimica Acta*, **535**, 161–168.

192 López Flores, J., Molina Díaz, A. and Fernández de Córdova, M.L. (2007) Development of a photochemically induced fluorescence-based optosensor for the determination of imidacloprid in peppers and environmental waters. *Talanta*, **72**, 991–997.

193 García Reyes, J.F., Llorent-Martínez, E.J., Ortega Barrales, P. and Molina Díaz, A. (2004) Multiwavelength fluorescence based optosensor for simultaneous determination of fuberidazole, carbaryl and benomyl. *Talanta*, **62**, 742–749.

194 Llorent Martínez, E.J., García Reyes, J.F., Ortega Barrales, P. and Molina Díaz, A. (2006) A multicommuted fluorescence-based sensing system for simultaneous determination of Vitamins B_2 and B_6. *Analytica Chimica Acta*, **555**, 128–133.

195 Yoshimura, K., Matsuoka, S., Inokura, Y. and Hase, U. (1992) Flow analysis for trace amounts of copper by ion-exchanger

phase absorptiometry with 4,7-diphenyl-2,9-dimethyl-1,10-phenanthroline disulphonate and its application to the study of karst groundwater storm runoff. *Analytica Chimica Acta*, **268**, 225–233.

196 Matsuoka, S., Yoshimura, K. and Tateda, A. (1995) Application of ion-exchanger phase visible light absorption to flow analysis. Determination of vanadium in natural water and rock. *Analytica Chimica Acta*, **317**, 207–213.

197 Ruedas Rama, M.J., Ruiz Medina, A. and Molina Díaz, A. (2003) Bead injection spectroscopic flow-through renewable surface sensors with commercial flow cells as an alternative to reusable flow-through sensors. *Analytica Chimica Acta*, **482**, 209–217.

198 Ruzicka, J. (1994) Discovering flow injection: journey from sample to a live cell and from solution to suspension. *Analyst*, **119**, 1925–1934.

199 Ruzicka, J. and Scampavia, L. (1999) Bead injection – a novel approach. *Analytical Chemistry*, **71**, 257A–263A.

200 Ruedas Rama, M.J., Ruiz Medina, A. and Molina Díaz, A. (2004) Implementation of flow-through multi-sensors with bead injection spectroscopy: fluorimetric renewable surface biparameter sensor for determination of berillium and aluminum. *Talanta*, **62**, 879–886.

201 Ruedas Rama, M.J., Ruiz Medina, A. and Molina Díaz, A. (2005) A flow injection renewable surface sensor for the fluorimetric determination of vanadium (V) with Alizarin Red S. *Talanta*, **66**, 1333–1339.

202 Zhu, H., Chen, H. and Zhou, Y. (2003) Determination of thiamine in pharmaceutical preparations by sequential injection renewable surface solid-phase spectrofluorometry. *Analytical Sciences*, **19**, 289–294.

203 Pascual-Reguera, I., Guardia-Rubio, M. and Molina-Díaz, A. (2004) Native fluorescence flow-through optosensor for the fast determination of diphenhydramine in pharmaceuticals. *Analytical Sciences*, **20**, 799–803.

204 Ortega-Algar, S., Ramos-Martos, N. and Molina-Díaz, A. (2004) Fluorimetric flow-through sensing of quinine and quinidine. *Microchimica Acta*, **147**, 211–214.

205 Fernández-Sánchez, J.F., Segura-Carretero, A., Costa-Fernández, J.M., Bordel, N., Pereiro, R., Cruces-Blanco, C., Sanz-Medel, A. and Fernández-Gutierrez, A. (2003) Fluorescence optosensors based on different transducers for the determination of polycyclic aromatic hydrocarbons in water. *Analytical and Bioanalytical Chemistry*, **377**, 614–623.

206 Ortega-Algar, S., Ramos-Martos, N. and Molina-Díaz, A. (2003) A flow-through fluorimetric sensing device for determination of α- and β-naphthol mixtures using a partial least-squares multivariate calibration approach. *Talanta*, **60**, 313–323.

207 López-Flores, J., Fernández de Córdova, M.L. and Molina-Díaz, A. (2007) Multicommutated flow-through optosensors implemented with photochemically induced fluorescence: determination of flufenamic acid. *Analytical Biochemistry*, **361**, 280–286.

208 Álava-Moreno, F., Díaz-García, M.E. and Sanz-Medel, A. (1993) Room temperature phosphorescence optosensor for tetracyclines. *Analytica Chimica Acta*, **281**, 637–644.

209 Guardia, L., Badía, R. and Díaz-García, M.E. (2007) Molecularly imprinted sol–gels for nafcillin determination in milk-based Products. *Journal of Agricultural and Food Chemistry*, **55**, 566–570.

210 Fernández-González, A., Badía-Laíño, R., Díaz-García, M.E., Guardia, L. and Viale, A. (2004) Assessment of molecularly imprinted sol–gel materials for selective room temperature phosphorescence recognition of nafcillin. *Journal of Chromatography B*, **804**, 247–254.

211 Guardia, L., Badía, R. and Díaz-García, M.E. (2006) Molecular imprinted

ormosils for nafcillin recognition by room temperature phosphorescence optosensing. *Biosensors and Bioelectronics*, **21**, 1822–1829.

212 Casado-Terrones, S., Fernández-Sánchez, J.F., Cañabate-Díaz, B., Segura-Carretero, A. and Fernández-Gutiérrez, A. (2005) A fluorescence optosensor for analyzing naphazoline in pharmaceutical preparations: comparison with other sensors. *Journal of Pharmaceutical and Biomedical Analysis*, **38**, 785–789.

213 Sánchez-Barragán, I., Costa-Fernández, J.M., Pereiro, R., Sanz-Medel, A., Salinas, A., Segura, A., Fernández-Gutiérrez, A., Ballesteros, A. and González, J.M. (2005) Molecularly imprinted polymers based on iodinated monomers for selective room-temperature phosphorescence optosensing of fluoranthene in water. *Analytical Chemistry*, **77**, 7005–7011.

214 Traviesa-Alvarez, J.M., Sánchez-Barragán, I., Costa-Fernández, J.M., Pereiro, R. and Sanz-Medel, A. (2007) Room temperature phosphorescence optosensing of benzo[a]pyrene in water using halogenated molecularly imprinted polymers. *Analyst*, **132**, 218–223.

215 Salinas-Castillo, A., Fernández-Sánchez, J.F., Segura-Carretero, A. and Fernández-Gutiérrez, A. (2005) Solid-surface phosphorescence characterization of polycyclic aromatic hydrocarbons and selective determination of benzo(a) pyrene in water samples. *Analytica Chimica Acta*, **550**, 53–60.

216 Casado-Terrones, S., Fernández-Sánchez, J.F., Segura-Carretero, A. and Fernández-Gutiérrez, A. (2005) The development and comparison of a fluorescence and a phosphorescence optosensors for determining the plant growth regulator 2-naphthoxyacetic acid. *Sensors and Actuators B*, **107**, 929–935.

217 Salinas-Castillo, A., Fernández-Sánchez, J.F., Segura-Carretero, A. and Fernández-Gutiérrez, A. (2004) A facile flow-through phosphorimetric sensing device for simultaneous determination of naptalam and its metabolite 1-naphthylamine. *Analytica Chimica Acta*, **522**, 19–24.

218 Fernández-Argüelles, M.T., Cañabate, B., Segura-Carretero, A., Costa-Fernández, J.M., Pereiro, R., Sanz-Medel, A. and Fernández-Gutiérrez, A. (2005) Flow-through optosensing of 1-naphthaleneacetic acid in water and apples by heavy atom induced–room temperature phosphorescence measurements. *Talanta*, **66**, 696–702.

219 Du, J., Sehn, L. and Lu, J. (2003) Flow injection chemiluminescence determination of epinephrine using epinephrine-imprinted polymer as recognition material. *Analytica Chimica Acta*, **489**, 183–189.

220 He, Y., Lu, J., Zhang, H. and Du, J. (2005) Molecular Imprinting – chemiluminescence determination of norfloxacin using a norfloxacin-imprinted polymer as the recognition material. *Microchimica Acta*, **149**, 239–244.

221 Xiong, Y., Zhou, H., Zhang, Z., He, D. and He, C. (2007) Flow-injection chemiluminescence sensor for determination of isoniazid in urine sample based on molecularly imprinted polymer. *Spectrochimica Acta Part A*, **66**, 341–346.

222 Zhou, H., Zhang, Z., He, D. and Xiong, Y. (2005) Flow through chemiluminescence sensor using molecularly imprinted polymer as recognition elements for detection of salbutamol. *Sensors and Actuators B*, **107**, 798–804.

223 Xiong, Y., Zhou, H., Zhang, Z., He, D. and He, C. (2006) Determination of hydralazine with flow injection chemiluminescence sensor using molecularly imprinted polymer as recognition element. *Journal of Pharmaceutical and Biomedical Analysis*, **41**, 694–700.

224 He, D., Zhang, Z., Zhou, H. and Huang, Y. (2006) Micro flow sensor on a chip for the determination of terbutaline in human serum based on chemiluminescence and a molecularly imprinted polymer. *Talanta*, **69**, 1215–1220.

225 Morais, I.P.A., Miró, M., Manera, M., Estela, J.M., Cerdá, V., Souto, M.R.S. and Rangel, A.O.S.S. (2004) Flow-through solid-phase based optical sensor for the multisyringe flow injection trace determination of orthophosphate in waters with chemiluminescence detection. *Analytica Chimica Acta*, **506**, 17–24.

226 Llorent-Martínez, E.J., Ortega-Barrales, P. and Molina-Díaz, A. (2006) Chemiluminescence optosensing implemented with multicommutation: determination of salicylic acid. *Analytica Chimica Acta*, **580**, 149–154.

227 Álava-Moreno, F., Valencia-González, M.J., Sanz Medel, A. and Díaz García, M.E. (1997) Oxygen sensing based on the room temperature phosphorescence intensity quenching of some lead–8-hydroxyquinoline complexes. *Analyst*, **122**, 807–810.

228 Costa-Fernández, J.M., Díaz-García, M.E. and Sanz-Medel, A. (1998) Sol–gel immobilized room-temperature phosphorescent metal-chelate as luminescent oxygen sensing material. *Analytica Chimica Acta*, **360**, 17–26.

229 Valencia-González, M.J., Liu, Y.M., Díaz-García, M.E. and Sanz Medel, A. (1993) Optosensing of D-glucose with an immobilized glucose oxidase minireactor and an oxygen room-temperature phosphorescence transducer. *Analytica Chimica Acta*, **283**, 439–446.

230 Paprovsky, D.B., Savitsky, A.P. and Yaropolov, A.I. (1990) Oxygen and glucose optical biosensors based on phosphorescence quenching of metalloporphyrins. *Journal of Analytical Chemistry*, **45**, 1441–1445.

231 Valencia-González, M.J. and Díaz-García, M.E. (1994) Enzymic reactor/room-temperature phosphorescence sensor system for cholesterol determination in organic solvents. *Analytical Chemistry,* **66**, 2726–2731.

232 Pererio García, R., Liu, Y.M., Díaz García, M.E. and Sanz Medel, A. (1991) Solid-surface room-temperature phosphorescence optosensing in continuous flow systems: an approach for ultratrace metal ion determination. *Analytical Chemistry,* **63**, 1759–1763.

233 Lu, J. and Zhang, Z. (1995) Determination of europium with solid-surface room-temperature phosphorescence optosensing. *Analyst,* **120**, 2585–2588.

234 Gong, Z. and Zhang, Z. (1997) Room temperature phosphorescence optosensing for gadolinium. *Microchimica Acta,* **126**, 117–121.

235 Gong, Z., Zhang, Z. and Zhang, A. (1996) Room Temperature Phosphorescence Optosensor for Terbium. *Analytical Letters,* **29**, 515–527.

236 Cuenca-Trujillo, R.M., Ayora-Cañada, M.J. and Molina-Díaz, A. (2002) Determination of ciprofloxacin with a room-temperature phosphorescence flow-through sensor based on lanthanide-sensitized luminescence. *Journal of AOAC International,* **85**, 1268–1272.

237 Traviesa-Alvarez, J.M., Costa-Fernández, J.M., Pereiro, R. and Sanz Medel, A. (2007) Direct screening of tetracyclines in water and bovine milk using room temperature phosphorescence detection. *Analytica Chimica Acta,* **589**, 51–58.

238 Llorent-Martínez, E.J., García-Reyes, J.F., Ortega-Barrales, P. and Molina-Díaz, A. (2005) Terbium-sensitized luminescence optosensor for the determination of norfloxacin in biological fluids. *Analytica Chimica Acta,* **532**, 159–164.

239 Álava-Moreno, F., Valencia-González, M.J. and Díaz-García, M.E. (1998) Room temperature phosphorescence optosensor for anthracyclines. *Analyst,* **123**, 151–154.

240 Aboul-Enein, H.Y., Stefan, R.I. and van Staden, J.F. (1999) Chemiluminescence-Based (Bio)Sensors – an overview.

Critical Reviews in Analytical Chemistry, **29**, 323–331.

241 Aboul-Enein, H.Y., Stefan, R.I., van Staden, J.F., Zhang, X.R., García-Campaña, A.M. and Baeyens, W.R.G. (2000) Recent developments and applications of chemiluminescence sensors. *Critical Reviews in Analytical Chemistry,* **30**, 271–289.

242 Huang, Y.M., Zhang, C., Zhang, X.R. and Zhang, Z.J. (1999) A novel chemiluminescence flow-through sensor for the determination of analgin. *Fresenius' Journal of Analytical Chemistry,* **365**, 381–383.

243 Zhang, S. and Li, H. (2001) Flow-injection chemiluminescence sensor for the determination of isoniazid. *Analytica Chimica Acta,* **444**, 287–294.

244 Lin, J.M., Qu, F. and Yamada, M. (2002) Chemiluminescent investigation of tris (2,2′-bipyridyl)ruthenium(II) immobilized on a cationic ion-exchange resin and its application to analysis. *Analytical and Bioanalytical Chemistry,* **374**, 1159–1164.

245 Lin, J.M., Sato, K. and Yamada, A. (2001) Hydrogen peroxide chemiluminescent flow-through sensor based on the oxidation with periodate immobilized on ion-exchange resin. *Microchemical Journal,* **69**, 73–80.

246 Li, B., Zhang, Z., Zhao, L. and Xu, C. (2002) Chemiluminescence flow-through sensor for ofloxacin using solid-phase PbO_2 as an oxidant. *Talanta,* **57**, 765–771.

247 Li, B., Zhang, Z., Zhao, L. and Xu, C. (2002) Chemiluminescence flow-through sensor for pipemidic acid using solid sodium bismuthate as an oxidant. *Analytica Chimica Acta,* **459**, 19–24.

248 Zhao, L., Li, B., Zhang, Z. and Lin, J.M. (2004) Chemiluminescent flow-through sensor for automated dissolution testing of analgin tablets using manganese dioxide as oxidate. *Sensors and Actuators B,* **97**, 266–271.

249 Collins, G.E. and Rose-Pehrsson, S.L. (1995) Chemiluminescent chemical

sensors for oxygen and nitrogen dioxide. *Analytical Chemistry*, **67**, 2224–2230.

250 Okabayashi, T., Fujimoto, T., Yamamoto, I., Utsomomiya, K., Wada, T., Yamashita, Y., Yamashita, N. and Nakagawa, M. (2000) High sensitive hydrocarbon gas sensor utilizing cataluminescence of γ-Al_2O_3 activated with Dy^{3+}. *Sensors and Actuators B*, **64**, 54–58.

251 Zhu, Y., Shi, J., Zhang, Z., Zhang, C. and Zhang, X.R. (2002) Development of a gas sensor utilizing chemiluminescence on nanosized titanium dioxide. *Analytical Chemistry*, **74**, 120–124.

252 Zhou, K., Ji, X., Zhang, N. and Zhang, X. (2006) On-line monitoring of formaldehyde in air by cataluminescence-based gas sensor. *Sensors and Actuators B*, **119**, 392–397.

253 Kiba, N., Itagaki, A., Fukumura, S., Saegusa, K. and Furusawa, M. (1997) Highly sensitive flow injection determination of glucose in plasma using an immobilized pyranose oxidase and a chemiluminometric peroxidase sensor. *Analytica Chimica Acta*, **354**, 205–210.

254 Kiba, N., Tachibana, M., Tani, K. and Miwa, T. (1998) Chemiluminometric branched chain amino acids determination with immobilized enzymes by flow injection analysis. *Analytica Chimica Acta*, **375**, 65–70.

255 Kiba, N., Ito, S., Tachinaba, M., Tani, K. and Koizime, H. (2001) Flow-through chemiluminescence sensor using immobilized oxidases for the selective determination of L-glutamate in a flow injection system. *Analytical Sciences*, **17**, 929–933.

256 Sekiguchi, Y., Nishikawa, A., Makita, H., Yamamura, A., Matsumoto, K. and Kiba, N. (2001) Flow-through chemiluminescence sensor using immobilized histamine oxidase from *Arthrobacter crystallopoietes* KAIT-B-007 and peroxidase for selective determination of histamine. *Analytical Sciences*, **17**, 1161–1164.

257 Zhang, R., Hirakawa, K., Seto, D., Soh, N., Nakano, K., Masadome, T., Nagata, K., Sakamoto, K. and Imato, T. (2005) Sequential injection chemiluminescence immunoassay for anionic surfactants using magnetic microbeads immobilized with an antibody. *Talanta*, **68**, 231–238.

258 Li, B.X., Zhang, Z.J. and Zhao, L.X. (2001) Chemiluminescent flow-through sensor for hydrogen peroxide based on sol–gel immobilized hemoglobin as catalyst. *Analytica Chimica Acta*, **445**, 161–167.

259 Lin, J.M. and Yamada, M. (2001) Chemiluminescent flow-through sensor for 1,10-phenanthroline based on the combination of molecular imprinting and chemiluminescence. *Analyst*, **126**, 810–815.

260 Zhou, H., Zhang, Z., He, D., Hu, Y., Huang, Y. and Chen, D. (2004) Flow chemiluminescence sensor for determination of clenbuterol based on molecularly imprinted polymer. *Analytica Chimica Acta*, **523**, 237–242.

261 Manz, A., Graber, N. and Widmer, H.M. (1990) Miniaturized total chemical analysis systems: a novel concept for chemical sensing. *Sensors and Actuators B*, **1**, 244–248.

262 Alivisatos, A.P. (1996) Semiconductor clusters, nanocrystals, and quantum dots. *Science*, **271**, 933–937.

263 Costa-Fernández, J.M., Pereiro, R. and Sanz-Medel, A. (2006) The use of luminescent quantum dots for optical sensing. *Trends in Analytical Chemistry*, **25**, 207–218.

264 Zhang, Z., He, D., Liu, W. and Lv, Y. (2005) Chemiluminescence micro-flow injection analysis on a chip. *Luminescence*, **20**, 377–381.

265 He, D., Zhang, Z., Huang, Y. and Hu, Y. (2007) Chemiluminescence microflow injection analysis system on a chip for the determination of nitrite in food. *Food Chemistry*, **101**, 667–672.

14
Enzymes in Flow Injection Analysis

Robert Koncki, Łukasz Tymecki, and Beata Rozum

14.1
Introduction

There are two main reasons for the incorporation of biocatalytic processes into analytical procedures. First is the significant expansion of the range of available analytes for conventional analytical methods. For example, electrochemically inactive analytes can be, in the course of some enzymatic reactions, converted into electro-active species. Secondly, in many cases the selectivity of such methods is significantly improved due to the high specificity of biocatalytic processes. Another bioanalytical purpose is the need to determine some enzyme activities. These analytical methods can be further applied for the indirect detection of compounds having an effect on enzyme activity like inhibitors, activators, coenzymes, and so on. These determinations can also be highly selective due to the biospecificity of such interactions.

The implementation of enzymatic methods into the FIA format exploits common advantages of this technique, including the possibility of analysis automation, decrease in (bio)reagents consumption, shortening of analysis time, and so on. Obviously, such an approach is rational in the case of multiple and frequently repeated analyses (mass-scale, routine analyses) as well as on-line real-time monitoring but there is no point in using the FIA development for single and occasional measurements. This field of analysis seems to be reserved rather for low-cost and easy-to-use disposable devices (e.g., strip tests, single-shot (bio)sensors, etc.) or for large central analytical laboratories equipped with highly specialized instrumentation.

The enzymatic methods of analysis in their essence have a kinetic character. The kinetics of a biocatalytic conversion is a function of reagent concentrations, enzyme activity, time, reaction conditions, and so on. Evidently, for the accurate determination of a substrate concentration or an enzyme activity all these parameters should be strictly controlled. FIA systems certainly offer such a possibility. The reaction time and transport conditions are well defined by the configuration of the FIA manifold (the length of reaction coil, flow rate, etc.). The high reproducibility of these conditions is an obvious and fundamental advantage of such systems.

Advances in Flow Analysis. Edited by Marek Trojanowicz
Copyright © 2008 WILEY-VCH Verlag GmbH & Co. KGaA, Weinheim
ISBN: 978-3-527-31830-8

Furthermore, the "chemical" conditions of the process (pH, buffer capacity, the concentration of the substrate or the enzyme activity, the presence of cofactors, activators, etc.) are well-defined by the composition of the carriers used. It should be pointed out, that FIA systems allow automation of all the steps of the analytical procedures (not only the enzymatic reaction). Additionally, each single step can be independently optimized. This means that well-designed FIA systems provide optimal conditions for sampling, removal/masking of interferents, (bio)chemical conversion(s) and finally detection. All these possibilities improve the comfort of the analyst's work as well as the quality of analysis, its accuracy, precision, reproducibility, and so on.

The main goal of this chapter is to report on the different ways enzymes are used in FIA systems and the application of the designed detection schemes in real analyses. Different analytical approaches: from the most common substrate determination (enzyme as a biocatalytic agent), through enzyme activity determination, to more sophisticated schemes, such as the indirect determination of inhibitors and cofactors are described. Some enzymatic applications become very difficult or even impossible to perform without flow analysis and instances of this are presented. Some examples are provided for the use of enzymes in flow analytical systems where soluble as well as immobilized enzymes are a part of the flow system as biosensors or bioreactors. In the above examples the authors focus on optical and electrochemical detection methods. Finally, some practical applications in the clinical, environmental and food analysis fields are reviewed. The chapter covers work reported in the analytical literature within the last five years (2003–2007).

14.2
Enzyme Substrates as Analytes

Analytical uses of enzymes are predominantly connected with the detection of respective substrates. Such applications necessitate the highest possible activity of biocatalytic systems in order to provide the highest possible conversion of an analyte into a detected product. The enzyme activity needs also to be constant, ensuring that the amount of product generated in the course of the reaction is dependent only on the substrate concentration and not on any other conditions. In these biodetection schemes enzymes are neither consumed nor deactivated, thus they can be used repeatedly. Hence, from strictly economical and practical points of view, enzymes in the FIA systems are mainly applied in their immobilized forms. FIA systems exploiting soluble enzymes are rarely reported in the literature and are developed merely for inexpensive enzymes easily available in the form of highly active forms, such as glucose oxidase. Additionally, the consumption of an enzyme is minimized by a suitable system configuration (i.e., microsystems or SIA systems working in the stop-flow mode) [1, 2]. In such approaches the enzyme activity required for a single analysis is significantly reduced. Generally, two predominant ways for the incorporation of immobilized enzymes into an FIA system are deployed. The enzyme can be used as a separate part of the system, in the form of a flow-through bioreactor.

Alternatively, it can be integrated with the detector, as its inseparable part playing the role of receptor committed to the biomolecular recognition process.

14.2.1
Biosensor-based FIA Systems

The integration of an enzyme with the detector simplifies the FIA system. Dispersion in manifolds with biosensors is significantly reduced, which aids the high sensitivity of the method, even when the enzymatic conversion of an analyte is not quantitative. Such systems are highly sensitive because changes in the concentrations of the bioreagents responsible for the generation of the analytical signal proceed directly on the detector surface. Finally, some biosensing schemes are impossible to realize in any other FIA configuration, for example, in the case of pH-metric detection, since the products of the biocatalytic reaction have protolytic properties. pH changes caused by an enzymatic reaction in the whole segment of the sample are mainly dependent on its buffer capacity. Therefore, the resulting pH changes are not connected to the detected substrate concentration. In the case of pH-based biosensors the buffer capacity of the samples influences the response, but the signal is still specific.

Unfortunately, in some cases FIA–biosensor integration also limits the possibility of biosensor optimization. It is known that, in some cases, the optimal operating conditions for a detector and an enzyme are different, hence measurements cannot be performed under the optimal conditions for both biosensor components and a compromise is needed, which obviously decreases the efficiency of the biodetection system. Another significant drawback is slowdown in the system response caused by biosensor dynamics (especially the biosensor "memory effect" leading to a long time to return to the baseline signal). This inconvenience may be minimized by a suitable choice of immobilization method, providing good penetration of the biocatalytic layer as well as fast species transport within the layer. On the other hand the applied immobilization method should provide a long operational lifetime of the biocatalytic layer(s) so that the biosensor can perform many analyses over a long period of time. The last disadvantage is that after destruction of the bioreceptor layer the whole detection element of the FIA system must be replaced.

Without doubt, amperometric enzymatic electrodes constitute the major group of biosensors applied in FIA systems. Numerous examples, differing mainly in the internal electrode employed, which enables the meaningful elimination of redox species interferences or effective enzyme immobilization, are gathered in Table 14.1.

Monoenzymatic electrodes are predominantly modified with enzymes belonging to the group of oxidases [3–19], dehydrogenases [20–23] or oxigenases [24–28]. Polyenzymatic sequential biosensors are rarely reported. These biosensors include a biocatalytic layer with a few enzymes catalyzing sequentially two or more conversions, leading to the formation of electroactive product from the detected substrate (e.g., biosensors for glycerol [18, 32], lactate [30, 31, 36], pyruvate [30, 36], and acetate [34, 36]). Another interesting approach to polyenzymatic sensitization is demonstrated by the bienzymatic glutamate biosensor, where the biosensing layer

Table 14.1 Selected FIA systems with amperometric biosensors as detectors.

Analyte(s)	Enzyme(s)	Biosensor construction	Sample(s)	Ref.
		(Monoenzyme electrodes)		
H_2O_2	Horseradish peroxidase	Enzyme cross-linked with the mediator (tetrathiafulvalene) using glutaral on 3-mercaptopropionic acid self assembled monolayers modified gold electrode	Hair dye, rainwater	[3]
H_2O_2	Horseradish peroxidase	Enzyme entrapped in the bulk of screen-printed graphite electrode	—	[4]
H_2O_2	Horseradish peroxidase	Enzyme entrapped in ormosil composite doped with ferrocene monocarboxylic acid-BSA conjugate and multiwall carbon nanotubes	—	[5]
Glucose	Glucose oxidase	Enzyme deposited with Nafion on RhO_2-doped-graphite screen-printed electrode	Instant tea, honey	[6]
Glucose	Glucose oxidase	Enzyme covalently bound to Prussian Blue-substituted polypyrrole composite film deposited on Pt electrode	Urine, serum, infusion fluids, soft drinks, wines	[7]
Glucose	Glucose oxidase	Enzyme immobilized on glassy carbon electrode modified electrochemically with Prussian Blue; Nafion film as protective layer	Instant coffee	[8]
Glucose	Glucose oxidase	Enzyme entrapped in porous titania sol–gel matrix on glassy carbon electrode to form enzyme/titania/Nafion; Nafion film as protective layer	Serum standards	[9]
Glucose	Glucose oxidase	Enzyme entrapped in electropolymerized *m*-phenylenediamine on platinum needle electrode	Rat brain dialysate	[10]
Glucose	Glucose oxidase	Enzyme immobilized with BSA by cross-linking with glutaral or entrapped inside oxysilane sol–gel matrix	Grape must	[11]
Galactose	Galactose oxidase	Enzyme covalently bound to single walled carbon nanotubes dispersed in chitosan and deposited on glassy carbon disk electrode; Nafion film as protective or/and intermediating layer	Blood plasma	[12]
Glycerol	Glycerol dehydrogenase	Enzyme co-immobilized with phenazine methosulfate or cross-linked onto Os-complex-modified poly(vinylimidazole) redox polymer using poly(ethyleneglycol) diglycidylether	Wine	[13]

Analyte	Enzyme	Method	Sample	Reference
Ethanol	Alcohol oxidase	Enzyme entrapped within pyrrole electropolymerized on gold electrode modified with overoxidized polypyrrole	Red wines	[14]
Ethanol	Alcohol oxidase	Enzyme and BSA crosslinked with glutaral on gold electrode modified with overoxidized polypyrrole	Red wines	[14]
Choline	Choline oxidase	Enzyme and BSA crosslinked with glutaral on platinum electrode modified with overoxidized polypyrrole	Milk powder, milk, soy lecithin hydrolysates	[15]
Lysine	L-lysine α-oxidase	Enzyme and BSA cross-linked with glutaral on gold electrode modified with electropolymerized m-phenylenediamine	Milk, soya milk, flour, pasta	[16]
Glutamate, β-ODAP	L-glutamate oxidase	Enzyme, BSA and Tween-20 cross-linked with glutaral on Prussian Blue modified glassy carbon electrode	—	[17]
NADH	NADH oxidase	Enzyme immobilized with Nafion and glutaral on screen-printed graphite electrode doped with Prussian Blue	Red and white wines	[18]
L-lactic acid	L-lactate oxidase	Enzyme cross-linked with polyethylenimine and glutaral onto a nylon membrane placed on the platinum electrode	Cell culture media	[19]
Glucose	D-glucose dehydrogenase	Enzyme entrapped with osphendione and NAD^+ in carbon paste electrode	Fruit juices, milk products	[20]
Fructose	D-fructose dehydrogenase	Enzyme entrapped with $Os(bpy)_2Cl_2$ and polyethylenimine in carbon paste electrode	Fruit juices, milk products	[20]
Lactose, Cellobiose	Cellobiose dehydrogenase	Enzyme adsorbed onto the graphite electrodes	—	[21]
Gluconate	Gluconate dehydrogenase	Enzyme cross-linked with mediator (tetrathiafulvalene) using glutaral on gold electrode modified with 3-mercaptopropionic acid self-assembled monolayers	Wines, wine musts	[22]
Histamine	Histamine dehydrogenase	Enzyme cross-linked with poly(ethylene glycol) diglycidyl ether and PVI-dmeos on a glassy carbon electrode	Extracts from tuna fish muscles	[23]
Phenol derivatives	Tyrosinase	Enzyme cross-linked with glutaral on gold electrode modified with 3-mercaptopropionic acid self-assembled monolayers	Refinery wastewaters	[24, 25]
Phenols	Laccase	Enzyme cross-linked with glutaral on glassy carbon electrode	Wines	[26]

(Continued)

Table 14.1 (*Continued*)

Analyte(s)	Enzyme(s)	Biosensor construction	Sample(s)	Ref.
Phenol derivatives	Laccase	Enzyme adsorbed on the graphite electrode	—	[27]
Trimethylamine	Flavin-containing monooxygenase	Enzyme photo-cross-linked with polyvinyl alcohol containing stilbazolium groups onto a dialysis membrane coupled with Clark-type oxygen electrode	Fish extracts (horse-mackerel)	[28]
Organo-phosphorus pesticides	Organophosphorus hydrolase	Enzyme bound to the self-assembled monolayer of cystamine using glutaral on the photolithographically microstructured gold electrodes	—	[29]
(Bienzyme electrodes)				
L-lactate pyruvate	L-lactate oxidase, Pyruvate oxidase	Enzymes and gelatin cross-linked with glutaral on platinum electrodes coated with electropolymerized 1,2-diaminobenzene	Rat brain dialysate, human serum standards	[30]
L-lactate	L-lactate dehydrogenase, Diaphorase	Enzymes and gelatin cross-linked with glutaral on platinum electrodes coated with electropolymerized 1,2-diaminobenzene	Beer, sake, red wine,	[31]
D-lactate	D-lactate dehydrogenase, Diaphorase	Enzymes and gelatin cross-linked with glutaral on platinum electrodes coated with electropolymerized 1,2-diaminobenzene	Beer, sake, red wine	[31]
Glycerol	NADH oxidase, Glycerol dehydrogenase	Enzymes immobilized with Nafion and glutaral on screen-printed graphite electrode doped with Prussian Blue	—	[18]
Glycerol	Glycerol dehydrogenase, Diaphorase	Enzymes entrapped in polycarbamoyl sulfonate on glassy carbon electrode	Anaerobic fermentation broth	[32]
L-glutamate	L-glutamate oxidase, L-glutamate dehydrogenase	Enzymes bound by cross-linking with glutaral on the polycarbonate membrane attached to sensing element of oxygen electrode	Soy and tomato sauces, chicken thai soup, gravy of chili chicken	[33]
(Trienzyme electrodes)				
Acetate	Acetate kinase, Pyruvate kinase, Pyruvate oxidase	Enzymes entrapped in membrane of poly(vinyl alcohol) with stilbazolium group; membrane photo-cross-linked to platinum electrode modified with poly(dimethylsiloxane)	Red wine	[34]

Analyte	Enzymes	Description	Samples	Ref
Acetate	Acetate kinase, Pyruvate kinase, Lactate dehydrogenase	Enzymes cross-linked with poly(ethylene glycol) diglycidyl ether on carbon rod electrode modified with mediator (Brilliant Cresyl Blue)	Wines, vinegar	[35]
Lactose	β-galactosidase, Mutarotase, D-glucose dehydrogenase	Enzymes entrapped with osphendione and NAD^+ in carbon paste electrode	Fruit juices, Milk derivatives	[20]
Sucrose	Invertase, Mutarotase, D-glucose dehydrogenase	Enzymes entrapped with osphendione and NAD^+ in carbon paste electrode	Fruit juices, Milk derivatives	[20]
Glucose, L-lactate, pyruvate	Glucose oxidase, L-lactate oxidase, Pyruvate oxidase	Enzyme and gelatin cross-linked with glutaral on platinum electrodes, modified with electropolymerized 1,2-diaminobenzene	Rat brain dialysate human serum standards,	[36]
Glycerol	Glycerol kinase, Glycerol-3-phosphate oxidase, Horseradish peroxidase	Enzymes entrapped in polycarbamoyl sulfonate on glassy carbon electrode	Anaerobic fermentation broth	[32]
o-phosphate	Maltose phosphorylase, Mutarotase, Glucose oxidase	Enzyme cross-linked with glutaral on platinum electrodes, modified with electropolymerized 1,2-diaminobenzene	—	[37]

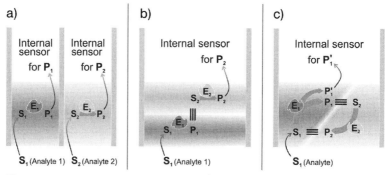

Figure 14.1 Biosensors containing two enzymes in the receptor part: (A) Configuration for simultaneous detection of two substrates (dual biosensor), (B) configuration for sequential biodetection of one substrate (bienzyme biosensor), (C) configuration for bioamplification of the signal for one substrate (bienzyme biosensor).

amplifies the analytical signal by redox recycling of the analyte [33]. Finally, it is worth mentioning FIA systems where the biosensor is a complex device consisting of two or three monoenzymatic electrodes sensitive to different bioanalytes. These so-called dual [30, 31] and triple [36] enzyme electrodes enable simultaneous multianalyte biodetection. Figure 14.1 illustrates three different sensing schemes for biosensors containing two enzymes in the biosensing layer.

Potentiometric biosensors are rarely applied under FIA conditions, the most used are those sensitive to urea [38–40] and creatinine [40–42]. Urea biosensors in the form of complete bioelectrochemical cells have been fabricated exclusively by means of screen-printing [38, 39]. In such microdevices both the internal electrode of the biosensor and the pseudoreference electrode are based on thick-film pH-sensitive ruthenium dioxide electrodes. Such strip biocells have been easily implemented into FIA manifolds dedicated to the analysis of urine and posthemodialysate fluids. Another example of a biosensing system working under a differential potentiometry regime is a FIA system with a solid-state ammonium ion-selective electrode modified with L-lysine oxidase, dedicated to amino acid detection [43]. Very stable creatinine and urea biosensors have been obtained by the covalent binding of creatinine deiminase or urease to the polymeric membranes of ammonium ion-selective electrodes [40, 41]. FIA systems based on such biosensors have been successfully applied for the analysis of urine, serum and postdialysate fluids. Another example of a potentiometric biosensor working under FIA conditions is a pH-membrane elec-trode modified with alkaline phosphatase [44]. Although the applied enzyme is not specific, this biosensor exhibits specific sensitivity towards monofluorphosphate. Finally, it is worth mentioning the potentiometric biosensor for paraoxon [29]. The enzyme applied for the biosensor construction catalyses the hydrolysis of this phosphoorganic compound. The acidic reaction products cause a change in the potential of the internal pH sensor. As *p*-nitrophenol generated in the course of the reaction is an electroactive product, amperometric detection is also possible. Thus the developed FIA system enables dual (potentiometric and amperometric) biodetection of selected phosphoorganic pesticides.

14.2.2
Bioreactor-based FIA Systems

A large majority of the above-mentioned biosensor disadvantages do not apply in FIA systems integrated with bioreactors. In such systems the operational conditions for detector and biocatalytic component can be optimized separately and there is no need for a compromise, as in the case of biosensors. This is especially important when more than one enzyme is involved in analyte bioconversion; the enzymes in the FIA system can then be separated and located in successive bioreactors, each of which can work under optimal conditions. A second significant advantage is the possibility to design FIA systems with replaceable bioreactors, which are selected with respect to the target analyte. Moreover, in such systems it is possible to place different bioreactors in parallel, giving the opportunity for the simultaneous determination of two or even more analytes with the same detector. A similar solution is possible using an array of biosensors, but unfortunately it requires the application of a multichannel measuring instrument. Finally, the bioreactor can be replaced without the need to change the detector as it is an independent part of the FIA system.

The wider applicability of bioreactors in FIA (in comparison with biosensors) is a result of one more advantage. Bioreactors can be used in FIA systems in conjunction with non-sensoric detectors, for example, based on the measurement of absorbance, fluorescence or chemiluminescence of a converted sample. As described above, such detection systems can be optimized independently of preceding enzymatic reactions and can even work under conditions which are unacceptable for enzymes. Several examples of bioreactor-based FIA systems reported in the literature within the last five years, where a variety of electrochemical [19, 37, 45–54] and optical [55–74] detection systems were applied, are collected in Table 14.2. The main disadvantage of FIA systems with incorporated bioreactors is the sample dispersion observed especially in the case of microcolumns packed with particles. To overcome this drawback flow bioreactors with high biocatalytic activity, providing fast and efficient analyte bioconversion are constructed. Alternatively, the dispersion could be significantly reduced by using "coated-tube bioreactors." Sample dispersion lengthens the determination time reducing the sample throughput of the FIA system. Finally, it should not be forgotten that the incorporation of any alternate components into an FIA system, increases its complexity and unfortunately, the failure frequency.

14.2.3
Additional Benefits Offered by FIA

Several examples compiled in Tables 14.1 and 14.2 clearly indicate the possibility of applying different enzymatic biosensors and bioreactors in FIA systems. Besides the evident advantages, which are a result of the approach mentioned in the introduction, it is worth remembering that the FIA concept can concern all steps of the analytical

Table 14.2 Selected FIA systems with bioreactors.

Analyte(s)	Enzyme(s)	Bioreactor(s)	Detection	Sample(s)	Ref
Glucose	Glucose oxidase	Enzyme covalently bound to the functionalized walls of fused silica column	Amperometric (H_2O_2) on platinum electrode	Human serum	[45]
Glucose	Glucose oxidase, Horseradish peroxidase	Enzymes immobilized separately on silica beads functionalized with tresyl; beads loaded to HPLC column	Amperometric (H_2O_2) on glassy carbon electrode	Human brain dialysate	[46]
Lactate	Lactate oxidase, Horse-radish peroxidase	Enzymes immobilized separately on silica beads functionalized with tresyl; beads loaded layer by layer onto HPLC column	Amperometric (H_2O_2) on glassy carbon electrode	Human brain dialysate	[46]
L-lactate, Glucose	Lactate oxidase, Glucose oxidase	Enzymes immobilized with glu-taral on CPG beads; each enzyme loaded into separate polymeric column	Amperometric (H_2O_2) on platinum electrode mod-ified with poly (1,2-diaminobenzene)	Rat brain dialysate, hu-man serum standards	[47]
L-malic acid	Malic enzyme	Enzyme immobilized on aminopropyl glass beads functionalized with glutaral; the enzyme-coated beads packed into a polymeric column	Amperometric (H_2O_2) on platinum electrode	Cell culture media	[19]
Phosphates	Acid phosphatase	Enzymes immobilized with glutaral on CPG beads; polymeric column	Amperometric (H_2O_2) on platinum electrode	—	[37]
Fructosyl pep-tides, Fructosyl amino acids	Fructosyl-amino acid oxidase, Fructosyl-peptide oxidases	Enzymes immobilized on Uni-port C beads functionalized with glutaral; each type of beads loaded into separate column	Amperometric (H_2O_2) on glassy carbon electrode	Protease-digested blood cells	[48, 49]

Analyte	Enzyme	Description	Detection	Sample	Ref.
Glycerol, Triacylglycerol	Glycerokinase, Glycerol-3-phosphate oxidase, Lipase	Enzymes bound covalently to the functionalized walls; glycerokinase and glycerol-3-phosphate oxidase coimmobilized, lipase in separate glass column	Amperometric (H_2O_2) on platinum electrode	Glycerol trioleate standards	[50]
Phenols	Laccase	Enzyme immobilized on ECH-Sepharose resin using carbodiimide method; gel packed in polymeric column	Amperometric (1,4-benzoquinone) On glassy carbon electrode	Olive oil mill wastewaters	[51]
Urea	Urease	Enzyme immobilized on CPG, silica gel or Poraver; functionalized with 3-amino-propyl-triethoxysilane and glutaral support support loaded onto polymeric column	Conductometric	Serum	[52]
Isocitrate	Isocitric dehydrogenase	Enzyme immobilized on aminopropyl glass beads functionalized with glutaral; the enzyme-coated beads packed into a Teflon column	Potentiometric (CO_2^{3-}) with ion selective electrode	Juices	[53]
Citrate	Citrate lyase, Oxaloacetate decarboxylase	Enzyme immobilized on aminopropyl glass beads functionalized with glutaral; the enzyme-coated beads packed into a Teflon column	Potentiometric (CO_2^{3-}) with ion selective electrode	Juices	[54]
H_2O_2	Peroxidase	Enzyme coupled to Amberlite IRA-173 resin with glutaral; resin with enzyme loaded into polymeric column	Photometric (H_2O_2) as antipyrilquinoneimine dye (505 nm)	Rainwater	[55]

(Continued)

Table 14.2 (*Continued*)

Analyte(s)	Enzyme(s)	Bioreactor(s)	Detection	Sample(s)	Ref
Glucose	Glucose oxidase	Enzyme coupled to Amberlite IRA-173 resin with glutaral; resin with enzyme loaded into polymeric column	Photometric (H_2O_2) as antipyrilquinoneimine dye (505 nm)	Whole human blood	[56]
Glucose	Glucose oxidase, Horseradish peroxidase	Enzymes cross-linked on functionalized CPG beads using glutaral; support loaded into a glass column	Photometric (H_2O_2) using Trinder method (490 nm)	Cell culture media	[57, 58]
Galactose	Galactose oxidase, Horseradish peroxidase	Enzymes cross-linked on functionalized CPG beads using glutaral; support loaded into a glass column	Photometric (H_2O_2) using Trinder method (490 nm)	Cell culture media	[58]
Ethanol	Alcohol oxidase, Horseradish peroxidase	Enzymes cross-linked on functionalized CPG beads using glutaral; support loaded into a glass column	Photometric (H_2O_2) using Trinder method (490 nm)	Cell culture media	[58]
Lactate	Lactate oxidase, Horseradish peroxidase	Enzymes cross-linked on functionalized CPG beads using glutaral; support loaded into a glass column	Photometric (H_2O_2) using Trinder method (490 nm)	Cell culture media	[58]
Amino acids	L-amino acid oxidase, Horseradish peroxidase	Enzymes cross-linked on functionalized CPG beads using glutaral; support loaded into a glass column	Photometric (H_2O_2) using Trinder method (490 nm)	Cell culture media	[58]
Cyanate	Cyanase	Enzyme immobilized on support (CPG) functionalized with 3-aminopropyltriethoxysilane and glutaral; then loaded to polymeric column	Photometric (NH_4^+) using Berthelot reaction (700 nm)	Exhausted electroplating liquids	[59]

Analyte	Enzyme	Detection method	Immobilization	Sample	Reference
Orthophosphate, Phytate	Phytase	Photometric (PO_4^{3-}) as molybdenum blue (650 nm)	Enzyme covalently bound to silanized controlled pore silica beads; beads loaded into a glass column	Corn, milk, soybean	[60]
Urea	Urease	Fluorimetric (isoindol derivatives, 485 nm)	Enzyme covalently immobilized onto CPG beads and loaded into polymer column	Alcoholic beverages	[61]
D-malate, L-malate	D-malate dehydrogenase, L-malate dehydrogenase	Fluorimetric (NADH or NADPH, 455 nm)	Enzymes bound to aminopropyl CPG beads; separate glass column for each enzyme	Fruit juices, soft drinks	[62]
Pyruvate	Pyruvate decarboxylase, Aldehyde dehydrogenase	Fluorimetric (NADH, 455 nm)	Enzymes bound to aminopropyl CPG with glutaral; beads loaded into glass column	—	[63]
Pyruvate, L-lactate.	L-lactate oxidase	Fluorimetric (NADH, 455 nm)	Enzymes bound to aminopropyl CPG with glutaral; beads loaded into glass column	—	[63]
Pyruvate, Acetate	Acetate kinase, Pyruvate kinase	Fluorimetric (NADH, 455 nm)	Enzymes bound to aminopropyl CPG with glutaral; beads loaded into glass column	—	[63]
Pyruvate, Citrate	Citrate lyase, Oxaloacetate decarboxylase	Fluorimetric (NADH, 455 nm)	Enzymes bound to aminopropyl CPG with glutaral; beads loaded into glass column	—	[63]
D-gluconate	Gluconate kinase, 6-phosphogluconate dehydrogenase	Fluorimetric (NADH, 455 nm)	Enzymes bound to aminopropyl CPG with glutaral; beads loaded into glass column	Honeys, vinegars, noble wines	[64]
Fumaric acid	Fumarase, Malic dehydrogenase	Fluorimetric (NADH, 440 nm)	Enzymes covalently bound to epoxy carrier beads; beads loaded into plastic column layer by layer	Cell culture media	[65]

(Continued)

Table 14.2 (*Continued*)

Analyte(s)	Enzyme(s)	Bioreactor(s)	Detection	Sample(s)	Ref
Succinic acid	Isocitrate lyase, Isocitrate dehydrogenase	Enzymes covalently bound to epoxy carrier beads; beads loaded into plastic column layer by layer	Fluorimetric (NADH, 440 nm)	Cell culture media	[66]
GABA, Glutamate	γ-aminobutyrate glutamate aminotransferase, L-glutamate oxidase, Catalase	Enzymes bound to aminopropyl CPG beads; separated glass column for coimmobilized L-glutamate oxidase/catalase and γ-aminobutyrate glutamate aminotransferase	Fluorimetric (NADPH, 455 nm)	Cell culture media	[67]
Glucose	Glucose oxidase, Horseradish peroxidase	Glucose oxidase immobilized onto aminopropyl glass beads with glutaral; beads loaded into a polymeric column; horseradish peroxidase in carrier solution	Chemiluminescence (luminol + H$_2$O$_2$, Catalysed by Co(II))	Human urine, low sugar drinks	[68]
Glucose	Glucose oxidase, Horseradish peroxidase	Enzymes bound to surface of silicon wet etched microchips modified with polyethylenimine & glutaral or glutaral & 3-aminopropyltriethoxysilane	Chemiluminescence (luminol + H$_2$O$_2$)	—	[69]
Ethanol	Alcohol oxidase, Horseradish peroxidase	Enzymes bound to surface of silicon wet etched microchips modified with polyethylenimine & glutaral or glutaral & 3-aminopropyltriethoxysilane	Chemiluminescence (luminol + H$_2$O$_2$)	—	[69]
Glycerol	Glycerokinase, Glycerol-3-phosphate oxidase	Enzyme bound to CPG beads (modified with 3-aminopropyl-triethoxysilane) with glutaral; beads loaded into glass column	Chemiluminescence (luminol + H$_2$O$_2$)	Triglycerides in blood serum (pretreated with lipase)	[70]

Choline, Acetylcholine	Enzymes immobilized separately and choline oxidase co-immobilized on tresylated surface of vinyl polymer beads; beads packed layer by layer into polymeric column	Chemiluminescence (luminol + H_2O_2)	Rabbit brain homogenates	[71]
Phosphatydylcholine	Enzymes co-immobilized onto aminopropyl-CPG beads functionalized with glutaral; loaded into polymeric column	Chemiluminescence (luminol + H_2O_2, Catalysed by Co(II))	Sediments	[72]
PO_4^{3-}	Enzymes immobilized onto NHS porous cellulose beads; beads loaded into stainless steel column	Chemoluminescence (luminol + H_2O_2)	Freshwater	[73]
PO_4^{3-}	Enzyme bound to CPG beads (modified with 3-aminopropyl-triethoxysilane) with glutaral; beads loaded into glass column.	Chemiluminescence (luminol + H_2O_2)	Waters	[74]

Phospholipase C, Alkaline phosphatase, Choline oxidase

Maltose phosphorylase, Mutarotase, Glucose-oxidase

Pyruvate oxidase

Choline oxidase, Peroxidase, Acetylcholinesterase

procedure, not only the analyte conversion and detection steps. The quality improvement of analytical procedures performed according to the FIA concept is also connected with the possibility of automation of such processes as sampling or the minimization of interferences.

Three strategies for the elimination of interferences can be applied: (i) the use of anti-interferent barriers, (ii) interferent masking and (iii) interferent removal. The common example of the first approach is the construction of amperometric biosensors coated with Nafion membranes that exclude effects from such electroactive sample components as paracetamol, ascorbic and uric acids. Another is the development of electrode materials providing a low redox reaction potential for the products of the bioconversion reaction. The masking of interferents can be easily realized in FIA systems by a suitable choice of carrier composition. An example of such an approach is the spiking of the carrier solution with chelators like EDTA, masking the traces of heavy metal ions that are able to inhibit enzymatic systems in FIA conditions. Another example is the use of carriers with high and well controlled levels of interfering species. However, such an approach is possible only in the case of samples with relatively high levels of target analyte. Finally, interferent removal, often a laborious step of conventional analytical procedures, can sometimes be easily performed under FIA conditions. A good example is the removal of endogenous ammonium and alkaline cations that strongly influence the response of potentiometric biosensors for creatinine and urea, based on poorly selective internal ammonium ion-selective electrodes [40, 41]. The scheme of the applied system is shown in Figure 14.2. The interfering ions can be removed from the sample segment using a cation exchanger microcolumn before the biosensor. Importantly, the implementation of the microcolumn into an appropriately designed FIA system allows not only the automation of interferent removal, but also the continual regeneration of the ion exchanger in the microcolumn [40].

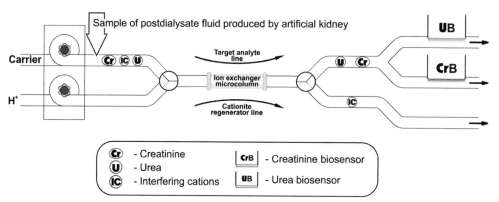

Figure 14.2 FIA system with interferent removal unit. The microcolumn with ion-exchanger retained cations influencing responses of urea and creatinine biosensors. The FIA system is used for the biparametric monitoring of hemodialysis treatments. Details are given in [40].

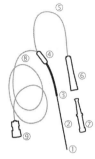

Parts of CMA 70

① Dialysis membrane
② Shaft
③ Liquid cross
④ Stopper
⑤ Outlet tube
⑥ Vial holder
⑦ Microvial
⑧ Inlet tube
⑨ Luer lock connection

Figure 14.3 Microsampling on-line unit used in the bioanalytical FIA system applied for the monitoring of glucose and lactate in the brain (Brain Microdialysis Catheter with Gold tip from CMA microdialysis, www.microdialysis.se). Details are given in [46].

Automation of the sampling process is another important advantage of biodetection realized in the FIA format. An interesting example is microsampling by means of dialysis. Such flow-through samplers can be applied either in the form of a microdialysis sampling tube [31] immersed in the liquid sample, or as microdialysis probes located directly in the analyzed environment [10, 30, 36, 46, 47]. Figure 14.3 shows such a microsampling probe applied in the bioreactor-based FIA system for on-line monitoring of glucose and lactic acid levels in the human brain during surgery [46].

14.2.4
Analytical Applications. FIA Systems as Monitors

It should be stressed, that the main goal of FIA development is its application for the analysis of real samples. The bioanalytical systems reviewed in this chapter fulfill this demand. The nature of enzymatic substrate recognition defines the area of practical analytical chemistry, where such systems can find wide application. As can be seen from Tables 14.1 and 14.2, the main fields where the systems are especially useful are: clinical/biomedical [7, 9, 10, 12, 19, 30, 36, 45–49, 52, 56, 68, 70, 71], food [4–8, 14–16, 18, 20, 26, 28, 31, 33–35, 54, 60–62, 64, 68] and environmental [3, 24, 25, 51, 55, 59, 72–74] analysis. At this point it is necessary to highlight one more advantage of the FIA concept and its practical consequences. FIA systems are suitable for the successive analysis of many samples and the measurements can be performed in real time. This means that FIA systems can play the role of analyzers as well as monitors. This attribute has a great influence on the development of FIA systems for the discrete monitoring of real processes. Several examples of FIA systems for the monitoring of clinical and biotechnological processes are widely reported in the analytical literature.

The already mentioned FIA systems with microdialysis sampling, based on both, biosensors [10, 30, 36] and bioreactors [47], were applied for the monitoring of glucose, lactate and pyruvate – the main brain nutrient and its metabolites. Striatal dialysate from rat brain was automatically sampled. A similar bioanalytical FIA system [46], with two parallel bioreactors and amperometric detectors, has been applied for the continuous monitoring of brain glucose and lactate during the course of surgery and during the neurointensive care of patients with head trauma. Another example of biomedical monitoring with enzymatic FIA systems is the control of hemodialysis treatment. The developed systems, based on thick-film urea biosensors [39] or ammonium ion-selective electrodes enzymatically sensitized for urea and creatinine [40], have been successfully applied for the monitoring of both these uremic marker levels in the spent fluids generated by an artificial kidney. Such monitoring is useful to ascertain whether the treatment is failure-free and efficient. Urea and creatinine profiles obtained from such on-line measurements are useful for the quantitative description of therapy as well as for its adequacy evaluation.

Enzyme-based FIA systems have also found several applications for the monitoring of advanced biotechnological processes. Two FIA systems [19] based on an amperometric L-lactate biosensor and a malic enzyme bioreactor coupled with an amperometric H_2O_2 detector have been applied for the monitoring of micro-malolactic fermentations (the biotechnological processes to improve the quality of wines by decreasing the malate level) in red wines induced by lactic acid bacteria. A fluorimetric FIA system with the bioreactor containing immobilized fumarase and malic dehydrogenase has been applied for the on-line monitoring of the culture of fumarate-producing microorganisms (*Rhizopus oryzae*) [65]. A similar FIA system with a bienzyme reactor containing co-immobilized isocitrate lyase and dehydrogenase has been applied for the monitoring of the biotechnological process of succinic acid production by immobilized *Escherichia coli* bacteria [66]. Several photometric FIA systems based on bienzymatic reactors with immobilized horseradish peroxidase and respective oxidases have been applied for the monitoring of glucose [57, 58], ethanol [58], lactate [58] and L-leucine [58] in *Escherichia coli* fermentation broths, as well as glucose [57, 58], galactose [58] and ethanol [58] in *Saccharomyces cerevisiae* fermentation broths. These bioanalytical systems have also been applied for the monitoring of glucose in the medium for rat Langerhans islets cell culture (bioartificial pancreas device) [57] as well as for the monitoring of lactate and leucine levels in the medium for mesenchymal stem cell culture [58]. A good example of an enzyme-based FIA device developed for environmental needs is a photometric system with immobilized cyanase for the monitoring of cyanate consumption in the bioremediation processes of waste waters from the jewellery industry [59].

14.3
Methods Based on Enzyme Activity Measurements

Enzymatic biodetection schemes for substrates reported in the previous section can be easily adapted for the detection of selected enzymes. Briefly, using the same detection system and properly chosen, constant substrate concentration, the analytical signals obtained in the FIA system, after a certain enzymatic reaction time should be (according to the Michaelis–Menten theory) proportional to the activities of the assayed enzyme. Bioanalytical procedures, oriented on the detection of enzyme activity, can be easily realized in the FIA format. Moreover, it should be noted that the enzyme activity detection gives the opportunity for further analytical applications connected with the indirect detection of substances influencing this activity. A survey of the reports on FIA systems for enzyme activity detection clearly indicates the main fields of their applications. The first large area is the control of either industrial enzymes production or their activity levels in consumer products. The second field is the analysis of enzyme activity in physiological fluids (mainly serum) for clinical diagnostics. The methods based on enzyme inhibition are mainly applied for environmental analysis. Rarely the systems are adapted for indirect detection of enzyme cofactors. General schemes for such detections are depicted in Figure 14.4. Some examples of real applications are reviewed briefly below.

14.3.1
Enzyme Activity Detection

Cutinases (lipolytic esterases) are promising catalysts in a variety of chemical and biotechnological applications (the dairy and oleochemical industries) due to their ability to operate in both aqueous and nonaqueous environments. The common chromogenic substrate enabling cutinase detection is *p*-nitrophenylbutyrate which is easily hydrolyzed to yellow *p*-nitrophenol. A photometric FIA system based on this reaction and detection has been additionally adjusted for the generation of micellar forms of this substrate [75]. This FIA system has been applied for the monitoring of bioproduction, preconcentration and purification of cutinase secreted by the *Saccharomyces cerevisiae* strain directly from whole fermentation broth [75, 76]. Industrial subtilisin-type protease enzymes like savinase and Purafect are mass-produced and widely used in the detergent industry. The main part of the FIA system developed for the activity detection of these enzymes is a column containing gelatin-Texas Red conjugate immobilized on the surface of sol–gel particles [77]. The exposure of the bioreactor with the immobilized fluorophore-labelled protein substrate to subsequent samples of subtilisin-type enzymes resulted in the formation of fluorescence peaks.

Another important industrial enzyme is α-amylase, widely used for starch degradation and oligosaccharides production by transglucosylation. Moreover, from the clinical point of view, serum and urine α-amylase activity determinations are useful for the diagnosis of acute pancreatitis. The amylase activity assays are based on the detection of maltose generated enzymatically from the selected oligosaccharides, like starch or maltopentaose. FIA systems for such purposes are based on the maltose bienzyme electrode [78] or on the glucosidase bioreactor connected with the amperometric glucose biosensor [79]. The latter system has been used for salivary α-amylase detection, considered as a factor associated with stress.

Two different FIA systems have been developed for the detection of alkaline phosphatase activity in human serum. This enzyme is considered to be indicative for bone and liver diseases. Bioanalysis in a semi-automated flow/bead-injection system is based on the photometric detection of *p*-nitrophenol released from its phosphate in the course of the enzymatic reaction [80]. The use of microbeads coated with wheat

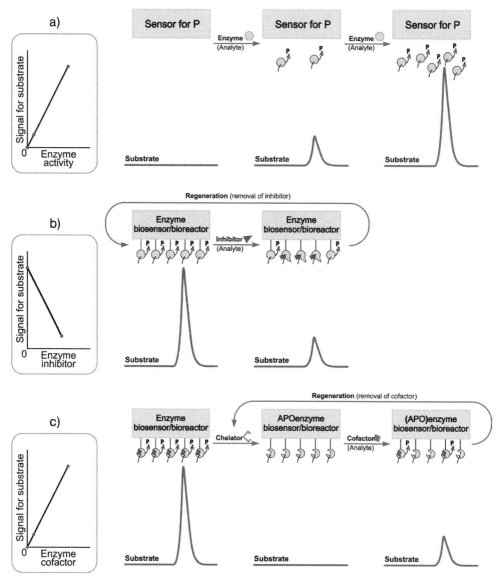

Figure 14.4 General schemes for the detection of (a) enzyme
activity, (b) enzyme inhibitor, (c) enzyme cofactor.

germ enables biospecific separation and detection of the bone-isoform of the
enzyme. A significantly simpler potentiometric FIA system based on a fluoride
ion-selective electrode utilizes the ability of the enzyme to catalyze monofluorophosphate hydrolysis [81]. The amount of fluoride ions generated from the substrate in
the course of the reaction is proportional to the enzyme activity. This FIA system is
useful for the determination of physiological and pathological levels of serum

alkaline phosphatase as well as the quantitative differentiation of bone and liver isoforms after the thermal pretreatment of serum samples.

14.3.2
Enzyme Inhibitor Detection

The biosensing schemes and FIA systems used for enzyme activity assays can be easily adapted for the indirect detection of the respective enzyme inhibitors. The most intensive investigations have been devoted to the detection of acetylcholinesterase (AChE) inhibitors. The common target analytes are phosphoorganic species. These analytes define the potential utility of such systems for environmental, agriculture and military analysis. FIA systems for AChE inhibitor determination differ in the forms of applied enzyme (biosensor, bioreactor or soluble enzyme) as well as in the detection methods. An amperometric biosensor has been fabricated using a layer-by-layer electrostatic self-assembly method, creating a sandwich-like structure where the enzyme is located between polyelectrolyte layers [82]. As internal sensor a nanotube modified glassy carbon electrode was applied. The developed biosensor-based FIA system has been proposed for the monitoring of organophosphate pesticides and nerve agents. For carbamate detection the FIA system based on a semi-disposable reactor with AChE immobilized on silica gel by covalent binding has been demonstrated [83]. The detection was based on measurements of pH and conductivity. The developed method was applied for the detection of carbaryl in waters without any sample preconcentration. A similar method for both organo-phosphate and carbamate pesticides determination by means of the inhibition of soluble AChE was developed using a pH-sensitive probe and highly sensitive fluorimetric detection [84]. Another spectrophotometric assay of organophosphoric insecticides utilizing an FIA system has been obtained using a reactor filled with AChE immobilized on aminated glass pearls filling. The optical detection of thiocho-line formed in the course of the enzymatic hydrolysis of thioacethylcholine using Ellman's reagent was applied for residue enzyme activity determination. The bioreactor was regenerated on-line after each inhibition using obidoxime [85]. The method has been applied for the analysis of real water samples. Many earlier publications concerning the determination of pesticides using AChE-based flow systems are reported in recently published reviews [86, 87].

Apart from acetylocholinesterase, other enzymes are not frequently applied for the indirect determination of inhibitors under FIA conditions. Recently, alkaline phos-phatase (in the soluble form) was applied for the determination of some selected inhibitors [88]. Monofluorophosphate and a fluoride ion-selective electrode were used as substrate and detector, respectively. It should be emphasized that this FIA system provides improvement in the selectivity of this inhibitive detection due to the kinetic discrimination of slower inhibition processes. Thanks to this approach, the system allows detection of beryllium and vanadate ions in the ppb range without significant influences from other inhibitors, including heavy metal ions. Another FIA system has been developed for the determination of inhibitors of xanthine oxidase [89]. All measurements were performed with soluble enzyme applying a

Clark-type oxygen electrode as a detector. The enzymatic reaction as well as inhibition was performed on-line in the FIA system. The system was applied for the inhibitive detection of quercetin, a compound playing an important role in the inhibition of many diseases related to reactive oxygen species. Under optimized conditions the reported FIA allows determination of quercetin in the submillimolar range of concentration.

14.3.3
Enzyme Cofactor Detection

Some enzymes are metalloproteins, having in their bioactive centers metal ions playing a cofactor role. The removal of the metal ion results in the formation of apoenzyme exhibiting no biocatalytic activity. Conversely, the addition of the cofactor to the apoenzyme causes the recovering of the enzyme activity. Such reversible conversions constitute a biosensing platform for the indirect determinations of either cofactor ion (regenerator) or chelator (inactivator). Both these reversible operations (inactivations and regenerations), seems to be especially easy to perform reproducibly under FIA conditions. The small number of publications concerning apoenzyme formation and regeneration under FIA conditions is surprising taking into account the potentialities of such an approach.

The apoenzyme-based recognition for the detection of calcium ions has been demonstrated using α-amylase [90]. The FIA system has been applied for the evaluation of the activity of an immobilized α-amylase, being a biorecognition element. The enzyme was immobilized in the flow-through minicolumn. Starch was applied as a substrate for the enzymatic reaction. The assay was possible using an FIA manifold containing additionally a double-layered column consisting of immobilized α-glucosidase and pyranose oxidase and finally an amperometric detector. The system was capable of detecting calcium(II) ions at millimolar levels.

Another, much simpler FIA system is based on a pH-electrode modified with a covalently bound monomolecular layer of alkaline phosphatase [44]. The cofactor role for this enzyme is played by zinc ions. Cysteine was applied as a chelating agent used for the generation of the immobilized apoenzyme. All steps of the analytical procedure, including the enzyme deactivation (the bioreceptor formation), the activity recovery (the analyte recognition) and the activity detection (the signal measurement for the substrate) were repeatedly performed on-line by alternating changes of the carrier composition. The reported potentiometric FIA system enables selective detection of zinc ions in the micromolar range.

14.4
Conclusions

Enzymes have been applied for analytical purposes for a long time. Nowadays enzymes play a significant role when coupled with flow injection analyses. They are considered as analytes and a large number of procedures for their activity determination is performed in the FIA format. Moreover, enzymes are responsible for biomolecular recognition, not only of substrates but also inhibitors and cofactors. Taking into account the variety of biomolecular recognition processes, the number of described enzyme applications for different assays is not surprising. New biodetection schemes using enzymes are still being developed. Moreover, for the already developed analytical systems, novel materials, technologies and instrumentation concepts, like nanoparticles applications, thin and thick film microdevices, as well as novel platforms for SIA and µTAS are being explored. Beyond such developments attention is being paid towards achieving lower detection limits and reagent consumption as well as reduction in sample volume and shortening of analysis time. All these demands could be fulfilled by various enzymes-based FIA systems. Finally, it should be remembered that FIA is inevitably connected with real analytical chemistry. As shown in this chapter, enzyme-based FIA systems perfectly affirm this statement.

Finally it should be mentioned that enzyme activity detection plays a significant role in a large group of bioassays not reported in this chapter. Some enzymes, like alkaline phosphatase, horseradish peroxidase, glucose oxidase and urease, are often used as biocatalytic markers in various immunochemical methods of analysis. The use of enzyme-labeled antibodies and other components of immunorecognition systems is a common and popular practice, because enzyme markers enable significant amplification of the analytical signal and hence lower the detection limits. It is worth noting that the labeling of biomolecules with enzymes offers the same advantages in the case of genoanalysis. All these bioaffinity-based analyte recognitions together with the enzyme activity detection could be effectively realized in the FIA format, significantly improving the quality of such bioassays.

References

1 Panoutsou, P. and Economou, A. (2005) Rapid enzymatic chemiluminescent assay of glucose by means of a hybrid flow injection/sequential-injection method. *Talanta*, **67**, 603–609.

2 Economou, A., Panoutsou, P. and Themelis, D.G. (2006) Enzymatic chemiluminescent assay of glucose by sequential-injection analysis with soluble enzyme and on-line sample dilution. *Analytica Chimica Acta*, **572**, 140–147.

3 Campuzano, S., Pedrero, M. and Pingarrón, J.M. (2005) A peroxidase-tetrathiafulvalene biosensor based on self-assembled monolayer modified Au electrodes for the flow injection determination of hydrogen peroxide. *Talanta*, **66**, 1310–1319.

4 Ledru, S., Ruillé, N. and Boujtita, M. (2006) One-step screen-printed electrode modified in its bulk with HRP based on direct electron transfer for hydrogen peroxide detection in flow injection mode. *Biosensors & Bioelectronics*, **21**, 1591–1598.

5 Tripathi, V.S., Kandimalla, V.B. and Ju, H. (2006) Amperometric biosensor for hydrogen peroxide based on ferrocene-bovine serum albumin and multiwall carbon nanotube modified ormosil composite. *Biosensors & Bioelectronics*, **21**, 1529–1535.

6 Kotzian, P., Brázdilová, P., Řezková, S., Kalcher, K. and Vytřas, K. (2006) Amperometric glucose biosensor based on rhodium dioxide-modified carbon ink. *Electroanalysis*, **18**, 1499–1504.

7 Derwińska, K., Miecznikowski, K., Koncki, R., Kulesza, P.J., Głąb, S. and Malik, M. (2003) Application of prussian blue based composite film with organic polymer to construction of enzymatic glucose biosensor. *Electroanalysis*, **15**, 1843–1849.

8 De Mattos, I.L. and Da Cunha Areias, M.C. (2005) Automated determination of glucose in soluble coffee using Prussian Blue-glucose oxidase-Nafion® modified electrode. *Talanta*, **66**, 1281–1286.

9 Yu, J., Liu, S. and Ju, H. (2003) Glucose sensor for flow injection analysis of serum glucose based on immobilization of glucose oxidase in titania sol-gel membrane. *Biosensors & Bioelectronics*, **19**, 401–409.

10 Osborne, P.G. and Hashimoto, M. (2004) Chemical polymerization of m-phenylenediamine, in the presence of glucose oxidase, produces an enzyme-retaining electrooxidisable polymer used to produce a biosensor for amperometric detection of glucose from brain dialysate. *Analyst*, **129**, 759–765.

11 Barsan, M.M., Klinčar, J., Batič, M. and Brett, C.M.A. (2007) Design and application of a flow cell for carbon-film based electrochemical enzyme biosensors. *Talanta*, **71**, 1893–1900.

12 Tkac, J., Whittaker, J.W. and Ruzgas, T. (2007) The use of single walled carbon nanotubes dispersed in a chitosan matrix for preparation of a galactose biosensor. *Biosensors & Bioelectronics*, **22**, 1820–1824.

13 Niculescu, M., Sigina, S. and Csöregi, E. (2003) Glycerol dehydrogenase based amperometric biosensor for monitoring of glycerol in alcoholic beverages. *Analytical Letters*, **36**, 1721–1737.

14 Carelli, D., Centonze, D., De Giglio, A., Quinto, M. and Zambonin, P.G. (2006) An interference-free first generation alcohol biosensor based on a gold electrode modified by an overoxidised non-conducting polypyrrole film. *Analytica Chimica Acta*, **565**, 27–35.

15 Pati, S., Quinto, M., Palmisano, F. and Zambonin, P.G. (2004) Determination of choline in milk, milk powder, and soy lecithin hydrolysates by flow injection analysis and amperometric detection with a choline oxidase based biosensor. *Journal of Agricultural and Food Chemistry*, **52**, 4638–4642.

16 Divritsioti, M.H., Karalemas, I.D., Georgiou, C.A. and Papastathopoulos, D.S. (2003) Flow injection analysis system for L-lysine estimation in foodstuffs using a biosensor based on lysine oxidase immobilization on a gold-poly (m-phenylenediamine) electrode. *Analytical Letters*, **36**, 1939–1963.

17 Varma, S., Yigzaw, Y. and Gorton, L. (2006) Prussian blue-glutamate oxidase modified glassy carbon electrode: A sensitive L-glutamate and b-N-oxalyl-a, b-diaminopropionic acid (β-ODAP) sensor. *Analytica Chimica Acta*, **556**, 319–325.

18 Radoi, A., Compagnone, D., Devic, E. and Palleschi, G. (2007) Low potential detection of NADH with Prussian Blue bulk modified screen-printed electrodes and recombinant NADH oxidase from *Thermus thermophilus*. *Sensors and Actuators B*, **121**, 501–506.

19 Esti, M., Volpe, G., Micheli, L., Delibato, E., Compagnone, D., Moscone, D. and Palleschi, G. (2004) Electrochemical biosensors for monitoring malolactic fermentation in red wine using two strains of Oenococcus oeni. *Analytica Chimica Acta*, **513**, 357–364.

20 Maestre, E., Katakis, I., Narváez, A. and Domínguez, E. (2005) A multianalyte flow electrochemical cell: Application to the simultaneous determination of carbohydrates based on bioelectrocatalytic detection. *Biosensors & Bioelectronics*, **21**, 774–781.

21 Harreither, W., Coman, V., Ludwig, R., Haltrich, D. and Gorton, L. (2007) Investigation of graphite electrodes modified with cellobiose dehydrogenase from the Ascomycete Myriococcum

thermophilum. *Electroanalysis*, **19**, 172–180.

22 Campuzano, S., Gamella, M., Serra, B., Reviejo, A.J. and Pingarrón, J.M. (2007) Integrated electrochemical gluconic acid biosensor based on self-assembled monolayer-modified gold electrodes. Application to the analysis of gluconic acid in musts and wines. *Journal of Agricultural and Food Chemistry*, **55**, 2109–2114.

23 Takagi, K. and Shikata, S. (2004) Flow injection determination of histamine with a histamine dehydrogenase-based electrode. *Analytica Chimica Acta*, **505**, 189–193.

24 Campuzano, S., Serra, B., Pedrero, M., De Villena, F.J.M. and Pingarrón, J.M. (2003) Amperometric flow injection determination of phenolic compounds at self-assembled monolayer-based tyrosinase biosensors. *Analytica Chimica Acta*, **494**, 187–197.

25 Serra, B., Reviejo, A.J. and Pingarrón, J.M. (2003) Flow injection amperometric detection of phenolic compounds at enzyme composite biosensors application to their monitoring during industrial waste waters purification processes. *Analytical Letters*, **36**, 1965–1986.

26 Gamella, M., Campuzano, S., Reviejo, A.J. and Pingarrón, J.M. (2006) Electrochemical estimation of the polyphenol index in wines using a laccase biosensor. *Journal of Agricultural and Food Chemistry*, **54**, 7960–7967.

27 Haghighi, B., Jarosz-Wilkołązka, A., Ruzgas, T., Gorton, L. and Leonowicz, A. (2005) Characterization of graphite electrodes modified with laccases from Trametes hirsuta and Cerrena unicolor and their use for flow injection amperometric determination of some phenolic compounds. *International Journal of Environmental Analytical Chemistry*, **85**, 753–770.

28 Mitsubayashi, K., Kubotera, Y., Yano, K., Hashimoto, Y., Kon, T., Nakakura, S., Nishi, Y. and Endo, H. (2004) Trimethylamine biosensor with flavin-containing monooxygenase type 3 (FMO3) for fish-freshness analysis. *Sensors and Actuators B*, **103**, 463–467.

29 Schöning, M.J., Krause, R., Block, K., Musahmeh, M., Mulchandani, A. and Wang, J. (2003) A dual amperometric/potentiometric FIA-based biosensor for the distinctive detection of organophosphorus pesticides. *Sensors and Actuators B*, **95**, 291–296.

30 Yao, T. and Yano, T. (2004) On-line microdialysis assay of L-lactate and pyruvate *in vitro* and *in vivo* by a flow injection system with a dual enzyme electrode. *Talanta*, **63**, 771–775.

31 Nanjo, Y., Yano, T., Hayashi, R. and Yao, T. (2006) Optically specific detection of D- and L-lactic acids by a flow injection dual biosensor system with on-line microdialysis sampling. *Analytical Sciences*, **22**, 1135–1138.

32 Katrlík, J., Mastihuba, V., Voštiar, I., Šefčovičova, J., Štefuca, V. and Gemeiner, P. (2006) Amperometric biosensors based on two different enzyme systems and their use for glycerol determination in samples from biotechnological fermentation process. *Analytica Chimica Acta*, **566**, 11–18.

33 Basu, A.K., Chattopadhyay, P., Roychudhuri, U. and Chakraborty, R. (2006) A biosensor based on co-immobilized L-glutamate oxidase and L-glutamate dehydrogenase for analysis of monosodium glutamate in food. *Biosensors & Bioelectronics*, **21**, 1968–1972.

34 Mizutani, F., Hirata, Y., Yabuki, S. and Iijima, S. (2003) Flow injection analysis of acetic acid in food samples by using trienzyme/poly(dimethylsiloxane)-bilayer membrane-based electrode as the detector. *Sensors and Actuators B*, **91**, 195–198.

35 Mieliauskiene, R., Nistor, M., Laurinavicius, V. and Csoregi, E. (2006) Amperometric determination of acetate with a tri-enzyme based sensor. *Sensors and Actuators B*, **113**, 671–676.

36 Yao, T., Yano, T. and Nishino, H. (2004) Simultaneous *in vivo* monitoring of glucose, L-lactate, and pyruvate concentrations in rat brain by a flow injection biosensor system with an on-line microdialysis sampling. *Analytica Chimica Acta*, **510**, 53–59.

37 Yao, T., Takashima, K. and Nanjyo, Y. (2003) Simultaneous determination of orthophosphate and total phosphates (inorganic phosphates plus purine nucleotides) using a bioamperometric flow injection system made up by a 16-way switching valve. *Talanta*, **60**, 845–851.

38 Tymecki, Ł., Zwierkowska, E. and Koncki, R. (2005) Strip bioelectrochemical cell for potentiometric measurements fabricated by screen-printing. *Analytica Chimica Acta*, **538**, 251–256.

39 Tymecki, Ł. and Koncki, R. (2006) Thick-film potentiometric biosensor for bloodless monitoring of hemodialysis. *Sensors and Actuators B*, **113**, 782–786.

40 Radomska, A., Koncki, R., Pyrzyńska, K. and Głąb, S. (2004) Bioanalytical system for control of hemodialysis treatment based on potentiometric biosensors for urea and creatinine. *Analytica Chimica Acta*, **523**, 193–200.

41 Radomska, A., Bodenszac, E., Głąb, S. and Koncki, R. (2004) Creatinine biosensor based on ammonium ion selective electrode and its application in flow injection analysis. *Talanta*, **64**, 603–608.

42 Rasmussen, C.D., Andersen, J.E.T. and Zachau-Christiansen, B. (2007) Improved performance of the potentiometric biosensor for the determination of creatinine. *Analytical Letters*, **40**, 39–52.

43 García-Villar, N., Saurina, J. and Hernández-Cassou, S. (2003) Flow injection differential potentiometric determination of lysine by using a lysine biosensor. *Analytica Chimica Acta*, **477**, 315–324.

44 Rozum, B., Koncki, R. and Tymecki, Ł. (2007) The potentialities of pH-electrode modified with alkaline phosphatase. *Sensors and Actuators B*, **127**, 632–636.

45 Ho, J., Wu, L., Fan, N.C., Lee, M.S., Kuo, H.Y. and Yang, C.S. (2007) Development of a long-life capillary enzyme bioreactor for the determination of blood glucose. *Talanta*, **71**, 391–396.

46 Jones, D.A., Parkin, M.C., Langemann, H., Landolt, H., Hopwood, S.E., Strong, A.J. and Boutelle, M.G. (2002) On-line monitoring in neurointensive care: Enzyme-based electrochemical assay for simultaneous, continuous monitoring of glucose and lactate from critical care patients. *Journal of Electroanalytical Chemistry*, **538–539**, 243–252.

47 Yao, T., Yano, T., Nanjyo, Y. and Nishino, H. (2003) Simultaneous determination of glucose and L-lactate in rat brain by an electrochemical *in vivo* flow injection system with an on-line microdialysis sampling. *Analytical Sciences*, **19**, 61–65.

48 Nanjo, Y., Hayashi, R. and Yao, T. (2006) Determination of fructosyl amino acids and fructosyl peptides in protease-digested blood sample by a flow injection system with an enzyme reactor. *Analytical Sciences*, **22**, 1139–1143.

49 Nanjo, Y., Hayashi, R. and Yao, T. (2007) An enzymatic method for the rapid measurement of the hemoglobin A_{1c} by a flow injection system comprised of an electrochemical detector with a specific enzyme-reactor and a spectrophotometer. *Analytica Chimica Acta*, **583**, 45–54.

50 Wu, L.C. and Cheng, C.M. (2005) Flow-injection enzymatic analysis for glycerol and triacylglycerol. *Analytical Biochemistry*, **346**, 234–240.

51 Vianello, F., Ragusa, S., Cambria, M.T. and Rigo, A. (2006) A high sensitivity amperometric biosensor using laccase as biorecognition element. *Biosensors & Bioelectronics*, **21**, 2155–2160.

52 Limbut, W., Thavarungkul, P., Kanatharana, P., Asawatreratanakul, P., Limsakul, C. and Wongkittisuksa, B. (2004) Comparative study of controlled pore glass, silica gel and Poraver for the immobilization of urease to determine urea in a flow injection conductimetric biosensor system. *Biosensors & Bioelectronics*, **19**, 813–821.

53 Kim, M. and Kim, M.J. (2003) Isocitrate analysis using a potentiometric biosensor with immobilized enzyme in a FIA system. *Food Research International*, **36**, 223–230.

54 Kim, M. (2006) Determining citrate in fruit juices using a biosensor with citrate lyase and oxaloacetate decarboxylase in a flow injection analysis system. *Food Chemistry*, **99**, 851–857.

55 Matos, R.C., Coelho, E.O., de Souza, C.F., Guedes, F.A. and Matos, M.A.C. (2006) Peroxidase immobilized on Amberlite IRA-743 resin for on-line spectrophotometric detection of hydrogen peroxide in rainwater. *Talanta*, **69**, 1208–1214.

56 De Oliveira, A.C.A., Assis, V.C., Matos, M.A.C. and Matos, R.C. (2005) Flow-injection system with glucose oxidase immobilized on a tubular reactor for determination of glucose in blood samples. *Analytica Chimica Acta*, **535**, 213–217.

57 Vojinovič, V., Calado, C.R., Silva, A.I., Mateus, M., Cabral, J.M.S. and Fonseca, L.P. (2005) Micro-analytical GO/HRP bioreactor for glucose determination and bioprocess monitoring. *Biosensors & Bioelectronics*, **20**, 1955–1961.

58 Vojinovič, V., Esteves, F.M.F., Cabral, J.M.S. and Fonseca, L.P. (2006) Bienzymatic analytical microreactors for glucose, lactate, ethanol, galactose and L-amino acid monitoring in cell culture media. *Analytica Chimica Acta*, **565**, 240–249.

59 Luque-Almagro, V.M., Blasco, R., Fernandez-Romero, J.M. and Lucue de Castro, M.D.L. (2003) Flow-injection spectrophotometric determination of cyanate in bioremediation processes by use of immobilised inducible cyanase. *Analytical and Bioanalytical Chemistry*, **377**, 1071–1078.

60 Carvalho Vieira, E. and Nogueira, A.R.A. (2004) Orthophosphate, phytate, and total phosphorus determination in cereals by flow injection analysis. *Journal of Agricultural and Food Chemistry*, **52**, 1800–1803.

61 Iida, Y., Ikeda, M., Aoto, M. and Satoh, I. (2004) Fluorometric determination of urea in alcoholic beverages by using an acid urease column-FIA system. *Talanta*, **64**, 1278–1282.

62 Tsukatani, T. and Matsumoto, K. (2005) Sequential fluorometric quantification of malic acid enantiomers by a single line flow injection system using immobilized-enzyme reactors. *Talanta*, **65**, 396–401.

63 Tsukatani, T. and Matsumoto, K. (2006) Flow-injection fluorometric quantification of pyruvate using co-immobilized pyruvate decarboxylase and aldehyde dehydrogenase reactor: Application to measurement of acetate, citrate and l-lactate. *Talanta*, **69**, 637–642.

64 Tsukatani, T. and Matsumoto, K. (2005) Fluorometric quantification of total D-gluconate by a flow injection system using an immobilized-enzyme reactor. *Analytica Chimica Acta*, **530**, 221–225.

65 Rhee, J.I. and Sohn, O.J. (2003) Flow injection system for on-line monitoring of fumaric acid in biological processes. *Analytica Chimica Acta*, **499**, 71–80.

66 Sohn, O.J., Han, K.A. and Rhee, J.I. (2005) Flow injection analysis system for monitoring of succinic acid in biotechnological processes. *Talanta*, **65**, 185–191.

67 Tsukatani, T. and Matsumoto, K. (2005) Sequential fluorometric quantification of γ-aminobutyrate and L-glutamate using a single line flow injection system with immobilized-enzyme reactors. *Analytica Chimica Acta*, **546**, 154–160.

68 Manera, M., Miró, M., Estela, J.M. and Cerdá, V. (2004) A multisyringe flow injection system with immobilized glucose oxidase based on homogeneous chemiluminescence detection. *Analytica Chimica Acta*, **508**, 23–30.

69 Davidsson, R., Genin, F., Bengtsson, M., Laurell, T. and Emnéus, J. (2004) Microfluidic biosensing systems part I. Development and optimisation of enzymatic chemiluminescent µ-biosensors based on silicon microchips. *Lab on a Chip*, **4**, 481–487.

70 Yaqoob, M. and Nabi, A. (2003) Flow injection chemiluminescent assays for glycerol and triglycerides using a co-immobilized enzyme reactor. *Luminescence*, **18**, 67–71.

71 Kiba, N., Ito, S., Tachibana, M., Tani, K. and Koizumi, H. (2003) Simultaneous determination of choline and acetylcholine based on a trienzyme chemiluminometric biosensor in a single line flow injection system. *Analytical Sciences*, **19**, 1647–1651.

72 Amini, N. and McKelvie, I. (2005) An enzymatic flow analysis method for the determination of phosphatidylcholine in sediment pore waters and extracts. *Talanta*, **66**, 445–452.

73 Nakamura, H., Hasegawa, M., Nomura, Y., Ikebukuro, K., Arikawa, Y. and Karube, I. (2003) Improvement of a CL-FIA system using maltose phosphorylase for the determination of phosphate-ion in freshwater. *Analytical Letters*, **36**, 1805–1817.

74 Yaqoob, M., Anwar, M. and Nabi, A. (2005) Determination of phosphate in freshwaters by flow injection with immobilized enzyme and chemiluminescence detection. *International Journal of Environmental Analytical Chemistry*, **85**, 451–459.

75 Almeida, C.F., Calado, C.R.C., Bernardino, S.A., Cabral, J.M.S. and Fonseca, L.P. (2006) A flow injection analysis system for on-line monitoring of cutinase activity at outlet of an expanded bed adsorption column almost in real time. *Journal of Chemical Technology and Biotechnology*, **81**, 1678–1684.

76 Almeida, C.F., Cabral, J.M.S. and Fonseca, L.P. (2004) Flow injection analysis system for on-line cutinase activity assay. *Analytica Chimica Acta*, **502**, 115–124.

77 Theaker, B.J. and Rowell, F.J. (2003) A rapid and sensitive fluorometric flow injection assay for subtilisin-type enzymes utilising sol-gel particles directly coated with gelatin-Texas Red substrate. *Analyst*, **128**, 1043–1047.

78 Zajoncová, L., Jílek, M., Beranová, V. and Peč, P. (2004) A biosensor for the determination of amylase activity. *Biosensors & Bioelectronics*, **20**, 240–245.

79 Yamaguchi, M., Kanemaru, M., Kanemori, T. and Mizuno, Y. (2003) Flow-injection-type biosensor system for salivary amylase activity. *Biosensors & Bioelectronics*, **18**, 835–840.

80 Hartwell, S.K., Somprayoon, D., Kongtawelert, P., Ongchai, S., Arppornchayanon, O., Ganranoo, L., Lapanantnoppakhun, S. and Grudpan, K. (2007) Online assay of bone specific alkaline phosphatase with a flow injection-bead injection. *Analytica Chimica Acta*, **600**, 188–193.

81 Ogończyk, D. and Koncki, R. (2007) Potentiometric flow injection system for determination of alkaline phosphatase in human serum. *Analytica Chimica Acta*, **600**, 194–198.

82 Liu, G. and Lin, Y. (2006) Biosensor-based on self-assembling acetylcholinesterase on carbon nanotubes for flow injection/amperometric detection of organophosphate pesticides and nerve agents. *Analytical Chemistry*, **78**, 835–843.

83 Suwansa-Ard, S., Kanatharana, P., Asawatreratanakul, P., Limsakul, C., Wongkittisuksa, B. and Thavarungkul, P. (2005) Semi disposable reactor biosensors for detecting carbamate pesticides in water. *Biosensors & Bioelectronics*, **21**, 445–454.

84 Jin, S., Xu, Z., Chen, J., Liang, X., Wu, Y. and Qian, X. (2004) Determination of organophosphate and carbamate pesticides based on enzyme inhibition using a pH-sensitive fluorescence probe. *Analytica Chimica Acta*, **523**, 117–123.

85 Dăneţ, A.F., Bucur, B., Cheregi, M.C., Badea, M. andSerban, S. (2003)

Spectrophotometric determination of organophosphoric insecticides in a FIA system based on AChE inhibition. *Analytical Letters*, **36**, 59–73.

86 Prieto-Simón, B., Campás, M., Andreescu, S. and Marty, J.L. (2006) Trends in flow-based biosensing systems for pesticide assessment. *Sensors*, **6**, 1161–1186.

87 Solé, S., Merkoci, A. and Alegret, S. (2003) Determination of toxic substances based on enzyme inhibition. Part II. Electrochemical biosensors for the determination of pesticides, using flow systems. *Critical Reviews in Analytical Chemistry*, **33**, 127–143.

88 Koncki, R., Rudnicka, K. and Tymecki, Ł. (2006) Flow injection system for potentiometric determination of alkaline phosphatase inhibitors. *Analytica Chimica Acta*, **577**, 134–139.

89 Lam, L.H., Sakaguchi, K., Ukeda, H. and Sawamura, M. (2006) Flow injection determination of xanthine oxidase inhibitory activity and its application to food samples. *Analytical Sciences*, **22**, 105–109.

90 Iida, Y., Sato, Y. and Satoh, I. (2003) Novel detection system for calcium(II) ions based on an apoenzyme reactivation method using an amylase column as a recognition element. *Electrochemistry*, **71**, 449–452.

15
Flow Potentiometry

M. Conceição B.S.M. Montenegro and Alberto N. Araújo

15.1
Introduction

Analytical chemistry departments provide the information for problem resolution in many areas related to human wellbeing and sustainable development. However, restricted financial resources are available for their activities and thus the automation of method and procedures are obvious solutions to maximize the ratio between the volume and nature of the determinations offered and their unitary cost. Through automation, labour intensive tasks become significantly reduced and high quality results are produced in short periods of time. A significant part of modern automation solutions is based on liquid flow techniques operated under controlled hydrodynamic conditions. Either chromatography or electromigration techniques are available to encompass sample separation or continuous flow analysis-based systems when automation of additional unit operations is intended. In the most recent developments, these tools, together with detector devices, are being integrated and miniaturized to reduce reagent and sample consumption in a more environmentally friendly approach. In the same way as the more common spectroanalytical instrumentation, potentiometry based on the use of ion selective electrodes has been shown to be well suited to fulfilling the measuring step in continuous flow systems. Besides providing detection in a non-destructive way over a broad range of analyte concentrations, ISEs present low energy consumption, miniaturization capability without significant loss of sensitivity and low acquisition cost. In this chapter, attention is devoted to a review of the main features of potentiometric detection in continuous flow systems: the recent developments, coupling configurations and practical applications as well as trends and future prospects.

15.2
Background Concepts

Common undergraduate textbooks define potentiometry as an analytical technique in which the free energy change that occurs to achieve chemical equilibrium inside

Advances in Flow Analysis. Edited by Marek Trojanowicz
Copyright © 2008 WILEY-VCH Verlag GmbH & Co. KGaA, Weinheim
ISBN: 978-3-527-31830-8

an electrochemical cell is evaluated through potential measurement in the absence of current flow. This classical definition conforms to the initial experiments developed by Luigi Galvani (1737–1798) and Walter Nernst in 1888, when trying to explain the potential origin of either a metal wire immersed in a solution of its own ions or an inert metal wire immersed in a solution of a redox pair of ions [1]. As a practical result of his experiments, Nernst developed the first electrode capable of measuring the acidity of solutions, albeit with application limited by its complex construction [2]. Just nine years later, Cremer observed that the potential of a galvanic cell in which its constitutive semi-elements were separated by a thin glass membrane was dependent on the difference of proton concentration on both sides of the membrane [3]. Hence, the glass electrode was really the first ion-selective electrode to be proposed almost a century ago [4], but erroneously its response was seen as governed by proton diffusion through the glass membrane [5]. Despite these earlier achievements, the use of non-traditional electrodes only became common in the second half of the last century. This came about from consolidation of the theoretical aspects [6, 7], the appearance of high impedance voltmeters [8, 9], the proposal of an electrode possessing a crystalline membrane selective for fluoride [10] and electrodes supporting highly viscous liquid membranes selective for calcium [11] and potassium [12, 13] and the expeditious procedures for their construction [14]. An important contribution was also from their large scale introduction in the clinical chemistry routine during the nineties to enable determination of sodium, potassium, chloride, calcium and proton. This was mainly due to their almost maintenance-free operation and the ability to perform direct measurements in whole blood, plasma, serum and urine [15, 16]. Nowadays, the search for new electroactive species to improve selectivity and/or to extend the use to new applications, and efforts towards miniaturization and coupling to automated equipment are becoming popular fields of activity in the worldwide scientific community.

Potentiometric detection is achieved by simultaneous immersion of indicator and reference electrodes in the solution to be assayed. Ion-selective electrodes are usually employed as indicator electrodes although on some occasions use of conventional electrodes based on metallic surfaces provides satisfactory results. The magnitude of the cell potential is dependent on both the nature of the electrodes used and the ionic mobility at the liquid junction between sample and reference solutions. Thus, observed variations in the cell potential from sample to sample arise mainly from a change in activity of the monitored species contacting the indicator electrode surface or membrane. Accumulated knowledge about ISE performance has evidenced that it is the partitioning of sample ions at the sample/membrane interface to achieve electrochemical equilibrium that determines the potentiometric response [17, 18]. This phase-boundary potential model enables prediction of the analytical response based on the assumptions that the organic phase in contact with the aqueous sample solution is in chemical equilibrium and that the diffusion potential inside the membrane can be neglected [19]. According to the model, the potential developed in the organic/aqueous phase boundary of an ISE is given by:

$$E = \frac{RT}{z_i F} \ln k_1 + \frac{RT}{z_i F} \ln \frac{a_i(\text{aq})}{a_i(\text{org})}$$

where a_i(aq) and a_i(org) are the activities of the uncomplexed monitored ion (with charge z_i) in the aqueous sample and in the contacting organic phase boundary, respectively. For a constant a_i(org), a nernstian response similar to that described for the classical electrodes, is expected:

$$E_I = E_I^0 + \frac{RT}{z_i F} \ln a_I$$

The first term in the equation represents a standard potential once k_1 is a function of the relative free energies of solvation of the considered ion in the sample and the membrane phases. In this context, one can understand the rationale for the proposal of new chemical entities capable of extracting and keeping constant, preferably at a low value, the activity of the target ion in the membrane. Useful membranes arise from a careful combination of a hydrophobic charged ion exchanger that pulls the target ion and not its counter ion into the membrane (permselective properties of the membrane) and/or from a selective hydrophobic complexing agent (called an ionophore) to buffer its activity [20] (Figure 15.1).

The qualitative and quantitative compositions of the membrane together with its structure determine the final characteristics of the electrode, such as durability, analytical range, slope, response time and selectivity. When considering the analysis of complex samples it is expected that other ions of the same signal charge or more lipophilic (respectively, different a_i(org) and k_i values, see above) may interfere to some extent with the final potential reading. In these conditions, the ISE potential is better described by the semi-empirical Nikolskii–Eisenmann equation [6, 7], valid for three or four activity decades of nernstian response to the target ion in the sample:

$$E_i = E_i^0 + S \log\left(a_i + \sum K_j a_j^{z_i/z_j}\right)$$

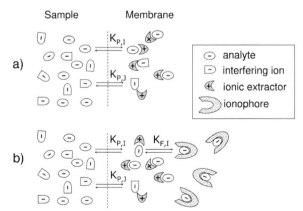

Figure 15.1 Equilibria mechanisms of ion-selective membranes based on the use of (a) ionic extractors; (b) ionic extractor plus ionophore. Counter ions on the sample solution side and equilibria between free and bonded main or interfering ions on the membrane side are not represented.

where E_i is the electrode potential to which the activity a_i of the monitored ion and a_j the activities of other (interfering) ions contribute. The contribution of each interfering ion is dependent on its actual activity in the sample, the ratio of the charges and of the potentiometric selectivity coefficient K_j. The coefficient value, determined according to IUPAC recommendations [21], is higher or lower than that for electrodes giving lower or higher potentiometric response for the target ion compared to the interfering ion. Finally, S represents the practical slope of the semi-logarithmic relationship between the electrode potential and the main ion activity. At 25 °C, positive or negative values of S of 59.2 mV decade^{-1} are obtained for main ions possessing, respectively, single positive or negative charges. This value decreases to net values around 29.6 mV decade^{-1} for double charged ions.

Albeit usually requiring simplified sample pre-treatment due to the selectivity characteristics and wide linear dynamic range, potentiometry-based measurements generally improve if made under flow conditions [22, 23]. The permanent liquid stream has a conditioning effect on the ISE membrane, leading to a better potential stability. The reproducible sample volume and transport timing contribute to reducing the analysis time and increasing the reproducibility and sensitivity of the potential readings when compared to those obtained using a conventional approach. Moreover, an improvement in selectivity potentiometric coefficients could be found compared to the conventional ones. Generally this improvement can be observed for interferences with stronger affinity to the membrane system than the main sensed ion, since both transient concentrations and the absence of thermodynamic equilibrium are usually employed for the determinations [24].

15.3
Electrode Developments and Detector Cell Designs

Whatever the actual application mode of potentiometric determinations, that is, batch or integrated in flow schemes, adequate sensitivity and selectivity depend mainly on the characteristics of the indicator electrode and on the detection cell design. The exploitation of conventional metallic electrodes is limited by their selectivity performance which hinders application to trace analysis or to samples with complex matrices. In contrast, polymer membrane ion-selective electrodes, with a selectivity profile that can be tailored according to their final use, are becoming a more attractive means to improve the analytical figures of potentiometric applications [25]. They can be generally described as a highly viscous membrane phase embedding an ion exchanger and a so-called "ionophore" or "ion carrier" which is a lipophilic complexing agent that chiefly dictates the final ISE selectivity behavior. These terms remain from the first used agents capable of transporting ions through lipophilic phases such as biological cell membranes or artificial model membranes [26, 27]. Theoretical modelling of this type of electrode is very recent [28–31] and yet has enabled prediction of how response characteristics could be improved. Though three or four decades of activity or concentration range should render attractive the use of potentiometry in several analytical fields its application is impaired by the micromolar limits of detection generally reported. This behavior arrives either from main ion counter diffusion with interfering ions present in the sample or from internal reference solution leaching across the membrane. These phenomena increase the main ion activity in the ESI/sample interface, impairing the measurement for dilute samples. The use of a high concentration of an interfering ion and a low concentration of the primary ion in the inner reference solution may improve the ESI detection limit and lead to unbiased thermodynamic selectivity coefficients. Highlighted by theory, a Pb^{2+}-selective electrode with a limit of detection in the picomolar range [32] was initially proposed, followed by new electrodes sensitive to silver [33], calcium [34], cadmium [35] and iodide [36], thus opening the opportunity to exploit their application in environmental analysis [37]. Future improvements will be guided by a deeper understanding of the basic principles and substantial reduction in experimental bias rather than through technological achievements like other analytical techniques [38, 39].

The routine use of potentiometry in laboratories is not enabled only by cost savings, simplicity or limits of detection achieved. Perhaps more important is the need to provide accurate results, which is dependent on the ISE membrane's ability to discriminate the analyte relative to other ions present in the sample. As stated before, this ability is evaluated by establishment of the potentiometric selectivity coefficients towards the main ion with respect to each of the probable interfering ions. Membrane preparation can be simply based on the use of high molecular weight lipophilic molecules, responsible for ion exchange through the membrane interface with the sample solution. Quaternary ammonium or phosphonium salts of the analyte are mostly used for the detection of anionic species and tetralkylborate derivatives or heteropolyacids for cationic-sensitive membranes [40–42]. However, a common selectivity pattern, similar to the Hofmeister series, is observed related to the hydrophobic character of both the ion exchanger and the measured ion. Numerous researchers have shown that the selectivity pattern can change and be improved by additional incorporation of an ionophore, even though the nature and relative proportions of the other membrane components still greatly determine their final electroanalytical properties [43, 44]. Ionophores are organic species with cationic or anionic selective recognition ability, developed in the supramolecular chemistry field. According to new theories, the main ion activity in the potentiometric bulk membrane is buffered by the ionophore with which it forms highly stable complexes [45]. Depending on the analyte nature and charge, successful ionophore classes have been identified, such as crown ethers containing bulky subunits or thiazoles [46–49], anion-selective species having hydrogen-bonding functional groups (e.g., urea, thiourea and guanidine) on rigid molecular frames [50–54], calixarenes with ion-recognizing pendants at the lower or upper rims [55–57], pyrrolic and polypyrrolic species [58, 59], metalloporphyrins and metallophtalocyanines [60], chiral anion recognising cyclodextrins and cholic and deoxycholic species [61]. However, this is a very open field of research and substantial increase in knowledge should be expected in the coming years, especially regarding anions where the size, shape, hydration and pH dependence variability is much larger than for cationic species. Still concerning the sensing membranes of ISE, one of the most described drawbacks is the progressive leaching with use of its embedded components which reflects on the selectivity loss and on the sensitivity. To overcome the limited but measurable partition coefficients of the components leading to their leakage, new membrane formulations based on the absence of plasticizer solvent have been proposed [62–66] as well as the covalent attachment of the ionophore to the membrane polymer [63, 65, 67], and the use of non-conventional polymers such as aminated or carboxylated PVC, poly(ethylene-co-vinyl)acetate-(EVA) [68], polymethylmetacrylate or polystyrene [69] and mixtures of electroactive polymers in solid contact electrodes [58].

Albeit the achievements reported before are mainly related to the construction of electrodes for batch analysis applications, some of them considered the use in continuous flow conditions where the elimination of the inner reference solution of the ISE, as will be described later, is of particular significance. Most of the works are focused on the determination of cationic species using electrodes formulated with ion-exchangers (e.g., heteropolyacids) incorporated in organic polymeric membranes or eventually in chemically modified carbon paste electrodes [70–90] (Table 15.1). The latter have the advantages of easier preparation/regeneration and possess low potential drifts and ohmic resistance. The reported improvement in the selectivity profile [81] seems more related to the kinetic discrimination provided by the flow conditions adopted. Our group has developed new ISE incorporating ionophores with specific metal–ligand interaction (metalloporphyrins) or ionophores that can form proton-bridged inclusion complexes with aliphatic and aromatic species (cyclodextrins). Besides presenting improved selectivity patterns, they yield more reproducible responses in the determinations of chloride [91], specific aliphatic carboxylates and aromatic species like diclofenac [92]. In the same way, by changing the PVC matrix membrane polymer to EVA [93] or by using a silica inorganic sol–gel support [94] prolonged use of the electrodes was observed, taking into consideration that shorter lifetimes were expected due to the continuous liquid stream over the membrane. The association of macrocyclic ionophores based on host–guest recognition with photocured polymers [95–97] has allowed increased potentiometric response robustness besides excellent adhesion to the conductor support. Significant advances were also achieved for continuous flow applications requiring potentiometric pH measurements. Prior approaches revealed it to be difficult to implement due to electrode design (quite obvious in the case of glass electrodes), narrow pH range and system coupling fragility. In this context flow set-ups were proposed using glass membranes [98], PVC or epoxyacrylate [99, 100], metal/metal oxide [101, 102], ISFETs [103] and stainless steel electrodes [104]. Both composite electrodes based on quinhydrone [105, 106] and an Fe_2O_3 graphite-epoxy resin tubular electrode obtained by filling the surface of a polyurethane resin block with that mixture [107] seem to present comparable characteristics to those of glass electrodes but are simpler to implement and maintain, to adapt to the flow mountings and to apply to pH measurements in emulsions.

The accumulated knowledge of new ISE performances and the robust coupling to liquid tubular conduits now perfectly established, make potentiometric detection a candidate for enhanced use in the different continuous-flow modes as well as in

Table 15.1 Membrane components of electrodes used recently (years 2000–2007) in flow injection systems for the determination of cationic organic species.

Main ion	Sensor immobilization	Membrane composition ionophore/plasticizer	Type of construction	Analytical application	Ref.
Phenylpropanolamine	Polymeric	PT/DOP and DBP	Conventional with inner reference solution	Pharmaceuticals	[70]
Terbutaline	Polymeric	PT/DOP	Conventional with inner reference solution	Pharmaceuticals	[71]
Piribedil	Chemically modified carbon paste	ST, SM, PT	Conventional without inner reference solution	Pharmaceuticals and Biological fluids (urine)	[72]
Hyoscyamine	Polymeric	TPB, PT/DOP	Conventional with inner reference solution	Pharmaceuticals	[73]
Chlorpromazine	Polymeric	TPB/o-NPOE	Tubular	Pharmaceuticals	[74]
Sildenafil	Polymeric	TP, Re/o-NPOE	Conventional with inner reference solution	Pharmaceuticals and Biological fluids (human serum)	[75]
Creatinine	Polymeric	PT, MP, PC/o-NPOE	Coated-wire and tubular	Biological fluids (human serum)	[76]
Dicyclomine	Polymeric	ST, SM, PT, MP, TPB/DBP	Conventional with inner reference solution	Pharmaceuticals Biological fluids (serum and urine) Milk	[77]
Meveberine	Polymeric	ST, SM, PT, MP/DBP	Conventional with inner reference solution	Pharmaceuticals	[78]
Amitriptyline	Polymeric	PT, MP and mixture of PT/MP/DOP	Conventional with inner reference solution	Biological fluids (serum and urine) Pharmaceuticals	[79]

(Continued)

Table 15.1 (*Continued*)

Main ion	Sensor immobilization	Membrane composition ionophore/plasticizer	Type of construction	Analytical application	Ref.
Tetracycline	Polymeric (PVC and EVA)	t-TCPB/BEHS	Tubular	Pharmaceuticals	[80]
Pipazethate	Polymeric	PT, MP and mixture of PT/MP/DOP	Conventional with inner reference solution	Pharmaceuticals	[81]
Alkylphenol polyethoxylates	Polymeric	TPB/o-NPOE	Tubular	—	[82]
Dipyridamole	Polymeric	TPB, Re/o-NPOE	Conventional with inner reference solution	Pharmaceuticals	[83]
Ambroxol	Polymeric	TPB/DOP	Conventional and Coated graphite	Pharmaceuticals	[84]
Chlormequat	Polymeric	TPB/o-NPOE, BEHS, DBP	Tubular	River waters	[85]
Cysteine	Polymeric	TPB, BTPPA/BEHS, o-NPOE	Tubular	Pharmaceuticals	[86]
Salbutamol	Polymeric	PT, MP, and mixture PT/MP/DOP	Conventional with inner reference solution	Pharmaceuticals	[87]
Ranitidine	Polymeric	TPB, PT/DOP, DBP	Conventional with inner reference solution	Pharmaceuticals	[88]
Drotaverine	Polymeric	ST, SM, PT, MPTPB/DBP, DOP, DOS, TCP, DINP	Conventional with inner reference solution	Pharmaceuticals	[89]
Chlordiazepoxide	Polymeric	PT, MP/DBP	Coated wire	Pharmaceuticals	[90]

PT, phosphotungstate; ST, silicotungstate; SM, silicomolybdate; TPB, tetraphenylborate; Re, reineckate; MP, molybdophosphate; PC, Picronolate, t-TCPB, tetrakis-(4-chlorophenylborate); BTPPA, bis (triphenylphos + phoranyleden)ammonium; DOP, dioctylphthalate; DBP, dibutylphthalate; o-NPOE, o-nitrophenyloctylether; BEHS, bis (2-ethylhexyl)sebacate; DOS, dioctylsebacate; TCP, tricresylphosphate; DINP, diisononylphtalate; PVC, polyvinylchloride; EVA, ethylene(vinylacetate).

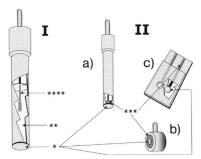

Figure 15.2 Ion-selective electrode configurations. I, liquid membrane; II (a) solid contact (b) tubular solid contact and (c) screen printed. *Polymeric sensitive membrane,** inner reference solution, *** conductive epoxy resin, metallic surface or conductive ink, **** inner reference electrode.

hyphenated techniques (with HPLC and capillary electrophoresis [108, 109]) and miniaturized flow systems like lab-on-chip [110] and lab-on-valve [111]. Coated wire configured electrodes or even solid wire contact electrodes (Figure 15.2) obtained by depositing the membrane over a platinum or glassy carbon substrate are preferred instead of liquid membrane inner solution ISEs (Figure 3.2). They are easy to assemble, can be made extremely sensitive and have one or two seconds response time by optimizing the difference in chemical potential for the analyte ion through the interface contacting the membrane [112, 113]. Their performance is superior to amperometric detection since the response is quasi-independent of the flow-regimen and rate adopted, an important issue in the new ultra-high flow rate HPLC systems. In this particular hyphenated technique a molecular recognition approach (host–guest chemistry) was used to optimize the difference in the interaction energies of the analyte and of the buffer ion. In this way the analysis of organic amines was achieved with sensitivities 20 times superior to that obtained using spectophotometric detection [114] and the procedure could be further extended to the determination of low molecular weight amino alcohols and alkylamines [115]. Furthermore, a quantitative structure–activity relationship (QSAR) approach was used for the first time to predict the potentiometric response and limits of detection of non-selective membrane electrodes for beta-adrenergic drugs and aliphatic amines on the basis of their lipophilic character and polarizability [116, 117]. The comparative use of electrodes based on ammonium salts with others based on podand urea derivatives with amine functionalities or macrocyclic polyamines revealed that the latter allowed the chromatographic analysis of carboxylic acids with better limits of detection due to lower leaching of the membrane components into the eluent. The linkage of HPLC to potentiometric detectors based on macrocyclic ionophores was shown also to be favorable in the analysis of organic acids of biochemical importance [113, 118–121], amines [116] and polyionic mono- and oligonucleotides [122] with very low limits of detection.

Potentiometric detection under flow-through conditions consists of monitoring, recording and processing a signal generated in the electroanalytical detection cell through an electronic circuit device. The detector is located downstream from the sampling point at a reduced distance if direct measurement is intended or at a greater distance if in-line sample pre-treatment is also required. Whatever the particular system considered, the flow-through detection cell monitors the concentration (mass)–time profile of the analyte in the flowing-stream. A plethora of flow-through detection cells could be found, each being characterized by length, diameter and the geometry of the channels in the cell, thus determining the main transport mechanism of sample transport to the membrane electrode interface. The cells are conceived with designs that assure an effective transport of the analyte towards the interface of the indicator electrode to fully exploit the response characteristics of the electrode. Depending on how the detector is probing the dispersed sample zone, the detection cell measures the average concentration over the channel cross-section in wall-jet configurations or the local concentration at the surface of the sensing probe in tubular and thin-cell configurations (Figure 15.3) where a tangential flow condition is kept [123]. Hence, the indicator electrode can be located at the center of the detection channel or at its wall. Potentiometric detectors can be characterized as surface detection devices which implies the adoption of reproducible radial concentration distribution. Whatever the case, robust electrodes are required in hydrodynamic methods of analysis. The materials used to construct a physically robust ISE surface vary from liquid membranes and conductive polymers to ceramic materials and metals. Their use was accomplished by several discussions among electrochemists as to whether the origin of the potential lay in ions being transported through the membrane, or was confined to its partition through the membrane interfaces. The first approach led to the proposal of detection flow-cells where liquid membrane electrodes with an inner solution of known concentration could be inserted, and the flowing analyte sample flushed the other side of the membrane in a wall-jet detection mode. In the last few years the new working models have clarified the discussion [27, 28, 44, 124] and seem to favor the latter approach. Consequently, good potentiometric signals were also obtained with indicator electrodes where the inner solution was omitted and the liquid

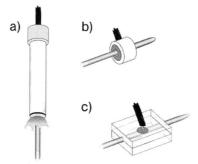

Figure 15.3 General coupling modes of ISE in continuous flow systems. (a) In wall-jet detection mode flowing sample zone strikes the electrode surface perpendicularly. (b) In tubular and (c) thin-cell electrodes the sample flows tangentially to the electrode surface.

membrane coats a conductive surface such as a metal, the so-called coated wire electrodes. As previously seen, coated wire ISEs are frequently associated with continuous flow procedures with the advantages of both being constructed with different configurations, namely tubular or planar, and being easily miniaturized for their use in LC, CE and µTAS systems [113]. Most coated wire electrodes, sometimes referred to as solid-contact and solid-state electrodes, are implemented with PVC membranes directly applied over the solid conductive surface. Poor adhesion is, however, a commonly encountered problem that leads to membrane detachment with the continual streaming of the carrier over its surface. Alternatively, electroactive conjugated polymers were proposed, working as ion-electron transducers when coated with the PVC membranes, or directly doped with the ion-recognition species. There are several examples of their use, mainly coupled to flow mountings in a planar microelectrode configuration [125–127], with significant gains in terms of durability and potential stability. The tubular configuration is also used successfully, where the membrane coats the inner walls of a longitudinal hole drilled in a conductive support. This type of configured ISE is adapted to flow mounting as an extension of its tubular conduits, thus reducing the inner dead volume of the detection cell and enabling measurement without disruption of the hydrodynamic transport conditions of the circulating sample plug. One of its main virtues is the possibility of further sample treatment downstream for speciation or multidetermination by other detection techniques, impaired in flow-through planar array sensors or wall-jet cells [128].

Potentiometry is a valuable analytical tool for various medical and environmental applications carried out remotely from the laboratory, fulfilling requirements of low-cost, simple design and small-size and even in a disposable format. Various planar technologies have been described to develop solid-contact sensors for flow methods [129, 130]. Thin and thick film technologies were shown to be well suited for production of disposable sensors. Screen-printing is especially useful due to the simplicity and efficiency in large scale production [127]. Through this fabrication procedure, miniaturization of electrodes is possible and hence the implementation of multisensoric systems in the form of sensor arrays [131]. As compared with field effect transistors (FET) the preparation has increased ruggedness and their integration in microfluidic systems is simple. The procedure also brought innovation to reference electrodes, with a planar liquid junction Ag/AgCl/KCl reference electrode being described [132]. The all-thick-film strip electrodes were constructed at the same time using a unique automated-screen-printing technology, without additional manual, chemical or electrochemical steps [133]. Simple, inexpensive, commercially available materials such as a plastic substrate and easily cured polymer-based pastes were used. The proposed construction procedure allowed mass and unified production of complete and fully integrated electrode systems for potentiometry consisting of the reference and indicator electrodes in one-strip as well as one-chip format. Screen-printing technology was also used in the preparation of miniaturized reference electrodes that could be integrated with indicator electrodes in a monolith where the flow analytical microsystem was produced using the low-temperature co-fired ceramic technology [134]. In this work the reference electrode is prepared by screen-printing over a ceramic tape containing an auxiliary channel through which a 0.1 M KCl solution continuously flowed to provide a constant reference potential.

15.4
Flow Analytical Techniques Based on Potentiometry

Mutual benefits from the coupling between different continuous flow concepts and potentiometric detection led to the publication of more than 200 papers where mainly single (Figure 15.4a) or two-channel flow injection mountings (Figure 15.4b) were adopted. In these basic preferred configurations fast and reliable concentration measurements are focused on the selectivity of the electrode used. The flow concept generally ensures the ionic strength and/or pH adjustment, the sample volume definition and its transport downstream. Being simple in design they confer system portability, required in environmental studies [135], reveal improved alternatives concerning highly polluting procedures [136] or enable exploitation of known chemical reactions difficult to implement using batch conditions. A singular approach was reported using a closed circulatory flow manifold to accomplish bromate determinations in the micromolar range [137]. The bromate present in the sample reacts with the bromide present in a Fe(II)–Fe(III) potential buffer used as carrier, generating labile bromine which is detected in a flow-through cell composed of a gold-plated electrode and an Ag/AСl reference electrode. Although flow injection analysis systems are still the preferred choice when potentiometric detection is used, increased use of other continuous flow systems employing microcomputer-controlled devices has been observed in recent years. Amongst these, sequential-injection analysis (SIA) (Figure 15.4c) was shown to be ideally suited for multiple determinations and, therefore, for the manipulation of samples containing different chemical forms of species in a matrix. In Table 15.2, a summary of recently developed procedures based on SIA is presented, where only procedures with real sample

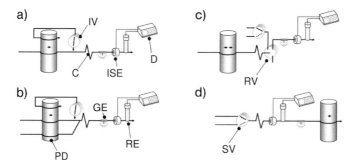

Figure 15.4 Typical continuous flow manifolds used with potentiometric detection: (a) single channel flow injection system; (b) double channel flow injection system; (c) sequential injection system with multitask ability; (d) multicom-mutated flow system, PD, propulsion device; IV, injection valve; SV, three-way solenoid valve; RV, stream selecting valve; C, coil; GE, grounding electrode; ISE, ion selective electrode; RE, reference electrode; D, decimillivoltmeter.

Table 15.2 Sequential-injection analysis with potentiometric detection applied in real samples determinations (years 2000–2007).

Analyte	Sample	Detection method	Linear range	Limit of detection	R.S.D. (%)	Sampling rate (h^{-1})	Ref.
Acetic acid	Vinegars	Titration with pH glass electrode	10–90 g L^{-1}	—	0.4	28	[138]
Acetylsalicylic acid	Aspirin tablets	Enzyme hydrolysis, salicylate electrode	5×10^{-5}–1×10^{-2} mol L^{-1}	5×10^{-5} mol L^{-1}	0.2	45	[139]
Acidity	Soft drinks	Titration with pH glass electrode	0.1–0.6% (w/v)	—	<0.5	45	[140]
Ammonium	Fertilisers	Neural network, array of NH$_4^+$, Na$^+$, K$^+$ and two generic electrodes	5×10^{-3}–4×10^{-2} mol L^{-1}	—	1.8–4.2	—	[141]
Barium	Waters (mineral)	Neural network, array of Ca electrodes	5–35×10^{-3} mg L^{-1}	—	1.8–4.2	—	[142]
Bicarbonate	Waters	Neural network plus Legendre decomposition, array of two Cl$^-$, two NO$_3^-$ and one generic electrodes	5×10^{-4}–5×10^{-3} mol L^{-1}	—	1.8–4.2	—	[143]
Calcium	Waters (mineral)	Neural network, array of Ca electrode	5–45×10^{-3} mg L^{-1}	—	1.8–4.2	—	[142]
Captopril	Pharmaceuticals	Titration with Ag/Ag$_2$S tubular electrode	2.5–10×10^{-4} mol L^{-1}	—	<1.3	—	[144]
Chloride	Electroplating baths	Tubular AgCl/Ag$_2$S electrode	0.1–1.0 mol L^{-1}	—	—	40	[145]
	Milk	Titration with Ag/Ag$_2$S tubular electrode	0.01–0.25 mol L^{-1}	—	<3.4	17	[146]
		Monosegmented, titration Ag/Ag$_2$S tubular electrode	8×10^{-4}–3×10^{-2} mol L^{-1}	—	<1	—	[147]
	Pharmaceuticals	Octaethyl(porphyrinato)In(III) based electrode	1×10^{-5}–1×10^{-2} mol L^{-1}	—	<0.64	60	[91]
	Waters	AgCl/Ag$_2$S electrode	10–3500 mg L^{-1}	1.6 mg L^{-1}	0.4	30	[148]

(Continued)

Table 15.2 (*Continued*)

Analyte	Sample	Detection method	Linear range	Limit of detection	R.S.D. (%)	Sampling rate (h^{-1})	Ref.
		Least squares multiple regression, AgCl/Ag$_2$S electrode	$10^{-4}-4\times10^{-1}$ mol L^{-1}	—	—	—	[149]
		Lab-at-valve concept, Ag/AgCl wire	$1\times10^{-4}-1.2\times10^{-1}$ mol L^{-1}	—	0.7–1.3	50	[150]
		Neural network plus Legendre decomposition, array of two Cl$^-$, two NO$_3^-$ and one generic electrode	$5\times10^{-3}-4.5\times10^{-2}$ mol L^{-1}	—	1.8	—	[143]
	Wines	Monosegmented, titration Ag/Ag$_2$S tubular electrode	$8\times10^{-4}-3\times10^{-2}$ mol L^{-1}	—	<1	—	[147]
Clavulanate	Pharmaceuticals	Clavulanate selective electrode	$2\times10^{-3}-1\times10^{-1}$ mol L^{-1}	—	0.6	53	[151]
Diclofenac	Pharmaceuticals	Cyclodextrin-based electrode	$5\times10^{-6}-1\times10^{-2}$ mol L^{-1}	2×10^{-6} mol L^{-1}	<1	33	[92]
Fluoride	Tap waters	LaF$_3$ crystal membrane electrode	$10^{-10}-10^{-6}$ mol L^{-1}	1.7×10^{-12} mol L^{-1}	0.3	30	[152]
	Toothpaste		$10^{-5}-10^{-1}$ mol L^{-1}	2×10^{-6} mol L^{-1}	0.3	30	
Gibberellic acid	Agriculture growth promoters	Tetraphenylporphyrinate Mn(III) tubular electrode	$5\times10^{-4}-8\times10^{-3}$ mol L^{-1}	3×10^{-4} mol L^{-1}	<0.4	30	[153]
Hydrochloric acid	Effluent streams	Titration with pH glass electrode	20 50 mmol L^{-1}	—	<0.22	30	[154]
	Acid production plant	Dilution, titration, pH glass electr.	5.93–8.99 mol L^{-1}	—	<0.4	30	[155]
Magnesium	Waters (mineral)	Neural network, array of Ca electr.	$5-40\times10^{-3}$ mg L^{-1}	—	1.8–4.2	—	[142]
Nitrate	Waters	Neural network, Legendre decomposition, array of two Cl$^-$, two NO$_3^-$ and one generic electrode	$5\times10^{-4}-5\times10^{-3}$ mol L^{-1}	—	1.8–4.2	—	[143]

Analyte	Sample	Method	Range				Ref.
	Waters	Multiple regression, tert-octylammo-nium bromide based electrode	5×10^{-4}–10^{-1} mol L^{-1}	—	—	—	[149]
Nitrogen	Silage	Ammonia electrode based on nonactin	10–120 mg L^{-1}	3 mg L^{-1}	<2	30	[156]
Penicillin-G	Pharmaceuticals	Tetraphenylporphyrinate Mn(III) tubular eletrode	2×10^{-4}–10^{-2} mol L^{-1}	1.5×10^{-4} mol L^{-1}	1.8	25	[157]
Potassium - free	Pharmaceuticals	K$^+$ tubular valinomycin electrode	2×10^{-3}–10^{-1} mol L^{-1}	—	0.5	53	[151]
- total	Table and port wines		39–3900 mg L^{-1}	0.8 mg L^{-1}	<2	12	[158]
	Fertilisers	After on-line microwave digestion Neural network, array of NH$_4^+$, Na$^+$, K$^+$ and two generic electrodes	5×10^{-3}–5×10^{-2} mol L^{-1}	—	1.8–4.2	—	[141]
pH	Electroplating baths	Tubular pH, tetrakis(p-chlorophenylborate) electrode	1–5 pH	—	—	40	[145]
Pyridoxine	Pharmaceuticals	Tubular pyridoxine electrode	5×10^{-5}–10^{-2} mol L^{-1}	—	8.6	20	[159]
Sodium	Fertilisers	Neural network, array of NH$_4^+$, Na$^+$, K$^+$ and two generic electrodes	5×10^{-3}–4.5×10^{-2} mol L^{-1}	—	1.8–4.2	—	[141]
Sucrose	Silage	Perchlorate electrode based on bis (triph-enylphosphoranilidene) ammonium	—	0.13%	<2	24	[160]
Valproate	Pharmaceuticals	Tubular PVC tetraphenylporphyrinate Mn(III) electrode	5×10^{-4}–10^{-2} mol L^{-1}	9×10^{-5} mol L^{-1}	1.2	30	[93]
Urea	Milk	Electrode with sol-gel membrane	1×10^{-3}–5×10^{-2} mol L^{-1}	9×10^{-4} mol L^{-1}	2.2	55	[161]
		Ammonia electrode based on non-actin after a gas diffusion unit	1×10^{-3}–10^{-2} mol L^{-1}	6×10^{-4} mol L^{-1}	1.9	20	

applications are considered. Comprehensive and detailed reviews on this subject have been published in the analytical journals [110, 162–164]. Potentiometric detection has also been recently described using the lab-on-valve concept [165] but requiring new configurations for the electrodes and detection cell design in order to enable injection of reduced volumes of sample, higher sampling rates and absence of a grounding electrode to reduce electrical noise [111]. Recent reports on both these flow concepts or other previously reported ones, such as flow injection analysis, monosegmented flow-analysis, multicommutated binary systems, (Figure 15.4d), aim to perform additional tasks. Thus, different approaches were able to achieve self-calibration of systems over the three or four orders of magnitude in which the nernstian detector response is observed. In one proposal [166] an elaborate set-up comprising two peristaltic pumps, two computer interface cards, a double three-way injection valve and a mixing chamber was used to prepare calibrating solutions inside a single flow injection system. Valve actuation enabled the recycling of solutions or their propelling in programmed volumes of 22 µL to 17 mL into the mixing chamber. Nevertheless, a rigorous previous calibration procedure is necessary due to the limited lifetime of the flexible pumping tubes. In a second proposal a miniaturized three-way solenoid valve actuated in a short series of on/off cycles is connected to a lateral port of the main valve of a sequential injection system [159]. In this approach only up to 100-fold dilution of the stock standard is possible, depending on the ratio between the on and off times of the solenoid valve. The accuracy of the calibration solutions is not compromised by the use of peristaltic pumps if a low sample dispersion condition is maintained.. By changing the chemical nature of the solutions it is also possible to implement fully automated assessment of interferences through separated and fixed interference methods [21], and perform determinations directly or through more elaborated protocols like standard additions or titration. The versatility of this new approach was assessed in the simultaneous determination of chloride and nitrate in waters using two tubular ISE [149]. After a previous assessment of the respective concentration values in each sample, automated in-line calibration of both electrodes proceeded in the vicinity of these values, varying the concentration of the main ion (chloride or nitrate) at the predicted concentration of the interfering ion (nitrate or chloride). The accuracy of the determinations measured by the relative root mean square error of prediction improved to 2.8% as compared with the value of 12.1% obtained for the standards addition method. A similar approach enabled Gran's plots based titration of chloride and iodide, both in the concentration range 1–100 µmol L^{-1} [167]. The flexibility of the approach enabled implementation of simple and robust procedures for real time accuracy assessment in the control of pharmaceuticals [91, 92, 144, 151]. This concept is based on the simultaneous evaluation of the same species by two quasi-independent techniques [168]. If the interferences are of different character the mean result achieved is more accurate. The development of e-tongues using a sequential-injection analysis to accelerate the calibration procedure was also proposed [169]. Electronic tongues comprise an array of chemical sensors with poor selectivity and cross-sensitivity to different species and an appropriate chemometric tool for data processing. A general sequential-injection system with a coupled mixing chamber on a lateral port of the main stream selecting valve allowed versatile sampling of different stock solutions of species and preparation of the training set for an artificial neural network or other chemometric algorithm. Using sets of tubular electrodes prepared using several previously described ion extractors and ionophores, multiparametric results with relative standard deviations less than 4% are predicted after integrated analysis of response profiles to the analytes, either as main ions or interfering species. Major benefits and drawbacks of these e-tongues were fully examined for the analysis of cationic species in water [141, 142, 170, 171], beverages [172], fertilisers [173] and clinical [174] samples and for anions in waters [147].

A method particularly suited to potentiometry due to its ability to respond to both broad activity or concentration variations and applicability to colored or turbid samples is titrimetry. Flow-injection titrations are the oldest gradient technique in flow injection analysis and are based on insertion of the sample in the titrant used as carrier. A concentration log dependence on the transient signal width measured between the equivalent points of the rising and falling edges of the signal is established after sample dispersion and reaction [138, 175]. Albeit fast, this procedure cannot be considered absolute since system calibration and sample dispersion reproducibility are required to get accurate results. New procedures have meanwhile been suggested to implement what the authors claim as "true titrations." These generally avoid calibrations by overcoming the influence of hydrodynamic fluctuations and consequently of the physical dispersion of the sample inside the tubular path. The principle of feedback-based flow ratiometry was applied in highly reproducible acid–base titrations lasting less than 15 s, with consumption of about 12 μL [176]. The flow-system needed up to three peristaltic pumps and string bead single packed reactors to guarantee good mixing conditions. The ratio between the titrant and titrand flow-rates increases linearly until the equivalence point is achieved. Then, that ratio is sequentially decreased and increased around the equivalence point by an electrical signal that controls the pump responsible for the titrant flow. Calculations are performed considering the mean value of the minimum and maximum electrical signals used in feedback control. The use of fast actuated solenoid three-way valves was proposed to perform titration procedures, this time based on simpler system configurations [177–179]. The concept is based on fast alternate injection of small volumes of titrant and titrand in the tubular path where complete mixing is favored. By step increasing the ratio between the on/off cycles, complete titration curves are gathered, minimizing solution consumption and allowing the use of traditional algorithms for results processing. The versatility of this approach increased once it was proven to be feasible also in a sequential injection mounting [147]. A recent monosegmented flow-system was proposed to achieve true titration of ascorbic acid or of Fe(II), with coulometric generation of the titrant, I_2 and Ce(IV), respectively [180]. Up to 100 μL of sample are intercalated between two 60 μL volumes of the titrant precursor solution, and carried towards the detection cell by an air stream. The detection cell comprised of three platinum electrodes and an Ag/AgCl reference electrode is limited on both sides by two opto-switches responsible for automatic reversal of the flow direction of the liquid monosegment. Increasing dispersion and reaction between both solutions enables generation of the titration curves. According to the authors, partial but progressive dispersion complies with the flow injection principles. The titration proceeds over one defined aliquot of titrand and calibration procedures are not necessary.

The use of packed reactors to enable automated sample cleaning, preconcentration, speciation or biochemical transformation has been shown to be compliant with potentiometric detection. In this context, simultaneous parallel urea and creatinine determination was proposed to accomplish real-time control of hemodialysis treatment, the setup being on-line coupled to the dyalisate spent channel of the hemodialysis equipment [181, 182]. The flow injection system required controlled timing of a peristaltic pump and of two injection valves. The first valve was located before the propelling device and enabled alternate dual injection of the 300 μL of dyalisate bloodless sample and 2.3 mol L^{-1} HCl regeneration solution, in a stream passing through a cationic exchange resin where the main cationic interfering species are retained. After passing the peristaltic pump, 100 μL of the cleaned sample are injected through the second valve in a Tris-HCl (pH 8.1) buffer solution which splits downstream to achieve the two detection cells based on ammonium-selective electrodes modified by chemical anchoring, respectively, urease and creatinine-deiminase. Use of packed columns proved effective in the accurate determination of fluoride in sera and urine 10-fold diluted samples in concentrations in the μmol L^{-1} range [183]. In this manifold an anion exchange resin was placed before the injection valve in order to reduce anion contaminants in the water used as sample carrier. The usually observed memory effect, drift and dissolution of lanthanum fluoride from the membrane crystal is overcome in a three channel flow injection manifold where acetate buffer (pH 5.3) with an ionic strength of 4.5 mol L^{-1} is added through a first confluence with the flowing sample to minimize fluctuation in fluoride activity caused by other ions present in the sample. Cationic interference was minimized by adding to the buffer trans-1,2-diaminocyclohexane-N,N,N,N-tetracetic acid as chelating agent. A 26.3 μmol L^{-1} (or 263 μmol L^{-1} for urine samples) fluoride solution containing phosphate at a concentration of 1 mmol L^{-1} was added through a second confluence to assure a stable baseline potential and minimize the interference of the anion on the recorded signals. Regarding the lower levels in water samples, a small column filled with cellulose fibers impregnated with zirconia was used to preconcentrate fluoride ions [184]. Adsorption occurred at pH 4.8 and the retained fluoride desorbed at pH 13. In this way it was possible to improve the detection limit up to 3×10^{-9} mol L^{-1} and provide results with a relative standard deviation better than 1.6%. For preconcentration and speciation of As(III) and As(V) in water samples it was proposed to use a packed column of amorphous oxyhydroxide embedded in silica gel and an As(V)-selective tubular electrode based on a film of the same material cast over a conducting graphite epoxy support [185]. The column was coupled to the injection valve thus being inserted on the sample or in the eluent/detector line. To achieve preconcentration the sample is continuously merged with a phosphate buffer solution (pH 12) with or without added iodine (As(V) or total As). After injection, a pH 7 phosphate buffer solution eluted the retained arsenic and allowed its transport to the detection cell. The preconcentration of 10 mL water samples provides results in the 40–500 μg L^{-1} range with a typical within batch precision of 2%. Determination of the total content of non-ionic surfactants containing hydrophilic chains between 6 and 18 ethoxylate units in river water samples was proposed with a flow injection system comprising a solid-phase extraction step on a C_{18} commercial resin placed on the loop of the injection valve to achieve sample enrichment and purification [186]. A tubular solid state electrode was used as detector, based on the tetraphenylborate salt of a barium complex of polyethoxylate non-ionic surfactant [187]. The procedure allowed a detection limit of 3×10^{-6} mol L^{-1} by eluting with 200 μL of 75% acetonitrile in water, if 40 mL of sample previously flowed through the resin. In a recent paper the authors extended the application to anionic surfactants using sodium dodecylsulfate as standard [188].

15.5
Trends and Future Prospects

Although not being exhaustive this review of recent achievements reveals that continuous flow-techniques with potentiometric detection are a fertile field for future research. Theoretical understanding of the response of new polymeric membrane electrodes with more robust configurations will guide the development of ISE with better selectivity characteristics and limits of detection. Instead of considering lack of selectivity a constraint in the quality of analysis, continuous flow systems have been used to demonstrate that electrode arrays and chemometric treatment of results constitute a new low cost multiparametric tool yielding good results at a modest cost. Fast and reproducible in-line preparation of solutions to fully exploit ISE response characteristics has become available, overcoming tedious laboratorial procedures like calibration over a wide concentration range, standard additions, titration and interferences testing. Separation techniques, to highly reproducible determinations near the limits of detection and biochemical assays are enabled by the use of packed in-line reactors. Furthermore, down-scaling of complete procedures for sample pre-treatment does not lead to loss of detection, physical robustness, or compromization of the response performance.

References

1 Nernst, W. and Loeb, K. (1988) Zur Kinetik der in Lösung befindlichen Körper. Zweite Abhanlung Überfuhrungszahlen und Leitvermogen einiger Silbersaze. *Zeitschrift fur Physikalische Chemie-International Journal of Research in Physical Chemistry & Chemical Physics*, **2** (12), 948–963.

2 Nernst, W. (1897) Die elektrolytische Zersetzung wässriger Lösungen. *Berichte der Deutschen Chemischen Gesellschaft*, **30** (12), 1547–1563.

3 Cremer, M. (1906) Über die Urasche der elektromotorischen Eigenschaften der Gewebe, zugleich ein Beitrag zur Lehre von den polyphasischen Elektrolytketten. *Zeitschrift für Biologie*, **347**, 562–608.

4 Haber, F. and Klemensiewicz, Z. (1909) On electrical interfacial potentials. *Zeitschrift fur Physikalische Chemie-Leipzig*, **67**, 385–431.

5 Donnan, F.G. (1911) Theory of membrane equilibrium and membrane potentials in the presence of nondialyzable electrolytes. A contribution to physical-chemical physiology. *Zeitschrift für Elektrochemie und physikalische Chemie*, **17**, 572–581.

6 Eisenman, G., Rudin, D.O. and Casby, J.U. (1957) Glass electrode for measuring sodium ion. *Science*, **126**, 831–834.

7 Nikolskii, B.P. and Schults, M.M. (1962) Some aspects of glass electrode theory. *Zhurnal Fizicheskoi Khimii*, **36**, 1327–1330.

8 Inzelt, G. (2005) Patent No. 2,058,761—or the beginning of electrochemical instrumentation. *Journal of Solid State Electrochemistry*, **9**, 181–182.

9 Radiometer Annual Report, Copenhagen, 1994/1995.

10 Frant, M.S. and Ross, J.W. (1966) Electrode for sensing fluoride ion activity in solution. *Science*, **154**, 1553–1555.

11 Ross, J.W. (1967) Calcium-selective electrode with liquid ion exchanger. *Science*, **156**, 1378–1379.

12 Stefanac, Z. and Simon, W. (1966) *In-vitro*-verhalten von makrotetroliden in membranen als grundlage fur hochselektive kationenspezifische elektrodensysteme. *Chimia*, **20**, 436.

13 Stefanac, Z. and Simon, W. (1967) Ion specific electrochemical behavior of macrotetrolides in membranes. *Microchemical Journal*, **12**, 125–132.

14 Moody, G.J., Oke, R.B. and Thomas, J.D.R. (1970) Calcium-sensitive electrode based on a liquid ion exchanger in a poly (vinyl-chloride) matrix. *Analyst*, **95**, 910–918.

15 Gunaratna, P.C., Koch, W.F., Paule, R.C., Cormier, A.D., D'Orazio, P., Greenberg, N., O'Connell, K.M., Malenfant, A., Okorodudu, A.O., Miller, R., Kus, D.M. and Bowers, G.N. (1992) Frozen human serum reference material for standardization of sodium and potassium measurements in serum or plasma by ion-selective electrode analyzers. *Clinical Chemistry*, **38**, 1459–1465.

16 Bakker, E., Diamond, D., Lewenstam, A. and Pretsch, E. (1999) Ion sensors: current limits and new trends. *Analytica Chimica Acta*, **393**, 11–18.

17 Guggenheim, E.A. (1929) The conceptions of electrical potential difference between two phases and the individual activities of ions. *Journal of Physical Chemistry*, **33**, 842–849.

18 Guggenheim, E.A. (1930) On the conception of electrical potential difference between two phases. II. *Journal of Physical Chemistry*, **34**, 1540–1543.

19 Bakker, E., Buhlmann, P. and Pretsch, E. (2004) The phase-boundary potential model. *Talanta*, **63**, 3–20.

20 Amemiya, S., Buhlmann, P., Pretsch, E., Rusterholz, B. and Umezawa, Y. (2000) Cationic or anionic sites? Selectivity optimization of ion-selective electrodes based on charged ionophores. *Analytical Chemistry*, **72**, 1618–1631.

21 Buck, R.P. and Lindner, E. (1995) Recommendations for nomenclature of ion-selective electrodes – (IUPAC recommendations 1994). *Pure and Applied Chemistry*, **66**, 2527–2536.

22 Frenzel, W. (1988) Enhanced performance of ion-selective electrodes in flow injection analysis – non-nernstian response, indirect determination, differential detection and modified reverse flow injection analysis. *Analyst*, **113**, 1039–1046.

23 Chudy, M., Wróblewski, W., Dybko, A. and Brzozka, Z. (2001) Multi-ion analysis based on versatile sensor head. *Sensors and Actuators B*, **78**, 320–325.

24 Trojanowicz, M., Szewczynska, M. and Wcislo, M. (2003) Electroanalytical flow measurements – recent advances. *Electroanalysis*, **15**, 347–365.

25 Krawczyk, T.K.V., Trojanowicz, M. and El-Murr, N. (2000) Enhancement of

selectivity of electrochemical detectors by kinetic discrimination in flow injection systems. *Laboratory Robotics and Automation*, **12**, 205–215.

26 Moore, C. and Pressmann, B.C. (1964) Mechanism of action of valinomycin on mitochondria. *Biochemical and Biophysical Research Communications*, **15**, 562–567.

27 Visser, H.C., Reinhoudt, D.N. and de Jong, F. (1994) Carrier-mediated transport through liquid membranes. *Chemical Society Reviews*, **23**, 75–81.

28 Sokalski, T., Zwickl, T., Bakker, E. and Pretsch, E. (1999) Lowering the detection limit of solvent polymeric ion-selective electrodes. 1. Modelling the influence of steady-state ion fluxes. *Analytical Chemistry*, **71**, 1204–1209.

29 Sokalski, T., Ceresa, A., Fibbioli, M., Zwickl, T., Bakker, E. and Pretsch, E. (1999) Lowering the detection limit of solvent polymeric ion-selective membrane electrodes. 2. Influence of composition of sample and internal electrolyte solution. *Analytical Chemistry*, **71**, 1210–1214.

30 Bakker, E., Pretsch, E. and Buhlmann, P. (2000) Selectivity of potentiometric ion sensors. *Analytical Chemistry*, **72**, 1127–1133.

31 Zwickl, T., Sokalski, T. and Pretsch, E. (1999) Steady-state model calculations predicting the influence of key parameters on the lower detection limit and ruggedness of solvent polymeric membrane ion-selective electrodes. *Electroanalysis*, **11**, 673–680.

32 Sokalski, T., Ceresa, A., Zwickl, T. and Pretsch, E. (1997) Large improvement of the lower detection limit of ion-selective polymer membrane electrodes. *Journal of the American Chemical Society*, **119**, 11347–11348.

33 Ceresa, A., Radu, A., Peper, S., Bakker, E. and Pretsch, E. (2002) Rational design of potentiometric trace level ion sensors. A Ag^+-selective electrode with a 100 ppt detection limit. *Analytical Chemistry*, **74**, 4027–4036.

34 Qin, W., Zwickl, T. and Pretsch, E. (2000) Improved detection limits and unbiased selectivity coefficients obtained by using ion-exchange resins in the inner reference solution of ion selective polymeric membrane electrodes. *Analytical Chemistry*, **72**, 3236–3240.

35 Ion, A.C., Bakker, E. and Pretsch, E. (2001) Potentiometric Cd^{2+}-selective electrode with a detection limit in the low ppt range. *Analytica Chimica Acta*, **440**, 71–79.

36 Malon, A., Radu, A., Qin, W., Qin, Y., Ceresa, A., Maj-Zurawska, M., Bakker, E. and Pretsch, E. (2003) Improving the detection limit of anion-selective electrodes: An iodide-selective membrane with a nanomolar detection limit. *Analytical Chemistry*, **75**, 3865–3871.

37 Ceresa, A., Bakker, E., Gunther, D., Hattendorf, B. and Pretsch, E. (2001) Potentiometric polymeric membrane electrodes for measurement of environmental samples at trace levels: New requirements for selectivities and measuring protocols, and comparison with ICPMS. *Analytical Chemistry*, **73**, 343–351.

38 Bakker, E. and Pretsch, E. (2005) Potentiometric sensors for trace-level analysis. *Trends Analytical Chemistry*, **24**, 199–207.

39 Morf, W.E., Badertscher, M., Zwickl, T., Rooij, N.F. and Pretsch, E. (2002) Effects of controlled current on the response behavior of polymeric membrane ion-selective electrodes. *Journal of Electroanalytical Chemistry*, **526**, 19–28.

40 Arnold, M.A. and Solsky, R.L. (1986) Ion-selective electrodes. *Analytical Chemistry*, **58**, R84–R101.

41 Yu, R.Q. (1986) Aspects of the develop-ment of liquid membrane anion sensitive electrodes. *Ion-Selective Electrode Review*, **3**, 153–172.

42 Wotring, V.J., Johnson, D.M. and Bachas, L.G. (1990) Polymeric membrane anion-selective electrodes based on diquaternary

ammonium-salts. *Analytical Chemistry*, **62**, 1506–1510.

43 Simon, W. and Carafoli, E. (1979) Design, properties, and applications of neutral ionophores. *Methods in Enzymology*, **56**, 439–448.

44 Bakker, E., Buhlmann, P. and Pretsch, E. (1997) Carrier-based ion-selective electrodes and bulk optodes. 1. General characteristics. *Chemical Reviews*, **97**, 3083–3112.

45 Beer, P.D. and Gale, P.A. (2001) Anion recognition and sensing: The state of the art and future perspectives. *Angewandte Chemie-International Edition in English*, **40**, 486–516.

46 Siswanta, D., Nagatsuka, K., Yamada, H., Kumakura, K., Hisamoto, H., Shichi, Y., Toshima, K. and Suzuki, K. (1996) Structural ion selectivity of thia crown ether compounds with a bulky block subunit and their application as an ion-sensing component for an ion-selective electrode. *Analytical Chemistry*, **68**, 4166–4172.

47 Suzuki, K., Siswanta, D., Otsuka, T., Amano, T., Ikeda, T., Hisamoto, H., Yoshihara, R. and Ohba, S. (2000) Design and synthesis of a more highly selective ammonium ionophore than nonactin and its application as an ion-sensing component for an ion-selective electrode. *Analytical Chemistry*, **72**, 2200–2205.

48 Kim, H.S., Park, H.J., Oh, H.J., Koh, Y.K., Choi, J.H., Lee, D.H., Cha, G.S. and Nam, H. (2000) Thiazole containing benzo crown ethers: A new class of ammonium selective ionophores. *Analytical Chemistry*, **72**, 4683–4688.

49 Benco, J.S., Nienaber, H.A. and McGimpsey, W.G. (2003) Synthesis of an ammonium ionophore and its application in a planar ion-selective electrode. *Analytical Chemistry*, **75**, 152–156.

50 Xiao, K.P., Buhlmann, P., Nishisawa, S., Amemiya, S. and Umezawa, Y. (1997) A chloride ion-selective solvent polymeric membrane electrode based on a hydrogen

bond forming ionophore. *Analytical Chemistry*, **69**, 1038–1044.

51 Hutchins, R.S., Bansal, P., Molina, P., Alajarin, M., Vidal, A. and Bachas, L.G. (1997) Salicylate selective electrode based on a biomimetic guanidinium ionophore. *Analytical Chemistry*, **69**, 1273–1278.

52 Amemiya, S., Buhlmann, P., Umezawa, Y., Jagessar, R.C. and Burns, D.H. (1999) An ion-selective electrode for acetate based an urea-functionalized porphyrin as a hydrogen-bonding ionophore. *Analytical Chemistry*, **71**, 1049–1054.

53 Fibbioli, M., Berger, M., Schmidtchen, F.P. and Pretsch, E. (2000) Polymeric membrane electrodes for monohydrogen phosphate and sulfate. *Analytical Chemistry*, **72**, 156–160.

54 Berrocal, M.J., Cruz, A., Badr, I.H.A. and Bachas, L.G. (2000) Tripodal ionophore with sulfate recognition properties for anion-selective electrodes. *Analytical Chemistry*, **72**, 5295–5299.

55 Giannetto, M., Mori, G., Notti, A., Pappalardo, S. and Parisi, M.F. (1998) Discrimination between butylammonium isomers by calix[5]arene-based ISE's. *Analytical Chemistry*, **70**, 4631–4635.

56 Zeng, X., Weng, L., Chen, L., Leng, X., Zhang, Z. and He, X. (2000) Improved silver ion-selective electrodes using novel 1,3-bis(2-benzothiazolyl)thioalkoxy-p-tert-butylcalix[4]arenes. *Tetrahedron Letters*, **41**, 4917–4921.

57 Mahajan, R.K., Kaur, I. and Kumar, M. (2003) Silver ion-selective electrodes employing Schiff base p-tert-butyl calix[4]arene derivatives as neutral carriers. *Sensors and Actuators B*, **91**, 26–31.

58 Bobacka, J. (2006) Conducting polymer-based solid-state ion-selective electrodes. *Electroanalysis*, **18**, 7–18.

59 Bobacka, J., Ivaska, A. and Lewenstam, A. (2003) Potentiometric ion sensors based on conducting polymers. *Electroanalysis*, **15**, 366–374.

60 Arvand, M., Pourhabib, A., Shemshadi, R. and Giahi, M. (2007) The potentiometric

behavior of polymer-supported
metallophthalocyanines used as
anion-selective electrodes. *Analytical
and Bioanalytical Chemistry*, **387**,
1033–1039.

61 Shim, J.H., Jeong, I.S., Lee, M.H., Hong,
H.P., On, J.H., Kim, K.S., Kim, H.S., Kim,
B.H., Cha, G.S. and Nam, H. (2004) Ion-
selective electrodes based on molecular
tweezer-type neutral carriers. *Talanta*, **63**,
61–71.

62 Heng, L.Y. and Hall, E.A.H. (2000)
Producing self-plasticizing ion-selective
membranes. *Analytical Chemistry*, **72**,
42–51.

63 Heng, L.Y. and Hall, E.A.H. (2000) One-
step synthesis of K$^+$-selective
methacrylic-acrylic copolymers
containing grafted ionophore and
requiring no plasticizer. *Electroanalysis*,
12, 178–186.

64 Heng, L.Y. and Hall, E.A.H. (2000) Taking
the plasticizer out of methacrylic-acrylic
membranes for K$^+$-selective electrodes.
Electroanalysis, **12**, 187–193.

65 Malinowska, E., Gawart, L.,
Parzuchowski, P., Rokicki, G. and
Brzózka, Z. (2000) Novel approach of
immobilization of calix[4]arene type
ionophore in 'self-plasticized' polymeric
membrane. *Analytica Chimica Acta*, **421**,
93–101.

66 Qin, Y., Peper, S. and Bakker, E. (2002)
Plasticizer-free polymer membrane ion-
selective electrodes containing a
methacrylic copolymer matrix.
Electroanalysis, **14**, 1375–1381.

67 Bereczki, R., Gyurcsányi, R.E., Ágai, B.
and Tóth, K. (2005) Synthesis and
characterization of covalently
immobilized bis-crown ether based
potassium ionophore. *Analyst*, **130**,
63–70.

68 Jammal, A.E., Bouklouse, A.A.,
Patriarche, G.J. and Christian, G.D.
(1991) Use of ethylene-vinyl-acetate as a
new membrane matrix for calcium ion-
selective electrode preparation. *Talanta*,
38, 929–935.

69 Thomas, J.D.R. (1986) Solvent polymeric
membrane ion-selective electrodes.
Analytica Chimica Acta, **180**, 289–297.

70 Badawy, S.S., Youssef, A.F. and Mutair,
A.A. (2004) Construction and
performance characterization of ion-
selective electrodes for potentiometric
determination of phenylpropanolamine
hydrochloride applying batch and flow
injection analysis techniques. *Analytica
Chimica Acta*, **511**, 207–214.

71 Rizk, M.S., Abdel-Ghani, N.T. and El
Nashar, R.M. (2001) Construction and
performance characteristics of
terbutaline plastic membrane electrode in
batch and FIA conditions. *Microchemical
Journal*, **70**, 93–101.

72 Ibrahim, H. (2005) Chemically modified
carbon paste electrode for the
potentiometric flow injection analysis of
piribedil in pharmaceutical preparation
and urine. *Journal of Pharmaceutical and
Biomedical Analysis*, **38**, 624–632.

73 Badawy, S.S., Issa, Y.M. and Mutair, A.A.
(2005) PVC membrane ion-selective
electrodes for the determination of
hyoscyamine in pure solution and in
pharmaceutical preparations under batch
and flow modes. *Journal of Pharmaceutical
and Biomedical Analysis*, **39**, 117–124.

74 Sales, M.G.F., Tomás, J.F.C. and
Lavandeira, S.R. (2006) Flow injection
potentiometric determination of
chlorpromazine. *Journal of Pharmaceutical and Biomedical Analysis*, **41**,
1280–1286.

75 Hassan, S.S.M., Elnemma, E.M.,
Mahmoud, W.H. and Mohammed,
A.H.K. (2006) Continuous potentiometric
monitoring of viagra (sildenafil) in
pharmaceutical preparations using novel
membrane sensors. *Journal of Applied
Electrochemistry*, **36**, 139–146.

76 Hassan, S.S.M., Elnemma, E.M. and
Mohammed, A.H.K. (2005) Novel
biomedical sensors for flow injection
potentiometric determination of
creatinine in human serum. *Electroanalysis*, **17**, 2246–2253.

77 Ibrahim, H., Issa, Y.M. and Abu-Shawish, H.M. (2005) Potentiometric flow injection analysis of dicyclomine hydrochloride in serum, urine and milk. *Analytica Chimica Acta*, **532**, 79–88.

78 Ibrahim, H., Issa, Y.M. and Abu-Shawish, H.M. (2005) Potentiometric flow injection analysis of mebeverine hydrochloride in serum and urine. *Journal of Pharmaceutical and Biomedical Analysis*, **36**, 1053–1061.

79 El-Nashar, R.M., Abdel-Ghani, N.T. and Bioumy, A.A. (2004) Flow injection potentiometric determination of amitriptyline hydrochloride. *Microchemical Journal*, **78**, 107–113.

80 Sales, M.G.F. and Montenegro, M.C.B.S.M. (2001) Tetracycline-selective electrode for content determination and dissolution studies of pharmaceuticals by flow injection analysis. *Journal of Pharmaceutical Sciences*, **90**, 1125–1133.

81 Abdel-Ghani, N.T., Shoukry, A.F. and El-Nashar, R.M. (2001) Flow injection potentiometric determination of pipazethate hydrochloride. *Analyst*, **126**, 79–85.

82 Martinez-Barrachina, S., del Valle, M., Matia, L., Prats, R. and Alonso, J. (2001) Potentiometric flow injection system for the determination of polyethoxylate nonionic surfactants using tubular ion-selective electrodes. *Analytica Chimica Acta*, **438**, 305–313.

83 Hassan, S.S.M. and Rizk, N.M.H. (2000) Potentiometric dipyridamole sensors based on lipophilic ion pair complexes and native ionic polymer membranes. *Analytical Letters*, **33**, 1037–1055.

84 Abdel-Ghani, N.T. and Hussein, S.H. (2003) Determination of ambroxol hydrochloride in pure solutions and some of its pharmaceutical preparations under batch and FIA conditions. *Il Farmaco*, **58**, 581–589.

85 Sales, M.G.F., Lino, N.F.M.C. and Paiga, P.C.B. (2003) Chlormequat selective electrodes: Construction, evaluation and application at FIA systems. *International Journal of Environmental Analytical Chemistry*, **83**, 295–305.

86 Sales, M.G.F., Pille, A. and Paigu, P.C.B. (2003) Construction and evaluation of cysteine selective electrodes for FIA analysis of pharmaceuticals. *Analytical Letters*, **36**, 2925–2940.

87 Abdel-Ghani, N.T., Rizk, M.S. and El-Nashar, R.M. (2002) Potentiometric flow injection determination of salbutamol. *Analytical Letters*, **35**, 39–52.

88 Issa, Y.M., Badawy, S.S. and Mutair, A.A. (2005) Ion-selective electrodes for potentiometric determination of ranitidine hydrochloride, applying batch and flow injection analysis techniques. *Analytical Sciences*, **21**, 1443–1448.

89 Ibrahim, H., Issa, Y.M. and Abu-Shawish, H.M. (2005) Potentiometric flow injection analysis of drotaverine hydrochloride in pharmaceutical preparations. *Analytical Letters*, **38**, 111–132.

90 Issa, Y.M., Abdel-Ghani, N.T., Shoukry, A.F. and Ahmed, H.M. (2005) New conventional coated-wire ion-selective electrodes for flow injection potentiometric determination of chlordiazepoxide. *Analytical Sciences*, **21**, 1037–1042.

91 Pimenta, A.M., Araújo, A.N., Montenegro, M.C.B.S.M., Pasquini, C., Rohwedder, J.J.R. and Raimundo, I.M., Jr (2004) Chloride-selective membrane electrodes and optodes based on an indium(III) porphyrin for the determination of chloride in a sequential injection analysis system. *Journal of Pharmaceutical and Biomedical Analysis*, **36**, 49–55.

92 Pimenta, A.M., Araújo, A.N. and Montenegro, M.C.B.S.M. (2002) Simultaneous potentiometric and fluorimetric determination of diclofenac in a sequential injection analysis system. *Analytica Chimica Acta*, **470**, 185–194.

93 Garcia, C.A., Júnior, L.R. and Neto, G.O. (2003) Determination of potassium ions in pharmaceutical samples by FIA using a potentiometric electrode based on ionophore nonactin occluded in EVA membrane. *Journal of Pharmaceutical and Biomedical Analysis*, **31**, 11–18.

94 Santos, E.M.G., Araújo, A.N., Couto, C.M.C.M. and Montenegro, M.C.B.S.M. (2005) Construction and evaluation of PVC and sol-gel sensor membranes based on Mn(III)TPP-Cl. Application to valproate determination in pharmaceutical H. Kahlert, J.R. Porksen, J. Behnert, F. Scholz, FIA acid-base titrations with a new flow-through pH detector. *Analytical and Bioanalytical Chemistry*, **382**, 1981–1986.

95 Farrell, J.R., Iles, P.J. and Dimitrakopoulos, T. (1996) Photocured polymers in ion-selective electrode membranes. Part 5: Photopolymerised sodium sensitive ion-selective electrodes for flow injection potentiometry. *Analytica Chimica Acta*, **334**, 133–137.

96 Farrell, J.R., Iles, P.J. and Dimitrakopoulos, T. (1996) Photocured polymers in ion-selective electrode membranes Part 6; Photopolymerized lithium sensitive ion-selective electrodes for flow injection potentiometry. *Analytica Chimica. Acta*, **335**, 111–116.

97 Alexander, P.W., Dimitrakopoulos, T. and Hibbert, D.B. (1997) A photo-cured coated-wire calcium ion selective electrode for use in flow injection potentiometry. *Talanta*, **44**, 1397–1405.

98 Edmonds, T.E. and Coutts, G. (1983) Flow injection analysis system for determining soil pH. *Analyst*, **108**, 1013–1017.

99 Hongbo, C., Ruzicka, J. and Hansen, E.H. (1985) Evaluation of critical parameters for measurement of pH by flow injection analysis determination of pH in soil extracts. *Analytica Chimica. Acta*, **169**, 209–220.

100 del Mundo, F.R., Cardwell, T.J., Catrall, R.W., Iles, P.J. and Hamilton, I.C. (1989) An ultraviolet-cured, pH-sensitive membrane electrode for use in flow injection analysis. *Electroanalysis*, **1**, 353–356.

101 Marzouk, S.A.M. (2003) Improved Electrodeposited Iridium Oxide pH Sensor Fabricated on Etched Titanium Substrates. *Analytical Chemistry*, **75**, 1258–1266.

102 Hassan, S.S.M., Marzouk, S.A.M. and Badawy, N.M. (2002) Solid State Iridium Oxide-Titanium Based Sensor for Flow Injection pH Measurements. *Analytical Letters*, **35**, 1301–1311.

103 de Roiij, N.F., and Vlekkert, H.H. (1991) Microstructured ISFETs. *Chemical Sensor Technology*, **3**, 213–231.

104 Zampronio, C.G., Rohwedder, J.J.R. and Poppi, R.J. (2000) Development of a potentiometric flow cell with a stainless steel electrode for pH measurements. Determination of acid mixtures using flow injection analysis. *Talanta*, **51**, 1163–1169.

105 Kahlert, H., Porksen, J.R., Behnert, J. and Scholz, F. (2005) FIA acid-base titrations with a new flow-through pH detector. *Analytical and Bioanalytical Chemistry*, **382**, 1981–1986.

106 Kahlert, H., Porksen, J.R., Isildak, I., Ardac, M., Yolau, M., Behnert, J. and Scholz, F. (2005) Application of a new pH-sensitive electrode as a detector in flow injection potentiometry. *Electroanalysis*, **17**, 1085–1090.

107 Teixeira, M.F.S., Ramos, L.A., Cassiano, N.M., Fatibello-Filho, O. and Bocchi, N. (2000) Evaluation of a Fe_2O_3-based graphite-epoxy tubular electrode as pH sensor in flow injection potentiometry. *Journal of the Brazilian Chemical Society*, **11**, 27–31.

108 Nagels, L.J. (2004) Potentiometric detection for high-performance liquid chromatography is a reality: Which classes of organic substances are the targets? *Pure and Applied Chemistry*, **76**, 839–845.

109 Bohets, H., Vanhoutte, K., de Maesschalck, R., Cockaerts, P., Vissers, B. and Nagels, L.G. (2007) Development of *in situ* ion selective sensors for dissolution. *Analytica Chimica Acta*, **581**, 181–191.

110 Prieto-Simon, B., Campas, M., Andreescu, S. and Marty, J.L. (2006) Trends in flow-based biosensing systems for pesticide assessment. *Sensors*, **6**, 1161–1186.

111 Amorim, C.G., Araújo, A.N. and Montenegro, M.C.B.S.M. (2007) Exploiting sequential injection analysis with lab-on-valve and miniaturized potentiometric detection: epinephrine determination in pharmaceutical products. *Talanta*, **72**, 1255–1260.

112 Nagels, L.G., Bazylak, G. and Zielinska, D. (2003) Designing potentiometric sensor materials for the determination of organic ionizable substances in HPLC. *Electroanalysis*, **15**, 533–538.

113 Nagels, L.G. and Poels, I. (2000) Solid state potentiometric detection systems for LC, CE and MTAS methods. *Trends in Analytical Chemistry*, **19**, 410–417.

114 Bazylak, G. and Nagels, L.J. (2002) Potentiometric detection of N,N'-diethylaminoethanol and lysosomotropic amino alcohols in cation exchange high-performance liquid chromatography systems. *Analytica Chimica Acta*, **472**, 11–26.

115 Poels, L. and Nagels, L.J. (2001) Potentiometric detection of amines in ion chromatography using macrocycle-based liquid membrane electrodes. *Analytica Chimica Acta*, **440**, 89–98.

116 Bazylak, G. and Nagels, L.J. (2002) Potentiometric detection of exogenic beta-adrenergic substances in liquid chromatography. *Journal of Chromatography. A*, **973**, 85–96.

117 Bazylak, G. and Nagels, L.J. (2003) A novel potentiometric approach for detection of beta-adrenergics and beta-adrenolytics in high-performance liquid chromatography. *Il Farmaco*, **58**, 591–603.

118 Picioreanu, S., Poels, I., Frank, J., van Dam, J.C., van Dedem, G.W.K. and Nagels, L.G. (2000) Potentiometric detection of carboxylic acids, phosphate esters, and nucleotides in liquid chromatography using anion-selective coated-wire electrodes. *Analytical Chemistry*, **72**, 2029–2034.

119 Poels, I., Nagels, L.G., Verreyt, G. and Geise, H.J. (1998) Conducting polymer based potentiometric detection applied to the determination of organic acids with narrow-bore LC systems. *Biomedical Chromatography*, **12**, 124–125.

120 Poels, I., Schasfoort, R.B.M., Picioreanu, P., Frank, J., van Dedem, G.W.K., van den Berg, A. and Nagels, L.G. (2000) An ISFET-based anion sensor for the potentiometric detection of organic acids in liquid chromatography. *Sensors and Actuators B*, **67**, 294–299.

121 Zielinska, D., Poels, I., Pietraszkiewicz, M., Radecki, J., Geise, H.J. and Nagels, L.J. (2001) Potentiometric detection of organic acids in liquid chromatography using polymeric liquid membrane electrodes incorporating macrocyclic hexaamines. *Journal of Chromatography. A*, **915**, 25–33.

122 Bao, Y., Everaert, J., Pietraszkiewicz, M., Pietraszkiewicz, O., Bohets, H., Geise, H.J., Peng, B.X. and Nagels, L.J. (2003) Behaviour of nucleotides and oligonucleotides in potentiometric HPLC detection. *Analytica Chimica Acta*, **550**, 130–136.

123 Toth, K., Stulik, K., Ktner, W., Feher, Z. and Lindner, E. (2004) Electrochemical detection in liquid flow analytical techniques: Characterization and classification (IUPAC Technical Report). *Pure and Applied Chemistry*, **76**, 1119–1138.

124 Pungor, E. (1998) The theory of ion-selective electrodes. *Analytical Sciences*, **14**, 249–256.

125 Gyurcsányi, R., Rangisetty, N., Clifton, S., Pendley, B.D. and Lindner, E. (2004) Microfabricated ISEs: critical comparison of inherently conducting polymer and hydrogel based inner contacts. *Talanta*, **63**, 89–99.

126 Gyurcsányi, R., Nyback, A.S., Toth, K., Nagy, G. and Ivaska, A. (1998) Novel polypyrrole based all-solid-state potassium-selective microelectrodes. *Analyst*, **123**, 1339–1344.

127 Tymecki, L., Glab, S. and Koncki, R. (2006) Miniaturized, planar ion-selective electrodes fabricated by means of thick-film technology. *Sensors*, **6**, 390–396.

128 Bohm, S., Olthius, W. and Bergveld, P. (2000) A generic design of a flow-through potentiometric sensor array. *Mikrochimica Acta*, **134**, 237–243.

129 Hassan, S.S.M., Sayour, H.E.M. and al-Mehrezi, S.S. (2007) A novel planar miniaturized potentiometric sensor for flow injection analysis of nitrates in wastewaters, fertilizers and pharmaceuticals. *Analytica Chimica Acta*, **581**, 13–18.

130 Toczylowska, R., Pokrop, R., Dybko, A. and Wróblewski, W. (2005) Planar potentiometric sensors based on Au and Ag microelectrodes and conducting polymers for flow-cell analysis. *Analytica Chimica Acta*, **540**, 167–172.

131 Ciosek, P., Brzózka, Z. and Wróblewski, W. (2006) Electronic tongue for flow-through analysis of beverages. *Sensors and Actuators B*, **118**, 454–460.

132 Tymecki, L., Zwierkowska, E. and Koncki, R. (2004) Screen-printed reference electrodes for potentiometric measurements. *Analytica Chimica Acta*, **526**, 3–11.

133 Yoon, H.Y., Shin, J.H., Lee, S.D., Nam, H., Cha, G.S., Strong, T.D. and Brown, R.B. (2000) Solid-state ion sensors with a liquid junction-free polymer membrane-based reference electrode for blood analysis. *Sensors and Actuators B*, **64**, 8–14.

134 Ibanez-Garcia, N., Mercader, M.B., Rocha, Z.M., Seabra, C.A., Gongora-Rubio, M.R. and Chamarro, J.A. (2006) Continuous flow analytical microsystems based on low-temperature co-fired ceramic technology. Integrated potentiometric detection based on solvent polymeric ion-selective electrodes. *Analytical Chemistry*, **78**, 2985–2992.

135 Watanabe, A., Kayanne, H., Nozaki, K., Kato, K., Negishi, A., Kudo, S., Kimoto,

H., Tsuda, M. and Dickson, A.G. (2004) A rapid, precise potentiometric determination of total alkalinity in seawater by a newly developed flow-through analyzer designed for coastal regions. *Marine Chemistry*, **85**, 75–87.

136 Dantan, N., Kroning, S., Frenzel, W. and Kuppers, S. (2000) Comparison of spectrophotometric and potentiometric detection for the determination of water using Karl Fischer method under flow injection analysis conditions. *Analytica Chimica Acta*, **420**, 133–142.

137 Ohura, H., Imato, T., Kameda, K. and Yamasaki, S. (2004) Potentiometric determination of bromate using an Fe(III)-Fe(II) potential buffer by circulatory flow injection analysis. *Analytical Sciences*, **20**, 513–518.

138 van Staden, J.F., Mashamba, M.G. and Stefan, R.-I. (2002) An on-line potentiometric sequential injection titration process analyser for the determination of acetic acid. *Analytical and Bioanalytical Chemistry*, **374**, 141–144.

139 Paseková, H., Sales, M.G.F., Montenegro, M.C.B.S.M., Araújo, A.N. and Polásek, M. (2001) Potentiometric determination of acetylsalicylic acid by sequential injection analysis (SIA) using a tubular salicylate-selective electrode. *Journal of Pharmaceutical and Biomedical Analysis*, **24**, 1027–1036.

140 van Staden, J.F., Mashamba, M.G. and Stefan, R.-I. (2005) Determination of the total acidity in soft drinks using potentiometric sequential injection titration. *Talanta*, **58**, 1109–1114.

141 Cortina, M., Gutés, A., Alegret, S. and del Valle, M. (2005) Sequential injection system with higher dimensional electrochemical sensor signals. Part 2. Potentiometric e-tongue for the determination of alkaline ions. *Talanta*, **66**, 1197–1206.

142 Calvo, D., Grossl, M., Cortina, M. and Del Valle, M. (2007) Automated SIA system using an array of potentiometric sensors

for determining alkaline-earth ions in water. *Electroanalysis*, **19**, 644–651.

143 Cortina, M., Duran, A., Alegret, S. and del Valle, M. (2006) A sequential injection electronic tongue employing the transient response from potentiometric sensors for anion multidetermination. *Analytical and Bioanalytical Chemistry*, **385**, 1186–1194.

144 Pimenta, A.M., Araújo, A.N. and Montenegro, M.C.B.S.M. (2001) Sequential injection analysis of captopril based on colorimetric and potentiometric detection. *Analytica Chimica Acta*, **438**, 31–38.

145 Silva, J.E., Pimentel, M.F., Silva, V.L., Montenegro, M.C.B.S.M. and Araújo, A.N. (2004) Simultaneous determination of pH, chloride and nickel in electro-plating baths using sequential injection analysis. *Analytica Chimica Acta*, **506**, 197–202.

146 Reis-Lima, M.J., Fernandes, S.M.V. and Rangel, A.O.S.S. (2004) Sequential injection titration of chloride in milk with potentiometric detection. *Food Control*, **15**, 609.

147 Vieira, J.A., Raimundo, I.M., Jr, Reis, B.F., Montenegro, M.C.B.S.M. and Araújo, A.N. (2003) Monosegemented flow potentiometric titration for the determination of chloride in milk and wine. *Journal of the Brazilian Chemical Society*, **14**, 259–264.

148 Andrade-Eiroa, A., Erustes, J.A., Forteza, R., Cerda, V. and Lima, J.L.F.C. (2002) Determination of chloride by multisyringe flow injection analysis and sequential injection analysis with potentiometric detection. *Analytica Chimica Acta*, **467**, 25–33.

149 Santos, E., Montenegro, M.C.B.S.M., Couto, C., Araújo, A.N., Pimentel, M.F. and Silva, V.L. (2004) Sequential injection analysis of chloride and nitrate in waters with improved accuracy using potentiometric detection. *Talanta*, **63**, 721–727.

150 Jakmunee, J., Patimapornlert, L., Suteerapataranon, S., Lenghor, N. and Grudpan, K. (2005) Sequential injection with lab-at-valve (LAV) approach for potentiometric determination of chloride. *Talanta*, **65**, 789–793.

151 Pimenta, A.M., Araújo, A.N. and Montenegro, M.C.B.S.M. (2002) A sequential injection analysis system for potassium clavulanate determination using two potentiometric detectors. *Journal of Pharmaceutical and Biomedical Analysis*, **30**, 931–937.

152 van Staden, J.F., Stefan, R.-I. and Birghila, S. (2000) Evaluation of different SIA sample-buffer configurations using a fluoride-selective membrane electrode as detector. *Talanta*, **52**, 3–11.

153 Santos, E.M.G., Couto, C.M.C.M., Araújo, A.N., Montenegro, M.C.B.S. and Reis, B.F. (2004) Determination of Gibberellic acid by sequential injection analysis using a potentiometric detector based on Mn(III)-Porphyrin with improved characteristics. *Journal of the Brazilian Chemical Society*, **15**, 701–707.

154 van Staden, J.F., Mashamba, M.G. and Stefan, R.-I. (2002) On-line determination of hydrochloric acid in process effluent streams by potentiometric sequential injection acid-base titration. *South African Journal of Chemistry*, **55**, 43–55.

155 van Staden, J.F., Mashamba, M.G. and Stefan, R.-I. (2002) On-line dilution and determination of the amount of concentrated hydrochloric acid in the final products from a hydrochloric acid production plant using a sequential injection titration system. *Talanta*, **58**, 1089–1094.

156 Silva, F.V., Nogueira, A.R.A., Souza, G.B. and Cruz, G.M. (2000) Determination of total, volatile and acid detergent insoluble nitrogen in silage by sequential injection. *Analytical Sciences*, **16**, 361–364.

157 Santos, E.M.G., Araújo, A.N., Couto, C.M.C.M., Montenegro, M.C.B.S.M., Kejzarova, A. and Solich, P. (2004) Ion selective electrodes for penicillin-G based on Mn(III)TPP-Cl and their application in pharmaceutical formulations control by sequential injection analysis. *Journal of Pharmaceutical and Biomedical Analysis*, **36**, 701–709.

158 Zárate, N., Araújo, A.N., Montenegro, M.C.B.S.M. and Pérez-Olmos, R. (2003) Sequential injection analysis of free and total potassium in wines using potentiometric detection and microwave digestion. *American Journal of Enology and Viticulture*, **54**, 46–49.

159 Fernandes, R.N., Sales, M.G.F., Reis, B.F., Zagatto, E.A.G., Araújo, A.N. and Montenegro, M.C.B.S.M. (2001) Multi-task flow system for potentiometric analysis: its application to the determination of vitamin B-6 in pharmaceuticals. *Journal of Pharmaceutical and Biomedical Analysis*, **25**, 713–720.

160 Silva, F.V., Souza, G.B. and Nogueira, A.R.A. (2001) Use of yeast crude extract for sequential injection determination of carbohydrates. *Analytical Letters*, **34**, 1377–1388.

161 Silva, F.V., Nogueira, A.R.A., Souza, G.B., Reis, B.F., Araujo, A.N., Montenegro, M.C.B.S.M. and Lima, J.L.F.C. (2000) Potentiometric determination of urea by sequential injection using Jack bean meal crude extract as a source of urease. *Talanta*, **53**, 331.

162 Perez-Olmos, R., Soto, J.C., Zarate, N., Araújo, A.N., Lima, J.L.F.C. and Saraiva, M.L.M.F.S. (2005) Application of sequential injection analysis (SIA) to food analysis. *Food Chemistry*, **90**, 471–490.

163 Perez-Olmos, R., Soto, J.C., Zarate, N., Araújo, A.N., Lima, J.L.F.C. and Montenegro, M.C.B.S.M. (2005) Sequential injection analysis using electrochemical detection: A review. *Analytica Chimica Acta*, **554**, 1–16.

164 Pimenta, A.M., Araújo, A.N., Montenegro, M.C.B.S.M. and Calatayud, J.M. (2006) Application of sequential injection analysis to pharmaceutical analysis. *J. Pharmaceutical and Biomedical Analysis*, **40**, 16–34.

165 Kikas, T. and Ivaska, A. (2007) Potentiometric measurements in sequential injection analysis lab-on-valve (SIA-LOV) flow-system. *Talanta*, **71**, 160–164.

166 Dorneanu, S.A. Coman, V., Popescu, I.C. and Fabry, P. (2005) Computer-controlled system for ISEs automatic calibration. *Sensors and Actuators B*, **105**, 521–531.

167 Araújo, A.N., Montenegro, M.C.B.S.M., Kousalova, L., Skleranova, H., Solich, P. and Perez-Olmos, R. (2004) Sequential injection system for simultaneous determination of chloride and iodide by a Gran's plot method. *Analytica Chimica Acta*, **505**, 161–166.

168 Oliveira, C.C., Sartini, R.P., Zagatto, E.A.G. and Lima, J.L.F.C. (1997) Flow analysis with accuracy assessment. *Analytica Chimica Acta*, **350**, 31–36.

169 Duran, A., Cortina, M., Velasco, L., Rodriguez, J.A., Alegret, S. and del Valle, M. (2006) Virtual instrument for an automated potentiometric e-tongue employing the SIA technique. *Sensors*, **6**, 19–24.

170 Gallardo, J., Alegret, S., de-Roman, M.A., Munoz, R., Hernandez, P.R., Leija, L. and del Valle, M. (2003) Determination of ammonium ion employing an electronic tongue based on potentiometric sensors. *Analytical Letters*, **36**, 2893–2908.

171 Gallardo, J., Alegret, S., Munoz, R., de-Roman, M., Leija, L., Hernandez, P.R. and del Valle, M. (2003) An electronic tongue using potentiometric all-solid-state PVC-membrane sensors for the simultaneous quantification of ammonium and potassium ions in water. *Analytical and Bioanalytical Chemistry*, **377**, 248–256.

172 Gallardo, J., Alegret, S. and del Valle, M. (2005) Application of a potentiometric electronic tongue as a classification tool in food analysis. *Talanta*, **66**, 1303–1309.

173 Gallardo, J., Alegret, S., Munoz, R., Leija, L., Hernandez, P.R. and del Valle, M. (2005) Use of an electronic tongue based on all-solid-state potentiometric sensors

for the quantitation of alkaline ions. *Electroanalysis*, **17**, 348–355.

174 Gutierrez, M., Alegret, S. and del Valle, M. (2007) Potentiometric bioelectronic tongue for the analysis of urea and alkaline ions in clinical samples. *Biosensors & Bioelectronics*, **22**, 2171–2178.

175 Kahlert, H., Porksen, J.R., Behnert, J. and Scholz, F. (2005) FIA acid-base titrations with a new flow-through pH detector. *Analytical and Bioanalytical Chemistry*, **382**, 1981–1986.

176 Dasgupta, P.K., Tanaka, H. and Jo, K.D. (2001) Continuous on-line true titrations by feedback based flow ratiometry: application to potentiometric acid-base titrations. *Analytica Chimica Acta*, **435**, 289–297.

177 Borges, E.P., Martelli, P.B. and Reis, B.F. (2000) Automatic stepwise potentiometric titration in a monosegmented flow system. *Mikrochimica Acta*, **135**, 179–184.

178 Almeida, C.M.V., Lapa, R.A.S. and Lima, J.L.F.C. (2001) Automatic flow titrator based on a multicommutated unsegmented flow system for alkalinity monitoring in wastewaters. *Analytica Chimica Acta*, **438**, 291–298.

179 Paim, A.P.S., Almeida, C.M.N.V., Reis, B.F., Lapa, R.A.S., Zagatto, E.A.G. and Lima, J.L.F.C. (2002) Automatic potentiometric flow titration procedure for ascorbic acid determination in pharmaceutical formulations. *Journal of Pharmaceutical and Biomedical Analysis*, **28**, 1221–1225.

180 Aquino, E.V., Rohwedder, J.J.R. and Pasquini, C. (2006) A new approach to flow-batch titration. A monosegmented flow titrator with coulometric reagent generation and potentiometric or biamperometric detection. *Analytical and Bioanalytical Chemistry*, **386**, 1921–1930.

181 Koncki, R., Radomska, A. and Glab, S. (2000) Bioanalytical flow injection system for control of hemodialysis adequacy. *Analytica Chimica Acta*, **418**, 213–224.

182 Radomska, A., Koncki, R., Pyrzynska, K. and Glab, S. (2004) Bioanalytical system for control of hemodialysis treatment based on potentiometric biosensors for urea and creatinine. *Analytica Chimica Acta*, **523**, 193–200.

183 Itai, K. and Tsunoda, H. (2001) Highly sensitive and rapid method for determination of fluoride ion concentrations in serum and urine using flow injection analysis with a fluoride ion-selective electrode. *Clinica Chimica Acta*, **308**, 163–171.

184 Hosseini, M.S. and Rahiminegad, H. (2006) Potentiometric determination of ultratrace amounts of fluoride enriched by zirconia in a flow system. *Journal of Analytical Chemistry*, **6**, 166–171.

185 Rodriguez, J.A., Barrado, E., Vega, M. and Lima, J.L.F.C. (2005) Speciation of inorganic arsenic in waters by potentiometric flow analysis with on-line preconcentration. *Electroanalysis*, **17**, 504–511.

186 Martinez-Barrachina, S., del Valle, M., Matia, L., Pratz, R. and Alonso, J. (2002) Determination of polyethoxylated non-ionic surfactants using potentiometric flow injection systems. Improvement of the detection limits employing an on-line pre-concentration stage. *Analytica Chimica Acta*, **454**, 217–227.

187 Jones, D.L., Moody, G.J. and Thomas, J.D.R. (1981) Potentiometry of alkoxylates. *Analyst*, **106**, 439–447.

188 Martinez-Barrachina, S. and del Valle, M. (2006) Use of a solid-phase extraction disk module in a FI system for the automated preconcentration and determination of surfactants using potentiometric detection. *Microchemical Journal*, **83**, 48–54.

16

Flow Voltammetry

Ivano G.R. Gutz, Lúcio Angnes, and Andrea Cavicchioli

16.1
Introduction

Electrochemistry gives flow analysis a large assortment of versatile detectors, from universal transducers to quite selective ones, capable of measuring concentrations or activities, evaluating metal lability, providing redox state speciation, preconcentrating analytes on the detector, investigating adsorption or even carrying out molecular recognition. Currently, many electrochemical detectors (ECDs) are marketed ready for use in FA or can be easily configured or adapted thereto from cells, electrodes, and instruments produced for general use or for HPLC. ECDs can also be constructed in the laboratory, as exemplified in this chapter with cells developed by the authors – including microfluidic ones.

Advanced ECDs fit perfectly the time–size–flow scales involved in conventional FA and micro-FA, as well as HPLC and CE. They present low dead-volume and fast response, require minute amounts of sample and cover various orders of magnitude of analyte concentration (up to twelve), with detection limits ranging between 10^{-4} and $10^{-14}\,\text{mol}\,\text{L}^{-1}$ or amounts down to femtograms. The electrodes plug into low-power solid-state instruments, based on modern analog and digital integrated circuits that render them simple, dependable, rugged, compact, portable, cheap, and directly interfaceable to computers. Compared with spectrometric FA detectors, ECDs are free of optical components, flames, torches, ovens, pumps, vacuum systems or any moving mechanical parts. Most ECDs present moderate selectivity and work undisturbed by colored, turbid or saline matrices. However, the direct physical contact of the electrodes with the sample matrix, reagents, and products makes virtually all ECDs – including the voltammetric ones – vulnerable to gradual inactivation or alteration of the response to the analyte. This drawback is mitigated in FA because exposure times are shorter than with batch operation, and compensated by recalibration up to the point where sensitivity loss makes electrode reactivation mandatory. Therefore, some of the advances produced by the most recent research efforts are mainly related to the improvement of the endurance of the

Advances in Flow Analysis. Edited by Marek Trojanowicz
Copyright © 2008 WILEY-VCH Verlag GmbH & Co. KGaA, Weinheim
ISBN: 978-3-527-31830-8

sensor's response (e.g., new or modified electrode materials, *in situ* reactivation by electrochemical or other means, shielding by membranes, sample pretreatment, disposable electrodes, etc.). With modified electrodes, especially amperometric biosensors, shelf life and leaching during operation are additional problems that need to be tackled. Advances continue to emerge in terms of selectivity, sensitivity, application spectrum and, for voltammetric detectors, widening the potential domain, conversion of batch voltammetric stripping methods to FA, extended use of techniques like square wave voltammetry and multi-pulse amperometry, development of simultaneous determinations, coupling of ECDs with other techniques, study of certain adsorption processes at electrodes approachable only with FA, and development of microfluidic systems.

However, electrochemistry can provide far more than detectors to FA. In-flow electrochemical operations include: preconcentration of analytes (by electroplating, adsorption, migration, amalgamation), purification of the carrier or sample (removal of electroactive interferents), phase transfer promotion, organic material destruction (oxidation, mineralization or "burning" at diamond electrode), surface polarity tuning, ion transport by migration, fluid displacement by electroosmotic pumping (described in detail in Chapter 5), sampling by electrodissolution of metals and alloys, coulometric generation of reagents (including unstable ones), electropolymerization, organic electrosynthesis, electrochemiluminescence, photoelectrocatalysis, spectroelectrochemistry, and so forth. Microfluidics (see Chapter 6) opens additional research opportunities to elegantly associate this rich repertoire of electrochemical mass transport and chemical transformation tools with ECDs and non-electrochemical detectors.

In order to illustrate the trends and advances in ECD-FA, a profuse, though not exhaustive, compilation of representative examples published mainly during the last decade have been included in this chapter, conceived for experienced users and newcomers to the field. For sake of brevity, heavy theoretical discussions and equations have been omitted, nonetheless these are available in the cited references.

16.2
Voltammetric/Amperometric Flow Analysis

16.2.1
Principles and Techniques

The foundations of electrochemistry were solidly established over the last two centuries and the theory, instrumentation, experimental practice, and application of electroanalytical chemistry are set down in an authoritative series of over 20 monographs [1], a recent handbook [2] and numerous books (e.g., [3–8]). Regarding FA with ECD, the most comprehensive treatise is still a book published in 1987, *Electroanalytical Measurements in Flowing Liquids* [9], covering principles, cell design, features, and applications. Meantime, more specific works on ECDs for one or another form of FA have appeared as book chapters [10–12] and reviews [13–15]. The ample literature on ECDs for chromatography [16–18] and capillary electrophoresis [19] is valuable for FA as well. Reliable tutorials on voltammetric and amperometric ECDs with flow cells, helpful for newcomers and practitioners, are available on the WWW (e.g., [20, 21]).

In flow analysis as a whole, detectors involving electrolysis, like the amperometric and voltammetric systems, are more disseminated than those free of electrolysis, mainly the potentiometric and conductometric ones. Activity sensing using potentiometric detectors (described in Chapter 15) relies on the measurement of the equilibrium potential of indicator electrodes (e.g., ion-selective electrodes) at essentially null current. In conductometric detection (extensively used as a flow detector in ion chromatography) electrolysis is circumvented during ion displacement under the effect of an electric field by alternating the excitation signal. At frequencies around one megahertz, direct contact of the electrodes with the sample is no longer required, due to capacitive coupling. This property allowed the recent introduction and growth of contactless conductivity detection in capillary electrophoresis by simply fitting twin metal rings around the capillary [22–24]. Contactless conductivity detection is of interest in FA as long as selectivity is assured by chemical means or in some other way [25]. Truly contactless detection cannot be extended to amperometric (or potentiometric) sensors but direct exposure to the whole sample matrix can sometimes be avoided, for example, for volatile or gaseous species, by interposition of a hydrophobic membrane as in the Clark cell, a very popular amperometric oxygen sensor applicable in FA.

Cyclic voltammetry with continuous (or sampled) current measurement during a linear (or staircase) scan of the difference in potential at a working electrode/solution interface under stationary conditions, with scan direction reversal at a defined potential (triangular wave), is used universally for redox characterization of ions and molecules. The resulting current–voltage curves, recorded at various scan rates ($v = dE/dt$), can provide information about the number of heterogeneous charge transfer processes, their conditional standard potentials, $E^{0'}$, the kinetics of the electron transfer itself and of coupled chemical reactions preceding or following the redox process, as well as on adsorption of reactants and/or products at the electrode surface [5]. The stability and lability of metal complexes and the prevailing oxidation state of an electroactive species in a sample (speciation) can also be evaluated voltammetrically.

Besides the faradaic current (i_f) flowing to or from the electrode to promote anodic (i_{an}) or cathodic (i_{cat}) processes, any change in the applied potential during voltammetry is accompanied by a capacitive current (i_c) that makes up for the charge spent for the rearrangement of neutral and charged chemical species in the diffuse and compact regions of the electric double layer at the electrode, including adsorption/desorption. The i_c is an important contributor to the background or residual current (i_r) that typically also includes a faradaic component, originated by traces of electroactive impurities in the solution, electrolyte or electrode, or surface oxidation/reduction of the electrode material itself.

The potential range available for voltammetry in the positive and negative directions is limited by the regions where exponential growth of i_r erupts as a result of massive oxidation of the electrode material, the electrolysis of the solvent or also of a component of the sample matrix or electrolyte added to handle the migration current and to adjust the medium. Conditions must be chosen so that the analyte (or a species generated or consumed by the analyte) is electroactive at a potential within this "clean" range.

The richness of cyclic voltammetry (CV) as a tool for thermodynamic, mechanistic, kinetic, and mass transport studies stems from the interplay of many factors that may pose difficulties in the selection of optimum operational conditions for establishing new ECD-FA methods, and background knowledge in electrochemistry is of value. For FIA, the CV scan must be synchronous with the injection and sufficiently fast to be executed during the passage of the analyte plug through the flow cell. However, the higher v the more significant is i_r even for sampled current modes, to the detriment of the detection limit. The advantages of voltammetry over amperometry at a fixed (or pulsed) potential, more widely used in FA, include multi-analyte quantification and better interference diagnosis/immunity (like FA with multi-λ versus single-λ spectrophotometric detection). A fast scan technique that is gaining acceptance in batch voltammetry and is promising but not much used in FA yet is square wave voltammetry (SWV) [6]. It resembles differential pulse voltammetry (DPV) in that, instead of sigmoid waves, peak shaped signals are recorded at the $E^{0'}$ of the analytes. But unlike DPV, scan rates are much higher and the recorded i values refer to $i_{an} - i_{cat}$ measured by double current sampling during each pulse (at the top and base potentials, after a short decay of i_c). On the other hand, SWV is strongly affected by the reversibility of the electrode processes.

All staircase and pulse voltammetric techniques, including DPV and SWV, explore the faster decay of the unwanted i_c, a component of i_r, over i_f, dictated by mass transport [11]. The interference is thus reduced by allowing some milliseconds before sampling the current after each increment or pulse of the potential. Nonetheless, with solid and paste electrodes, the minimization of i_r takes longer than with liquid mercury, so that the best signal to background current ratio in FA is reached only at constant potential operation. This behavior and simplicity are reasons why, whenever a single electroactive component with a well-defined and interference-free voltammetric wave is to be quantified by FA, the technique of choice, as a rule, is DC amperometric detection (or pulse amperometry). The potential (or pulse width) is simply "frozen" at a value judiciously chosen in the voltammogram, while the current is monitored. Once the high i_r observed when switching on the system decays to a low and stable baseline, determinations can start; peak-shaped i_f records are obtained, similar to the absorbance peaks observed at single wavelength detection, with heights directly proportional to concentration (area integration can be used instead, but this is unusual). The (pulsed) amperometric operation mode is, thus, highly favorable for FIA, BIA, SIA, continuous FA and HPLC in terms of detection limit [16].

Simultaneous amperometric determination of multiple electroactive analytes requires various detectors in series at increasingly positive (or negative) potentials. In amperometric measurements, only a small fraction of the electroactive analyte is electrolyzed. Consequently, the most easily oxidizable (or reducible) species will produce a current not only at the first but at all working electrodes, and the algebraic subtraction of contributions may jeopardize accuracy. Coulometric cells with a high ratio of electrode area to sample flow (porous or tubular working electrodes or certain low-flow thin-layer geometries), overcome this problem by depleting one species at each detection stage.

Although not often used in FA, polarographic detection is an interesting alternative, at least for systems presenting severe "electrode fouling" problems. Polarography is the only technique that employs, as the electrode, a mercury drop that is frequently renewed (every 0.3 to 3 s) during each linear, staircase or pulsed potential scan. With modern valved automatic static mercury drop electrodes, the amount of (recyclable) mercury for a scan is of the order of 50 drops of 1 μL each. Drop renewal causes enough convection to homogenize the solution so that forward and reverse scans define the same sigmoid curve for an electroactive species. There are thousands of polarographic (batch) methods in the literature, adaptable to FA with the help of a flow adaptor that fits to commercial mercury drop electrodes, as will be seen.

As mentioned, flow voltammetry and amperometry benefit from convective transport of the analyte to the electrode (and also the removal of soluble product), reducing the diffusion boundary layer thickness with consequent increase in i_f. This is why sigmoidal curves (like polarographic ones) are also obtained with constant area electrodes in flow cells under continuous sample admission or recirculation; under diffusive/convective mass transfer control and low v, forward and reverse scans become indistinguishable, at least for electrode reaction products that are soluble and non-adsorbable. Equations relating the measured current with the electrode and flow cell geometry are available for the most common designs [9, 10]. The theory and numerical simulation of the voltammetric response in a flow cell with channel electrodes (a rectangular strip crossing the flow channel, common also in micro-fluidic cells) received contributions and was reviewed by Cooper and Compton [26]. Working curves are available on line and a wide range of electrode widths from 10 mm down to microns, and flow rates from 10^{-4} to over 10^{-1} cm^3 s^{-1} are covered by this study, as are different electrode process mechanisms.

Further enhancement in mass transport and sensitivity has been obtained in BIA by employing a fast rotating disk electrode [27]. The cell, also applicable to FIA or SIA, is however, less simple to construct and operate. A comparable gain in current densities (i_f/a) is also reachable by reduction of the exposed area of the electrode to micrometric size, in order to promote radial diffusion [28]. Despite the advantageous improvement in the i_f/i_c ratio, the absolute value of i_f for such minute electrodes is very low, demanding higher electronic amplification which results in a less favorable signal/noise ratio. Interconnection of a matrix or array of microelectrodes or use of band or wire electrodes seem to offer a good compromise [29].

The capability for *in situ* accumulation of the analytes at the electrode interface under flow conditions, followed by its voltammetric determination, generally known as anodic or cathodic stripping analysis (ASV or CSV), is decisive in reaching very low detection limits, comparable to sophisticated spectroscopic techniques [30]. Experimental conditions for stripping analysis have to be carefully chosen because when various metals are simultaneously accumulated on solid electrodes or in amalgams there is a chance of formation of intermetallic compounds presenting ASV waves at potentials differing from the pure elements. CSV is applied mostly after adsorptive accumulation of electroactive organic molecules or of metal complexes with adsorption induced by the ligands. The preconcentration process follows an adsorption isotherm and involves molecular interaction and surface saturation at the upper limit, which implies that a calibration curve will be linear over a limited range. Competitive surface saturation with molecules more strongly adsorbable than the analyte also poses problems for analytical applications. This is one of the reasons why samples with organic ligands are submitted to photolysis, digestion or mineralization before metal trace analysis, with the exception of speciation and lability studies.

Minimum instrumentation for amperometric detection consists of a potentiostat, a recorder or computer for data acquisition and a flow cell with a working, a reference and an auxiliary electrode. Cells and electrodes will be discussed in the next sections. Simple potentiostats can be built in the laboratory with low cost operational amplifiers [3, 5, 11] and interfaced with computers using data acquisition cards. A wide range of potentiostats and electrochemical systems is commercially available, many of them overspecified for FA-ECD and with a very broad repertoire of techniques. Representative manufacturer brands are: Ecochemie, BAS, PAR, Metrohm, Pine, Radiometer, PamlSens, Solartron, Uniscan, Gamry, EDAQ, and Bank (Wenking). These companies also produce an assortment of cells and electrodes. Technical data are available at the sites of the companies on the internet. Instruments designed for flow voltammetry/amperometry are sold by some of the companies that provide flow cells, cited in the forthcoming section.

Regardless of the potentiostat used in FA, current signals recorded with normal-sized electrodes under diffusive/convective mass transport control present fluctuations that are originated by the flow pulsation inherent to the operation of peristaltic and solenoid pumps. Microelectrodes under radial diffusion regimen are rather insensitive to this pulsation. Other ways to attenuate such disturbance are: pneumatic damping (a chamber with an air bubble followed by a flow restriction); analog or digital signal filtering, current sampling in synchrony with the pulsation; "pulseless" flow delivered by motor-driven syringe pumps (common in SIA but not in FIA), motorized pipetters (typical for BIA) or pressurization of the carrier (and reagents) by gravity or compressed gas action (compatible with FIA). An inexpensive membrane pump, of the type used to insufflate air in aquaria, is fine for this purpose [31].

16.2.2
Electrode Materials

The most common materials for voltammetric working electrodes are mercury (renewable drop, film or amalgam), noble metals (mainly Pt and Au) and a variety of electron-conducting carbon allotropes (e.g., carbon paste, graphite, carbon fiber, glassy carbon, reticulated vitreous carbon, highly orientated pyrolytic graphite, C60-fullerene, carbon nanotubes, and doped diamond) [30]. Synthetically produced diamond films, doped with boron to attain conductivity, have received great attention in recent literature [30, 32, 33] due to a more favorable potential window, lower background current, and higher chemical and mechanical resistance than glassy carbon.

Flat electrodes, usually circular, with areas in the range 1 to 20 mm^2, tightly embedded in insulating polymeric blocks, are representative of the geometry adopted in most flow cells because the blocks can be easily disassembled and polished. Other shapes, sizes, and mountings are not uncommon, including printed or sputtered electrodes, tubular electrodes, porous column electrodes, and arrays of microelectrodes.

Specific adsorption processes and changes in the solid electrode's morphology or surface oxidation are not always reversible, so that replicated voltammograms with electrodes other than renewable drops or films of mercury may exhibit certain displacement or distortion of the waves and reduction in their current intensities. New or repolished solid (or paste) electrodes require a conditioning period at a working potential until the background current decays to a low and stable value; excursions to regions where hydrogen and/or oxygen evolution starts, or other sequences of pulses, were found effective for electrode activation in many cases [16]. Eventually, voltammetric detectors work unattended for months [11]. Mechanical polishing of the electrode surface is the customary *modus operandi* when other activation procedures fail. Alternatives include sonication, chemical treatment with aggressive solutions (e.g., *piranha* solution), heat or laser treatments.

The renewable mercury drop electrode, whenever applicable, offers the best way to obtain perfectly reproducible voltammograms and is still unbeatable when it comes to redox processes at negative potentials vs. NHE. Commercial mercury drop electrode generators are automatic, with drop changes triggered by a computer, manually or by the voltammetric analyzer itself. Renewal can be made before each sample injection (typical) or after a series of injections. Faster drop change (e.g., every second) may be required for analytes that rapidly block the electrode [34].

Besides solid amalgams, bismuth electrodes or Bi-plated electrodes have been proposed as an alternative to mercury electrodes, because of their non-toxic nature and favorable hydrogen overvoltage, although not as high as that of Hg [35]. Bi electrodes have been successfully applied, for example, in ASV-SIA of metals [36, 37]. With the main purposes of improving selectivity and durability, modified electrodes have also been widely applied in FA, as will be seen at the end of this chapter.

16.2.3
Commercial Flow Cells

Amperometric flow cells, designed to be used as detectors in liquid chromatography, have been on the market since 1974 [11]. Nowadays, due to wider use of ECD in HPLC for the determination of easily oxidizable and reducible compounds, an assortment of flow cells is offered by producers of electrochemical instrumentation in general and manufacturers of HPLC equipment in special. For example, BAS (Lafayette, USA) offers various thin layer flow cells with one to four electrodes in series or parallel [38]; Waters produces the 2465 ECD [39]; Antec Leyden sells the VT-03 electrochemical flow cell, and a reactor cell for electrochemical derivatization [40]; Shimadzu manufactures the L-ECD-6A EQD [41]; Dionex commercializes the ICS-3000 ECD [42]; the flow cell 66-CL010 from Cypress Systems is an all-purpose one, indicated for FIA [43]; a flow cell with biosensor chips is produced by TRACE [44].

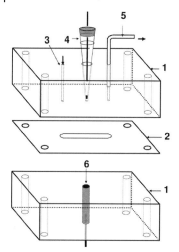

Figure 16.1 General purpose voltammetric/amperometric flow cell made in the authors' laboratory: (1) upper and lower Plexiglas (or PTFE) blocks; (2) spacer gasket; (3) entrance of the flowing solution; (4) reference electrode, miniature Ag/AgCl [48] or other type with adapter; (5) stainless steel (hypodermic needle) outlet tube also serving as counter electrode; (6) disk working electrode, for example, a Pt, Au or glassy carbon cylinder tightly embedded in the lower block, or a brass machine screw, recessed and with carbon paste filling on top. The blocks are pressed together with (not shown) fixing screws or clamps.

Most of the mentioned devices are of the cross-flow thin-layer type, with the flow lines parallel to the surface of a flat block with an embedded platinum or glassy carbon disk working electrode. A cell constructed in the authors' laboratory, depicted in Figure 16.1, exemplifies this basic design. Some manufacturers prefer the wall-jet design (flow inlet perpendicular to the center of the electrode disk). Detectors for HPLC are constructed from stainless steel and polymers like Teflon and Kel-F, inert to almost all solvents, resistant to considerably high pressures and presenting response volumes of 1 μL or less, for microbore HPLC. Although appropriate for FIA and SIA, they are rather overspecified and overpriced for work in aqueous solution at almost atmospheric pressure.

Coulometric flow cells with a porous working electrode (glassy carbon grains) are provided by ESA (Chelmsford, USA) [45]. They provide ultimate sensitivity, but not necessarily the best S/N ratio [11] because of increased background current. The application of Faraday's laws is sufficient to obtain the concentration of the analyte with such cells, rendering the calibration procedures with standard solutions superfluous. In a flow system with two or more coulometric cells, multiple analytes can be determined; the first can also be used for the removal of traces of electroactive contaminants from the electrolyte.

Cells with multiple (up to 65) interdigitated microelectrodes (2 to 20 μm wide, 2 to 3 mm long bands, Au or Pt) are also commercially available from at least two suppliers, IJ Cambria Scientific (Burry Port, UK) and Abtech Scientific (Richmond, USA) [46, 47].

16.2.4
Adaptors for Commercial Batch Cells

Although the interfacing of a flow system with a voltammetric or amperometric detector can be implemented by means of a flow-through cell, like one of those described in the previous paragraph, adaptors for batch cells can also be used. The distinction refers to two different strategies that can be employed to allow the flow stream to reach the working electrode surface, where the actual redox processes involved in the detection step take place. Strict-sense flow cells have lower dead-volume and incorporate the three-electrode system within the same space in close proximity, whereas cell adaptors are devices for the impingement of the flow solution onto the working electrode dipped in a conventional electrolytic bath, typically with several mL volume, in which are also immersed the reference and the auxiliary electrodes. In practical terms, the use of one or the other approach has its own advantages and drawbacks. Flow cells are more elaborate and require the possibility of precisely fitting miniaturized electrodes into a tiny and compact housing, which requires some skill, including in the assembly of the reference electrode [48–51]. On the other hand, adaptors can be effectively mounted in any commercial cell and are somewhat easier to make. However, they generally have larger dead volumes and, depending on the actual design, are subjected to recontamination from the electro-lytic bath by already-disposed analyte and pose some problems of reproducible adjustment of their position when removed and reinstalled.

A context in which flow adaptors are particularly convenient is the coupling of flow manifolds and detectors using the mercury drop as a working electrode (MDEs). MDEs still remain an interesting type of electrode on account of the important advantages associated with their use, like high hydrogen overvoltage, smooth and uniform surface, and the fact that, in contemporary equipment, the electrode surface is easily renewed *in situ* by an automatic knocker. Clearly, in this case the design of both flow cells and adaptors must contemplate the outflow of one dislodged mercury drop (1 µL) per determination. A review of several FIA interfaces for MDEs from the literature up to 2003 and a critical comparison of the performance of four models designed for amperometric and stripping voltammetry detection in flow systems can be found in the literature [52]. In that paper, aspects like sensitivity, flow rate dependence, response time, drop stability, tolerance to the passage of bubbles, extent of uncompensated resistance and recontamination were taken into account in the assessment of each model, shown in Figure 16.2.

Despite the restrictions derived from the use of mercury, the interest in the use of flow adaptors and flow cells for MDEs is still quite strong, as reflected by the variety of new models that have been proposed over the last years. A flow cell consisting of a cylindrical acrylic block with an axial channel of 0.7 mm diameter and with a hole in its longitudinal center to fit the mercury capillary perpendicularly to the flow port, has been exploited in several investigations in which both the use of mercury electrodes and FIA mode of analysis are quite critical [53–57]. Still in environmental analysis, a similar device was developed for the amperometric determination of U^{III} in real water samples [58] whereas a cylindrical acrylic adaptor, in which the incoming flow

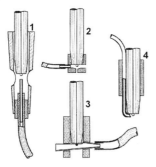

Figure 16.2 Four designs of simple flow adapters for commercial batch cells with automatic mercury drop electrodes (HMDE and SMDE) [52] with permission. (1) Upflow adapter made from a pipette tip, pressure fitted to the glass capillary of the electrode; (2) lateral flow adaptor, made of acrylic, pressure fitted; (3) lateral flow adapter, made from a section of a pipette tip with an upper V-shaped cut to accommodate the drop, held in place by perforated silicon tube; (4) L-shaped PTFE tube, held in place by a silicone ring. Copyright (2004) Wiley-VCH.

was split prior to entering the electrode compartment, was adopted in the FIA determination of atrazine in soil extracts [59] and of Cd^{II}, Pb^{II} and Cu^{II} in water and wastewater [60, 61] on HMDE. In a very straightforward solution, an L-shaped PTFE tube was attached with a silicone ring to the glass capillary that delivers the mercury drop and used in such a way as to direct the impinging liquid onto the drop [62]. Of course, a miniaturized batch cell fed under flowing conditions with a magnetic stirrer at its base and the glass capillary at the top can be used for such a purpose [63, 64]. The use of mercury hemisphere ultramicroelectrodes as microjet electrodes, that is, in which a fine jet of solution impinges from a nozzle on the actual electrodic surface under high mass transfer rates, has also been suggested [65]. A number of issues must be borne in mind when designing interfaces for MDEs, such as the necessity of allowing effective detachment of the mercury drop by the original solenoid or pneumatic driven knocker, free outflow of the drops to the bottom of the cell, the hydrodynamics of the system and how this affects the mass transfer in accumulation/stripping voltammetry, the immunity to the passage of gas bubbles, proper sealing of the adaptor and inflow tubing from oxygen (electroactive at negative potentials vs. ECS), among others.

For solid electrodes, the use of adaptors is less common since, in this case, a flow cell represents a more practical solution, the most common ones being specially designed cells for batch injection analysis, a class of cells detailed in Chapter 4. Nonetheless, there are examples of adaptors, like the one given by Walcarius *et al.*, where amperometric detection of ammonium, alkali, and alkaline earth ions is carried out on a zeolite-modified carbon paste electrode [66, 67], or that of Bergelin and Wasberg [68]. This paper presents an adaptor configuration for the working electrode and the liquid outlet based on the partial submersion of the interface zone resulting in the formation of a meniscus between solid electrode and flow outlet so that a shielding between the WE and the cell electrolyte is created.

16.2.5
Specially Designed Flow Cells and Systems

The construction of voltammetric microcells was reviewed in 1997 [69]. Since then, a number of novel designs have appeared in the literature. Gun *et al.* expanded the possibilities of an adaptor previously developed by Hua *et al.* made of a mercury-coated gold microwire transversally fitted into a length of silicone tubing (3 cm × 5 mm o.d. × 2 mm i.d.) [70, 71].

A general purpose model of a cell, adequate for any flat solid electrode, including inexpensive (disposable) gold electrodes obtained from CD-Rs (compact disks for computers) proposed by Angnes and coworkers [72, 73], is reproduced in Figure 16.3. The cell performs almost as well as expensive commercial ones and can also be adapted for sandwiching arrays of microelectrodes or multiple band electrodes or screen printed electrodes, if assembled in/on a solid, flat dielectric material (ceramic, polymeric or vitreous).

Screen-printed (disposable) electrodes are being increasingly exploited in flow voltammetric detection on account of the continuous development of accessible technologies for their production [74–76]. An interesting example is the manufacture of electrochemical filters, that is, electrolytic micro-reactors for the elimination of electrochemically active components from biofluids prior to an in-series amperometric biosensor [77]. Disposable detectors for flow analysis consisting of

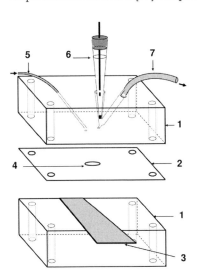

Figure 16.3 Laboratory-made voltammetric/amperometric flow cell for flat electrodes: (1) Upper and lower Plexiglas blocks, pressed together with fixing screws (not shown); (2) spacer gasket; (3) working electrode, for example, a slice of a CD-R of the gold sputtered type (Mitsui Gold) [72, 73], a Pt or glassy carbon plate, a chip with a matrix of microelectrodes [97] or interdigitated electrodes [98]; (4) hole in the spacer (defines the electrode area and the flow channel geometry); (5) entrance tubing of the flowing solution; (6) miniature Ag/AgCl reference electrode [48]; (7) stainless steel (hypodermic needle) outlet tube, serving as counter electrode.

a polyester-carbon fiber channel structure made by laser etching techniques have been recently proposed [78]. Alternative solutions to immobilize the enzyme acetyl-cholinesterase onto screen-printed electrodes for monitoring, under FIA conditions, organophosphate pesticides in water samples have also been described [79–81]. Gao *et al.* developed immunoelectrode strips based on the screen-printed carbon electrode for a flow injection enzyme-linking immunoassay system with electrochemical detection [82]. In the same work, more examples of enzyme immunoassays coupled with flow injection systems and amperometric detection are cited.

The importance of thin-layered flow cells (flow channel thickness lower than 100 μm) is increasing, stimulated by the advances in microdesign and micro-manufacture of electrochemical systems [83, 84], with possible mass production. This trend leads to a reduction in the overall cell volume and solution fluxes, thus allowing minimum sample and reagents consumption. Bratten *et al.* [85] produced, for the purpose of analyzing the response of single cells *in vitro*, a micromachined 200-μm diameter × 20-μm depth electrochemical cell using lithographic procedures. Although their work was not intended for flow analysis, it constituted the basis for subsequent development of thin-layer flow cells [86]. Thin-layer flow cells are particularly useful in the wall-jet ring disk electrode configuration, either for analytical determinations or for kinetic studies. Toda *et al.* [87] micromachined a flow cell with a depth as low as 5 μm for application in a wall-jet ring disk electrode. Such a thin-layer cell was proposed to optimize the collection, by the ring working electrode, of the derivative product formed on the disk electrode in an application for the detection of chloramphenicol by FIA-amperometry [88]. A micrometric electrode was developed and tested with a capillary FIA system in the determination of galactose [89]. The idea of a dual thin-layer flow cell with high disk generation/ring collection efficiency was also implemented [90, 91]. Gutz and Daniel [92] obtained a 19-μm deep cell using laser-printed toner masks. The cell, depicted in Figure 16.4 worked at the same time as a TiO_2-mediated photodegradation reactor (illuminated with UV radiation from a light emitting diode, with the photocatalyst immobilized onto a gold electrode that served as an electron-scavenger interface) and as a voltammetric detector for Cu^{II}. The thin layer was essential to rapidly promote the

 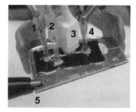

Figure 16.4 Microfluidic voltammetric cell with on-chip photoelectrocatalytic digestion of organic matter (schematic diagram [92], with permission, Copyright (2007) Elsevier). (1) Cell inlet; (2) miniaturized Ag/AgCl reference electrode [48]; (3) UV-LED radiation source; (4) stainless steel tube serving as auxiliary electrode and cell outlet; (5) cell with gold electrodes prepared as described in Ref. [98].

photocatalytic destruction of organic matter in the entire volume of liquid flowing over the electrode surface.

Although not using a physically enclosed thin-layer cell, a hydrodynamic system has been described in which the liquid is expelled from a nozzle placed at 20 μm from a planar substrate holding a ring electrode, so as to flow radially over it, that is, virtually as if within a 20-μm channel [93]. Electrochemical detectors are particularly well suited for miniaturization and, in recent years, microfluidic voltammetric cells have undergone remarkable advances leading to an increased tendency to assemble entire flow systems in the form of labs-on-chips (see Chapter 6).

Microelectrodes with at least one of the dimensions smaller than the diffusion layer thickness (micro-wire or micro-band or micro-disk electrodes) present increased mass transport by radial diffusion (while linear diffusion prevails for larger electrodes), as amply demonstrated in the literature [94]. Convective transport produces a similar current increase for electrodes of any size. Higher current densities (by radial diffusion and/or convection) are advantageous because the i_f/i_r ratio increases for a given area-dependent double-layer capacity. For microelectrodes, this advantage is partially outweighed by the need for higher amplification of the weak signal, resulting in more noise. However, arrays of microelectrodes sufficiently apart one from another can be interconnected to retain only the advantages. In flow systems, microelectrodes are less sensitive to current modulation by pulsation originated by a peristaltic pump. Also, for reversible analytes, interdigitated electrodes in the flow path can originate signal amplification by redox recycling [95]. Conversely, each electrode of an array can be operated at different potentials and/or submitted to distinct surface modification, thus favoring simultaneous determination of various analytes [96]. We have proposed simple ways to produce arrays of individually addressable gold electrodes [97] as well as of multi-band and interdigitated electrodes [98].

16.3
Strategies for Improving Selectivity, Sensitivity, and Durability

16.3.1
Preconcentration

The increasing necessity for detection and quantification of numerous species at ever lower concentrations in a variety of samples stimulates the development of preconcentration strategies, alongside with improvement in modern instruments and techniques. In the literature, there are an impressive number of studies involving the association of flow techniques with preconcentration processes, aiming not only at attaining better detection limits, but also at eliminating interferences in studies involving the most different kinds of samples. The majority of the preconcentration schemes are based on the reversible chemical bonding of the analyte (e.g., on ion exchange resins), adsorptive processes (e.g., on solid materials or electrodes) or a redox process subsequent to the deposition on conductive surfaces (i.e., an electrochemical electrode).

Flow injection analysis involving on-line separation and/or preconcentration, with emphasis on the environmental monitoring of metal ions has recently been reviewed by Miro *et al.* [99]. The main focus of this work was on the association of flow injection analysis with spectroscopic techniques. In the same direction, a second review of solid-phase extraction (SPE) was carried out by Camel [100], also covering many flow injection applications. The progress in the methodology and practical applications of FIA in conjunction with atomic absorption was also reviewed [101]. Recently, the association of miniaturized columns and selective procedures, denominated lab-on-valve hyphenated with electrothermal atomic absorption spectrometry, have been proposed for the quantification of ultra-trace levels of metals [102, 103].

The association of FA-ECD with preconcentration at the electrode presents many advantages. The analyte is accumulated on the sensor surface and this feature can dramatically improve the detection limit. By choosing the most favorable potential the preconcentration can be optimized while interferences are minimized. The first study describing the association of continuous-flow and anodic stripping voltammetry was published by Zirino and Liberma even before the advent of flow injection analysis [104]. Processes involving FIA plus differential pulse [105] or stripping potentiometry of metals [106] were developed for determination of thallium in soils and mercury in oceanic and tap water. Automated systems to perform metals preconcentration on-line in the laboratory [107] or even on a ship on the ocean [108] have been described. Mercury is the preferred electrode material in FIA-ASV; the same pre-plated Hg film on glassy carbon can be reused many times for the determination of Cd(II), Cu(II), Pb(II) and Zn(II) in wastewater after off-line UV digestion [109]. Masini's group applied SI-ASV to the determination of the same elements in river sediments [110] and wastewater from the coatings industry, after off-line acid digestion [111].

Adsorptive preconcentration of organic species was also extensively explored in many applications, such as in the analysis of flavonoids in wines, tea or tomato juices [112] or for the determination of paraxetine in pharmaceuticals [113] and riboflavin [114]. For the quantification of low concentrations of cobalt in high purity iron samples, the coupling of electrochemical preconcentration with chemiluminescence was a useful procedure, since cobalt can be selectively stripped back to the solution [115].

16.3.2
Medium Exchange

The potentiality of flow analysis in managing solutions has so far been scarcely explored for medium exchange in stripping voltammetry. The electrode is exposed to the acidified carrier and the metal-ion-containing sample during the deposition phase, while a more suitable medium (buffered complexing solution) is chosen for the stripping step. The ligand can provide some extra selectivity, by anticipating the stripping of one metal in relation to another. Medium exchange was explored for trace analysis of metals in complex samples such as brine [116] or soils [117] or to quantify organic compounds in complex samples [118–120]. The association of modified electrodes and medium exchange was also explored with advantage [117, 119, 121, 122] to attain elevated sensitivity. The inherent advantages presented by SIA for medium exchange were also demonstrated by Ivaska and Kubiak [123] for the quantification of copper in tap water.

Sawamoto pioneered, in the 1990s, the use of FA and medium exchange for fundamental studies involving adsorption/desorption at electrodes that cannot be accessed otherwise [124, 125]. With the evolution of these studies, it was even possible to measure the adsorption of a second species on top of a previously formed adsorption monolayer [126]. In another study, the reorientation of quinoline on a mercury electrode was demonstrated [127]. The reversible or irreversible adsorption of surfactants on mercury was also studied using flow analysis and differential capacity–time curves [128].

16.3.3
Oxygen Removal

Special attention should be given, in voltammetric measurements at negative potentials, both in stationary and flowing mode of operation, to the presence of dissolved oxygen. On account of its electroreduction to H_2O_2 or H_2O at potentials lower than approximately the NHE, it generally interferes in determination procedures and needs to be removed (techniques that respond mainly to reversible electrode processes, like SWV, are less affected). In batch analysis, this is usually performed by time-consuming previous bubbling of the solution with inert gas. This is obviously not practical for flow analysis and to date various alternative solutions have been proposed. One possibility is to extract the dissolved oxygen from the solution through the walls of a permeable narrow bore silicone tubing coiled within a small chamber maintained at a reduced pressure of air or of an inert gas, for example N_2 [129]. The silicone tubing can be replaced by a microporous capillary like Oxyphan, a polypropylene hollow fiber, fluxed with N_2 at room pressure.

Other membrane materials, as well as O_2-scavengers in a degassing chamber, have been compared [130]. Even an enzymatic cross-linked O_2 removing gel was considered [131]. Furthermore, Reinke and Simon proposed the use of chromatomembrane-based flow cells, consisting of polytetrafluorethylene micropores and macropores. In this approach, the gas phase is selectively eliminated through the micropores exploiting the inability of the liquid solution to penetrate them because of capillary forces [132].

16.3.4
Catalytic Electrode Processes

An intelligent approach to enhance the faradaic current to gain sensitivity in voltammetric and amperometric methods is to explore catalytic electrode processes in which the analyte is cycled repeatedly, the simplest example being the chemical reoxidation of a metal ion that is reduced from a higher to a lower oxidation state at the electrode. Other mechanisms are presented in a review [133] with over 100 catalytic voltammetric methods; many of them take advantage also of the accumulation of the analyte at the electrode interface by adsorption to reach extra sensitivity [134]. Most methods rely on mercury electrodes (drop or film) and show excellent detection limits, in the region of 10^{-7} to 10^{-14} mol L^{-1}. Selectivity is usually good, but samples with organic material must be digested first, because surfactants may interfere by adsorption at the electrode. Some catalytic voltammetric methods have been implemented in flow systems, although many more are suitable. The catalytic FIA of uranium exemplifies the concept. At the mercury drop electrode, U(IV) is reduced to U(III). In an electrolyte containing nitrate in large excess, this anion oxidizes the U(III) back to U(IV), making it available for a new electroreduction step. This catalytic cycle is fast and occurs repeatedly at the mercury drop under normal FIA conditions, greatly enhancing the measured current and the detection limit [58].

16.3.5
Spectroelectrochemistry

The amount of information extracted from a sample can be expanded by *in situ* observation of the electrode processes through spectroscopic techniques. Cells of different designs were proposed for FA spectroelectrochemistry and applied in fundamental studies and analytical methods, with gains mainly in selectivity. Measurement of light absorption and/or emission at/near electrodes in the UV–Visible region of the spectrum is perhaps the simplest approach. For example, a long optical path (10 mm) flow cell with twin gold working electrodes fitting into a standard size quartz, glass or acrylic cuvette was proposed (Figure 16.5a) [135]. The cell was inserted in the unmodified cuvette holder of a diode array spectrophotometer and applied to the spectroelectrochemical determination of prometazine and chlorpromazine [136, 137]. Absorbance spectra versus potential 3-D surfaces can be generated (Figure 16.5b) for mechanistic studies and for the selection of optimum conditions for analytical methods with amperometric and single or multi-wavelength detection. A standard quartz cuvette was used years earlier as a stationary electrochemical cell during invention of the technique of cyclic fluorovoltammetry [138]. By using the same concept shown in Figure 16.5a, this cell can be converted to a flow cell for fluorovoltammetry and fluoroamperometry. A recent book on electrochemiluminescence [139] covers its association with FA, inspiring many analytical applications [140]. For example, the inhibition of the RuO_2^{2+} electrogenerated chemiluminescence by chloramphenicol is the basis of a FIA method for this drug [141]. The trend of scaling-down electrochemiluminescence detection to microchip electrophoresis, presented in a review by Du and Wang [19], with over 200 references, is extensible to µFIA too [142].

Surface-enhanced resonance Raman spectroscopy (SERRS) was coupled to ECD-FA [143] and applied to the determination of iron. In the injected sample, the metal ion is complexed as $Fe(phen)^{2+}$, preconcentrated by adsorption on a silver-plated glassy carbon electrode and irradiated with a laser beam for acquisition of the scattering spectrum and stripped off afterwards, together with the silver layer. The spectrum allows selective determination of the complex down to the femtomol level.

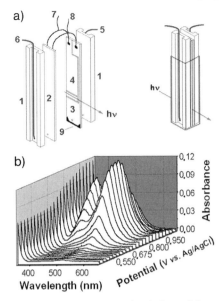

Figure 16.5 (a) Long-optical-path flow cell for spectroelectrochemistry (reproduced with permission from [135], Copyright (2001) Wiley-VCH). (1) Acrylic plates; (2) polycarbonate back plane of the Au-CDtrode; (3) Au-CDtrode working electrode area [72]; (4) auxiliary electrode area; (5) solution inlet tube; (6) salt bridge tube to reference electrode [48] (not shown); (7) electric interconnection of the twin auxiliary electrodes; (8) electric interconnection of the twin working electrodes; (9) silicone rubber spacer, 200 μm thick. (b) 3-D spectrovoltammograms of promethazine electroxidation in $0.1\,mol\,L^{-1}$ H_2SO_4, obtained in cell (a) (reproduced with permission from [136] Copyright (2003) Elsevier).

For routine analytical applications the lack of reasonably priced, ready to use, Raman spectrometer and spectroelectrochemical cell is presently a limitation.

16.3.6
Modified Electrodes

The development of modified electrodes opened possibilities to create sensors for an impressive number of applications. Most of these modified electrodes can advantageously be used as detectors in FA. Films or coatings, sometimes multilayered, have been applied to the bare electrodes by adsorption, dip coating, brush painting, spray painting, spin coating, Langmuir–Blodgett thin-film deposition, electropolymerization, co-electrodeposition and molecular self-assembly. Dispersion of modifiers in carbons pastes or in screen printed inks is also widely employed. The main aims of applying one or more modifiers are: (i) to favor the electrolysis of the targeted species by mediating or catalyzing their charge transfer; (ii) to favor the preconcentration of the target analyte on the sensor, improving the detection limit; (iii) to protect the electrode surface against interfering species and/or of passivation agents; (iv) to immobilize or avoid leaching of other modifiers. A review of stripping voltammetry presented by Brainina and coworkers [30] covers the modification of electrodes by *in situ* and *ex situ* procedures, distinguishing surface coating from in-bulk insertion and presenting hundreds of references on such sensors and their applications in environmental and food analysis, most of them still carried out in the batch mode.

The association of modified electrodes with flow injection analysis has increased significantly in recent years, as has the range of materials used to modify their surface. Electrodes modified with biological materials (predominantly enzymes) still dominate the scenario [144–147]. Additional information on the use of enzymes in FA can be found in Chapter 14. Electrode modification with compounds like Prussian blue and other similar salts [148, 149], macrocycles (phthalocyanins, porphyrins, porphyrazins) [150, 151], SAMs [152, 153] and many other materials is increasing rapidly. The preparation of composites [154, 155], the utilization of new electrodes such as doped diamonds [30, 156], carbon nanotubes [157], conducting polymers [158] or the association of different materials in the same sensor [159–162] are gaining increasing importance.

Another approach to the modification of electrodes is the utilization of polymeric films on their surface. This kind of strategy can provide high selectivity and stability, two mandatory characteristics to meet the growing demand for rapid and direct analysis of components in complex matrices. These requirements have stimulated researchers to explore polymeric membranes, which can be very effective for separations. The selectivity can be based on electrostatic effects (ion exchanger, e.g., Nafion), molecular size (e.g., dialysis membranes) and/or hydrophobic/hydrophilic characteristics (e.g., PTFE) of the membrane. Membrane application on the electrode surface is typically made by drop casting and electropolymerization [158, 163, 164]. The placement of membranes between two flowing streams should be cited at this point as an alternative to electrode modification; it has been widely used for separations based on dialysis of ions [165–167] or the diffusion of gaseous species [168–170].

16.4
Trends and Perspectives

Examination of the literature on flow analysis over the last 15 years shows an increase of about 5% in the annual number of publications dealing with amperometric detection, roughly the same growth rate as for spectrophotometric detection, which is more disseminated, while the use of voltammetry, notably ASV and CSV, is growing twice as fast (10% annually).

The advent of new electrode materials, synthetic analogs of active centers of enzymes, nanomaterials for surface modification and the improvement of techniques for electrode modification reflect favorably on flow analysis inasmuch as they provide amperometric sensors and biosensors with greater selectivity and durability. The broader acknowledgement of advantages like multi-analyte capabilities of fast voltammetric scanning (e.g., SWV) and the perfect compatibility of highly sensitive, reasonably selective voltammetric stripping analysis with FIA, BIA or SIA will probably sustain the intensified growth in research activity with such detection methods for a long period.

The availability of commercial instrumentation (electrodes, cells, potentiostats, and multipotentiostats) at moderate or low cost contributes to this trend. Many of such instruments can be controlled by a computer. More flexible and user friendly software for simultaneous control of the flow analysis system (valves, pumps, sample changer, etc.) and the voltammetric detector (unlimited user programmable pulse and scan sequences, current sampling schemes, electrode activation procedures and synchronization with sample injection) would be welcomed by practitioners with more complex needs and not interested in writing their own programs.

The development of portable FA-ECD systems for personal health care, use in the field or remote places with unattended operation is certainly of growing interest. Since voltammetric detectors easily fulfill requirements like low cost and size, battery operability, ruggedness and automation, the challenges reside in aspects like electrode durability (system redundancy can be considered for such low-cost devices), auto-calibration and reagents/electrolyte supply to allow long-lasting unsupervized operation.

The trend towards miniaturization of FA-ECD is apparent and feasible, and the progressive integration of more steps like sampling, sample treatment by digestion, extraction, reaction, followed by preconcentration, medium exchange, stripping and so forth, culminating with µTASs, is on the horizon.

References

1 Bard, A.J. and Rubinstein, I. (eds) (1966–2004) *Electroanalytical Chemistry: a Series of Advances*, Vols. 1–22, Marcel Dekker, New York.

2 Zoski, C.G. (2007) *Handbook of Electrochemistry*, Elsevier, Boston.

3 Sawyer, D.T., Sobkowiak, A. and Roberts, J.L. (1995) *Electrochemistry for Chemists*, John Wiley & Sons, New York.

4 Brett, C.M.A. and Brett, A.M.O. (1998) Electroanalysis, Oxford University Press, New York.

5 Bard, A.J. and Faulkner, L.R. (2001) *Electrochemical Methods: Fundamentals and Applications*, Wiley, New York.

6 Scholz, F. (2002) *Electroanalytical Methods: Guide to Experiments and Applications*, Springer, Berlin.

7 Bard, A.J., Stratmann, M. and Unwin, P.R. (eds) (2003) *Instrumentation and Electro-analytical Chemistry*, Vol. 3, Wiley-VCH, Weinheim.

8 Wang, J. (2006) *Analytical Electrochemistry*, Wiley-VCH, Weinheim.

9 Stulik, K. and Pacakova, V. (1987) *Electroanalytical Measurements in Flowing Liquids*, Ellis Horwood, Chichester.

10 Gunasingham, H. and Fleet, B. (1990) Hydrodynamic Voltammetry in Continuous-Flow Analysis, in *Electroanalytical Chemistry: a Series of Advances*, 16, Marcel Dekker, New York, pp. 89–180.

11 Kissinger, P.T. and Heineman, H. (1996) *Laboratory Techniques in Electroanalytical Chemistry*, Marcel Dekker, New York.

12 Trojanowicz, M. (2000) *Flow Injection Analysis: Instrumentation and Applications*, World Scientific, Singapore.

13 Trojanowicz, M., Szewczynska, M. and Wcislo, M. (2003) Electroanalytical flow measurements – recent advances. *Electroanalysis*, 15, 347–365.

14 Pérez-Olmos, R., Soto, J.C., Zarate, N., Araújo, A.N. and Montenegro, M.C.B.S.M. (2005) Sequential injection analysis using electrochemical detection: a review. *Analytica Chimica Acta*, **554**, 1–16.

15 Quintino, M.S.M. and Angnes, L. (2004) Batch injection analysis: an almost unexplored powerful tool. *Electroanalysis*, **16**, 513–523.

16 LaCourse, W.R. (1997) *Pulsed Electrochemical Detection in High-Performance Liquid Chromatography*, Wiley-Interscience, New York.

17 Flanagan, R.J., Perrett, D. and Whelpton, R. (2005) *Electrochemical Detection in HPLC*, Royal Society of Chemistry, Cambridge.

18 LaCourse, W.R. and Modi, S.J. (2005) Microelectrode applications of pulsed electrochemical detection. *Electroanalysis*, **17**, 1141–1152.

19 Du, Y. and Wang, E.K. (2007) Capillary electrophoresis and microchip capillary electrophoresis with electrochemical and electrochemiluminescence detection. *Journal of Separation Science*, **30**, 875–890.

20 http://epsilon-web.com/Lc/manuals/Principles/principles.html, last accessed on 10/Aug/2007.

21 http://www.chem.agilent.com/scripts/LiteraturePDF.asp?iWHID=20644, last accessed on 10/Aug/2007.

22 Zemann, A.J., Schnell, E., Volgger, D. and Bonn, G.K. (1998) Contactless conductivity detection for capillary electrophoresis. *Analytical Chemistry*, **70**, 563–567.

23 Fracassi da Silva, J.A. and do Lago, C.L. (1998) An oscillometric detector for capillary electrophoresis. *Analytical Chemistry*, **70**, 4339–4343.

24 Kubáň, P. and Hauser, P.C. (2004) Contactless conductivity detection in capillary electrophoresis: a review. *Electroanalysis*, **16**, 2009–2021.

25 Hoherčáková, Z. and Opekar, F. (2005) A contactless conductivity detection cell for flow injection analysis: determination of total inorganic carbon. *Analytica Chimica Acta*, **551**, 132–136.

26 Cooper, J.A. and Compton, R.G. (1998) Channel electrodes – a review. *Electroanalysis*, **10**, 141–155.

27 Tur'yan, Y.I., Strochkova, E.M., Kuselman, I. and Shenhar, A. (1996) Microcell for anodic stripping voltammetry of trace metals. *Fresenius' Journal of Analytical Chemistry*, **354**, 410–413.

28 Macpherson, J.V., Simjee, N. and Unwin, P.R. (2001) Hydrodynamic ultramicroelectrodes: kinetic and analytical applications. *Electrochimica Acta*, **47**, 29–45.

29 Nascimento, V.B., Augelli, M.A., Pedrotti, J.J., Gutz, I.G.R. and Angnes, L. (1997) Arrays of gold microelectrodes made from split integrated circuit chips. *Electroanalysis*, **9**, 335–339.

30 Brainina, K.Z., Malakhova, N.A. and Stojko, N.Y. (2000) Stripping voltammetry in environmental and food analysis. *Fresenius' Journal of Analytical Chemistry*, **368**, 307–325.

31 Matos, R.C., Gutz, I.G.R., Angnes, L., Fontenele, R.S. and Pedrotti, J.J. (2001) A versatile and pulsation-free pneumatic impeller for flow analysis systems. *Quimica Nova*, **24**, 795–798.

32 Fujishima, A., Einaga, Y., Rao, T.N. and Tryk, D.A. (2005) *Diamond Electrochemistry*, Elsevier, Amsterdam.

33 Fujishima, A. (2007) *Electrochemistry of Diamond*, BKS, Tokyo.

34 Hidalgo, P. and Gutz, I.G.R. (2001) Determination of low concentrations of the flotation reagent ethyl xanthate by sampled DC polarography and flow injection with amperometric detection. *Talanta*, **54**, 403–409.

35 Wang, J., Lu, J.M., Kirgöz, U.A., Hocevar, S.B. and Ogorevc, B. (2001) Insights into the anodic stripping voltammetric behavior of bismuth film electrodes. *Analytica Chimica Acta*, **434**, 29–34.

36 Kefala, G. and Economou, A. (2006)
Polymer-coated bismuth film electrodes
for the determination of trace metals by
sequential-injection analysis/anodic
stripping voltammetry. *Analytica Chimica
Acta*, **576**, 283–289.

37 Economou, A. and Voulgaropoulos, A.
(2007) On-line stripping voltammetry of
trace metals at a flow-through bismuth-
film electrode by means of a hybrid flow
injection/sequential-injection system.
Talanta, **71**, 758–765.

38 http://www.bioanalytical.com/products/
lc/flowcells.html, last accessed on
10/Aug/2007.

39 http://www.waters.com/WatersDivision/
ContentD.asp?watersit=JDRS-5LWJX8,
last accessed on 10/Aug/2007.

40 http://www.antecleyden.com/products/
vt03cell.shtml, last accessed on
10/Aug/2007.

41 http://www.ssi.shimadzu.com/products/
product.cfm?product=l-ecd-6a, last
accessed on 10/Aug/2007.

42 www1.dionex.com/en-us/
ins_dxprn29100.html, last accessed on
10/Aug/2007.

43 http://www.cypresssystems.com/Cells/
FIA.html, last accessed on 10/Aug/2007.

44 http://www.trace.de/en/onlinemessung/
fliesszelle-bild.htm, last accessed on 10/
Aug/2007.

45 http://www.esainc.com/products/
type/instruments/HPLC/
specialty_detectors/coularray, last
accessed on 10/Aug/2007.

46 http://www.ijcambria.com/
ALS_printed_electrodes.htm, last
accessed on 10/Aug/2007.

47 http://www.abtechsci.com/labproducts.
html, last accessed on 10/Aug/2007.

48 Pedrotti, J.J., Angnes, L. and Gutz, I.G.R.
(1996) Miniaturized reference electrodes
with microporous polymer junctions.
Electroanalysis, **8**, 673–675.

49 Keller, O.C. and Buffle, J. (2000)
Voltammetric and reference
microelectrodes with integrated
microchannels for flow through

microvoltammetry. 1. The microcell.
Analytical Chemistry, **72**, 936–942.

50 Hashimoto, M., Upadhyay, S., Kojima, S.,
Suzuki, H., Hayashi, K. and Sunagawa, K.
(2006) Needle-type Ag/AgI reference
electrode with a stagnant electrolyte layer
and an active liquid junction. *Journal of the
Electrochemical Society*, **153**, 155–160.

51 Polk, B.J., Stelzenmuller, A., Mijares, G.,
MacCrehan, W. and Gaitan, M. (2006) Ag/
AgCl microelectrodes with improved
stability for microfluidics. *Sensors and
Actuators B-Chemical*, **114**, 239–247.

52 Cavicchioli, A., Daniel, D. and
Gutz, I.G.R. (2004) Critical comparison
of four simple adaptors for flow
amperometric and stripping voltammetry
with mercury drop electrodes in batch
cells. *Electroanalysis*, **16**, 391–398.

53 Colombo, C., van den Berg, C.M.G. and
Daniel, A. (1997) A flow cell for on-line
monitoring of metals in natural waters by
voltammetry with a mercury drop
electrode. *Analytica Chimica Acta*, **346**,
101–111.

54 Colombo, C. and van den Berg, C.M.G.
(1998) Determination of trace metals (Cu,
Pb, Zn and Ni) in soil extracts by flow
analysis with voltammetric detection.
*International Journal of Environmental
Analytical, Chemistry*, **71**, 1–17.

55 Al-Farawati, R. and van den Berg, C.M.G.
(1999) Metal-sulfide complexation in
seawater. *Marine Chemistry*, **63**, 331–352.

56 Al-Farawati, R. and van den Berg, C.M.G.
(2001) Thiols in coastal waters of the
western North Sea and English Channel.
Environmental Science & Technology, **35**,
1902–1911.

57 Achterberg, E.P., van den Berg, C.M.G.
and Colombo, C. (2003) High resolution
monitoring of dissolved Cu and Co in
coastal surface waters of the Western
North Sea. *Continental Shelf Research*, **23**,
611–623.

58 Aguiar, M.A.S., Marquez, K.S.G. and
Gutz, I.G.R. (2000) Determination of
traces of uranium in environmental
samples using a flow injection system

with amperometric catalytic detection. *Electroanalysis*, **12**, 742–746.

59 Abate, G., Lichtig, J. and Masini, J.C. (2002) Construction and evaluation of a flow-through cell adapted to a commercial static mercury drop electrode (SMDE) to study the adsorption of Cd(II) and Pb(II) on vermiculite. *Talanta*, **58**, 433–443.

60 dos Santos, L.B.O., Abate, G. and Masini, J.C. (2005) Application of sequential injection-square wave voltammetry (SI-SWV) to study the adsorption of atrazine onto a tropical soil sample. *Talanta*, **68**, 165–170.

61 dos Santos, L.B.O., Abate, G. and Masini, J.C. (2006) Developing a continuous flow-square wave voltammetry method for determination of atrazine in soil solutions using the hanging mercury drop electrode. *Journal of the Brazilian Chemical Society*, **17**, 36–42.

62 Pedrotti, J.J. and Gutz, I.G.R. (2003) Ultra-simple adaptor to convert batch cells with mercury drop electrodes in voltammetric detectors for flow analysis. *Talanta*, **60**, 695–705.

63 Korolczuk, M. and Grabarczyk, M. (1999) Voltammetric determination of Cr(VI) in a flow system in the presence of diethylenetriaminepentaacetic acid (DTPA) following its deposition in the metallic state. *Analytica Chimica Acta*, **387**, 97–102.

64 Korolczuk, M. (1999) Voltammetric determination of Cr(VI) in natural water in the presence of bipyridine following its deposition to the metallic state. *Electroanalysis*, **11**, 1218–1221.

65 Macpherson, J.V. and Unwin, P.R. (1997) Characterization and application of a mercury hemisphere microjet electrode. *Analytical Chemistry*, **69**, 5045–5051.

66 Walcarius, A., Mariaulle, P., Louis, C. and Lamberts, L. (1999) Amperometric detection of nonelectroactive cations in electrolyte-free flow systems at zeolite modified electrodes. *Electroanalysis*, **11**, 393–400.

67 Walcarius, A., Vromman, V. and Bessiere, J. (1999) Flow injection indirect amperometric detection of ammonium ions using a clinoptilolite-modified electrode. *Sensors and Actuators B-Chemical*, **56**, 136–143.

68 Bergelin, M. and Wasberg, M. (1998) The impinging jet flow method in interfacial electrochemistry: an application to bead-type electrodes. *Journal of Electroanalytical Chemistry*, **449**, 181–191.

69 Tur'yan, Y.I. (1997) Microcells for voltammetry and stripping voltammetry. *Talanta*, **44**, 1–13.

70 Hua, C., Jagner, D. and Renman, L. (1987) Automated-determination of total arsenic in sea-water by flow constant-current stripping analysis with gold fiber electrodes. *Analytica Chimica Acta*, **201**, 263–268.

71 Gun, J., Salaun, P. and van den Berg, C.M.G. (2006) Advantages of using a mercury coated, micro-wire, electrode in adsorptive cathodic stripping voltammetry. *Analytica Chimica Acta*, **571**, 86–92.

72 Angnes, L., Richter, E.M., Augelli, M.A. and Kume, G.H. (2000) Gold electrodes from recordable CDs. *Analytical Chemistry*, **72**, 5503–5506.

73 Munoz, R.A.A., Matos, R.C. and Angnes, L. (2001) Amperometric determination of dipyrone in pharmaceutical formulations with a flow cell containing gold electrodes from recordable compact discs. *Journal of Pharmaceutical Sciences*, **90**, 1972–1977.

74 Nascimento, V.B. and Angnes, L. (1998) Screen-printed electrodes. *Quimica Nova*, **21**, 614–629.

75 Neufeld, T., Eshkenazi, I., Cohen, E. and Rishpon, J. (2000) A micro flow injection electrochemical biosensor for organophosphorus pesticides. *Biosensors & Bioelectronics*, **15**, 323–329.

76 Hsu, C.T., Chung, H.H., Lyuu, H.J., Tsai, D.M., Kumar, A.S. and Zen, J.M. (2006) An electrochemical cell coupled with disposable screen-printed electrodes

for use in flow injection analysis. *Analytical Sciences*, **22**, 35–38.

77 Okawa, Y., Kobayashi, H. and Ohno, T. (1995) Direct and simultaneous determination of uric-acid and glucose in serum with electrochemical filter/ biosensor flow injection analysis system. *Analytica Chimica Acta*, **315**, 137–143.

78 Kilbey, G., Karousos, N.G., Eglin, D. and Davis, J. (2006) Laser etched carbon fibre composites: disposable detectors for flow analysis applications. *Electrochemistry Communication*, **8**, 1315–1320.

79 Rippeth, J.J., Gibson, T.D., Hart, J.P., Hartley, I.C. and Nelson, G. (1997) Flow-injection detector incorporating a screen-printed disposable amperometric biosensor for monitoring organophosphate pesticides. *The Analyst*, **122**, 1425–1429.

80 Shi, M.H., Xu, J.J., Zhang, S., Liu, B.H. and Kong, J.L. (2006) A mediator-free screen-printed amperometric biosensor for screening of organophosphorus pesticides with flow injection analysis (FIA) system. *Talanta*, **68**, 1089–1095.

81 Law, K.A. and Higson, S.P.J. (2005) Sonochemically fabricated acetylcholinesterase micro-electrode arrays within a flow injection analyser for the determination of organophosphate pesticides. *Biosensors & Bioelectronics*, **20**, 1914–1924.

82 Gao, Q., Ma, Y., Cheng, Z.L., Wang, W.D. and Yang, X.R. (2003) Flow injection electrochemical enzyme immunoassay based on the use of an immunoelectrode strip integrate immunosorbent layer and a screen-printed carbon electrode. *Analytica Chimica Acta*, **488**, 61–70.

83 Pernaut, J.M. and Matencio, T. (1999) Thin layer electrochemical cells, principle and application. *Quimica Nova*, **22**, 899–902.

84 Toda, K. (2004) Development of miniature key devices for flow analysis and their applications. *Bunseki Kagaku*, **53**, 207–219.

85 Bratten, C.D.T., Cobbold, P.H. and Cooper, J.M. (1997) Micromachining sensors for electrochemical measurement in subnaliter volumes. *Analytical Chemistry*, **69**, 253–258.

86 Muck, A., Jr, Wang, J. and Barek, J. (2003) Microfluidic platform for FIA with electrochemical detection. *Chemicke Listy*, **97**, 957–960.

87 Toda, K., Oguni, S., Takamatsu, Y. and Sanemasa, I. (1999) A wall-jet ring disk electrode fabricated within a thin-layered micromachined cell. *Journal of Electroanalytical Chemistry*, **479**, 57–63.

88 Liao, C.Y., Chang, C.C., Ay, C. and Zena, J.M. (2007) Flow injection analysis of chloramphenicol by using a disposable wall-jet ring disk carbon electrode. *Electroanalysis*, **19**, 65–70.

89 Kovalcik, K.D., Kirchhoff, J.R., Giolando, D.M. and Bozon, J.P. (2004) Copper ring-disk microelectrodes: fabrication, characterization, and application as an amperometric detector for capillary columns. *Analytica Chimica Acta*, **507**, 237–245.

90 Jusys, Z., Kaiser, J. and Behm, R.J. (2004) A novel dual thin-layer flow cell double-disk electrode design for kinetic studies on supported catalysts under controlled mass-transport conditions. *Electrochimica Acta*, **49**, 1297–1305.

91 Paixão, T.R.L.C., Matos, R.C. and Bertotti, M. (2003) Design and characterisation of a thin-layered dual-band electrochemical cell. *Electrochimica Acta*, **48**, 691–698.

92 Daniel, D. and Gutz, I.G.R. (2007) Microfluidic cell with a TiO_2-modified gold electrode irradiated by an UV-LED for *in situ* photocatalytic decomposition of organic matter and its potentiality for voltammetric analysis of metal ions. *Electrochemistry Communication*, **9**, 522–528.

93 Macpherson, J.V. and Unwin, P.R. (1998) Radial flow microring electrode: development and characterization. *Analytical Chemistry*, **70**, 2914–2921.

94 Wittstock, G., Gründig, B., Strehlitz, B. and Zimmer, K. (1998) Evaluation of

microelectrode arrays for amperometric detection by scanning electrochemical microscopy. *Electroanalysis*, **10**, 526–531.

95 Wittstock, G. (2002) Sensor arrays and array sensors. *Analytical and Bioanalytical Chemistry*, **372**, 16–17.

96 Matos, R.C., Augelli, M.A., Lago, C.L. and Angnes, L. (2000) Flow injection analysis-amperometric determination of ascorbic and uric acids in urine using arrays of gold microelectrodes modified by electrodeposition of palladium. *Analytica Chimica Acta*, **404**, 151–157.

97 Augelli, M.A., Nascimento, V.B., Pedrotti, J.J., Gutz, I.G.R. and Angnes, L. (1997) Flow-through cell based on an array of gold microelectrodes obtained from modified integrated circuit chips. *The Analyst*, **122**, 843–847.

98 Daniel, D. and Gutz, I.G.R. (2005) Microfluidic cells with interdigitated array gold electrodes: fabrication and electrochemical characterization. *Talanta*, **68**, 429–436.

99 Miro, M., Estela, J.M. and Cerda, V. (2004) Application of flowing stream techniques to water analysis Part III. Metal ions: alkaline and alkaline-earth metals, elemental and harmful transition metals, and multielemental analysis. *Talanta*, **63**, 201–223.

100 Camel, V. (2003) Solid phase extraction of trace elements. *Spectrochimica Acta Part B-Atomic Spectroscopy*, **58**, 1177–1233.

101 Begak, O.Y. and Borodin, A.V. (1998) Flow-injection analysis in atomic absorption spectrometry (review). *Industrial Laboratory*, **64**, 371–381.

102 Long, X.B., Miro, M. and Hansen, E.H. (2005) An automatic micro-sequential injection bead injection Lab-on-Valve (μSI-BI-LOV) assembly for speciation analysis of ultra trace levels of Cr(III) and Cr(VI) incorporating on-line chemical reduction and employing detection by electrothermal atomic absorption spectrometry (ETAAS). *Journal of Analytical Atomic Spectrometry*, **20**, 1203–1211.

103 Long, X.B., Miro, M., Jensen, R. and Hansen, E.H. (2006) Highly selective micro-sequential injection lab-on-valve (μ SI-LOV) method for the deter-mination of ultra-trace concentrations of nickel in saline matrices using detection by electrothermal atomic absorption spectrometry. *Analytical and Bioanalytical Chemistry*, **386**, 739–748.

104 Zirino, A. and Lieberma, S. (1973) Continuous-flow anodic stripping voltammetry of Zn, Cd, and Pb in San Diego bay water. *Journal of the Electrochemical Society*, **120**, C254–C254.

105 Lukaszewski, Z. and Zembrzuski, W. (1992) Determination of thallium in soils by flow injection differential pulse anodic stripping voltammetry. *Talanta*, **39**, 221–227.

106 Richter, E.M., Augelli, M.A., Kume, G.H., Mioshi, R.N. and Angnes, L. (2000) Gold electrodes from recordable CDs for mercury quantification by flow injection analysis. *Fresenius' Journal of Analytical Chemistry*, **366**, 444–448.

107 Wang, J., Setiadji, R., Chen, L., Lu, J.M. and Morton, S.G. (1992) Automated-system for online adsorptive stripping voltammetric monitoring of trace levels of uranium. *Electroanalysis*, **4**, 161–165.

108 Daniel, A., Baker, A.R. and van den Berg, C.M.G. (1997) Sequential flow analysis coupled with ACSV for on-line monitoring of cobalt in the marine environment. *Fresenius' Journal of Analytical Chemistry*, **358**, 703–710.

109 Suteerapataranon, S., Jakmunee, J., Vaneesorn, Y. and Grudpan, K. (2002) Exploiting flow injection and sequential injection anodic stripping voltammetric systems for simultaneous determination of some metals. *Talanta*, **58**, 1235–1242.

110 da Silva, C.L. and Masini, J.C. (2000) Determination of Cu, Pb, Cd, and Zn in river sediment extracts by sequential injection anodic stripping voltammetry with thin mercury film electrode. *Fresenius' Journal of Analytical Chemistry*, **367**, 284–290.

111 Santos, A.C.V. and Masini, J.C. (2006) Development of a sequential injection anodic stripping voltammetry (SI-ASV) method for determination of Cd(II), Pb(II) and Cu(II) in wastewater samples from coatings industry. *Analytical and Bioanalytical Chemistry*, **385**, 1538–1544.

112 Volikakis, G.J. and Efstathiou, C.E. (2005) Fast screening of total flavonols in wines, tea-infusions and tomato juice by flow injection/adsorptive stripping voltammetry. *Analytica Chimica Acta*, **551**, 124–131.

113 Nouws, H.P.A., Delerue-Matos, C., Barros, A.A. and Rodrigues, J.A. (2006) Electroanalytical determination of paroxetine in pharmaceuticals. *Journal of Pharmaceutical and Biomedical Analysis*, **42**, 341–346.

114 Kubiak, W.W., Latonen, R.M. and Ivaska, A. (2001) The sequential injection system with adsorptive stripping voltammetric detection. *Talanta*, **53**, 1211–1219.

115 Economou, A., Clark, A.K. and Fielden, P.R. (2001) Determination of Co(II) by chemiluminescence after *in situ* electrochemical pre-separation an a flow-through mercury film electrode. *The Analyst*, **126**, 109–113.

116 Romanus, A., Muller, H. and Kirsch, D. (1991) Application of adsorptive stripping voltammetry (AdSV) for the analysis of trace-metals in brine. 2. Development and evaluation of a flow injection system. *Fresenius' Journal of Analytical Chemistry*, **340**, 371–376.

117 Vazquez, M.D., Tascon, M.L. and Debran, L. (2006) Determination of Pb(II) with a dithizone-modified carbon paste electrode. *Journal of Environmental Science and Health Part A,-Toxic/Hazardous Substances & Environmental Engineering*, **41**, 2735–2746.

118 Han, J., Chen, H. and Gao, H. (1991) Alternating-current adsorptive stripping voltammetry in a flow system for the determination of ultratrace amounts of folic-acid. *Analytica Chimica Acta*, **252**, 47–52.

119 Khodari, M. (1993) Voltammetric determination of the antidepressant trimipramine at a lipid-modified carbon-paste electrode. *Electroanalysis*, **5**, 521–523.

120 Villar, J.C.C., Garcia, A.C. and Blanco, P.T. (1992) Determination of mitoxantrone using phase-selective ac adsorptive stripping voltammetry in a flow system with selectivity enhancement. *Analytica Chimica Acta*, **256**, 231–236.

121 Guo, S.X. and Khoo, S.B. (1999) Highly selective and sensitive determination of silver(I) at a poly(8-mercaptoquinoline) film modified glassy carbon electrode. *Electroanalysis*, **11**, 891–898.

122 Thomsen, K.N., Kryger, L. and Baldwin, R.P. (1998) Voltammetric determination of traces of nickel(II) with a medium exchange flow system and a chemically modified carbon paste electrode containing dimethylglyoxime. *Analytical Chemistry*, **60**, 151–155.

123 Ivaska, A. and Kubiak, W.W. (1997) Application of sequential injection analysis to anodic stripping voltammetry. *Talanta*, **44**, 713–723.

124 Sawamoto, H. and Gamoh, K. (1990) Desorption studies at a hanging mercury drop electrode by a flow injection method. *Journal of Electroanalytical Chemistry*, **283**, 421–424.

125 Sawamoto, H. (1993) Adsorption-desorption studies of alcohols at a hanging mercury drop electrode by a flow injection method. *Journal of Electroanalytical Chemistry*, **361**, 215–220.

126 Sawamoto, H. (1994) Adsorption of methanol, 1-propanol and 1-octanol on a mercury-electrode and adsorption of 1-octanol on the adsorbed layer of methanol at a mercury-electrode. *Journal of Electroanalytical Chemistry*, **375**, 391–394.

127 Sawamoto, H. (1997) The study of adsorption and reorientation of quinoline at mercury electrodes by measuring differential capacity-potential and differential capacity-time curves.

Journal of Electroanalytical Chemistry, **432**, 153–157.

128 Sawamoto, H. (2003) Reversible and irreversible adsorption of Surfactants at a hanging mercury drop electrode. *Analytical Sciences*, **19**, 1381–1386.

129 Pedrotti, J.J., Angnes, L. and Gutz, I.G.R. (1994) A fast, highly efficient, continuous degassing device and its application to oxygen removal in flow injection analysis with amperometric detection. *Analytica Chimica Acta*, **298**, 393–399.

130 Colombo, C. and van den Berg, C.M.G. (1998) In-line deoxygenation for flow analysis with voltammetric detection. *Analytica Chimica Acta*, **377**, 229–240.

131 Tercier-Waeber, M.L. and Buffle, J. (2000) Submersible online oxygen removal system coupled to an *in situ* voltammetric probe for trace element monitoring in freshwater. *Environmental Science & Technology*, **34**, 4018–4024.

132 Reinke, R. and Simon, J. (2002) The online removal of dissolved oxygen from aqueous solutions used in voltammetric techniques by the chromatomembrane method. *Analytical and Bioanalytical Chemistry*, **374**, 1256–1260.

133 Bobrowski, A. and Zarebski, J. (2000) Catalytic systems in adsorptive stripping voltammetry. *Electroanalysis*, **12**, 1177–1186.

134 Czae, M.Z. and Wang, J. (1999) Pushing the detectability of voltammetry: how low can we go? *Talanta*, **50**, 921–928.

135 Daniel, D. and Gutz, I.G.R. (2001) Long-optical-path thin-layer spectroelectrochemical flow cell with inexpensive gold electrodes. *Electroanalysis*, **13**, 681–685.

136 Daniel, D. and Gutz, I.G.R. (2003) Flow injection spectroelectroanalytical method for the determination of promethazine hydrochloride in pharmaceutical preparations. *Analytica Chimica Acta*, **494**, 215–224.

137 Daniel, D. and Gutz, I.G.R. (2005) Spectroelectrochemical determination of chlorpromazine hydrochloride by flow injection analysis. *Journal of Pharmaceutical and Biomedical Analysis*, **37**, 281–286.

138 Rubim, J.C., Gutz, I.G.R. and Sala, O. (1985) Cyclic-fluorovoltammetry as a technical tool in the study of passivating films generated on electrode surfaces. *Journal of Electroanalytical Chemistry*, **190**, 55–63.

139 Bard, A.J. (2004) *Electrogenerated Chemiluminescence*, CRC Press, New York.

140 Fähnrich, K.A., Pravda, M. and Guilbault, G.G. (2001) Recent applications of electrogenerated chemiluminescence in chemical analysis. *Talanta*, **54**, 531–559.

141 Lindino, C.A. and Bulhões, L.O.S. (2004) Determination of chloramphenicol in tablets by electrogenerated chemiluminescence. *Journal of the Brazilian Chemical Society*, **15**, 178–182.

142 Al-Gailani, B.R.M., Greenway, G. and McCreedy, T. (2007) A miniaturized flow injection analysis (µFIA) system with on-line chemiluminescence detection for the determination of iron in estuarine water. *International Journal of Environmental Analytical, Chemistry*, **87**, 637–646.

143 Gouveia, V.J.P., Gutz, I.G.R. and Rubim, J.C. (1994) A new spectroelectrochemical cell for flow injection analysis and its application to the determination of Fe(II) down to the femtomol level by surface-enhanced resonance Raman-scattering (SERRS). *Journal of Electroanalytical Chemistry*, **371**, 37–42.

144 Gorton, L. (1995) Carbon-paste electrodes modified with enzymes, tissues, and cells. *Electroanalysis*, **7**, 23–45.

145 Mello, L.D. and Kubota, L.T. (2002) Review of the use of biosensors as analytical tools in the food and drink industries. *Food Chemistry*, **77**, 237–256.

146 Baeumner, A.J. (2003) Biosensors for environmental pollutants and food contaminants. *Analytical and Bioanalytical Chemistry*, **377**, 434–445.

147 Prieto-Simon, B., Campas, M., Andreescu, S. and Marty, J.L. (2006) Trends in flow-based biosensing systems for pesticide assessment. *Sensors*, **6**, 1161–1186.

148 Koncki, R. (2002) Chemical sensors and biosensors based on Prussian blues. *Critical Reviews in Analytical Chemistry*, **32**, 79–96.

149 Jayasri, D. and Narayanan, S.S. (2007) Manganese(II) hexacyanoferrate based renewable amperometric sensor for the determination of butylated hydroxyanisole in food products. *Food Chemistry*, **101**, 607–614.

150 Ozoemena, K.I. and Nyokong, T. (2006) Novel amperometric glucose biosensor based on an ether-linked cobalt(II) phthalocyanine-cobalt(II) tetraphenylporphyrin pentamer as a redox mediator. *Electrochimica Acta*, **51**, 5131–5136.

151 da Rocha, J.R.C., Angnes, L., Bertotti, M., Araki, K. and Toma, H.E. (2002) Amperometric detection of nitrite and nitrate at tetraruthenated porphyrin-modified electrodes in a continuous-flow assembly. *Analytica Chimica Acta*, **452**, 23–28.

152 Wang, J., Wu, H. and Angnes, L. (1993) Online monitoring of hydrophobic compounds at self-assembled monolayer modified amperometric flow detectors. *Analytical Chemistry*, **65**, 1893–1896.

153 Pedrosa, V.A., Lowinsohn, D. and Bertotti, M. (2006) FIA determination of paracetamol in pharmaceutical drugs by using gold electrodes modified with a 3-mercaptopropionic acid monolayer. *Electroanalysis*, **18**, 931–934.

154 Zacco, E., Pividori, M., Llopis, X., Del Valle, M. and Alegret, S. (2004) Renewable protein a modified graphite-epoxy composite for electrochemical immunosensing. *Journal of Immunological Methods*, **286**, 35–46.

155 Brahim, S., Narinesingh, D. and Guiseppi-Elie, A. (2002) Polypyrrole-hydrogel composites for the construction of clinically important biosensors. *Biosensors & Bioelectronics*, **17**, 53–59.

156 Chailapakul, O., Siangproh, W. and Tryk, D.A. (2006) Boron-doped diamond-based sensors: a review. *Sensor Letters*, **4**, 99–119.

157 Arribas, A.S., Bermejo, E., Chicharro, M., Zapardiel, A., Luque, G.L., Ferreyra, N.F. and Rivas, G.A. (2006) Analytical applications of a carbon nanotubes composite modified with copper microparticles as detector in flow systems. *Analytica Chimica Acta*, **577**, 183–189.

158 Nagels, L.J. and Staes, E. (2001) Polymer (bio) materials design for amperometric detection in LC and FIA. *Trends in Analytical Chemistry*, **20**, 178–185.

159 Ricci, F. and Palleschi, G. (2005) Sensor and biosensor preparation, optimisation and applications of Prussian Blue modified electrodes. *Biosensors & Bioelectronics*, **21**, 389–407.

160 Trojanowicz, M. (2005) Electroanalytical flow measurements. *Annali di Chimica*, **95**, 421–435.

161 Hart, J.P., Crew, A., Crouch, E., Honeychurch, K.C. and Pemberton, R.M. (2004) Some recent designs and developments of screen-printed carbon electrochemical sensors/biosensors for biomedical, environmental, and industrial analyses. *Analytical Letters*, **37**, 789–830.

162 Vidal, J.C., Espuelas, J., Garcia-Ruiz, E. and Castillo, J.R. (2004) Amperometric cholesterol biosensors based on the electropolymerization of pyrrole and the electrocatalytic effect of Prussian-Blue layers helped with self-assembled monolayers. *Talanta*, **64**, 655–664.

163 Wang, J. (1992) in *Biosensors and Chemical Sensors* (eds P. Edelman and J. Wang), America Chemical Society, Washington, DC.

164 Guerrieri, A., Lattanzio, V., Palmisano, F. and Zambonin, P.G. (2006)

Electrosynthesized poly(pyrrole)/poly(2-naphthol) bilayer membrane as an effective anti-interference layer for simultaneous determination of acethylcholine and choline by a dual electrode amperometric biosensor. *Biosensors & Bioelectronics*, **21**, 1710–1718.

165 Ortuno, J.A., Sanchez-Pedreno, C. and Gil, A. (2005) Flow-injection pulse amperometric detection based on ion transfer across a water-plasticized polymeric membrane interface for the determination of verapamil. *Analytica Chimica Acta*, **554**, 172–176.

166 Ortuno, J.A., Sanchez-Pedreno, C., Hernandez, J. and Oliva, D.J. (2005) Flow-injection potentiometric determination of triiodide by plasticized poly(vinyl chloride) membrane electrodes and its application to the determination of chlorine-containing disinfectants. *Talanta*, **65**, 1190–1195.

167 Silva, I.S., Richter, E.M., do Lago, C.L., Gutz, I.G.R., Tanaka, A.A. and Angnes, L.

(2005) FIA-potentiometry in the sub-Nernstian response region for rapid and direct chloride assays in milk and in coconut water. *Talanta*, **67**, 651–657.

168 Amini, N., Cardwell, T.J., Cattrall, R.W. and Kolev, S. (2005) Determination of mercury(II) at trace levels by gas-diffusion flow injection analysis with amperometric detection. *Analytica Chimica Acta*, **539**, 203–207.

169 Oshima, M., Wei, Y.L., Yamamoto, M., Tanaka, H., Takayanagi, T. and Motomizu, S. (2001) Highly sensitive determination method for total carbonate in water samples by flow injection analysis coupled with gas-diffusion separation. *Analytical Sciences*, **17**, 1285–1290.

170 Santos, J.C.C. and Korn, M. (2006) Exploiting sulphide generation and gas diffusion separation in a flow system for indirect sulphite determination in wines and fruit juices. *Mikrochimica Acta*, **153**, 87–94.

17
Affinity Interaction Profiling of Protein–Protein and Protein–Ligand Interactions Using Flow Analysis

J. Kool, N.P.E. Vermeulen, H. Lingeman, R.J.E. Derks, and H. Irth

17.1
Introduction

During the last decade, advances in the pharmaceutical industry have fuelled developments in analytical chemistry. Impressive improvements in separation sciences and spectroscopic methods have facilitated research in key areas of drug discovery. In recent years the implementation of biochemical assays (BCAs) as a detection method in analytical chemistry has become very popular. Two main strategies can be distinguished, that is, high throughput screening (HTS) techniques based on microtiter-plate formats and continuous-flow BCAs in combination with flow analysis (FA) and high-performance liquid chromatography (HPLC) analysis. In this chapter we give an overview of continuous-flow BCA methodologies and their integration with key analytical techniques such as mass spectrometry (MS).

MS has become one of the most popular analytical techniques, mainly due to the increase in performance in terms of resolution and sensitivity, increasing compatibility with matrices of biological origin and the advent of new LC–MS interfaces. The use of MS in combination with BCAs is rather new, as most BCAs are based on either fluorescence or radiometric detection. While fluorescence and radiometric BCAs typically offer outstanding sensitivity and high reproducibility, the need to synthesize a fluorescent probe or substrate that retains significant affinity for receptor or enzyme often hampers the speed of BCA development. Another striking feature of the novel MS-based approaches is the technical set-up. While conventional HTS BCAs are typically performed in microtiter-plates, the novel MS-based BCAs are carried out in a continuous-flow environment. Moreover, these formats enable direct coupling of the BCA to separation techniques such as HPLC, thus enabling analysis of highly complex mixtures from biological or synthetic origin. Another interesting application of FA-MS is the analysis of noncovalent protein complexes, protein–ligand interactions and other biologically relevant protein–protein interactions, which evolved rapidly with the development and commercial availability of high performance mass spectrometers. This area has now become known as native MS.

Mainly for the reason that BCAs in flow injection analysis (FIA) mode can be coupled to powerful separation techniques, many different FIA technologies based on fluorescence have been developed lately. Apart from the fluorescence- and MS-based technologies, other detection methodologies are described in the literature that deal with BCA formats for which MS- or fluorescence-based detection strategies are not easily developed. When BCA formats rely on MS readout, these BCAs must be developed in flow (injection) analysis in order to introduce the tracer molecules or proteins to the MS. The FIA BCA formats, however, are compromised in terms of throughput (30–60 samples per hour) as compared to microtiter-plate formats (50–2000 samples per hour). Also, FIA BCAs with real on-line biochemical incubations in post-injection reaction coils are restricted to small reaction times (<5 min) as a result of band broadening in these coils. Apart from the fact that the BCA readout in compound mixtures only represents the biological activity present in the whole sample, no information is obtained about the identity of the compound(s) in the sample causing this biological activity. Also, no information about the concentration of these biologically active compounds is obtained. Hence the question: was the biological activity caused by a potent compound at low concentration or a weak compound at high concentration? Another problem which can occur when analyzing (very) complex samples is the interference of sample components with the biological interactions of the BCAs, resulting in inaccurate readouts. Analytical separation techniques can be used to overcome these problems. Several different approaches have been used to combine biochemical interactions with analytical separation techniques. These biochemical interactions were combined before, during and after analytical separations. An example of using biochemical interactions before the analytical separation is the use of immobilized affinity protein columns to preconcentrate the compound of interest and to "clean-up" the sample before the analytical separation step [1, 2]. Affinity chromatography is a separation technique based on biospecific interactions and an example of using biochemical interactions during the analytical separation step [1, 3]. Both approaches increase the selectivity of the analysis, but no real enhancement of the sensitivity is observed. A third approach is to combine the biochemical interactions after the analytical separation. This approach offers several advantages over the other two approaches. First, the detection principles of the BCA are still the same as in the batch BCA. Hence, similar sensitivity should be expected. Secondly, by separating the (complex) sample and removing interfering matrix components the selectivity of the method is also increased. The easiest way to do this is to collect fractions of the sample after the analytical separation, remove the solvents, redissolve the fractions into a BCA-compatible buffer and perform the BCA for each fraction. This approach has been widely used in many different application fields such as bio-analysis, environmental analysis, natural product screening and metabolic profiling [4–7]. However, this approach has several disadvantages. The chromatographic fractionation and the off-line measurement of all the fractions are laborious but a more serious disadvantage is the risk of losing the active compounds during the removal of the solvents after the fractionation step and the redissolving step. By coupling the biochemical detection (BCD) and an analytical separation step on-line these disadvantages are circumvented.

This chapter presents the various FA-MS-based BCA formats that can be employed in the affinity interaction profiling of protein–protein and protein–ligand interactions and the screening of bioactive compounds. It also focuses on other, mainly fluorescence-based, BCA formats in flow (injection) analysis format that were developed for coupling on-line to analytical separation technologies (gradient HPLC).

17.2
Profiling of Noncovalent Protein–Protein and Protein–Ligand Interactions Based on Mass Spectrometry Flow Analysis

For FA-MS-based BCAs, several fundamentally different approaches can be chosen. The interaction of bioactive compounds with proteins can be monitored by measuring

- the formation of noncovalent protein–ligand complexes;
- the concentration change of reporter molecules, that is, known active compounds upon interaction of the target protein with the analyte;
- analytes dissociated from the protein, after the protein–analyte complex has been separated from unbound compounds.

Figure 17.1 summarizes the different interactions involved; the species detected by MS are highlighted in gray boxes. In the case of enzyme inhibitor screening, the interaction of inhibitors with the enzyme can also be detected by measuring the change of either a substrate concentration or an enzymatic product concentration.

17.2.1
Flow Injection and Continuous Infusion Mass Spectrometry of Noncovalent Complexes Using Electrospray Ionization

17.2.1.1 Introduction
Electrospray ionization (ESI) and matrix-assisted laser desorption/ionization (MALDI) are the most suitable methods for generating gas phase ions of large biomolecules. Both ionization methods are so-called "soft ionization" techniques.

ESI as well as MALDI have been used to study various forms of complexes. Noncovalent bond energies are about an order of magnitude lower than covalent bond energies. This makes noncovalent complexes hard to detect with MS.

17.2.1.2 Electrospray Ionization Mass Spectrometry of Noncovalent Complexes
In the last few decades ESI has proven itself to be a very good ionization technique for detecting noncovalent complexes with MS. Katta *et al.* and Ganem *et al.* were the first groups to report the use of ESI-MS for the detection of noncovalent complexes in their studies on the globin–heme interaction of myoglobin [8] and the receptor–ligand complex [9]. Examples of noncovalent complexes studied are: receptor–ligand, enzyme–substrate, host–guest, intact multimeric proteins, DNA

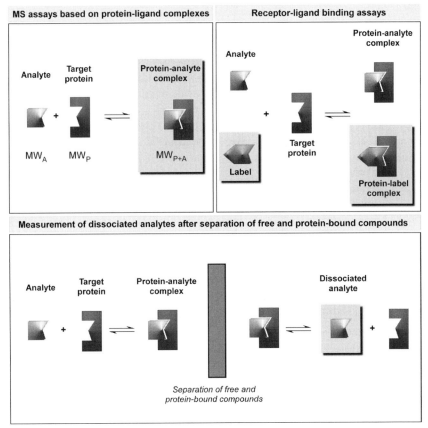

Figure 17.1 Different BCA formats employed in combination with mass spectrometry.

duplex and quadruplex species, oligonucleotide complexes with drugs and proteins and protein–drug complexes [8, 10–17]. The rapidly expanding field of native MS was thoroughly reviewed a few years ago by van den Heuvel and Heck [18] and since then this field of research is still increasing.

17.2.1.3 Limitations of Electrospray Ionization Mass Spectrometry for the Study of Noncovalent Complexes

A very important requirement for detecting noncovalent complexes with MS is that the solution used is MS compatible. Unfortunately, the typical solvent conditions used in ESI-MS to achieve maximum sensitivity are not always optimal for maintaining an intact noncovalent complex. Generally, a phosphate buffer and salts (e.g., sodium chloride) are used to mimic the physiological conditions needed to keep a noncovalent complex intact. In addition, a blocking agent (e.g., Tween 20) is often added to prevent non-specific binding of the noncovalent complex (and protein) to the surface. However, non-volatile additives in the eluent, such as phosphate buffer and blocking reagents, which can cause ion suppression, are not compatible with MS detection. The percentage of organic modifier (i.e., methanol, acetonitrile or isopropanol) has to be kept as low as possible to prevent dissociation of the noncovalent complex or even denaturation. However, ESI prefers an acidic (for positive ESI) environment and a certain amount of organic modifier. Keeping all the above in mind, a (delicate) balance is required between providing sufficient heating/ activation for ion desolvation and keeping interface conditions soft enough to preserve the noncovalent complex during the ESI process.

17.2.2
Flow Injection Mass Spectrometry Assays Using Reporter Molecules

17.2.2.1 Introduction
There are several possibilities to couple MS to a BCA. One possibility is coupling the MS in parallel to a (fluorescence-based) BCA. During recent years multiple applications have been developed for this technique [19–23]. This approach differs significantly from the direct MS-based method. Instead of using MS for the detection of bioactivity, biochemical readouts based on fluorescence are used (analytical scheme, see Figure 17.2). In this way biochemical and chemical information are obtained simultaneously (this will be discussed in detail later).

The traditional fluorescence BCA methods typically offer outstanding sensitivity and high reproducibility. Despite these advantages, however, the need to synthesize a fluorescent probe or substrate or ligand that retains significant affinity for receptor or

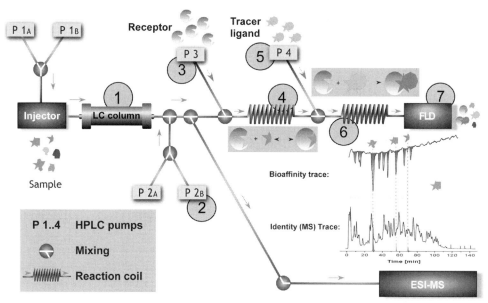

Figure 17.2 Configuration of an on-line receptor BCA. (1) Sample injection followed by gradient HPLC; (2) counter gradient followed by a flow split to MS and the BCA; (3) reagent pump for receptor solution; (4) reaction coil; (5) reagent pump for tracer ligand solution; (6) reaction coil; (7) fluorescence detector (FLD).

enzyme often hampers BCA development. Instead of developing a fluorescently labeled substrate or ligand, any substrate, formed product(s) or ligand used in an on-line BCA can also be monitored by MS in a direct MS-based approach.

17.2.2.2   Requirements for Mass Spectrometry-Based Biochemical Assays

In contrast to the requirements of the continuous-flow BCA, the MS prefers high organic modifier contents. The organic modifier content should therefore be carefully optimized in order to obtain good reaction kinetics in the BCA and good detectability/sensitivity in the MS. A second important requirement is the compatibility of the buffer, and buffer additives, used in the continuous-flow BCA with MS. Often in batch BCAs and in continuous-flow BCAs, which are not coupled to MS, physiological buffers and additives/salts can be used. Generally, a phosphate buffer and salts (e.g., sodium chloride) are used to mimic the physiological conditions needed in a BCA. In addition, blocking agents are often also added to the eluent to prevent sticking of the protein or protein complexes to the surface of the reaction coils or tubing. All these buffer salts, additives, and blocking agents can cause ion suppression and are therefore not suitable for continuous-flow BCAs coupled to MS. Often the phosphate buffer can be replaced by volatile buffers like an ammonium formate or ammonium acetate solution. Ammonium formate or ammonium acetate solutions at a pH of approximately 6–7.5 are a good compromise for the BCAs and are MS compatible.

17.2.2.3   Flow Injection Ligand-Binding Assays Using Mass Spectrometry as Readout

An example of a competitive MS-based BCA was first described by Hogenboom *et al.* [24]. They developed two different competitive binding BCAs to obtain proof-of-principle. Streptavidin and fluorescein-biotin were respectively selected as the model molecular target and reporter ligand for the BCA. As the affinity interaction between biotin and streptavidin is one of the strongest currently known in nature $(K_a = 0.6 \times 10^{15}\,L\,mol^{-1})$, the affinity complex is therefore likely to be preserved during the ionization process. A fluorescent reporter ligand was used to demonstrate that MS-based BCA formats generated similar results as their fluorescence-based counterparts. In order to demonstrate proof-of-principle for weaker affinity interactions anti-digoxigenin and digoxin were chosen as the target protein and reporter ligand for the second MS-based BCA $(K_a = \sim 10^9\,L\,mol^{-1})$.

17.2.2.4   Flow Injection Enzyme Assays Using Mass Spectrometry as Readout

An MS-based enzymatic BCA format was recently reported by De Boer *et al.* [25]. In this case, the protease cathepsin B was used as the model molecular target. During the biochemical reaction cathepsin B converted the substrate into two products, respectively, Z-FR and AMC, which were then continuously monitored by MS. If no inhibitors were injected, the ion currents of the products Z-FR and AMC reached a maximum level. Upon injection of inhibitors, however, the enzymatic conversion rate was reduced, causing a decrease in the concentration of both Z-FR and AMC. A slightly different, but important approach to screen for products of enzymatic conversions is the so-called pulsed ultrafiltration MS [26]. This technology, which can also be applied in receptor–ligand interaction screening, can be used for medium throughput screening of enzymatic products. In an early application, ligands contained in a library mixture were bound to a macromolecular receptor, followed by purification of the ligand–receptor complexes by ultrafiltration and finally dissociation using methanol to elute the ligands into the electrospray mass spectrometer for detection [27]. Later, the same group developed similar methodologies to evaluate the metabolic stability of drugs [28].

17.2.3
Reporter-Free Assays after Dissociation of Protein–Ligand Complexes

In addition to the continuous-flow BCA formats several MS-based techniques have been developed recently which enable detection of ligands without the need for reporter molecules [29–32]. These so called reporter-free MS-based approaches have been used to perform "ligand fishing" in mixtures of synthetic compounds or in samples of biological origin. Fully automated LC-UV/MS systems have been reported by Hsieh *et al.*, Kaur *et al.* and Lenz *et al.* [33–35]. In general, the protein targets are incubated with compound libraries followed by separation of affinity-bound and unbound ligands. A similar approach was reported recently in which two sequentially configured restricted access columns were used to isolate ligands present in complex matrices [29]. Figure 17.3 shows the flow scheme used. This approach shows many resemblances to the ultrafiltration based approaches described in Section 17.2.2.4. Another reporter-free system for measuring protein–protein and protein–ligand interactions is surface plasmon resonance (e.g., the Biacore system). Although this is not an MS-based flow analysis methodology, it is one of the most important

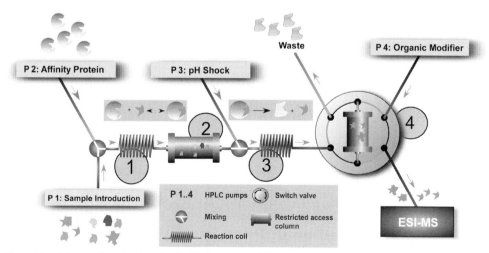

Figure 17.3 Scheme of the MS-based BCA. (1) Reaction coil for association; (2) trapping of unbound ligands by restricted-access column; (3) reaction coil for dissociation; (4) trapping of bioactive molecules by restricted-access column.

technologies for measuring real affinities and association and dissociation rates of biomolecular interactions. As the system is more or less mass selective, it is most suitable for large ligand interactions, however with the latest improvements in technology small ligand binding studies also become in reach [36].

17.3
Integration of Flow Analysis and High-Performance Liquid Chromatography for the Bioaffinity Screening of Mixtures

17.3.1
Introduction

Compound activity screening is common practice in drug discovery and it is often performed in high throughput fashion to facilitate the screening of large numbers of compounds [37]. Compound screening is usually performed in a broad variety of different BCAs. While radioligand-based affinity BCAs were the primary BCAs in the past, the use of radiolabeled tracer molecules and the heterogenicity of the BCAs render them less suitable for high throughput automatization than, for example, fluorescence-based BCAs. This has resulted in the development of fluorescence-based BCAs in all areas of drug discovery. Examples such as time-resolved fluorescence resonance energy transfer, fluorescence anisotropy, fluorescence correlation spectroscopy, fluorescence fluctuation spectroscopy and normal fluorescence-based BCAs can be named in this regard [38]. A primary bottleneck in compound screening is the identification of the compound or compounds that are responsible for biological activities in complex mixtures, such as combinatorial and natural compound libraries. This also holds true for the screening of metabolic mixtures [39]. The activities and/or affinities of metabolites, which usually closely resemble the structure of the mother compound, are extremely important determinants in the final actions of a drug. Cumbersome labor-intensive and costly dereplication processes usually have to be performed in order to separate the compounds in the mixtures (usually in 96- to 1536-well plates) before the separated compounds can be resubmitted to the original BCAs [4, 40, 41]. An example is shown in Figure 17.4.

Receptor screening technologies [42, 43] do allow affinity screening of individual components, but they are usually not capable of detecting individual compounds in mixtures. Some receptor affinity screening strategies concern the binding of active metabolites to a receptor followed by off-line centrifugal ultrafiltration [27, 44] or restricted access column isolation [29], thus resulting in the collection of bound ligands and allowing characterization by LC–MS. These technologies, however, lack the ability of efficient trapping of low affinity ligands in the presence of high affinity ligands or in the presence of high concentrations of parent compound in the case of metabolic mixtures. This major bottleneck is circumvented by direct on-line post-column connection of the desired flow analysis BCAs to HPLC [45], indicated as flow analysis BCD in HPLC mode (FA-BCD-HPLC). When performed in parallel with MS measurements, chemical information on the individual compounds could be linked

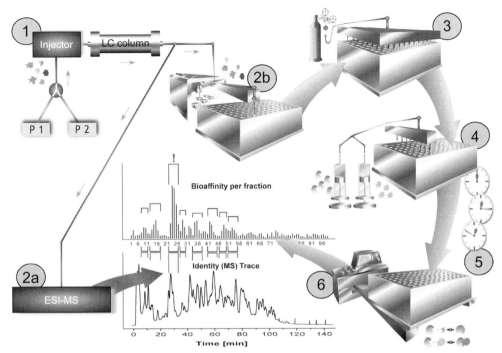

Figure 17.4 Active metabolite profiling using an HPLC-based dereplication format with 96-well plates. BCA-guided detection of active metabolites with characterization by MS allows identification of active metabolites in metabolic mixtures. Metabolite mixtures are separated by HPLC (1) followed by splitting to MS (2a) and a fraction collector for fractionation into 96-well plates (2b). The fractions are then dried (3), mixed with BCA components (4), incubated (5) and screened for activity with a plate reader (6). Finally, a correlation is made between the fractionated bioaffinities and the MS trace.

directly to their biological activities [21, 22]. This section reviews the evolution and current state of FA-BCD-HPLC. Principles and applications are also discussed and advantages and drawbacks are evaluated.

17.3.2
On-line Coupling of Flow Biochemical Assays to HPLC

The term FA-BCD-HPLC dates from the 1990s and was introduced with the development of the first truly on-line post-column BCA [45]. With the advent of FA-BCD-HPLC, novel approaches arose in screening methodologies, thereby opening the way for the rapid and efficient screening of complex mixtures. The ease of parallelization allowed the simultaneous screening towards several targets [23]. A general scheme of a FA-BCD-HPLC setup which employs BCA parallelization is shown in Figure 17.5. The following section gives a detailed description of the different BCA formats.

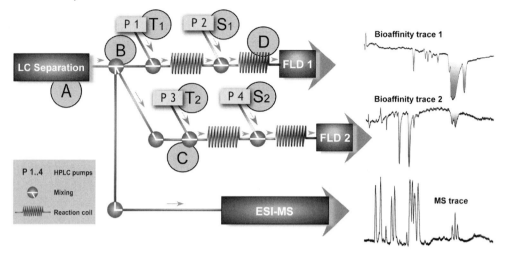

Figure 17.5 General scheme of a FA-BCD-HPLC set-up with two parallelized BCAs. After gradient HPLC with post-column make-up gradient (A), the total flow is split (B) to the two parallelized BCAs. Via subsequent BCA T-unions (C), the target proteins (T_1 and T_2) are added to the split parts of the HPLC effluent in reaction coils (D). Substrate or ligand (S_1 and S_2) is then added to the following reaction coils. After the on-line bioreactions, a fluorescence readout is commonly employed (FLD). Parallel MS readout allows identification of active compounds. A parallel BCA format provides selectivity data when two similar targets are used (e.g., ER-α and ER-β).

17.3.2.1 General Principles

The main characteristic of FA-BCD-HPLC-based BCAs is the continuous mixing of target protein and tracer molecules, usually from different reservoirs, the effluent of a separation technique, such as HPLC [46]. After mixing in a reaction chamber, which commonly consists of coiled or non-coiled tubing, the reaction products go directly to a detection unit (homogeneous format) or first pass some kind of separation step to separate bound or reacted tracer molecules from non-bound or non-reacted tracer molecules (heterogeneous format). To maintain relatively stable target and tracer solutions over time, they are usually stored at appropriate temperatures (0–4 °C) and introduced into the reaction unit by superloops [47], which are actually large volume syringes. When eluting compounds interact with the target screened for, temporary changes in detector signal are reported as they elute.

Basically, FA-BCD-HPLC assay formats can be divided into two major classes, namely heterogeneous and homogeneous BCAs. In heterogeneous assay formats, tracer ligand that is not bound to the BCA target is trapped prior to detection (e.g., with on-line restricted access or affinity columns). The main advantages of heterogeneous BCA formats are their intrinsically high sensitivity due to low background fluorescence and trapping of potentially interfering compounds. An example is given by Oosterkamp *et al.*, who simultaneously developed a heterogeneous and homogeneous BCA format for the estrogen receptor (ER) [48]. BCAs that rely mainly on homogeneous formats are enzymatic BCAs in which the inhibition of the conversion of a non-fluorescent substrate into a fluorescent product by test compounds is a measure of enzyme affinity [49]. While heterogeneous BCAs usually prove to be less stable due to clogging risks of the restricted access columns used, as is the case with the ER, the removal of the background fluorescence usually renders the BCA more sensitive. A FA-BCD-HPLC-based receptor BCA in heterogeneous format is almost identical to the homogeneous FA-BCD-HPLC scheme discussed in Figure 17.2. In the heterogeneous case, the following parts are distinguished: (1) Effluent from HPLC, (2) counter gradient followed by a flow split to MS and the BCA, (3) addition of ER, (4) reaction coil, (5) addition of the fluorescent ligand coumestrol, (6) reaction coil. The second reaction coil is then followed by a restricted access column (not shown in the figure). (7) Fluorescence detector. When coupling gradient HPLC on-line to BCAs, the well-known incompatibility of most biochemical targets with organic modifiers is slightly compensated by the relatively short post-column BCA reaction times. Another important aspect in this regard is the increasing concentration of organic modifier when gradient HPLC separation is applied. This problem can efficiently be solved by the addition of a counteracting post-column gradient (Figure 17.2; point 2), allowing an effluent with a constant concentration of organic modifier to enter the BCAs [22].

17.3.2.2 On-line Ligand-Binding Flow Assays Coupled to HPLC

The first post-column on-line BCD system was developed by Przyjazny et al. [50]. This proof of principle set-up made use of the high affinity interactions of avidin with biotin and biocytin. Avidin with an affinity dye was continuously mixed with the HPLC effluent. Upon biotin or biocytin elution, the dye was displaced from the active site of avidin, resulting in optical changes which were measured by UV. Although this on-line BCA showed novel possibilities for rapid screening of biochemical interactions in mixtures, the UV-based detection methodology rendered the BCA format less useful for the screening of unknown mixtures as many compounds in mixtures show absorbance at the wavelength used. Subsequently, Przyjazny et al. developed a similar post-column homogeneous fluorophore-linked BCA [51]. While this set-up largely overcame the UV-associated problems, this BCA format posed problems for biotin-like compounds that did not result in fluorescence enhancement when bound in the active site. This general problem was circumvented with the development of another on-line biodetection strategy [45]. This strategy made use of immobilised digoxin and fluorescently labeled anti-digoxigenin FAB fragments for the detection of digoxin-like molecules. Eluting digoxins can react with the fluorescent antibodies, thereby occupying their active sites and rendering them inactive for further binding to an affinity column with immobilized digoxin placed upstream. The non-retained fluorescent antibodies can subsequently be detected with a fluorescence detector.

Table 17.1 lists most of the currently developed FA-BCD-HPLC assay formats and their BCA characteristics. All these FA-BCD-HPLC assays are discussed in this chapter. From the FA-BCD-HPLC assays developed, most BCAs are based on fluorescence detection. In the last few years, however, FA-BCD-HPLC assays have been developed which use a mass spectrometer for BCD. One of the few exceptions is a FA-BCD-HPLC assay developed by Neungchamnong et al. For the detection of anti-oxidants in mixtures, they developed a FA-BCD-HPLC strategy capable of on-line

Table 17.1 Currently developed on-line flow BCA formats sorted in assay format.

Year	Author	Target	Analytes	Refs.
Optical changes of ligand when bound to target or by substrate conversion				
1990	Przyjazny	Avidin	Biotin-like	[50]
2000, 2005	Ingkaninan, Fabel	AChE	AChE inhibitors	[68, 69, 86]
2005	Nuengchamnong	Free radicals	Antioxidants	[52]
Fluorescence enhancement of ligand when bound to target				
1993	Przyjazny	Avidin	Biotin-like	[51]
1996, 2001, 2004,	Oosterkamp, Schobel, van Elswijk,	Estrogen Receptor	(Xeno)-Estrogens	[22, 23, 48,
2006	Kool, van Liempd			63, 64]
2003	Schenk	Fluorophore-labeled phosphate binding protein	Phosphate consuming or releasing enzymes	[20]
Fluorescent product formation				
2003	Rhee	Acetylcholine esterases (AChE)	AChE inhibitors	[49]
2003	Schenk	Phosphodiesterases (PDEs)	PDE inhibitors	[21]
2005, 2006	Kool	Cytochromes P450 (CYPs)	CYP inhibitors	[77, 78]
2007	Kool	GSTs	GST inhibitors	[79]
2007	Kool	ROS targeted biomolecules	ROS producing compounds and antioxidants	[85]
Time resolved FRET based bioaffinity screening				
2004	Hirata	Tyrosine kinase	Tyrosine kinase inhibitors	[70]
Heterogeneous assay format based on immobilized antigen/ligand column				
1993, 1994	Irth, Oosterkamp	Fluorescein-labeled antibodies	Digoxin-like	[45, 46]
1996	Miller	Fluorescein-labeled antibodies	GCSF-like	[55]
1999, 2000	Van Bommel	Enzyme-labeled streptavidin	Biotin-like	[58–60]
2001	Schenk	Fluorescein -abeled antibodies	Cytokines	[56]

Year	Author	Column / probe	Analyte	Reference
Heterogenous assay format based on immobilized antibodies/target protein column				
1996	Oosterkamp	Antibodies	Leukotrienes	[57]
1998	Oosterkamp	Fluorescein-labeled urokinase receptor	Urokinase receptor ligands	[67]
Heterogenous assay format based on flow cytometry				
2003	Schenk	Fluorescein-labeled antibodies	Digoxin-like	[47]
Heterogeneous assay format based on hollow membrane filter				
1996	Lutz	Antibodies	Biotin-like	[61]
Heterogeneous assay format based on free flow electrophoresis				
2000	Mazereeuw	Streptavidin	Biotin-like	[62]
Theoretical concepts of on-line screening				
1997	Oosterkamp	Estrogen Receptor	Estradiol	[53]
1997	Oosterkamp	Biotin based affinity resin	Biotin	[54]
Binding to optical biosensors				
2000	Haake	Antibodies	Isoproturon pesticides	[32]
MS based affinity interaction				
2001, 2003	Hogenboom, Derks	Antibodies	Biotin and digoxin	[24, 87]
2003, 2005, 2006	Krabbe	Iron(III)methyl calcein blue	Analytes (e.g., phosphorylated peptides) with high affinity for metal ions.	[88–90]
2007	Krabbe	Iron(II)methyl calcein blue	Metal–ligand complexes	[91]
MS based enzyme product formation				
2003, 2006	Van Elswijk, de Jong	Angiotensin converting enzyme (ACE)	ACE inhibitors	[19, 92]
2004, 2005	De Boer	Cathepsin B	Cathepsin B inhibitors	[25], [72], [80]
2007	Bruyneel	Protein identification	Post-column digested peptides	[93]

post-column detection of anti-oxidants in mixtures based on UV–Vis detection [52]. On-line post-column chemical oxidation of a model compound resulted in the continuous formation of a colored product, which can be monitored spectrophotometrically.

When transferring batch-like BCA formats into on-line systems, one of the important questions is whether the biochemical characteristics of the target protein in the batch format resemble the characteristics in the on-line format. Because multiple binding sites on the target molecule might complicate the interactions, Oosterkamp *et al.* used theoretical models to show that biochemical interactions are preserved in on-line systems that employ labeled ligands [53] or when labeled proteins are used, also if multiple binding sites are present [54]. The theoretical models were ultimately validated with a (fluorescence-enhancement-based) receptor affinity detection BCA [48] and fluorescein-labeled streptavidin and biotin, respectively.

FA-BCD-HPLC assay formats based on antibodies are usually very stable and sensitive due to the inherent stability and high affinity of the antibodies. Therefore, these BCA formats are used frequently for proof of principle studies on new FA-BCD-HPLC methodologies. An on-line BCA format, based on high affinity antibody interactions, was the on-line post-column methionyl granulocyte colony stimulating factor analysis [55]. The development by Schenk *et al.* of an on-line post-column BCA sensitive to cytokines and based on immunochemistry created the possibility of rapidly determining relative concentrations of cytokines, thereby allowing the measurement of important cytokine profiles in biological matrices [56]. Another similar on-line BCA relied on the passage of fluorescence-labeled leukotrienes through an immobilized antibody column bound to antibodies in solution. Eluting leukotrienes compete with the fluorescence-labeled leukotrienes for the limited amount of antibodies present in solution, thereby altering the amount of detected fluorescence-labeled leukotrienes [57]. Enzymatic signal amplification was examined by van Bommel *et al.* [58, 59]. Compared to the strategy described by Oosterkamp *et al.* [48], which employed fluorescein-labeled streptavidin and biotin, here the fluorescein label on the streptavidin was replaced by a peroxidase. This enzyme-based signal amplification did afford a more sensitive BCA than the fluorescence-based on-line BCA. A similar system, which was developed for the determination of proteins, was also described by van Bommel *et al.* [60]. With the development of a hollow-fiber-based on-line BCA, new possibilities for separating free and bound tracer in continuous flow immunochemical detection arose [61]. Moreover, the use of this methodology was also demonstrated with small enzymatic labels. Another immunoaffinity-based BCA was the on-line coupled label-free optical biosensor based on highly cross-reactive antibodies against the pesticide isoproturon [32]. Yet another technology, in which free flow electrophoresis was used, employs an electric field to generate a laminar flow perpendicular to the on-line BCA flow [62]. All biochemical reagents undergo this laminar flow with different mobilities because of their different electrophoretic characteristics. This allows bound and non-bound tracer to be separated, after which optical detection took place through the transparent electrophoresis cell. Other novel and recent FA-BCD-HPLC-based approaches are the use of flow cytometry for on-line post-column detection [47]. With this technology, fluorescence interferences of eluting compounds and unbound fluorescent immunoreagents are not detected. The principle is based on the cytometer's sole measurement of fluorescent immunoreagents bound to antigen-labeled beads. In this respect, the BCA principle is similar to that described by Oosterkamp *et al.* [46] and could also be designated as a heterogeneous BCA in which the immobilized support is continuously pumped through. This results in a BCA format having both the stability characteristics of a homogeneous BCA and the advantages of a heterogeneous BCA, such as fluorescent background elimination and enhanced sensitivity [48]. The antibody- or avidin-based BCA formats, described in this section, were used in most cases to demonstrate a new FA-BCD-HPLC methodology. Most of the mentioned methodologies, however, are restricted to use with high affinity interactions. As enzyme systems generally show much lower affinities compared to antibody- or avidin-based systems, they are usually not suited for these strategies. This also holds true for receptor systems, although receptor systems generally have higher affinities than enzymes. For receptor systems, the generally lower receptor concentrations and complex biological behavior make them less suitable for these BCA formats. Moreover, many enzymes and receptors are membrane bound and will therefore probably clog on-line heterogeneous BCA formats, in which free and bound ligand are to be separated prior to detection.

A post-column BCA for the human ERα was the first on-line receptor affinity BCA [48]. This on-line BCA has already been discussed in more detail in Section . The BCA showed its value in the analysis of mixtures containing multiple estrogens, such as environmental mixtures with contaminating estrogens. Applications of this methodology were reported by Schobel *et al.* [23] and van Elswijk *et al.* [22]. They used the technology, in which ERα and β were used in parallel with MS, to rapidly screen for ERα and/or ERβ selective compounds in natural extracts. In 2006, this methodology was used to demonstrate on-line ERα affinity screening and MS identification of individual metabolites in metabolic mixtures of model drugs like tamoxifen and raloxifen [63, 64]. Figure 17.6 shows tamoxifen (TM) metabolites formed during microsomal incubations (MS traces) and the corresponding ERα-affinity traces with the post-column BCA. It must be stated, however, that ERα-affinity responses can be measured very rapidly due to fast receptor–ligand kinetics, hence, resulting in a successful FA-BCD-HPLC methodology [48]. Also, the high stability towards organic modifiers of the ERα that was used in the study renders the BCA format very suitable for FA-BCD-HPLC purposes. In this regard, one can think of less suitable receptor systems, like GPCRs, which need very delicate BCA formats [65] that prevent loss of activity due to receptor destruction or solubilization from membranes.

The urokinase receptor plays an important role in cell adhesion, migration, proliferation and crucial matrix degradations and is involved in intracellular degradation mediated diseases and effects proteolitic cascade systems of plasminogen activations [66]. This receptor was labeled with fluorescein for a heterogeneous on-line BCA format similar to most heterogeneous immunoaffinity BCAs [67]. For receptor-based FA-BCD-HPLC assays, receptor concentrations in the BCA have to be

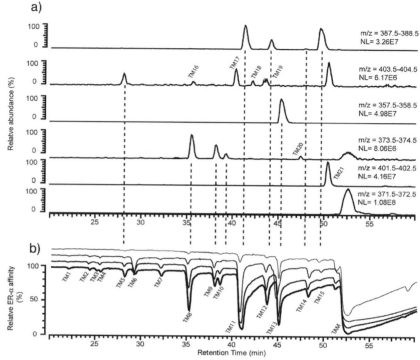

Figure 17.6 TIC chromatograms of mass ranges corresponding to tamoxifen metabolites with or without ERα-affinity (A). ERα affinity traces at increasing incubation time. From top to bottom the affinity traces correspond to 0, 7, 17, and 26 min microsomal incubation (B). TIC chromatograms are correlated with ERα affinity traces of the on-line ERα affinity BCA.

relatively high in order to obtain good signal to noise ratios in short reaction times. Also, the readout of many receptor-based batch BCAs is based on second messenger formation. As the formation of second messengers in these (often adherent-cell-based) BCAs usually occurs in time spans unsuitable for the relatively short post-column BCA times in FA-BCD-HPLC assays, monitoring receptor interactions this way is not an option. FA-BCD-HPLC assays, therefore, can only be used for a limited number of receptor systems with the current assay formats and biochemical detection methodologies available.

17.3.2.3 On-line Enzyme Flow Assays Coupled to HPLC

The first on-line post-column enzyme affinity BCA was described by Ingkaninan *et al.* [68]. This colorimetric BCA for acetylcholine esterase (AChE) activity used the common strategy in which a substrate is converted into a colored product. Enzyme inhibition due to eluting ligands resulted in a change in the UV baseline due to temporarily decreased formation of product. Ingkaninan *et al.* later used this methodology in combination with parallel MS to rapidly examine plant extracts for individual active compounds with simultaneous identification [69]. In 2003, Rhee *et al.* improved the on-line AChE inhibitor BCA with the replacement of the colorimetric probe with the substrate 7-acetoxy-1-methyl quinolinium iodide [49]. A generic phosphate consuming BCA developed for on-line post-column purposes allowed the screening of mixtures for phosphate consuming or releasing enzymes [20]. Phosphodiesterases, which play important roles in cell signalling, for example, by hydrolyzing second messengers like cAMP and cGMP, were the targets in an on-line post-column detection method described by Schenk *et al.* [21]. The conversion of the fluorescent substrate Mant-cGMP by phosphodiesterases into the highly fluorescent Mant-GMP was key to this on-line BCA. A similar strategy was exploited in a FA-BCD-HPLC system for angiotensin converting enzymes (ACE) [19]. With this methodology, bioactive peptides found in hydrolyzed milk samples were simultaneously identified with MS. While most fluorescence-based enzymatic BCAs rely on straightforward fluorescence changes upon product formation, a more delicate approach, namely time resolved fluorescence resonance energy transfer (TR-FRET), was used by Hirata *et al.* [70]. This on-line FIA BCA used TR-FRET-dependent emission of light from a fluorescently labeled substrate bound to a fluorescently labeled antibody when phosphorylated by tyrosine kinase. In a FRET-based follow-up study by Hirata *et al.*, another separation strategy than the normally used HPLC separation, namely size-exclusion chromatography, was applied [71]. Here, Hirata *et al.* described the separation of a mixture of compounds containing protease inhibitors followed by on-line post-column screening of inhibitors towards the protease Subtilisin Carlsberg. This BCA is based on a substrate peptide which is labeled with two fluorescent compounds. After digestion by the protease, the substantial intramolecular FRET quenching is reduced, leading to an increase in fluorescence. For screening of expensive targets, miniaturization is very desirable. In order to reduce target consumption, a continuous flow microfluidic system capable of performing on-line BCAs has been developed [72]. With a total flow rate of only 4 μl min^{-1}, considerable reductions in costs compared to traditional on-line BCA formats were made. With the cysteine protease cathepsin B as model enzyme, however, lower sensitivities were obtained compared to conventional on-line BCAs, but non-specific binding was minimized.

Biotransformation enzymes play a crucial role in the metabolism of both endogenous compounds and xenobiotics. Cytochromes P450 (CYPs) are the most important phase I metabolism enzymes and play a central role in oxidative metabolic reactions of drugs and other xenobiotics [73, 74]. Affinity of drugs and metabolites towards CYPs can result in unwanted drug–drug interactions at the level of drug metabolism. Phase II metabolism can also result in adverse drug reactions. The phase II metabolic enzymes, among which are the glucuronosyl transferases, sulfotransferases and glutathione *S*-transferases (GSTs), add substituents to functional groups in compounds, usually resulting in strongly increased hydrophilicity and facilitated excretion. As metabolism of drugs can result in the formation of pharmacologically active, drug–drug interacting and/or reactive metabolites, it renders them essential for early consideration in drug discovery and development programs [75]. The screening process of compounds often employs HTS methodologies [37]. However, most HTS-methodologies used do not allow the identification of individual ligands in metabolic mixtures. These mixtures have to be separated chromatographically before affinity screening of individual compounds can occur [39, 76]. With the on-line coupling of a novel CYP enzyme affinity detection (EAD) system to gradient HPLC, for the first time CYP affinity screening of individual compounds in mixtures was enabled by Kool *et al.* [77]. By a parallelized configuration of different CYP EAD systems, a FA-BCD-HPLC methodology was devised which is capable of simultaneous affinity screening of individual compounds in mixtures towards a panel of up to three relevant CYPs [78]. In addition to the present FA-BCD-HPLC methodologies for screening CYP interactions, GST interaction screening with similar methodologies can aid the drug discovery and development process as well. Since GSTs scavenge reactive intermediates and electrophilic compounds, a FA-BCD-HPLC system for GSTs might find use in identifying glutathione-conjugates and electrophilic compounds in mixtures. Recently, a parallel FA-BCD-HPLC methodology for screening for iso-enzyme specific GST inhibitors in mixtures was described [79]. As an example, a parallel EAD chromatogram (measured with a parallel FA-BCD-HPLC configuration with post-column split to a cytosolic GSTs BCA and to a human GST Pi BCA) obtained from a test mixture is shown in Figure 17.7. One of the problems seen with this FA-BCD-HPLC methodology is the severe tailing of apolar compounds that elute at higher concentrations of organic modifier (see Figure 17.7). This is probably caused by precipitation of apolar compounds after mixing the post-column counteracting gradient. However, compounds with affinity that elute shortly after each other appear superimposed and can still be distinguished. Other separation methodologies

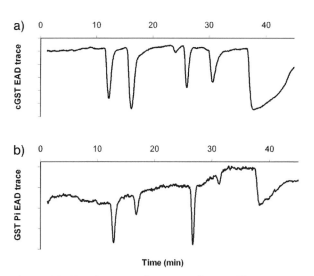

Figure 17.7 GST EAD traces of a mixture of six compounds injected in the total FA-BCD-HPLC system for GSTs. Injected compounds are: acrolein (3600 μM; 12.5 min), crotonaldehyde (2900 μM; 16.5 min), 2-chloro-5-nitropyridine (2400 μM; 24.0 min), cinnumaldehyde (1500 μM; 26.5 min), 4-chloro-3-nitro-benzoic acid butyl ester (300 μM; 33.0 min) and ethacrynic acid (300 μM; 38.5 min). (A) cGSTs EAD trace. (B) GST Pi EAD trace.

like high temperature liquid chromatography (HTLC) or size exclusion chromato-graphy (SEC) can be used to reduce organic modifier concentrations in the on-line BCAs, but the resolutions obtained are still inferior to conventional HPLC [71, 80].

ROS-producing compounds, sometimes generated upon bio-activation, are gen-erally regarded as hazardous compounds. As the products of ROS-producing compounds can react with antioxidants in the same mixtures, false negatives might result from traditional batch BCAs [81, 82] usually conducted with mixtures. In principle, on-line HPLC-based BCAs that measure individual antioxidants in mix-tures already exist [83, 84], however, these BCAs are not suitable for the simultaneous screening of both pro-oxidants and antioxidants in mixtures. With the development of a FA-BCD-HPLC methodology for ROS-producing compounds and antioxidants, the rapid screening of such compounds in mixtures can be performed in a single run [85].

When the rate of formation of fluorescent product in EAD systems is low, the resolution of detected ligands also declines as longer reaction times (thus longer reaction coils) are needed to create sufficient signal to noise ratios. Therefore, enzymatic- or receptor-based FA-BCD-HPLC systems giving rapid and sensitive responses [19, 22] are essential for FA-BCD-HPLC systems to be competitive with HTS systems. Attempts to allow longer on-line post-column reaction times with minimum reaction coil band broadening, however, have been reported by Fabel et al. [86]. This was done by applying gas-segmented flows in these reaction coils. Since the need for fluorescent targets or ligands may seriously complicate BCA development if none are easily available, MS-based on-line BCAs were developed to overcome this problem [24, 25, 87]. In these BCAs, the tracer was not monitored by its fluorescent characteristics, but by its characteristic molecular mass [24]. One of the main advantages of MS, namely the ease of measuring different ligand–target interactions in one BCA, was demonstrated by Derks et al. [87]. De Boer et al. [25] used ESI-MS as a detection method for the identification of cathepsin B inhibitors by measuring changes in the formation of enzymatic products (AMC and Z-FR) of particular cathepsin B substrates. Another use of MS for on-line post-column BCD involved the detection of analytes with a high affinity for metal ions [88]. Selective ion monitoring of a tracer ligand (methylcalcein blue) that is released upon ligand-exchange of the metal–ligand complex (iron(III)methylcalcein blue) by eluting analytes was used as the basis for this type of analysis. This system was able to detect phosphorylated peptides as competing affinity compounds. Non-phosphory-lated and phosphorylated peptides were separated by HPLC prior to the on-line BCD. A modified format of this BCA allowed the screening of ligands with affinity for certain metal ions after an on-line ligand exchange reaction [89, 90]. The same author also used ESI-MS to investigate complex formation of different metal complexes in a continuous-flow ligand-exchange reactor [91]. In another study, an on-line BCA in which enzymatic product formation was monitored with MS was used to screen for acetylcholine esterase (AChE) inhibitors [92]. This BCA monitored the hydrolysis of acetylcholine with MS. Inhibitors eluting from HPLC temporarily inhibit AChE, thereby altering the acetylcholine hydrolysis, which is monitored by MS. The MS simultaneously measured the mass spectrum of the eluting inhibitor. Analytical methodologies for the identification of proteins typically include a digestion step, often using trypsin as the proteolytic enzyme. In the majority of cases, off-line and on-line digestion methods are implemented prior to an LC-MS analysis system in order to obtain high sequence coverages for unambiguous protein identification. For proteins with a strong overlap in amino acid sequence, for example, therapeutic proteins and their metabolites, it is essential to separate proteins prior to digestion and the subsequent electrospray MS analysis of marker peptides. Bruyneel *et al.* [93] demonstrated an on-line post column solution-phase digestion methodology that is based on the continuous infusion of the proteolytic enzyme pepsin downstream to a nano-C18 reversed-phase column. Proteins are identified based on their retention time in combination with the detection of specific marker peptides formed in the post column digest.

FA-BCD-HPLC assays based on MS detection are a possible option for enzymatic BCA systems if fluorescence-based BCA formats are not available. These BCA formats, however, are seriously hampered by the fact that optimal BCA conditions usually employ buffer and co-factor conditions that are not compatible with MS. For these BCAs, in which enzymatic product formation is followed, MS compatible BCA conditions have to be chosen. This most probably results in BCA conditions far from optimal or BCA conditions that render the BCA unsuitable for on-line BCA formats. When looking at MS-based FA-BCD-HPLC assays in which binding interactions are monitored, high affinity interactions are needed to obtain good signal to noise ratios. Therefore, these BCA formats can probably not be used for most receptor systems that generally have lower affinity ranges as compared to antibody and avidin-based systems. Also, when lower receptor concentration ranges are used as compared to antibody and avidin-based systems, this usually results in low signal to noise ratios. Moreover, MS compatible BCA conditions are in most cases unsuitable for receptor BCAs and membrane-bound receptor systems will most probably give MS related problems. The obvious choice for MS-based FA-BCD-HPLC formats, therefore, are non-membrane-bound enzyme systems which rapidly convert a substrate to its product (which is measured by MS) and for which no fluorescent enzymatic BCA is available.

17.4
Conclusions

The combination of flow injection analysis or liquid chromatography with biochemical assays and mass spectrometric detection have proven to be an excellent approach to increase both the selectivity and sensitivity as well as sample throughput during affinity interaction profiling of protein–protein and protein–ligand interactions. The traditional fluorescence or radiometric detection modes using biochemical assays possess the disadvantages that a fluorescent probe or substrate must be synthesized that retains significant affinity for the receptor or the enzyme. Another limitation is that the labeling reaction can be laborious and time-consuming. Mass spectrometric detection, on the other hand, only requires the use of a reporter ligand – a chemical compound that binds to the receptor protein – and that can be easily detected by the mass spectrometer. This means that only the ionization characteristics of the reporter ligand are of importance, with the result that a wide variety of compounds is available.

In order to combine a flow system with a biochemical assay three approaches can be applied: using a biochemical interaction before the analytical separation, using a biochemical interaction during the analytical separation or using a biochemical interaction after the analytical separation. All three procedures have there own pros and cons with respect to the selectivity (e.g., influence of interfering matrix components) and the sensitivity (e.g., potential sample losses during fractionation). By using on-line systems, high throughputs and reproducibilities can be obtained. In addition this type of procedure allows the simultaneous determination of both chemical (e.g., quantitative and qualitative information) and biological (e.g., IC50 values) information.

References

1 Hage, D.S. (1998) Survey of recent advances in analytical applications of immunoaffinity chromatography. *Journal of Chromatography. B, Biomedical Sciences and Applications*, **715**, 3–28.

2 Hennion, M.C. and Pichon, V. (1000) Immuno-based sample preparation for trace analysis. *Journal of Chromatography. A*, **2003**, 29–52.

3 Hage, D.S. (2002) High-performance affinity chromatography: a powerful tool for studying serum protein binding. *Journal of Chromatography. B, Analytical Technologies in the Biomedical and Life Sciences*, **768**, 3–30.

4 Phillipson, D.W., Milgram, K.E., Yanovsky, A.I., Rusnak, L.S., Haggerty, D.A., Farrell, W.P., Greig, M.J., Xiong, X. and Proefke, M.L. (2002) High-throughput bioassay-guided fractionation: a technique for rapidly assigning observed activity to individual components of combinatorial libraries, screened in HTS bioassays. *Journal of Combinatorial Chemistry*, **4**, 591–599.

5 Hewitt, L.M. and Marvin, C.H. (2005) Analytical methods in environmental effects-directed investigations of effluents. *Mutation Research*, **589**, 208–232.

6 Trubetskoy, O.V., Gibson, J.R. and Marks, B.D. (2005) Highly miniaturized formats for *in vitro* drug metabolism assays using vivid fluorescent substrates and recombinant human cytochrome P450 enzymes. *Journal of Biomolecular Screening*, **10**, 56–66.

7 Reineke, N., Bester, K., Huhnerfuss, H., Jastorff, B. and Weigel, S. (2002) Bioassay-directed chemical analysis of River Elbe surface water including large volume extractions and high performance fractionation. *Chemosphere*, **47**, 717–723.

8 Katta, V. and Chait, B.T. (1991) Observation of the heme-globin complex in native myoglobin by electrospray-ionization mass spectrometry. *Journal of the American Chemical Society*, **113**, 8534.

9 Ganem, B., Li, Y. and Henion, J.D. (1991) Detection of noncovalent receptor-ligand complexes by mass spectrometry. *Journal of the American Chemical Society*, **113**, 6294–6296.

10 Chowdhury, S.K., Katta, V. and Chait, B.T. (1990) Probing conformational changes in proteins by mass spectrometry. *Journal of the American Chemical Society*, **112**, 9012–9013.

11 Craig, T.A., Veenstra, T.D., Naylor, S., Tomlinson, A.J., Johnson, K.L.M.S., Juranic, N. and Kumar, R. (1997) Zinc binding properties of the DNA binding domain of the 1,25-dihydroxyvitamin D3 receptor. *Biochemistry*, **36**, 10482–10491.

12 Kheterpal, I., Cook, K.D. and Wetzel, R. (2006) Hydrogen/deuterium exchange mass spectrometry analysis of protein aggregates. *Methods in Enzymology*, **413**, 140–166.

13 Whitelegge, J., Halgand, F., Souda, P. and Zabrouskov, V. (2006) Top-down mass spectrometry of integral membrane proteins. *Expert Review of Proteomics*, **3**, 585–596.

14 Evers, T.H., van Dongen, J.L., Meijer, E.W. and Merkx, M. (2007) Ligand-induced monomerization of Allochromatium vinosum cytochrome c′ studied using native mass spectrometry and fluorescence resonance energy transfer. *Journal of Biological Inorganic Chemistry*.

15 Mazon, H., Gabor, K., Leys, D., Heck, A.J., van der Oost, J. and van den Heuvel, R.H. (2007) Transcriptional activation by CprK1 is regulated by protein structural changes induced by effector binding and redox state. *Journal of Biological Chemistry*, **282**, 11281–11290.

16 van Duijn, E., Bakkes, P.J., Heeren, R.M., van den Heuvel, R.H., van Heerikhuizen, H., van der Vies, S.M. and Heck, A.J. (2005) Monitoring macromolecular complexes involved in the chaperonin-assisted protein folding cycle by mass spectrometry. *Nature Methods*, **2**, 371–376.

17 Bovet, C., Wortmann, A., Eiler, S., Granger, F., Ruff, M., Gerrits, B., Moras, D. and Zenobi, R. (2007) Estrogen receptor-ligand complexes measured by chip-based nanoelectrospray mass spectrometry: an approach for the screening of endocrine disruptors. *Protein Science*, **16**, 938–946.

18 van den Heuvel, R.H. and Heck, A.J. (2004) Native protein mass spectrometry: from intact oligomers to functional machineries. *Current Opinion in Chemical Biology*, **8**, 519–526.

19 van Elswijk, D.A., Diefenbach, O., van der Berg, S., Irth, H., Tjaden, U.R. and van der Greef, J. (1020) Rapid detection and identification of angiotensin-converting enzyme inhibitors by on-line liquid chromatography-biochemical detection, coupled to electrospray mass spectrometry. *Journal of Chromatography A*, **2003**, 45–58.

20 Schenk, T., Appels, N.M., van Elswijk, D.A., Irth, H., Tjaden, U.R. and van der Greef, J. (2003) A generic assay for phosphate-consuming or -releasing enzymes coupled on-line to liquid chromatography for lead finding in natural products. *Analytical Biochemistry*, **316**, 118–126.

21 Schenk, T., Breel, G.J., Koevoets, P., van den Berg, S., Hogenboom, A.C., Irth, H., Tjaden, U.R. and van der Greef, J. (2003) Screening of natural products extracts for the presence of phosphodiesterase inhibitors using liquid chromatography coupled online to parallel biochemical detection and chemical characterization. *Journal of Biomolecular Screening*, **8**, 421–429.

22 van Elswijk, D.A., Schobel, U.P., Lansky, E.P., Irth, H. and van der Greef, J. (2004) Rapid dereplication of estrogenic compounds in pomegranate (Punica granatum) using on-line biochemical detection coupled to mass spectrometry. *Phytochemistry*, **65**, 233–241.

23 Schobel, U., Frenay, M., van Elswijk, D.A., McAndrews, J.M., Long, K.R., Olson, L.M., Bobzin, S.C. and Irth, H. (2001) High resolution screening of plant natural product extracts for estrogen receptor alpha and beta binding activity using an online HPLC-MS biochemical detection system. *Journal of Biomolecular Screening*, **6**, 291–303.

24 Hogenboom, A.C., de Boer, Λ.R., Derks, R.J. and Irth, H. (2001) Continuous-flow, on-line monitoring of biospecific interactions using electrospray mass spectrometry. *Analytical Chemistry*, **73**, 3816–3823.

25 de Boer, A.R., Letzel, T., van Elswijk, D.A., Lingeman, H., Niessen, W.M. and Irth, H. (2004) On-line coupling of high-performance liquid chromatography to a continuous-flow enzyme assay based on electrospray ionization mass spectrometry. *Analytical Chemistry*, **76**, 3155–3161.

26 Johnson, B.M., Nikolic, D. and van Breemen, R.B. (2002) Applications of pulsed ultrafiltration-mass spectrometry. *Mass Spectrometry Reviews*, **21**, 76–86.

27 van Breemen, R.B., Huang, C.R., Nikolic, D., Woodbury, C.P., Zhao, Y.Z. and Venton, D.L. (1997) Pulsed ultrafiltration mass spectrometry: a new method for screening combinatorial libraries. *Analytical Chemistry*, **69**, 2159–2164.

28 Geun Shin, Y., Bolton, J.L. and van Breemen, R.B. (2002) Screening drugs for metabolic stability using pulsed ultrafiltration mass spectrometry. *Combinatorial Chemistry & High Throughput Screening*, **5**, 59–64.

29 van Elswijk, D.A., Tjaden, U.R., van der Greef, J. and Irth, H. (2001) Mass spectrometry-based bioassay for the screening of soluble orphan receptors. *International Journal of Mass Spectrometry*, **210**, 625–636.

30 Annis, D.A., Nazef, N., Chuang, C.C., Scott, M.P. and Nash, H.M. (2004) A general technique to rank protein-ligand binding affinities and determine allosteric versus direct binding site competition in compound mixtures. *Journal of the American Chemical Society*, **126**, 15495–15503.

31 Annis, D.A., Athanasopoulosa, J., Currana, P.J., Felscha, J.S., Kalghatgia, K., Leea, W.H., Nasha, H.M., Orminatia, J.-P.A., Rosnera, K.E., Shipps, G.W., Jr., Thaddupathyb, G.R.A., Tylera, A.N., Vilenchek, L., Wagner, C.R. and Wintner, E. (2004) An affinity selection-mass spectrometry method for the identification of small molecule ligands from self-encoded combinatorial libraries. Discovery of a novel antagonist of *E. coli* dihydrofolate reductase.

32 Haake, H.M., de Best, L., Irth, H., Abuknesha, R. and Brecht, A. (2000) Label-free biochemical detection coupled on-line to liquid chromatography. *Analytical Chemistry*, **72**, 3635–3641.

33 Hsieh, Y.F., Gordon, N., Regnier, F., Afeyan, N., Martin, S.A. and Vella, G.J. (1997) Multidimensional chromatography coupled with mass spectrometry for target-based screening. *Molecular Diversity*, **2**, 189–196.

34 Kaur, S., McGuire, L., Tang, D., Dollinger, G. and Huebner, V. (1997) Affinity selection and mass spectrometry-based strategies to identify lead compounds in combinatorial libraries. *Journal of Protein Chemistry*, **16**, 505–511.

35 Lenz, E., Taylor, S., Collins, C., Wilson, I.D., Louden, D. and Handley, A. (2002) Flow injection analysis with multiple on-line spectroscopic analysis (UV, IR, 1H-NMR and MS). *Journal of Pharmaceutical and Biomedical Analysis*, **27**, 191–200.

36 Rich, R.L., Hoth, L.R., Geoghegan, K.F., Brown, T.A., LeMotte, P.K., Simons, S.P., Hensley, P. and Myszka, D.G. (2002) Kinetic analysis of estrogen receptor/ligand interactions. *Proceedings of the National Academy of Sciences of the United States of America*, **99**, 8562–8567.

37 Schuffenhauer, A., Popov, M., Schopfer, U., Acklin, P., Stanek, J. and Jacoby, E. (2004) Molecular diversity management strategies for building and enhancement of diverse and focused lead discovery compound screening collections. *Combinatorial Chemistry & High Throughput Screening*, **7**, 771–781.

38 Liu, B., Li, S. and Hu, J. (2004) Technological advances in high-throughput screening. *American Journal of PharmacoGenomics*, **4**, 263–276.

39 Fura, A., Shu, Y.Z., Zhu, M., Hanson, R.L., Roongta, V. and Humphreys, W.G. (2004) Discovering drugs through biological transformation: role of pharmacologically active metabolites in drug discovery. *Journal of Medicinal Chemistry*, **47**, 4339–4351.

40 van Rhee, A.M., Stocker, J., Printzenhoff, D., Creech, C., Wagoner, P.K. and Spear, K.L. (2001) Retrospective analysis of an experimental high-throughput screening data set by recursive partitioning. *Journal of Combinatorial Chemistry*, **3**, 267–277.

41 Fura, A. (2006) Role of pharmacologically active metabolites in drug discovery and development. *Drug Discovery Today*, **11**, 133–142.

42 Ohno, K., Fukushima, T., Santa, T., Waizumi, N., Tokuyama, H., Maeda, M. and Imai, K. (2002) Estrogen receptor binding assay method for endocrine disruptors using fluorescence polarization. *Analytical Chemistry*, **74**, 4391–4396.

43 de Boer, T., Otjens, D., Muntendam, A., Meulman, E., van Oostijen, M. and Ensing, K. (2004) Development and validation of fluorescent receptor assays based on the human recombinant estrogen receptor subtypes alpha and beta. *Journal of Pharmaceutical and Biomedical Analysis*, **34**, 671–679.

44 Lim, H.K., Stellingweif, S., Sisenwine, S. and Chan, K.W. (1999) Rapid drug metabolite profiling using fast liquid chromatography, automated multiple-stage mass spectrometry and receptor-binding. *Journal of Chromatography A*, **831**, 227–241.

45 Irth, H., Oosterkamp, A.J., van der Welle, W., Tjaden, U.R. and van der Greef, J. (1993) On-line immunochemical detection in liquid chromatography using fluorescein-labelled antibodies. *Journal of Chromatography*, **633**, 65–72.

46 Oosterkamp, A.J., Irth, H., Beth, M., Unger, K.K., Tjaden, U.R. and van de Greef, J. (1994) Bioanalysis of digoxin and its metabolites using direct serum injection combined with liquid chromatography and on-line immunochemical detection. *Journal of Chromatography. B, Biomedical Applications*, **653**, 55–61.

47 Schenk, T., Molendijk, A., Irth, H., Tjaden, U.R. and van der Greef, J. (2003) Liquid chromatography coupled on-line to flow cytometry for postcolumn homogeneous biochemical detection. *Analytical Chemistry*, **75**, 4272–4278.

48 Oosterkamp, A.J., Villaverde Herraiz, M.T., Irth, H., Tjaden, U.R. and van der Greef, J. (1996) Reversed-phase liquid chromatography coupled on-line to receptor affinity detection based on the human estrogen receptor. *Analytical Chemistry*, **68**, 1201–1206.

49 Rhee, I.K., Appels, N., Luijendijk, T., Irth, H. and Verpoorte, R. (2003) Determining acetylcholinesterase inhibitory activity in plant extracts using a fluorimetric flow assay. *Phytochemical Analysis*, **14**, 145–149.

50 Przyjazny, A., Kjellstrom, T.L. and Bachas, L.G. (1990) High-performance liquid chromatographic postcolumn reaction detection based on a competitive binding system. *Analytical Chemistry*, **62**, 2536–2540.

51 Przyjazny, A., Hentz, N.G. and Bachas, L.G. (1993) Sensitive and selective liquid chromatographic postcolumn reaction detection system for biotin and biocytin using a homogeneous fluorophore-linked assay. *Journal of Chromatography A*, **654**, 79–86.

52 Nuengchamnong, N., de Jong, C.F., Bruyneel, B., Niessen, W.M., Irth, H. and Ingkaninan, K. (2005) HPLC coupled on-line to ESI-MS and a DPPH-based assay for the rapid identification of anti-oxidants in Butea superba. *Phytochemical Analysis*, **16**, 422–428.

53 Oosterkamp, A.J., Irth, H., Villaverde Herraiz, M.T., Tjaden, U.R. and van der Greef, J. (1997) Theoretical concepts of on-line liquid chromatographic-biochemical detection systems. I. Detection systems based on labelled ligands. *Journal of Chromatography A*, **787**, 27–35.

54 Oostercamp, A.J., Irth, H., Tjaden, U.R. and van der Greef, J. (1997) Theoretical concepts of on-line liquid chromatographic-biochemical detection systems. II. Detection systems based on labelled affinity

proteins. *Journal of Chromatography. A,*
787, 37–46.

55 Miller, K.J. and Herman, A.C. (1996)
Affinity chromatography with
immunochemical detection applied to the
analysis of human methionyl granulocyte
colony stimulating factor in serum.
Analytical Chemistry, **68**, 3077–3082.

56 Schenk, T., Irth, H., Marko-Varga, G.,
Edholm, L.E., Tjaden, U.R. and van der
Greef, J. (2001) Potential of on-line micro-
LC immunochemical detection in the
bioanalysis of cytokines. *Journal of
Pharmaceutical and Biomedical Analysis,*
26, 975–985.

57 Oosterkamp, A.J., Irth, H., Heintz, L.,
Marko-Varga, G., Tjaden, U.R. and van
der Greef, J. (1996) Simultaneous determination of cross-reactive leukotrienes in
biological matrices using on-line liquid
chromatography immunochemical
detection. *Analytical Chemistry,* **68**,
4101–4106.

58 van Bommel, M.R., de Jong, A.P., Tjaden,
U.R., Irth, H. and van der Greef, J. (1999)
Enzyme amplification as detection tool in
continuous-flow systems. I. Development
of an enzyme-amplified biochemical
detection system coupled on-line to flow
injection analysis. *Journal of
Chromatography A,* **855**, 383–396.

59 van Bommel, M.R., de Jong, A.P., Tjaden,
U.R., Irth, H. and van der Greef, J. (1999)
Enzyme amplification as detection tool in
continuous-flow systems. II. On-line
coupling of liquid chromatography to
enzyme-amplified biochemical detection
after pre-column derivatization with
biotin. *Journal of Chromatography A,* **855**,
397–409.

60 van Bommel, M.R., de Jong, A.P., Tjaden,
U.R., Irth, H. and van der Greef, J. (2000)
High-performance liquid chromatography
coupled to enzyme-amplified biochemical
detection for the analysis of hemoglobin
after pre-column biotinylation. *Journal of
Chromatography A,* **886**, 19–29.

61 Lutz, E.S., Irth, H., Tjaden, U.R. and van
der Greef, J. (1996) Applying hollow fibres

for separating free and bound label in
continuous-flow immunochemical
detection. *Journal of Chromatography A,*
755, 179–187.

62 Mazereeuw, M., de Best, C.M., Tjaden,
U.R., Irth, H. and van der Greef, J. (2000)
Free flow electrophoresis device for
continuous on-line separation in analytical
systems. An application in biochemical
detection. *Analytical Chemistry,* **72**,
3881–3886.

63 Kool, J., Ramautar, R., van Liempd, S.M.,
Beckman, J., de Kanter, F.J., Meerman,
J.H., Schenk, T., Irth, H., Commandeur,
J.N. and Vermeulen, N.P. (2006) Rapid on-
line profiling of estrogen receptor binding
metabolites of tamoxifen. *Journal of
Medicinal Chemistry,* **49**, 3287–3292.

64 van Liempd, S.M., Kool, J., Niessen, W.M.,
van Elswijk, D.E., Irth, H. and Vermeulen,
N.P. (2006) On-line formation, separation,
and estrogen receptor affinity screening of
cytochrome P450-derived metabolites of
selective estrogen receptor modulators.
Drug Metabolism and Disposition, **34**,
1640–1649.

65 Cooper, M.A. (2004) Advances in
membrane receptor screening and
analysis. *Journal of Molecular Recognition:
JMR,* **17**, 286–315.

66 Ragno, P. (2006) The urokinase receptor: a
ligand or a receptor? Story of a sociable
molecule. *Cellular and Molecular Life
Sciences: CMLS.*

67 Oosterkamp, A.J., van der Hoeven, R.,
Glassgen, W., Konig, B., Tjaden, U.R., van
der Greef, J. and Irth, H. (1998) Gradient
reversed-phase liquid chromatography
coupled on-line to receptor-affinity
detection based on the urokinase receptor.
*Journal of Chromatography B, Biomedical
Sciences and Applications,* **715**, 331–338.

68 Ingkaninan, K., de Best, C.M., van der
Heijden, R., Hofte, A.J., Karabatak, B.,
Irth, H., Tjaden, U.R., van der Greef, J. and
Verpoorte, R. (2000) High-performance
liquid chromatography with on-line
coupled UV, mass spectrometric and
biochemical detection for identification of

acetylcholinesterase inhibitors from natural products. *Journal of Chromatography. A*, **872**, 61–73.

69 Ingkaninan, K., Hazekamp, A., de Best, C.M., Irth, H., Tjaden, U.R., van der Heijden, R., van der Greef, J. and Verpoorte, R. (2000) The application of HPLC with on-line coupled UV/MS-biochemical detection for isolation of an acetylcholinesterase inhibitor from narcissus 'Sir Winston Churchill'. *Journal of Natural Products*, **63**, 803–806.

70 Hirata, J., de Jong, C.F., van Dongen, M.M., Buijs, J., Ariese, F., Irth, H. and Gooijer, C. (2004) A flow injection kinase assay system based on time-resolved fluorescence resonance energy-transfer detection in the millisecond range. *Analytical Chemistry*, **76**, 4292–4298.

71 Hirata, J., Chung, L.P., Ariese, F., Irth, H. and Gooijer, C. (1081) Coupling of size-exclusion chromatography to a continuous assay for subtilisin using a fluorescence resonance energy transfer peptide substrate: testing of two standard inhibitors. *Journal of Chromatography A*, **2005**, 140–144.

72 de Boer, A.R., Bruyneel, B., Krabbe, J.G., Lingeman, H., Niessen, W.M. and Irth, H. (2005) A microfluidic-based enzymatic assay for bioactivity screening combined with capillary liquid chromatography and mass spectrometry. *Lab-on-a-Chip*, **5**, 1286–1292.

73 Guengerich, F.P. (2001) Common and uncommon cytochrome P450 reactions related to metabolism and chemical toxicity. *Chemical Research in Toxicology*, **14**, 611–650.

74 Nebert, D.W. and Gonzalez, F.J. (1987) P450 genes: structure, evolution, and regulation. *Annual Review of Biochemistry*, **56**, 945–993.

75 Vermeulen, N.P. (2003) Prediction of drug metabolism: the case of cytochrome P450 2D6. *Current Topics in Medicinal Chemistry*, **3**, 1227–1239.

76 Strege, M.A. (1999) High-performance liquid chromatographic-electrospray

ionization mass spectrometric analyses for the integration of natural products with modern high-throughput screening. *Journal of Chromatography B, Biomedical Sciences and Applications*, **725**, 67–78.

77 Kool, J., van Liempd, S.M., Ramautar, R., Schenk, T., Meerman, J.H., Irth, H., Commandeur, J.N. and Vermeulen, N.P. (2005) Development of a novel cytochrome p450 bioaffinity detection system coupled online to gradient reversed-phase high-performance liquid chromatography. *Journal of Biomolecular Screening*, **10**, 427–436.

78 Kool, J., van Liempd, S.M., van Rossum, H., van Elswijk, D.A., Irth, H., Commandeur, J.N. and Vermeulen, N.P. (2007) Development of three parallel cytochrome P450 enzyme affinity detection systems coupled on-line to gradient high-performance liquid chromatography. *Drug Metabolism and Disposition*, **35**, 640–648.

79 Kool, J., Eggink, M., van Rossum, H., van Liempd, S.M., van Elswijk, D.A., Irth, H., Commandeur, J.N., Meerman, J.H. and Vermeulen, N.P. (2007) Online biochemical detection of glutathione-S-transferase P1-specific inhibitors in complex mixtures. *Journal of Biomolecular Screening*, **12**, 396–405.

80 de Boer, A.R., Alcaide-Hidalgo, J.M., Krabbe, J.G., Kolkman, J., van Emde Boas, C.N., Niessen, W.M., Lingeman, H. and Irth, H. (2005) High-temperature liquid chromatography coupled on-line to a continuous-flow biochemical screening assay with electrospray ionization mass spectrometric detection. *Analytical Chemistry*, **77**, 7894–7900.

81 Sugita, O., Ishizawa, N., Matsuto, T., Okada, M. and Kayahara, N. (2004) A new method of measuring the antioxidant activity of polyphenols using cumene hydroperoxide. *Annals of Clinical Biochemistry*, **41**, 72–77.

82 Manzocco, L., Calligaris, S. and Nicoli, M.C. (2002) Assessment of pro-oxidant activity of foods by kinetic analysis of

crocin bleaching. *Journal of Agricultural and Food Chemistry*, **50**, 2767–2771.

83 Cardenosa, R., Mohamed, R., Pineda, M. and Aguilar, M. (2002) On-line HPLC detection of tocopherols and other antioxidants through the formation of a phosphomolybdenum complex. *Journal of Agricultural and Food Chemistry*, **50**, 3390–3395.

84 Cano, A., Alcaraz, O., Acosta, M. and Arnao, M.B. (2002) On-line antioxidant activity determination: comparison of hydrophilic and lipophilic antioxidant activity using the ABTS* + assay. *Redox Report*, **7**, 103–109.

85 Kool, J., Van Liempd, S.M., Harmsen, S., Schenk, T., Irth, H., Commandeur, J.N. and Vermeulen, N.P. (2007) An on-line post-column detection system for the detection of reactive-oxygen-species-producing compounds and antioxidants in mixtures. *Analytical and Bioanalytical Chemistry*, **388**, 871–879.

86 Fabel, S., Niessner, R. and Weller, M.G. (1099) Effect-directed analysis by high-performance liquid chromatography with gas-segmented enzyme inhibition. *Journal of Chromatography A*, **2005**, 103–110.

87 Derks, R.J., Hogenboom, A.C., van der Zwan, G. and Irth, H. (2003) On-line continuous-flow, multi-protein biochemical assays for the characterization of bioaffinity compounds using electrospray quadrupole time-of-flight mass spectrometry. *Analytical Chemistry*, **75**, 3376–3384.

88 Krabbe, J.G., Lingeman, H., Niessen, W.M. and Irth, H. (2003) Ligand-exchange detection of phosphorylated peptides using liquid chromatography electrospray mass spectrometry. *Analytical Chemistry*, **75**, 6853–6860.

89 Krabbe, J.G., Lingeman, H., Niessen, W.M. and Irth, H. (1093) Screening for metal ligands by liquid chromatography – ligand-exchange – electrospray mass spectrometry. *Journal of Chromatography A*, **2005**, 36–46.

90 Krabbe, J.G., Gao, F., Li, J., Ahlskog, J.E., Lingeman, H., Niessen, W.M. and Irth, H. (2006) Selective detection and identification of phosphorylated proteins by simultaneous ligand-exchange fluorescence detection and mass spectrometry. *Journal of Chromatography A*, **1130**, 287–295.

91 Krabbe, J.G., de Boer, A.R., van der Zwan, G., Lingeman, H., Niessen, W.M. and Irth, H. (2007) Metal-complex formation in continuous-flow ligand-exchange reactors studied by electrospray mass spectrometry. *Journal of the American Society for Mass Spectrometry*, **18**, 707–713.

92 de Jong, C.F., Derks, R.J., Bruyneel, B., Niessen, W. and Irth, H. (2006) High-performance liquid chromatography-mass spectrometry-based acetylcholinesterase assay for the screening of inhibitors in natural extracts. *Journal of Chromatography A*, **1112**, 303–310.

93 Bruyneel, B., Hoos, J.S., Smoluch, M.T., Lingeman, H., Niessen, W.M. and Irth, H. (2007) Trace analysis of proteins using postseparation solution-phase digestion and electrospray mass spectrometric detection of marker peptides. *Analytical Chemistry*, **79**, 1591–1598.

18
Atomic Spectroscopy in Flow Analysis

José L. Burguera and Marcela Burguera

18.1
Introduction

The most critical and time-consuming step of any analytical method is the sample pretreatment. The analyst must often deal with a wide variety of biological, environmental and geological samples, occurring in all three states (solid, liquid, and gas), which require application of the most appropriate preparative techniques, such as dilution, filtration, dissolution, digestion, extraction, and so on. Most of these sample treatment procedures are still performed manually. Their implementation on-line includes all procedures from sample preparation to data management and plays a major role in the delicate stage of analyte transfer to the detector. With this purpose, continuously flowing systems, including flow injection (FI) have been easily interfaced to atomic spectroscopic (AS) techniques such as: atomic absorption spectrometry (AAS) either with flame (FAAS) or electrothermal (ET AAS) atomization, inductively coupled plasma optical emission spectrometry (ICP-OES) or ICP-mass spectrometry (ICP-MS) and atomic fluorescence spectrometry (AFS).

A literature search reveals that there are entire chapters in a number of books [1–4] and full reviews [5–13] dedicated to on-line sample preparation techniques coupled to AS, which report critically on the most important developments published up to the year 2001. The annual reviews published regularly in the *Journal of Analytical Atomic Spectrometry* (one example is given in the reference section) cover this matter to a great extent [14]. This chapter reviews the different and most relevant on-line-AS sample preparation techniques described in the literature during the last six years, emphasizing on-line separation/preconcentration procedures like chemical vapor generation (CVG), sorption on columns or on knotted reactors (KR) and some of the hyphenated techniques used for speciation, like high performance liquid chromatography (HPLC), capillary electrophoresis (CE) or microdialysis. Publications related to ICP-OES and ICP-MS are treated in a separate chapter of this book. The selection of the papers cited here was based on the criteria of: (i) dealing with on-line separation/preconcentration systems and (ii) being published during the period

Advances in Flow Analysis. Edited by Marek Trojanowicz
Copyright © 2008 WILEY-VCH Verlag GmbH & Co. KGaA, Weinheim
ISBN: 978-3-527-31830-8

2001–2007, in analytical journals of high impact factor. We also avoided citing papers with little innovation in the analytical procedure. It is interesting to state that on-line microwave-assisted sample digestion is no longer a subject of research as it was during the nineties [6–8, 11, 12], probably because it has become a routine technique, broadly available, or because it was difficult to implement for complex, solid matrices.

18.2
Flame Atomic Absorption Spectrometry

FAAS coupled to FI systems is still one of the most popular methods for the routine determination of many inorganic elements in a wide variety of samples. This may be due to its low cost and the ease to interface any FI system to the direct aspiration produced by the suction effect of the pneumatic nebulizer for sample introduction into the flame. Despite its uni-elemental features, FAAS is a robust technique and is available in most clinical and analytical laboratories for elemental analysis at $mg\,L^{-1}$ level. Therefore, for many samples, FAAS does not meet the specific requirements in sensitivity and various on-line separation and/or preconcentration procedures are often used to improve the performance. Among them, CVG after separation on specific columns has been the most widely used recently, so that the analyte ions are on-line concentrated in the column, while potential interferents are discarded. As the same instrumentation used for hydride-forming elements determination is adequate for cold vapor generation, we included selected papers related to mercury determination in this section, although the flame is not used for atomization.

18.2.1
Preconcentration/Separation Systems Using FAAS Detection

18.2.1.1 Chemical Vapor Generation
CVG is a versatile technique, generally used for the separation of certain elements from complex matrices, but also for their preconcentration and, under controlled conditions, for speciation purposes. The on-line connection of cold vapor or hydride generation (HG) manifolds with AS detectors after on-line separation/preconcentration of the analyte on minicolumns packed with a specific sorbent material, has become a selective and sensitive approach for the determination of elements like Hg [15, 16], As [17, 18] or Bi [19]. In principle, the procedures developed so far are very similar: the corresponding inorganic ions are on-line preconcentrated and/or separated and eluted with an acidic solution (mainly HCl and/or HNO_3) which mixes with a reducing agent ($SnCl_2$ or $NaBH_4$) to generate the specific volatile species which are stripped to the QC atomizer by a stream of nitrogen or argon. Electrochemically (EC) generated hydrides could also be preconcentrated directly on a heated QC [20] or on a minicolumn [21]. Such systems were exclusively designed with the purpose of increasing sensitivity and removing the rest of the matrix and some of

Table 18.1 Selected applications of on-line chemical vapor generation after preconcentration on minicolumns for FAAS determinations.

Element determined	Sample type	Sorption material	Retained species	Eluent	DL (ng L^{-1})	Ref.
Total Hg	Tap water and hair	Activated carbon	Hg^{2+}	HNO$_3$	10	[15]
Hg(II)	Lake and deep well water	Cyanex 923	Hg^{2+}	HNO$_3$	0.2	[16]
As(III)	Natural waters	PTFE turnings	As-PDC	HCl	20	[17]
Total As	Mineral, drinking, river and spring waters	Activated alumina	As^{3+}	HCl	150	[18]
Bi	Urine	Amberlite	BiCl$_4^-$	HCl	225	[19]
Total Sb (EC-HG)	Drinking water, soil/sediment	Heated QC	SbH$_3$	Heat + H$_2$	53	[20]
Total Sb (EC-HG)	Compost, marine sediments	Chelex-100	Fe(II)	HCl	—	[21]

DL = Detection Limit.

their characteristics are shown in Table 18.1 [15–21]. In these cases, the selection of sample type is conditioned by the sample consumption volume, which might be of 25 mL or more, depending, obviously, on the sample flow rate used in the system and on the preconcentration to be achieved. The enrichment factor (EF) reported, also shows wide variations (from 10- to 100-fold) and depends on the preconcentration time which could be varied at the expenses of the sample throughput, which merely reaches 25 h^{-1} or less. The main difference between the developed procedures resides in the type of packing material and the nature of the real samples determined, although most of the applications deal with natural waters. The detection limits (DL) reach, in all cases, ng L^{-1} levels.

There are also some other CVG developments which do not use any sorbent material for analyte separation/preconcentration, but are worth mentioning because of their novel procedures. For instance, a 7 cm-long silica tubing located between the gas liquid separator (GLS) and the QC was used to trap, and thus to preconcentrate, the Pb hydride generated in a conventional HG-FI system at 500 °C [22]. The analyte species is released when the trap is heated further to 750 °C. Hydrogen is required for re-volatilization and oxygen must also be present. The DL achieved for a 60 s trapping time was 19 ng L^{-1}, but the overall efficiency was only 49%. This principle was recently applied to trap bismuthine directly into a modified QC atomizer which was externally heated to 900 °C [23]. In this case, the collection/volatilization efficiency was 100 ± 2.5% with a preconcentration ratio of 530 for a collection time of 300 s, corresponding to a 20 mL sample volume. Under these conditions, the DL was 3.9 ng L^{-1}. The authors claim that the same approach is analytically useful also for stibine but not for arsine.

With few exceptions, the systems mentioned above have no ability to discriminate between the different species of such analytes and never speciate the respective organic compounds.

Other experimental arrangements, however, incorporate separation techniques like HPLC prior to CVG. Hyphenation of HPLC with the conventional on-line HG-AAS systems has proved adequate for such speciation studies. Most of these coupled approaches have been used for As speciation [24–27] although Se [27] in samples containing their respective organic and inorganic species were also separated. Species-independent quantifications were carried out after submission of samples containing arsenic species to alkaline peroxodisulfate oxidation in a microwave oven [24]. Such applications were directed to toxicologically important species in environmental and biological samples.

18.2.1.2 Preconcentration of Trace Elements by Solid-phase Extraction

Given the low sensitivity of FAAS, numerous on-line sorbent extraction/preconcentration systems have been optimized for the determination of trace elements present in samples of various natures at $\mu g\,L^{-1}$ levels. For that purpose, natural (kaolinite, zeolites, alumina, rice husks) and synthesized materials (amberlite, silica gel, polyurethane foam) were tested recently as adsorptive materials after being chemically immobilized or functionalized with different compounds, in order to separate the metal ions from the matrix and preconcentrate them for subsequent determination by FAAS. The adsorbed complex is then eluted with a diluted acid or an organic solvent and carried to the detector. Analytical figures of merit, some specific characteristics and current applications of selected publications related to this matter are given in Table 18.2 [28–72]. In this case, the applications were directed to more complex matrices like food, biological and environmental samples. For preconcentration times of 60–90 s, sampling frequency rarely exceeded $40\,h^{-1}$.

In parallel to sorption on columns, the retention of complexing agents on KR also allowed the selective retention of an analyte. This approach involves the precoating of the KR with a chelating reagent to adsorb the analyte, followed by the introduction of a fixed small volume of eluent for the quantitative elution of the adsorbed chelate. The progress in FI on-line separation and preconcentration systems for AS techniques employing KR as a sorption medium for organometallic complexes has been reviewed by Cerutti *et al.* [73], where the determination of different elements utilizing diverse organic and inorganic reagents in different samples was discussed.

18.2.1.3 Other Preconcentration Systems for FAAS Detection

Other preconcentration approaches have also been implemented for the preconcentration of metal ions in different real samples before being determined by FAAS. Among them, precipitation/dissolution [74] and solvent extraction [75, 76] are worth mentioning. For example, Cd was preconcentrated 32-fold by precipitation as an ion pair between tetraiodocadmate and quinine, followed by dissolution of the precipitate with ethanol [74]. The method was applied to the determination of Cd in mussel samples at the $\mu g\,g^{-1}$ level. Also Fe(III) and total Fe were sequentially determined in table wines by sequential injection (SI)-FAAS at a sampling rate of $18\,h^{-1}$ [75].

Table 18.2 On-line FAAS column or KR sorption/preconcentration systems for the determination of trace elements.

Element	Sample type	Sorption material	Complexing agent	Eluent	EF	DL (µg L^{-1})	Ref.
Cd, Pb	CRM: pig kidney, beech leaves	Rice husks modified	NaOH	HNO$_3$	72.4 46	1.14 14.1	[28]
Co	Water, Vit B$_{12}$, B-complex ampoules	Surfactant-coated alumina	2-nitroso-1-naphthol	Ethanol	125	0.02	[29]
Ag	Tap, well, rain, sea waters, radiology film	Surfactant-coated alumina	Ag-DDTC. pH 3–4	Ethanol	125	1.7	[30]
Cu, Pb	Water, vitamin tablets, alloys	Surfactant-coated alumina	Salen I. pH 9	HNO$_3$	100, 75	0.3, 2.6	[31]
Cr(VI) Cr(III)	Sewage waters	Activated alumina	—	NH$_4$OH HNO$_3$	— —	42 81	[32]
Mn	Complex materials	Kaolinite	5-Br-PADAP pH 8.5–10.0	H$_2$SO$_4$	—	4.3	[33]
Cu, Zn	Water, CRM	Natrolite zeolite	5-Br-PADAP pH 7.5–9.5	HNO$_3$	—	0.03, 0.006	[34]
Pb	Shrimp, oyster, crab, fish, mussel	Polyurethane foam	BTAC pH	HCl	26	1.0	[35]
Pb	Wine	Polyurethane foam	BTAC, pH 7	HCl	26	1.0	[36]
Cu, Pb, Cr(VI)	Natural waters Biological material	Polyurethane foam	Me-PDC	MIBK	170 131 28	0.2 1.8 2.0	[37]
Ga	Aluminum alloys, natural waters, urine	Polyurethane foam	Ga chloride complex	MIBK	40	6	[38]

(*Continued*)

Table 18.2 (*Continued*)

Element	Sample type	Sorption material	Complexing agent	Eluent	EF	DL ($\mu g\ L^{-1}$)	Ref.
Zn	Natural waters	Polyurethane foam	Me-BTABr. pH 6.5–9.2	HCl	23	0.37	[39]
Zn	Drinking water, saline waste from oil refinery	Polyurethane foam	PAR. pH 8.3	HCl	91.23	0.28	[40]
Pb	Drinking water, saline waste from oil refinery	Polyurethane foam	PAR. pH 8.3	HCl	51	0.4	[41]
Cd	Drinking water	Polyurethane foam	PAR pH 8.2	HCl	158	0.02	[42]
Cr(VI)	Natural waters, sediments	PTFE turnings	Cr-PDC. pH 0.8–1.4	MIBK	80	0.8	[43]
Pb	Water, sediments, fish tissue	PTFE turnings	Pb-PDC. pH 1.4–3.2	MIBK	330	0.8	[44]
Cu, Cd	CRM and water	Silica gel modified	Nb(V) oxide	HNO$_3$	34.2 / 33	0.4 / 0.1	[45]
Zn	Urine, blood plasma, erythrocites	Silica gel modified	Nb(V) oxide	HCl	77	0.77	[46]
Cr(III) Cr(VI)	Water	Silica gel modified	Zr(IV) phosphate Zr(IV) oxide	HNO$_3$ THAM	20.8 / 24.9	1.9 / 2.3	[47]
Cu	Aqueous solutions	Silica gel	3(1-imidazolyl)propyl groups	HNO$_3$	19.5–25.8	0.4	[48]
Cu, Pb	Waters	Polychlorotrifluoroethy-lene beads	Me-DDPA	MIBK	250	0.07, 2.7	[49]
Cr(VI) Pb	Waters	Polychlorotrifluoroethy-lene beads	Me-PDC. pH 1–1.6 for Cr(VI). pH 1.5–3.2 for Pb	MIBK	94 / 220	0.4 / 1.2	[50]

Pb	Surface, ground, effluent water, soil	Merrifield chloromethylated resin beads	Me-DDTC pH 8.0–9.0	Acidified methanol	48	1.3	[51]
Pd	Synthetic geological, pellet-type used car catalyst	Polyamine Metalfix-Chelamine resin	Pd^{2+}	Thiourea in HCl	20	9	[52]
Mn	Water, CRM	Amberlite IRA-904	5,10,15,20-tetrakis(4-carboxyphenil) porphyrin	HNO_3	30	12	[53]
Cd, Mn, Ni, Zn, Fe	Aqueous solutions	Amberlite XAD-4 loaded with pyrocatechol violet	Benzyl dimetyl tetradecyl ammonium chloride pH 7.4–8.2	HNO_3	85	—	[54]
Pb	Biological	Amberlite XAD-2 modified	BTAC. pH 6.5–8.5	HCl	27	3.7	[55]
Ni	Rice flour, spinach, orchard and peach leaves	Amberlite XAD-2 loaded	BTAC pH	HCl	30	1.1	[56]
Cu	Rice flour, spinach and apples leaves	Amberlite XAD-2 loaded	2-(2-thiazolylazo)-5-dimethilamino-phenol	HCl	62	0.23	[57]
Cu	Rice flour and starch	Amberlite XAD-2 functionalized	3,4-dihydroxybenzoic acid	HCl	33	0.27	[58]
Cd, Co, Cu, Ni	Rice flour, spinach, black tea	Amberlite XAD-2 anchored with pyrocatechol	None	HNO_3	36–69	0.31–1.64	[59]
Cr	Parenteral solution	Amberlite-XAD-16	TAR	Ethanol	50	0.02	[60]
Cd	Biological	Amberlite XAD-4	Me-DDTP	HCl	20	5–1	[61]

(Continued)

Table 18.2 (Continued)

Element	Sample type	Sorption material	Complexing agent	Eluent	EF	DL (μg L^{-1})	Ref.
Cd	Seawater	Chelite P	Aminomethyl phosphonic groups	HCl	1556	2.7	[62]
Cr(VI), Co, Ni, Cu	Aqueous solutions	KR	Me-PDC	Ethanol	44, 78 65, 75	0.40, 0.33 0.31, 0.26	[63]
Cd	Aqueous solutions	KR	Me-PAN	HCl	18	0.1	[64]
Pb	Aqueous solutions	KR	Me-PAN	HCl	26.5	0.43	[65]
Cu	Urine, blood serum	Metalfix Chelamine	Cu^{2+}	HNO$_3$	—	35, 0.67	[66]
Cu	Aqueous solutions	Styrene divinyl benzene resin	(S)-2-[hydroxyl-bis-(4-vinyl-phenyl)-methyl]-pyrrolidine-1-carboxylic acid ethyl ester pH 9	HCl	43	1.1	[67]
Co, Cu, Ni	Food (black tea, rice flour)	Amberlite XAD-2	Me-BTAC	HCl	19, 12 12	—	[68]
Cd, Cu	Potable, lake and sea waters	Chitosan microspheres	5-sulphonic acid 8-hydroxiquinoline pH 7 (Cd), pH 10 (Cu)	HNO$_3$	19.1, 13.9	0.2, 0.3	[69, 70]
Pd	Street/fan blade dust and rocks	Exfoliated graphite	Pd-DADC	Methanol	—	1.0	[71]
Pb	Aqueous standards	Multiwall carbon nanotubes	Pb(II) oxidized with HNO$_3$ sorbed at pH 4.7	HNO$_3$	44.2	2.6	[72]

EF = enrichment factor; DL = detection limit; THAM = tris(hydroxymethyl)amine; Salen I = N,N'-bis(salicylidene) ethylene diamine; 5-Br-PADAP = 2-(5-Br-2-pyridylazo)-5-diethylaminophenol; BTAC = 2-(2-benzothiazolylazo)-p-cresol; TAR = 4-(2-thiazolylazo)-resorcinol; PAR = 4-(2-pyridylazo)-resorcinol; DADC = diethyl ammonium dithio carbamate.

The complex formed between Fe(III) and thiocyanate was extracted with MIBK, while total Fe was determined in an aliquot without solvent extraction. A newly designed gravitational phase separator was used to extract on-line the Cd-PDC complex into MIBK with a sampling frequency of $33\,h^{-1}$, an enrichment factor of 155 and a DL of $20\,ng\,L^{-1}$ [76]. An on-line continuous ultrasound-assisted extraction was used for the determination of Cu [77] and Mn [78] in seafood samples by FAAS with DL of 0.06 and $0.4\,\mu g\,g^{-1}$ respectively. Such systems are not very popular in the recent literature.

18.2.1.4 Indirect Determinations

Based on specific chemical properties of certain anionic species like cyanide, iodide, nitrite, and so on, their indirect determination by FAAS was also possible.

A simple FI-FAAS system was used for the indirect determination of cyanide in industrial wastewaters after passing the sample solution through a column packed with cadmium carbonate suspended on silica gel beads [79]. Sodium hydroxide solution (pH 10), used as eluent, carried the analyte as a cadmium cyanide complex to the flame. The absorbance is proportional to the cyanide concentration in the sample. The system provided a DL of $0.2\,mg\,L^{-1}$, a relative standard deviation of 1.22% and a sampling rate of $72\,L^{-1}$. Nitrite was also indirectly determined in foodstuffs and wastewaters by FAAS after its on-line oxidation in a microcolumn filled with lead(IV) dioxide [80]. The flow of sample reduces the PbO_2 solid phase reagent to Pb(II) which is measured by FAAS with a DL of $0.11\,mg\,L^{-1}$ for 0.4 ml of sample injected and a sampling rate of about $80\,h^{-1}$.

Besides the use of solid-phase extraction, the DL in all the above cases was in the $mg\,L^{-1}$ range, which is still too high to allow applications to environmental or biological samples. However, the reported sampling rate was $>70\,h^{-1}$.

18.3
Sample Dilution

To date, the AS techniques coupled with FI manifolds for on-line dilution are those with continuous detecting systems like FAAS [1, 2, 81, 82] and ICP-MS [83] as well as ET AAS [84]. Generally, the purpose of the dilution is to increase sample throughput and precision, to extend the analytical dynamic range of FI-FAAS systems, to bring the analyte concentration within the instrumental working range and to reduce the sample pretreatment steps, thus minimizing the risk of contamination.

In such systems, the on-line dilution was achieved in manifolds based on different principles, like zone penetration and dispersion of micro-volumes, implementation of the cascade method or using different devices, like dilution chambers, recirculating loops, and so on. The precision, which is an important criterion for a dilution system, depends strongly on the performance of the propulsion devices and on the geometry of the transport conduits. Improved precision is obtained when piston pumps are used instead of peristaltic pumps (which present impaired pulsations), and when the conduits are knotted or coiled instead of straight, thus favoring better radial mixing of the sample zone with the carrier. By choosing appropriate pumps and sampling times, a wide range of dilution degrees can be obtained (2 to 130 000). According to the aims of the constructed manifold, sampling frequency also show wide variations (60 to 200 h^{-1}), the repeatability, under optimized conditions, remains <3% and some of the systems are highly compatible with routine applications.

18.4
Electrothermal Atomic Absorption Spectrometry

ET AAS is a robust analytical technique, which is available in most laboratories and is sufficiently sensitive to be routinely applied to the determination of most trace elements at $\mu g\,L^{-1}$ or $\mu g\,g^{-1}$ levels. Despite the advances in ET AAS instrumentation, complex samples cannot be directly processed owing to severe matrix interferences, which have not been minimized despite the development of efficient background correction devices.

The implementation of on-line sample processing systems coupled to ET AAS has been limited by the difficulty in coupling a discontinuous technique like ET AAS to any FI system. The development of sophisticated interfaces has allowed on-line ET AAS to become a fast developing technique, which merited the publication of a few reviews providing the main developments and recent trends in the various on-line ET AAS systems [5, 7, 9–11, 13, 85]. Sample processing systems developed so far are: precipitation and coprecipitation, sorption on columns, solvent extraction, microdialysis, EC reactions, CVG, emulsification and dilution.

18.4.1
Analyte Preconcentration for ET AAS Detection

18.4.1.1 Precipitation or Coprecipitation/Dissolution Reactions
Precipitation or coprecipitation processes are aimed at preconcentrating an analyte or separating it from a cumbersome matrix. The following dissolution process is used to dissolve the precipitate containing the analyte or the other matrix components in order either to introduce the analyte into the atomizer or to divert selected components to waste. The advantages of using on-line precipitation and coprecipitation reactions were the subject of a series of articles published before 2001, which were described in reviews by Vereda et al. [10] and Burguera and Burguera [11]. Since then, the development of such systems has decreased, probably because they are not as versatile as those based on solid-phase extraction.

Nakajima et al. [86] described the determination of Pb in seawater, first by on-line coprecipitation with Fe(III) hydroxide, followed by the dissolution of the precipitate in HNO$_3$ and the separation of the analyte from Fe by solid-phase extraction with a Pb-selective resin (Pb-Spec). The sorbed Pb was eluted with an EDTA solution. The 30-µl portion of the eluate, corresponding to the highest analyte concentration zone, was injected into the graphite furnace.

Burguera *et al.* [87] described a precipitation/dissolution system for the determination of Mo in blood serum and whole blood after a complete on-line flow injection MW-assisted mineralization. After the on-line exposure of the sample to MW radiation, the Mo was precipitated with potassium ferrocyanide. The precipitate of molybdenyl ferrocyanide was dissolved with diluted NaOH and a sub-sample was then collected in a capillary of a sampling arm assembly. Finally, an aliquot of this solution was introduced into the atomizer by positive air displacement. The method was applied over the range 0.2–20.0 $\mu g\,L^{-1}$ Mo with a DL of 0.1 $\mu g\,L^{-1}$. This system has the advantage of being fully controlled by the software, without modification of the instrument software.

Chemically modified beads of C_{18} were used for collection of a Cd hydroxide precipitate in a lab-on-valve (LOV) system coupled with ET AAS [88]. An enrichment factor of 28 and a DL of 1.7 $ng\,L^{-1}$, along with a sampling frequency of 13 h^{-1} were obtained after eluting the precipitate with HNO_3. The enrichment factor could be further enhanced to 44 by increasing the sample volume from 600 to 1200 μL.

The most salient advantage of the above described papers has been their high tolerance to coexisting interferences, which permitted the determination of analytes in rather complex matrices, such as seawater [86], serum and whole blood [87], river sediment, lettuce and frozen cattle blood [88]. However, the on-line manipulation of a relatively large amount of precipitate in a FI system poses some difficulties in producing robust systems. This can be circumvented by the use of LOV systems [88], which provide controlled collection capacities in well-defined tubing and more versatile processes.

18.4.1.2 Sorption on Columns

The most commonly used FI ET AAS systems involve the incorporation of packed microcolumns in the FI manifold, which either replaced or were placed in the tip of the ET AAS autosampler arm, although neither design is clearly better or worse than the other as regards the analytical performance obtained. Then, the desired volume is introduced into the atomizer by air displacement or by using a selection valve.

Various sorbent materials have been used to pack minicolumns for preconcentration purposes, such as hydrophobic PTFE turnings [89], activated carbon [90, 91], silica gel [92–97], newly developed chelating resins [98, 99], although the most widely used have been the commercially available resins of the Amberlite [93] Dowex [100, 101] and Muromac [102] series. The results of coupling FI sorption on a KR with ET AAS demonstrated that the column-in-loop systems can be substituted for different analytical purposes. The volumes of the KR are larger than the column, which causes a serious problem with the limited capacity of the atomizer. This problem was solved by either utilizing column reactors packed with PTFE beads or elution of discrete zones segmented by air. In this way, the eluted volume to be introduced into the atomizer is confined between air bubbles and its dispersion is minimized to obtain better enrichment factors and to improve the reproducibility of the results. The most recent papers dealing with these approaches are summarized in Table 18.3 [89–106].

Table 18.3 On-line sorption preconcentration in packed columns with ET AAS detection.

Analyte	Sample	Sorbing material	Eluent	EF	DL (ng L^{-1})	Ref.
Co	Waters, mussels	PTFE turnings	MIBK	87-fold?	4	[89]
Co	Drinking water	Activated carbon (pH 9.5)	HNO$_3$	190-fold	5	[90]
Cr(III) Cr(VI)	Drinking water	Activated carbon (pH 5)	HNO$_3$	35	3	[91]
Cr(III), Cr(VI)	Natural waters	Amberlite or functionalized silica gel	HNO$_3$	7.4, 5.6	140, 80	[92]
Cd	Seawater	Functionalized silica gel	HNO$_3$	2	60	[93]
Sb(III)	Environmental	Immobilized silica gel	HNO$_3$	—	300	[94]
Rh	Catalyst	Silica gel	HNO$_3$	—	300	[95]
Pt	Catalyst	Silica gel	HNO$_3$	40–42	800	[96, 97]
Pt	Aerosol	Chelating reagent	Ethanol	42	100	[98]
Pb	Seawater	Chelating resin	Ethanol	20.5	140	[99]
Se(IV), Se(VI)	Drinking water	Dowex	HCl	82	10	[100]
Rh	Environmental	Dowex loaded with PSTH	HNO$_3$	—	300	[101]
Bi, Cd, Pb	Urine	Muromak	HNO$_3$		13, 2, 4.5	[102]
Hg(II) Organic Hg	Fish tissue	Cigarette filters	Ethanol	75	6.8 3.4 ng g^{-1}	[103]
Pd	Blood, dust	KR (ion pair: K$^+$-18-crown-6 and Pd(SCN)$_4{}^{2-}$)	Methanol	29	16	[104]
As(III)	Seawater	KR (As(III)-PDC)	Ethanol	44	8	[105]
Co	Standards	KR (hydroquinones)	—	10–34	—	[106]

EF = enrichment factor; DL = detection Limit; PSTH = 1,5-bis(2-pyridyl)-3-sulfophenyl methylene thiocarbonohydrazide.

Despite the low DL assessed with the above-mentioned preconcentration procedures (Table 18.3), they were still inadequate when the sample had a complex matrix. For this reason, Hansen's group [107–114] used bead-packed microcolumns and the SI-LOV schemes with renewable reversed phase surfaces, which enhanced the potential of SI-ET AAS for the analysis of more complex samples. These authors claim that the reversal of the flow direction of SI for the sorption step increases the analyte–sorbent material interaction and improves the enrichment factors, as the sample plugs are stacked one after the other to allow the physical operation to take place.

The renewable reversed-phase surface involves the use of: (i) poly(styrene-divinylbenzene) beads containing pendant octadecyl moieties (C_{18}-PS/DVB) [109, 113] impregnated with a selective organic metal chelating agent, (ii) poly (tetrafluoroethylene) (PTFE) – granular Algoflon beads containing ammonium diethyl dithiophosphate (ADDP) [110], and (iii) hydrophilic chelating Sepharose [111, 112] prior to manipulation of the beads in the microbore conduits of the LOV unit. The size homogeneity and perfect sphericity of the beads improved the reproducibility, retention efficiency, enrichment factor and linear dynamic range [109]. The potential of these SI-LOV schemes was demonstrated by determining Cd [108, 110, 111], Cr(III) [112], Cr(VI) [112, 113], Pb(II) [111], Ni(II) [111] and Bi [114] in environmental (natural and/or sea waters [109, 112], and river sediment [113]), and a biological sample (urine) [111] without requiring any dilution step. In general, wider dynamic ranges, high enrichment factors (above 20) and sample throughputs of ≥ 10 were readily obtained.

18.4.1.3 Solvent Extraction

In some cases, manually conducted liquid–liquid extraction processes mean a large number of successive extractions. An alternative approach has been the on-line mass transfer between two liquid phases, with which an efficient extraction can be conducted with a minimum of sample volume and extracting solvent when using an appropriate set-up.

Arsenic in steel has been determined based on the reaction of As(III) with iodide ion in a concentrated HCl medium to produce AsI_3, which is extracted into benzene using a gravitational phase separator, and back-extracted into water [115]. The absolute DL was 0.2 µg As. The same research team [116] used this extraction system for the determination of As or Sn in steels but used a H_2SO_4 solution for the back-extraction. In this case, the DL were 0.2 and 0.1 µg for As and Sn, respectively.

Cadmium in aqueous solutions [117] and in natural waters and urine [118] was first complexed with APDC and ADDP, respectively, and the corresponding chelate extracted into MIBK which was separated from the aqueous phase by means of a gravitational phase separator. The Cd-PDC complex was back extracted into dilute HNO_3 containing Hg(II) ions as stripping agent, while an aliquot of the Cd-DDP complex was injected directly into the graphite furnace. Enrichment factors of 21.4 and 24.6 and DL of 2.7 and 2.8 ng L^{-1}, along with sampling frequencies of 13 and 30 h^{-1}, respectively, were obtained.

A dynamic two-dimensional admicelles (by cetyltrimethylammonium bromide as a cation surfactant) solvent extraction coupled to ET AAS for the determination of Cr(VI) in drinking water was developed by Nan and Yan [119]. The analyte was on-line complexed with APDC and was solubilized in the admicelles of a microcolumn packed with silica gel. The column was then eluted with acetonitrile for determination by ET AAS. The formation of admicelles demonstrated reversibility, which allowed 22 replicate determinations. The enhancement factor was 32 with a sampling throughput of 31 h^{-1} and a DL of 3.0 ng L^{-1}.

Correctly assessing the analytical potential of the above described liquid–liquid FI-ET AAS systems, the enrichment factors are quite modest relative to other preconcentration approaches. Ensuring correct performance and reliable analytical results requires sound operator experience, particularly in the mass transfer step (which may affect the sensitivity), and the kind of phase separators used (which affect the reproducibility). In particular, the use of gravitational phase separators [115–118] for a smooth operation in routine analyses is difficult owing to their minimal ruggedness. In order to minimize these drawbacks, the following alternatives have been implemented: (i) back-extraction into aqueous solutions [115–117] and (ii) solubilization of the organic complex in admicelles of a microcolumn packed with silica gel [119].

18.4.2
Analyte Separation Prior to ET AAS Detection

In some cases, the analyte needs only to be separated (not necessarily preconcentrated) from the matrix before its determination by ET AAS. There are several novel approaches which assess the separation using microdialysis membranes, EC processes or chemical vapor generation.

The outstanding feature of flow-through dialysis-based units is the high-resolution separation of analyte from matrix constituents via the membrane. In FI, miniaturized systems are required for low reagent consumption and to increase throughput. An on-line microdialysis sampling coupled with FI-ET AAS was developed by Tseng *et al.* [120] for *in vivo* monitoring of extracellular diffusible Mn in brains of living rats. Microdialysed samples perfuse through implanted probes and are collected with a sample loop of an on-line injection valve and directly introduced into the atomizer by a FI system. The DL was of $0.43\,ng\,Mn\,L^{-1}$, enabling monitoring of Mn in anesthetized rats after administration of $MnCl_2$. A temporal resolution of 25 min was possible. The same methodology was applied to the *in vivo* monitoring of diffusible Mg in the blood of living rabbits after $MgSO_4$ administration with a temporal resolution of 1.5 min [121].

The DL was $0.53\,mg\,L^{-1}$. A similar approach for *in vitro* determination of bioavailability of Fe by simulated gastrointestinal digestion has been described by Promchan and Shiowatana [122]. The method used to estimate the mineral bioavailability was based on gastric digestion in a batch system and in a dynamic continuous-flow intestinal digestion, which was performed in a dialysis bag placed inside a chamber containing a flowing stream of dialysing solution. Mineral concentration and dialyzed minerals in the intestinal digestion stage were used to estimate the dialysability of Fe in different kinds of milk. Recently, Tseng *et al.* [123] undertook continuous multi-element (Cu, Mn, Ni, and Se) monitoring in saline and cell suspension using on-line microdialysis coupled with SIMAAS. As an ultrapure NaCl solution was used as the perfusate, the samples contained a high concentration of salt in the matrix. The use of $Pd-Mg(NO_3)_3$ as the matrix modifier improved the pyrolysis temperature, preventing the target elements from undergoing evaporation before the NaCl. Miró *et al.* [124] presented a novel concept for on-line microsampling and continuous monitoring of metal ions in soils with minimum disturbance of the sampling site. It involved a hollow-fiber microdialyser that was implanted in the soil body as a miniaturized sensing device. The idea behind microdialysis in this application was to mimic the function of a passive sampler to predict the actual, rather than potential, mobility and bioavailability of trace metals. Although almost quantitative dialysis recoveries were obtained for lead ($\geq$98%) from aqueous model solutions with sufficiently long capillaries ($l \geq$ 30 mm, 200 μm i.d.) at perfusion rates of 2.0 μL min^{-1} the resistance of an inert soil matrix was found to reduce metal uptake by 30%. Preliminary investigations of the potential of the microdialysis analyser for risk assessment of soil pollution, and for metal partitioning studies, were performed by implanting the dedicated probe in a laboratory-made soil column and hyphenating it to ET AAS so that minute, well-defined volumes of clean microdialysates were injected on-line into the atomizer. A noteworthy feature of the implanted microdialysis-based device is the capability to follow the kinetics of metal release under simulated natural scenarios or anthropogenic actions.

The EC techniques connected on-line with ET AAS have scarcely been used, probably due to the complexity of synchronization of the on-line electrodissolution process with ET AAS. To our knowledge, there are only two papers describing this approach in the last five years. Knápek *et al.* [125] determined Cd in seawater by ET AAS using EC separation in a microcell. The method comprised four steps: (i) electrolysis in the microcell, (ii) electrodeposition of analyte on the graphite electrode under conditions of controlled current in a flow-through mode, (iii) dissolution of the analyte into the eluent (diluted HNO_3 solution) and (iv) direct injection of the whole volume (40 (L) into the atomizer. The DL was 25 ng L^{-1}. A method using on-line bi-directional electrostacking was used for As speciation and preconcentration in natural waters with ET AAS as detector [126]. A DL of 0.35 μg L^{-1} and a preconcentration factor of 4.8 were achieved when a voltage of 750 V was applied for 20 min. $KMnO_4$ was used to oxidize all As species to As(V) in order to determine total As.

The on-line CVG approach connected to ET AAS can provide lower DL down to the ng L^{-1} level, matrix modification, sensitivity enhancement, selective determinations and suppression of interferents [13]. The transfer line between the CVG unit and the atomizer should be as short as possible in order to minimize losses of analyte during transport.

Ma *et al.* [127] trapped the generated Au volatile species from an ore sample digest solution on the inner wall of the graphite furnace. A detection limit of 0.8 ng mL^{-1} was obtained with a 40 s trapping time and a sampling frequency of 50 h^{-1}.

Chuachuad and Tyson [128] undertook the first FI-CVG of Cd using a tetrahydroborate-form anion-exchanger for its determination in natural waters, wine, human saliva, orange juice and human hair. The DL was in the range 40–50 ng L^{-1} for all samples. This system was later improved using the in-atomizer trapping technique [129]. In this case, an Amberlite IRA-400 column was first loaded with a borohydride solution of the sample and then a flow of diluted HNO_3 was passed through the column. After the elution, the analyte was trapped in the graphite tube coated with Zr–Ir. The signals were enhanced in the presence of L-cysteine, thiourea and cobalt. The DL was greatly improved to 3–15 ng L^{-1} for natural waters and 90–180 ng L^{-1} for the other samples.

Recently, Petrov *et al.* [130] described a procedure for the determination of toxicologically relevant arsenic species in urine by FI-HG with collection of generated inorganic and methylated hydrides on an integrated platform of a transversely-heated graphite atomizer. In this case, a large (20 cm^3) custom-made GLS, which exhibited a good tolerance to flooding and aerosol formation, was used. A quartz pipette tip was adjusted to deliver hydrides about 1.3 mm above the platform treated with Zr and Ir, which served as both an efficient hydride sequestration medium and permanent chemical modifier, allowing up to 700–800 firings. The characteristic mass, characteristic concentration and DL were 39 pg, 0.078 μg L^{-1} and 0.038 μg L^{-1}, respectively, with 10- and 25-fold dilutions and the sample throughput rate was 25 h^{-1}.

From the above, it can be concluded that the *in situ* trapping approach was applied in all cases probably because, in this way, atomization is independent of minor changes in the atomizer surface and the composition of the gaseous medium used to transport the volatile species. While, L-cysteine was used to enhance the sensitivity [128–130], an anion-exchanger permitted the determination of the analyte in rather complex matrices [129].

18.4.3
Miscellaneous

An emulsion is a mixture of two or more liquids in which one is present in droplets of microscopic or ultramicroscopic size [131, 132]. The use of emulsions has led to various on-line FI-ET AAS automated determinations of metals in highly viscous samples [132].

The SI technique was applied by Burguera *et al.* [131] for the on-line preparation of an "oil in water" microemulsion for the determination of Al in new and used lubricating oils. A mixture of ethoxylated nanylphenol surfactants, sample and co-surfactant (*sec*-butanol) solutions were sequentially aspirated to a holding coil. The sonication and repetitive change of flow direction improved the stability of the emulsion. The calibration graph was linear from 7.7 to 120 μg L^{-1} Al with a DL of 2.3 μg L^{-1}. The optimum phase behavior formulation of surfactant/oil/water methodology with regard to a reduction in the time of preparation of these highly viscous samples was greatly simplified. Additionally, the analyses of sample solutions which differ significantly in viscosity from that of the standards is solved in a very elegant manner, as the microemulsion reduces the viscosity effect on the accuracy of determinations.

A SI-ET AAS method was recently described for the comprehensive determination of Zn in human saliva by ET AAS after on-line dilution of samples without changing the sample transport conduits during the experiments [84]. The dilution procedure and the injection of solutions into the atomizer are computer controlled and synchronized with the operation of the temperature program. The precision, tested by repeated analysis of real saliva samples, was better than 3% and the DL was 0.35 μg L^{-1}. The potential of this approach is just emerging, as the controlled dilution of stock solutions for calibration purposes can lead to the implementation of the standard addition method, and other standardization procedures with a significant reduction in time and in the number of firings.

18.5
Atomic Fluorescence Spectrometry

Atomic fluorescence spectrometry is a detection system used on a routine basis for HG and CV determinations in samples of environmental and toxicological importance. However, it presents a lower running cost and similar DL when compared to ET AAS and ICP-OES. Its enormous potential of application in clinical practice only needs the symbiosis between the physicians and analytical chemists.

18.5.1
Preconcentration of the Analyte for AFS Detection

AFS has been extensively applied to the determination of hydride forming elements after their separation from the matrix and preconcentration by retention on packed columns or KR, by coprecipitation, selective extraction or complex formation. Most of these systems allowed speciation studies and achievement of low DL, although the applications, with few exceptions, are limited to environmental samples. Some analytical characteristics of selected systems are given in Table 18.4 [133–139].

Inorganic As species have been determined in soil samples using sequential extraction combined with FI-HG-AFS [140]. The sequence of extractants was: water, mono-potassium phosphate, HCl and KOH and As(III) was analyzed in citric acid medium, while As(V) was reduced on-line with L-cysteine in order to determine total As. The DL for As(III) and As(V) were 0.11 and 0.07 $\mu g\,L^{-1}$, respectively and the RSD was 1.43% ($n = 11$) at the $10\,\mu g\,L^{-1}$ level. Following the same idea, Dong and Yan [141] developed an on-line sequential extraction coupled to HG-AFS for rapid and automatic fractionation of arsenic in soils. The soil sample, packed into a microcolumn, was dynamically extracted by continually pumping individual extractants (water, KOH and HCl solutions) through the column. The extracted arsenic was merged with a potassium persulfate solution for the on-line oxidation of all arsenic species to As(V). The total extracted As was quantitated using an on-line standard addition calibration strategy. The developed methodology was successfully applied to the fractionation of As in certified soil reference materials.

The HG-AFS technique has also been extended to elements like Cd [139] and Au [142], which are not among the hydride-forming elements. Using a programmable intermittent flow reactor, volatile species of Au were generated at room temperature by reduction with KBH_4 in HCl medium, followed by the determination by AFS in geological certified reference materials [142]. The presence of micro amounts of DDTC greatly enhanced the generation efficiency of the volatile species, while the flow rate of the carrier gas had a significant impact on the signal intensity of Au. In addition, a high surface/volume ratio and rapid GLS are essential for the generation

Table 18.4 On-line preconcentration systems for AFS detection.

Element	Sample	Sorption on column	Eluent	DL (ng L^{-1})	EF	Samples (h^{-1})	RSD (%)	Ref.
Se(IV)	River, lake, sea waters	Se(IV)-PDC on PTFE fiber	NaBH$_4$	4	67%	26	1.5	[133]
	Waters	Se(IV) coprecipitated with La(OH)$_2$ on PTFE beads	HCl	5	11%	38	1.2	[134]
	Waters	Se(IV) coprecipitated with La(OH)$_2$ on KR	HCl	14	18	24	2.5	[135]
Ge	Aqueous solutions	Ge(IV) coprecipitated with Ni(OH)$_2$ on KR	Phosphate	110	11	24	5.6	[136]
Hg(II), MeHg	Waters	Sorption on silica gel-2-mercapto benzimidazol after sample decomposition with UV radiation	KCN, HCl	0.07, 0.05	13, 24	—	8.8, 10	[137]
	Waters	Hg(II)-DDP and MeHg-dithizone on KR	HCl	3.6, 2.0	13, 24	30, 20	2.2, 2.8	[138]
Cd	Tea	Sorption on Cyanex 923	H$_2$SO$_4$	10.8	—	40	0.97	[139]

DL = detection limit; EF = enhancement factor; RSD = relative standard deviation.

of the volatile gold species. Under optimized conditions, a DL of $0.23 \, \mu g \, L^{-1}$ was obtained and the precision of the measurements for $20 \, \mu g \, L^{-1}$ of Au was 1.7%. The Cd volatile species generation efficiency could be greatly enhanced in the presence of Co^{2+} ions.

As in the case of FAAS, the above-mentioned systems at most are able to differentiate between two valence states of the same element (e.g., As(III) and As(V)). For the speciation of organic compounds containing hydride-forming elements, powerful separation techniques are connected to the HG manifold before AFS detection.

18.5.2
Hyphenated Techniques for Speciation Studies with AFS Detection

Coupling any HG-AFS system with separation techniques like HPLC or reversed phase liquid chromatography (RPLC) has extended the speciation studies to complex organic-containing molecules like selenoamino acids, monomethylarsonic acid (MMAA), dimethylarsinic acid (DMAA), arsenobethaine (AsB), arsenocholine (AsC), methyl mercury (MeHg), ethylmercury (EthHg) or phenylmercury (PhHg) compounds.

Selected selenoaminoacids like selenocystine (SeCys), selenomethionine (SeMet), selenoethionine (SeEt) as well as Se(IV) were separated by on-line coupling HPLC and HG-AFS using ammonium acetate buffer (pH 4) containing methanol and DDAB as eluent [143]. The DL obtained for SeCys, SeMet, SeEt and Se(IV) were 18, 70, 96, and $16 \, \mu g \, L^{-1}$, respectively. SeMet was also determined in breast and formula milk after chiral separation using HPLC on a glycopeptide teicoplanin-based chiral stationary phase (Chirobiotic T) coupled to AFS and ICP-MS [144]. Such instrumental combinations require on-line post-column analyte treatment, like microwave-assisted digestion, and a severe sample clean-up for elimination of fat and proteins by centrifugation and ultrafiltration. Underivatized DL-SeMet enantiomers were completely resolved in 10 min using unbuffered water mobile phase. The good selectivity and sensitivity (DL 3.1 and $3.5 \, ng \, mL^{-1}$ as Se for L-SeMet and D-SeMet, respectively, together with the method robustness and simplicity as well as the low cost of the AFS detector, make it suitable for infant milk routine analysis.

An HPLC-HG-AFS set-up was developed for the speciation of selenium in cow's milk obtained after the supplementation of cow forages with different selenium species (organic selenium as selenized yeast and inorganic selenium as sodium selenite) [145]. The separation was carried out in a μBondapack C_{18} column using tetraethylammonium chloride (as positively charged ion-pairing agent) in the mobile phase. The pre-reduction of the different selenium compounds in the eluted fractions was improved with UV radiation and a heating block. The milk samples obtained after organic supplementation of feeding presented SeCys, SeMet and Se(IV), while only SeCys and Se(IV) were present in milk samples obtained after an inorganic supplementation of feeding.

Ion-exchange LC coupled on-line to AFS through HG was used for the determination of inorganic (As(III) and As(V)) and organic (MMAA and DMAA) arsenic compounds in orthophosphoric acid extracts of soils, sediments and sewage sludge [146]. The method showed a good potential for routine As speciation in environmental solids, with very low DL, ranging from 0.02 to 0.04 mgAs kg^{-1} for all species in all matrices. The efficiency of the extraction procedure and the stability of the extracts were studied in detail. Similarly, 12 inorganic and organic arsenic compounds were separated in a Dionex AS7 column using an acetate buffer and nitric acid as mobile phases [147]. After separation, the compounds were oxidized on-line by UV irradiation, volatilized by HG and carried to the detector by a stream of argon. The DL varied in the range 4–22 pg and the method was applied to marine samples. AsB was detected as the main species in all samples. The chromatographic conditions proposed to separate AsB, AsC, trimethylarsine oxide and tetramethylarsonium ion within 20 min were concomitantly published in another journal [148].

Four mercury species (Hg^{2+}, MeHg, EthHg, and PhHg chlorides) were baseline separated within 13 min in a RP C$_{18}$ column, using a mobile phase of methanol containing tetrabutyl ammonium bromide and sodium chloride [149]. In order to improve the conversion efficiency of the organic mercury compounds into inorganic mercury, a post-column microwave digestion, in the presence of acidified potassium persulfate, was applied. To avoid water vapor and methanol entering into the AFS detector, an ice–water mixture bath was used to cool the digest. The proposed method was accurately applied to the speciation of mercury in seafood samples. Improved DL was obtained for Hg^{2+} and MeHg, simultaneously determined by AFS after HG or ethylation cryofocusing gas chromatographic separation [150]. Also based on RPC coupled on-line with CV-AFS, analysis of certain metallothioneines [151], phytochelatins (PCs) [152] and volatile thiols, including hydrogen sulfide [153] was possible. First, the thiolic proteins were pre-column denatured and derivatized in phosphate buffer containing urea and *p*-hydroximercurybenzoate (*p*-HHgB), and then the corresponding complexes were separated on a C4 Vydac reverse phase column. Post column on-line reaction of the derivatized denatured proteins with bromine, generated *in situ* by KBr/KBrO$_3$ in HCl medium, allowed the fast conversion of the free and proteins-bound *p*-HHgB to inorganic Hg, which is selectively detected by AFS after reduction to Hg^0. Using such a methodology, it was possible to determine the number of –SH groups containing metallothioneins from rabbit liver capable of being complexed by *p*-HHgB. In the case of phitochelatins, synthetic solutions (apo-PCs and Cd^{2+}-complexed PCs) were analyzed: (i) by size exclusion chromatography (SEC) and (ii) by the derivatization of PC-SH groups in the SEC fractions with *p*-HHgB, separation of complexes by RPC followed by their indirect determination by CV-AFS. PCs were determined in extracts of cell cultures from *Phaeodactylum tricornutum* grown in Cd-containing nutrient solutions. Volatile sulfur compounds (H$_2$S, methanethiol, ethanthiol, and propanethiol) from contaminated air were trapped and preconcentrated in alkaline solutions containing *p*-HHgB to form strong Hg–SH covalent bonds, which are stable for at least 12 h at room temperature and for 3 months if stored frozen at −20 °C. Thus different sulfur-containing compounds were indirectly determined by AFS in biogas and the air of a plant for fractional distillation of crude oil. The DL obtained with the CV-AFS detection were respectively 9.7, 13.7, and 17.7 µg L^{-1} for

CH$_3$SH, C$_2$H$_5$SH, and C$_3$H$_7$SH. Exactly the same pre- and post-column derivatization procedures were used for speciation and direct determination of inorganic and organic mercury (MeHg, EthHg, and PhHg) in the form of cysteine, penicillamine, and glutathione complexes by CV-AFS [154]. The separated mercury-thiol complexes were also oxidized with bromine to inorganic mercury in less than 2.5 s, at room temperature, in a 30 cm knitted coil. Under optimized complexation and elution conditions, the DL for Hg(II), MeHg, EthHg, and PhHg were, respectively, 16, 18, 18, and 20 pg (as mercury).

A novel approach interfaced a microfluidic chip-based CE to a volatile species generation system prior to the determination of As(III) and As(V) in locally collected water samples by AFS [155]. Species separation was achieved within 54 s using a mixture of H$_3$BO$_3$ and CTAB (pH 8.9) as buffer. The DL were 76 and 112 µg L^{-1} for As(III) and As(V), respectively, and the precision (RSD% for $n = 5$) varied in the range 1.4–207% at the 3.0 mg L^{-1} level (total As). The same system was used to determine inorganic (Hg(II)) and organic (MeHg(I)) mercury species [156], which were separated as their L-cysteine complexes within 64 s with a precision (RSD for $n = 5$) of 0.7 and 0.9% for 2 mg L^{-1} Hg(II) and 4 mg/L MeHg(I), respectively and a detection limit of 53 and 161 µg/L for Hg(II) and MeHg(I), respectively. With small variations in the system design, the same work was published recently by the same authors, but in a different journal [157]. Curiously, the figures of merit are identical, with the exception of the DL which greatly deteriorated.

18.6
Conclusions and Further Developments

The further development of fully automated manifolds involving on-line coupling of sample pretreatment systems with AS techniques should increasingly offer the advantages of low sample consumption together with a higher frequency of sample analysis per unit time. Microsample introduction systems, preconcentration and, chemical vapor generation procedures will be used to improve the tolerance to high dissolved solids, organic solvents and highly viscous samples. In this context, future studies should involve the use of ultrasonic and microwave radiation to remove some of the matrix components, while the available preconcentration manifolds for sensitivity enhancements, require further study.

The efficiency of sorption columns used both for the elimination of matrix components and preconcentration of the analyte(s) should be further improved as it is not possible from the huge amount of papers published to say which is the most appropriate material for a specific application. Perhaps it might be advisable to compile all the available information about the so-called "home-made" microcolumns and, using available certified reference materials, try to standardize methodologies for specific purposes, in terms of reproducibility and long-term stability. SI and LOV units will allow chemical reaction(s) and/or physical operations to take place in ever miniaturized units, which could eventually be used for field studies.

Many relevant problems related to environmental and biological systems could be better understood if sufficient speciation information were available. Given the complexity of the biological systems, data on speciation is still scarce when compared with that existing for environmental systems, especially natural waters. Validation of the speciation results requires the availability of adequate standard reference materials. For instance, dialysis manifolds coupled to AS techniques permitted the *in situ* and *in vitro* monitoring of diffusible species in order to know more about their bioavailability in biosystems. Stimulus-response schemes adapted from neurochemical applications and pharmacokinetic studies could be extended to other highly specific separation techniques, like electrochemical processes or capillary electrophoreses, which are easily connected on-line, providing that sensitivity limitations are circumvented by coupling to sufficiently sensitive techniques like ET AAS and ICP-MS. Such arrangements would impact heavily on future clinical and environmental monitoring programs of unknown element species at trace levels.

Finally, it is hoped that in the next phase, the use of on-line AS systems in clinical and environmental laboratories will increase, particularly in routine analysis.

References

1 Burguera, J.L. (ed.) (1989) *Flow Injection Atomic Spectroscopy*, Marcel Dekker, New York.

2 Fang, Z. (1995) *Flow Injection Atomic Spectrometry*, John Wiley, New York.

3 Sanz-Medel, A. (ed.) (1999) *Flow Analysis with Atomic Spectrometric Detectors*, Elsevier, Amsterdam.

4 Ruzucka, J. and Hansen, E.H. (1988) *Flow Injection Analysis*, 2nd edn, Wiley, New York.

5 Fang, Z. and Tao, G. (1996) New developments in flow injection separation and preconcentration techniques for electrothermal atomic absorption spectrometry. *Fresenius' Journal of Analytical Chemistry*, **355**, 576–580.

6 Burguera, J.L. and Burguera, M. (1997) Flow injection for automation in atomic spectrometry. *Journal of Analytical Atomic Spectrometry*, **12**, 643–651.

7 Burguera, J.L. and Burguera, M. (1998) On-line sample pre-treatment systems interfaced to electrothermal atomic absorption spectrometry. *Analyst*, **123**, 561–569.

8 Burguera, M. and Burguera, J.L. (1998) Microwave-assisted sample decomposition in flow analysis. *Analytica Chimica Acta*, **366**, 63–80.

9 Fang, Z.L. (1998) Trends and potentials in flow injection on-line separation and preconcentration techniques for electrothermal atomic absorption spectrometry. *Spectrochimica Acta Part B-Atomic Spectroscopy*, **53**, 1371–1379.

10 Vereda, A., García de Torres, A. and Cano Pavón, J.M. (2001) Flow injection on-line electrothermal atomic absorption spectrometry. *Talanta*, **55**, 219–232.

11 Burguera, J.L. and Burguera, M. (2001) Flow injection-electrothermal atomic absorption spectrometry configurations: recent developments and trends. *Spectrochimica Acta Part B-Atomic Spectroscopy*, **56**, 1801–1829.

12 Burguera, J.L. and Burguera, M. (2001) On-line flow injection-atomic spectroscopic configurations: road to practical environmental analysis. *Quím Analítica*, **20**, 255–273.

13 Burguera, J.L. and Burguera, M. (2001) Volatile species generation in flow injection for the on-line determination of species in environmental samples by ET AAS. *Journal of Flow Injection Analysis*, **18**, 5–12.

14 Evans, E.H., Day, J.A., Palmer, C., Price, W.J., Smith, C.M.M. and Tyson, J.F. (2006) Analytical spectrometry update. Advances in atomic emission, absorption and fluorescence spectrometry, and related techniques. *Journal of Analytical Atomic Spectrometry*, **21**, 592–625.

15 Ferrúa, N., Cerutti, S., Salonia, J.A., Olsina, R.A. and Martinez, L.D. (2007) On-line preconcentration and determination of mercury in biological and environmental samples by cold vapor-atomic absorption spectrometry. *Journal of Hazardous Materials*, **141**, 693–699.

16 Duan, T., Song, X., Xu, J., Guo, P., Chen, H. and Li, H. (2006) Determination of Hg(II) in waters by on-line preconcentration using Cyanex 923 as a sorbent-cold vapor atomic absorption spectrometry. *Spectrochimica Acta Part B-Atomic Spectroscopy*, **61**, 1069–1073.

17 Anthemidis, A.N. and Martavaltzoglou, E.K. (2006) Determination of arsenic(III) by flow injection solid-phase extraction coupled with on-line hydride generation atomic absorption spectrometry using PTFE turnings-packed micro column. *Analytica Chimica Acta*, **57**, 413–418.

18 Bartoleto, G.G. and Cadore, S. (2005) Determination of total inorganic arsenic in water using on-line preconcentration and hydride generation atomic absorption spectrometry. *Talanta*, **67**, 169–174.

19 Carrero, P., Gutierrez, L., Rondón, C., Burguera, J.L., Burguera, M. and Petit de Peña, Y. (2004) Flow injection determination of bismuth in urine by successive retention of Bi(III) and tetrahydroborate(III) on an anion-exchange resin and hydride generation atomic absorption spectrometry. *Talanta*, **64**, 1309–1316.

20 Menemenlioglu, I., Korkmaz, D. and Yavuz Ataman, O. (2007) Determination of antimony by using a quartz atom trap and electrochemical hydride generation atomic absorption spectrometry. *Spectrochimica Acta Part B-Atomic Spectroscopy*, **62**, 40–47.

21 Bolea, E., Arroyo, D., Laborda, F. and Castillo, J.R. (2006) Determination of antimony by electrochemical hydride generation atomic absorption spectrometry in samples with high iron content using chelating resins as on-line removal system. *Analytica Chimica Acta*, **569**, 227–233.

22 Korkmaz, D.K., Ertas, N. and Yavuz Ataman, O. (2002) A novel silica trap for lead determination by hydride generation atomic absorption spectrometry. *Spectrochimica Acta Part B-Atomic Spectroscopy*, **57**, 571–580.

23 Kratzer, J. and Dedina, J. (2006) *In situ* trapping of bismuthine in externally heated quartz tube atomizers for atomic absorption spectrometry. *Journal of Analytical Atomic Spectrometry*, **21**, 208–210.

24 Villa-Lojo, M.C., Alonso-Rodriguez, E., López-Mahía, P., Muniategui-Lorenzo, S. and Prada-Rodriguez, D. (2002) Coupled high performance liquid chromatography–microwave digestion hydride generation-atomic absorption spectrometry for inorganic and organic arsenic speciation in fish tissue. *Talanta*, **57**, 741–750.

25 Sur, R. and Dunemann, L. (2004) Method for the determination of five toxicologically relevant arsenic species in human urine by liquid chromatography-hydride generation atomic absorption spectrometry. *Journal of Chromatography*, **807**, 169–176.

26 Tseng, W.C., Cheng, G.W., Lee, C.F., Wu, H.L. and Huang, Y.L. (2005) On-line coupling of microdialysis sampling with high performance liquid chromatography and hydride generation atomic absorption spectrometry for continuous *in vivo*

monitoring of arsenic species in the blood of living rabbits. *Analytica Chimica Acta*, **543**, 38–45.

27 Niedzielski, P. (2005) The new concept of hyphenated analytical systems: simultaneous determination of inorganic arsenic(III), arsenic(V), selenium(IV) and selenium(VI) by high performance liquid chromatography-hydride generation-(*fast sequential*) atomic absorption spectrometry during single analysis. *Analytica Chimica Acta*, **551**, 199–206.

28 Tarley, C.R.T., Ferreira, S.L.C. and Arruda, M.A.Z. (2004) Use of modified rise husks as a natural solid adsorbent of trace metals: characterization and development of an on-line preconcentration system for cadmium and lead determination by FAAS. *Microchemical Journal*, **77**, 163–175.

29 Aji Shabani, A.M., Dadfarnia, S. and Dehghan, K. (2003) On-line preconcentration and determination of cobalt by chelating microcolumns and flow injection atomic spectrometry. *Talanta*, **59**, 719–725.

30 Dadfarnia, S., Haji Shabani, A.M. and Gohari, M. (2004) Trace enrichment and determination of silver by immobilized DDTC microcolumn and flow injection flame atomic absorption spectrometry. *Talanta*, **64**, 682–687.

31 Dadfarnia, S., Haji Shabani, A.M., Tamaddon, F. and Rezaei, M. (2005) Immobilized salen (*N*,*N'*-bis(salicylidene) ethylenediamine) as a complexing agent for on-line, sorbent extraction/preconcentration and flow injection-flame atomic absorption spectrometry. *Analytica Chimica Acta*, **539**, 69–75.

32 Marquéz, M.J., Morales-Rubio, A., Salvador, A. and de la Guardia, M. (2001) Chromium speciation using activated alumina microcolumns and, sequential injection analysis- flame atomic absorption spectrometry. *Talanta*, **53**, 1229–1239.

33 Afzali, D., Taher, M.A., Mostafavi, A. and Mobarakeh, S.Z.M. (2005) Thermal modified Kaolinite as useful material for separation and preconcentration of trace amounts of manganese ions. *Talanta*, **65**, 476–480.

34 Mostafavi, A., Afzali, D. and Taher, M.A. (2006) Atomic absorption spectrometric determination of traces amounts of copper and zinc after simultaneous solid-phase extraction and preconcentration onto modified natrolite zeolite. *Analytical Sciences*, **22**, 849–853.

35 Lemos, V.A. and Ferreira, S.L.C. (2001) On-line preconcentration system for lead determination in seafood simples by flame atomic absorption spectrometry using polyurethane foam loaded with 2-(2-benzothiazolylazo)-2-p-cresol. *Analytica Chimica Acta*, **441**, 281–289.

36 Lemos, V.A., de la Guardia, M. and Ferreira, S.L.C. (2002) An on-line system for preconcentration and, determination of lead in wine samples by FAAS. *Talanta*, **58**, 475–480.

37 Anthemidis, A.N., Zachariadis, G.A. and Stratis, J.A. (2002) On-line preconcentration and determination of copper, lead and chromium(VI) using unloaded polyurethane foam packed column by flame atomic absorption spectrometry in natural waters and biological samples. *Talanta*, **58**, 831–840.

38 Anthemidis, A.N., Zachariadis, G.A. and Stratis, J.A. (2003) Gallium trace on-line preconcentration/separation and determination using polyurethane foam minicolumn and flame atomic absorption spectrometry. *Talanta*, **60**, 929–936.

39 Lemos, V.A., dos Santos, W.N.L., Santos, J.S. and de Carvalho, M.B. (2003) On-line preconcentration system using a minicolumn, of polyurethane foam loaded with Me-BTABr for zinc determination by flame atomic absorption spectrometry. *Analytica Chimica Acta*, **481**, 283–290.

40 dos Santos, W.N.L., Santos, C.M.C. and Ferreira, S.L.C. (2003) Application of three-variables Doehlert matrix for optimization o fan on-line preconcentration system for zinc determination in natural water samples by flame atomic absorption spectrometry. *Microchemical Journal*, **75**, 211–221.

41 dos Santos, W.N.L., dos Santos, C.M.M., Costa, J.L.O., Andrade, H.M.C. and Ferreira, S.L.C. (2004) Multivariate optimization and validation studies in on-line preconcentration system for lead determination in drinking water and saline waste from oil refinery. *Microchemical Journal*, **77**, 123–129.

42 dos Santos, W.N.L., Costa, J.L.O., Araujo, R.G.O., de Jesus, D.S. and Costa, A.C.S. (2006) An on-line preconcentration system for determination of cadmium in drinking water using FAAS. *Journal of Hazardous Materials*, **137**, 1357–1361.

43 Anthemidis, A.N., Zachariadis, G.A., Kougoulis, J.S. and Stratis, J.A. (2002) Flame atomic absorption spectrometric determination of chromium(VI) by on-line preconcentration system using a PTFE packed column. *Talanta*, **57**, 15–22.

44 Zachariadis, G.A., Anthemidis, A.N., Bettas, P.G. and Stratis, J.A. (2002) Determination of lead by on-line solid-phase extraction using a PTFE microcolumn and flame atomic absorption spectrometry. *Talanta*, **57**, 919–927.

45 da Silva, E.L., Ganzarolli, E.M. and Carasek, E. (2004) Use of Nb_2O_5-SiO_2 in an automated on-line preconcentration system for determination of copper and cadmium by FAAS. *Talanta*, **62**, 727–733.

46 Dutra, R.L., Maltez, H.F. and Carasek, E. (2006) Development of an on-line preconcentration system for zinc determination in biological samples. *Talanta*, **69**, 488–493.

47 Maltez, H.F. and Carasek, E. (2005) Chromium speciation and preconcentration using Zr(IV) oxide and Zr(IV) phosphate chemically

immobilized onto silica gel surface using a flow system and FAAS. *Talanta*, **65**, 537–542.

48 da Silva, E.L., Martins, A.O., Valentini, A., de Favere, V.T. and Carasek, E. (2004) Application of silica gel organofunctionalized with 3(1-imidazolyl)propyl in an on-line preconcentration system for the determination of copper by FAAS. *Talanta*, **64**, 181–189.

49 Anthemidis, A.N. and Ioannou, K.I.G. (2006) Evaluation of polychloro trifluoroethylene as sorbent material for on-line solid-phase extraction systems: determination of copper and lead by flame atomic absorption spectrometry in water samples. *Analytica Chimica Acta*, **575**, 126–132.

50 Anthemidis, A.N. and Koussoroplis, S.J.V. (2007) Determination of chromium(VI) and lead in water samples by on-line sorption preconcentration coupled with flame atomic absorption spectrometry using a PCTFE-beads packed column. *Talanta*, **71**, 1728–1733.

51 Praveen, R.S., Naidu, G.R.K. and Prasada Rao, T. (2007) Dithiocarbamate functionalized or surface sorbed Merrifield resin beads as column materials for on-line flow injection-flame atomic absorption spectrometry determination of lead. *Analytica Chimica Acta*, **600**, 205–213.

52 Iglesias, M., Antico, E. and Salvado, V. (2003) On-line determination of trace level of palladium by flame atomic absorption spectrometry. *Talanta*, **59**, 651–657.

53 Knap, M., Kilian, K. and Pyrzynska, K. (2007) On-line enrichment system for manganese determination in water samples using FAAS. *Talanta*, **71**, 406–410.

54 Mostafavi, A., Afzali, D. and Taher, M.A. (2006) Simultaneous separation and preconcentration of traces amounts of Cd^{+2}, Mn^{+2}, Ni^{+2}, Zn^{+2} and Fe^{+2} onto modified amberlite XAD-4 resin loaded with, pyrocatechol violet. *Asian Journal of Chemistry*, **18**, 2303–2309.

55 Ferreira, S.L.C., Lemos, V.A., Santelli, R.E., Ganzarolli, E. and Curtius, A.J. (2001) An automated on-line flow system for the preconcentration and determination of lead by flame atomic absorption spectrometry. *Microchemical Journal*, **68**, 41–46.

56 Ferreira, S.L.C., dos Santos, W.N.L. and Lemos, V.A. (2001) On-line preconcentration system for nickel determination, in food simples by flame atomic absorption spectrometry. *Analytica Chimica Acta*, **445**, 145–151.

57 Ferreira, S.L.C., Bezerra, M.A., dos Santos, W.N.L. and Neto, B.B. (2003) Application of Doehlert designs for optimization o fan on-line preconcentration system for copper determination by flame atomic absorption spectrometry. *Talanta*, **61**, 295–303.

58 Lemos, V.A., Baliza, P.X., Yamaki, R.T., Rocha, M.E. and Alves, A.P.O. (2003) Synthesis and application of a functionalized resin in on-line system for copper preconcentration and determination in food by flame atomic absorption spectrometry. *Talanta*, **61**, 675–682.

59 Lemos, V.A., da Silva, D.G., de Carvalho, A.l., de Andrade Santana, D., dos Santos Novaes, G. and dos Passos, A.S. (2006) Synthesis of amberlite XAD-2-PC resin for, preconcentration and determination of trace elements in food simples by flame atomic absorption spectrometry. *Microchemical Journal*, **84**, 14–21.

60 Wuilloud, G.M., Wuilloud, R.G., de Wuilloud, J.C.A., Olsina, R.A. and Martinez, L.D. (2003) On-line preconcentration and determination of chromium, in parenteral solutions by flow injection-flame atomic absorption spectrometry. *Journal of Pharmaceutical and Biomedical Analysis*, **31**, 117–124.

61 dos Santos, E.J., Herrmann, A.B., Ribeiro, A.S. and Curtius, A.J. (2005)

Determination of Cd in biological simples by flame AAS following on-line preconcentration by complexation with *O, O*-diethyldithiophosphate and solid-phase extraction with Amberlite, XAD-4. *Talanta*, **65**, 593–597.

62 Yebra-Biurrun, M.C., Moreno-Cid, A. and Puig, L. (2004) Minicolumn field preconcentration and flow injection flame atomic absorption spectrometric determination of cadmium in seawater. *Analytica Chimica Acta*, **524**, 73–77.

63 Li, Y., Jiang, Y. and Yan, X.P. (2004) Further study on a flow injection on-line multiplex sorption preconcentration coupled with flame atomic absorption spectrometry for trace element determination. *Talanta*, **64**, 758–765.

64 Souza, A.S. dos Santos, W.N.L. and Ferreira, S.L.C., (2005) Application of Box-Behnken design in the optimization, of an on-line preconcentration system using knotted reactor for cadmium determination by flame atomic absorption spectrometry. *Spectrochimica Acta Part B-Atomic Spectroscopy*, **60**, 737–742.

65 Souza, A.S., Brandao, G.C., dos Santos, W.N.L., Lemos, V.A., Ganzarolli, E.M., Bruns, R.E. and Ferreira, S.L.C. (2007) Automatic on-line preconcentration system using, knotted reactor for the FAAS determination of lead in drinking water. *Journal of Hazardous Materials*, **141**, 540–545.

66 Lopes, C.M.P.V., Almeida, A.A., Santos, J.L.M. and Lima, J.L.F.C. (2006) Automatic flow system for the sequential determination of copper in serum and urine by flame atomic absorption spectrometry. *Analytica Chimica Acta*, **555**, 370–376.

67 Cassella, R.J., Magalhaes, O.I.B., Couto, M.T., Lima, E.L.S., Neves, M.A.F.S. and Cautinho, F.M.B. (2005) Synthesis and application of a functionalized resin for flow injection/FAAS copper determination in waters. *Talanta*, **67**, 121–128.

68 Lemos, V.A., David, G.T. and Santos, L.N. (2006) Synthesis and application of XAD-2/Me-BTAP resin for n-line solid-phase extraction and determination of trace metals in biological simples by FAAS. *Journal of Brazilian Chemical Society*, **17**, 697–704.

69 Martins, A.O., da Silva, E.L., Carasek, E., Laranjeira, M.C.M. and de Favere, V.T. (2004) Sulphoxine immobilized onto chitosan microspheres by, spray drying: application for metal ions preconcentration by flow injection analysis. *Talanta*, **63**, 397–403.

70 Martins, A.O., da Silva, E.L., Carasek, E., Goncalves, N.S., Laranjeira, M.C.M. and de Favere, V.T. (2004) Chelating resin from functionalization of chitosan with, complexing agent 8-hydroxyquinoline: application for metal ions on-line preconcentration system. *Analytica Chimica Acta*, **521**, 157–162.

71 Praveen, R.S., Daniel, S., Prasada Rao, T., Sampath, S. and Sreenivasa Rao, K. (2006) Flor injection on-line solid phase extractive preconcentration of palladium(II) in dust and rock samples using exfoliated graphite packed microcolumns and determination by flame atomic absorption spectrometry. *Talanta*, **70**, 437–443.

72 Barbosa, A.F., Segatelli, M.G., Pereira, A.C., de Santana Santos, A., Kubota, L.T., Luccas, P.O. and Tarley, C.R.T. (2007) Solid-phase extraction system for Pb(II) ions enrichment, based on multiwall carbon nanotubes coupled on-line to flame atomic absorption spectrometry. *Talanta*, **71**, 1512–1519.

73 Cerutti, S., Martínez, L.D. and Wuilloud, R.G. (2005) Knotted reactors and their role in flow injection on-line preconcentration systems coupled to atomic spectrometry-based detector. *Applied Spectroscopy Reviews*, **40**, 71–101.

74 Yebra, M.C., Enriquez, M.F. and Cespon, R.M. (2000) Preconcentration and flame atomic absorption spectrometry determination of cadmium in mussels by an on-line continuous precipitation-dissolution flow system. *Talanta*, **52**, 631–636.

75 de Campos Costa, R.C. and Araujo, A.N. (2001) Determination of Fe(III) and total Fe in wines by sequential injection analysis and flame atomic absorption spectrometry. *Analytica Chimica Acta*, **438**, 227–233.

76 Anthemidis, A.N., Zachariadis, G.A., Farastelis, C.G. and Stratis, J.A. (2004) On-line liquid–liquid extraction system using a new phase separator for flame atomic absorption spectrometric determination of ultra-trace cadmium in natural waters. *Talanta*, **62**, 437–443.

77 Moreno-Cid, A. and Yebra, M.C. (2002) Flow injection determination of copper in mussels by flame atomic absorption spectrometry after on-line continuous ultrasound-assisted extraction. *Spectrochimica Acta Part B-Atomic Spectroscopy*, **57**, 967–974.

78 Yebra, M.C. and Moreno-Cid, A. (2003) On-line determination of manganese in solid seafood samples by flame atomic absorption spectrometry. *Analytica Chimica Acta*, **477**, 149–155.

79 Naroozifar, M., Khorasani-Motlagh, M. and Hosseini, S.N. (2005) Flow injection analysis- flame atomic absorption spectrometry system for indirect determination of cyanide using cadmium carbonate as a new solid-phase reactor. *Analytica Chimica Acta*, **528**, 269–273.

80 Naroozifar, M., Khorasani-Motlagh, M., Taheri, A. and Homayoonfard, M. (2006) Indirect determination of nitrite by flame atomic absorption spectrometry using lead (IV) dioxide oxidant microcolumn. *Bulletin of the Korean Chemical Society*, **27**, 875–880.

81 Tao, G., Fang, Z.-L., Baasner, J. and Welz, B. (2003) Flow injection on-line dilution for flame atomic absorption spectrometry by micro-sample introduction and dispersion using syringe pumps. *Analytica Chimica Acta*, **481**, 273–281.

82 López García, I., Kozak, J. and Hernández Córdoba, M. (2007) Use of membrane micropumps for introducing the simple solution in flame atomic absorption spectrometry. *Talanta*, **71**, 1369–1374.

83 Wang, J., Hansen, E.H. and Gammelgaard, B. (2001) Flow injection on-line dilution for multielement determination in human urine with detection by ICP-MS. *Talanta*, **55**, 117–126.

84 Burguera-Pascu, M., Rodriguez-Archilla, A., Burguera, J.L., Burguera, M., Rondón, C. and Carrero, P. (2007) Flow injection on-line dilution for zinc determination in human saliva with ETAAS detection. *Analytica Chimica Acta*, **600**, 214–220.

85 Wang, J. and Hansen, H.E. (2005) Trends and perspectives of flow injection/ sequential injection on-line sample-pretreatment schemes coupled to ETAAS. *Trends in Analytical Chemistry*, **24**, 1–8.

86 Nakajima, J., Hirano, Y. and Oguma, K. (2003) Determination of lead in seawater by flow injection on-line preconcentration-electrothermal atomic absorption spectrometry after coprecipitation with iron(III) hydroxide. *Analytical Science*, **19**, 585–588.

87 Burguera, J.L., Burguera, M. and Rondón, C. (2002) An on-line flow injection microwave-assisted mineralization and a preconcentration/dissolution system for the determination of molybdenum in blood serum and whole blood by electrothermal atomic absorption spectrometry. *Talanta*, **58**, 1167–1175.

88 Wang, Y., Wang, J.-H. and Fang, Z.-L. (2005) Octadecyl immobilized surface for precipitate collection with a renewable microcolumn in a lab-on-valve coupled to an electrothermal atomic absorption spectrometer for ultratrace cadmium determination. *Analytical Chemistry*, **77**, 5396–5401.

89 Anthemidis, A.N., Zachariadis, G.A. and Stratis, J.A. (2002) Cobalt ultra-trace on-line preconcentration and determination using a PTFE turnings packed column

and electrothermal atomic absorption spectrometry. Applications in natural waters and biological samples. *Journal of Analytical Atomic Spectrometry*, **17**, 1330–1334.

90 Cerutti, S., Moyano, S., Gásquez, J.A., Stripeikis, J., Olsina, R.A. and Martínez, L.D. (2003) On-line preconcentration of cobalt in drinking water using a minicolumn packed with activated carbon coupled to electrothermal atomic absorption spectrometric determination. *Spectrochimica Acta Part B-Atomic Spectroscopy*, **58**, 2015–2021.

91 Gil, R.A., Cerutti, S., Gásquez, J.A., Olsina, R.A. and Martinez, L.D. (2006) Preconcentration and speciation of chromium in drinking water samples by coupling of on-line sorption on activated carbon to ETAAS determination. *Talanta*, **68**, 1065–1070.

92 Cordero, M.T.S., Alonso, E.I.V., García de Torres, A. and Pavón, J.M.C. (2004) Development of a new system for the speciation of chromium in natural waters and human urine samples by combining ion exchange and ETA-AAS. *Journal of Analytical Atomic Spectrometry*, **19**, 398–403.

93 Vereda Alonso, E.I., Gil, L.P., Siles Cordero, M.T., García de Torres, A. and Cano Pavón, J.M. (2001) Automatic on-line column preconcentration system for determination of cadmium by electrothermal atomic absorption spectrometry. *Journal of Analytical Atomic Spectrometry*, **16**, 293–295.

94 Ojeda, C.B., Rojas, F.S., Pavón, J.M.C. and Martín, L.T. (2005) Use of 1,5-bis(di-2-pyridyl)methylene thiocarbohydrazide immobilized on silica gel for automated preconcentration and selective determination of antimony(III) by flow injection electrothermal atomic absorption spectrometry. *Analytical and Bioanalytical Chemistry*, **382**, 513–518.

95 Sánchez Rojas, F., Bosch Ojeda, C. and Cano Pavón, J.M. (2005) Application of flow injection on-line electrothermal

atomic absorption spectrometry to the determination of rhodium. *Annali di Chimica*, **95**, 437–445.

96 García, M.M., Rojas, F.S., Ojeda, C.B., García de Torres, A. and Pavón, J.M.C. (2003) On-line ion-exchange preconcentration and determination of traces of platinum by electrothermal atomic absorption spectrometry. *Analytical and Bioanalytical Chemistry*, **375**, 1229–1233.

97 Bosch Ojeda, C., Sánchez Rojas, F., Cano Pavón, J.M. and García de Torres, A. (2003) Automated on-line separation preconcentration system for platinum determination by electrothermal atomic absorption spectrometry. *Analytica Chimica Acta*, **494**, 97–103.

98 Limbeck, A., Rudolph, E., Hann, S., Koellensperger, G., Stingeder, G. and Rendl, J. (2004) Flow injection on-line pre-concentration of platinum coupled with electrothermal atomic absorption spectrometry. *Journal of Analytical Atomic Spectrometry*, **19**, 1474–1478.

99 Vereda, A.E., Siles Cordero, M.T., Garcia de Torres, A. and Cano Pavon, J.M. (2006) Lead ultra-trace on-line preconcentration and determination using selective solid-phase extraction and electrothermal atomic absorption spectrometry. Applications in seawaters and biological samples. *Analytical and Bioanalytical Chemistry*, **385**, 1178–1185.

100 Stripeikis, J., Pedro, J., Bonivardi, A. and Tudino, M. (2004) System coupled to electrothermal atomic spectrometry with permanent chemical modifiers. *Analytica Chimica Acta*, **502**, 99–105.

101 Sánchez Rojas, F., Bosch Ojeda, C. and Cano Pavón, J.M. (2004) On-line preconcentration of rhodium on an anion-exchange resin loaded with 1,5-bis (2-pyridyl)-3-sulphophenyl methylene thiocarbonohydrazide and its determination in environmental samples. *Talanta*, **64**, 230–236.

102 Sung, Y.-H. and Huang, S.-D. (2003) On-line preconcentration system coupled to electrothermal atomic absorption spectrometry for the simultaneous determination of bismuth, cadmium, and lead in urine. *Analytica Chimica Acta*, **495**, 165–176.

103 Yan, X.-P., Li, Y. and Jiang, Y. (2003) Selective measurement of ultratrace methylmercury in fish by flow injection on-line microcolumn displacement sorption preconcentration and separation coupled with electrothermal atomic absorption spectrometry. *Analytical Chemistry*, **75**, 2251–2255.

104 Dimitrova, B., Benkhedda, K., Ivanova, E. and Adams, F. (2004) Flow injection on-line preconcentration of palladium by ion-pair adsorption in a knotted reactor coupled with electrothermal atomic absorption spectrometry. *Journal of Analytical Atomic Spectrometry*, **19**, 1394–1396.

105 Herbello-Hermelo, P., Barciela-Alonso, M.C., Bermejo-Barrera, A. and Bermejo-Barrera, P. (2005) Flow on-line sorption preconcentration in a knotted reactor coupled with electrothermal atomic absorption spectrometry for selective As (III) determination in sea-water samples. *Journal of Analytical Atomic Spectrometry*, **20**, 662–664.

106 Tsakovski, S., Benkhesdda, K., Ivanova, E. and Adams, F.C. (2002) Comparative study of 8-hydroxyquinoline derivatives as chelating reagents for flow injection preconcentration of cobalt in a knotted reactor. *Analytica Chimica Acta*, **453**, 143–154.

107 Wang, J., Hansen, E.H. and Miró, M. (2003) Sequential injection-bead injection-lab-on-valve schemes for on-line solid-phase extraction and preconcentration of ultra-trace levels of heavy metals with determination by electrothermal atomic absorption spectrometry and inductively coupled plasma mass spectrometry. *Analytica Chimica Acta*, **499**, 139–147.

108 Wang, J. and Hansen, E.H. (2002) Sequential injection on-line matrix

removal and trace metal preconcentration using a PTFE beads packed column as demonstrated for the determination of cadmium with detection by electrothermal atomic absorption spectrometry. *Journal of Analytical Atomic Spectrometry*, **17**, 248–252.

109 Miró, M., Jóncyk, S., Wang, J. and Hansen, E.H. (2003) Exploiting the bead-injection approach in the integrated sequential injection lab-on-valve format using hydrophobic packing materials for on-line matrix removal and preconcentration of trace levels of cadmium in environmental and biological samples via formation of non-charged chelates prior to EETAAS detection. *Journal of Analytical Atomic Spectrometry*, **18**, 89–98.

110 Long, X., Chomchoei, R., Gala, P. and Hansen, E.H. (2004) Evaluation of a novel PTFE material for use as a means for separation and preconcentration of trace levels of metal ions in sequential injection (SI) and sequential injection lab-on-valve (SI-LOV) systems: Determination of cadmium(II) with detection by electrothermal atomic absorption spectrometry (ETAAS). *Analytica Chimica Acta*, **523**, 279–286.

111 Long, X., Hansen, E.H. and Miró, M. (2005) Determination of trace metal ions via on-line separation and preconcentration by means of chelating Sepharose beads in a sequential injection lab-on-valve (SI-LOV) system coupled to electrothermal atomic absorption spectrometric detection. *Talanta*, **66**, 1326–1332.

112 Long, X., Miró, M. and Hansen, E.H. (2005) An automatic micro-sequential injection bead injection Lab-on-Valve, (μSI-BI-LOV) assembly for speciation analysis of ultra trace levels of Cr(III) and Cr(VI) incorporating on-line chemical reduction and employing detection by electrothermal atomic absorption spectrometry (ETAAS). *Journal of*

Analytical Atomic Spectrometry, **20**, 1203–1211.

113 Long, X., Miró, M. and Hansen, E.H. (2005) Universal approach for selective trace metal determinations via sequential injection-bead injection-lab-on-valve using renewable hydrophobic bead surfaces as reagent carriers. *Analytical Chemistry*, **77**, 6032–6040.

114 Wang, J. and Hansen, E.H. (2001) On-line ion exchange preconcentration in a sequential injection lab-on-valve micro-system incorporating a renewable column with ETAAS for trace level determination of bismuth in urine and river sediments. *Atomic Spectroscopy*, **22**, 312–318.

115 Sakuragawa, A., Taniai, T. and Uzawa, A. (2003) Determination of arsenic in steel by metal furnace-AAS and flow injection analysis based on method with on-line iodide trace extraction system. *Tetsu-To-Hagane/Journal of the Iron and Steel Institute of Japan*, **89**, 927–934.

116 Taniai, T., Sakuragawa, A. and Uzawa, A. (2004) Determination of arsenic or tin in steels by the automated extraction system with a recycled solvent and an improved gravity phase separation column. *ISIJ Interntional*, **44**, 1852–1858.

117 Wang, J. and Hansen, E.H. (2002) Development of an automated sequential injection on-line solvent extraction-back extraction procedure as demonstrated for the determination of cadmium with detection by electrothermal atomic absorption spectrometry. *Analytica Chimica Acta*, **456**, 283–292.

118 Anthemidis, A.N., Zachariadis, G.A. and Stratis, J.A. (2003) Development of on-line solvent extraction system for electrothermal atomic absorption spectrometry utilizing a new gravitational phase separator. Determination of cadmium in natural waters and urine samples. *Journal of Analytical Atomic Spectrometry*, **18**, 1400–1403.

119 Nan, J. and Yan, X.-P. (2005) On-line dynamic two-dimensional admicelles solvent extraction coupled to electrothermal atomic absorption spectrometry for determination of chromium(VI) in drinking water. *Analytica Chimica Acta*, **536**, 207–212.

120 Tseng, W.-C., Sun, Y.C., Yang, M.-H., Chen, T.-P., Lin, T.-H. and Huang, Y.-L. (2003) On-line microdialysis sampling coupled with flow injection electrothermal atomic absorption spectrometry for *in vivo* monitoring extracellular manganese in brains of living rats. *Journal of Analytical and Atomic Spectrometry*, **18**, 38–43.

121 Tseng, W.-C., Sun, Y.C., Lee, C.-F., Chen, B.-H., Yang, M.-H., Chen, T.-P. and Huang, Y.-L. (2005) On-line microdialysis sampling coupled with flame atomic absorption spectrometry for continuous *in vivo* monitoring of diffusible magnesium in the blood of living animals. *Talanta*, **66**, 740–745.

122 Promchan, J. and Shiowatana, J. (2005) A dynamic continuous-flow dialysis system with on-line electrothermal atomic-absorption spectrometric and pH measurements for in-vitro determination of iron bioavailability by simulated gastrointestinal digestion. *Analytical and Bioanalytical Chemistry*, **382**, 1360–1367.

123 Tseng, W.-C., Chen, P.-H., Tsay, T.-S., -H-Chen, B. and Huang, Y.-L. (2006) Continuous multi-element (Cu, Mn, Ni, se) monitoring in saline and cell suspension using on-line microdialysis coupled with simultaneous electrothermal atomic absorption spectrometry. *Analytica Chimica Acta*, **576**, 2–8.

124 Miró, M., Jimoh, M. and Frenzel, W. (2005) A novel dynamic approach for automatic microsampling and continuous monitoring of metal ion release from soils exploiting a dedicated flow-through microdialyser. *Analytical and Bioanalytical Chemistry*, **382**, 396–404.

125 Knàpek, J., Komàrek, J. and Kràsensky, P. (2005) Determination of cadmium by electrothermal atomic absorption spectrometry using electrochemical

separation in a microcell. *Spectrochimica Acta Part B-Atomic Spectroscopy*, **60**, 393–398.

126 Coelho, L.M., Coelho, N.M.M., Arruda, M.A.Z. and de la Guardia, M. (2007) On-line bi-directional electrostacking for As speciation/preconcentration, using ET AAS. *Talanta*, **71**, 353–358.

127 Ma, H., Fan, X., Zhou, H. and Xu, S. (2003) Preliminary studies on flow injection *in situ* trapping of volatile species of gold in graphite furnace and atomic absorption spectrometric determination. *Spectrochimica Acta Part B-Atomic Spectroscopy*, **58**, 33–41.

128 Chuachuad, W. and Tyson, J.F. (2004) Determination of cadmium by electrothermal atomic absorption spectrometry with flow injection chemical vapor generation from a tetrahydroborate-form anion-exchanger and in-atomizer trapping. *Canadian Journal of Analytical Science and Spectroscopy*, **49**, 362–373.

129 Chuachuad, W. and Tyson, J.F. (2005) Determination of cadmium by flow injection atomic absorption spectrometry with cold vapor generation by a tetrahydroborate-form-anion-exchanger. *Journal of Analytical Atomic Spectrometry*, **20**, 273–281.

130 Petrov, P.K., Serafimovski, I., Dtafilov, T. and Tsalev, D.L. (2006) Flow injection hydride generation electrothermal atomic absorption spectrometric determination of toxicologically relevant arsenic in urine. *Talanta*, **69**, 1112–1117.

131 Burguera, J.L., Burguera, M., Anton, R.-E., Salager, J.L., Arandia, M.A., Rondón, C., Carrero, P., Petit de Peña, Y., Brunetto, R. and Gallignani, M. (2005) Determination of aluminum by electrothermal atomic absorption spectroscopy in lubricating oils emulsified in a sequential injection analysis system. *Talanta*, **68**, 179–186.

132 Burguera, J.L. and Burguera, M. (2004) Analytical applications of organized assemblies for on-line spectrometric determinations: present and future. *Talanta*, **64**, 1099–1108.

133 Lu, C.-Y., Yan, X.-P., Zhang, Z.-P., Wang, Z.-P. and Liu, L.-W. (2004) Flow injection on-line sorption preconcentration coupled with hydride generation atomic fluorescence spectrometry using a polytetrafluoroethylene fiber packed microcolumn for determination of Se(IV) in natural waters. *Journal of Analytical Atomic Spectrometry*, **19**, 277–281.

134 Tang, X., Xu, Z. and Wang, J. (2005) A hydride generation fluorescence spectrometric procedure for selenium determination after flow injection on-line coprecipitation preconcentration. *Spectrochimica Acta Part B-Atomic Spectroscopy*, **601**, 580–1585.

135 Wu, H., Jin, Y., Shi, Y. and Bi, S. (2007) On-line organoselenium interference removal for inorganic selenium species by flow injection coprecipitation preconcentration coupled with hydride generation atomic fluorescence spectrometry. *Talanta*, **71**, 1762–1768.

136 Jianbo, S., Zhiyong, T., Chunhua, T., Quan, C. and Zexiang, J. (2002) Determination of trace amounts of germanium by flow injection hydride generation atomic fluorescence spectrometry. *Talanta*, **56**, 711–716.

137 Bagheri, H. and Gholami, A. (2001) Determination of very low levels of dissolved mercury(II) and methylmercury in river waters by continuous flow with on-line UV decomposition and cold vapor atomic fluorescence spectrometry after preconcentration on a silica gel-2-mercaptobenzimidazol sorbent. *Talanta*, **55**, 1141–1150.

138 Wu, H., Jin, Y., Han, W., Miao, Q. and Bi, S. (2006) Non-chromatographic speciation analysis of mercury by flow injection on-line preconcentration in combination with chemical vapor generation atomic fluorescence spectrometry. *Spectrochimica Acta Part B-Atomic Spectroscopy*, **61**, 831–840.

139 Duan, T., Song, X., Jin, D., Li, H., Xu, J. and Chen, H. (2005) Preliminary results on the determination of ultratrace amounts of cadmium in tea samples using a flow injection on-line solid-phase extraction separation and preconcentration technique to couple with a sequential injection hydride generation atomic fluorescence spectrometry. *Talanta*, **67**, 968–974.

140 Bo Shi, J., Tang, Z.-Y., Jin, Z.-X., Chi, Q., He, B. and Jiang, G.-B. (2003) Determination of As(III) and As(V) in soils using sequential extraction combined with flow injection hydride generation atomic fluorescence spectrometry. *Analytica Chimica Acta*, **477**, 139–147.

141 Dong, J.-M. and Yan, X.-P. (2005) On-line coupling of flow injection sequential extraction to hydride generation atomic fluorescence spectrometry for fractionation of arsenic in soils. *Talanta*, **65**, 627–631.

142 Li, Z. (2006) Studies on the determination of trace amounts of gold by chemical vapor generation non-dispersive atomic fluorescence spectrometry. *Journal of Analytical Atomic Spectrometry*, **21**, 435–438.

143 Ipolyi, I., Stefanka, Z. and Fodor, P. (2001) Speciation of Se(IV) in the selenoaminoacids by high performance liquid chromatography-direct hydride generation atomic fluorescence spectrometry. *Analytica Chimica Acta*, **435**, 367–375.

144 Gómez-Ariza, J.L., Bernal-Daza, V. and Villegas-Portero, M.J. (2004) Comparative study of the instrumental couplings of high performance liquid chromatography with microwave-assisted digestion hydride generation atomic fluorescence spectrometry and inductively coupled plasma mass spectrometry for chiral speciation of selenomethionine in breast milk. *Analytica Chimica Acta*, **520**, 229–235.

145 Muñiz-Naveiro, O., Dominguez-Gonzalez, R., Bermejo-Barrera, A.,

Bermejo Barrera, P., Cocho, J.A. and Fraga, J.M. (2007) Selenium speciation in cow milk obtained after supplementation with different selenium forms to the cow feed using liquid chromatography coupled with hydride generation-atomic fluorescence. *Talanta*, **71**, 1587–1593.

146 Gallardo, M.V., Bohari, Y., Astruc, A., Potin-Goutier, M. and Astruc, M. (2001) Speciation analysis of arsenic environmental solids reference materials by high-performance liquid chromatography-hydride generation-atomic fluorescence spectrometry following orthophosphoric acid extraction. *Analytica Chimica Acta*, **441**, 257–268.

147 Simón, S., Tran, H., Pannier, F. and Potin-Gautier, M. (2004) Simultaneous determination of twelve inorganic and organic arsenic compounds by liquid chromatography-ultraviolet irradiation-hydride generation atomic fluorescence spectrometry. *Journal of Chromatography*, **1024**, 105–113.

148 Simón, S., Lobos, G., Pannier, F., de Gregori, I., Pinochet, H. and Potin-Gautier, M. (2004) Speciation analysis of organoarsenical compounds in biological matrices by coupling ion chromatography to atomic fluorescence spectrometry photooxidation and hydride generation. *Analytica Chimica Acta*, **521**, 99–108.

149 Liang, L.-N., Jiang, G.-B., Liu, J.-F. and Hu, J.-T. (2003) Speciation analysis of mercury in seafood by using high-performance liquid chromatography on-line coupled with cold vapor atomic fluorescence spectrometry via a post column microwave digestion. *Analytica Chimica Acta*, **477**, 131–137.

150 Stoichev, T., Rodriguez, R.C., Tessier, E., Amouroux, D. and Donard, O.F.X. (2004) Improvement of analytical performances for mercury speciation by on-line derivatization, cryofocusing and atomic fluorescence spectrometry. *Talanta*, **62**, 433–438.

151 Bramanti, E., Lomonte, C., Galli, A., Onor, M., Zamboni, R., Raspi, G. and D'Ulivo, A. (2004) Characterization of denatured metallothioneins by reverse phase coupled with on-line chemical vapor generation and atomic fluorescence spectrometric detection. *Journal of Chromatography*, **1054**, 285–291.

152 Bramanti, E., Toncelli, D., Morelli, E., Lampugnani, L., Zamboni, R., Miller, K.E., Zemewtra, J. and D'Ulivo, A. (2006) Determination and characterization of phytochelatins by liquid chromatography coupled with on-line chemical vapor generation and atomic fluorescence spectrometric detection. *Journal of Chromatography*, **1133**, 195–203.

153 Bramanti, E., D'Ulivo, L., Lomonte, C., Galli, A., Onor, M., Zamboni, R., Raspi, G. and D'Ulivo, A. (2006) Determination of hydrogen sulfide and volatile thiols in air samples by mercury probe derivatization coupled with liquid chromatography-atomic fluorescence spectrometry. *Analytica Chimica Acta*, **579**, 38–46.

154 Bramanti, E., Lomonte, C., Galli, A., Onor, M., Zamboni, R., D'Ulivo, A. and Raspi, G. (2005) Mercury speciation by liquid chromatography coupled with on-line chemical vapor generation and atomic fluorescence spectrometric detection. *Talanta*, **66**, 762–768.

155 Li, F., Wang, D.-D., Yan, X.-P., Su, R.-G. and Lin, J.-M. (2005) Speciation analysis of inorganic arsenic by microchip capillary electrophoresis coupled with hydride generation atomic fluorescence spectrometry. *Journal of Chromatography*, **1081**, 232–237.

156 Li, F., Wang, D.-D., Yan, X.-P., Lin, J.-M. and Su, R.-G. (2005) Development of a new hybrid technique for rapid speciation analysis by directly interfacing capillary electrophoresis system to atomic fluorescence spectrometry. *Electrophoresis*, **26**, 2261–2268.

157 Wang, D.-D., Li, F. and Yan, X.-P. (2006) On-line hyphenation of flow injection miniaturized capillary electrophoresis and atomic fluorescence spectrometry for high throughput speciation analysis. *Journal of Chromatography*, **1117**, 246–249.

19
Flow Injection Mass Spectrometry

Maria Fernanda Giné

19.1
Introduction

19.1.1
The Role and Importance of Flow Injection Analysis Mass Spectrometry

Flow injection analysis presents versatility for on-line management of solutions allowing the performance of different unit operations and processes under computer-controlled conditions [1]. These features are of great value when coupled to analytical techniques with high detection power, such as mass spectrometry (MS) [2].

The injection of discrete sample volumes in a carrier solution produces a transient passage through the FIA-MS interface, which allows methodical base line recovery, and avoids carryover between injected solutions. The efficiency of FIA coupled to MS allows the injection of samples, standards, or spikes of isotopic solutions in sequence, thus simplifying quantification and minimizing systematic errors and contaminations [3, 4]. The capability of FIA to run small volumes (e.g., <100 µl) of samples, standards and reagents in a highly reproducible manner has generated numerous applications in MS [5]. Flow injection systems have been employed for molecular and elemental analyses with MS detection. These applications have enhanced the efficiency of sample introduction into the ionization sources and thus have overcome matrix effects and provided the convenience of performing on-line calibrations.

19.1.2
Mass Spectrometry

MS is an instrumental technique with a powerful capacity for separation of ions, identification and quantification. The MS systems comprise a sample introduction device, an ionization source followed by a compartment for transferring the ions to the mass analyzer where they are separated by their mass to charge ratio (m/z) and sent to a detector system. The compartments of a generalized mass spectrometer are

Advances in Flow Analysis. Edited by Marek Trojanowicz
Copyright © 2008 WILEY-VCH Verlag GmbH & Co. KGaA, Weinheim
ISBN: 978-3-527-31830-8

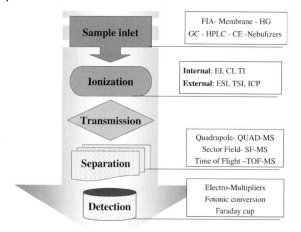

Figure 19.1 Mass spectrometry steps. Sample inlet and external ionization sources are at atmospheric pressure, the other components operate under vacuum.

presented in the scheme of Figure 19.1. Sample introduction can combine FIA with membrane probes which are permeated by organic species by pervaporation. Transient analyte species separations by gas chromatography (GC) [6] and liquid chromatography (LC) [7], capillary electrophoresis (CE) [8], field flow fractionation (FFF) [9] have been interfaced with MS. On-line vapor generation or, more specifically, hydride generation (HG) as well as different types of nebulizers have been employed for sample introduction [10].

Molecular or elemental ions from organic or inorganic samples in the gaseous phase are analyzed by MS. The production of ions occurs by different mechanisms, involving transfer of energy to form gas-phase molecular or elemental ions, depending on the analytical purpose. The ionization sources may be located inside the MS, under vacuum, or outside the vacuum and connected through an interface. For most molecular ions the ionization processes involve electron impact (EI), chemical ionization (CI), or thermal ionization (TI) in an internal chamber under vacuum [11]. For molecular analysis combined sample nebulization and ionization can be accomplished at atmospheric pressure by electrospray (ESI) [12] and thermospray ionization (TSI) [13]. Direct ionization techniques in open-air, such as desorption electrospray ionization (DESI) [14] and direct analysis in real time (DART) [15], have become one of the fastest growing fields in organic mass spectrometry. The inductively coupled plasma is a powerful atmospheric pressure ionization source for multi-elemental inorganic analysis [16]. For ICP-MS, the sample is normally introduced as a solution by a nebulizer.

Representative atomic and molecule ions from the ion sources are guided through electric and magnetic fields under vacuum in the mass analyzer and collected with a detector system. Some characteristics of mass analyzers commonly used for detecting ions from transient FIA are described in subsequent sections.

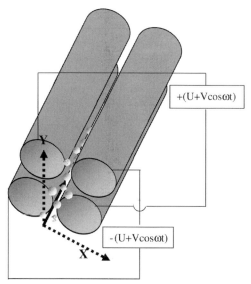

+(U+Vcosωt)

-(U+Vcosωt)

Figure 19.2 Scheme of a quadrupole mass analyzer, showing direct and alternate currents applied to opposite rods. The arrow shows the ion entrance direction.

19.1.2.1 Quadrupole Mass Spectrometers (QMS)

A quadrupole mass spectrometer (QMS) is an array of four rod electrodes, to which a combination of alternate (rf) and direct current voltages are applied, as shown in Figure 19.2 [16–19]. The electrode rods define an internal hyperbolic electrically charged space, and ions are introduced into the center of the charged field space after being processed by appropriate ion optics following the ion source. Ions meeting the criterion for stable passage follow a well-defined path through the quadrupole to the detector. Other ions are unstable and are lost. The stable trajectory of a defined m/z ion is dependent on the charged field parameters, which define the electrical combinations for transmission of a defined mass to charge ion. The transmission of ions is scanned through the mass range in fast sequences by changing the applied quadrupole voltage parameters. Quadrupoles are characterized by low resolution (e.g., $R = m/\Delta m = 300$), and their mass range is up to 4000 uma.

19.1.2.2 Sector Field Mass Spectrometers

Sector field mass spectrometers (SFMS) include a magnetic and/or an electrostatic analyzer in a number of configurations [17–19].

Ions entering SFMS are accelerated by a constant voltage V which, depending on their charge z and the potential, is given by:

$$E = zV$$

which is converted into kinetic energy, then:

$$zV = \frac{1}{2}mv^2 \tag{1}$$

Ions entering with velocity v into the magnetic sector field B acquire a trajectory perpendicular to both, owing to the Lorentz force $F = Bzv$ resulting in a circular trajectory because of centrifugal force:

$$F = \frac{mv^2}{R}$$

Comparing both forces affecting the trajectory R yields

$$Bzv = \frac{mv^2}{R} \text{ and } BR = \frac{mv}{z} \tag{2}$$

The m/z dependence on the applied accelerated voltage V is deduced from relations (1) and (2), resulting in Equation (3):

$$\frac{m}{z} = \frac{B^2 R^2}{2V} \tag{3}$$

One or several detectors positioned in the focal plane can collect ions that travel through the magnetic and/or electrostatic field.

The electrostatic sector is constructed with two curved plates separated by a distance d and with curvature ratio r. On applying a voltage difference ΔE, typically from 0.5 to 1.0 kV, to the plates, an ion with m/z and velocity v will be focused at the exit window if:

$$\frac{mv^2}{z} = \frac{Er}{d} \tag{4}$$

For ions entering the electrostatic sector the stable trajectory $r = 2v/E$ is given for those ions with energy in the range from E_1 to E_2 which are focused at the exit slit. The capability for discriminating ion energies has been used in MS with a double sector arrangement to attain high resolving power (e.g., $R = m/\Delta m \sim 400$–10 000). The SF-MS presents the capability of separating masses in a wide range (up to 15 000 uma) with the possibility of mass separation at low, medium and high resolution (e.g., $R = 400$, 3000, 10 000).

19.1.2.3 Multicollector Mass Spectrometer

The multicollector mass spectrometer (MCMS) is an arrangement of a double focusing mass analyzer combining a pole-based mass separation with a magnetic sector MS which is interfaced to the ICP ionization source allowing simultaneous multiple detection with improved detection limits [19, 21]. The mounting coupled to the ICP ionization source is designed for attaining precise isotope ratios for numerous elements, which is adequate for determining the natural variations of radiogenic or stable isotopes. The main applications have been focused on earth and geological science. A good performance is commonly obtained for continuous sample introduction once the isotope ratio measurements in transient signals were affected by drifts and biased measurements [20].

19.1.2.4 Time of Flight Mass Spectrometers (TOFMS)

Time of flight (TOF) separation is achieved by a pulsing electrostatic field which accelerates packages of ions through an evaluated field-free linear tube. The flight time depends upon the m/z of the ions, which are recorded with a fast electronic detection system [19, 22].

Ions entering the TOF spectrometer are accelerate by a pulsing voltage, typically of 2 kV lasting about 5 ns, and sent into a field-free vacuum inside a cylindrical tube. The acceleration potential pulse is transformed into kinetic energy (Equation (1)). Packages of ions fly very rapidly inside the vacuum where ions with different mass to charge ratio are discriminated by their velocity. From Equation (1)

$$zV = \frac{1}{2}mv^2$$

The velocity to cross a linear distance d of the tube in a defined time t allows the substitution:

$$\frac{m}{z} = \frac{2V}{d^2}t^2 \tag{5}$$

The time interval for detecting ions with different m/z accelerated by the same pulse is calculated by:

$$t_1 - t_2 = \sqrt{\frac{d^2}{2zV}}(\sqrt{m_2} - \sqrt{m_1}) \tag{6}$$

Two ICP-TOFMS configurations with different ICP interface alignment have been described [23]. The trajectories of the ions in the orthogonal and on-axis ICP-TOFMS are shown in Figure 19.3. An electrode plate in front of the extraction region (repeller electrode) in the orthogonal arrangement facilitates the extraction of the ions. The

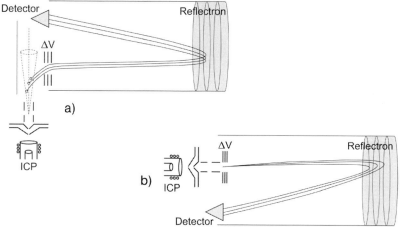

Figure 19.3 Scheme of the ICP-TOF mass spectrometer: orthogonal (a) and axial (b) configurations.

deflection trajectories are separated, resulting in higher resolution when compared with the on-axis scheme where the extraction electrode is coaxial with the flight tube. Additionally, modern TOF spectrometers include a reflecting ion lens (reflectron) at the end of the linear drift tube. The ions are reflected by an electric field, and the cloud of ions with the same m/z is compacted, improving resolution. Resolution for Pb isotopes in the on-axis TOFMS is 1400–1500 and that of the orthogonal design can reach 2300.

Ions of different m/z are detected in fast sequences. For example, consider a 1 m flight-tube and a 2 kV pulse, the time for detecting H^+ is 1.7 μs and for a molecular ion with m/z 2500 is 50 μs. The fast acquisition (about 20 000–30 000 mass spectra per second) of ions carried in packages accelerated by the pulsing voltage permit the detection of several points along transient signals, minimizing the spectral skew error. This error is a distortion when detection is performed by scanning a variable time – concentration peak. Figure 19.4 helps to understand the spectral skew error. The TOFMS presents excellent performance for transient-sampling technology (e.g., laser ablation, electrothermal vaporization) and FIA [10, 23–25]. In some situations time resolution eliminates direct mass spectral overlaps [24].

19.2
FIA-MS Sample Introduction Devices

The application of FIA for elemental and organic analysis by mass spectrometry requires interfacing sample introduction devices with ionization sources and mass spectrometers. The instrumental devices already described for coupling FIA-MS for elemental and molecular analyses are summarized in Figure 19.5.

The capability for monitoring transient signals from an interfaced FIA-MS depends on the dynamics and efficiency of transference of liquid analytes to the ionization source and the rate of conversion to gaseous ions, the efficiency and

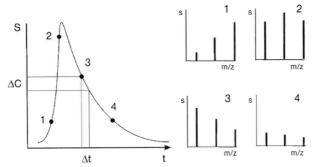

Figure 19.4 Illustration of the skew effect error. The transient profile of the signal (S) corresponds to the temporal variations in concentration. Points 1–4 indicates different possibilities between 3 m/z abundances. (From Ref. [10], with permission.)

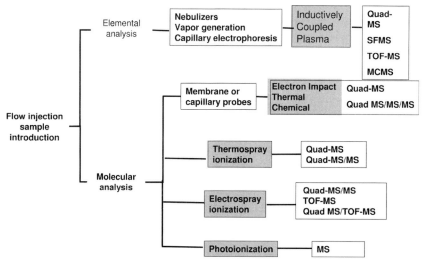

Figure 19.5 Flow chart of the instrumentation described for coupling FIA-MS for elemental and molecular analysis. The gray shadow areas indicate the ionization techniques.

dynamics of the ions sampling and rate of transmission to the spectrometer, the mass separation and the speed of the m/z detection.

19.2.1
Transient Sample Introduction into the MS Ionization Chamber

Classical mass spectrometers with internal ionization sources (i.e., EI, TI and CI) were characterized by the introduction and analysis of only gaseous samples. The gaseous compounds were introduced as steady-state streams by different procedures, such as differential diffusion, direct insertion through a probe, a heated capillary and a septum. The introduction of transient gaseous species into the ionization chamber was initiated with the first gas chromatography mass spectrometer described in 1957 [6]. Part of the gases emerging from the chromatographic column at elevated pressure was conveyed through a capillary to the MS ionization chamber in a mild vacuum. The spectrum was monitored by an oscilloscope. Thereafter several GC-MS interface devices such as membrane probes, a heated capillary, a capillary probe and diffusion jet were devised to introduce gaseous and liquid samples into the MS ionization source. Looking for enhanced MS detection sensitivity and selectivity, different approaches were proposed to couple liquid chromatography to MS [11]. Discrete sample solutions were transferred to a probe tip where gaseous compounds produced by flash evaporation under vacuum were introduced into the ionization chamber. In LC-MS, the post column solutions were subjected to flash evaporation before gas phase separation in a three concentric silicon membrane tubes device. The molecular analytes were transmitted through the membranes in increased vacuum conditions [7].

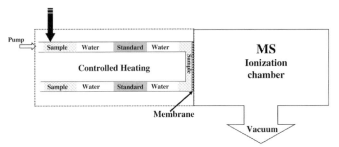

Figure 19.6 Scheme of sample and standards introduction by membrane permeation FIA-MS. (From Ref. [27], with permission.)

Following from GC-MS and LC-MS interfaces FIA-MS with membrane permeation was described for the introduction of volatile organic compounds into the MS ionization chamber. The membrane introduction mass spectrometer (MIMS) was modified as a direct membrane probe (DIMP) and introduced into the MS ionization chambers under vacuum [26]. The molecular ions of the permeated volatile compounds are produced by electron impact, thermal ionization and/or chemical ionization. The mass analyzers typically used are QMS and tandem MS in two or three stages to obtain information about fragments.

Advantages of FIA with highly selective and specific membranes are the on-line segmentation by water for cleaning the membrane and the promotion of baseline recovery [27]. The manifold presented in Figure 19.6 allows the injection of sequences of sample/water/standards for on-line quantification.

The water segmentation avoids carryover from sample to standards by cleaning the membrane surfaces. Pumping 200 µl of organic sample segmented by water into a direct insertion membrane probe allows injection of just the volatile species into the MS ionization chamber. The separation by pervaporation requires the selective adsorption of analytes from the high pressure side of a heated (70–95 °C) membrane, permeation and desorption into the vapor state under vacuum. The reproducible permeation of organic compounds from the FIA segmented flow system to the ionization chamber has been demonstrated. The dynamics of analyte transportation in a FIA-MS using membrane introduction has been modeled to study the effect of flow parameters over the mass transfer profiles across the film [28].

Membranes and capillary probes were useful in implementing FIA-MS for on-line monitoring of volatile products from fermentation and bioreactors.

Thin membranes were tested in flow conditions for on-line monitoring processes when high throughput is required [29]. Other FIA-MS applications were associated with the on-line monitoring of petrochemical derivates with the advantage of processing real-time chemical reaction such as photolysis, providing automated feedback control [30].

Different combinations of mass spectrometers such as single quadrupole QMS and quadrupoles in tandem (QMS/MS/MS) have been employed for FIA-MS. The well-known liquid–liquid dispersion profiles characteristic of FIA suffer an enlargement in the path through the gaseous environment and an efficient hyphenation arrangement is needed to avoid contamination. This condition was attained for on-line monitoring of bioreactions in ethanol production plants for selectively controlling major products and volatile metabolites of fermentation compounds [31]. However, the analytical application to other kinds of samples is restricted by the specificity of the membranes.

19.2.2
FI Sample Introduction to External Ionization Sources – MS

The interfaces coupling FIA-MS are intended to enhance the management of sample solutions before mass spectrometry detection. Classical MS sample preparation was restricted by the requirement to produce volatile or gaseous compounds, and this limited the MS analysis of several materials, mainly for elemental analysis. However, different external ionization sources have been developed to facilitate the production of representative molecular and atom ions from liquid and solid samples.

Most FIA-MS applications were developed by combining the versatility of on-line liquid management procedures with the introduction of efficient external MS ionization sources. Liquid solutions were directly admitted into external ionization sources, such as plasmas [32], thermospray [33], and electrospray [12, 34]. The applicability of FIA-ICP-MS for elemental analysis was demonstrated for on-line dilutions, calibrations, additions, analyte separations, preconcentrations and other chemical procedures [2].

Coupling FIA to sample introduction/ionization sources, such as ESI [34], TSI [35], atmospheric pressure chemical ionization (APCI) [36], atmospheric pressure photoionization (APPI) [37] and laser induced multiphoton or pulsed sample introduction interface (PSI) [38] for molecular analysis have been reported. The reasons for using FIA sample introduction are high sample throughput without contamination, on-line dilution, and the possibility of standardization. The wide linear range of calibration using FIA with TSI and APCI was two orders of magnitude [39]. The quantification of non-biodegradable pollutants in sewage treatment plant effluents by standard additions using FIA-MS was 25–30 times faster than LC-MS [33]

19.3
Flow Systems Coupled to External Ionization Sources – MS

19.3.1
FIA-ICP-MS

Extensive use of FIA-MS has been described for elemental analysis with ICP-MS. Highly precise trace element determinations were achieved by coupling FIA with the multi-isotopic capability, high detection power, fast acquisition speed and compatibility for sample introduction of ICP-MS. The convenience of on-line procedures coupling FIA-ICP-MS for several applications has resulted in about 300 papers. Most of the research coupling FIA to ICP-MS targeted improving limits of detection for trace elements, either by on-line matrix separation, analyte preconcentration, or vapor generation. Quantification by standard additions and isotope dilution was improved by the precise on-line analyte addition or isotopic spiking. The arrangements of the plasma source for ion production and the interface for sampling ions and guiding them to the spectrometer are described in the next sections.

19.3.1.1 The ICP Source

The ICP is an atmospheric pressure ionization source commonly produced in argon subjected to an oscillator magnetic field maintained by a radio frequency (RF) power supply flowing through an inductive coil. The RF operates at power levels between 1.0 and 1.5 kW at 27 and 40 MHz. The plasma is formed in the end of an assembly of concentric quartz tubes (i.e., a torch) at atmospheric pressure, releasing heat, radiation and ions. The torch comprises three concentric cylindrical tubes with $12–15\ L\,min^{-1}$ of argon flowing through the external section to maintain the plasma. The ICP is not in thermodynamic equilibrium and is characterized by an annular distribution of heat by the Joule effect, attaining the highest temperatures (10 000 K) in the plasma core, in the induction coil region. The plasma electron density is variable from 10^{13} to 10^{15} (electrons cm^{-3}) provided by about 2% of the ionized argon [18, 19]. The liquid sample is commonly introduced as an aerosol produced by a nebulizer. The homogeneity of the aerosol depends on several parameters, such as the droplet size and distribution and the discrimination of drop size by the spray chamber. The efficiency of sample transport depends on the cloud homogeneity which is affected by the liquid surface tension, the density of the solution, the nebulizer gas flow-rate, and the spray chamber geometric factors. The injected sample volume transported by a liquid carrier in FIA is characterized by laminar dispersion, presenting a typical peak profile when detected by a flow-through detector. This behavior is completely altered by the turbulent aerosol transport by nebulizers, mainly due to the loss of the major part of the analytes inside the spray chamber. This produces transient peaks band broadening close to the Gaussian distribution profile.

The plasma energy is transferred to the sample and results in drying, evaporation, dissociation, atomization, and ionization of the sample components. The efficiency of ion production depends on the sample matrix, the RF power supplied, the sample aerosol production, the transportation flow-rate, and the plasma gas composition. The energy available in the argon plasma is 15.7 eV which is enough to ionize most of the elements of the periodic table.

The direct injection nebulizer system (DIN) for small sample volume introduction is similar to a FIA system [40]. The manifold consists of a three-channel solenoid valve with exchangeable loops to inject different sample volumes. The injected volumes are pumped by a liquid pressurized gas displacement pump (GDP) through a capillary that is inserted near the end of the central tube of the torch. The injected sample forms an aerosol injected directly into the plasma where it is vaporized, atomized and ionized. The transient peaks produced present a flat maximum and area depending on the injected volume and pumping pressure of the carrier solution. The peak format with a large plateau reveals a small dispersion effect, mainly at the extremes of the injected sample plug and a highly efficient introduction into the plasma, minimizing dead volume.

The FIA liquid sample introduction into the ICP can be by vapor generators [41], thermospray (TS) [42] and hydride generators (HG) [43]. The analyte conversion from liquid to vapor state increases the efficiency of transportation to the plasma and presents the convenience of matrix separation, which results in lower limits of detection.

19.3.1.2 FIA-ICP-MS for Reducing Matrix Effects

Matrix effects are non spectroscopic interferences associated with complex samples that present characteristics out of the range of the working instrumental conditions. The sample introduction nebulizers for ICP are commonly affected by matrix effects (e.g., physical matrix effects from solvent viscosity) and devices coupling nebulizers to desolvation units have been efficient in reducing solvent loading of the plasma discharge.

In ICP-MS besides the nebulizer's drawbacks, the interface for ion sampling and transmission to the mass spectrometer is also affected by the sample matrix. The interface for sampling ions is composed of two metallic cones (i.e., sampler and skimmer cones), placed in sequence with small central holes of 1 and 0.7 mm diameter, respectively. The sampler cone hole is subject to clogging by salts deposition. This restricts sample solids contents to less than 2%.

After the sampling cone most of the neutral components are discarded by the vacuum (10^{-3} bar) between the cones, while ions are attracted by a low potential applied to the skimmer cone. Ions crossing the skimmer cone are guided through cylindrical lenses by electric and magnetic fields under vacuum (10^{-4} bar) to the spectrometer compartment.

Some instrumental devices and compartments are not dimensioned to manage complex solutions and ions can be affected by mass discrimination. The instrumental mass bias is corrected by analyzing reference solutions with certified isotopic composition, such as boric acid prepared by the National Institute of Standards NIST- 951[19]. However, the conduction of light isotopes can be affected by mass bias due to sample matrix effects during their passage through the lens region. The heavy ions are preferentially transmitted along the center axis and light ions can be lost by coulomb repulsion. This effect is dependent on the relative concentration of heavy to light ions and their ionization potentials and is considered the major cause of matrix effects in ICP-MS. The isotopic mass discrimination in the transmission of light isotopes affects the isotope ratios of B and Li in sample matrices with heavy elements. The mass bias effect due to the sample matrix can be corrected by applying the algorithm of common analyte internal standardization [44].

The first experiments coupling FIA-ICP-MS presented promising isotopic data for multielement determination of trace metals when introducing microliter volumes of solutions [45]. Those results anticipated the advantages of the FIA-ICP-MS in reducing matrix effects. The injection of defined sample volumes in between a carrier solution provides on-line dilution in a reproducible manner, allowing fast baseline recovery and less risk of salt deposits on the cones.

Besides the advantages of FIA for sample introduction other facilities for managing solutions before mass spectrometric determination have been improved. A flow system coupled ICP-MS was used with different types of nebulisers, such as a microflow nebulizer for introduction of nL volumes [46]. FIA coupled with highly efficient nebulizers has been useful for injection of small volumes of sample containing low analyte concentrations [47]. Matrix effects were overcome using FIA systems with on-line matrix removal [48–50], solvent extraction/back extraction preconcentration or liquid–solid-phase extraction [51], and addition of an internal standard [52]. The diagnosis of matrix interferences can be carried out by gradient dilution in a FIA-ICP-MS similarly to the gradient calibration method [53]. The arrangement of gradient elution was performed in FIA-ICP-TOFMS to characterize matrix interferences [54]. FIA designs to perform quantification by on-line standard additions have been described to lessen matrix effects. The application of the standard addition method (SAM) requires measuring the analyte in a sample before and after addition of known amounts of standards to the sample to obtain a calibration curve free from matrix effects. Several schemes to perform the on-line additions in flowing systems coupled to atomic spectrometers have been described [55].

FIA configurations using confluent streams by adding the standard solution to the dispersed sample zone continuously have been used. Also, the merging zones flow scheme where the dispersed sample and standard zones merge downstream by confluence has been coupled to ICP-MS. A simple procedure involving the injection of just one volume of standard overlaying the sample peak was enough to apply the SAM. The exploitation of the concentration–time profile proposed earlier [56] was recently improved for FIA-ICP-MS by a peak fitting program for calculating the areas under the sample peak before and after the additions [57].

Evaluation of calibration techniques using a modular FIA manifold for on-line sample dilution and addition of standards coupled to ICP-MS was applied for direct analysis of wine and urine [58]. Lead quantification by standard additions, isotope dilution, and external calibration was compared in terms of accuracy and also flexibility. The external calibration with Tl on-line addition as internal standard presented simplicity for applications in large scale analysis, but the accuracy is highly dependent on the matrix. Isotope dilution and standard additions presented accurate results for Pb but with less flexibility. The comparison of the same calibration techniques was studied in FIA-ICP-MS with a TS interface for Pt determination [59]. The comparison of FIA transient peaks by using pneumatic nebulization or TS sample introduction presented interesting features, the analyte transportation using TS is more efficient but less precise than the nebulizer introduction. The FIA design promoted high sample dilution and owing to the low Pt concentrations the choice for the best calibration was not conclusive.

19.3.1.3 FIA Systems to Perform Isotope Dilution (ID)

Isotope dilution (ID) is an absolute method of quantification and uses mass ratio measurements as the primary metrological procedure [60]. It is an accurate quantification method applicable to those elements with at least two isotopes (i.e., A and B) not exhibiting spectral interferences in a sample. A defined sample volume with unknown mass M_s of the element is spiked with a known and exact amount m_{sp} of the spike with high atom% of isotope A. The isotopic composition of the sample (A_s and B_s) and spike (A_{sp} and B_{sp}) are previously known. The spiked sample is mixed to attain isotopic equilibrium after which the isotope ratio A/B is measured. The isotope ratio measured by ICP-MS is the resultant of the isotopic mass balance after equilibrium, as shown in the following relationship:

$$R = \frac{M_s A_s + m_{sp} A_{sp}}{M_s B_s + m_{sp} B_{sp}}$$

The development of flow injection procedures facilitates the process of spiking the samples accurately, with less contamination risk and reduction in the sample and spike volumes. Different FIA arrangements have been applied for adding the aqueous spike to the sample solutions in a reproducible way while being transported to measure both isotopes involved in the dilution process by ICP-MS. On-line ID by FIA-ICP-MS was first reported in 1989 [61] using time-based injection of sample and spikes in a merging zone flow configuration. The authors demonstrated the feasibility of Pb determination by FIA-ID-ICP-MS for routine analysis of blood, dust, paints, marine and soil sediments [61, 62].

The r-FIA or reverse FIA arrangement to perform ID consists of the injection of the sample volume over a constant flow of the spike solution. A reverse flow injection with programmable time-based injections of the sample on the continuously flowing spike solution was used to attain optimum isotope ratio [63]. Different sample to spike ratios are easily programmed by changing the dilution rates. The main drawbacks of the r-FIA configuration arise from the continuous pumping of the spike solution. When a defined sample volume is injected over the spike, the transient dilution peak is obtained in the m/z of the enriched isotope spike and an increase in the other isotope, depending on the analyte concentration in the sample. The application requires the preliminary analysis of the samples and is practical when the analyte concentrations are in a straight concentration range [64]. Incorrect prediction of the analyte concentration leads to low dilution. In those cases one of the isotope measurements could not satisfy counting statistics, and the isotope ratio is affected by poor precision. The sandwich arrangement considered the injection of a spike volume in between two injections of sample. In this case the spike to sample zones are dispersed before reaching the ICP-MS. Configurations of flowing systems for on-line performance of standard additions and isotope dilution and their applications are summarized in Table 19.1.

The exactness of results obtained by ID is dependent on the precision of the isotope ratio measurements. The precision of isotope ratio measurements using ICP-MS is dependent on the mass spectrometer and other instrumental factors from the sample introduction device, sample matrix and data acquisition parameters [65].

However improvements in precision of the isotope ratio measurements are dependent on the relative precision of measurements on both isotopes σ_A and σ_B. Precise isotope ratio measurements reflect accurate determination of the element in the sample. The highest precision of isotope ratio measurements depends on the

Table 19.1 Applications of FIA-ICP-MS with on-line standard additions and isotope dilution.

Description and analytical purpose	Analytes	Samples	Ref.
n-FIA in ICP- SFMS	U, Pu	urine, waters	[46]
On-line desalter minimize interferences in Cr speciation	Cr(VI) total Cr	saline solutions	[50]
Flow strategies to control matrix effects: gradient dilution	Co, Ni, Cd, Pb	saline solutions	[54]
FIA standard additions clibration for transient signals in ICP-MS	Mn, Co, Se, Mo, Cd, Pb, Sb, Tl	urine	[57]
Comparison of calibration techniques: external calibration, standard additions, isotope dilution.	Pb	wines, urine	[58]
FIA standard additions and ID with a TS interface	Pt	corn leaves, mouse liver, $Ca_3(PO_4)$	[59]
Performing ID using confluent streams of spike and sample solutions	Pb	biological sample sediments	[61, 62]
Reverse flow injection for spiking the sample and a standard to minimize isotopic carryover	Ni, Mo	river and saline waters	[63]
Sandwich injection of the spike creating a concentration gradient of the spike solution	Mo	saline water	[64]
Programmable introduction of sample and spike solutions to attain precise isotope ratios.	Cd	aqueous solutions	[68]
On-line dialysis, ID	CU, Zn, Se, Pb	serum	[69]
nanoHPLC-ID	S	peptides	[72]
ID for quantification of the electrolytic dissolved mass	Pb	Cu	[78]
Cold vapor generation and ID	Hg	water	[79]
Merging sample and spike before cation exchange chromatography	Pd	urine	[80]
Spike of ^{26}Mg to apply pseudo ID measuring ^{26}Mg/^{23}Na	Na	serum	[81]
Direct analysis of organic matrix-ID with ultrasonic nebulizer	Ag, Cd, Co, Cr, Fe, Mn, Ni, Pb	fuel ethanol	[82]
Comparison of ID with and without matrix separation	Cu, Cd, Pb	urine	[83]
Electrolytic on-line matrix separation	B	steels	[84]
Size exclusion chromatography Metallothioneins bounded metals ID	Cu-Cd-Zn	kidney citosol of carps	[86]

following conditions: (i) measurements on both isotopes A and B satisfying the counting statistic characterized by low standard deviations; (ii) the counting level must be corrected by the detector dead time factor, and (iii) similarity of the precision of both isotopes measurements [60].

The precision of isotope ratio measurements in ICP-MS is dependent on the sample introduction, plasma stability, ions sampling, ions transmission, and mass spectrometer. The efficiency of the sample introduction system and the matrix effects affect the plasma stability and the representative ions transmission to the mass analyzer. The measurements of both isotopes in sequential spectrometers such as ICP-QMS are affected by plasma fluctuations, and the precision of isotope ratios are characterized by RSD > 0.5%. This precision is attained by fast measurements of both isotopes with small dwell times [57]. In transient analysis the main limitations for measuring isotope ratios (IR) with good precision are the time-dependent profile and the limited peak width. The IR measurements of B by injecting 50 μL using DIN- ICP-QMS were characterized by precision <1.4%. The plant digested solutions contained $300 \, \mu g \, L^{-1}$ of B. The DIN transient signals in Figure 19.7 correspond to the injection of different volumes of a solution of $100 \, \mu g \, L^{-1}$ whose peak steady state maximum was convenient to attain improved IR measurements [66].

Spectrometer designs have been improved to attain simultaneous detection of a few isotopes with high resolution in the multi-collector (MC) spectrometer [20] or the real time ICP ions analysis with TOFMS [10]. Peak profiles of the triplicate simultaneous determination of 22 rare earth elements by FIA-ICP-TOFMS are presented in Figure 19.8 [67]. The coupled FIA- ICP-TOFMS allowed detection of IR with precision <0.5% for FIA signals lasting 6 s. High precision was also achieved for IR measurements in transient peaks obtained by coupling CE and HPLC with ICP-TOFMS for organometallic speciation analysis [24].

Designed to improve the performance of ID, computer-controlled FIA manifolds can perform programmable, on-line spikes for application of ID with ICP-MS. FIA systems include a set of solenoid valves for introducing defined volumetric ratios of

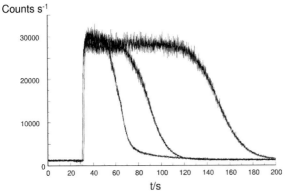

Figure 19.7 Transient peaks (*m/z* 10) obtained by injecting 20, 50 and 100 μL of a $100 \, \mu g \, L^{-1}$ solution by DIN-ICP-MS. (From Ref [66], with permission.)

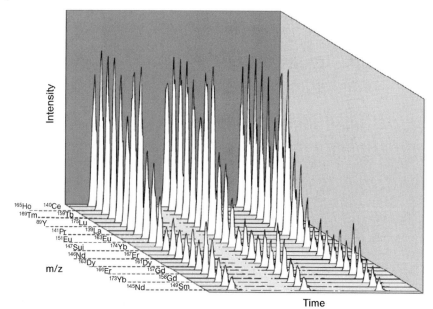

Figure 19.8 Transient peaks obtained by FIA-TOFMS for 22 rare earth elements (in triplicate). (From Ref [67].)

sample and spike solutions, guaranteeing their on-line mixing before reaching the nebulizer. Programmed on-line spikes with different analytes in concentrations estimated for attaining precise IR have been described [68]. The adjusted sample to spike volumes for each sample presented good correlation between the programmed and found IR.

19.3.1.4 FIA Applications of ID-ICP-MS

Isotope dilution is a quantification procedure that is applicable in processes with partial analyte recoveries; in these cases the spike can be considered a perfect internal standard. The application is limited to processes without mass discrimination. Some artifacts such as desalters and dialysers were incorporated for on-line decreasing of sample matrix elements. This strategy is especially useful to decrease Na, K with more than 90% efficiency and Ca, Mg from samples [69].

FIA-ID is very useful for multi-elemental analysis in small sample volumes. The application of ID allowed the determination of several elements in saliva [47], and B in subcellular fractions in coffee leaves [66], and in body fluids [70]. Samples were injected and introduced by DIN-ICP-MS.

The isotope dilution quantification of metals in different oxidation states or bonded to diverse organic molecules can be performed by the species specific or the species-unspecific spiking modes [71].

The species-specific ID mode consists in spiking the sample with the labeled-species before starting the analytical procedures. Spiking samples before digesting or sample matrix separations by different processes, such as dialysis, solid-phase extraction and HG allows accurate quantification of analytes although their recovery may not be quantitative. No volumetric assessment is required, which avoids contamination. This procedure was performed in a serum sample incubated with spiked metals before being dialyzed by FIA-ICP-MS [69].

The species-unspecific on-line ID mode has been applied for quantification in speciation analysis by HPLC or CE [85]. The post-column separated species are spiked while flowing to the ICP-MS and the synchronized IR measurements allowed quantification of the elements. The spike solution is continuously pumped to the plasma and reverse transient peaks of analyte are obtained. The reverse ID transient peaks are affected by the same effects described earlier [62]. Recently a pre-column spike ID was demonstrated for quantification of cysteine and methionine peptides via S isotope dilution. The instrumental system to perform pre-column ID includes the possibility of on-line preconcentration of analytes and compensation for instrumental drifts [72].

ID is widely applicable in on-line processes with partial recovery, such as removing matrix elements or analyte retention/elution using different on-line artifacts. These procedures are usually necessary for geological and environmental samples, especially for seawater samples [73]. The determinations of trace elements in small sample volumes of blood serum, brain and drug metabolites have been performed by FIA-ICP-MS with ID.

19.3.1.5 FIA Coupled with Hyphenated Techniques to ICP-MS

One of the frequently reported applications of FIA is the inclusion of flow-through packed columns for on-line matrix/analyte separation [47, 48] and preconcentration [66] or in knotted reactors [74]. The main applications of using flow-through devices for matrix separation, analyte preconcentration and element speciation are presented in Table 19.2.

Several calibration strategies have exploited FIA facilities for automation such as, internal standardization, standard additions and isotope dilution with high reproducibility [54]. The incorporation in flow systems of vapor generator devices enhances the sample introduction efficiency into the MS [43]. The FIA facilities for managing small sample volumes were emphasized in specific applications, such as identification of metals relevant to virus infection [75], or B in cellular fractions of biological samples [66]. The flow injection of nano-volumes associated with a total consumption nebulizer and a desolvating system efficiently reduced the oxides formation for trace determinations of lanthanides in digested biological tissues [76].

A system coupling FIA-CE-ICP-MS was described for on-line speciation of Cr(VI) and Cr(III). Sample injection using FIA over the electrolyte solution facilitates the sample introduction in capillary electrophoresis (CE) avoiding changing tubes for replacement of solutions [77]. The determination of Pb impurities in electrolytic copper was performed by sample electrodissolution performed on-line with ICP-MS. ID was employed for quantifying the electrodissolved Pb [78]. A combined FIA-HG-ICP-TOFMS was proposed for multi-element determination of Ge, Sb, Sn, Hg and Bi in urine samples [10]. Five peak profiles recorded for independent injections by scanning the Hg isotopes in a $200\,ng\,mL^{-1}$ solution are presented in Figure 19.9.

Table 19.2 Applications of FIA ICP-MS for on-line matrix separation and/or analyte preconcentration.

Description and analytical purpose	Analytes	Samples	Ref.
Matrix removal with 8-HQS column	Pb	seawater	[73]
Knotted reactor-IDQ	Ag, Cd, Co, Ni, Pb, U	natural waters	[74]
Removal of Ca, Na, Mg in an imidioacetate resin column	Cu, Ni, Zn, Co, Pb, Cd, Fe	seawater fish otholits	[48]
Removal of anions to avoid matrix interferences	B, Al, Cu, Mn, Fe, Cr, Cd, Pb, S.	plants	[49]
Solvent/back extraction DIN	Cu, Pb	urine, water, sediments,	[51]
Preconcentration in a knotted reactor on-line with ICP-TOFMS	REE	seawater	[67]
Preconcentration from the sample plus EDTA at pH 10.6 in Ln resin (HDEHP)	Ra-226	waters	[98]
Preconcentration/matrix separation method, using TRU resin	Am-241	urine	[96]
Preconcentration in anion exchange resin and use of CO in the reaction cell	Pu	vegetation	[88]
Extraction chromatography resin	U, Th	blood urine	[90]
In vitro micro dialysis-solid-phase extraction	Cu, Zn, Mn	brain extracellular fluid	[91]
Column with modified carbon nanofibers	REE	biological	[87]
Immobilized 8-hydroquinolinol	V, Mn, Co, Ni, Cu, Zn, Mo, Cd, Pb, U	seawater	[94]
Iminodiacetate resin, Muromac A-1	REE	seawater	[95]
Solid phase extraction of Ag-PDC	Pd	rocks	[89]
Knotted reactor Diethyldithiourea complexes	Pt, Rh, Pd	urine serum road dust	[92]
SIA-chelating resin	Rh, Pt, Pd	urine	[93]
FIA- Hydride generation to avoid matrix interferences	As As, Ge, Se		[43, 10]
Preconcentration in resin, desolvation nebulization using a highly sensitive device.	Pu	urine	[97]
SF-ICP-MS ICP-TOF-MS. On-line pre-reduction for cold vapor generation	Hg	water	[79]

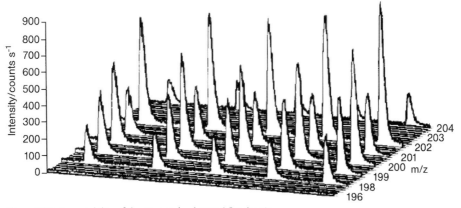

Figure 19.9 Repeatability of the FI signals obtained for the six mercury isotopes (198, 199, 200, 201, 202 and 204) by FI-HG-ICP-TOF-MS. (From Ref. [10].)

A FIA system was used to determine Hg in water at the ultra-trace level by generating the cold vapor in flowing conditions. The generated Hg^0 was separated in a gas–liquid device, purified in a liquid trap, and amalgamated in a gold foil connected to the ICP-MS. The transient Hg signal produced when heating the gold column was quantified by ID. The performance of an ICP-SFMS was compared with an orthogonal ICP-TOFMS for the determination of Hg. The direct analysis of the Hg cold vapor generated, performed under identical conditions, presented high precision of IR measurements by ICP-SFMS. Meanwhile, when the Hg was preconcentrated by amalgamation in gold, the isotope ratios measurements in the transient signal by ICP-TOFMS provided higher precision, and the calculated LOD was $0.9\,\mathrm{pg\,g^{-1}}$ [79].

19.3.2
FIA-Thermospray MS

The ionization by thermospray (TSI) is attained by generating an ultrasonic aerosol of microdrops emerging from a capillary tip of diameter less than 0.1 mm into a heated chamber. The drops containing the solvated analyte molecules begin losing the solvent while crossing the heated chamber until they become dried particles of solutes. The solute particles are ionized by collision with ions through ion–molecule or charge transfer mechanisms. Alternative ionization devices are used such as electrical discharges, electron bombardment, and fission particles bombardment. An alternative way is to add electrolytes, such as $0.1\,\mathrm{mol\,L^{-1}}$ ammonium acetate, to the liquid sample in order to promote charge transfer in the liquid phase. In this case the ions are produced in solution, and the ionic evaporation occurs during the droplet drying step. Typically ionic species MH^+ are produced. The ions produced by TSI are guided through an ionic path to the spectrometer.

A fully automated FIA-TSI-MS was used for determination and characterization of agricultural chemicals. The FIA-TSI coupled to tandem QMS/MS was employed to determine herbicide and pesticides [99]. The comparison of FIA-TSI-MS/MS with LC-TSI-MS/MS for sulfonamides in meat and blood presented good sensitivity and acceptable detection limits. FIA gave the capability of analyzing 50–70 extracts per hour with no contamination [100]. FIA-TSI-MS and LC-TSI-MS were compared for the determination of nonbiodegradable polar compounds from biological wastewater treatment plants [101].

19.3.3
Electrospray Ionization (ESI – MS)

19.3.3.1 Electrospray Ionization
The ionization process by electrospray transforms the aerosol of charged liquid droplets, containing ions in solutions to ions in the gas phase. The electrospray device consists of a metal capillary and a counter electrode connected to a 2–3 kV power supply. The charge separation process in the liquid media subjected to a continuous electrical current is a mechanism similar to electrophoresis and produces charged droplets at the capillary tip. The liquid sample droplets emerge from the metallic capillary tip through a 0.1 mm i.d. hole and form an aerosol of charged droplets which are subjected to gradual solvent evaporation in a heated environment. The drops become smaller and highly charged. The increasing charge repulsion eventually overcomes the solvent surface tension when gas-ions are liberated. The electric field around the capillary tip presents high electric density $10^6\,V\,m^{-1}$. The counter electrode contains a 0.3–0.5 mm hole for sampling ions to a low pressure region. After the hole free jet expansion occurs, and neutrals are removed by the vacuum pump. The skimmer cone has a small orifice (0.6–1.4 mm) in the apex, and ions passing through the hole are guided by magnetic fields to the mass spectrometer [12].

19.3.3.2 FIA-ESI-MS
The FIA systems are easily connected into the capillary of the ESI with carrier flow-rates below $1\,mL\,min^{-1}$. The sample is injected in a polar solvent mixture such as methanol, ethanol:water or acetone, acetonitrile, or dimethylsulfoxide. FIA is commonly used for introducing small sample volumes in applications of FIA-ESI-MS.

The flow injection addition of an isotopically labeled internal standard thiabenda-zole-^{13}C to fungicides allows their ionization and determination by FIA-ESI-MS/MS. The method is an alternative to LC and GC but is easily performed since tedious sample clean-up is not required [102]. The same hyphenated arrangement was used to quantify ^{10}B in antitumoral compounds used in boron neutron capture therapy. The ^{10}B-containing drug compounds were quantitatively determined in urine and plasma by injecting just 1 μL [103]. Some applications of FIA-ESI-MS are presented in Table 19.3.

The hyphenated FIA-CE-ESI-MS was used for determination of biogenic amines in wines. Samples were filtered before injecting into the CE for amine separations. The coupling required the choice of a CE electrolyte ensuring a stable electrospray with the ESI. In the CE-ESI-MS interface a sheath solution was added to close the CE circuit and carry the separated analytes to the electrospray. The advantage of MS

Table 19.3 Applications of flow injection sample introduction in electrospray ionization mass spectrometry (FIA-ESI-MS).

Description and analytical purpose	Analyte	Samples	Ref.
FIA-ESI-MS	fungicides	citrus	[102]
FIA-ESI-MS/MS Borotherapy	B-10 molecular compounds	biological samples	[103]
FIA-CE-ESI-MS Toxicology	biogenic amines	wines	[104]
SIA-CE-ESI-MS Sample pre-treatment for CE separation	several amines	water	[105]
FIA-ESI-MS Biochemical markers	guanidinoacetate and creatinine	dried blood	[106]
FIA-ESI-MS/MS Medical purposes	succinylacethone	urine, dried blood spots	[107]
FIA ESI-MS Increase sample throughput	underivatized amino acids	yeast	[108]
FIA-HPLC ESI – TOF-MS Post-column standardization	fragments and impurities	drugs	[109]
FIA-ESI-MS Substitute for chromatographic separation	saccharides	beer	[110]
HPLC-FIA-ESI-MS Standardization	alkaloids	natural health products	[111]
Direct introduction	caffein creatine	pharmaceutical preparation	[112]
On-line photo derivatization FIA-MSpharmaceutica	indole	sulfa drugs	[113]
FIA-ESI-MS Reproducible and fast	bacterial	crude cell extracts	[114]
FIA-ESI-MS Mobility	Th-siderophore	environment	[115]

compared with UV detection was the unequivocal identification of different amines with high sensitivity [104].

An automated sequential injection analysis (SIA) was coupled to CE-ESI-MS for the analysis of several amines in water samples. The SIA system allowed the addition of reagents for sample pre-conditioning and clean up before separation by CE [105].

The analysis of extracts from dried blood spots for determination of biochemical markers (guanidinoacetate and creatine) of primary creatine disorders in newborns was performed by ID in FIA-ESI-MS/MS. The FIA allowed management of the small sample volume, and both compounds were determined in 1 min [106]. Similarly, the metabolite succinylacetone (SA), a diagnostic of hepatorenal disorder, was determined as a Girard T derivate, which formed ionic species by ESI. Isotope dilution using ^{13}CSA and FIA-ESI-MS-MS was used for quantification in blood spots and urine samples [107].

A FIA-ESI-MS procedure for accurate mass measurements of compounds separated by LC was proposed [108]. A post-column FIA transient standard solution is introduced to superimpose the reference eluted peak and partially overlap the eluted analyte. Ionization by TSI occurs independently without an ion suppression effect, and the FIA analyte serves as a reference for mass measurements.

Hydrogen peroxide was determined in a FIA-ESI-MS system. The FIA system allowed formation of a peroxide adduct by promoting the contact of hydrogen peroxide with a dinuclear heptadentate Fe(III) complex. The reaction product was analyzed by ESI-MS. Two adduct fragments were monitored at m/z 251.5 and 240.5, and the hydrogen peroxide LOD was 10^{-7} mol L^{-1} [109].

19.4
Conclusions

Flow injection analysis has been coupled to different mass spectrometers for organic and inorganic analytical purposes. Flowing systems present a major contribution in facilitating sample and standards management before MS detection. The main applications converge in performing on-line calibration procedures by injecting samples and standards in sequence to overcome instrumental drift effects on quantification. Calibration procedures by on-line standardization, either by performing standard additions or by isotope dilution by using different MS instrumental arrangements proved to be a valuable contribution of FIA systems towards attaining accurate results. On-line procedures for matrix removal or analyte preconcentration constitute an important part of the FIA-MS applications. The insertion of solid phase resin columns or vaporization devices are definitive to avoid matrix effects and to attain lower detection limits. Flow injection using miniaturized sample introduction devices coupled to efficient ionization sources allows trace analysis in small sample volumes with reduced interferences.

References

1 Ruzicka, J. and Hansen, E.H. (1988) *Flow Injection analysis,* 2nd edn, Wiley, New York.

2 Luque de Castro, M.D. and Tena, M. (1995) Hyphenated flow injection systems and high discrimination instruments. *Talanta,* **42,** 151–169.

3 Bier, M.E. and Cooks, R.G. (1987) Membrane interface for selective introduction of volatile compounds directly into the ionization chamber of a mass spectrometer. *Analytical Chemistry,* 59, 597–601.

4 Wang, X., Viczian, M., Lasztity, A. and Barnes, R. (1988) Lead hydride generation for isotope analysis by inductively coupled plasma mass spectrometry. *Journal of Analytical Atomic Spectrometry,* **3,** 821–827.

5 Hansen, E.H. (2005) Use of flow injection and sequential injection analysis schemes for the determination of trace-level concentrations of metals in complex matrices by ETAAS and ICPMS. *Journal of Environmental Science and Health Part A-Toxic/Hazardous Substances & Environmental Engineering,* 40, 1507–1524.

6 Holmes, J.C. and Morrell, F.A. (1957) Oscillographic mass spectrometric monitoring of gas chromatography. *Applied Spectroscopy,* **2,** 86–88.

7 Jones, P.R. and Yang, S.K. (1975) A liquid chromatography/mass spectrometer interface. *Analytical Chemistry*, **47**, 1000–1003.

8 Smith, R.D., Olivares, J.A., Nguyen, N.T. and Udseth, H.R. (1988) Capillary zone electrophoresis mass spectrometry using an electrospray ionization interface. *Analytical Chemistry*, **60**, 436–441.

9 Beckett, R. (1991) Field flow fractionation ICP-MS a powerful new analytical tool for characterizing macromolecules and particles. *Atomic Spectroscopy*, **12**, 228–232.

10 Centineo, G., Montes Bayon, M., De la Campa, R.F. and Sanz-Medel, A. (2000) Flow injection analysis with inductively coupled plasma time-of-flight mass spectrometry for the simultaneous determination of elements forming hydrides and its application to urine. *Journal of Analytical Atomic Spectrometry*, **15**, 1357–1362.

11 Schröder, H.Fr. (1997) Mass spectrometric detection and identification of polar pesticides and their degradation products a comparison of different ionization methods. *Environmental Monitoring and Assessment*, **44**, 503–513.

12 Cole, R.B. (1997) *Electrospray Ionization Mass Spectrometry Fundamentals, Instrumentation and Applications*, John Wiley & Sons, New York.

13 Vestal, M.L. (1983) Studies of ionization mechanisms involved in thermospray LC-MS. *International Journal of Mass Spectrometry*, **46**, 193–196.

14 Takáts, Z., Wiseman, J.M., Gologan, B. and Cooks, R.G. (2004) Mass spectrometry sampling under ambient conditions with desorption electrospray ionization. *Science*, **306**, 471–473.

15 Cody, R.B., Laramée, J.A. and Durst, D. (2005) Versatile new ion source for the analysis of materials in open air under ambient conditions. *Analytical Chemistry*, **77**, 2297–2302.

16 Yinon, J. and Klein, F.S. (1971) The quadrupole and its applications in vacuum technology and mass spectrometry. *Vacuum*, **21**, 379–383.

17 Wollnik, H. (1999) Ion optics in mass spectrometry. *Journal of Mass Spectrometry*, **34**, 991–1006.

18 Jarvis, K.E., Gray, A.L. and Houk, R.S. (1992) *Handbook of Inductively Coupled Plasma Mass Spectrometry*, Blackie, Glasgow.

19 Montaser, A. (1998) *Inductively Coupled Plasma Mass Spectrometry*, Wiley-VCH, New York.

20 Günther-Leopold, I., Wernli, B., Kopajtic, Z. and Günther, D. (2004) Measurement of isotope ratios on transient signals by MC-ICP-MS. *Analytical and Bioanalytical Chemistry*, **378**, 241–249.

21 Rehkämper, M., Schönbächler, M. and Stirling, C.H., (2001) Multicollector ICP-MS: introduction to instrumentation, measurement techniques and analytical capabilities. *Geostandards Newsletter*, **25**, 23–40.

22 Price, D. and Milnes, G.J. (1990) The renaissance of time-of-flight mass spectrometry. *International Journal of Mass Spectrometry and Ion Processes*, **99**, 1–39.

23 Ray, S.J. and Hieftje, G.M. (2001) Mass analyzers for inductively coupled plasma time-of-flight mass spectrometry. *Journal of Analytical Atomic Spectrometry*, **16**, 1206–1216.

24 Ray, S.J., Andrade, F., Gamez, G., McClenathan, D.M., Rogers, D., Schilling, G., Wetzel, W. and Hieftje, G.M. (2004) Plasma-source mass spectrometry for speciation analysis: state-of-the-art. *Journal of Chromatography A*, **1050**, 3–34.

25 Carrión, M.C., Andrés, J.R., Rubí, J.A.M. and Emteborg, H. (2003) Performance optimization of isotope ratio measurements in transient signals by FI-ICP-TOFMS. *Journal of Analytical Atomic Spectrometry*, **18**, 437–443.

26 Bier, M.E. and Cooks, R.G. (1987) Membrane interface for selective introduction of volatile compounds

directly into the ionization chamber of a mass spectrometer. *Analytical Chemistry,* **59**, 597–601.

27 Hayward, M.J., Kotiaho, T., Lister, A.K., Cooks, R.G., Austin, G.D., Narayan, R. and Tsao, G.T. (1990) On-line monitoring of bioreactions of bacillus *Polymyxa* and *Klebsiella oxytoca* by membrane introduction tandem mass spectrometry with flow injection analysis sampling. *Analytical Chemistry,* **62**, 1798–1804.

28 Srinivasan, N., Johnson, R.C., Kasthurikrishnan, N., Wong, P. and Cooks, R.G. (1997) Membrane introduction mass spectrometry. *Analytica Chimica Acta,* **350**, 257–271.

29 Tsai, G.-J., Austin, G.D., Syu, M.J., Tsao, G.T., Hayward, M.J., Kotiaho, T. and Cooks, R.G. (1991) Theoretical analysis of probe dynamics in flow injection/ membrane introduction mass spectrometry. *Analytical Chemistry,* **63**, 2460–2465.

30 Bier, M.E., Kotiaho, T. and Cooks, R.G. (1990) Direct insertion membrane probe for selective introduction of organic compounds into a mass spectrometer. *Analytica Chimica Acta,* **231**, 175–190.

31 Srinivasan, N., Kasthurikrishnan, N., Cooks, R.G., Krishnan, M.S. and Tsao, G.T. (1995) On-line monitoring with feedback control of bioreactors using a high ethanol tolerance yeast by membrane introduction mass spectrometry. *Analytica Chimica Acta,* **316**, 269–276.

32 Thompson, J.J. and Houk, R.S. (1986) Inductively coupled plasma mass spectrometry detection for multielement flow injection analysis and elemental speciation by reversed-phas liquid chromatography. *Analytical Chemistry,* **58**, 2541–2548.

33 Schröder, H.Fr. (1993) Surfactants: non-biodegradabls, significant pollutants in seawage treatment plant effluents Separation, identification and quantification by liquid chromatography,

flow injection analysis-mass spectrometry and tandem mass spectrometry. *Journal of Chromatography,* **647**, 219–234.

34 Barco, M., Planas, C., Palacios, O., Ventura, F., Rivera, J. and Caixach, J. (2003) Simultaneous quantitative analysis of anionic, cationic and nonionic surfactants in water by electrospray ionization mass spectrometry with flow injection analysis. *Analytical Chemistry,* **75**, 5129–5136.

35 Kristiansen, G.K., Brock, R. and Bojesen, G. (1994) Comparison of Flow injection/ thermospray MS/MS and LC/ Thermospray MS/MS methods for determination of sulfonamides in meat and blood. *Analytical Chemistry,* **66**, 3253–3258.

36 Schröder, H.Fr. and Meesters, R.J.W. (2005) Stability of fluorinated surfactants in advanced oxidation processes A follow up of degradation products using flow injection mass spectrometry, liquid chromatography mass spectrometry and liquid chromatography multiple stage mass spectrometry. *Journal of Chromatography A,* **1082**, 110–119.

37 Gómez-Ariza, J.L., Arias-Borrego, A. and García-Barrera, T. (2006) Use of flow injection atmospheric pressure photoionization quadrupole time-of-flight mass spectrometry for fast olive oil fingerprinting. *Rapid Communications in Mass Spectrometry,* **20**, 1181–1186.

38 Wang, A.P.L. and Li, L. (1992) Pulsed sample introduction interface for combining flow injection analysis with multiphoton ionization time-of-flight mass spectrometry. *Analytical Chemistry,* **64**, 769–775.

39 Koeber, R., Niesser, R. and Bayona, J.M. (1997) Comparison of liquid chromatography-mass spectrometry interfaces for the analysis of polar metabolites of benzopyrene. *Fresenius' Journal of Analytical Chemistry,* **359**, 267–273.

40 Wiederin, D.R. and Houk, R.S. (1991) Measurements of aerosol-particle sizes

from a direct injection nebulizer. *Applied Spectroscopy*, **45**, 1408–1412.

41 Bouyssiere, B., Ordóñez, Y.N., Lienemann, C.P., Schaumlöffel, D. and Łobiński, R. (2006) Determination of mercury in organic solvents and gas condensates by (flow injection-inductively coupled plasma mass spectrometry using a modified total consumption micronebulizer fitted with single pass spray chamber. *Spectrochimica Acta Part B-Atomic Spectroscopy*, **61**, 1063–1068.

42 Koropchak, J.A. and Veber, M. (1992) Thermospray sample introduction to atomic spectrometry. *Critical Reviews in Analytical Chemistry*, **23**, 113–141.

43 Machado, L.F.R., Jacintho, A.O., Menegário, A.A., Zagatto, E.A.G. and Giné, M.F. (1998) Electrochemical and chemical processes for hydride generation in flow injection ICP-MS: determination of arsenic in natural waters. *Journal of Analytical Atomic Spectrometry*, **13**, 1343–1346.

44 Al-Ammar, A.S. and Barnes, R.M. (2001) Improving isotope ratio precision in inductively coupled plasma quadrupole mass spectrometry by common analyte internal standardization. *Journal of Analytical Atomic Spectrometry*, **16**, 327–332.

45 Houk, R.S. and Thompson, J.J. (1983) Trace-metal isotopic analysis of microliter solution volumes by inductively coupled plasma mass-spectrometry. *Biomedical Mass Spectrometry*, **10**, 107–112.

46 Schaumlöffel, D., Giusti, P., Zority, M.V., Pickhardt, C., Szpunar, J., Łobiński, R. and Becker, J.S. (2005) Ultratrace determination of uranium and plutonium by nano-volume flow injection double-focusing sector field inductively coupled plasma mass spectrometry (nFI-ICP-SFMS). *Journal of Analytical Atomic Spectrometry*, **20**, 17–21.

47 Menegário, A.A., Packer, A.P. and Giné, M.F. (2001) Determination of Ba, Cd, Cu and Zn in saliva by isotope dilution direct injection inductively coupled plasma mass spectrometry. *Analyst*, **126**, 1363–1366.

48 Willie, S.N., Lam, J.W.H., Yang, L. and Tao, G. (2001) On-line removal of Ca, Na and Mg from imidioacetate resin for the determination of trace elements in seawater and fish otoliths by flow injection ICP-MS. *Analytica Chimica Acta*, **447**, 143–152.

49 Menegário, A.A. and Giné, M.F. (1997) On-line removal of anions for plant analysis by inductively coupled plasma mass spectrometry. *Journal of Analytical Atomic Spectrometry*, **12**, 671–674.

50 Sun, Y.C., Lin, C.Y., Wu, S.F. and Chung, Y.T. (2006) Evaluation of on-line desalter-inductively coupled plasma-mass spectrometry system for determination of Cr(III), Cr(VI), and total chromium concentrations in natural water and urine samples. *Spectrochimica Acta Part B-Atomic Spectroscopy*, **61**, 230–234.

51 Wang, J.H. and Hansen, E.H. (2002) FI/SI on-line solvent extraction/back extraction preconcentration coupled to direct injection nebulization inductively coupled plasma mass spectrometry. *Journal of Analytical Atomic Spectrometry*, **17**, 1284–1289.

52 Klinkenberg, H., Beeren, T. and Borm, W.V. (1993) The use of an enriched isotope as an on-line internal standard in inductively coupled plasma mass spectrometry: a reference method for a proposed determination of tellurium in industrial waste water by means of graphite furnace atomic absorption spectrometry. *Spectrochimica Acta Part B-Atomic Spectroscopy*, **48**, 649–661.

53 Sperling, M., Fang, Z. and Welz, B. (1991) Expansion of dynamic working range and correction for interferences in flame atomic absorption spectrometers using flow injection gradient ratio calibration with a single standard. *Analytical Chemistry*, **63**, 151–159.

54 McClenathan, D.M., Ray, S.J. and Hieftje, G.M. (2001) Novel flow injection

strategies for study and control of matrix interferences by inductively coupled plasma time-of-flight mass spectrometry. *Journal of Analytical Atomic Spectrometry,* **16**, 987–990.

55 Giné, M.F., Reis, B.F., Krug, F.J., Jacintho, A.O., Bergamin Filho, H. and Zagatto, E.A.G. (1994) Flow injection plasma spectrometry. *ICP Information Newsletter,* **20**, 413–417.

56 Araújo, M.C.U., Pasquini, C., Bruns, R.E. and Zagatto, E.A.G. (1985) A fast procedure for standard additions in flow injection analysis. *Analytica Chimica Acta,* **171**, 337–343.

57 Antler, M., Mazwell, E.J., Duford, D.A. and Salin, E.D. (2007) On-line standard additions calibration of transient signals for inductively coupled plasma mass spectrometry. *Analytical Chemistry,* **79**, 688–694.

58 Goossens, J., Moens, L. and Dams, R. (1994) Determination of lead by flow injection inductively coupled plasma mass spectrometry comparing several calibration techniques. *Analytica Chimica Acta,* **293**, 171–181.

59 Parent, P.M., Vanhoe, H., Moens, L. and Dams, R. (1996) Evaluation of a flow injection system combined with an inductively coupled plasma mass spectrometry with a thermospray nebulization for the determination of trace level of platinum. *Analytica Chimica Acta,* **320**, 1–10.

60 Magnusson, B., Trešl, I. and Haraldsson, C. (2005) Isotope dilution with ICP-MS simplified uncertainty estimation using a robust procedure based on a higher target value of uncertainty. *Journal of Analytical Atomic Spectrometry,* **20**, 1024–1029.

61 Lasztity, A., Viczián, M., Wang, X. and Barnes, R.M. (1989) Sample analysis by on-line isotope dilution inductively coupled plasma mass spectrometry. *Journal of Analytical Atomic Spectrometry,* **4**, 761–766.

62 Viczián, M., Lasztity, A., Wang, X. and Barnes, R.M. (1990) On-line isotope

dilution and sample dilution by flow injection and inductively coupled plasma mass spectrometry. *Journal of Analytical Atomic Spectrometry,* **5**, 125–133.

63 Beauchemin, D. and Specht, A.A. (1997) On-line isotope dilution analysis with ICPMS using reverse flow injection. *Analytical Chemistry,* **69**, 3183–3187.

64 Specht, A.A. and Beauchemin, D. (1998) Automated on-line isotope dilution analysis with ICP-MS using sandwich flow injection. *Analytical Chemistry,* **70**, 1036–1040.

65 Furuta, N. (1991) Optimization of the mass scanning rate for the determination of lead isotope ratio using an inductively coupled plasma mass spectrometry. *Journal of Analytical Atomic Spectrometry,* **6**, 199–203.

66 Bellato, A.C.S., Menegário, A.A. and Giné, M.F. (2003) Boron Isotope Dilution in cellular fractions of coffee leaves evaluated by Inductively Coupled Plasma Mass Spectrometry with Direct Injection nebulization (DIN-ICP-MS). *Journal of the Brazilian Chemical Society,* **14**, 269–273.

67 Benkhedda, K., Infante, H.G. and Addams, F.C. (2002) Inductively coupled plasma mass spectrometry for trace analysis using flow injection on-line preconcentration and time-of-flight mass analyser. *Trends in Analytical Chemistry,* **21**, 332–342.

68 Packer, A.P.C., Giné, M.F., Reis, B.F. and Menegário, A.A. (2001) Micro flow system to perform programmable isotope dilution for inductively coupled plasma-mass spectrometry. *Analytica Chimica Acta,* **438**, 267–272.

69 Giné, M.F., Bellato, A.C.S. and Menegário, A.A. (2004) Determination of trace elements in serum samples by isotope dilution inductively coupled plasma mass spectrometry using on-line dyalisis. *Journal of Analytical Atomic Spectrometry,* **19**, 1252–1256.

70 Bellato, A.C.S., Giné, M.F. and Menegário, A.A. (2004) Determination of B in body fluids by isotope dilution

inductively coupled plasma mass spectrometry with direct injection nebulization. *Microchemical Journal*, 77, 119–123.

71 Heumann, K.G. (2004) Isotope dilution ICP-MS for trace element determination and speciation: from a reference method to a routine method? *Analytical and Bioanalytical Chemistry*, 378, 318–329.

72 Schaumlöffel, D., Giusti, P., Preud'Homme, H., Szpunar, J. and Łobiński, R. (2007) Pre-column isotope dilution analysis in nano HPLC-ICPMS for absolute quantification of sulfur containing peptides. *Analytical Chemistry*, 79, 2859–2868.

73 Huang, Z.Y., Chen, F.R., Zhuang, Z.X., Wang, X.R. and Lee, F.S.C. (2004) Trace lead measurement and on-line removal of matrix interferences in seawater by isotope dilution coupled with flow injection and ICP-MS. *Analytica Chimica Acta*, 508, 239–245.

74 Dimitrova-Koleva, B., Benkhedda, K., Ivanova, E. and Adams, F. (2007) Determination of trace elements in natural waters by inductively coupled plasma time of flight mass spectrometry after flow injection preconcentration in a knotted reactor. *Talanta*, 71, 44–50.

75 DeNicola Cafferky, K., Thompson, R.L., Richardson, D.D. and Caruso, J.A. (2007) Determination, by inductively coupled plasma mass spectrometry, of changes in cellular metal content resulting from herpes simplex virus (HSV-1) infection. *Analytical and Bioanalytical Chemistry*, 387, 2037–2043.

76 Dressler, V.L., Pozebon, D., Matusch, A. and Becker, J.S. (2007) Micronebulization for trace analysis of lanthanides in small biological specimens by ICP-MS. *International Journal of Mass Spectrometry*, 266, 25–33.

77 Giné, M.F., Gervasio, A.P.G., Lavorante, A.F., Miranda, C.E.S. and Carrilho, E. (2002) Interfacing flow injection with capillary electrophoresis and inductively coupled plasma mass spectrometry for Cr speciation in water samples. *Journal of Analytical Atomic Spectrometry*, 17, 736–738.

78 Packer, A.P., Gervasio, A.P.G., Miranda, C.E.S., Reis, B.F., Menegário, A.A. and Giné, M.F. (2003) On-line electrolytic dissolution for lead determination in high purity copper by isotope dilution inductively coupled plasma mass spectrometry. *Analytica Chimica Acta*, 485, 145–153.

79 Yang, L., Willie, S. and Sturgeon, R. (2005) Ultra-trace determination of mercury in water by cold-vapor generation isotope dilution mass spectrometry. *Journal of Analytical Atomic Spectrometry*, 20, 1226–1231.

80 Falta, T., Limbeck, A., Stingeder, G. and Hann, S. (2007) Ultra-trace determination of palladium in human urine samples via flow injection coupled with ICP-MS. *Atomic Spectroscopy*, 28 (3), 81–89.

81 Long, S.E. and Vetter, T.W. (2002) Determination of sodium in blood serum by inductively coupled plasma mass spectrometry. *Journal of Analytical Atomic Spectrometry*, 17 (12), 1589–1594.

82 Saint'Pierre, T.D., Tormen, L., Frescura, V.L.A. and Curtius, A.J. (2006) The direct analysis of fuel ethanol by ICP-MS using a flow injection system coupled to an ultrasonic nebulizer for sample introduction. *Journal of Analytical Atomic Spectrometry*, 21, 1340–1344.

83 Lu, P.-L., Huang, K.-S. and Jiang, S.-J. (1993) Determination of traces of copper cadmium and lead in biological and environmental samples by flow injection isotope dilution inductively coupled plasma mass spectrometry. *Analytica Chimica Acta*, 284, 181–188.

84 Coedo, A.G., Dorado, M.T. and Padilla, I. (2005) Evaluation of different sample introduction approaches for the determination of boron in unalloyed steels by inductively coupled plasma mass spectrometry. *Spectrochimica Acta Part B-Atomic Spectroscopy*, 60, 73–79.

85 Koellensperger, G., Hann, S., Nurmi, J., Prohaska, T. and Stingeder, G. (2003) Uncertainty of species unspecific quantification strategies in hyphenated ICP-MS analysis. *Journal of Analytical Atomic Spectrometry*, **18**, 1047–1055.

86 Infante, H.G., Van Campenhout, K., Schaumloffel, D., Blust, R. and Adams, F.C. (2003) Multi-element speciation of metalloproteins in fish tissue using size-exclusion chromatography coupled "on-line" with ICP-isotope dilution-time-of-flight-mass spectrometry. *Analyst*, **128**, 651–657.

87 Chen, S., Xiao, M., Lu, D. and Zhan, X. (2007) Use of a microcolumn packed with modified carbon nanofibers coupled with inductively coupled plasma mass spectrometry for simultaneous on-line preconcentration and determination of trace rare earth elements in biological samples. *Rapid Communications in Mass Spectrometry*, **21**, 2524–2528.

88 Epov, V.N., Benkhedda, K. and Evans, R.D. (2005) Determination of Pu isotopes in vegetation using a new on-line FI- ICP-DRC-MS protocol after microwave digestion. *Journal of Analytical Atomic Spectrometry*, **20**, 990–992.

89 Fang, J., Liu, L.-W. and Yang, X.-P. (2006) Minimization of matrix interferences in quadrupole inductively coupled plasma mass spectrometric (ICP-MS) determination of Palladium using a flow injection mass displacement solid-phase extraction protocol. *Spectrochimica Acta Part B-Atomic Spectroscopy*, **61**, 864–869.

90 Tolmachyov, S.Y., Kuwabara, J. and Noguchi, H. (2004) Flow injections extraction with ICP-MS for thorium and uranium determination in human body fluids. *Journal of Radioanalytical and Nuclear Chemistry*, **261**, 125–131.

91 Sun, Y., Lu, Y. and Chung, Y. (2007) Online in-tube solid-phase extraction coupled to ICP-MS for *in vivo* determination of the transfer kinetics of trace elements in the brain extracellular fluid of anesthetized

rats. *Journal of Analytical Atomic Spectrometry*, **22**, 77–83.

92 Benkhedda, K., Dimitrova, B., Infante, H.G., Ivanova, E. and Adams, F.C. (2003) Simultaneous on-line preconcentration and determination of Pt, Rh and Pd in urine, serum and road dust by flow injection combined with inductively coupled plasma time-of-flight mass spectrometry. *Journal of Analytical Atomic Spectrometry*, **18**, 1019–1025.

93 Lopes, C.M.P.V., Almeida, A.A., Saraiva, M.L.M.F.S. and Lima, J.L.F.C. (2007) Determination of Rh, Pd and Pt in urine samples using a preconcentration sequential injection analysis system coupled to a quadrupole-inductively coupled plasma-mass spectrometer. *Analytica Chimica Acta*, **600**, 226–232.

94 Hirata, S., Kajiya, T., Takano, N., Aihara, M., Honda, K., Shikino, O. and Nakayama, E. (2003) Determination of trace metals in seawater by on-line column preconcentration inductively coupled plasma mass spectrometry using metal alkoxide glass immobilized 8-quinolinol. *Analytica Chimica Acta*, **499**, 157–165.

95 Hirata, S., Kajiya, T., Aihara, M., Honda, K., Shikino, O. and Nakayama, E. (2002) Determination of rare earth elements in seawater by on-line column preconcentration inductively coupled plasma mass spectrometry. *Talanta*, **58**, 1185–1194.

96 Epov, V.N., Benkhedda, K., Brownell, D., Cornett, R.J. and Evans, R.D. (2005) Comparative study of three sample preparation approaches for the fast determination of americium in urine by flow injection ICP-MS. *Canadian Journal of Analytical Sciences and Spectroscopy*, **50**, 14–22.

97 Epov, V.N., Benkhedda, K., Cornett, R.J. and Evans, R.D. (2005) Rapid determination of plutonium in urine using flow injection on-line preconcentration and inductively coupled plasma mass spectrometry. *Journal of*

Analytical Atomic Spectrometry, **20**, 424–430.

98 Benkhedda, K., Lariviere, D., Scott, S. and Evans, D. (2005) Hyphenation of flow injection on-line preconcentration and ICP-MS for the rapid determination of Ra-226 in natural waters. *Journal of Analytical Atomic Spectrometry*, **20**, 523–528.

99 Geerdink, R.B., Berg, P.J., Kienhuis, P.G.M. and Brinkman, U.A.T. (1996) Flow-injection analysis thermospray tandem mass spectrometry of triazine herbicides and some of their degradation products in surface water. *International Journal of Environmental Analytical Chemistry*, **64**, 265–278.

100 Kristiansen, G.K., Brock, R. and Bojesen, G. (1994) Comparison of flow injection/ thermospray MS/MS and LC/ thermospray MS/MS methods for determination of sulfonamides in meat and blood. *Analytical Chemistry*, **66**, 3253–3258.

101 Schröder, H.Fr. (1995) Polar organic pollutants in the Elbe river — liquid-chromatographic mass-spectrometric and flow injection analysis mass-spectrometric analyses demonstrating changes in quality and concentration during the unification process in germany. *Journal of Chromatography A*, **712**, 123–140.

102 Ito, Y., Goto, T., Oka, H., Matsumoto, H., Miyasaki, Y., Takahashi, N. and Nakazawa, H. (2003) Simple and rapid determination of thiabendazole, imazalil and *o*-phenylphenol in citrus fruit using flow injection electrospray ionization tandem mass spectrometry. *Journal of Agricultural and Food Chemistry*, **51**, 861–866.

103 Basilico, F., Sauerwein, W., Pozzi, F., Wittig, A., Moss, R. and Mauri, P.L. (2005) Analysis of B-10 antitumoral compounds by means of flow injection into ESI-MS/ MS. *Journal of Mass Spectrometry*, **40**, 1546–1549.

104 Santos, B., Simonet, B.M., Ríos, A. and Valcárcel, M. (2004) Direct automatic determination of biogenic amines in wine by flow injection-capillary electrophoresis-mass spectrometry. *Electrophoresis*, **25**, 3427–3433.

105 Santos, B., Simonet, B.M., Lendl, B., Ríos, A. and Valcárcel, M. (2006) Alternatives for coupling sequential injection systems to commercial capillary electrophoresis-mass spectrometry equipment. *Journal of Chromatography A*, **1127**, 278–285.

106 Carducci, C., Santagata, S., Leuzzi, V., Carducci, C., Artiola, C., Giovanniello, T., Battini, R. and Antonozzi, I. (2006) Quantitative determination of guanidinoacetate and creatine in dried blood spot by flow injection analysis-electrospray tandem mass spectrometry. *Clinica Chimica Acta*, **364**, 180–187.

107 Johnson, D.W., Gerace, R., Rainieri, E., Trihn, M. and Fingerhut, R. (2007) Analysis of succinylacetone, as a Girard T derivative, in urine and dried bloodspots by flow injection electrospray ionization tandem mass spectrometry. *Rapid Communications in Mass Spectrometry*, **21**, 59–63.

108 Charles, L. (2003) Flow injection of the lock mass standard for accurate mass measurement in electrospray ionization time-of-flight mass spectrometry coupled with liquid chromatography. *Rapid Communications in Mass Spectrometry*, **17**, 1383–1388.

109 McCooeye, M. and Mester, Z. (2006) Comparison of flow injection analysis electrospray mass spectrometry and tandem mass spectrometry and electrospray high-field asymmetric waveform ion mobility mass spectrometry and tandem mass spectrometry for the determination of underivatized amino acids. *Rapid Communications in Mass Spectrometry*, **20**, 1801–1808.

110 Mauri, P., Minoggio, M., Simonetti, P., Gardana, C. and Pietta, P. (2002) Analysis of saccharides in beer samples by flow injection with electrospray mass spectrometry. *Rapid Communications in Mass Spectrometry*, **16**, 743–748.

111 McCooeye, M., Ding, R., Gardner, G.J., Fraser, C.A., Lam, J., Sturgeon, R.E. and Mester, Z. (2003) Separation and quantitation of the stereoisomers of ephedra alkaloids in natural health products using flow injection-electrospray ionization high field asymetric waveform ion mobility spectrometry-mass spectrometry. *Analytical Chemistry*, **75**, 2538–2542.

112 Wade, N. and Miller, K. (2005) Determination of active ingredient within pharmaceutical preparations using flow injection mass spectrometry. *Journal of Pharmaceutical and Biomedical Analysis*, **37**, 669–678.

113 Numan, A. and Danielson, N.D. (2002) On-line photo derivatization with flow injection and liquid chromatographic-atmospheric pressure electrospray mass spectrometry for the identification of indoles. *Analytica Chimica Acta*, **460**, 49–60.

114 Vaidyanathan, S., Kell, D.B. and Goodacre, R. (2002) Flow-injection electrospray ionization mass spectrometry of crude cell extracts for high throughput bacterial identification. *Journal of the American Society for Mass Spectrometry*, **13**, 118–128.

115 Keith-Roach, M.J., Buratti, M.V. and Worsfold, P.J. (2005) Thorium complexation by hydroxamate siderophores in perturbed multicomponent systems using Flow Injection Electrospray Ionization Mass Spectrometry. *Analytical Chemistry*, **77**, 7335–7341.

III

Applications

Advances in Flow Analysis. Edited by Marek Trojanowicz
Copyright © 2008 WILEY-VCH Verlag GmbH & Co. KGaA, Weinheim
ISBN: 978-3-527-31830-8

20
Environmental Applications of Flow Analysis

Shoji Motomizu

20.1
Introduction

Environmental science is one of the widest areas requiring chemical analyses. In environmental analyses, there are often a very large number of samples, and the analysis must be carried out as quickly as possible, or carried out on site. Further, the samples are very complex and the matrix must be eliminated before measurement; analyte concentration is often quite low and therefore some enrichment procedures are required.

To overcome such problems, a computer-controlled flow method, which is constructed by coupling conventional flow injection analysis (FIA) and sequential injection analysis (SIA) possessing pretreatment functions, must be developed. Computer-controllable pumps, valves and on-line pretreatment devices are now available, and therefore sophisticated analysis systems can be constructed easily at reasonable cost.

20.2
Analysis of the Aquatic Environment by Flow Methods

20.2.1
Substances Related to Eutrophication

Eutrophication may occur in lakes, coastal and inland seas, due to the influx of domestic and industrial wastewaters, where a red tide sometimes happens, and fish and shellfish can no longer live. The main substances involved in eutrophication are nitrogen and phosphorus compounds, both inorganic and organic. Amongst inorganic nitrogen compounds are nitrite, nitrate and ammonia, and amongst inorganic phosphorus compounds, orthophosphate and condensed phosphorus compounds, such as pyrophosphate and triphosphate. In general, with respect to eutrophication, the total nitrogen and phosphorus contents must be determined in wastewaters, river and lake waters, and seawater. The determination is based on the oxidative decomposition of nitrogen and phosphorus compounds to nitrate and orthophosphate, respectively, with peroxodisulfate at high temperature and pressure. Therefore, methods for the determination of nitrate and orthophosphate are important for eutrophication analysis.

20.2.1.1 Nitrogen Compounds

Nitrite and Nitrate Determination In general, the total concentrations levels of nitrite and nitrate are about 10^{-5}–10^{-4} M in an aquatic environment, and the concentration ratio of nitrite to nitrate is about 1 : 10. Therefore, the determinable range and limit of detection (LOD) for nitrite must be at the concentration levels of 10^{-6} M and 10^{-7} M, respectively.

Photometric detection can be used for the determination of nitrite by a flow injection technique. The detection reaction used is the diazotization of nitrite with sulfanilamide (SA), followed by coupling the diazonium ion with N-(1-naphthyl)-ethylenediamine (NEDA) [1]. The product can be measured at 540 nm.

This diazotization-coupling reaction can be applied to the determination of nitrate after reducing nitrate to nitrite with appropriate reducing agents. In a convenient reducing method for nitrate to nitrite, copperized cadmium (Cd/Cu) particles (0.5–1.0 mm) are packed into a glass column (2 mm i.d. × 10 cm) and installed in a carrier line, where a sample solution is introduced by a six-way switching valve, as shown in Figure 20.1(a). The reagent solution is prepared as follows: 0.1% SA and 0.01% NEDA (dihydrochloride) in 0.1 M HCl, and the carrier contains EDTA in order to refresh the surface of the Cd/Cu particles [2].

For the determination of nitrite, the same flow diagram can be used by modifying the system: a carrier is replaced by water and the reducing column is uninstalled.

In Figure 20.1(b), flow signals for nitrite and nitrate are shown. It can be seen from that the peak height of nitrate is almost identical to that of nitrite, which indicates that the reduction of nitrate to nitrite is almost 100%.

The methods can be favorably applied to river and sea waters: the correlation between the results obtained by the proposed methods and obtained by the batchwise standard methods (Japanese Industrial Standards) is quite good [2].

To reduce nitrate to nitrite, a low-pressure mercury lamp can be used [3]; a reaction tubing (PTFE: 0.8 mm i.d., 1.6 mm o.d. × 3 m) is wound around the lamp (14 mm o.d. × 134 mm, 4 W), and is wrapped with aluminum foil. As a carrier, the mixed solution of 0.1 M KH_2PO_4 and 1×10^{-3} M EDTA is used. The lamp is installed just after the sample injection valve: the flow diagram is similar to Figure 20.1.

As a simple flow injection system, phloroglucinol (1,3,5-trihydroxybenzene) can be used for the detection of nitrite; the product of the nitrosation reaction can be spectrophotometrically detected at 312 nm [4]. Similarly, N,N-bis(2-hydroxypropyl) aniline can be used, the detection wavelength then being 500 nm [5].

Highly sensitive detection of nitrite and nitrate can be carried out by fluorophotometry based on diazotization of C acid (3-amino-1,5-naphtalenedisulfonic acid) in an acidic medium, followed by addition of NaOH to the reaction solution to produce a

a)

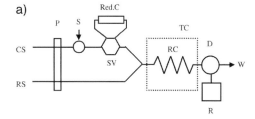

b)

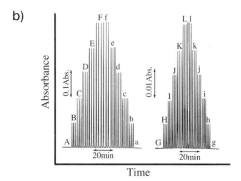

Figure 20.1 (a) Flow diagram for nitrate determination with a reducing column [2]. CS: carrier (EDTA, pH 8.2); RS: reagent solution (SA + NEDA); P: double-plunger pump; S: sample injector(100 μL); SV: 6-way switching valve; Red. C: reducing column; TC: temperature-controlled bath(40 °C); RC: reaction column

(0.5 mm i.d. ×2 m); D: detector(540 nm); R: recorder; W: waste. (b) Flow profiles for nitrite and nitrate. [N–NO$_2^-$]/ppm: A, 0; B, 0.2; C, 0.4; D, 0.6; E, 0.8; F, 1.0. [N–NO$_3^-$]/ppm: a, 0; b, 0.2; c, 0.4; d, 0.6; e, 0.8; f, 1.0. [N–NO$_2^-$]/ppb: G, 0; H, 20; I, 40; J, 60; K, 80; L, 100. [N–NO$_3^-$]/ppb: g, 0; h, 20; i, 40; j, 60; k, 80; l, 100.

deprotonated species of the diazoic acid: $\lambda_{ex} = 365$ nm, $\lambda_{em} = 470$ nm. The method can be applied to river and sea waters [6, 7]

A sequential-injection/spectrophotometric method can be applied to the determination of nitrite with the same reagents as in the flow injection technique. However, the sensitivity is not sufficient for determination of nitrite at concentrations of 10^{-6} M or lower; it is sufficient for the determination of nitrate at concentrations of 10^{-5} M or higher but the refreshment of the Cd/Cu column is tedious and not reproducible.

A multi-syringe and a solenoid pumping system can be used for nitrite/nitrate determination [8] by spectrophotometric detection.

Ammonia Determination   Ammonia in aqueous solutions can be determined by a spectrophotometric method based on an Indophenol Blue reaction, where phenols and hypochlorite are used for the coloration. Ammonia can react with hypochlorite to form monochloramine (NH_2Cl) in the presence of sodium nitroprusside as a catalyst, which in turn reacts with phenols to form Indophenol Blue. Various kinds of phenol have been proposed for the determination of ammonia from which salicylic acid is recommended because of the stability of the phenol (salicylic acid) and high

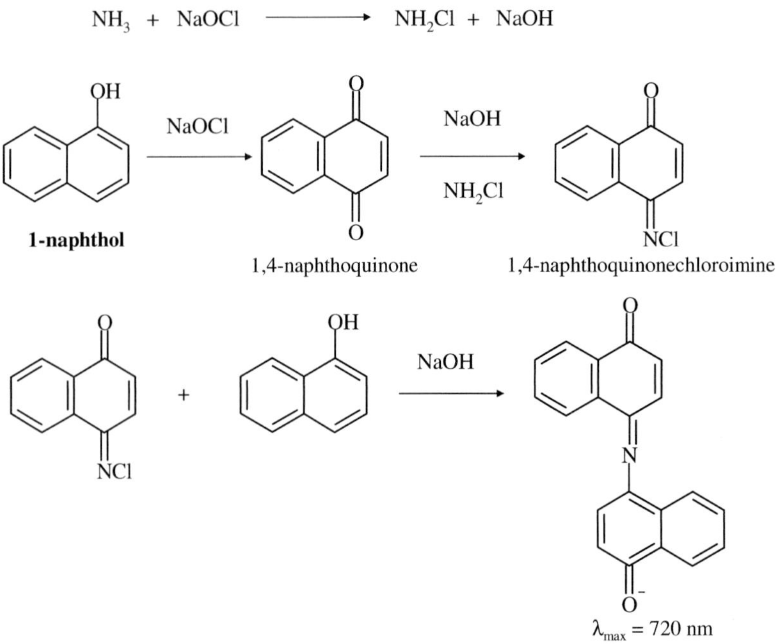

$$NH_3 + NaOCl \longrightarrow NH_2Cl + NaOH$$

1-naphthol

1,4-naphthoquinone

1,4-naphthoquinonechloroimine

$\lambda_{max} = 720$ nm

Scheme 20.1 Detection reaction for the determination of ammonia by 1-naphthol method.

sensitivity [9]. However, nitroprusside is a toxic substance, and great attention must be paid to the treatment of the experimental waste.

One of the most interesting and sensitive Indophenol Blue methods is based on the reaction involving 1-naphthol and hypochlorite in an alkaline solution without nitroprusside as a catalyst [10]. In this method, acetone is used to improve the sensitivity [11], and the solution of sodium isocyanurate can be used in a similar manner as in the method with sodium hypochlorite: isocyanurate can be converted to hypochlorite in an alkaline solution. The reaction may occur as shown in Scheme 20.1.

In Figure 20.2(a) and (b), the flow diagram and the signal profile, respectively, for ammonia determination are shown. By the proposed method, ammonia at concentrations of about 10^{-7} M can be determined.

One of the alternative methods is based on gas diffusion of ammonia in an alkaline solution through a hydrophobic membrane into an absorption solution containing coloration reagents. Ammonium ion in aqueous samples is converted to gaseous ammonia in an alkaline carrier stream, and the gaseous ammonia can pass through a gas-permeable membrane and be absorbed into the reagent solution containing an acid–base indicator. As a gas permeable membrane, microporous PTFE tubing can be used [12–14]. In Figure 20.3(a) and (b), a gas diffusion apparatus of a double-tubing type and a flow diagram with the gas diffusion apparatus are shown [15].

In the gas diffusion apparatus, the hydrophobic membrane may become less hydrophobic and the efficiency of the permeation decreased after a long-time flow of aqueous solutions. In order to avoid the decrease in the hydrophobicity of the

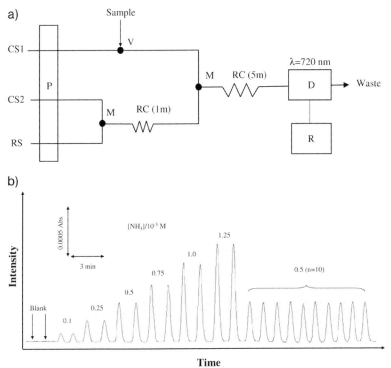

Figure 20.2 (a) Flow diagram for the determination of ammonia CS1: purified water; CS2: 0.01 M $C_3Cl_2N_3NaO_3$, 0.1 M NaOH; RS: 0.5% 1-naphthol in 35% acetone; V: injection valve; M: mixing joint; RC: reaction coil; D: detector; R: recorder. (b) Flow profile for the calibration graph of ammonia determination RSD: 0.81% ($n = 10$); LOD: 0.79×10^{-7} M; sample throughput: 45 h^{-1}.

membrane, the aqueous solutions in the inner and the outer tubing may be replaced by air after experiments by turning additional switching valves for air aspiration and aspirating air into the tubings with a water aspirator or a vacuum pump for several minutes [15].

In general, the permeation efficiency of gaseous substances, such as ammonia and carbon dioxide, is about 10%, and therefore LOD may become worse than the direct detection with an indophenol blue method. However, almost all of the effect of interfering substances existing in real samples can be eliminated. The determinable range and LOD of NH_3-N are 0–1.0 and 0.01 ppm, respectively [15].

Organic Nitrogen Compounds Total nitrogen contents in water samples can be measured by spectrophotometry. After the oxidative decomposition of nitrogen compounds in the samples with potassium peroxodisulfate at 120 °C in an autoclave, nitrate can be determined directly by spectrophotometry at 220 nm [16] or can be determined by the spectrophotometric method based on the reduction of nitrate to nitrite and the diazotization-coupling reaction.

a)

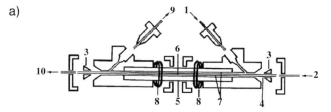

b)

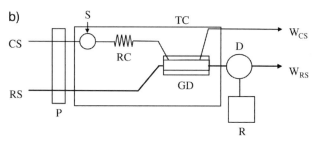

Figure 20.3 (a) Gas diffusion apparatus with double tubings [15]. 1, carrier in and out; 2, reagent solution in and out; 3, ferrule; 4, porous PTFE tubing; 5, glass tubing; 6, reagent stream; 7, carrier stream; 8, O-ring; 9, carrier waste; 10, reagent in and out, to detector (b) Flow diagram for ammonia determination by gas diffusion method CS: carrier stream (0.02 M NaOH); RS: reagent stream (4×10^{-4} M HEPES, 1.25×10^{-4} M Cresol Red, pH 7.0); P: pump (1.0 ml min^{-1}); S:sample (200 μl); RC: reaction coil(0.5 mm i.d. $\times$ 1 m); GD: gas diffusion apparatus (gas diffusion: 5 cm); TC: temperature-controlled bath (40 °C); D: detector (550 nm); R: recorder; W$_{CS}$: carrier waste; W$_{RS}$: reagent waste.

20.2.1.2 Phosphorus Compounds

Phosphorus is an essential nutrient for all life including phytoplankton in an aquatic system and, therefore, excessive inputs of phosphorus compounds from domestic and industrial wastes, and terrestrial soils can lead to the eutrophication of lakes, coastal and inland seas, which may cause a red tide. The red tide is often accompanied by abnormal growth of toxic algae which sometimes leads to damage to fish.

Orthophosphate (PO_4^{3-} and its protonated forms) is the most bioavailable of all of the phosphorus forms. The criterion for phosphorus concentrations in non-eutrophic waters is considered to be less than 20 ppb [17], which means the lowest determinable concentration of phosphorus in an aquatic environment must be several ppb.

Orthophosphate One of the conventional spectrophotometric methods for the determination of orthophosphate (PO_4^{3-}) is based on a classical Molybdenum Blue formation reaction with molybdate in an acidic solution in the presence of a reducing agent. Orthophosphate can react with molybdate to form a yellow heteropoly acid, molybdophosphate or $H_3PMo_{12}O_{40}$, which is reduced with a reducing agent, such as ascorbic acid and stannous chloride, to form Molybdenum Blue: molar absorptivity is $(1-2) \times 104$ L mol^{-1} cm^{-1}. One of the disadvantages of the Molybdenum Blue method in a batch-wise method is the lack of reproducibility of the reaction, especially the reducing reaction. However, this disadvantage can be completely overcome by using flow injection techniques [18]. The flow diagram is a conventional double-line system: one for a carrier stream, the other for a reagent stream. For example, the carrier solution may be just water, while the reagent solution is prepared by mixing an equal volume of each of the solutions (A) and (B): (A) contains 5.5 g of ammonium molybdate (tetrahydrate), 0.25 g of potassium antimonyl tartrate and concentrated sulfuric acid in 1000 mL of aqueous solution, and (B) contains 3.0 g of ascorbic acid and 1.0 g of sodium dodecyl sulfate in 1000 mL aqueous solution [19].

It is very interesting with respect to sensitivity improvement for phosphate determination that Malachite Green(MG), one of the cationic triphenylmethane dyes, reacts with molybdophosphoric acid in an acidic medium to form a colored ion associate [20]. By using this reaction with MG, a more sensitive spectrophotometric method for orthophosphate determination can be developed. This coloration reaction is based on the ion association reaction of molybdophosphoric acid with a yellow protonated Malachite Green, HMG^{2+}. This coloration reaction can proceed in an acidic medium as follows:

$$\underset{(\text{yellow})}{H_3PMo_{12}O_{40}} + \underset{(\text{yellow, } \lambda_{max}=446 \text{ nm})}{HMG^{2+}} \rightarrow \underset{(\text{green, } \lambda_{max}=650 \text{ nm})}{(MG^+)(H_2PMo_{12}O_{40}^-)} + 2H^+ \tag{1}$$

$$(MG^+)(H_2PMo_{12}O_{40}^-) + HMG^{2+} \rightarrow (MG^+)_2(HPMo_{12}O_{40}^{2-}) + 2H^+ \tag{2}$$

$$(MG^+)_2(HPMo_{12}O_{40}^{2-}) + HMG^{2+} \rightarrow (MG^+)_3(PMo_{12}O_{40}^{3-}) + 2H^+ \tag{3}$$

The final product of (3 : 1) ion associate formed as in Equation (3) can precipitate in an aqueous solution, which can be extracted into an organic phase [21], or can be adsorbed on a hydrophobic solid phase, such as a cellulose nitrate membrane filter [22, 23]. By adding poly(vinyl alcohol) (PVA) to the reaction solution, the ion association reactions (2) and (3) are stopped, and the ion associate can be soluble in the solution. The molar absorptivity in the aqueous solution was about 8×10^4 l mol^{-1} cm^{-1} at 650 nm: this coloration reaction is called the MG method. The advantages of the MG method are: (i) fast coloration reaction without heating, (ii) higher sensitivity than for Molybdenum Blue methods and (iii) longer maximum wavelength than for the Molybdenum Yellow methods.

The MG method coupled with FIA can be applied to the determination of phosphate in water samples: the LOD is 1 ppb of phosphorus at 650 nm. The MG method can be applied to the determination of total phosphorus in industrial waste waters by on-line digestion/spectrophotometric FIA [24].

A fluorophotometric method can be used for the sensitive determination of orthophosphate using a double-line FIA system: the carrier is a water stream, and the reagent stream contains molybdate, Rhodamine 6G and hydrochloric acid [25]. The method is based on the fluorescence quenching of a cationic dye due to the formation of an ion associate of molybdophosphoric acid with Rhodamine 6G: the LOD is 0.1 ppb of phosphorus, and 0.5 M NaCl does not interfere with the determination. The method can be applied to the determination of orthophosphate in real waters, such as river and sea waters. Instead of Rhodamine 6G, Rhodamine B (RB) can be used in the presence of poly(vinyl alcohol) [26]: the LOD is about 1×10^{-8} M.

In the RB/fluorescence quenching method improved by re-examining the experimental conditions, the LOD is much improved to 5×10^{-9} M [27].

Total Phosphorus In general, the Molybdenum Blue method can be applied to total phosphorus determination in real water samples after the decomposition of organic and inorganic phosphorus compounds, organic and inorganic condensed phosphorus compounds, and solid and dissolved phosphorus compounds to orthophosphate by heating at high temperature (120–160 °C) in the presence of potassium peroxodisulfate and sulfuric acid [28, 29], by heating at high temperature with a platinum wire in a reaction coil [24], or by irradiating aqueous samples containing peroxodisulfate with UV light [19, 30]. By using a reaction coil, which is wound on two low-pressure mercury lamps (14 mm o.d., 134 mm length, 4 W germicidal use), most organic phosphorus compounds were decomposed to orthophosphate at 70 °C: the LOD was 1 ppb P at 830 nm [19].

The oxidation reaction of thiamine with molybdophosphoric acid can be applied to the determination of total dissolved phosphorus in water samples, in which phosphorus compounds can be decomposed using a simple UV photoreactor in the presence of peroxodisulfate: the LOD is 1×10^{-8} M [31]. Though the thiamine method is very sensitive, the reaction is less selective to orthophosphate: oxidizing agents may interfere with the determination and cause positive errors.

A luminol chemiluminescence detection system can be used: the method is based on the oxidation reaction of luminol with molybdophosphoric acid, and the LOD is 30 ppt of phosphorus [32]. The method is very sensitive and practical: it can be applied to the determination of trace amounts of phosphorus in aquatic environmental samples.

20.2.2
Organic Compounds Related to Water Pollution

20.2.2.1 Surfactants
Vast amounts of various kinds of surfactant have been produced world-wide, and they are now requisite for industrial as well as household use. Some parts of the surfactants used, especially those in household use, are discharged as domestic waste into rivers, lakes and sea and can contaminate them. Of the surfactants, anionic surfactants, which are often used for household washing, are serious causes of water pollution.

In general, anionic surfactants can be determined by solvent extraction with a cationic dye as an ion associate. In FIA, a PTFE porous membrane, which is permeable to a hydrophobic organic solvent, is used for phase separation [33].

In Figure 20.4(a) and (b), a flow diagram and signal profile for the determination of anionic surfactants are shown. In this system, Methylene Blue is used as a pairing ion for anionic surfactants, and *o*-dichlorobenzene replaces chloroform because of its lower volatility and toxicity [34]. A simple phase separator, composed of two blocks made of poly(chlorotrifluoroethylene)(CTFE) and PTFE membrane (pore size: 0.8 μm) [34], is installed. A double-membrane phase separator [35] and

a)

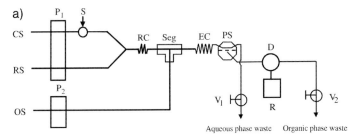

b)

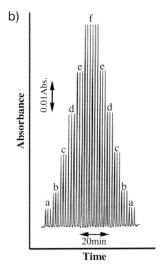

Figure 20.4 (a) Flow diagram of solvent extraction/FIA for the determination of anionic surfactants. CS: carrier; RS: reagent solution; OS: extraction solvent; P_1 P_2: pumps; S: sample injection; RC: reaction coil; Seg: segmentor; EC: extraction coil; PS: phase separator; D: detector; R: recorder; V_1, V_2: needle valves. (b) Flow profiles for the determination of anionic surfactant [lauryl sulfate]/10^{-6} M: (a) 0.5; (b) 1; (c) 2; (d) 3; (e) 4; (f) 5.

a cylindrical cavity-type phase separator [36] can be used for better phase separation in a solvent extraction.

Anionic and cationic surfactants can be determined in an aqueous phase on the basis of the ion association reactions/color change with an anionic dye and a cationic surfactant without solvent extraction [37]. The flow system and the reagent composition are very simple compared with a solvent extraction system. However, interferences from naturally occurring substances, such as fumic acids, must be taken into account with environmental samples; they may interfere with the determination of anionic surfactants with positive errors.

20.2.2.2 Chemical Oxygen Demand (COD)
Chemical oxygen demand (COD) is one of the indexes of pollution of organic matter, and is defined as the amount of oxygen necessary to oxidize organic substances

with a certain oxidizing agent, such as permanganate, chromate or cerium(IV). The oxidizing agents used for measuring COD in aquatic-environment samples have different standard methods. In general, a long reaction coil and a higher temperature are necessary for improvement of the oxidation of organic substances; therefore a double-plunger pump is necessary to propel the two solutions under high pressure.

In a FIA system for COD with chromate [38], a two-line FIA system can be used: one is for the carrier stream (H_2O), the other for the reagent stream ($K_2Cr_2O_7$ in H_2SO_4). Other conditions are as follows; a reaction coil: 50 m (PTFE, 0.5 mm) in a 120 °C bath, a back-pressure coil: 0.25 mm i.d. × 3 m PTFE; sample size: 100 μL; wavelength: 445 nm with a 10-mm path quartz cell; each flow rate: 0.3 mL min^{-1}; sample throughput: 15 h^{-1}; LOD: 5 mg L^{-1} as D-glucose.

In a FIA system with permanganate [39], platinum tubing (0.5 mm i.d. × 1 m or 3 m) can be used as a reaction coil in a hot bath (95 °C). Platinum can act as a catalyst for the oxidation with permanganate in an acid solution of 0.9 M sulfuric acid solution and 0.1% H_3PO_4.

A cyclic FIA for the repetitive determination of COD with $KMnO_4$ as an oxidizing agent can be used; samples (20 μL) are injected into the reagent solution (0.2 M $KMnO_4$ in a solution of 0.8 M H_2SO_4 and 0.001 M HIO_4), which is circulating in PTFE tubing from a 50 mL reservoir to a flow cell (5-mm path, 9 μL) via a reaction coil (0.5 mm × 2 m) in a hot bath (70 °C), a cooling coil (0.5 mm i.d. × 1.5 m) and a backpressure coil (0.5 mm × 10 m) to the reservoir; flow rate: 0.75 mL min^{-1}; wavelength: 525 nm [40].

20.2.3
Organic Compounds Related to Toxic/Hazardous Problems

Analysis for various kinds of pesticide, herbicide and endocrine disrupting compounds in aquatic environments is one of the most important problems in environmental and analytical science. The difficulties in such analyses are: (i) detection concentrations of the methods must be very low, because analyte concentrations are often quite low, and (ii) the selectivity of the detection methods must be good, because various kinds of interfering substance may coexist in samples at high concentrations. Usually, in measuring such compounds, any pretreatments for the improvement of sensitivity and selectivity are performed prior to the measurement by FIA, similarly as in a batchwise measuring method.

Conventional pretreatment procedures by solid-phase extraction (SPE) with a column and a cartridge can be applied to the enrichment of analytes and the removal or separation of interfering/matrix substances from analytes for a batchwise pretreatment prior to FIA measurement or on-line pretreatment. For the collection/enrichment of organic analytes or interfering substances, hydrophobic adsorbents, such as ODS, are used for non-ionic organic substances, while ion exchangers are used for ionic substances.

Trace amounts of phenols in water samples can be preconcentrated on an Amberlite XAD-4 column (pH 2) installed in an FIA system, being separated from the sample matrix. The phenols, enriched on the column, can be eluted with an alkaline solution (pH 13), and are measured by a flow injection technique with the 4-aminoantipyrine (4-AAP) method at 510 nm [41].

For the fluorometric determination of 17β-estradiol, one of the endocrine disruptors, enrichment can be carried out on a molecularly imprinted polymer (MIP) packed in a microcolumn [42]. A molecularly imprinted polymer/SPE (MIP/SPE) combined with FI chemiluminescence can be used for the determination of tetracycline [43].

20.2.4
Metals and Metal Compounds Related to Water Pollution and Toxic/Hazardous Problems

20.2.4.1 Spectroscopic Detection
FIA coupled with spectroscopic detection methods, such as atomic absorption, ICP-AES and ICP-MS, are much more convenient and versatile systems for metal analysis in aquatic environmental samples than spectrophotometric or fluorophotometric detection methods. In a classical FIA method coupled with atomic absorption spectrometry, a single-line system (carrier stream) was used, and the sample was injected into the carrier. The main aim of such a system is to automate the measurement and improve sample throughput. In the analysis of water pollutants, however, analyte concentrations are often very low, and therefore some preconcentration and/or clean-up procedures are necessary to improve detection sensitivity in AAS and ICP-AES and remove the sample matrix in ICP-MS.

In Table 20.1, an example of the minicolumn pretreatment for sensitivity improvement by ICP-AES is summarized; trace and ultratrace metal ions at concentrations of several or several tens of ppt can be detected [44]. The on-line pretreatment system is shown in Figure 20.5, where on-line pretreatment is carried out by using a sequential-injection system.

In Table 20.2, an example of the application of the on-line system to the analysis of standard reference materials is shown: most trace and ultratrace metal ions in river water samples can be determined by ICP-AES.

In the on-line system coupled with a multi-detection system, such as ICP-AES and ICP-MS, a versatile chelating resin possessing iminodiacetate function groups for multi-element collection, such as Chelex 100 and Muromac A-1, is useful.

On-line preconcentration/AAS coupled with FIA can be used for trace and ultratrace metals; for example, lead can be determined by flame-AAS [45, 46], and bismuth, cadmium and lead can be determined by electrothermal (graphite furnace) atomization-AAS (ET-AAS) [47].

In ET-AAS, an eluent zone, in which the concentration of an analyte is the highest, must be held just at the tip of the injection nozzle of the auto-sampler of the ET-AAS, which can be easily and reproducibly repeated with a computer-controlled syringe pump system of SIA.

ICP-AES coupled with an on-line pretreatment system, as in Figure 20.5, can also be used for single metal analysis: for example, trace and ultratrace amounts of lead can be determined [48, 49].

Table 20.1 Analytical characteristics for the collection/
concentration of metal ions with a mini-column using an
automated pretreatment system.

						LOD, ng mL^{-1}	
	Wavelength,	Linear range,		Enrichment			Direct
Elements	nm^a	ng mL^{-1}	Linearity	factor^b	%RSD^c	This work^{b,d}	ICP-AES^e
Ba	493.408	0.1–10	0.9973	5	4.6	0.02	0.4
Be	313.042	0.001–10	0.9946	10	5.1	0.001	0.4
Cd	226.502	0.01–10	0.9991	16	6.7	0.018	0.6
Co	228.615	0.1–10	0.9987	9	9.6	0.10	1.5
Cr	205.560	0.1–10	0.9998	14	4.6	0.09	1.3
Cu	324.754	0.1–10	0.9909	15	2.1	0.08	1.5
Fe	259.940	0.1–10	0.9968	13	2.7	0.05	1.3
Mn	257.610	0.01–10	0.9997	10	4.4	0.008	0.4
Ni	231.604	0.05–10	0.9971	12	6.7	0.16	2.4
Pb	220.353	0.1–10	0.9980	16	5.5	0.18	4.4
Sc	361.383	0.01–10	0.9916	19	4.5	0.01	0.3
V	292.401	0.1–10	0.9986	17	2.9	0.09	1.4
Zn	213.856	0.1–10	0.9955	12	8.7	0.02	1.3

^{a}Emission wavelengths used: these are based on Method 200.7 US-EPA.
^{b}5 mL of the sample solutions were used.
^{c}Obtained by using 5 mL of the mixed standard solution, the concentration of each metal ion used
was 0.5 ppb ($n = 7$).
^{d}Limit of detection, corresponding to 3 (S/N).
^{e}Instrumental detection limit, corresponding to 3 SD (standard deviation) of 0.01 M HNO$_3$
($n = 10$).

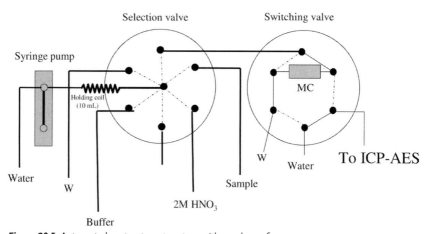

Figure 20.5 Automated pretreatment system with a column for
collection/concentration of metal ions. Column size:
40 mm × 2 mm i.d.

Table 20.2 Analytical results for standard river waters.

Elements	SLRS-4[a]		JSAC 030 1-1[b]		JSAC 0302[c]	
	Certified values, ng mL⁻¹	This work, ng mL⁻¹	Certified values, ng mL⁻¹	This work ng mL⁻¹	Certifed values, ng mL⁻¹	Tills work ng mL⁻¹
Ba	12.2 ± 0.6	OR[d]	0.60 ± 0.02	0.55 ± 0.03	0.60 ± 0.01	0.58 ± 0.01
Be	0.007 ± 0.002	0.009 ± 0.001	—	(0.006 ± 0.001)[f]	0.99 ± 0.04	1.02 ± 0.01
Cd	0.012 ± 0.002	0.014 ± 0.005[e]	0.0023 ± 0.0007	0.005 ± 0.002[e]	1.01 ± 0.01	0.92 ± 0.06
Co	0.033 ± 0.006	0.040 ± 0.012[e]	—	(0.040 ± 0.010)[e,f]	—	(0.032 ± 0.013)[e,f]
Cr	0.33 ± 0.02	0.33 + 0.02	0.15 ± 0.01	0.18 ± 0.02	10.1 ± 0.2	9.8 ± 0.2
Cu	1.81 ± 0.08	1.92 ± 0.13	0.57 ± 0.07	0.61 ± 0.05	10.3 ± 0.2	10.7 ± 0.3
Fe	103 ± 5	OR[d]	4.7 ± 0.3	4.6 ± 0.5	56 ± 1	OR[d]
Mn	3.37 ± 0.18	3.33 ± 0.16	0.125 ± 0.007	0.13 ± 0.03	5.0 ± 0.1	4.9 ± 0.1
Ni	0.67 ± 0.08	0.65 ± 0.02	—	(0.082 ± 0.013)[e,f]	9.9 ± 0.2	9.4 ± 0.3
Pb	0.086 ± 0.007	0.092 ± 0.002	(0.005)[f]	ND[d]	10.1 ± 0.2	9.8 ± 0.6
Sc	—	(0.026 ± 0.003)[f]	—	(0.036 ± 0.008)[f]	—	(0.015 ± 0.004)[f]
V	0.32 ± 0.03	0.34 ± 0.04	—	(0.47 ± 0.04)[f]	—	(0.49 ± 0.04)[f]
Zn	0.93 ± 0.10	0.92 ± 0.08	0.19 ± 0.03	0.17 ± 0.03	10.2 ± 0.3	9.8 ± 0.3

[a]River water reference material for trace metals issued by National Research Council Canada.
[b]River water reference material for trace metals issued by The Japan Society for Analytical Chemistry (unspiked).
[c]River water reference material for trace metals issued by The Japan Society for Analytical Chemistry (spiked), diluted 2 times with 0.01 M HNO₃.
[d]OR, over range; ND, cannot be detected.
[e]Sample volume: 15 mL. LODs of Cd, Co, Ni and Pb are 0.005 ng mL⁻¹, 0.03 ng mL⁻¹, 0.05 ng mL⁻¹, 0.05 ng mL⁻¹ and 0.06 ng mL⁻¹, respectively when 15 mL was used.
[f]Figures in parentheses were information values.

For the speciation of metals, two or more mini-columns can be installed in FIA, which is coupled with AAS or ICP-AES. The speciation of Cr(III) and (VI) can be carried out using small-sized thin solid phase (STSP) columns: one packed with cation exchanger for Cr(III) collection and the other packed with anion exchanger for CrO_4^{2-} collection [50]; the detection of both species is carried out by ICP-AES. The speciation of selenite and selenate at the sub-μg L^{-1} level can be carried out by using mini-columns packed with anion exchanger by ICP-MS. At pH 1.5, selenate is present as $HSeO_4^-$ and selenite is present as an uncharged H_2SeO_3 and, therefore, only anionic selenate can be collected on the first column and selenite is collected on the second column at higher pH [51].

20.2.4.2 Other Detection Methods

In general, trace and ultratrace analyses of metals in an aquatic environment are carried out by using spectroscopic methods. However, in special cases, spectro-photomeric, fluorophotometric and chemiluminescence methods are much more sensitive than spectroscopic methods. Trace amounts of iron and copper in river water samples can be determined by FIA coupled with spectrophotometric detection using a catalytic reaction [52–54]; in the presence of N,N-dimethyl-1,4-penylenedia-mine (DPA) and hydrogen peroxide, iron and copper can act as a catalyst for the oxidation of DPA, and the LOD is 0.02 ppb of Fe [52], the LODs are 0.01 and 0.07 ppb for iron and copper, respectively.

Chemiluminescence detection with luminal coupled with FIA can be used for ultratrace determination of iron in sea water: the LOD of total iron (Fe(II) and (III)) is 40 p.m. In the system, an 8-quinolinol chelating column is installed for in-line matrix elimination/preconcentration [55].

20.3
Analysis of an Atmospheric Environment by Flow Methods

Trace and ultratrace amounts of analytes in air samples can be determined by FIA after the collection/concentration of analytes in a certain absorbing solution by a batchwise or an on-line method. In a classical method, the mass transfer of analytes from a gaseous phase to a liquid phase can be done by bubbling sample gas into the absorbing solution. Such methods are very simple in principle and the collection efficiency is almost 100%. However, the collection procedure needs a long time to attain a high enrichment factor and is difficult to miniaturize or install in an FIA system. To overcome the disadvantages of the classical gas bubbling method, several gas–liquid extraction apparatuses for on-line collection/concentration, which are installed in FIA, are proposed; they are a denuder (DN), a gas diffusion scrubber (GDS) and a chromatomembrane cell (CMC).

20.3.1
Denuder (DN) and Gas Diffusion Scrubber (GDS)

In a usual DN device, such absorbing materials as viscous liquids or solids are coated on the inner wall of a cylindrical tube into which a gas sample flows; the transfer of analytes occurs from a gas phase into a liquid or a solid phase by gas diffusion. After the sampling is finished, the inner wall of the tube is washed with a suitable eluent, and the analytes are recovered in the eluent. The collection method with a DN device is not suitable for installing in a flow analysis system.

In a GDS device, an absorbing solution is kept at one side of a gas-permeable membrane, while a gas sample flows over the other side of the membrane. The gas sample can flow continuously over the absorbing layer in the GDS device, which can be recovered and used for continuous detection of an analyte. Therefore, the GDS device can be used in a continuous flow measurement method.

GDS is one of the most useful devices for on-line gas–liquid extraction, which can be miniaturized and installed in various kinds of flow analysis system. Since 1984, a number of GDS-based analytical systems have been developed and applied to air analysis by Dasgupta and his colleagues [56–62].

In the collection of analytes in gaseous samples into absorbing solutions, the collection efficiency with a DN and a GDS device is usually 10–20%, and therefore standard gas samples are requisite for the preparation of calibration graphs.

20.3.2
Chromatomembrane Cell (CMC)

A CMC device is a different type of gas collector from the DN and the GDS device [62]. Since 1994, the CMC device has been used for the concentration of analytes in aqueous solutions into organic solutions [63, 64] and for the collection/concentration of gas analytes in the atmosphere [65–68].

The membrane block of the CMC is made of biporous PTFE, which has two kinds of pore, micropores (0.1–0.5 µm) and macropores (250–500 µm); a schematic illustration of the CMC device and its collection mechanism are shown in Figures 20.6 and 20.7, respectively. The specific interface area is about $65 \, cm^2 \, cm^{-3}$ [69]. In the membrane block, a polar liquid is filled into the macropores, while the micropores can be used only for a non-polar phase, such as organic solvents or gas, because the capillary pressure of polar liquids prevents their penetration into the micropores. Therefore, the non-polar phase, such as air or organic solvents can go through the membrane block, even if the polar phase, such as aqueous solutions, fills the macropores. While the non-polar phase goes through the CMC membrane block, in which the macropores are filled with a polar phase, an analyte in the non-polar phase can distribute between the two phases, and mass transfer is carried out from the non-polar phase to the polar phase, similar to the distribution in a column packed with some packings; therefore, in the membrane, the peak shapes areas in Figure 20.7(b) [70].

In practical use for the collection/concentration of an analyte in an air sample into an absorbing aqueous solution, first the absorbing solution is filled into the macropores as a stationary phase, and then the air sample is flowed continuously into the CMC membrane, where the gas phase goes through the micropores and out of the CMC membrane as in Figure 20.6. The mass transfer of the analytes in the gaseous sample occurs while the sample is flowing through the biporous PTFE block. In the CMC membrane, the complete transfer of the analytes can be easily attained quickly

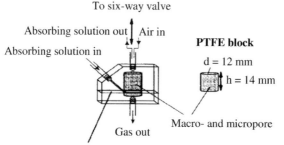

Figure 20.6 Chromatomembrane cell (CMC) with biporous PTFE membrane. CMC: three-hole chromatomembrane cell.

because the interface between the two phases is very large. Furthermore, highly efficient enrichment of the analyte can be achieved, because only a small volume of the absorbing solution is filled in the macropores as the stationary phase, while a large volume of the gas sample flows through the CMC membrane; the volume ratio of the gas sample to the absorbing solution is very large, and usually several 100-fold enrichment can be achieved. Furthermore, the collection efficiency of the analyte from the air sample is almost 100% by flowing the sample in the membrane at a reasonable flow rate. As a result, no gaseous standards are necessary for the preparation of a calibration graph and, usually, standard aqueous solutions can be used for the calibration [70].

On-line collection/concentration of trace and ultratrace analytes in atmospheric samples can be realized by coupling the CMC device with a conventional or micro-FIA. An automated CMC system, which consists of a six-way switching valve, a liquid pump for filling an absorbing solution into the CMC membrane, a gas pump for aspirating an air sample and delivering it to the CMC membrane, a degassing device,

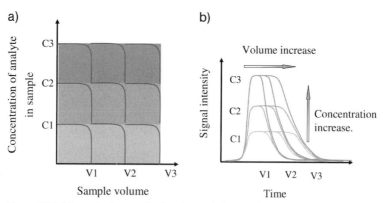

Figure 20.7 Schematic diagram for distribution behavior of analyte in a CMC membrane. (a) chromatogram in the CMC membrane, (b) peak profiles according to concentration of analyte and sample volume.

Flow injection system

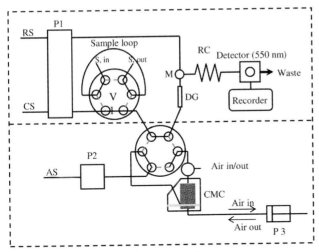

Absorption/concentration system

Figure 20.8 FIA system coupled with a three-hole CMC. RS: 5×10^{-5} M pararosaniline $+ 4.5 \times 10^{-2}$ M formaldehyde at pH 1.4; CS/AS: absorbing solution ($2 \, g \, L^{-1}$ triethanolamine (TEA)), P1: double plunger pump (each flow rate 0.2 mL min^{-1}), P2: peristaltic pump (flow rate: 0.6 mL min^{-1}), P3: syringe type pump (5 mL min^{-1}), SL: standard sample loop (300 (l), DG: degas unit; V1 and V2: six-way valves, RC: reaction coil (0.5 mm i.d. $\times 200$ cm).

a three-way switching valve and a CMC device packed with biporous PTFE block (12 mm o.d. $\times$ 14 mm), as shown in Figure 20.8. In the system, the CMC device is connected to the switching valve, and two ports of the valve are connected to the carrier stream of a FIA system. A six-way switching valve can be used to introduce standard solutions for the preparation of a calibration graph. The CMC system coupled with FIA can be applied to the determination of nitrogen dioxide in air (LOD: 0.9 ppbv) [70–72], sulfur dioxide in air (0.5 ppbv) [73] and formaldehyde in air (0.03 ppbv) [74]. In Table 20.3, the experimental conditions for the determination of nitrogen dioxide, sulfur dioxide and formaldehyde are summarized.

A flow injection method coupled with the CMC device can be applied to the determination of other gaseous analytes, such as ammonia, ozone, hydrogen peroxide and hydrogen sulfide in the atmosphere.

20.3.3
Simple Batchwise Collection/Concentration Method for Substances in Air

In the analysis of gaseous samples, including the atmosphere, one of the most important procedures is the collection of analytes in an absorbing solution. The GDS and the CMC device are very useful for this. However, sometimes an analytical instrument cannot be used on site for various reasons. In such cases, a batchwise

Table 20.3 Experimental conditions for the determination of sulfur dioxide, nitrogen dioxide and formaldehyde in air.

Conditions	SO$_2$ determination	NO$_2$ determination	HCHO determination
Carrier solution	2 g l^{-1} triethanolamine solution		purified water
Reagent solution	5 × 10^{-5} M pararosaniline + 4 × 10^{-2} M CH$_2$O at pH 1.25	20 g l^{-1} sulfanilamide + 0.5 g l^{-1} NED + 25 ml l^{-1} HCl	0.03 M acetylacetone + 2 M of ammonium acetate buffer at pH 5.8–6.0
Wavelength	550 nm	525 nm	412 nm
Each flow rate	0.20 ml min^{-1}		0.40 ml min^{-1}
Standard	300 µl		
Mixing coil length	200 cm		500 cm
Air sample taken	20 ml		
Flow rate of air sample	5–6 ml min^{-1}		

method is needed; the method must be a simple procedure, a simple apparatus and low cost; furthermore the collection efficiency should be quantitative and the enrichment reasonable.

A plastic syringe, of volume 50 mL, can be applied to the collection/concentration of substances in air. The procedure is very simple and does not need special techniques; the procedure for ammonia collection is shown in Figure 20.9. The method can be applied to the collection/concentration of formaldehyde [75] and ammonia [11] in air, followed by determination by a conventional FIA. The total volume of the syringe is 62.2 ± 0.2 mL, when the air is filled to the point where a piston stops, as shown in Figure 20.9 [2].

The syringe collection method is useful and convenient for on-site collection/concentration, and the measurement can often be done on-site, or in a laboratory.

20.4
Analysis of the Geosphere Environment by Flow Methods

In the analysis of the geosphere environment, solid samples must first be converted to aqueous solutions by a conventional method, such as the dissolution of an analyte from the samples, acid digestion of the sample with or without heating and a microwave, followed by pretreatment for clean-up and enrichment with or without a column and a cartridge. The aqueous sample solutions thus prepared are used for measuring a target analyte by conventional flow methods, in a similar manner as in aqueous samples.

The most important procedure in the analysis of geosphere samples is the same as in the analysis of atmospheric samples; that is, the preparation of pretreated

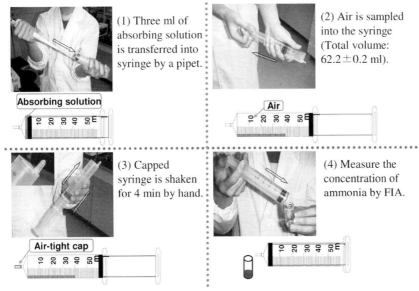

Figure 20.9 Simple batchwise collection/concentration method for gaseous substances in air.

aqueous samples is a requisite for reliable analyses. Suitable methods necessary for reliable sample preparations are to be found in manuals issued as standard methods, recommended methods and so forth.

20.5
Future of Environmental Analysis

In general, in environmental analysis a number of samples must be analyzed in a short time and therefore flow-based methods are required. Further, environmental samples are sometimes very complex, and therefore interfering effects must always be taken into account. As is in the case of dioxin analysis, analyte concentrations are often very low, and an effective enrichment method or a sophisticated and very high-cost instrument may be necessary. Also, the consumption of samples, reagents, consumables, energy, labor and time must be lowered.

In order to overcome such problems, a computer-assisted flow chemical analysis system (CAFCA system), which can carry out complex procedures according to fully refined programs, is requisite for environmental sample analysis. The CAFCA system can be assembled by using computer-controllable pumps, selection valves, switching valves, solenoid pumps and valves, and special devices for on-line pretreatment and a computer to control the system including auto-sampling and data acquisition/feedback. However, reliable sampling and preparation of pretreated samples are essential and fundamental for analysis. Analysts must always take responsibility for the reliability of the analytical data obtained.

References

1 Anderson, L. (1979) Simultaneous spectrophotometric determination of nitrite and nitrate by flow injection analysis. *Analytica Chimica Acta*, **110**, 123–128.

2 Inoue, A., Higuchi, K. and Tamanouchi, H. (2000) Spectrophotometric determination of nitrite and nitrate in water samples by flow injection techniques. *Journal of Flow Injection Analysis*, **16** (Supplement), 57.

3 Motomizu, S. and Sanada, M. (1995) Photo-induced reduction of nitrate to nitrite and its application to the sensitive determination of nitrate in natural waters. *Analytica Chimica Acta*, **308**, 406–412.

4 Burakham, R., Oshima, M., Grudpan, K. and Motomizu, S. (2004) Simple flow-injection system for the simultaneous determination of nitrite and nitrate in water samples. *Talanta*, **64**, 1259–1265.

5 Motomizu, S., Rui, S.C., Oshima, M. and Toei, K. (1987) Spectrophotometric determination of trace amounts of nitrite based on the nitrosation reaction with *N*,*N*-bis(2-hydroxypropyl)aniline and its application to flow injection analysis. *Analyst*, **112**, 1261–1263.

6 Motomizu, S., Mikasa, H. and Toei, K. (1987) Fluorimetric determination of nitrate in natural waters with 3-amino-1,5-naphthalenedisulphonic acid in a flow injection system. *Analytica Chimica Acta*, **193**, 343–347.

7 Motomizu, S., Mikasa, H. and Toei, K. (1987) Fluorometric determination of nitrite in natural waters with 3-aminonaphthalene-1,5-disulphonic acid by flow injection analysis. *Talanta*, **33**, 729–732.

8 Joichi, Y., Lenghor, N., Takayanagi, T., Oshima, M. and Motomizu, S. (2006) Development of computer-controlled flow injection instruments and its application to determination of nitrate, nitrite, and ammonium ions in

environmental samples. *Bunseki Kagaku*, **55**, 707–713.

9 Muraki, H., Higuchi, K., Sasaki, M., Korenaga, T. and Tôei, K. (1992) Fully automated system for the continuous monitoring of ammonium ion in fish farming plant sea water by flow injection analysis. *Analytica Chimica Acta*, **261**, 345–349.

10 Morita, Y. and Kogure, Y. (1963) Spectrophotometric determination of nitrogen with hypochlorite and a-naphthol. *Nippon Kagaku Zasshi*, **84**, 816–823.

11 Suekane, T., Oshima, M. and Motomizu, S. (2005) Determination of trace amounts of ammonia in air using batchwise collection/concentration method by spectrophotometry. *Bunseki Kagaku*, **54**, 953–957.

12 Aoki, T., Uemura, S. and Munemori, M. (1983) Continuous flow fluorometric determination of ammonia in water. *Analytical Chemistry*, **55**, 1620–1622.

13 Motomizu, S., Toei, K., Kuwaki, T. and Oshima, M. (1987) Gas-diffusion unit with tubular microporous poly (tetrafluoroethylene) membrane for flow-injection determination of carbon dioxide. *Analytical Chemistry*, **59**, 2930–2932.

14 Sanada, M., Oshima, M. and Motomizu, S. (1993) Assembly of a new gas-diffusion unit and its application to the determination of total carbonate and ammoniacal nitrogen by FIA. *Bunseki Kagaku*, **42**, T123–T128.

15 Higuchi, K., Inoue, A., Tsuboi, T. and Motomizu, S. (1999) Development of a new gas-permeation system and its application to the spectrophotometric determination of ammonium ion by FIA. *Bunseki Kagaku*, **48**, 253–259.

16 Goto, M., Murobushi, S. and Ishii, D. (1988) Continuous monitoring method of total nitrogen in wastewater using

continuous microflow analysis. *Bunseki Kagaku*, **37**, 47–51.

17 Hori, T., Kanada, Y. and Fujinaga, T. (1983) Fractionation of total phosphorus occurring in Lake Biwa by sensitive indirect phosphate determination. *Bunseki Kagaku*, **31**, 592–597.

18 Ruzicka, J. and Hansen, E.H. (1975) Flow injection analyses. Part I. A new concept of fast continuous flow analysis. *Analytica Chimica Acta*, **78**, 145–157.

19 Higuchi, K., Tamanouchi, H. and Motomizu, S. (1998) On-line photo-oxidative decomposition of phosphorus compounds to orthophosphate and its application to flow injection spectrophotometric determination of total phosphorus in river water and waste waters. *Analytical Sciences*, **14**, 941–946.

20 Itaya, K. and Ui, M. (1966) A new micromethod for the colorimetric determination of inorganic phosphate. *Clinica Chimica Acta*, **14**, 361–366.

21 Motomizu, S., Wakimoto, T. and Toei, K. (1984) Solvent extraction-spectrophotometric determination of phosphate with molybdate and malachite green in river water and sea water. *Talanta*, **31**, 235–240.

22 Matsubara, C., Yamamoto, Y. and Takamura, K. (1987) Rapid determination of trace amounts of phosphate and arsenate in water by spectrophotometric detection of their heteropoly acid-Malachite Green aggregates following preconcentration by membrane filtration. *Analyst*, **112**, 1257–1260.

23 Susanto, J.P., Oshima, M. and Motomizu, S. (1995) Determination of micro amounts of phosphorus with Malachite Green using a filtration-dissolution preconcentration method and flow injection-spectrophotometric detection. *Analyst*, **120**, 187–191.

24 Aoyagi, M., Yasumasa, Y. and Nishida, A. (1988) Rapid spectrophotometric determination of total phosphorus in industrial wastewaters by flow injection analysis including a capillary digestor. *Analytica Chimica Acta*, **214**, 229–237.

25 Motomizu, S., Mikasa, H., Oshima, M. and Toei, K. (1984) Continuous flow method for the determination of phosphorus using the fluorescence quenching of rhodamine 6G with molybdophosphate. *Bunseki Kagaku*, **33**, 116–119.

26 Motomizu, S., Oshima, M. and Katsumura, N. (1995) Fluorimetric determination of phosphate in sea water by flow injection analysis. *Analytical Science & Technology*, **8**, 843–848.

27 Ii, Z., Oshima, M., Sabarudin, A. and Motomizu, S. (2005) Trace and ultratrace analysis of purified water samples and hydrogen peroxide solutions for phosphorus by flow injection method. *Analytical Sciences*, **21**, 263–268.

28 Korenaga, T. and Okada, K. (1984) Automated system for total phosphorus in wastewaters by flow injection analysis. *Bunseki Kagaku*, **33**, 683–686.

29 Goto, M., Nishimura, M., Tominaga, T. and Ishii, D. (1988) Continuous monitoring method of total phosphorus in wastewater using continuous microflow analysis. *Bunseki Kagaku*, **37**, 52–55.

30 Vlessidis, A.G., Kotti, M.E. and Evmiridis, N.P. (2004) A Study for the validation of spectrophotometric methods for detection, and of digestion methods using a flow injection manifold, for the determination of total phosphorus in wastewaters. *Journal of Analytical Chemistry*, **59**, 77–85.

31 Perez-Ruiz, T., Martinez-Lozano, C., Tomas, V. and Martin, J. (2001) Flow-injection spectrofluorimetric determination of dissolved inorganic and organic phosphorus in waters using on-line photo-oxidation. *Analytica Chimica Acta*, **442**, 147–153.

32 Yaqoob, M., Nabi, A. and Worsfold, P.J. (2004) Determination of nanomolar concentrations of phosphate in freshwaters using flow injection with luminol chemiluminescence

detection. *Analytica Chimica Acta*, **510**, 213–218.

33 Kawase, J., Nakae, A. and Yamanaka, M. (1979) Determination of anionic surfactants by flow injection analysis based on ion-pair extraction. *Analytical Chemistry*, **51**, 1640–1643.

34 Motomizu, S., Oshima, M. and Kuroda, T. (1988) Spectrophotometric determination of anionic surfactants in water after solvent extraction coupled with flow injection. *Analyst*, **113**, 747–753.

35 Sakai, T., Chung, Y.S., Ohno, N. and Motomizu, S. (1993) Double-membrane phase separator for liquid–liquid extraction in flow injection analysis. *Analytica Chimica Acta*, **276**, 127–131.

36 Motomizu, S. and Korechika, K. (1989) Modified cylindrical cavity-type phase separator for liquid/liquid extraction in flow injection system. *Analytica Chimica Acta*, **220**, 275–280.

37 Motomizu, S., Oshima, M. and Hosoi, Y. (1992) Spectrophotometric determination of cationic and anionic surfactants with anionic dyes in the presence of nonionic surfactants, Part II: Development of batch and flow injection methods. *Microchimica Acta*, **106**, 67–74.

38 Korenaga, T. and Ikatsu, H. (1982) The determination of chemical oxygen demand in wastewaters with dichromate by flow injection analysis. *Analytica Chimica Acta*, **141**, 301–309.

39 Tsuboi, T., Hirano, Y., Kinoshita, K., Oshima, M. and Motomizu, S. (2004) Rapid and simple determination of chemical oxygen demand (COD) with potassium permanganate as an oxidizing agent by a flow injection technique using a platinum-tube reactor. *Bunseki Kagaku*, **53**, 309–314.

40 Zenki, M., Fujiwara, S. and Yokoyama, T. (2006) Repetitive determination of chemical oxygen demand by cyclic flow injection analysis using on-line regeneration of consumed permanganate. *Analytical Sciences*, **22**, 77–80.

41 Sakai, T., Fujimoto, S., Higuchi, K. and Teshima, N. (2005) Determination of phenols at low levels in water samples using automatic flow injection analysis coupled with on-line solid-phase extraction. *Bunseki Kagaku*, **54**, 1183–1188.

42 Bravo, J.C., Fernandez, P. and Durand, J. (2005) Flow injection fluorimetric determination of β-estradiol using a molecularly imprinted polymer. *Analyst*, **130**, 1404–1409.

43 Xiong, Y., Zhou, H., Zhang, Z., He, D. and He, C. (2006) Molecularly imprinted on-line solid-phase extraction combined with flow injection chemiluminescence for the determination of tetracycline. *Analyst*, **131**, 829–834.

44 Katarina, Rosi K., Lenghor, Narong. and Motomizu, Shoji. (2007) On-line preconcentration method for the determination of trace metals in water samples using a fully automated pretreatment system (Auto-Pret AES System) coupled with ICP – AES. *Analytical Sciences*, **23**, 343–350.

45 Seki, T., Hirano, Y. and Oguma, K. (2002) On-line preconcentration and determination of traces of lead in river-water and seawater by flow injection-flame atomic absorption spectrometry and ICP-mass spectrometry. *Analytical Sciences*, **18**, 351–354.

46 Martin, A.O., Ruiz da Silva, E., Laranjeira, M.C.M. and Favere, V.T. (2005) Application of chitosan functionalized with 8-hydroxyquinoline: determination of lead by flow injection flame atomic absorption spectrometry. *Microchimica Acta*, **150**, 27–33.

47 Sung, Y.H. and Huang, S.D. (2003) On-line preconcentration system coupled to electrothermal atomic absorption spectrometry for the simultaneous determination of bismuth, cadmium, and lead in urine. *Analytica Chimica Acta*, **495**, 165–176.

48 Zougagh, M., Garcia de Torres, A., Alonso, E.V. and Pavon, J.M.C. (2004) Automatic on line preconcentration and determination of lead in water by ICP-AES using a TS-microcolumn. *Talanta*, **62**, 503–510.

49 Sabarudin, A., Lenghor, N., Yu, L.-P., Furusho, Y. and Motomizu, S. (2006) Automated online preconcentration system for the determination of trace amounts of lead using Pb-selective resin and Inductively coupled plasma–atomic emission spectrometry. *Spectroscopy Letters*, **39**, 669–682.

50 Motomizu, S., Jitmanee, K. and Oshima, M. (2003) On-line collection/concentration of trace metals for spectroscopic detection via use of small-sized thin solid phase(STSP) column resin reactors Application to speciation of Cr(III) and Cr(VI) *Analytica Chimica Acta*, **499**, 149–155.

51 Jitmanee, K., Teshima, N., Sakai, T. and Grudpan, K. (2007) DRC ICP-MS coupled with automated flow injection system with anion exchange minicolumns for determination of selenium compounds in water samples. *Talanta*, **73**, 352–357.

52 Lunvongsa, S., Oshima, M. and Motomizu, S. (2006) Determination of total and dissolved amount of iron in water samples using catalytic spectrophotometric flow injection analysis. *Talanta*, **68**, 969–973.

53 Lunvongsa, S., Tsuboi, T. and Motomizu, S. (2006) Sequential determination of trace amounts of iron and copper in water samples by flow injection analysis with catalytic spectrophotometric detection. *Analytical Sciences*, **22**, 169–172.

54 Lunvongsa, S., Takayanagi, T., Oshima, M. and Motomizu, S. (2006) Determination of ultratrace amounts of iron in concentrated acids by flow injection spectrophotometric method based on the catalytic effect of iron ion on the oxidation reaction of *N,N*-dimethyl-*p*-phenylenediamine with hydrogen peroxide. *Journal of Flow Injection Analysis*, **23**, 25–28.

55 Bowie, A.R., Achterberg, E.P., Fauzi, R., Mantoura, C. and Worsfold, P.J. (1998) Determination of sub-nanomolar levels of iron in seawater using flow injection with chemiluminescence detection. *Analytica Chimica Acta*, **361**, 189–200.

56 Dasgupta, P.K. (1994) A diffusion scrubber for the collection of atmospheric gases. *Atmospheric Environment*, **18**, 1593–1599.

57 Dasgupta, P.K., McDowell, W.L. and Rhee, J. (1986) Porous membrane-based diffusion scrubber for the sampling of atmospheric gases. *Analyst*, **111**, 87–90.

58 Philps, D.A. and Dasgupta, P.K. (1987) A diffusion scrubber for the collection of gaseous nitric acid. *Separation Science and Technology*, **22**, 1255–1267.

59 Dasgupta, P.K., Dong, S., Hwang, H., Yang, H. and Genfa, Z. (1988) Continuous liquid-phase fluorometry coupled to a diffusion scrubber for the real-time determination of atmospheric formaldehyde, hydrogen peroxide and sulfur dioxide. *Atmospheric Environment*, **22**, 949–963.

60 Lindgren, P.F. and Dasgupta, P.K. (1989) Measurement of atmospheric sulfur dioxide by diffusion scrubber coupled ion chromatography. *Analytical Chemistry*, **61**, 19–24.

61 Dasgupta, P.K. and Lindgren, P.F. (1989) Inlet pressure effects on the collection efficiency of diffusion scrubbers. *Environmental Science and Technology*, **23**, 895–897.

62 Zhang, G. and Dasgupta, P.K. (1992) Determination of gaseous hydrogen peroxide at parts per trillion levels with a Nafion membrane scrubber and a single-line flow injection system. *Analytica Chimica Acta*, **260**, 57–64.

63 Moskvin, L.N. (1994) Chromatomembrane method for the continuous separation of substances. *Journal of Chromatography A*, **669**, 81–87.

64 Moskvin, L.N. and Simon, J. (1994) Flow injection analysis with the chromatomembrane – a new device for gaseous/liquid and liquid/liquid extraction. *Talanta*, **41**, 1765–1769.

65 Moskvin, L.N., Simon, J., Loffer, P., Michaolova, N.V. and Nicolaevna, D.N. (1996) Photometric determination of anionic surfactants with a flow injection analyzer that includes a chromatomembrane cell for sample preconcentration by liquid–liquid solvent extraction. *Talanta*, **43**, 819–824.

66 Erxleben, H., Moskvin, L.N., Nikitina, T.G. and Simon, J. (1998) Determination of small quantities of nitrogen oxides in air by ion chromatography using a chromatomembrane cell for preconcentration. *Fresenius' Journal of Analytical Chemistry*, **361**, 324–325.

67 Loffler, P., Simon, J., Katruzov, A. and Moskvin, L.N. (1995) Separation and determination of traces of ammonia in air by means of chromatomembrane cells. *Fresenius' Journal of Analytical Chemistry*, **352**, 613–614.

68 Moskvin, L.N. and Rodinkov, O.V. (1996) Continuous chromatomembrane headspace analysis. *Journal of Chromatography A*, **725**, 351–359.

69 Erxleben, H., Simon, J., Moskvin, L.N., Vladimirovna, L.O. and Nikitina, T.G. (2000) Automized procedures for the determination of ozone and ammonia contents in air by using the chromatomembrane method for gas–liquid extraction. *Fresenius' Journal of Analytical Chemistry*, **366**, 322–325.

70 Wei, Y., Oshima, M., Simon, J. and Motomizu, S. (2002) The application of the chromatomembrane cell for the absorptive sampling of nitrogen dioxide followed by continuous determination of nitrite using a micro-flow injection system. *Talanta*, **57**, 355–364.

71 Wei, Y., Oshima, M., Simon, J. and Motomizu, S. (2001) Application of chromatomembrane cell to flow injection analysis of trace pollutant in ambient air. *Analytical Sciences*, **17** (Suppl), a325–a328.

72 Wei, Y., Oshima, M., Simon, J., Moskvin, L.N. and Motomizu, S. (2002) Absorption, concentration and determination of trace amounts of air pollutants by flow injection method coupled with a chromatomembrane cell system: application to nitrogen dioxide determination. *Talanta*, **58**, 1343–1355.

73 Sritharathikhun, P., Oshima, M., Wei, Y., Simon, J. and Motomizu, S. (2004) On-line collection/concentration and detection of sulfur dioxide in air by flow injection spectrophotometry coupled with a chromatomembrane cell. *Analytical Sciences*, **20**, 113–118.

74 Sritharathikhun, P., Oshima, M. and Motomizu, S. (2005) On-line collection/concentration of trace amounts of formaldehyde in air with chromatomembrane cell and its sensitive determination by flow injection technique coupled with spectrophotometric and fluorometric detection. *Talanta*, **67**, 1014–1022.

75 Sritharathikhun, P., Suekane, T., Oshima, M. and Motomizu, S. (2004) On-site analysis of trace amounts of formaldehyde in ambient air using batchwise collection/concentration method and portable flow injection system. *Journal of Flow Injection Analysis*, **21**, 53–58.

21
Flow Methods in Pharmaceutical Analysis

J. Martínez Calatayud and J.R. Albert-García

21.1
Introduction

Strictly speaking, pharmaceutical analyses are restricted to pharmaceutical and veterinary formulations and include the determination of active principles, excipients and impurities. Obviously, determining contents in this context also entails assessing their stability and hence that of any degradation intermediates and end products. Because industrial drug production processes include the control of raw materials, pharmaceutical analyses also involve the determination of other parameters including content uniformity, solubility or dissolution rate. In fact, drug analyses cannot be restricted to formulations as they often entail determining specific drugs in a variety of much more complex matrices including, but not limited to, animal foods, drinks, cattle feed and cosmetics, in addition to many clinical, forensic and veterinary samples spanning a variety of matrix types such as blood, urine or tissue. At present, the analysis of pharmaceuticals and its metabolites is a relevant topic for environmental samples. *Analysis of pharmaceuticals* is maybe a more accurate term than *pharmaceutical analysis*.

The use of flow techniques in drug analysis is as old as FIA or even continuous segmented flow analysis. Ever since the former technique was developed, a ceaseless stream of procedures for determining drugs has been reported that can be expected to continue growing by virtue of the usefulness of flow analysis as a practical *analytical tool*. The inception of SIA and its miniaturized, automated version called Lab-on-valve and other, more recent methodologies such as multicommutated analysis and multisyringe analysis signaled the start of a new stage in the development and dissemination of flow analysis methods. A book was published at the time to describe both the foundation of FIA and its operational modes that revolved around the achievements in relation to drug analysis. As expected, flow techniques have been used with all types of detectors, flow systems and physico-chemical processes of both the homogeneous and heterogeneous type.

Advances in Flow Analysis. Edited by Marek Trojanowicz
Copyright © 2008 WILEY-VCH Verlag GmbH & Co. KGaA, Weinheim
ISBN: 978-3-527-31830-8

The discussion below is organized around the analytical chemical process involved and hence around the type of detector used. Further distinction in terms of the specific continuous flow methodology used in each case would have been difficult and unnecessary.

21.2
Analysis of Pharmaceutical Formulations

21.2.1
Spectrophotometry (UV–vis and IR). Homogeneous Systems

UV–vis absorption detectors are by far the most widely used with flow techniques. In fact, roughly one-third of all FIA work has been conducted with them. Diode array spectrophotometers, which are in wide use under computer control, have enabled the development of new types of uses and strongly facilitated others. The ability to expeditiously record the absorbance at several wavelengths, or even full spectra, in order to obtain averaged data for a selected wavelength range, and the absence of moving parts in instruments has resulted in substantially increased reproducibility with respect to conventional spectrophotometers.

Multicomponent determinations based on UV–vis measurements are especially difficult with substances present at different concentrations in the sample and having partly or completely overlapped spectra. The flow assembly-diode array spectrophotometry couple provides a useful tool for resolving mixtures such as those in pharmaceutical formulations. Electronic derivation, spectral derivation and other chemometric methods provide highly useful tools for the simultaneous determination of two or three drugs. Thus, FIA has enabled the simultaneous resolution of mixtures of two, three or even four active principles with completely overlapped spectra (viz. etafedrine hydrochloride, phenylephrine hydrochloride, doxylamine succinate and theophylline) in a commercial formulation by using an integration time of 0.4 s.

21.2.1.1 Applications of Flow-Based Molecular Absorption Spectrophotometry to the Determination of Drugs

Molecular absorption spectrophotometry allows drugs to be determined by using reactionless procedures (viz. from their native spectra). This simply entails measuring the absorbance at a given wavelength, as one would measure any other physical variable. With solids, it suffices to grind the sample, extract it with an aqueous solution at an appropriate pH or an organic solvent such as methanol or ethanol, dilute to the required concentration and read the solution absorbance. This procedure has been used in FIA to determine proflavin, 3,6-diaminoacridine and stilboestrol phosphate (fosfestrol).

The precision is always higher than with methods involving some chemical pretreatment and so, obviously, is the operational simplicity. However, the sensitivity and selectivity are usually so low that some chemical derivatization is required. Drug analyses invariably require derivatization of the analyte, whether chemical or physical (e.g., photochemical, by irradiating the sample with a UV light source). Chemical derivatization processes span a variety of reactions of the redox, color-forming or photodegradation type, for example.

Especially commonplace in this context are the strong inorganic oxidants Ce(IV), potassium hexacyanoferrate (II), potassium permanganate and potassium dichromate. Insoluble oxidants such as MnO_2 and PbO_2 immobilized in solid-phase reactors – whether by the effect of their own solubility or with the help of some support – have also been frequently used in flow techniques, and so have milder oxidants such as Fe (III) and organic reactants including chloramine T and 2-iodyl benzoate.

Potassium dichromate in sulfuric acid has been used for the FIA oximetric determination of various members of the phenothiazine family. Cerium(IV) has been used as oxidant in both solid-phase reactors and homogeneous systems included in single-channel FIA assemblies where the sample was injected into a carrier-reagent stream consisting of an acid solution of the metal. In this way, cerium was used for the FIA determination of drugs such as procainamide, trimeprazine tartrate and diphenylhydramine.

The high oxidizing power of alkaline hexacyanoferrate(III) has been used for the determination of isoprenaline. In combination with 4-aminotantipyrine, this oxidant gives an oxidation-condensation reaction that facilitates the determination of phenol compounds including some drugs such as terbutaline sulfate. A spectrophotometric method for paracetamol and *N*-acetyl-*p*-aminophenol exists, based on their oxidation with hexacyanoferrate(III) and, for the latter, the subsequent reaction of *N*-acetyl-*p*-benzoquinoneimine with phenol in an ammoniacal medium at 80 °C to give *N*-(*p*-hydroxyphenyl)-*p*-benzoquinoneimine, which is blue-colored.

Paracetamol in pharmaceutical formulations has been determined by SIA following oxidation with potassium hexacyanoferrate(III) and reaction with phenol in an ammoniacal medium. The ferricyanide-aminoantipyrine reaction has been proposed for the SIA determination of fenoterol in drug formulations.

Other spectrophotometrically useful redox agents include vanadate, the reduction from V(V) to (IV) in acid medium of which has been proposed for the determination of some phenothiazines. Also, vanadium(V) forms an amber-coloured 1 : 1 complex with isoniazid or isonicotinic acid hydrazide. Excess reagent in the medium causes the complex to slowly decompose into isonicotinic acid, V(V) and gaseous nitrogen; although the reaction is very slow, it can be accelerated by using Os(VIII) as catalyst.

Oxidants such as periodate, occasionally supplemented with a small amount of hydrogen peroxide, have been used to determine emetine, ergonovine and ergotamide, and, in the presence of a catalyst, reserpine. This last drug is oxidized by periodate ion in the presence of Mn(II) as catalyst in an acid medium; under these conditions, the reaction stops at 3,4-didehydroreserpine, which is yellow-colored, rather than proceeding to 3,4,5,6-tetradehydroreserpine, which is reddish-brown. Hydrogen peroxide, an oxidant widely used in traditional drug analyses, was avoided in the early times of FIA owing to its forming abundant bubbles and to prevent its absorption from blocking the UV region. The problem was solved by destroying excess H_2O_2 with a pyrolusite reactive bed to decompose it into water and oxygen, and using a debubbler to remove oxygen bubbles.

Chloramine T facilitates the detection of bromide ions in pharmaceutical formulations such as clidinium, glycopyrronium, homatropine, neostignine, propantheline, pyridostigmine and scopolamine. This oxidant has been used in indirect methods including that for *p*-aminobenzoic acid based on its reaction with acid hypochlorite and subsequent determination of residual chlorine with *o*-tolidine.

The catalytic effect of iodine on the redox reaction between Ce(IV) and As(III) has been used to determine iodine in drug formulations. Calibration was done by monitoring the absorbance of Ce(IV) as a function of the iodine concentration.

SIA methodology has been used to examine the oxidation kinetics of ascorbic acid by Fe(III), using 1,10-phenanthroline as indicator. The reaction was proposed for the determination of vitamin C. Subsequently, the vitamin was determined by spectrophotometric titration, using sulfuric Ce(IV) as oxidant and measuring the decrease in the absorbance at 410 nm of ceric ion. A very similar subsequent method for vitamin C involves its oxidation with potassium permanganate and monitoring the absorbance decrease in the oxidant.

Ferric ion immobilized in an SIA flow cell has also been used, among others, to determine vitamin C with the redox indicator tris(1,10-phenanthroline)-iron(II) impregnated in a Nafion membrane.

Colored complexes. Although little difference in foundation exists – the underlying chemical process is identical – the operational conditions to be used depend on whether the metal ion is used as the color-forming reagent or, much less frequently, is contained in the active principles of the formulation concerned.

Ferric ion has been proposed as a reagent for obtaining colored complexes, and so have Co(II) and Ni(II). *N*-Acetylcysteine in injectables can be readily determined by formation of Ni(II) complexes; samples are injected into an Ni(II) solution in 0.1 M NH_3/NH_4Cl buffer and measured spectrophotometrically at 415 nm. The analyte is quantified from its peak height and width (i.e., by pseudo-titration), using a dispersion tube inserted in the FIA manifold. An identical assembly has been used to determine *N*-penicillamine in capsules, but with Co(II) in 0.2 M ammonium acetate as oxidant.

The formation of chelates between drugs and Fe(III) is a classical reaction in pharmaceutical analysis. Salicylate, salicylamide and methyl salicylate ions react with Fe(III) via their phenol groups – following hydrolysis in the case of methyl salicylate. This has been used to determine oxytetracycline, norfloxacin and ciprofloxacin in drug formulations. Hydroxy aminolysis of the β-lactam ring of penicillins and the formation of colored complexes with ferric ion have also been used for the FIA determination of penicillins.

Salicylic and acetylsalicylic acids have been determined in a double-injection flow system by injecting one sample aliquot into a NaOH stream that was hydrolyzed and forced to travel a longer path than the other. One of two FIA signals obtained for the Fe (III)-salicylic acid complex was proportional to the content in acetylsalicylic acid and the other (viz. the one for the hydrolysed stream) to the combination of both analytes.

Ferric ion has been used as an oxidant and the resulting ferrous ion employed to form coloured complexes for spectrophotometric monitoring. One case in point is the determination of paracetamol by oxidation with Fe(III), Fe(II) released being used to form a complex with 2,4,6-tripyridyl-S-triazine. Ascorbic acid has been determined similarly; the sample was oxidized upon injection into a stream of acid Fe(III) and the reaction product merged with a stream of 1,10-phenanthroline, the Fe(II) coloured complex thus formed being monitored at 508 nm. The anxiolytic sedative bromazepam was determined by using hydrochloric Fe(II) rather than ferric ion as colour-forming reagent.

Trimeprazine and perphenazine have been determined by monitoring the formation of their palladium chelates in a hydrochloric medium. An SIA method for the simultaneous determination of captopril was developed in two variants with a view to improving accuracy; one involved the formation of Pd chelates in an acid medium and the other potentiometric titration of the analyte with Ag(I).

Metal ions as active principles of formulations. Metals present as active principle ingredients in pharmaceutical formulations are usually determined by formation of colored chelates as above; there is one exception, iron is quantified as ferrous rather than ferric ion, which requires a prior oxidation or reduction reaction.

Iron is present in various forms in pharmaceutical formulations. The procedure repeatedly used to determine it usually involves using a flow-through dialyser on-line coupled to an SIA manifold in order to remove colored interferents or turbidity in the samples; following dialysis, Fe(III) is complexed with Tiron under photometric monitoring at 667 nm. This entails previously oxidizing Fe(II) in the sample to Fe(III). Alternatively, total iron is determined as Fe(II) by inserting a column packed with Cd granules to reduce ferric ion to ferrous ion, which is then reacted with 1,10-phenanthroline and the reaction monitored at 515 nm. Ferrous ion has also been chelated with 2,2-bipyridyl for this purpose.

Recently, Fe(II) and Fe(III) in drug formulations have been successfully speciated by chelating the latter with Tiron and monitoring the resulting complex; this is followed by oxidation of all iron to Fe(III) with hydrogen peroxide and its chelation.

Captopril has been determined by using FIA and SIA manifolds; the drug was oxidized with ferric ion and the resulting ferrous ion chelated with 2,2N-dipyridyl-2-pyridylhydrazone to obtain the well-known red-colored complex.

Additional determinations in this area include that of calcium by chelation with cresolphthalein or zinc in diet supplements with Xylenol Orange in the presence of sodium thiosulfite.

Especially common among FIA spectrophotometric determinations of drugs are those based on the formation of ion-pairs of low or no solubility in water. The underlying methodology has been used with a variety of techniques including liquid–liquid extraction and precipitation. These have in fact been the most frequent choices for the FIA spectrophotometric analysis of drugs and their uses are explained in detail in the turbidimetry and liquid–liquid extraction sections.

Other derivatization methods. While important, these have been less frequently used in this context than have the previous ones. Such is the case with the reactions involving nitrite ion. The classical Bratton–Marshall reaction, by which a free aryl primary amine is converted into a diazonium salt under the action of nitrous acid, is used to obtain an azo dye by coupling with a chromogenic agent following removal of excess reagent with sulfamic acid. This classical method was adapted to FIA methodology with a view to quantifying various sulfonamides and also for their dissolution testing.

Subsequent methods have used this chemical procedure in that the nitrite reagent has been prepared *in situ* by reducing a nitrate solution in a solid-phase reactor filled with copperized cadmium. This avoids the need to store unstable nitrite solutions under strictly controlled solutions and discard them on a frequent basis.

The formation of azo dyes has been used in combination with *S*-nitrosation as a general assay for thiols in both batch and FIA systems. The method involves the formation of *S*-nitrosothiols by reaction with nitrous acid and their hydrolysis with mercuric ion following destruction of excess acid; nitrous acid released in the latter step forms the azo dye to be measured spectrophotometrically. Obopindol was determined by SIA, using the color-forming reaction of the indole group in Ehrlich's reagent in an acid medium. Paracetamol has also been determined by nitrosation with sodium nitrite in an acid medium followed by addition of sodium hydroxide to improve the stability of the resulting compound.

FIA pseudo-titrations have been exploited to determine drugs with the chemical system previously employed to develop the method based on spectrophotometric readings at a single wavelength. Flow assembly used for pseudo-titration, however, included a dilution or dispersion unit in the form of a stirring chamber or a long, wide tube. While the conventional method is simpler and more expeditious, and uses smaller sample volumes, the pseudo-titration method provides wider determination ranges and slightly better accuracy. Because it enables the measurement of higher concentrations, the pseudo-titration method can be a better choice for analyzing pharmaceutical formulations with minimal or no need for dilution.

21.2.1.2 UV–vis Heterogeneous Systems (Turbidimetry, Solid–Liquid and Liquid–Liquid)

Turbidimetry Ion-pairing can be a simple, cost-effective alternative to liquid–liquid extraction for the determination of drugs. One can obviously expect precipitates to adhere to tubing and cell walls, and preclude the use of turbidimetry in routine flow analyses as a result. In fact, the earliest FIA turbidimetric determinations involved inserting a flushing stream after each sample, which halved the throughput. Flocculation of precipitates can be avoided by using an effective antiflocculant, which will additionally ensure homogeneity in the suspension. Fast, cost-effective procedures for control analyses of pharmaceutical formulations currently exist based on the formation of ion association compounds (with inorganic or organic counter-ions) and their turbidimetric detection. The presence of a surfactant usually suffices to ensure proper, reproducible operation, with little or no need for flushing. Such is the case with the determination of chlorhexidine with Bromocresol Green, using Triton X-100 as colloid protector. In some cases, the hydrodynamic properties of the continuous flow (viz. microstreams in an FIA manifold) suffice to ensure reproducible measurements without any antiflocculants or periodic flushing. Some authors have successfully determined various drugs without the need for any suspension stabilizer, as in the determinations of the anthelminthic levamisole with

$HgI_4{}^2$; phenformin with wolframate ion; thiamine with silicotungstic acid; diphenhydramine and promethazine hydrochlorides with Bromophenol Blue; and amitriptyline with Bromocresol Purple. The required procedures are quite simple as samples are usually injected directly into a carrier-reagent stream and rarely is the carrier used as the reaction medium and merged with a reagent stream.

Especially original are drug enantiomeric purity tests based on measurements of crystal growth inhibition. The turbidimetric method is based on the inhibition of crystal growth in L- and D-glutamic acid by the presence of paracetamol, the effect being monitored via absorbance measurements. The sample, containing L-glutamic acid, is merged with a propan-2-ol stream and the resulting solution injected into a stream carrying the same alcohol and monitored at 550 nm while equilibrium is reached. Pure water is used as a blank and flushing solution. The procedure is repeated using D-glutamic acid instead of the L-isomer.

The ion-pair formation approach has a variant involving extracting the product formed by the analyte and counterion, which is insoluble in water, with an organic solvent. The procedure is gradually declining in popularity as surfactants now dispense with the need for solvents (or even for solid–liquid extraction), but was rather a common choice in early FIA systems. A segmenter was used at the point of merging of the water carrier and organic solvent, and a separator was required to drive the organic solvent to the spectrophotometer and the aqueous solution to waste after the product was extracted. These early manifolds were subsequently replaced with simpler, more effective ones where extraction was done by dialysis across membranes; to this end, a device accommodating two parallel channels with an intervening membrane intended to bring the carrier (donor) and solvent (acceptor) into contact was used. These membrane-based systems facilitated the determinations of enoxacin, cisapride, sparfloxacin and trazodone using chloroform as extractant; and those of chloroquine and pyrimethamine with methylene chloride.

21.2.1.3 Infrared Absorption

Quantitation by IR absorption spectroscopy is usually less accurate and precise than with UV–vis spectrophotometry. In addition, deviations from Beer's law are much more common and cells are very narrow, impractical and scarcely reproducible. FIA IR systems are also essentially similar to FIA UV–vis configurations without chemical derivation, except for the need to use solvent-resistant tubes in the former. This, however, poses no special problem in practice at present. The earliest FIA-IR determination used carbon tetrachloride as solvent, phenyl isocyanate as carrier and a stainless steel piston pump as the propulsion system. Ibuprofen was subsequently determined with carbon tetrachloride as carrier and sample solvent to avoid the potential interference of excipients, which were insoluble in it. Additional uses of FIA- IR spectrophotometry include the simultaneous determination of acetylsalicylic acid and caffeine with dichloromethane.

Fourier transform IR spectrophotometry is especially useful with a view to obtaining a high sensitivity and data acquisition rate, and also simultaneously monitoring changes in specific absorption bands. The first use of this technique in FIA drug analysis involved supercritical fluids and its developers proposed specific mobile phases as the best choices for monitoring some solutes in the IR spectral region. The FIA-FTIR spectroscopy combination has also been used to determine choline compounds in aqueous solutions, using a cylindrical internal reflectance cell and a ZnSe crystal as reflecting element.

This methodology not always been used without the need for chemical derivatization of the sample. Thus, an FIA-FTIR method for the determination of acetaminophen (paracetamol) in pharmaceutical preparations based on a previous FIA UV–vis method required alkaline hydrolysis of the analyte to *p*-aminophenol prior to measurement.

One other sophisticated flow system integrating several derivatization processes was proposed for the determination of antimony in pharmaceuticals by hydride generation in combination with FTIR spectrophotometry. The process included on-line mineralization/oxidation of organic antimonials present in the sample and pre-reduction of Sb(V) to Sb(III) with $K_2S_2O_8$ and KI, respectively, prior to stibine production.

A prototype FIA system for the characterization of compounds like lidocaine by using a combination of diode array UV, H-NMR and FTIR spectroscopies, in addition to time-of-flight mass spectrometry, was tested on a number of pharmaceuticals and related compounds as models. This instrument combination allowed the on-line acquisition of UV, H-NMR, IR and mass spectra in addition to composition data, thereby enabling nearly complete structural characterization of the target compounds.

21.2.1.4 Flame Atomic Absorption Spectrometry

Because this technique is dealt with at length in another chapter, this section is exclusively concerned with its application to drug analysis. The significance of the FIA-AAS association is clearly apparent from the fact that it was the subject matter of the first book on detection methods used in FIA.

Atomic spectroscopy affords direct measurement of a large number of metal elements; this makes it highly suitable for clinical and forensic analysis, but hardly for drug analysis. As in many other circumstances, however, one can always use an indirect method to facilitate the detection of a variety of drugs. In any case, the AAS technique has never been among the favorite detection techniques for pharmaceutical analysis; possibly, its large number of uses in FIA was a result of the *rush* to automate virtually all types of analytical processes as no similar rush has ever occurred in SIA, a more recent technique. Also, few FIA drug analyses in this field have been conducted with variants other than flame AAS.

Metal ions are quite commonly used as reagents for drugs in classical analytical chemistry for purposes such as precipitation, chelation or redox reaction. In fact, metal ions have allowed a variety of pharmaceutical and non-pharmaceutical organic compounds to be determined via appropriate reactions.

The most common FIA AAS methods fall into either of two categories, namely:

1. Methods with solid–liquid extraction involving some precipitation reaction and measurement of the resulting precipitate after dissolution or the reagent excess.

2. Methods using solid–liquid extraction with immobilized reagents (e.g., metal ions in solid-phase reactors) to form soluble complexes with the target drug or, alternatively, an oxidative or reductive reactor.

21.2.1.5 Liquid–Liquid (Extraction) Systems Involving the Formation of Ion-Pairs or Neutral Chelates with the Analyte

Methods Involving Precipitate Formation

1. *With dissolution of the precipitate.* Methods based on the formation of an insoluble precipitate in a flow manifold differ in whether or not the precipitate is dissolved. In the former case, the precipitate retained on the filter is washed and dissolved, the metal ion in the resulting solution being the monitored species. The determination of chlorhexidine requires the use of a manifold including several valves to switch between various streams in order to dispense appropriate solutions or reagents; the drug is inserted into a carrier containing ammoniacal Cu(II) to obtain a precipitate that is retained in a filtering unit. Then, an ammonia washing solution is passed and the washed precipitate dissolved with a nitric acid stream and driven to the detector. A reversed FIA method has allowed levamisole to be determined by inserting an aliquot of reagent (HgI_4^{2-}) solution into the sample carrier, the resulting precipitate being retained on a filter for washing, dissolution and detection.

2. *Without dissolution of the precipitate.* These methods can be implemented in simpler manifolds. The precipitate is allowed to remain on the filter and the metal ion in it determined from its concentration decrease in the solution reaching the flame.

Lidocaine, procaine and tetracaine can be determined by precipitation with Co(II). To this end, a cobalt solution is inserted into a distilled water carrier in order to obtain a positive peak and then into the sample. A similar system has been used with Cu(II) or Ag(I) at pH 6–7 as precipitant for the determination of five sulfonamides. Nitrofurantoin reacts quantitatively with $AgNO_3$ in ammonia to form a precipitate that can be measured by flow injection on-line filtration. An FIA-AAS system has enabled the determination of urapidil based on the reaction of iodine with BiI_4^- in an acid medium to form an insoluble ion-pair.

In flow systems including a solid-phase reactor, the sample is passed through a reactive bed and the metal ion released by reaction with the analyte gives a transient peak that is used as the analytical signal for measurement. The reactions are usually of the chelation or redox type and reactive beds can be used in simple, robust flow systems typically consisting of a carrier channel for inserting the sample and a column located between the injection valve and detector or, for slow reactions, accommodated within the injection valve.

The aminoacid glycine has been determined by reaction with copper carbonate in the form of finely divided powder that is immobilized by natural means and constitutes the reactor. Natural immobilization can only be used with the few reagents that are sufficiently insoluble, reactive and mechanically stable under a continuous flow. More often, immobilizing the reagent entails using some type of support.

Soluble complexes have been used as the basis for the determination of aminoacids with immobilized copper carbonate, as have redox reactants for that of isoniazid with manganese dioxide or metamizole and ondasentron with lead dioxide. Some redox-based determinations (e.g., those of methadone, chloramphenicol and chlordiazepoxide) use reducing metals such as cadmium or zinc.

Flow-injection graphite furnace AAS has been used for the determination of platinum in salmeterol xinafoate and calcium folinate using oxime, sulfoxime and 2,2N-diaminodiethylamine cellulose microcolumns to preconcentrate the metal following reduction with iodide or sulfite ions.

An FIA AAS method for determining copper in multivitamin tablets uses on-line preconcentration by deposition of the analyte onto a microcolumn packed with 1,5-diphenylcarbazone immobilized on surfactant-coated alumina and subsequent elution by injection of a small volume of hydrobromic acid.

Determining some active principles and ingredients requires separation from their matrix using liquid–liquid extraction. A portion of organic phase containing the extracted analyte is injected into a water carrier to obtain the transient output.

One of the practical requirements with these methods is to adapt the flow rate of the isolated phase to that of the detector nebulizer. The most usual choice for this purpose involves continuously inserting the sample through a channel merging with the carrier-reagent in order to obtain an ion-pair or chelate in a relatively fast reaction along a short piece of tubing that is merged with an organic solvent at a segmenter. Once the two phases are separated, one is driven to the detector. The continuous flow of organic phase is passed through an injection valve and led to the detector with the aid of a water carrier. This type of system has been used for the determination of secondary amines, amphetamine and methylamphetamine by reaction with carbon disulfide in aqueous ammonia, the resulting dialkyldithiocarbamic derivatives forming chelates with copper, nickel or zinc that are extracted with methyl isobutyl ketone (MIBK) along the tubing portion between the segmenter and phase separator.

Bromazepam has been determined by forming an ion-pair between the $(bromazepam)_3Cu^{2+}$ chelate and perchlorate ion and extracting it with methyl isobutyl ketone. The ion-pairing approach has also been used to determine cocaine with various unidentate chelates.

Alternatively to the previous systems and as described above, extraction can be done across a membrane.

21.2.2
Luminescence

Recently, multicommutation assemblies have proved to be very useful for some novel purposes such as the theoretical prediction of the behavior of target compounds in molecular connectivity studies, which rely on the mathematical procedures of molecular topology. Specific equations for predicting the chemiluminescent behavior of organic compounds (drugs and pesticides mainly) with strong oxidants in solution have been developed. Predictions for drugs and pesticides were accurate in 92.7% of cases and the approach was subsequently applied to specific drug and pesticide families with success rates as high as 100% (e.g., for polyphenols). Further applications involved the prediction of photoinduced fluorescence and chemiluminescence, and the effect of potassium permanganate.

21.2.2.1 Flow-Fluorimetry in Drug Analysis

Traditionally, fluorimetry has been among the most frequently used techniques for determining both therapeutic and abuse drugs. Its use has no doubt been fostered by its excellent selectivity and the ability to detect very small amounts of some substances. Also, the fluorimetric technique is the recommended choice for quantifying the purity of active principles.

The proper use of fluorimetry requires considering the potential effects of interactions with the solvent(s), which can alter the strength and position of emission maxima. Such variables as polarity, viscosity and the presence or absence of heavy atoms are extremely important and require careful selection and optimization of the solvent or solvent mixture to be used. Thus, an aqueous solution of quinine emits strongly upon addition of sulfuric acid. Also, the determination of 9-aminoacridine is improved greatly by using methanol instead of water as solvent; in addition, the alcohol is an especially effective solvent for creams and suppositories. This is also the case with proflavin, which is much more readily determined in the presence of dimethylformamide than in pure water. In fact, a number of flow methods allow drugs to be quantified without the need for derivatization, simply by optimizing the influence of various factors on the emission. Thus, FIA has provided good limits of detection for 9-aminoacridine with insertion of the analyte into a 0.1 M HCl solution. Also, fluorescence measurements at two different pH values (6 and 11) have enabled the FIA determination of various coumarins. One method, based on the native fluorescence of ergotamine, has also been used to obtain the dissolution profile for a formulation of this drug in an SIA system. In a limited number of cases, an alternative medium is required owing to the low water-solubility of the analyte.

Similar considerations can be made about the factors influencing the solvent properties. Thus, organized media such as surfactants and cyclodextrin shield fluorophores from unwanted quenching effects of their ionic or molecular environment; in addition, they reduce the need for organic solvents. Warfarin forms an inclusion complex in β-cyclodextrin that raises the luminescence emission and has been used to develop an SIA method for its determination.

One other way of protecting an emissive species from interactions with its environment and increasing its structural rigidity is by immobilizing (adsorbing) the excited species on a solid surface; this is the basis for the solid surface photoluminescence approach, which is obviously more complex than using a homogeneous system in solution. Because the number of substances that are amenable to reversible retention (i.e., ready elution) is relatively small, this technique possesses increased selectivity. Applications to pharmaceutical analysis include the determination of tetracyclines by merging a buffered carrier containing the antibiotic with an Eu^{3+}

solution in the same buffer and retaining the resulting chelate in a sorbent bed packed in the flow cell.

While not widely in use, laser-based excitation sources could be employed to expand the detection potential of fluorimetric drug determinations inasmuch as the fluorescence emission intensity depends on the excitation intensity and the highly collimated beam facilitates control of the excited zone. Flow laser-based fluorimetry has been used to determine compounds such as riboflavin, tryptophan and some vitamins.

As with other techniques, determination methods usually require some prior derivatization of the analyte (e.g., chelation, ion-pairing, redox, enzymatic catalysis, condensation). This frequently entails the use of various approaches including photodegradation, solid or liquid extraction or electrochemical derivatization. Chemical derivatization in this context usually involves the use of an inorganic oxidant. Such a strong reductant as ascorbic acid can be oxidized to dehydro-L-ascorbic acid with mercuric chloride for chelation with o-phenylenediamine; the resulting quinoloxaline is highly fluorescent. The Cu(II)–H_2O_2 couple has been used to determine isoniazid as a solution of cupric ion has been used to determine cysteine and cystine.

A large number of methods use strong oxidants. Thus, hexacyanoferrate(III) immobilized in a solid-phase reactor has been used to determine paracetamol. Also, Ce(IV) and Tl(III) ions have enabled a number of indirect determinations based on the fluorescence of their reduced forms [Ce(III) and Tl(I), respectively]. Ceric ion has been used both in solution and immobilized in solid-phase reactors for the indirect determination of the antihypertensive captopril and the antihistamine diphenhydramine in addition to promethazine, trimeprazine and trifluoperazine. Thallic ion has enabled the joint determination of the aminoacids L-cysteine and L-cystine by their disparate rate of oxidation.

Thiamine has been used for the determination of proteins in serum by FIA, the sample was injected into a buffered carrier containing sodium hypochlorite and the chloroproteins thus formed were merged with a stream of thiamine and sodium nitrite. Following destruction of excess hypochlorite, thiamine was oxidized to thiochrome by the chloroproteins.

Careful selection of the strong oxidant and medium to be used, and elucidation of the influence of the concentration and reaction time, has enabled multicomponent determination based on differences in reaction rate between the analytes. In a recent study, a stopped-flow system allowed up to five tricyclic antidepressants to be determined. The oxidant of choice was potassium periodate at pH 4 and measurement times ranged from 129 to 490 s.

Although fluorescence-based flow methods using a redox reaction are probably the most common for drug analysis, there are effective alternatives based on virtually all types of reactions. Thus, FIA and SIA have been used to determine linosipril with o-phthalaldehyde in the presence of mercaptoethanol at pH 10. Also, boric acid has been determined by chelation with chromotropic acid in an SIA system.

Fluorescent derivatives obtained by methods other than UV irradiation (e.g., electrochemically) have rarely been used in flow analysis. One of the few available examples is the oxidation of thiamine to thiochrome at a potential of +0.4 V against Ag^+/Ag electrode in a flow cell accommodating the required electrodes.

Appropriate irradiation can facilitate the degradation of organic compounds by breaking their bonds via different mechanisms. This constitutes a highly useful procedure for drug analysis as it provides photofragments with properties that differ from those of the parent compound (e.g., higher or lower absorptivity, fluorescence, chemiluminescence, increased electroanalytical activity); this increases not only the sensitivity of the process, but also its selectivity. The use of this technique in flow methods provides substantial advantages as it affords the reproducible detection of intermediate and short-lived photodegradation products. Incorporating a reaction of this type into a flow system is always very easy: it suffices to place the light source into a coiled reactor. No monochromatic lamps or monochromator need be used as continuous-spectrum lamps are quite effective for the intended purpose. A substantial number of fluorimetric and chemiluminescent methods – and a smaller number of UV–vis methods – rely on the prior photodegradation of the analyte to a luminescent species. Some examples are the determination of folic acid by irradiation in an aqueous medium and carbonate/bicarbonate buffer; diclofenac in an SIA system; and, the FIA determination of emetine by irradiation in the presence of barium peroxide.

The advantages of irradiation as a replacement for chemical reagents have promoted an increase in its use. Such advantages include the use of fewer reagents (clean chemistry), the high stability of the light source, an increased simplicity in the flow assemblies required and no need to remove excess reagents. For these reasons, irradiation has been widely used in fluorimetric, chemiluminescent and electrochemical determinations.

While not exclusively associated to fluorimetric determinations, solid–liquid heterogeneous systems are widely used in this context, usually in the form of a solid-phase reactor intercalated at some point in the manifold. Using a solid-phase reactor to confine an insoluble or immobilized reagent somewhere in a manifold and circulating the sample-carrier solution through it has some analytical and operational advantages over the use of reagents in solution.

Such a purpose can simply be the pretreatment of samples, as in the determination of L-cysteine and L-cystine by oxidation with Tl(III). Because L-cystine is very slowly oxidized, the FIA system included a copperized cadmium column intended to effect the oxidation prior to injecting the sample into the carrier.

Solid-phase reactors can also be used to purify reagents *in situ* (with ion-exchange or sorbent columns), generate unstable reagents at the time of use or even dilute a reagent solution as required. Typical examples include the determination of sulfadiazine, sulfamerazine, sulfametoxazol, sulfametoxypyridazine and sulfamidothiazol with *in situ* prepared nitrite reagent obtained by reducing a nitrate solution on a copperized cadmium minicolumn. Using a reactor to dilute solutions avoids most of the imprecision in successive dilutions from the parent solution. This is especially useful in determinations involving the use of some catalyst such as (i) the determination of metamizol by passage through an immobilized Pb(II) reactor, released lead ions subsequently catalysing the reaction of potassium persulfate with Pyrogallol

Red; and (ii) the determination of aminoacids by which an amount of immobilized Cu(II) proportional to that of analyte is released to catalyze the reaction between ferric ion and sodium thiosulfate.

Placing the reactor (whether a single one or several arranged in series or parallel) between the sample injection point and the detector is the usual choice for derivatizing analytes with a view to determining a variety of compounds, and also for speciation and multicomponent determination purposes. The use of immobilized Ce(IV) has enabled the indirect fluorimetric determination of a number of drugs, including fluphenazine, thioridazine and promethazine, by monitoring the fluorescence of the resulting Ce(III).

Finally, derivatization and detection can be integrated by accommodating the reactor in the flow cell for increased sensitivity and selectivity. In this way, pyridoxal has been determined by retaining its complex with beryllium in a conventional flow cell packed with C_{18} silica. Subsequently, the procedure has been used for the joint determination of pyridoxic acid, pyridoxal and pyridoxal 5 N-phosphate.

21.2.2.2 Phosphorimetry

Phosphorescent light was first used for analytical purposes in 1957. The phosphorimetric technique has gained little acceptance except in a few, highly specific fields including drug and pesticide analysis. However, once the need to use very low temperatures was overcome, the popularity of room-temperature phosphorescence in solid supports and organized media has never ceased to grow. The variant using solid supports is especially suitable for flow work by virtue of the ease with which a flow cell can be packed with an appropriate support in order to retain the analytes for measurement. Phosphorimetry is subject to no interference from background or scattered light, nor from Rayleigh or Raman effects. This results in improved signal-to-noise ratios and hence in better limits of detection.

One example of increased selectivity is the determination of nafcillin; although few β-lactam antibiotics are phosphorescent, nafcillin possesses a naphthalene group at position 6 that confers luminescent properties.

The use of micelle-stabilized room-temperature phosphorescence in a stopped-flow system has enabled the kinetic determination of various drugs including nabumetone; naphthylacetic acid in formulations, soil and fruits; and that of nafronyl in formulations. The respective analytical systems were optimized for the most influential variables including surfactant, pH adjusting reagent and emission enhancer. The determinations were based on the respective kinetic curves, and the maximum slopes of the curves were used as analyte signals.

21.2.2.3 Chemiluminescence

Chemiluminescence can be defined as a phenomenon in which electromagnetic radiation is produced by a chemical (usually redox) reaction where some product is in an excited electronic state and releases part or all of its excitation energy as radiant energy upon return to its ground state.

Possibly, the principal advantage of chemiluminescence methods over other emission-based methods is that it requires no excitation lamp and is thus free of background noise. This results in increased simplicity, robustness and cost-effectiveness.

Early chemiluminescence determinations were developed by using a few reagents (luminol, lucigenin, lophine, oxalate esters) that were made chemiluminescent by oxidation with a strong oxidant. Rather than a reactant, the analyte played the role of an inhibitor, enhancer or catalyst for the chemiluminescent reaction. An increasing trend exists at present, however, to use the analyte as the species producing the chemiluminescence upon oxidation. Direct methods usually provide a higher sensitivity and lower detection limits than do indirect methods. Both have been exploited widely in drug analysis (particularly in FIA and SIA).

The direct CL methods involve reacting the analyte with a strong, usually inorganic, oxidant such as potassium permanganate – which is the most common and efficient of all –, potassium ferricyanide, Ce(IV), tris(2,2*N*-bipyridine)ruthenium(III), hydrogen peroxide, oxygen or *N*-bromosuccinimide or, less often, bromide or sodium hypochlorite. The emitted light intensity is strongly influenced by experimental variables including temperature, pH and the presence of organized media (surfactants, cyclodextrins).

Chemiluminescence-based determinations involving oxidation with potassium permanganate in a strongly acidic (sulfuric or polyphosphoric) medium exceed all others jointly in number. This is a result of the emitting species being one of the reduced forms of permanganate; so far, such a species has been hypothesized to be Mn(II), some intermediate complex with the analyte, Mn(III), Mn(IV) [or even Mn (VII) in phosphate solutions]. Whether sulfuric or polyphosphoric acid is the better choice is also a controversial subject; comparative studies have provided no clear-cut conclusions in this respect. Polyphosphoric acid appears to play a dual role in the process; thus, in addition to providing the acid strength needed, it stabilizes intermediate species formed in the reduction of permanganate, similarly to polyphosphate ion in a sulfuric medium.

Salbutamol has been determined by using an SIA system with integrated solid–liquid extraction. The analyte was adsorbed in silica gel modified with a carboxylic acid to facilitate separation from the matrix and subsequently rendered chemiluminescent by oxidation.

An SIA study of the kinetics of oxidation of procaine, benzocaine and tetracaine by potassium permanganate in aqueous sulfuric acid exposed differences in the respective emission-time profiles. A similar SIA system showed the sensitivity of the determination of sulfonamides with permanganate to be increased by the presence of glutaraldehyde. Other pharmaceutical active principles determined with permanganate in SIA include trimethoprim and promethazine.

An extensive scan conducted with a view to developing FIA CL-based methods for determining pharmaceuticals led to a total of 97 drugs of diverse molecular structure being subjected to on-line photodegradation tests with the lamp on and off. The results revealed the ability to determine a variety of drugs including sulfonamides, thiazides, nicotinamide, nortriptyline, levamisole and phenylbarbituric acid, among others, in a direct manner (i.e., without irradiation), and also to allow other drugs possessing no native chemiluminescence (e.g., chloramphenicol, dextromethorphan, riboflavin, ephedrine, piperazinamide, chlortrimazol, theophylline) to be made chemilumines-cent by irradiation.

A direct CL-based procedure for the determination of hydroquinone in various types of samples including pharmaceutical formulations, has been developed by using the emerging flow methodology known as multicommutation. The manifold consisted of a set of three channels and three solenoid valves, and the determination was performed at 60 °C. The chemical process was the oxidation of hydroquinone with sulfuric potassium permanganate. Light emission was clearly increased by the presence of quinine sulfate and benzalkonium chloride.

The oxidant tris(2,2N-bipyridine)ruthenium(III), which is an unstable species requiring fresh preparation from tris(2,2N-bipyridine)ruthenium(II), is frequently used in CL-based determinations. A variety of methods have been proposed to obtain active Ru(bipy)$_3^{3+}$ by oxidation of Ru(bipy)$_3^{2+}$, whether chemical [with acid Ce(IV) or permanganate], photochemical or electrochemical. The Ru(bipy)$_3^{3+}$ chelate is used to oxidize the analyte and produce the emitting species, the excited species probably being [Ru(bipy)$_3^{2+}$]*.

Very recently, indomethacin in semi-solid dosage forms was determined by oxidation with tris(2,2N-bipyridine)ruthenium(III) generated on-line by oxidation with sulfuric Ce(IV).

Irradiating Ru(bipy)$_3^{2+}$ in the presence of peroxydisulfate ions in an SIA system produces the oxidized species Ru(bipy)$_3^{3+}$, which can in turn oxidize L-cysteine and L-cystine. Because the product of the latter is non-chemiluminescent, a reductive column is used to previously covert it into L-cysteine and measure the combined signal of the two analytes.

The antibiotic cefadroxil has been determined by reaction with electrochemically produced Ru(bipy)$_3^{3+}$ in an SIA-CL system. The flow-cell contained a Pt working electrode, an AgCl reference electrode and a steel needle as auxiliary electrode. The oxidation of Ru(bipy)$_3^{2+}$ produced unstable Ru(bipy)$_3^{3+}$ on the surface of the Pt electrode which was used to oxidize the analyte.

Other direct CL-based oxidation systems include an FI-CL method for the determination of tryptophan that uses the strong chemiluminescence of the ami-noacid in a medium containing hydrogen peroxide, nitrite and sulfuric acid. The chemiluminescence produced was assigned to the peroxidation and epoxidation of tryptophan by peroxynitrous acid, and subsequent decomposition of the resulting dioxethane.

The chemiluminescent reaction of pentoxyverine citrate with the NaClO/H$_2$O$_2$ system was studied in a flow injection assembly. A new FI-CL system for the determination of propranolol hydrochloride by reaction with Ce(IV)/rhodamine has also been reported. Sodium metamizol has been determined by using the enhancing effect of formaldehyde on the chemiluminescence produced by oxidizing the analyte with dissolved Mn(IV). A method for the determination of pipemidic acid based on energy transfer from the excited state of peroxynitrous acid – which was synthesized on-line by mixing acid hydrogen peroxide with nitrite in a flow system – to the analyte has also been reported; the chemiluminescence was found to be produced by two excited states of pipemidic acid. The reaction of Rhodamine 6G with Ce(IV) in a sulfuric acid medium containing puerarin produces strong chemiluminescence which was used to develop a sensitive, selective method for determining the drug.

Worth special note here are electrochemiluminescence-based analyses, which exploit a chemiluminescent reaction occurring in the vicinity of an electrode to obtain an active oxidant from a passive precursor. This methodology has some disadvantages, including easy fouling of the electrode, poor repeatability, too short linear ranges because of the small electrode area involved, and the need for a complex flow cell design. The electrogenerated oxidant obtained can be used to oxidize well-known chemiluminescent reagents such as luminol or hydrogen peroxide, or employed to generate oxidants for the analyte. Various unstable oxidants including Mn(III), Ag(II), Co(III), $[Cu(HIO_6)_2]^{5-}$ have so far been *in situ* generated from $MnSO_4$, $AgNO_3$, $CoSO_4$ and $Cu(NO_3)_2$ in KIO_4, respectively.

The widest used indirect CL-based method is based on the oxidation of 5-aminophthalylhydrazide (luminol) in an alkaline medium. Luminol can be oxidized to 3-aminophthalate (the excited species) with various oxidants including potassium permanganate, periodate, dichromate, persulfate, chlorate, hydrogen peroxide, *N*-bromosuccinimide (or *N*-chlorosuccinimide), dichlorocyanurate (or trichlorocyanuric acid) and electrogenerated hypobromite; however, the most common choice by far for this purpose is potassium ferricyanide. This oxidation reaction is catalyzed by a large number of metal ions and organic compounds. The analyte (drug) can also play various roles including those of inhibitor, enhancer and catalyst chelating agent. For example, Cu(II) chelates have been used for the SIA determination of methimazole and carbamizole, and so have Co(II) chelates for that of diethylstilbestrol – with cobalamin, it suffices to acidify the medium in order to release cobalt ion (the catalyst).

A number of FIA systems have been used to determine an active principle by its inhibitory or enhancing effect on luminol chemiluminescence. Thus, a reversed FIA-CL system was recently used to determine dopamine hydrochloride from its strong quenching effect on the reaction between luminol and hexacyanoferrate(III) in an alkaline medium. Analgin was also found to sensitively inhibit the chemiluminescent reaction of luminol with alkaline $K_3Fe(CN)_6$; the decrease in CL intensity was proportional to the concentration of analgin and used to develop a new FIA method for the drug. A simple, sensitive method for determining amikacin sulfate uses CL emitted in the Cu(II)-catalyzed oxidation of luminol by H_2O_2; the drug forms a stable complex with the catalyst and thus interacts with the analytical reaction.

Chemiluminescence enhancers have also been widely used in flow techniques. Thus, light emission in the oxidation of luminol by potassium ferricyanide in an alkaline medium was found to be enhanced by phentolamine. Also, an FI-CL method for determining dopamine hydrochloride exists based on the chemiluminescent reaction of luminol with the KIO_4–tannic acid system. A CL-based method for the determination of acetylspiramycin has been developed from the greatly enhancing effect of the drug on the chemiluminescent reaction of luminol with hydrogen peroxide.

A sensitive FI CL-based method for the antipsychotic risperidone uses its catalytic effect on the reaction of luminol with hydrogen peroxide. The drug increases the emitted light intensity in proportion to its concentration.

An FI-CL system has been used for the simultaneous determination of ascorbic acid and L-cysteine with partial least-squares calibration. The ensuing method relies on the ability of both compounds to quantitatively reduce Fe^{3+} to Fe^{2+} at a different rate. The resulting ferrous ion is detected with the luminol/O_2 system; thus, the amount of light emitted at different times of reaction of ferric ion with ascorbic acid and cysteine is measured and processed using partial least-squares methodology.

Luminol can be oxidized electrochemically, e.g., for the determination of isoniazid and norfloxacin with dissolved oxygen forms a superoxide radical which can oxidize luminol (e.g., in the determination of β-glucose); the extent of oxidation is strongly dependent on the reaction conditions.

The use of liquid–liquid and solid–liquid heterogeneous systems has proved effective with a view to increasing the selectivity of the luminol reaction. One case in point is the extraction of chlorpromazine with dichloromethane following ion-pairing with tetrachloroaurate(III) in an organized medium (cetylmethylammonium bromide micelles). Solid–liquid extraction has been used to retain luminol and its oxidant (ferricyanide, dichromate, periodate) separately in two different resins packed jointly in a minicolumn placed on-line in front of the detector flow cell.

The number of SIA-CL and, especially, FIA-CL methods reported to date is overwhelmingly greater than that of other CL-based flow methods. Among the latter few is *multipumping* for chemiluminometric inhibition. The hypoglycemic drug metformin has been determined by its inhibitory (scavenging) effect on Cu(II) ions, which catalyze the chemiluminescent reaction between luminol and hydrogen peroxide.

Also worth special note here are analytical systems based on the oxidation of sulfite ion with strong oxidants. The reaction produces light of low intensity which, however, can be increased by oxidation with an organic compound; the reduced species is probably sulfur dioxide which acts as an intermediate by transferring its excitation energy to a neighboring chromophore. Quinolone grapefloxacin produces low-intensity chemiluminescence that is dramatically enhanced by the presence of a lanthanide; the excitation energy of sulfur dioxide is absorbed by the quinolone-lanthanide complex formed.

As in fluorimetric determinations, the use of on-line light sources has proved highly effective in chemiluminescence-based determinations. The assemblies used for this purpose are quite simple and have enabled a variety of determinations such as that of sulfonamides, the chemiluminescence of which with potassium permanganate is strongly increased by the presence of a on-line source; that of chloramphenicol with the previous oxidant; and that of cephalosporins with permanganate in the presence of glyoxal to accelerate the reaction. The ensuing method has also been used to obtain dissolution profiles for commercially available formulations of orally administered solid sulfamethoxaloze. Lactate has also been determined in this way and so has vitamin K_3, both using the luminol reaction.

The potential of molecular imprinting processes (MIP) in continuous flow analytical methodologies has been examined recently with a view to improving the selectivity and robustness of CL-based determinations. The process involves the copolymerization of functional and cross-linking monomers in the presence of the target analytes, the subsequent removal of the latter leaving binding sites that are complementary to the target analytes in the resulting molecularly imprinted polymer. In FI systems, particles of the MIP thus obtained are used for packing into a flow cell. This technique is deemed a simple, robust choice for specific molecular recognition, and has been used to determine adrenalin among other compounds.

21.2.3
Electrochemistry

21.2.3.1 Conductimetry

Flow conductimetric determinations, which are not very common – and even less so in pharmaceutical analysis – rely on various types of cells. The few exceptions include a method for the determination of ascorbic acid in which a portion of sample was injected into a 5 mM NH_3 carrier stream and the mixture passed through a flow cell in order to measure the resulting increase in conductivity. In one other method, portions of acetylcholine standard solutions were injected into a carrier of phosphate buffer at pH 7.4 and passed through a polyethylene tube containing acetylcholinesterase immobilized on glass beads. The solution was then merged with a stream of H_2SO_4 and passed through a PTFE membrane diffusion cell. Acetic acid diffusing through the membrane was driven to a conductimetric cell in a water carrier. The ensuing method was used to determine acetylcholine in a pharmaceutical formulation.

21.2.3.2 Potentiometry

Potentiometric techniques provide various advantages, including not only instrumental simplicity, but also selectivity, sensitivity, responsiveness and the virtual absence of chemical reactions preceding analytical detection.

Unlike batch potentiometric detection, the analyte–electrode contact time in continuous flow potentiometry is constant and precision invariably better as a result. Also, the kinetic nature of flow measurements results in improved selectivity by virtue of the ability to measure the response of the analyte with no interference from other active principles or excipients. In addition, the short contact time between the analyte and electrode surface minimizes the risk of poisoning and prolongs the life span of electrodes.

Ion-selective electrodes are effective alternatives to other, expensive, labour-intensive choices for drug analysis in pharmaceutical formulations and complex matrices. These electrodes require no complicated sample treatments – the sample color or turbidity has no adverse effect. Frequently, all they require is adjusting the pH and ionic strength of the medium. More detailed information about potentiometric detectors is given in Chapter 15.

Ion-selective electrodes can be used in two different modes depending on whether they are sensitive to the analyte or to some other species (i.e., whether the target compound can be determined in a direct or indirect manner).

Direct determinations. A tubular electrode consisting of epoxy resin and graphite has been proposed for the determination of barbital and phenobarbital in tablets and solutions. The determination method is based on the retention of a quaternary ammonium salt dissolved in *o*-nitrophenyl ether immobilized on PVC. A similar system was used to determine benzoate in five different pharmaceutical formulations. Other electrodes based on membrane-supported benzoate ion-pairs have also been used for this purpose (e.g., one based on a tridodecylmethylammonium salt to determine cholinesterase in blood serum and another using tetraheptylammonium ion for amides).

Vitamins B_1 and B_6 have been determined by using a tubular selective electrode with no reference solution. The electrode membranes were prepared from vitamin-tetra(2-chlorophenyl)borate dissolved in *o*-nitrophenyl octyl ether for immobilization onto PVC. Vitamin B_6 in drug formulations has also been determined by using an ion-selective electrode in a multicommutated flow system. The manifold included a three-way solenoid valve coupled to the switching valve and facilitated the automation of various procedures using potentiometric detectors.

Potassium clavulanate in formulations was determined in an SIA system including two potentiometric detectors one for clavulanate and the other for potassium. This configuration allowed the accuracy of the results and clavulanate degradation to be assessed, and also isolated faults in the flow system to be detected. The underlying principle was subsequently applied to the determination of chloride ion by using an electrode and optrode based on the same ionophore.

Dicyclomine hydrochloride in pharmaceutical preparations has been determined by using FIA in combination with various plastic membrane electrodes consisting of dicyclominium-silicotungstate, silicomolybdate, phosphotungstate, phosphomolybdate and tetraphenylborate ion associations dispersed in PVC matrices with dibutyl phthalate plasticizer that were characterized in terms of composition, life span, usable pH range, working concentration range and temperature. A similar approach was performed to determine drotaverine hydrochloride using drotaverinium-silicotungstate, silicomolybdate, phosphotungstate, phosphomolybdate or tetraphenylborate ion associations dispersed in a PVC matrix with dibutyl phthalate plasticizer. The approach employed for the determination of drotaverinium ion was also employed in the determination of mebeverine hydrochloride with mebeverinium-silicotungstate, silicomolybdate, phosphotungstate or phosphomolybdate ion associations dispersed in the same plasticizer.

Selective electrodes in combination with multicommutation were used for the determination of acetaminophen in various formulations. The manifold included an enzymatic reactor to hydrolyze acetaminophen to 4-aminophenol. The ensuing method provides several advantages over existing choices for the determination of this analyte in complex matrices such as human serum – which pose no interference with selective electrodes.

Indirect determinations. An extremely large number of selective electrodes for determining drugs is currently available. However, drugs can also be determined by derivatizing the analyte and monitoring some species – whether a reactant or reaction product – other than the target drug. This is a widely used approach in classical pharmaceutical analysis.

1-Fluoro-2,4-dinitrobenzene (FDBN) is a selective reagent for various functional groups including those in amines, aminoacids, phenols, thiols, hydrazines, hydrazides and azides; this has facilitated the determination of a number of drugs by using kinetic potentiometric methods. Detection is based on a fluoride-selective electrode that is used to measure the amount of fluoride ion released in the reaction between FDBN and the drug concerned.

Isoxsuprine was determined in an FIA system also by using the reaction of the drug with FDBN and monitoring the F^- ions released with a fluoride-selective electrode. The reaction was catalyzed by cetylmethylammonium bromide.

A carbon paste electrode for drotaverine hydrochloride was constructed and fully characterized for composition, life span, usable pH range, response time and temperature prior to use in the FIA potentiometric determination of drotaverinium ions in pharmaceutical preparations and biological fluids. The electrode was based on a mixture of two ion exchangers (viz. drotaverinium-silicotungstate and drotaverinium-tetraphenylborate dissolved in tricresyl phosphate as pasting liquid). This modified carbon paste electrode was used for end-point indication in the potentiometric titration of drotaverinium ions with various titrants.

Worth special note among potentiometric analyses are those involving the monitoring of pH changes with a glass electrode. In the determination of penicillin V with a solid-phase reactor containing immobilized penicillinase, the hydrolysis of the β-lactam ring produces an acid that is detected with a glass electrode; quantitation is based on the difference between the signals provided by a sample reaching the detector as such and another previously passed through the reactor.

Glass electrodes can be modified for use as enzyme electrodes by immobilizing penicillinase; to this end, the enzyme is cross-linked with a thin film of glutaraldehyde that is sprayed onto the surface of two different glass electrodes.

Total penicillin V and penicilloic acid have been determined jointly by using three different procedures based on (i) the reduction of molybdoarsenic acid to Molybdenum Blue by penicilloic acid in the presence of mercuric chloride and spectrophotometric detection of the dye formed; (ii) an FIA adaptation of the classical procedure for measuring penicilloic acid where iodine is reduced to iodide ion and the absorbance of the iodine-starch compound decreased as a result; and (iii) measuring the pH change caused by the hydrolysis of penicillin to penicilloic acid. In the three procedures, penicillin is quantitatively converted into penicilloic acid in a solid-phase reactor containing β-lactamase.

Finally, metal electrodes have also been used for potentiometric measurements. Thus, pre-oxidized nickel electrodes have facilitated the determination of alcohols, aminoacids and carbohydrates, among others.

21.2.3.3 Polarography

Historically, voltammetry was assumed to derive *polarography*, an electroanalytical technique developed by J. Heyrovsky in 1920. At present, polarography is regarded simply as a branch of voltammetry. The main trait and greatest difference from other electroanalytical techniques is that it uses a mercury drop electrode as a working electrode. Also, it differs in the absence of convective mass transfer and, hence, in the fact that limiting currents are exclusively diffusion-controlled.

Polarographic measurements are strongly dependent on temperature and the presence of dissolved oxygen. As a result, the simplest possible polarographic system for use in continuous flow work requires a highly precise thermostat. The oxidation of oxygen can be shifted to more negative potentials by using an appropriate surfactant or a differential technique can be used to suppress the unwanted signal.

The mercury drop electrode was difficult to integrate in an FIA system; hence the small number of FIA polarographic determinations developed to date, which includes those of penicilloic acid, isosorbide dinitrate and allopurinol in pharmaceutical formulations.

In the polarographic determination of allopurinol, the oxidation of metal mercury is accompanied by the formation of highly insoluble chelates of general formula $HgL_x)^{2-xn}$, where x can be 1 or 2. The polarogram for allopurinol in phosphate buffer contains two signals, one of which is due to the formation of a thin film of insoluble material that inhibits the electrode reaction. The sample treatment required is very simple: it suffices to powder tablets, dissolve the powder in an alkaline medium, filter and dilute for injection into a borate buffer carrier. Benzylpenicilloic acid is a degradation product of penicillins; therefore, it can be used to determine penicillin purity (penicillin is polarographically inert whereas penicilloic acid gives an anodic wave). Based on the polarograms for various penicilloates from the hydrolysis of ampicillin, cloxacillin, carbenicillin and phenoxymethylpenicillin, all give a similar anodic wave with a slightly different half-wave potential. Because none of these penicillins is polarographically active, the response obtained from their hydrolyzed solutions can only be ascribed to the presence of their respective penicilloic acids. Based on this knowledge, a continuous flow assembly was used to deliver mercury drops frontally to the carrier and sample inlet. The combined use of a solid-phase reactor and polarographic detection was the basis for an FIA determination of ascorbic acid. The manifold included a reactor containing ascorbate oxidase immobilized on Sepharose that was used to measure the amount of oxygen produced by enzymatic oxidation of the acid.

An SIA system for determining vitamin B_2 in tablets based on its adsorption on a mercury drop electrode has been developed and additionally used to monitor the photodegradation of the vitamin in aqueous solutions.

21.2.3.4 **Amperometry**

Voltammetry encompasses a body of electroanalytical techniques which provide analytical information from current–potential curves. Careful examination of such curves has enabled large families of electroanalytical determinations which can be classified into two broad categories according to whether they rely on potential scans (*voltammetries*) or use a constant potential (*amperometries*). Most electroanalytical work in continuous flow systems has focused on cell design, miniaturization and configuration, and, especially, on the development of new, innovatively shaped, miniaturized electrodes with surfaces consisting of special materials. Most FIA and SIA electroanalytical systems are very simple and require no derivatization of the sample prior to detection. Some, however, do require chemical derivatization; for greater details please see Chapter 16. Carbon-based electrodes, which encompass both carbon paste and glassy carbon electrodes, are among the most popular amperometric detectors, judging by the large number of uses their highly reproducible operation have provided.

Cyanide and hydroxylamine are two decomposition products of formulations containing pralidoxime salts, which are used against anticholinesterase poisoning. Both products have been determined by using a cell accommodating a silver electrode, Ag/AgCl as reference and glassy carbon as auxiliary electrode. Following separation of HCN by diffusion, the acid is driven to an amperometric detector via a separate channel. Hydroxylamine is determined by oxidation with triiodine, the resulting nitrite being measured colorimetrically.

Metal electrodes are less common than carbon electrodes here. Aminoacids and peptides can be determined by oxidation in two ways with a copper electrode. In a neutral or slightly alkaline solution at about 0.0 V vs. Ag/AgCl, the reaction involves the formation of aminoacid-Cu(II) chelates with the ions from the anodic solution of copper on the electrode surface.

Metal electrodes include those based on other species (oxides) and combinations of several metals. An antimony-doped tin oxide electrode for the electrochemical detection of tricyclic antidepressants (imipramine and desipramine) was constructed and tested in FIA systems with cyclic voltammetric and differential pulse voltammetric detection. Other authors have used a nickel oxide electrode to determine caffeine and salicylic acid. The initial electrode was a Ni wire, onto which a layer of nickel oxide was electrolytically formed.

So-called *chemically modified electrodes* have helped alleviate some classical problems in amperometry and expand the use of electrochemical detectors in flow systems as a result. These electrodes possess specific chemical functions that are deliberately bonded to their surfaces. A number of polyhydroxyl compounds undergo electrocatalytic oxidation on an electrode prepared by depositing a cuprous salt on a glassy carbon surface. This type of electrode has been tested with a variety of substances including carbohydrates (monosaccharides, disaccharides and oligosaccharides), amino sugars, alditols, antibiotics (streptomycin sulfate, kanamycin sulfate, digitoxin and erythromycin), and aldonic, uronic and aldaric acids. The significance of chemically modified electrodes for flow detection derives from the phenomenon of electrocatalysis. In fact, these electrodes can catalyze the oxidation or reduction of solutes exhibiting high overpotentials against conventional electrodes. This is facilitated by the ability of some surface-bound redox mediators to accelerate the electron transfer and lower the working potential. One of the most common ways of modifying the activity of conventional electrodes is by incorporating the biocatalyst into a carbon paste matrix. Carbon paste electrodes modified by addition of cobalt phthalocyanine have been found to decrease by several hundred millivolts the potential required for the electro-oxidation of various species and proved useful for determining sulfydryl compounds such as cysteine, *N*-acetylcysteine and homocysteine. An immobilization approach incorporates the enzyme into silicone grease that is then used to fill the micropores of a graphite surface; this provides a layer of silicone lying very close to sensitive sites in the graphite. Tyrosinase immobilized in this way has been used in the FIA determination of various phenol compounds.

One common way of binding an enzyme to an electrode is by placing a fixed membrane in the vicinity of the electrode surface. Thus, theophylline oxidase entrapped on a platinum disk electrode enabled the amperometric determination of theophylline in clinical samples.

The electrode selectivity and sensitivity can be further improved by expanding the previous approach with an additional variable: *permselectivity*. To this end, the electrode is coated with an appropriate membrane capable of selecting the specific components to be allowed to reach the electrode or hindering rapid diffusion of the analyte. In addition to being scarcely prone to poisoning, coated electrodes typically exhibit wider useful ranges than their uncoated counterparts, as shown by a comparative study of biological fluids containing a mixture of analytes such as acetaminophen, dopamine and perphenazine. A glassy carbon electrode coated with a diacyanobis(1,10-phenanthroline)iron(II) membrane enabled the FIA amperometric determination of acetaminophen in physiological samples (urine) with no interference from ascorbic or uric acid. As regards permselectivity, Nafion facilitates the exclusion of anions usually present in urine (e.g., ascorbate, urate). Nafion coatings have also been used to study electrode selectivity and stability for cationic drugs with promethazine as model compound.

21.2.3.5 Continuous Flow Voltammetry

Continuous flow voltammetric studies have been conducted in FIA since 1980. For example, dopamine was determined in the presence of ascorbic acid by inserting samples into a coulometric flow-through detector furnished with a porous reticulated vitreous carbon working electrode, where the flow was stopped in order to facilitate trapping of the products of the initial electrolysis by the electrode and regeneration of dopamine to obtain the analytical signal. The efficiency of this procedure relies on the irreversible oxidation of ascorbic acid and on the quasi-reversible behavior of dopamine.

An SIA system for adsorptive stripping voltammetric measurements has allowed the determination of various substances such as riboflavin by absorption at a mercury electrode in a simple manifold using a sample aspiration device.

A voltammetric method has enabled the determination of nineteen drugs from ten different families. The analytes were subjected to cyclic voltammetry and the results compared with those obtained with amperometric measurements, the latter proving more sensitive and reliable for determining traces of the analytes.

Because background currents (baseline) can form by the effect of the release of hydrogen, reduction of dissolved oxygen or oxidation of other solvents, among others, avoiding them entails deaerating samples prior to analysis. This approach has been assessed in terms of reproducibility, linearity and detection limit in the determination of various drugs including chlorpromazine, acetaminophen and norepinephrine.

The metal ions Cu and Zn in a multivitamin formulation were determined by dissolving the tablets in nitric acid. The voltammetric scan was performed following electrodeposition at 0–0.8 V with HNO_3 or KNO_3 as the supporting electrolyte.

21.2.3.6 Continuous Flow Amperometry

The amperometric technique measures the current intensity produced by the oxidation or reduction of an analyte upon application of an appropriate voltage. Amperometry is therefore especially similar to coulometry, the sole major difference between the two being the proportion of analyte that is oxidized or reduced. Amperometric measurements require the use of a cell accommodating the working, reference and auxiliary electrode, the reaction taking place in the first. The potential of choice is selected from the voltammogram; in order to maximize the sensitivity and selectivity, the chosen current usually coincides with the limiting current for the analyte. The auxiliary electrode avoids the passage of current through the reference electrode and hence unwanted changes in it or the applied voltage. Ensuring that the detector will not respond to other sample components entails carefully choosing the operating conditions. Thus, if the analyte is the sole electroactive (electro-oxidizable or electro-reducible) species present in the sample (e.g., in drug analyses), measurements will be virtually specific. However, if any electroactive species other than the analyte is present and its oxidation or reduction potential falls within the applied potential window, then the target species cannot be determined free of interferences.

Amperometry is a very common choice for detection in continuous flow systems by virtue of its providing low detection limits in addition to substantial selectivity as long as an appropriate working potential is used. One of the earliest uses of amperometric detection in FIA was for the determination of ascorbic acid, epinephrine and L-dopa with a reticulated glassy carbon flow-through electrode that was operated in both the amperometric and coulometric modes. The amperometric determination of ascorbic acid was conducted at $+0.19$ V vs. SCE in acetate buffer at pH 5.5. In one other application, ascorbic and uric acids were determined with spectrophotometric detection at 293 nm for the former and amperometric detection at $+0.6$ V for both.

An SIA system has enabled the simultaneous determination of L-thyroxine, D-thyroxine and L-triiodothyronine, the former with an amperometric biosensor and the latter two with immunosensors constructed from chemically modified carbon paste containing them.

Penicillins have been determined by *pulsed amperometry* on a gold electrode, using Ag/AgCl as reference and Pt as counter-electrode. The sensitivity was maximal in a strong acid medium, where, however, chemical stability was inadequate. An acetic-acetate buffer provided improved stability at the expense of slightly lower sensitivity.

Biamperometry is an amperometric mode involving the use of two polarized electrodes to measure the current intensity passing through them upon application of a low voltage (10–500 mV). In flow systems, this can be accomplished by using two identical Pt electrodes immersed in the flow. If the solution contains the oxidized and reduced forms of a reversible (or quasi-reversible) redox couple at the applied potential, then the reduced form can be oxidized and the oxidized form reduced. This behavior has been observed in redox couples including Br_2/Br^-, I_2/I^-, Ce(IV)/Ce(III), Ti(IV)/Ti(III), VO_3^-/VO_2^+, Fe(III)/Fe(II) and $Fe(CN)_6^{3-}/Fe(CN)_6^{4-}$. Thus, the $Fe(CN)_6^{3-}/Fe(CN)_6^{4-}$ couple has been used to determine reducing sugars in syrups by oxidizing the analyte with hexacyanoferrate(III) in an alkaline medium at a high temperature. Detection was done with a cell furnished with two Pt electrodes polarized at 200 mV. Sucrose required prior hydrolysis to a carbohydrate by injecting 1 M HCl; once the sugar was hydrolyzed, the acid was merged with the oxidant in a strongly alkaline medium.

The reversibility of the Fe(III)/Fe(II) couple in an HCl medium has been exploited for the FIA biamperometric determination of the iron content in multivitamin formulations with Pt electrodes.

Indirect biamperometric methods rely on a redox reaction of the analyte, excess reagent and the reaction product providing the oxidized and reduced form, respectively, of the reversible redox couple needed. The indirect biamperometric determination of promethazine and thioridazine revealed Fe(III) to be a more effective oxidant for the intended purpose than were $Fe(CN)_6^{3-}$, Ce(V), VO_3^-, Ti(IV), I_2 and Br_2. Detection was performed at 20 °C, using a voltage of 150 mV between the two working electrodes.

The iodine-azide reaction is readily induced by elemental sulfur and sulfur-containing compounds. Induction times vary strongly between compounds; thus, while some compounds have induction times as short as 30 s, or even less, in others such as disulfides (cystine, vitamin B_1, sulfathiazole) the induced reaction only starts after the azide slowly succeeds in cleaving C−S or S−S bonds in the presence of triiodide and takes several hours to complete. Compounds with short induction times can be determined in an FIA system by injecting the sample solution into a carrier containing the iodine-azide mixture and monitoring iodine consumption biamperometrically.

21.3
Flow Process Analyzers in the Pharmaceutical Industry

Chemical and biochemical reactions can be monitored via a number of variables. Flow methods provide an ideal means for monitoring industrial production and quality control processes. The pharmaceutical industry has devoted much effort to developing control procedures – mainly of the FIA and SIA types – for obtaining useful information in real time with a view to improving consistency, quality and throughput in production processes. Flow methodologies avoid, or at least minimize, the need for sample preparation. This facilitates the rapid acquisition of the data needed to assess the status of a production process at any time. Also, the ability to simultaneously determine several variables allows greatly improved information relative to officially endorsed methods to be obtained. Ever since continuous flow methodologies were devised, the interest in developing robust, efficient systems allowing samples to be directly treated in the reaction medium and several analytes to be determined at once has never ceased to grow.

21.3.1
Process Analysis in Pharmaceutical Production

As can be seen in Table 21.1, flow systems have also been used to monitor the production processes for some drugs. Thus, SIA has proved effective for monitoring

Table 21.1 Evaluation of the production process.

Analyte	Sample	Detection	Linear range R.S.D. (%)	Throughput (h^{-1})/assay interval (h)	Ref.
Penicillin, glucose, Lactic acid	*P. chrysogenum*	CL	$0.01–1.200\,g\,L^{-1}$ $0.01–7.000\,g\,L^{-1}$ $0.005–5.000\,g\,L^{-1}$ 2.06; 2.56; 2	30/160 20/160 20/160	[1]
Penicillin, glucose	*P. chrysogenum*	CL UV–vis spectrophotometry	$0.1–1.8\,g\,L^{-1}$ $0.01–7.000\,g\,L^{-1}$	$-/425$ $-/350$	[2]
Morphine	Aqueous extracts of *P. somniferum*	CL (250–600 nm)	$2.5 \times 10^{-6}–3 \times 10^{-6}\,M$ 1.4	100/144	[3]
Morphine	No-aqueous extracts of *P. somniferum*	CL (250–600 nm)	0.001–0.1% w/v 6	120/144	[4]
Cimetidine Bromazepam Diclofenac and residues	Residual water samples	Mass spectrometry	—	—	[5]

the fermentation process involved in the industrial production of penicillin. The specific system used was devised for monitoring three major parameters (viz. glucose, penicillin and lactic acid) during cultivation of the filamentous fungus *Penicillium chrysogenum*. Glucose, which acts as a source of carbon and energy, must be maintained at controlled levels during the production stage. Lactic acid is a major controlling factor for the penicillium culture inasmuch as the target product is the antibiotic. Glucose and lactic acid were determined by using the enzymes glucose oxidase (GOD) and lactate oxidase (LOD) immobilized in solid-phase enzymatic reactors in order to catalyze the oxidation of β-D-glucose and lactic acid to β-D-glucono-δ-lactone and pyruvate, respectively. Both reactions produce hydrogen peroxide that was detected by the chemiluminescence generated upon addition of luminol and $K_3FE(CN)_6$ to the reaction medium. Penicillin was determined by hydrolysis to penicillinic acid in a reactor containing immobilized penicillase as catalyst. The resulting penicillinic acid was reacted with iodine, which decreased the chemiluminescence generated in the reaction between iodine and luminol. The original system was subsequently modified for the on-line determination of glucose and penicillin using the previous indirect methods in addition to the reaction of penicillinic acid with iodine and subsequent measurement of the absorbance decrease in the iodine-starch complex for the antibiotic. The system was reliable enough to allow the accurate determination of penicillin after more than 400 h of fermentation.

Two SIA systems have been proposed for monitoring and controlling the industrial production process for morphine, which involves a series of extractions of the plant *Papaver somniferum* with aqueous and non-aqueous solvents. Concentrations were determined by using the chemiluminescent reaction of the drug with potassium permanganate in an acid medium containing sodium hexametaphosphate, emitted light being transferred from the flow cell to a photomultiplier via fiber optics.

Flow methods have also been used to control pharmaceutical industrial waste water. Thus, a flow system was used to determine several substances and their metabolites in waste water from a pharmaceutical plant. The determination method included a pretreatment step involving liquid–liquid extraction of the solid phase prior to insertion into the flow manifold and detection by mass (FIA MS) or tandem mass spectrometry (FIA MS–MS).

21.3.2
Automated Drug Dissolution and Drug Release Testing

Dissolution testing is important for many reasons, one of which is that it facilitates drug quality control. Deriving comprehensive information about the dynamics of the dissolution processes can facilitate the development of new dosage forms. Dissolution tests officially endorsed by pharmacopoeias involve withdrawing samples from the dissolution medium at preset times. The analytical information thus obtained is inadequate to describe the dissolution kinetics and precludes the development and assessment of new dosing forms. Also, official methods exhibit poor sample collection efficiency and are impractical and difficult or even impossible to implement when several multi-dissolution vessels (usually six) are operated simultaneously. For this reason, continuously monitoring active principles in dissolution media entails using an automated system.

Ideally, automated systems provide a number of advantages in this context including the following: (i) they enable fast determinations and sample pretreatments, which facilitates real-time monitoring; (ii) they exhibit a high throughput (and also high resolution as a result); (iii) they use modest amounts of sample and cause minimal volume changes in the dissolution medium, especially at high sampling rates; (iv) they afford the simultaneous control of several dissolution tests with a single detector and also the simultaneous determination of several components in the same formulation; (v) they allow samples to be pretreated in a stable manner and continuous detection over long periods; (vi) they afford continuous monitoring of the baseline and on-line recalibration of the detection system and (vii) they feature low reagent consumption.

Flow systems possess the previous features. Recently, the ability to collect samples directly from the dissolution medium – by on-line filtration –, conduct simultaneous tests in multi-dissolution vessels and use chemometric methods to for multi-component determinations was assessed in flow systems (see Table 21.2).

The earliest attempts at accomplishing the above ideal objectives date from the beginning of FIA. Thus, new analytical procedures for some active principles were used not only to determine the target analyte, but also to obtain its dissolution profile.

Table 21.2 Dissolution assays and drug release testing on pharmaceutical formulations. The reviewed period spans from 1998 to 2007.

Dissolution profiles of pharmaceutical formulations			
Pharmaceutical	Flow-system detection	Linear range R.S.D. (%)	Ref.
Ibuprofen	SIA UV–vis spectrophotometry (222 nm)	0–$0.12\,\mathrm{g\,L^{-1}}$ 0.5	[6]
Acetylsalicylic acid-caffeine-phenacetin	SIA UV–vis spectrophtometry (220–310 nm)	44–$220\,\mathrm{mg\,L^{-1}}$ 30–$1500\,\mathrm{mg\,L^{-1}}$ 10–$50\,\mathrm{mg\,L^{-1}}$ 0.4–0.5	[7]
Reserpine	FIA photo-induced fluorimetry ($\lambda_{exc} = 386\,\mathrm{nm}$, $\lambda_{em} = 490\,\mathrm{nm}$)	0.01–$0.75\,\mathrm{mg\,L^{-1}}$ 1.4	[8]
Isoniazid	FIA CL	5×10^{-7}–$1\times10^{-4}\,\mathrm{mg\,L^{-1}}$ 3.0	[9]
Bumetanide	SIA fluorimetry ($\lambda_{exc} = 314\,\mathrm{nm}$; $\lambda_{em} = 370\,\mathrm{nm}$)	0.05–$10.0\,\mathrm{mg\,L^{-1}}$ 0.46	[10]
Acetylsalicylic acid	SIA Selec. electr.	0.05–$10\,\mathrm{mM}$ 0.20	[11]
Ergotamine tartrate	SIA fluorimetry ($\lambda_{exc} = 236\,\mathrm{nm}$, $\lambda_{em} = 390\,\mathrm{nm}$)	0.03–$0.61\,\mathrm{mg\,L^{-1}}$ <0.86	[12]
Sulfadiazine-trimethoprim and amitriptyline-perphenazine	FIA UV–vis spectrophotometry reactionless Two simultaneous profiles		[13]
Rutin trihydrate-ascorbic acid	SIA UV–vis spectrophotometry (262 nm)	2–$20\,\mathrm{mg\,L^{-1}}$ 0–$100\,\mathrm{mg\,L^{-1}}$ 0.4–0.7	[14]
Sulfamethoxazole	FIA photo-induced chemiluminescence	0–$8\,\mathrm{mg\,L^{-1}}$ 1.4	[15]
Sulfamethoxazole – trimethoprim	FIA UV–vis spectrophotometry. Reactionless. Two simultaneous profiles		[16]
Bromhexine-amoxicilline and amoxicilline-clavulanic acid	FIA UV–vis spectrophotometry. Three simultaneous profiles		[17]
Hydrochlorothiazide-captopril	FIA multicommutated UV–vis spectrophotometry (273 and 250 nm). Three simultaneous profiles	10–$50\,\mathrm{mg\,L^{-1}}$ 5–$25\,\mathrm{mg\,L^{-1}}$	[18]

(Continued)

Table 21.2 *(Continued)*

Dissolution profiles of pharmaceutical formulations			
Pharmaceutical	Flow-system detection	Linear range R.S.D. (%)	Ref.
Sulfamethoxazole-trimetoprim	FIA multicommutated reactionless. Three simultaneous profiles UV–vis spectrophotometry (257 and 247 nm)	0–200 mg L^{-1} 0–90 mg L^{-1}	[18]
Prazosin hydrochloride	SIA fluorimetry ($\lambda_{exc} = 244$ nm; $\lambda_{em} = 389$ nm)	0.02–2.43 mg L^{-1} 1.9	[19]
Chloroquine	FIA Selec. electr.	0.01–100 mM	[20]
Famotidine	FIA UV–vis spectrophotometry (265 nm)	20–60 mg l^{-1} 1.0%	[21]
Carbamazapine	FIA UV–vis spectrophotometry (288 nm)	1.08×10^{-5}–6.48×10^{-5} M 1.92	[22]
Bumetanide	FIA fluorimetry ($\lambda_{ex} = 314$ nm and $\lambda_{em} = 370$ nm)		[23]
Propranolol hydrochloride	SIA dissolution and permeation assembly		[24]
Drug release testing			
Pharmaceutical	Flow-system detection	Ref.	
Acetyl salicylic acid	FIA	[25]	
Indometacin	SIA	[26]	
Salicylic acid		[27]	
Salicylic acid	SIA fluorimetry Simultaneous release tests	[28]	
Lidocaine and prilocaine	SIA UV–vis spectrophotometry	[29]	
Triamcinolone acetonide and salicylic acid	SIC (SIA + L chromatography)	[30]	

The result was a vast number of disperse procedures using various types of detectors that hindered the automatization of dissolution tests. Such procedures were the subject of a review. There were, however, subsequent attempts at improving the procedures for obtaining dissolution profiles by simultaneously determining the *overall* profile using the recommended standard procedures and up to two *individual* profiles by using derivative spectrophotometry.

The first SIA system used for dissolution testing was developed to assess the dissolution profiles for ibuprofen tablets and capsules. Samples were collected following on-line filtration and the analyte determined by monitoring the absorbance. Subsequently, the high flexibility of the system was used to simultaneously monitor the dissolution of acetylsalicylic acid, phenacetin and caffeine in aspirin tablets, using partial least-squares calibration. UV–vis absorption, selective electrodes, fluorescence, biamperometry, chemiluminescence and so on, have been proposed as detectors for flow systems dealing with the dissolution profiles of one or several drugs simultaneously. FIA, SIA and multicommutation assemblies have also been exploited for such applications.

Multicommutation methodology has also enabled the obtainment of the dissolution profiles for individual drugs in sulfamethoxazole–trimethoprim and hydrochlorothiazide–captopril mixtures in two different formulations by using a manifold containing one and two solenoid valves, respectively, in combination with derivative spectrophotometry. The use of a multicommutated system resulted in substantially reduced reagent consumption (up to 25 times relative to the conventional flow procedure).

21.3.3
Membrane Diffusion

The search for new drug release mechanisms has relied on their ability to penetrate tissues and diffuse through blood vessels. This has required the development of effective methods for understanding, optimizing and predicting the release- and adsorption-related properties of the drugs prior to their general use. Such methods tend to be tested *in vitro* rather than *in vivo* in order to facilitate strict, reproducible control. However, *in vitro* results are only valid if they are accurately correlated with every aspect of the corresponding *in vivo* process. This has fostered research aimed at developing experimental designs capable of mimicking as closely as possible the way a molecule penetrates the human body *in vivo*. The *in vitro* method in widest use at present involves the simulated permeation of a membrane for adsorption tests. Experiments usually require the use of various membranes mimicking epithelial tissue, properly adjusted environments, a variety of protocols and diffusion cells to simulate the conditions inside and outside the body.

A diffusion or membrane permeation cell typically consists of two compartments containing a donor and an acceptor that are separated by a membrane mimicking the skin or some other tissue of interest. The donor compartment can be of the open, *single-chamber cells*, or closed type *two-chamber cells*. The latter have the advantage that the donor can be used at an infinite rate, which facilitates the monitoring of permeation profiles for analytes with high permeability coefficients.

Based on the way the fluid is delivered to the acceptor compartment, diffusion cells can be of the batch or flow type. In batch cells, the acceptor medium is continuously stirred in its compartment; however, the medium is not replaced, but only replenished to offset the volume losses caused by sample withdrawals. The acceptor compartment has a relatively large volume in order to facilitate homogenization and dilution of the analyte. In flow cells, however, the acceptor medium is circulated continuously through the acceptor compartment and the medium is continuously renewed as a result. For this reason, flow cells are better at simulating blood flow

Table 21.3 Membrane diffusion studied on pharmaceutical formulations.

Analyte	Detection	Linear interval	R.S.D. (%)	Number of samples (h^{-1})/ assay interval (h)	Ref.
Caffeine, aminophilline	UV–vis spectrophotometry (272 nm)	0.4–20 mg L^{-1}	—	170/2	[31]
Lidocaine, Prilocaine	UV–vis spectrophotometry (212 nm)	2.5–200 mg L^{-1}	1.89 2.44	5/4	[32]
Indomethacin	Fluorimetry ($\lambda_{exc} = 330$ nm; $\lambda_{em} = 385$ nm	0.05–10 mg L^{-1}	2.3	120/6	[33]
Salicylic acid	UV–vis spectrophotometry (385 nm)	0.05–10 mg L^{-1}	0.52	120/6	[34]

across the skin. Unlike batch cells, however, flow cells require the volume of the acceptor stream to be as low as possible. Therefore, the chamber volume and flow rate are critical in a flow system; in fact, the volume of both the acceptor chamber and collected fraction should be as low as possible, and the opposite holds for the flow rate. In any case, these variables will have no effect on a permeation profile if mixing occurs under appropriate conditions.

Table 21.3 lists selected references to studies on drug diffusion through various types of membranes.

Multicommutated methodology was used to insert the acceptor solution in a study on the diffusion of caffeine and aminophylline across an artificial lipophilic membrane. The membrane consisted of a cellulose nitrate support impregnated with N and S$_2$ lipids. The potential effects of surfactants on the diffusion process were studied. The drugs were monitored by using a flow cell inserted in the path of the acceptor solution. The large number of measurements obtained in each test (in the region of 340 nm) could be useful with a view to facilitating kinetic studies on artificial membranes and elucidating the influence of excipients on the diffusion process.

A Franz cell coupled to an SIA system was used to determine the extent of release of active principles in semi-solid formulations, using membranes of various materials (polycarbonate, cellulose mixed esters, Teflon, silicone, polyvinylidene fluoride) and variable pore size to compare the liberation profiles for a gel containing 1% indomethacin, and 2.5% lidocaine and prilocaine. A polycarbonate membrane of 0.4 μm pore size was selected in order to compare the release rate for the previous drugs in gels and pomades using various procedures. The release profiles for indomethacin in three semi-solid preparations were successfully monitored over a period of 6 h, and those for lidocaine and prilocaine for 4 h, with no human intervention. Recently, the same methodology was used to assess the release of salicylic acid in three semi-solid formulations containing the active principle in a 3% proportion. No operator control was required during the tests. The ability to use up to six Franz cells make this system also highly useful for release testing as part of quality control analyses, as well as for monitoring pre- and post-changes in product properties, batch monitoring and product development.

21.3.4
Functional Cellular Assays for Screening Potential Drugs

The mere binding of a ligand or reactant to a cell receptor does not necessarily elicit a physiological response. Identifying and characterizing new drugs requires confirming the efficiency of candidate compounds by using functional tests in order to classify substances as agonists or antagonists, depending on whether they elicit or inhibit a biological response. If binding of a drug to a receptor does elicit a response, then the response can be assessed by monitoring some parameters such as the release of cytosolic calcium, changes in intracellular pH, the release or extrusion of acids, glucose or oxygen uptake or even changes in membrane potentials. Manual tests of this type frequently involve the continuous exposure of a biological material to variable drug concentrations rather than the use of a new sample at each concentration. This can influence the response by reducing the sensitivity of the receptor or even destroying the biological material. Biological tests were recently automated by using SIA in combination with appropriate reagents for sequential injection-bead injection; the system used cell cultures and measured various cell response variables (see Table 21.4). Cells were adsorbed onto the surface of microdrops and a small volume of the resulting suspended particles was used as a representative sample for analysis. Because each portion of stimulant inserted reacted with a fresh portion of cells over a controlled exposure period, the response was highly reproducible and the typical biological variability encountered when cells are cultured in the traditional manner suppressed. Also, bead injection methodology provides kinetic information that cannot be extracted when stimulants are directly added to the culture medium by hand. Based on the kinetic behavior of the

Table 21.4 Cellular flow-assays.

Analyte	Sample	Detection	Linear range, Assay interval (s)	Ref.
Intracellular calcium	CHO M1 cells treated with acethylcoline, pilocarpine and atropine	Fluorimetry with fura-2-am	−500	[35]
Oxygen	CHO M1 cells treated with amobarbital	Fluorimetry Pt-porphyrin complex	180	[36]
pH intracellular	CHO M1 cells treated with carbachol	Fluorimetry with BCECF-AM	700	[37]
Glucose	TABX2S cells with azide	UV–vis spectrophotometry (340 nm)	$0.1–5.6\,mmol\,L^{-1}$ 120	[38]
Glucose lactate	TABX2S cells with azide	UV–vis spectrophotometry (340 nm)	$0.1–5.6\,mmol\,L^{-1}$ 10 $0.05–1.00\,mmol\,L^{-1}$ 30	[39]

interaction with the receptor, and on the duration and maximum strength of the response, one can discriminate between specific antagonists and assess their efficiency.

Chinese hamster ovary cells containing rat muscarinic receptors were used in a functional test for antagonists. Cells were cultured on microdrops loaded with fluorescence-sensitive intracellular calcium (fura-2-am) and suspended in buffering solution. An SIA-bead injection system was used to automatically aspirate a small volume of suspension into a jet-ring chamber. Then, cells captured on the micro-column were allowed to interact with the target drugs. The cellular responses elicited by the drugs were assessed by monitoring the transient release of cytosolic calcium (a result of the receptor being stimulated by the drug). The system was also used to examine the effect of muscarinic antagonists of the response on intracellular calcium by analyzing a series of samples consisting of a mixture of acetylcholine at a constant concentration and the antagonist at a variable concentration.

One other similar system has been used to measure cell stimulation *via* the oxygen uptake. Oxygen plays a central role in the aerobic metabolism of cells, where it acts as a final electron acceptor in the aerobic oxidation pathway; therefore, measuring the oxygen uptake can help investigate biochemical events involving an aerobic mechanism. The ensuing methodology was used to examine the effect of amobarbital and acetylcholine by monitoring the oxygen uptake. The sensor measured the decrease caused by the presence of oxygen in the phosphorescence of a Pt-porphyrin complex immobilized on microdrops which also served as substrate for the cell culture. Stimulating muscarinic receptors in cells was found to increase the oxygen uptake, as reflected in the metabolic change caused by some drugs.

The developers of the previous system proposed a similar SI-BI configuration for measuring the response of cells to a drug *via* the intracellular and extracellular pH. All living cells produce acids as part of their energy-related processes. Therefore, by monitoring intracellular and extracellular pH one can study the events potentially interfering with cell metabolism (e.g., during the stimulation of cells by a chemical agent). The system in question allowed both types of pH to be measured. For intracellular pH measurements, cells were cultured on microdrops and dyed with a fluorescent pH indicator. The resulting changes in fluorescence intensity were proportional to those in intracellular pH. Extracellular pH was measured by using an insoluble fluorescent indicator immobilized on microdrop surfaces, cells being cultured in a subsequent step. The microdrops were inserted into the bead injection system and extracellular pH monitored via the fluorescence intensity. No physical attraction was required in either determination, the sole difference between the two being the way the indicator was used inside the cells to measure intracellular pH or adhered to an external membrane to measure extracellular pH. The study was conducted using carbachol as stimulant. Exposure to the stimulant resulted in the release of acid and acidification of cytosol. Both changes were proportional to the carbachol concentration, which allowed a concentration–response curve to be constructed.

The lab-on-valve (LOV) concept could be used with bead injection methodology to assess the glucose uptake by living cells *in situ*. Glucose is the main source of carbon and energy for cells and its uptake can be used to assess their metabolic status upon chemical stimulation or to study metabolic effects before, during and after a hypoxic episode in cells. In the earliest use of LOV in biological tests, hepatocyte cells (TABX2S) were cultured on microdrops and subsequently packed and embedded in a microcolumn containing a microbioreactor at a controlled temperature. The amount of glucose used was measured in the liquid embedding the microdrops *via* an enzymatic process occurring in a flow cell furnished with fiber optics. Glucose was determined by a two-step enzymatic reaction the reaction product of which being NADH. Tests were also performed in the presence of azide, which is an inhibitor for cytochrome C oxidase (an enzyme involved in the aerobic photodegradation of glucose). In the presence of azide, glucose could only be degraded anaerobically, so cells increased their consumption of this source in order to offset the energy deficiency resulting from this degradation pathway.

An LOV-bead injection combined system was designed for the determination of the glucose uptake and lactate extrusion by cultured cells. The lactate extrusion rate, in combination with the cellular glucose uptake, provides a means for identifying the metabolic regime prevailing in a cell culture. By comparing these two parameters, one can estimate the fraction of energy that is produced aerobically or anaerobically. Similarly, metabolic regimes can be used to discriminate between primary and secondary cultures – primary cultures tend to be aerobic – or even to elucidate the effect of some chemicals on living cells. The effects of chemicals on cell metabolism can be assessed by comparing glucose uptake and lactate extrusion in their presence. This approach can also be used to control environmental effects potentially inducing metabolic alterations. For example, an ischemic event (e.g., a bump, cardiac arrest) can be simulated by degassing the carrier. The system proposed by Schulz *et al.* was used to compare the metabolic regime of two types of hepatocyte cells, namely: TABXS2, which produces excess Bcl-xL (a mitochondrial membrane protein regulating the permeability to major metabolites) and TABX1A, which produces no Bcl-xL. Cells were cultured, packed and embedded in a microcolumn containing a microbioreactor, as in the previous study. The metabolic regime of the cells was established by monitoring glucose uptake and lactate extrusion in the receptor liquid, using quantitative tests based on the same enzyme assay (viz. NADH production). Tests were also performed in the presence of azide, which resulted in increased glucose uptake and lactate extrusion.

References

1 Min, R.W., Nielsen, J. and Villadsen, J. (1995) *Analytica Chimica Acta*, **312**, 149.

2 Min, R.W., Nielsen, J. and Villadsen, J. (1996) *Analytica Chimica Acta*, **320**, 199.

3 Barnett, N.W., Lewis, S.W. and Tucker, D.J. (1996) *Analytical Chemistry*, **355**, 591.

4 Barnett, N.W., Lenehan, C.E., Lewis, S.W., Tucker, D.J. and Essery, K.M. (1998) *Analyst*, **4**, 601.

5 Schröder, H.F. (1999) *Waste Management*, **19**, 111.

6 Liu, X.Z. and Fang, Z.L. (1998) *Analytica Chimica Acta*, **358**, 103.

7 Liu, X.Z., Liu, S.S., Wu, J.F. and Fang, Z.L. (1999) *Analytica Chimica Acta*, **392**, 273.

8 Chen, H. and He, Q. (2000) *Talanta*, **53**, 463.

9 Li, B., Zhang, Z. and Liu, W. (2001) *Talanta*, **54**, 697.

10 Solich, P., Polydorou, C.K., Koupparis, M.A. and Efstathiou, C.E. (2001) *Analytica Chimica Acta*, **438**, 1331.

11 Paseková, H., Sales, M.G., Montenegro, M.C., Araujo, A.N. and Polasek, M. (2001) *Journal of Pharmaceutical and Biomedical Analysis*, **24**, 1027.

12 Legnerová, Z., Sklenárová, H. and Solich, P. (2002) *Talanta*, **58**, 1151.

13 Moreno Galvez, A., García Mateo, J.V. and Martínez Calatayud, J. (2002) *Journal of Pharmaceutical and Biomedical Analysis*, **27**, 1027.

14 Legnerová, Z., Satinský, D. and Solich, P. (2003) *Analytica Chimica Acta*, **497**, 165.

15 Catala Icardo, M., Garcia Mateo, J.V., Fernandez Lozano, M. and Martínez Calatayud, J. (2003) *Analytica Chimica Acta*, **499**, 57.

16 Moreno Galvez, A., Gomez Benito, C. and Martínez Calatayud, J. (2003) *Journal of Flow Injection Analysis*, **20**, 8.

17 Vranic, E., Catalá Icardo, M. and Martínez Calatayud, J. (2003) *Journal of Pharmaceutical and Biomedical Analysis*, **33**, 1039.

18 Tomšů, D., Catala Icardo, M. and Martínez Calatayud, J. (2004) *Journal of Pharmaceutical and Biomedical Analysis*, **36**, 549.

19 Legnerová, Z., Huclová, J., Thun, R. and Solich, P. (2004) *Journal of Pharmaceutical and Biomedical Analysis*, **34**, 115.

20 Saad, B., Zihn, Z.M., Jab, M.S., Rahman, I.A., Saleh, M.I. and Mahsufi, S. (2005) *Analytical Sciences*, **21**, 521.

21 Tzanavaras, P.D., Verdoukas, A. and Balloma, T. (2006) *Journal of Pharmaceutical and Biomedical Analysis*, **41**, 437.

22 Çomoglu, T., Gönül, N., Sener, E., Dal, A.G. and Tunel, M. (2006) *Journal of Liquid Chromatography and Related Techniques*, **29**, 2677.

23 Tzanavaras, P.D. and Themelis, D. G. (2007) *Analytica Chimica Acta*, **588**, 1.

24 Motz, S.A., Klimundová, J., Schaefer, U.F., Balbach, S., Eichinger, T., Solich, P. and Lehr, C.-M. (2007) *Analytica Chimica Acta*, **581**, 174.

25 Solich, P., Ogrocká, E. and Schaefer, E. (2001) *Pharmazie*, **56**, 787.

26 Solich, P., Sklenárová, H., Huclová, J., Stínsky, D. and Schaefer, U.F. (2003) *Analytica Chimica Acta*, **499**, 9.

27 Klimundová, J., Sklenárová, H., Schaefer, U.F. and Solich, P. (2005) *Journal of Pharmaceutical and Biomedical Analysis*, **37**, 893.

28 Klimundová, J., Mervartová, K., Sklenářová, H., Solich, P. and Polášek, M. (2006) *Analytica Chimica Acta*, **366**, 573.

29 Klimundová, J., Šatinský, D., Sklenářová, H. and Solich, P. (2006) *Talanta*, **69**, 730.

30 Chocolouš, P., Holik, P., Šatinský, D. and Solich, P. (2007) *Talanta*, **72**, 854.

31 Sales, M.G.F., Reis, B.F. and Montenegro, M.C.B.S.M. (2001) *Journal of Pharmaceutical and Biomedical Analysis*, **26**, 103.

32 Klimundová, J., Satinský, D., Sklenárová, H. and Solich, P. (2005) *Talanta*, **69**, 730.

33 Solich, P., Sklenárová, H., Huclová, J., Satinský, D. and Schaefer, U.F. (2003) *Analytica Chimica Acta*, **499**, 9.

34 Klimundová, J., Sklenárová, H., Schaefer, U.F. and Solich, P. (2004) *Journal of Pharmaceutical and Biomedical Analysis*, **37**, 893.

35 Hodder, P.S. and Ruzicka, J. (1999) *Analytical Chemistry*, **71**, 1160.

36 Lahdesmaki, I., Scampavia, L., Beeson, C. and Ruzicka, J. (1999) *Analytical Chemistry*, **71**, 5248.

37 Lahdesmaki, I., Beeson, C., Christian, G.D. and Ruzicka, J. (2000) *Talanta*, **51**, 497.

38 Schulz, C.M. and Ruzicka, J. (2002) *Analyst*, **127**, 1293.

39 Schulz, C.M., Scampavia, L.S. and Ruzicka, J. (2002) *Analyst*, **127**, 1583.

Further Reading

Abdulrahman, L.K., Al-Abachi, A.M. and Al-Qaissy, M.H. (2005) Flow injection-spectrophotometeric determination of some catecholamine drugs in pharmaceutical preparations via oxidative coupling reaction with p-toluidine and sodium periodate. *Analytica Chimica Acta*, **538**, 331–335.

Catalá Icardo, M., Fernandez Lozano, M. and Martinez Calatayud, J. (2003) Enhanced flow injection–chemilumino- metric determination of sulphonamides by on-line photochemical reaction. *Analytica Chimica Acta*, **499**, 57–69.

Catalá Icardo, M., Lahuerta Zamora, L., Antón Fos, G.M., Martínez Calatayud, J. and Duart, M.J. (2005) Molecular connectivity as a relevant new tool for predicting analytical behavior: A survey of chemiluminescence and chromatography. *Trends in Analytical Chemistry*, **24**, 782–791.

Gallignani, M., Ayala, C., Brunetto, M.R., Burguera, M. and Burguera, J.L. (2003) Flow analysis-hydride generation-Fourier transform infrared spectrometric determination of antimony in pharma-ceuticals. *Talanta*, **59**, 923–934.

Gerardi, R.D., Barnett, N.W. and Lewis, S.W. (1999) Analytical applications of tris (2,2′-bipyridyl)ruthenium(III) as a chemiluminescent reagent. *Analytica Chimica Acta*, **378**, 1–41.

Gilbert López, B., Llorent-Martínez, E.J., Ortega-Barrales, P. and Molina Díaz, A. (2007) Development of a multicommuted flow-through optosensor for the determination of a ternary pharmaceutical mixture. *Journal of Pharmaceutical and Biomedical Analysis*, **43**, 515–521.

Gómez-Taylor, B., Palomeque, M., García Mateo, J.V. and Martínez Calatayud, J. (2006) Photoinduced chemiluminescence of Pharmaceuticals. *Journal of Pharma-ceutical and Biomedical Analysis*, **41**, 347–357.

Grudpan, K., Kamfoo, K. and Jakmunee, J. (1999) Flow injection spectrophotometric or conductometric determination of ascorbic acid in a vitamin C tablet using permanganate or ammonia. *Talanta*, **49**, 1023–1026.

Martinez Calatayud, J. (1996) *Flow Injection Analysis of Pharmaceuticals. Automation in the Laboratory*, Taylor and Francis, London.

Martínez Calatayud, J. and Catalá Icardo, M. (2005) Flow injection analysis. Clinical and pharmaceutical applications. *Encyclopedia of Analytical Sciences*, 2nd edn, Academic Press, 0127641009.

Martínez Calatayud, J. (2005) *Spectrophotometry. Pharmaceutical Applications*. Encyclopedia of Analytical Sciences, 2nd edn, Academic Press, 0127641009.

Murillo Pulgarin, J.A., Alañon Molina, A. and Perez-Olivares Nieto, G. (2004) Determination of hydrochlorothiazide in pharmaceutical preparations by time resolved chemiluminescence. *Analytica Chimica Acta*, **518**, 37–43.

Pimenta, A.M., Montenegro, M.C.B.M.S., Araujo, A.N. and Martinez Calatayud, J. (2006) Application of Sequential Injection Techniques to flow analysis. *Journal of Pharmaceutical and Biomedical Analysis*, **40**, 16–34.

Ruedas Rama, M.J., Ruiz Medina, A. and Molina Diaz, A. (2004) Bead injection spectroscopy-flow injection analysis (BIS-FIA): an interesting tool applicable to pharmaceutical analysis: Determination of promethazine and trifluoperazine. *Journal of Pharmaceutical and Biomedical Analysis*, **35**, 1027–1034.

Sales, M.G.F., Tomás, J.F.C. and Lavandeira, S.R. (2006) Flow injection potentiometric determination of chlorpromazine. *Journal of Pharmaceutical and Biomedical Analysis*, **41**, 1280–1286.

22
Industrial and Environmental Applications of Continuous Flow Analysis

Kees Hollaar and Bram Neele

22.1
Introduction

The strength of continuous flow analyzers (CFA) or segmented flow analyzers (SFA) is the equal treatment and analysis conditions of all samples, standards and system quality checks. The possibility of sample pretreatment makes them suitable to handle many different applications in the industrial and environmental fields. Their robustness, not expected because of the technically complex looks of the system, is well proven and used in many laboratories with large sample flows, such as agricultural and environmental laboratories. The ability of the technique, introducing complex sample manipulations, has proven to be also a non-disputed strength in the automation of complex chemistries. Also, the ever-growing need to analyze more samples is translated into efforts for a high degree of automation in complex manual chemistries. Explicit examples are the analysis of total cyanide with in-line UV digestion and in-line distillation; the analysis of total nitrogen based on in-line UV digestion followed by a nitrate determination with cadmium reduction. Also, extractions with organic solvents can take place on a microscale using minimal volumes of extracting fluid. A typical example for this extraction process is the determination of bitterness in beer. Besides these entire sample treatment possibilities, simultaneous analysis of a number of parameters of the same sample is a very attractive perspective to the users by saving on analyses' turnaround time.

By applying these automated methods, significant cost reduction is achieved in terms of minimizing human resources in the laboratory. It is interesting to see that nowadays the shortage of laboratory staff in certain areas is even a justification for automation. It is also important to know that most of the studies to eliminate human participation, show that automation of analysis does not necessarily result in redundancy of staff as the growing number of problems encountered in our environment require additional new sections in the existing laboratories [1]. Costs related to the use and disposal of chemicals are also reduced as the automated procedures work with much smaller volumes. Transcription errors by human

Advances in Flow Analysis. Edited by Marek Trojanowicz
Copyright © 2008 WILEY-VCH Verlag GmbH & Co. KGaA, Weinheim
ISBN: 978-3-527-31830-8

handling will be avoided by the introduction of automated barcode readings on samples throughout the sample handling process. This automation is possible because of the digital era and its huge digital potential. With modern techniques in digitalized detection accuracy is improved by at least a factor of 2. Full automated system control is expanding the amount of analysis to be done, also by a factor >2. Daytime office hours are used for preparation of samples, while automated analysis is done during out of laboratory hours without operator interference.

The trend towards normalization of analysis methods is growing because of the commercialization of analysis in general. Comparability of results is being even more demanded when different commercial laboratories make the analyses. Breaches in trends are not welcome at any organization. Therefore analytical standards in methods and methodologies are a must and these are produced by, for example, EPA, ISO, Standard Methods of the examination of water and wastewater, MEBAK, ASBC, EBC, and many more related organizations nationwide. As a result of automation and standardization high analytical quality and comparability of the results produced will be achieved.

22.2
Overview of Environmental and Industrial Fields

It is impossible to give a detailed and particularly complete overview of all application fields and their specific applications. Only major application fields and their specific topics and parameters are listed, together with known advantages and disadvantages. The lists with applications are a limited overview to give an impression of automated applications in a particular application field. A variety in detection ranges and methods is available.

22.2.1
Environmental Applications

A vast amount of automated applications is available for the environmental laboratory (Table 22.1). The simultaneous analysis of nitrate + nitrite, nitrite, ammonium, and phosphate is in use in many laboratories worldwide. Often chloride and/or sulfate are added to make a five or six-channel simultaneous system. From different suppliers, these systems are able to perform up to 60 or 80 samples per hour per application, in this configuration producing 300–500 analytical results each hour. More and more complex applications like total cyanide, phenol index or Methylene Blue active substances (MBAS) are analyzed in the environmental laboratories. A more modern approach is to analyze total nitrogen and total phosphate in water samples without pretreatment of the samples by off-line digestion. Now digestion is carried out by ultra violet radiation and high temperature heating in-line during the analysis. This results in fast and low cost determinations compared with the previous (semi-) automated off-line digestion procedures which were followed by manual or (semi-) automated detection. A big advantage of these fully automated applications systems

Table 22.1 Overview of environmental applications.

Acidity (total)	Conductivity	Nitrate + Nitrite
Alkalinity (total)	Copper	Nitrite
Aluminum (hydrolyzed)	Cyanide (total, free, WAD)	Nitrogen (total)
Amino acids (total)	Dissolved organic carbon	Nonionics
Ammonia	Methanol	Phenol (total)
Boron	Fluoride	Phosphate (ortho, total, hydrolyzable)
Bromide	Formaldehyde	Potassium permanganate value (COD)
Calcium	Hardness (total)	Silicate
Carbonate (total)	Hydrazine	Sodium/Potassium
Chemical oxygen demand (COD)	Iron (II)/(III)	Sulfate
Chloride	Magnesium	Sulfide
Chlorine	Manganese	Sulfite
Cholinesterase inhibition	Methylene Blue active substances	Total organic carbon
Chromium(VI)	Molybdenum	Turbidity
Color	Nickel	Urea

is the lack of high concentrations of strong acids and toxic catalysts, which are used for digestion. The disadvantage of these automated flow analysis systems is that they are incapable of handling large particles. Thorough homogenizing and mixing of the samples prior to suction into the automated system, will partly overcome this disadvantage. However, the amount of particles should never exceed $100\,mg\,l^{-1}$ expressed as dry matter. Regulations in several countries around the world are demanding total nitrogen instead of Kjeldahl nitrogen as the governing authorities found out that not only the Kjeldahl part of the nitrogen determination is harmful to the environment, but also the nitrogen present as nitrate. Laboratories analyzing seawater samples are a group of users that nowadays use fully automated flow analysis techniques. Very low concentrations of nutrients at the sub $\mu g\,l^{-1}$ level can be detected; Kérouel and Aminot reported detection levels of $1.5\,nmol\,l^{-1}$ for ammonium by a fluorimetric method [2].

22.2.2
Plant and Soil Applications

Extracts from soil or plant samples are often colored by nature: in soil there are always medium to high concentrations of colored organics in the form of humic acids, especially in soils analyzed for fertilization recommendations. If the pH of the sample extract is changed during the analytical process, humic acids starts to coagulate and precipitate from solution. This may cause problems with the flow analysis. These organics can also disturb reaction processes like nitrate reduction by activated cadmium. Therefore a strong argument for using a flow analyser is the possibility to include a dialyzer. Dialyzers will block the humic acids and background color from entering the reaction processes. The group of applications which can be analysed simultaneously is different from that of the environmental users: here the nitrogen-containing compounds like ammonium and nitrate are of importance, together with phosphate and potassium (NPK). However, the wide variety of sample extraction procedures and the ratio between the weight of the soil/plant sample and the extracting liquid yield large differences in analytical ranges. Compare the results of a 1 : 60 ratio soil to water extract for "phosphate in water" (PW) and a 1 : 10 ratio soil to calcium chloride (0.01 M) extract for phosphate: completely different! A major difference will be found in doing analysis for agricultural purposes or for environmental purposes. The latter always request "total" determinations, demanding other extracting procedures or even digestion prior to analyses, while the agricultural field requests results representing analytes "available to plants". Important applications in the field of environmental investigations on soil samples are total cyanides, phenol index, total nitrogen, and total phosphate. An extension to analysis in soil and plants is the analysis of fertilizer and fertilizer raw material. Precise and accurate results must be obtained because inaccuracy in the final concentration of analytes in fertilizer will affect costs dramatically as thousands of tons of product are involved. Very important applications for the fertilizer industry are ammonia, nitrate, phosphate and potassium. Precision is all that matters in fertilizers. The list of analytes determined is shown in Table 22.2.

22.2.3
Pharmaceutical Applications

The use of continuous flow analysis equipment in pharmaceutical analyses is a growing need. Traceability and data integrity (21 CFR Part 11) Specific SOPs assure consistent operation and are a must in pharmaceutical laboratories. Total automation of analysis and the accuracy of digital data archiving is essential. Exactly these

Table 22.2 Overview of applications in plants and soil

Acetic acid	Cyanide (total, free)	Pectin
Aluminum	Dissolved organic carbon	Phenol (total)
Amino acids (total)	Fluoride	Phosphate (ortho, total, hydrolyzable.)
Amino nitrogen (free)	Galactose (D+)	Protein (soluble)
Ammonia	Gluconic acid (D)	Reducing sugars (total)
Boron	Glucose/Fructose/Saccharose.	Silicate
Bromide	Iron(II) + (III)	Sodium/Potassium
Calcium	Lactic acid (L/D)	Starch
Carbohydrates total	Lysine (L)	Sugars (total)
Carbon	Magnesium	Sulfate
Carbonate total	Malic acid (L)	Sulfide
Chloride	Manganese	Total organic carbon
Citric acid	Nitrate + Nitrite	Urea
Conductivity	Nitrite	
Copper	Nitrogen (total)	

requirements are met when using flow techniques. The complete analysis run can be validated during the run and afterwards. Samples are analyzed under identical conditions. Validated results and all digital data remain available by archiving. Whether it is determining glucose in a fermentation broth or glucose in readymade pills and powders, automation and the possibility of controlling all steps of the analytical procedure, are of importance to pharmaceutical laboratories. 21 CFR Part 11 is a well known credo in these laboratories. But, typically, in other application fields they will not even recognize the name. The use of flow systems is somewhat diffuse in terms of applications because pharmaceutical companies do not all produce the same products. As always in all laboratories, the automation of time and work consuming determinations is often well accepted. The answer to the question of sample analyses capacity is making the choice between semi-automated methods, manual methods, or applying a continuous flow analyzer. The latest topic in pharmaceutical applications is the analysis of vitamins or pre-vitamins. A strong point in favour of continuous flow analysis is the complete automation of the different steps (e.g., extraction) necessary to analyze these parameters. The list of determined analytes is shown in Table 22.3.

22.2.4
Beer and Wine Applications

Production control and an almost static quality are very important for beer analyses. Taste is what brewers all over the world sell, which means analyzing several parameters all having to do with taste. Determinations like bitterness, free amino nitrogen, polyphenols and diacetyl in beer or diastatic power, α-amylase and β-glucan in malt are in widespread use. The methodologies are not always from the conventional photometric types of detection, but include detection in the UV range and fluorimetric detection. An important advantage over manual determinations, for example, is the use

Table 22.3 Overview of pharmaceutical applications.

Acetic acid	Formaldehyde	Nitrite
Amino acids (total)	Formic acid	Nitrogen (total)
Ammonia	Galactose (D+)	Orcinol (sugar value)
Boron	Gluconic acid (D)	Penicillin
Calcium	Glucose/Fructose/Saccharose.	pH
Calcium pantothenate	Glutamic acid (L)	Phenylalanine
Carbohydrate	Glycerol	Phosphate (ortho, total, hydrolyzable)
Carbon	Hexuronic acid	Protein
Carbonate (total)	Lactic acid (L/D)	Pyridoxin
Cephalosporins	Lactose	Reducing sugars (total)
Chloride	Lysine (L)	Sodium/Potassium
Citric acid	Malic acid (L)	Starch
DNA	Morphine	Sugars (total)
Ethanol	Nitrate + Nitrite	Trimethoprim
Fluoride		

Table 22.4 Overview of applications in analysis of beer and wine.

Acetic acid	Ethanol	Protein soluble
Acidity (total)	Glucan (beta)	Reducing sugars (total)
Amino Nitrogen (free)	Glucanase (beta)	Sorbic acid
Anthocyanogen	Glucose/Fructose/Saccharose	Starch
Benzoic acid	Glycerol	Sugars (total)
Bitterness	Hydroxymethylfurfural	Sulfur dioxide (total, free)
Calcium	Iron (total, free)	Tartaric acid
Carbon dioxide	Lactic acid (L/D)	Thiobarbituric acid value
Citric acid	Malic acid (L)	Turbidity
Color	Nitrate + Nitrite	Viscosity
Cyanide (total)	pH	Volatile acidity
Density	Phosphate (ortho, total)	
Diacetyl	Polyphenols	

of small volumes of solvent in bitterness analyses. This automated application speeds up the sample throughput dramatically, which alone is very convenient, but also saves much cost in extracting solvent volume. Return of investment also plays a major role here. A parameter gaining interest nowadays in relation to European regulation, is the analysis of total sulfur dioxide in beer. The European regulation demands a maximum concentration of sulfur dioxide, which must be obeyed. Almost all automated applications (Table 22.4) are in accordance with EBC (European Brewery Convention), ASBC (American Society of Brewing Chemists) standards or MEBAK (Mitteleuropäischen Brautechnischen Analysenkommission).

In wine applications, a very important advantage is the use of simultaneous multiparameter systems. Analyzing eight parameters simultaneously is quite common. A single sample output includes results for glucose, fructose, saccharose, malic acid, citric acid, total reducing sugars, total sulfur dioxide, free sulfur dioxide, and several others like volatile acidity. These parameters could also easily be used in the analysis of fruit juices and other beverages. pH and density are also measured simultaneously, not because of the complexity of these measurements but only for the advantage of additional automation together with other parameters. Doing only these two applications manually and all the others automatically is obviously not sensible.

22.2.5
Tobacco Applications

By their nature, tobacco applications can be seen as plant determinations only these particular tobacco plants are treated with some flavors and other additives to be "tasty" in smoking. Nevertheless, these tobacco leaves are still plants. Extraction procedures and analysis are mostly carried out on the raw material prior to the production of cigarettes or cigars or other smoking material. In general, routine analyses on the flow analyzer are performed for ammonia, nitrate + nitrite, reducing sugars and total alkaloids (nicotine). Using an "all in one" extract and with each application built up with a dialyzer because of the strongly colored extracts a high sample throughput

Table 22.5 Overview of applications in analysis of tobacco.

Acetic acid	Glucose/Fructose/Saccharose	Protein
Alkaloids (total)	Lactic acid (L/D)	Reducing sugars (total)
Amino acids (total)	Malic acid (L)	Sorbic acid
Amino nitrogen (free)	Menthol	Starch
Ammonia	Nicotine	Sugars (total)
Carbohydrates (total)	Nitrate + Nitrite	Potassium
Chloride	Nitrogen (total)	Volatile acidity
Citric acid	Pectin	Volatile base
Cyanide (total)	Phosphate (ortho, total, hydrolyzable)	

is realized. Application methodologies are generally used according to Coresta (Cooperation Centre for Scientific Research Relative to Tobacco) and ISO. Different extraction procedures can be handled and automated. Determinations like cyanide, starch and malic acid can also be applied for specific analysis purposes. Alongside the analysis of tobacco itself, these analyses are also performed on the paper (used to produce the cigarettes) and tobacco smoke (Table 22.5).

22.2.6
Food Applications

As a start, in milk powder, concentrations of nitrate and nitrite are well monitored to be sure that infants especially do not get too high doses of these compounds. Separation of proteins and fat present in the milk samples is done by lengthy dialysis as the concentrations to be detected for nitrite are less than $2\,\mu mol\,l^{-1}$. Protein analysis, measured as total nitrogen after classic Kjeldahl digestion, is the most frequent contribution to flow analysis in the food sector. Although techniques like DUMAS are competitive for protein determinations, the DUMAS method is almost solely used in solid samples as the digestion step can be omitted with combustion analyzers. Kjeldahl nitrogen in solid food and beverages is determined by off-line wet digestion, followed by photometric detection. See the full list of determined analytes in Table 22.6.

22.3
Applications and Their Ranges

In this chapter several applications are discussed: total nitrogen, total phosphate, total cyanide, and phenol index for environmental analysis; total reducing sugars for the tobacco industry and bitterness for the beer industry will also be discussed.

Demands for new automated applications are endless. In particular there is pressure for time-consuming, and hence labor-consuming, applications to be automated. Complex applications, which include, for example, digestion, distillation, and/or extraction, are mainly required. The detection ranges of individual applications are also subject to improvement. These applications demand mostly hardware changes and the design of new components. In many laboratories workers develop

Table 22.6 Overview of applications in food analysis.

Acetic acid	Glycerol	Protein
Amino acids (total, free)	Histamine	Pyridoxine
Ammonia	Hydroxyproline	Pyruvate
Ascorbic acid	Iodine	Reducing sugars (total)
Benzoic acid	Iron (II) + (III)	Riboflavin
Bicarbonate	Lactic acid (L/D)	Sodium/Potassium
Caffeine	Lactose	Sorbic acid
Calcium	Lysine (L)	Starch
Carbohydrates (total)	Malic acid (L)	Sugars (total)
Carbon dioxide	Niacin	Sulfur dioxide
Chloride	Niacinamide	Thiamine
Citric acid	Nicotinic acid	Total organic carbon
Creatine/Creatinine	Nitrate + Nitrite	Vitamin B1
Cyanide (total)	Nitrite	Vitamin B2
Density	Nitrogen (total)	Vitamin B3
Ethanol	Orcinol (sugar value)	Vitamin B5
Fluoride	Pectin	Vitamin B6
Galactose ($D+$)	Phenylalanine	Vitamin C
Glucose/Fructose/Saccharose	Phosphate (ortho, total, hydrolyzable)	Vitamin PP
Glutamic acid (L)	Polyphenols	Volatile acidity

systems for their individual and local needs and, if published, others will copy these developments for their needs. However, it often happens that even then changes must be made. Standardization of changed hardware or newly developed hardware is needed. The applications as discussed further in this chapter are meant to be subject to standardization. To understand the basic principle of working of the CFA system, extended explanation can be read in the preceding chapters in this book.

The automated analysis system is built up in a modular way to provide the utmost flexibility to configure it to local user requirements (Figure 22.1). With limited hardware one is able to changeover to different applications easily. The central heart of the system is the applications which are built in.

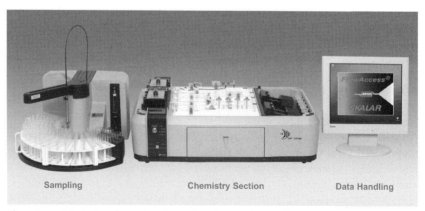

Figure 22.1 A typical configuration of an automated continuous flow system.

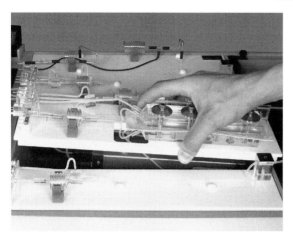

Figure 22.2 A typical design of an automated application.

The automated application is an individual module within the analysis system (Figure 22.2). From this configuration the analyzer can easily be handled as a single channel or a multi-channel in a simultaneous analysis system.

22.3.1
Total Nitrogen

The automated procedure for the determination of total nitrogen is based on the following reactions [4–6]; to digest the complex nitrogen compounds the sample is first mixed with a potassium peroxodisulfate/sodium hydroxide solution. Then a borax buffer is added and the flow is lead through a UV digester. On exiting the UV digester, the flow is heated to 107 °C. The time taken in the UV digester and during heating is approximately 20 min, in order to ensure complete digestion of the nitrogen compounds. To catalyze the digestion process titanium ions can be added to the digestion reagent. As all nitrogen is oxidized to nitrate, the next step is to determine this nitrate with the aid of a classic method: the modified Griess reaction. This method will first reduce the nitrate to nitrite with the addition of copper activated cadmium. This nitrite will then react with sulfanilamide and α-napthylethylenedia-mine dihydrochloride. The developed reddish-purple colored complex is measured at 540 nm and is proportional to the nitrogen originally present in the sample.

Addition of a dialyzer (see Figure 22.3) to the chemistry module is required to avoid excess peroxydisulfate destroying the activity of the cadmium reduction column.

This analysis procedure is applicable to all kinds of water samples, like surface water, ground water, wastewater, drinking water, seawater, and also for soil extracts. The minimum detection limit is better then $1 \, \mu g \, N \, l^{-1}$, which is sufficiently low for all sample types. The application is robust and will not be affected by acidic and seawater sample matrices. The flow diagram in Figure 22.3 shows the hardware components required for the automated application module to determine total nitrogen.

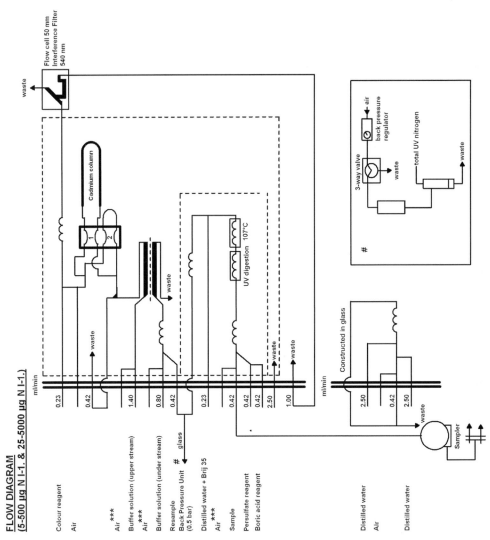

Figure 22.3 Typical flow diagram for automated nitrogen analysis.

Extra flexibility is provided by addition of a dilution loop prior to the UV oxidation process. In this way high ranges (>200 mg N l^{-1}) can be determined on a basic application layout of for example, 1–500 µg N l^{-1}. This automated method is highly sensitive and, in combination with digital detectors, suitable for a wide detection range.

An additional feature of this chemistry module is the possibility to use it for direct nitrate and nitrite analysis. By disconnecting the digestion part of the hardware, the sample follows the modified Griess reaction part of the module. This "double" use of hardware brings extra cost saving. The linear respondse of the curve is excellent from low nitrogen concentration, µg l^{-1}, to high, mg l^{-1}, levels. As the reagents are stable for a long time, calibration is carried out only once at the beginning of the analysis run. The stability, signal to noise ratio, is also excellent for a complex application like this. Initially, the calibration curve is built with a standard series, then samples follow batch wise, with quality controls in a user definable sequence. Long test runs, of more than 10 h, with many samples are no exception and are routinely made.

22.3.2
Total Phosphate

The automated procedure for the determination of total phosphate is based on the following reactions [7–10]: to hydrolyze the complex phosphate compounds to orthophosphate the sample is mixed with an acidic potassium peroxodisulfate solution. This flow is led through a UV digester. The organic phosphate complexes are decomposed to orthophosphates by the UV radiation and oxygen present. Then sulfuric acid is added to the stream and the flow is heated to 107 °C. Inorganic phosphate complexes are hydrolyzed to orthophosphate within that part of the application. The time for the total hydrolysis process, UV and high temperature, is approximately 15 min. After this hydrolysis the orthophosphate is analyzed with the aid of the well-established and reliable method using ammonium heptamolybdate. As this method is pH sensitive, the acidic sample flow coming out of the high temperature heater is neutralized with sodium hydroxide solution before addition of the ammonium heptamolybdate solution. The orthophosphate present in the flow then reacts and forms a phospho-molybdic acid complex, catalyzed by potassium antimony(III) oxide tartrate. This complex is reduced to an intensely blue colored complex by L(+) ascorbic acid. The blue complex is measured at 880 nm and is specific for phosphate. No interferences from silicate will be measured as the pH of the color reaction is kept below pH 1. Interference from extreme high background color of wastewater samples is avoided by the addition of a dialyzer.

This analysis procedure is applicable to all kinds of water samples, like surface water, ground water, wastewater, drinking water, seawater, and also for soil extracts. As this automated application is very robust its minimum detection limit is better then 1 µg P l^{-1} and so it is also very attractive for the analysis of seawater and soil samples.

As also mentioned for the total nitrogen application, the quality of the signal is excellent. It makes it possible to analyze at the sub-µg level. This application can be

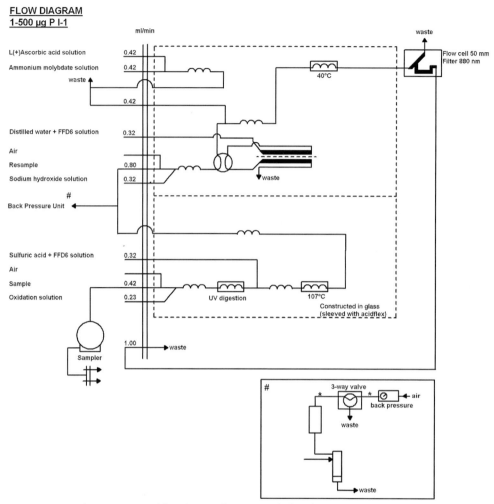

Figure 22.4 Typical flow diagram for automated phosphate analysis.

run simultaneously with the total nitrogen. In practice most samples have to be analyzed for both analytes and combining these applications in one system will double the sample throughput without operator interference. For both applications the sample conservation at pH 2 is according to international standards [12].

A typical flow diagram for automated phosphate analysis is shown in Figure 22.4.

22.3.3
Cyanides

Total cyanide is the sum of free cyanide ions, all organically bound cyanide, complex and simple metal cyanides, and those of gold and platinum, with the exception of the cyanide bound in cobalt complexes and thiocyanate. The cobalt complexes cannot be determined completely due to the catalytic decomposition of cyanide at high temperatures in acid medium, the cobalt ion catalyzes this decomposition.

Free cyanide is determined as easily liberatable cyanide. It is the sum of free cyanide ions and the cyanide bound in simple metal cyanides. Organic cyanides are not determined.

Weak Acid Dissociable Cyanide is basically the free cyanides like CN^- and HCN, together with weak metal-cyanocomplexes such as $[Cd(CN)_4]^{2-}$, $[Mn(CN)_6]^{3-}$, and $[Ni(CN)_4]^{2-}$, as well as mercury cyanide $Hg(CN)^2$. Iron complexes are not included. Acid dissociable cyanide, as determined in this way, is equivalent to cyanide amenable to chlorination (CATC) and available cyanide.

22.3.3.1   Total Cyanide

The automated procedure for the determination of total cyanide is based on the following reaction [11, 12]; complex bound cyanide is decomposed, in a continuously flowing stream at a pH of 3.8, by the radiation of UV light. A UV-B lamp and a decomposition coil of borosilicate glass (only permitting light from approximately 310 nm and upwards) is used to block UV light with a wavelength of less than 290 nm, thus preventing the conversion of thiocyanate into cyanide, which is decomposed to cyanide at that wavelength energy and below. After the decomposition, an in-line distillation unit, running at 125 °C, separates the released hydrogen cyanide from the digestion reagents. To distil with a constant flow a small vacuum is created in the distillation coil. In the distillate The hydrogen cyanide in the distillate is then determined spectrophotometrically. To develop a specific color reaction with the cyanide a few reagent solutions are added in the following order: chloramine-T, with which cyanogen chloride is formed. Then pyridine-4-carbonic acid and 1,3-dimethyl-barbituric acid are added, which develop a red-colored complex with the cyanogen chloride. The color is measured at 600 nm.

22.3.3.2   Free Cyanide

The automated procedure for the determination of free cyanide is based on the following reactions while using the same hardware configuration but with the UV lamp switched off. A zinc sulfate solution is added to the sample flow to precipitate any iron cyanides present in a zinc cyanoferrate complex. The free hydrogen cyanide in the sample is separated by the in-line distillation unit at 125 °C at a pH of 3.8 and under vacuum conditions. The hydrogen cyanide in the distillate is then determined spectrophotometrically, as described in the procedure for total cyanide.

Figure 22.5 shows a typical flow diagram for automated total cyanide analysis.

22.3.4
Phenol Index

The phenol index represents a group of phenolic compounds like phenol, ortho-, meta- substituted phenols and para-substituted phenols with a carboxl, halogen,

FLOW DIAGRAM

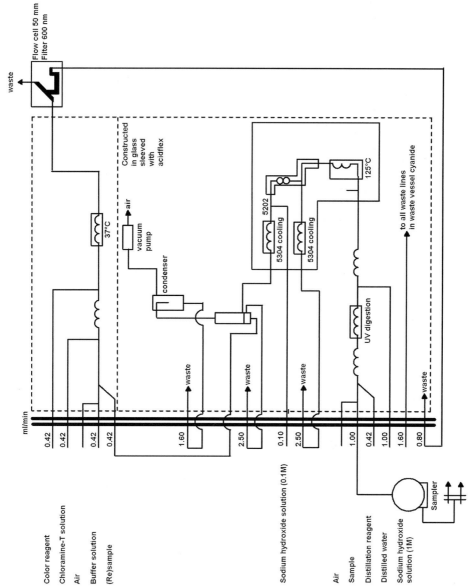

Figure 22.5 Typical flow diagram for automated total cyanide analysis.

methoxy or sulfonic acid group. These compounds are distillable, however, the volatility of these compounds is different. If a phenolic compound is of low volatility, the distillable amount is less than 100% and not fully taken into account in the total result. The measuring wavelength is normally set at 505 nm.

The automated procedure for the determination of ortho- and meta substituted phenols is based on the following reactions [13–16]: to determine the phenol index the sample is distilled, via an in-line distillation unit, in an acidic medium. The distillate is resampled into a buffer solution to control the pH at 10.0. Then an alkaline iron cyanide and 4-aminoantipyrine solution are added. The phenolic compounds present, will react with these chemicals to form a reddish-brown colored complex, which is measured at 505 nm.

To establish a stable signal the environment in which the analysis system is located plays an important role. The purity of chemicals and the reagent solutions prepared must be of analytical grade. Preparation must be carried out with care to clean glassware and pipettes. In particular, the hexacyanoferrate solution has to be protected from light. The measurement of the colored complex has been under discussion for a long time. Absorbance can be measured from 460 to 520 nm. Routinely, 505 nm is widely accepted. The absorbance of the detection range, in which normal values are to be expected, is low. $100\,\mu g$ phenol l^{-1} will yield an absorbance of approximately 0.035 AU. To detect in the low $\mu g\,l^{-1}$ level, detector resolution and stability of the signal must be high. In that case detection limits <1 μg phenol l^{-1} can be achieved. It also results in a coefficient of variation of ±0.3%.

Figure 22.6 shows a typical flow diagram for automated phenol index analysis.

22.3.5
Total Reducing Sugars

Reducing sugars, or reducing carbohydrates, present in samples can be determined in tobacco after extraction. As the name suggests, the sugars are not specifically determined themselves, but as a group, based on their reducing capacity. Saccharose, belonging to the group of reducing sugars, needs to be hydrolyzed prior to the determination.

The automated procedure for the determination of total sugars is based on the following reactions [17, 18]; after a separate extraction, the sample is mixed with an acetic acid solution and passed over a dialyzer membrane. If saccharose is present, an in-line hydrolysis step at 95 °C with hydrochloric acid is carried out first to convert the saccharose to glucose and fructose, both reducing sugars. After dialysis the sample is mixed with *p*-hydroxybenzoic acid hydrazine (PAHBAH) in a basic medium. To develop a yellow-colored osazone complex the flow is heated to 85 °C. The colored complex thus formed is measured at 410 nm.

In the flow diagram (Figure 22.7) the hydrolysis step for saccharose is included within the automated application. The hydrolysis can be done by addition of acid or an enzyme, for example, invertase or β-glucosidase. The hydrolysis must be performed prior to dialysis. This is because of the different speed in dialysis of different molecule

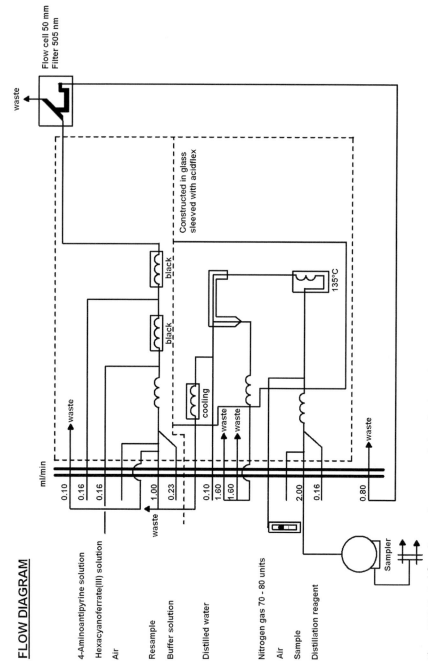

Figure 22.6 Typical flow diagram for automated phenol index analysis.

FLOW DIAGRAM

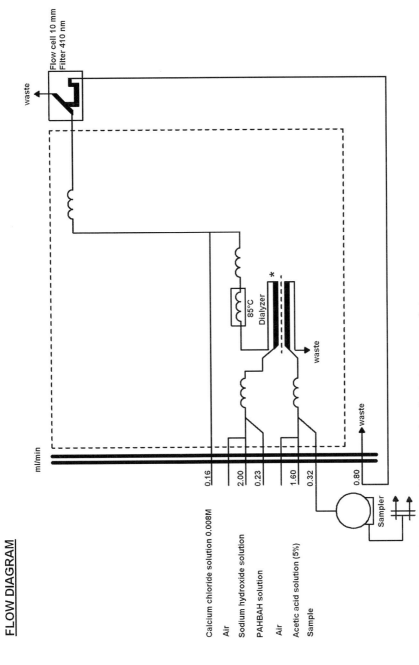

Figure 22.7 Typical flow diagram for automated total reducing sugar analysis.

sizes. If saccharose determination is not required a simple short cut enables exclusion of the hydrolysis step. The sample tube from the sampler is directly connected to the resample tube in the diagram.

It will be clear that reaction of various reducing sugars, like glucose and fructose, with the color reagent must yield equal color intensity. In the calibration the standards can be made up either from glucose or fructose. A quality check of the saccharose hydrolysis step is made with the addition of a pure saccharose standard. The theoretical concentrations of glucose and fructose to be formed are checked against pure glucose calibration standards. This check is of particular importance in the case of the use of enzymes.

22.3.6
Bitterness

Bitterness is a descriptive term in beer tasting. The bitter taste is formed by a group of iso-α-acids, among other taste affecting ingredients, which come from the hop into the beer during the brewing process.

The automated procedure for the determination of bitterness is based on an extraction method [19–21]: after acidification the sample is extracted in isooctane. The absorption of the iso-α-acids is measured at 275 nm

The phase separation, to separate the organic solvent isooctane is built in-line in the application. It is a small device which is treated inside with a coating to obtain a smooth continuous separation. As extractions are temperature dependent, the extraction coils are thermostatted. If tap water with a more or less constant temperature of 15 °C to a maximum of 20 °C is available this can be used. If not, a water circulating bath with internal cooling may be necessary. Pumping solvents does require special pump tubing, the classical available tubing is not resistant to solvents. A so called Acidflex (type of Viton) pump tube is used to convey the solvents.

Figure 22.8 shows a typical flow diagram for automated bitterness analysis.

It is hard to find artificial calibration standards. In practice this means that laboratories use known samples. These samples are once analyzed manually and are stable for at least 3 months at 4 °C. It is also possible to divide the samples into small portions and freeze them at −20 °C. When needed, simply defrost the portion for use.

As the iso-α-acids come from the hops, it is possible to use these purified acids as standard as well. It will be clear that the strength of these acids compared to the bitterness measured in beer is not known. Therefore these iso-α-acids need a separate quantification with the addition of a manual procedure.

Nowadays more and more variation in bitterness in beer is observed. The determination range, mostly between 20 and 35 bitterness units (BU), extends from 5 to 40 BU. Exceptional concentrations of bitterness in beer, up to 80 BU, will still occur. In those cases standardization around that concentration will be established to provide reliable results. The chemistry application does not need to be changed and can handle from low to high concentrations.

FLOW DIAGRAM

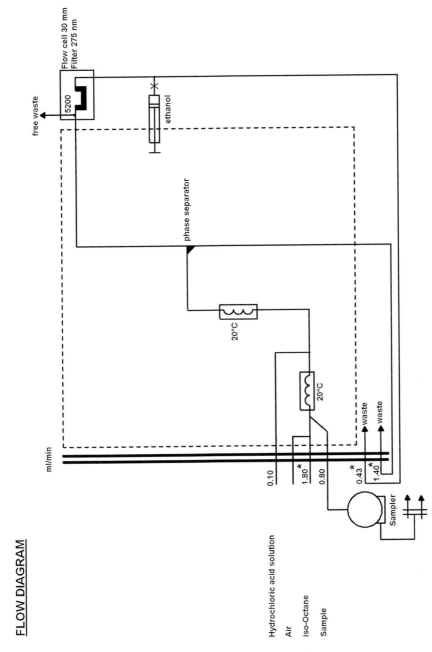

Figure 22.8 Typical flow diagram for automated bitterness analysis.

22.4
Development of Flow Analysis Applications

Automated flow analysis applications are replacing manual applications. They are justified when they increase sample throughput and decrease labor, laboratory space, use of chemicals and wastes. They must be efficient, accurate, and reliable. In general an amount of 10 samples per day for an application with distillation can already be sufficient to automate (e.g., cyanide).

An automated application determines analytical results from samples with a minimum of operator interference. It is mostly based on wet chemistry, followed by colorimetric measurement. Excluded from automated applications are sample preservations, extractions from solid samples to produce a liquid. All steps to automate an application must be known and understood. Any change from the original sample can be identified in a step; dilutions, additions of reagent, reaction time, temperature, extraction, distillation, and so on. Each step includes a condition, for example, for dilution – a dilution factor, for addition of reagent – a volume.

Listing all these steps will show the possibilities and impossibilities (not forgetting challenges). When steps are impossible to automate, all steps before that cannot be used in that automation either as it breaks the "flowing" process. The possibilities will give a list, which starts at a certain point in the analytical procedure and continues till the measurement in the detector. This list is the basis for automation of the manual procedure.

For translation of a manual step to an automated one must know very clearly the functions and features of a flow analyzer [3]. Below a simple example is given.

Steps of manual procedure to automate ammonia in soil samples:

1. Weigh out 10 g of dried soil
2. Add 50 ml distilled water
3. Mix for 1 h
4. Filter over paper filter
5. Pipette 5 ml sample filtrate into a 50 ml measuring flask
6. Add 10 ml buffer solution
7. Mix for 5 min
8. Add 5 ml reagent R1
9. Add 10 ml reagent R2
10. Add 5 ml reagent R3
11. Mix
12. Place flask in waterbath at 40 °C for 15 min
13. Cool and fill up to 50 ml with distilled water;
14. Measure the absorbance at 660 nm in a 10 mm cuvette
15. Measure a blank by replacing R3 with distilled water.

Steps to automate are as follows. The start of automation must be from point 5. Filtration of the soil slurry cannot be done in the flow analyzer and must be skipped. Till the end all steps are possibilities in automation. As we excluded the first four steps from automation it is clear that the final result to report is corrected for the sample intake in the extraction. A remarkable point is point 15, the blank measurement.

Manually it requires quite some extra work but using the feature of a dialyzer (a challenge) in the automated application, the blank does not require any extra effort. The dialyzer avoids the blank color of the sample influencing the measurement. The volumes used in the manual procedure are extensive compared to the volumes normally used in flow analysis. A total end flow is $1.6–2.0\,ml\,min^{-1}$. To convert volumes it is important to keep the ratio constant as much as possible, if this is not completely possible then modification of the reagent concentrations may be required. Reaction times are converted to the lengths ($=$ volumes) of the mixing coils and the flow running through. As mixing conditions are very intense and so reactions are optimized, it is not always necessary to take the full time as in the manual process. In addition, all samples, calibration standards and QCs are treated under identical circumstances. Temperature requirements are solved by heating (or cooling) reaction coils. This is possible at any place in the automated application. Finally, it is very important to understand the chemistry used to determine the analyte. Buffer capacity, pH of reactions, sensitivity to temperature, and light, and so on, play an important role in the manual procedure and certainly also in the automated procedure.

Figure 22.9 shows the final design of a flow diagram to analyze ammonia in soil.

22.5
Trends in Continuous Flow Analysis

The strength of analyses with continuous flow analyzers are the consistent quality and sample throughput. Although alternative automated analyzers are on the market, such as discrete analyzers that can do some simple straightforward applications, they can determine only a single parameter at a time. The simultaneous analysis of five, six or more parameters is an almost unbeatable advantage of continuous flow analyzers to gain speed in analyses throughput. Some disadvantages of flow injection analysis are slowing the implementation of flow systems in all application fields. These are technical problems of blockages in the small internal diameter tubing and inhandling samples with various matrices, particles and sample background color. This also has to be partly taken into account by discrete analyzers, which additionally suffer from low detection limits.

All of these issues need hardly be taken into account by CFA. For lengthy and complex applications with different analysis steps, like extraction, digestion and distillation, CFA will always be in favor for implementation of these techniques to configure a robust and reliable analyzer. The use of digital detectors brings extended range detection and a high level of accurate readings. Digital detectors nowadays use 20 bit and up analog to digital converters, resulting in detector resolution of less then 0.0003 AU. Compared with the early days with an 8 bit AD converter, the systems have improve significantly in a few decades of detection. Automated sample dilution and system intelligence for off-scale dilutions will lower the amount of labor put into the job. Integration of data acquisition and analyzer control programs into the laboratory network enables features for remote data access and analyzer control, even by the operator from home.

FLOW DIAGRAM

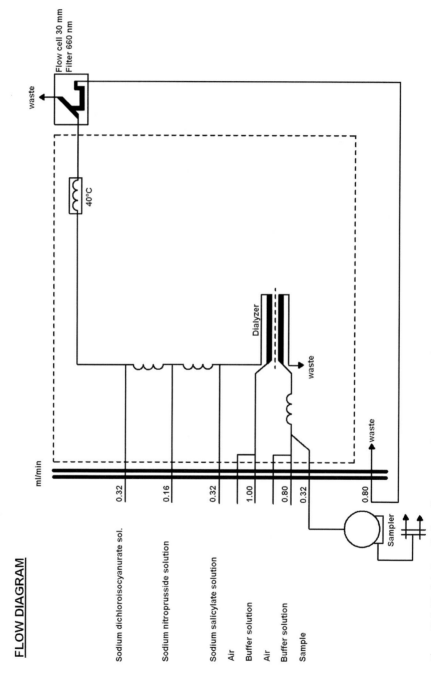

Figure 22.9 Final design of a flow diagram to analyze ammonia in soil.

Acknowledgements

The authors wish to acknowledge the company Skalar for its kind cooperation in making time available to time write this chapter.

References

1 Valcárcel, M. and Luque de Castro, M.D. (1988) *Automatic Methods of Analysis*, Elsevier, Amsterdam, pp. 143–149.

2 Kérouel, R. and Aminot, A. (1997) Fluorimeric determination of ammonia in sea and estuarine waters by direct segmented flow analysis. *Marine Chemistry*, **57**, 265–275.

3 Furman, W.B. (1976) *Continuous Flow Analysis, Theory and Practice*, Marcel Dekker, New York.

4 Houba, H.J.G., Novozamsky, I., Uittenbogaard, J. and van der Lee, J.J. (1987) Automatic determination of total soluble nitrogen in soil extracts. *Landwirtschaft Forschung*, **40**, 295–302.

5 Kroon, H. (1993) Determination of nitrogen in water; comparison of a continuous flow method with on-line UV digestion with the original Kjeldahl method. *Analytica Chimica Acta*, **276**, 287–293.

6 Skalar Automated N&P Analyser, Skalar Publication No. 0711293.

7 Standard Methods for Examination of Water and Waste Water. 15th edition 1980 APHA-AWWA-WPCF, pp. 410–425.

8 Boltz, D.F. and Mellon, M.G. (1948) Spectrophotometric determination of phosphate as molybdiphosphoric acid. *Analytical Chemistry*, **20**, 749–751.

9 Walinga, I., van Vark, W., Houba, V.J.G. and van der Lee, L.L. (1989) Plant analysis Procedures, Part 7, Department of Soil Science and Plant Nutrition, Wageningen Agricultural University, Syllabus, pp. 138–141.

10 ISO 15681-2. Determination of orthophosphate and total phosphorus contents by flow analysis, Part 2: Method by continuous flow analysis (CFA).

11 ISO 14403:2002. 1st edition 2002-03-01. Water quality - Determination of total cyanide and free cyanide by continuous flow analysis.

12 ISO 5667-3:2003. Water quality – Sampling – part 3: Guidance on the preservation and handling on water samples.

13 Standard method for the examination of water and wastewater. 21st edition, Chapter 5, Method 5530 Phenols, 2005.

14 ISO 14402. Water quality – Determination of phenol index by flow analysis.

15 EPA, method 420.2.

16 ASTM, Standard Test Methods for the Phenolic Compounds in Water, D1783-01.

17 Davis, R.E. (1976) A combined automated procedure for the determination of reducing sugars and nicotine alkaloids in tobacco products using a new reducing sugar method. *Tobacco Science*, **32**, 39–44.

18 ISO 15154. Tobacco, Determination of the content of reducing carbohydrates; Continuous flow analysis method.

19 Cooper, A.H. and Hudson, J.R. (1961) *Automated Analysis Applied to Brewing*, Brewing Industry Research Foundation, pp. 436–438.

20 European Brewery Convention, Analytica EBC, 5th edition, 2006 method 9.8.

21 The American Society of Brewing Chemists, Methods of Analysis of the ASBC, 2004, Beer Bitterness method 23 D.

Index

a

absorption techniques   244
acetylcholine esterase (AChE) inhibitors
    415, 501
acetylsalicylic acids   242, 604, 607
AChE-based flow systems   415
acid-base titrations   441
active operated devices   172
– real-time sample conditioning   172
– self-optimization of the flow system   172
adsorptive strtipping voltammetry   116
advanced calibration methods   203
advanced research projects agency network
    (ARPANET)   321
affinity chromatography   484
affinity microcolumn   94
aliphatic amines   433
alkaline phosphatase   160
amines   233
– derivatization of   233
ammonium ion-selective electrodes   412
amperometric biosensor   415
amperometric detectors   412
amperometric technique   625
angiotensin converting enzymes (ACE)
    499
aniline   240
– derivatization of   240
anionic surfactants   374
– CL determination of   374
anodic stripping analysis (ASV)   459
antiretroviral drugs   252
– didanosine   252
– zidovudine   252
artificial neural networks (ANNs)   250
– models   250
ascorbic acid   241, 605, 624
– binary determination of   241

at line, robotic arm   275ff.
atmospheric pressure chemical ionization
    (APCI)   553
atmospheric pressure photoionization
    (APPI)   553
atomic absorption spectrophotometry (AAS)
    50
atomic fluorescence spectrometry   527, 608
automated manifolds   531
automated-screen-printing technology   435
auxiliary reagents   244

b

bacterial cells   156
– lysis of   156
batch cells   107
batch injection analysis   107, 119
– capillary   119
beads   82
– analysis   80
– injection   60
– packed column   69
– suspensions   88
Beer's law   607
beta-adrenergic drugs   433
bienzymatic glutamate biosensor   397
bioassay system   155
– integration of   155
biocatalytic process   395
– specificity   395
biochemical assays (BCAs)   483
– immunoaffinity-based   496
– implementation   483
– methodologies   483
– technique   162
biological fluids   190
bioluminescence   365
biosensor chips   461

Advances in Flow Analysis. Edited by Marek Trojanowicz
Copyright © 2008 WILEY-VCH Verlag GmbH & Co. KGaA, Weinheim
ISBN: 978-3-527-31830-8

BIS-FIA fluorimetric methodology 371
bismuth electrodes 461
Bratton–Marshall reaction 605

c
calibration curve method 203, 253
calibration graph(s) 205, 212, 214
– multi-line two-point 212
calibration methods 247
– multivariate 247
calibration strategy, *see* standard addition
method (SAM)
capacity-time curves 469
capillary chromatography 139
capillary electrophoresis (CE) 142, 134,
274ff., 546, 561
– systems 53, 134
– techniques 135
carbamate pesticides 415
carbon-based electrodes 622
carbon paste 116
catalytic electrode processes 469
cathodic stripping analysis (CSV) 459
cation exchange polymers 115
cell-based assay 164
cell design 110ff.
cell suspension 156
chelating reagents 153
chemical operations 228
– liquid-liquid extraction 228
– solid-phase extraction 228
chemical oxygen demand (COD) 585
chemical processing (CFCP) 149
chemical reaction 157
– real-time monitoring 157
chemical vapor generation 512
chemically modified electrodes 623
chemiluminescence 51, 614
– advantage of 614
– based determinations 615
– cerium(IV)-based CL systems 359
– detector 51
– direct 360
– electrogenerated 360
– luminol 359
– methods 590, 614
– permanganate-based 360
– peroxyoxalate 359
– reagents 374
– sensors 374
– solid-phase 372
– *Tris*(2,20-bipyridyl)ruthenium(II) 359
chemiluminescent reaction 617, 618
chemiluminometric methods 244

chemometric methods 232, 250, 252
– performance of 250
chemometric techniques 245
– data analysis 245
chip-based slide valve 160
chip technology 152
chromatomembrane cell (CMC) 590, 591
– device 592
– membrane 592
– system 592
chronoamperometry 109
– potential-step 109
Clark-type oxygen electrode 416
combined injection methods 70
comma separated variable (CSV) 323
commutation 168
– commuter 168
– rotary valve 168
– sliding bars 168
– solidary commutation (mechanical) 168
components of flow systems
– contribution to dispersion 21
– detection 24
– injection 23
– transport 25
computer-controlled burettes 54
computer-controlled flow method 577
continuous flow analysis 44, 346, 639, 659
– technique 44
– trends 659
continuous flow analyzers 639
– configurations 44, 45
– system 646
– technique 44
continuous flow voltammetry 624
controlled dispersion 176
controlled variable volume injection 54
coulometric flow cells 462
cumulative injection 55
current-voltage curves 457
cyanide amenable to chlorination (CATC) 651
cystine 240

d
data sets 245
– data augmentation 246
– one-way data 245
– three-way data 246
– two-way data 246
– zero-way data 245
derivatization 240, 350ff., 351, 612
– electrochemical 612
– hydrolysis reactions 350
– liquid extraction 612

– photochemically induced fluorescence
  (PIF)   350ff.
– processes   603
– reactions   293
desorption electrospray ionization
  (DESI)   546
detection   112
– amperometric   112, 114
– calorimetry   114
– optical   114
– potentiometric   112, 114
– voltammetric   114
deuterated solvents   158
dialysis   228, 274
– ASTED   274
– derivatization   228
– gas-diffusion   228
– microdialysis   274
– process   50
diazotization-coupling reaction   581
diffusion-convection equation   9
diffusion layer   109
diffusion limited current   108
digital detectors   659
dimensionless numbers in flow systems   11
DIN transient signals   559
diode array spectrophotometers   602
direct analysis in real time (DART)   546
direct injection nebulizer system (DIN)   554
dispersion in flow injection analysis   7
disposable syringe   53
dissolution reactions   520
distribution of times of residence   10
double-beam spectrophotometer   242
double-injection flow system   604
drug analysis   611
– flow-fluorimetry   611
drug dissolution   628
drug release testing   628
drugs   605
– FIA spectrophotometric analysis   605
– FIA spectrophotometric determinations
  605
dual sequential injection analysis   71
DUMAS method   645

e
EAD systems   501
ECD-FA methods   458
Ehrlich's reagent   606
electrokinetic (EK) injection   142
electrical double layer (EDL)   129
electrochemical detectors (ECDs)   89, 107,
  455

electrode arrays   443
electrode fouling   459
electrodes   108
– wall-jet   108
– wall-tube   108
electromagnetic radiation   614
electromigration techniques   425
electroosmosis   127
– based pumps   129
– phenomenon   127
electroosmosis-driven flow   129
electroosmotic flow (EOF)   129
– based pump   131, 140, 143
– channel(s)   137
– driven flow analysis system   134, 136, 142,
  143
– electrokinetic injection   142
– flow analysis system   145
– flow rate   139
– hydrodynamic/pressure injection   142
– microchannel-based   138
– pH-dependent   139
– principles   129
– pumped SIA system   143, 145ff.
– valve injection   142
electrospray ionization mass spectrometry
  (ESI–MS)   485, 486
electrothermal atomic absorption
  spectrometry   468, 520
emission-based methods   614
– techniques   244
enzymatic cofactors   243
– detection   416
enzymatic signal amplification   496
enzyme activity   412, 415
– assays   415, 635
– measurements   412
enzyme affinity detection (EAD) system   500
enzyme flow assays   498
enzyme inhibitor detection   415
enzyme-linked immunosorbent assay   162
– microchip-based   162
enzyme-linking immunoassay system   466
enzyme substrates   396
enzymes, immobilization   373, 374
Escherichia coli   412
ETAAS detection   93
Extensible Markup Language (XML)   340
extrapolative version of multivariate
  calibration   219

f
fast-switching solenoid valve   63
ferricyanide-aminoantipyrine reaction   603

FIA-FTIR method 608
FIA-FTIR spectroscopy 608
FIA-ICP-MS 556
FIA-ICP-TOFMS 556
FIA-MS methods 245
field effect transistors (FET) 435
first-order methods 250
– partial least square regression (PLS) 250
– principal component regression (PCR) 250
flame atomic absorption spectrometry 244, 512, 608
flame emission spectrometry 208
flow-based molecular absorption spectrophotometry 602
flow analysis applications
– database 323
– environmental applications 577
– techniques 43
flow biochemical assays 491
– on-line coupling 491
flow conductimetric determinations 619
flow injection analysis (FIA) 43, 46, 63, 107, 203, 205, 218, 227, 244, 321, 395, 484, 556, 560, 577
– applications 560
– calibration methods 203, 209, 210, 220
– configurations 556
– drug analysis 607, 608
– electrochemical detection 244
– enzyme assays 488
– enzyme-based 412, 417
– fluorescence-based methods 351
– gradient concept 217
– immobilized reagents 237
– immunoassay 162
– ligand-binding assays 488
– manifold 220, 556
– methods 221, 231, 587, 609
– multicomponent 227, 234, 237
– multi-syringe 208
– polarographic determinations 622
– principles 46
– procedure 209, 210, 214
– pseudo-titrations 606
– rotary injection valve 64
– systems 159, 160, 162, 213
– titrations 441
flow injection mass spectrometry 487, 545
– analysis 545
– automated 641
– bioreactor-based 403, 411
– biosensor-based 397

– importance 545
– photometric 413
– potentiometric 414
– role 545
– signals 604
– turbidimetric determinations 606
– UV–vis configurations 607
– UV–vis method 608
flow injection multicomponent 232, 237
– analysis 252
– chemiluminescence 243
– determinations 237
– immunoassay-based 237
– manifold 232
– methods 229, 230
– optimization process 229, 230
flow potentiometry 425
flow processing devices 265ff.
flow systems 205
– batch flow injection analysis 6
– classification and fundamentals 4
– continuous flow analysis 4
– feature 205
– flow injection analysis 4
– multicommutation 5
– sequential injection analysis 4
– stopped flow 6
flow voltammetry 455
fluorescence-based flow methods 612
fluorescence 349, 350
– complexation 351
– laser induced 349
– native, 347ff.
– solid phase 350ff.
– solid surface 350
– spectrometry 346
fluoride-selective electrode 621
fluorimetric assay 60
food analysis 182
FTIR spectrophotometry 608
fully rotary valve (FRV) 206
fused silica capillaries (FSCs) 129

g
galactic network 321
gas diffusion 50, 57
– scrubber 590
gas displacement pump (GDP) 554
gas-liquid device 563
gas-segmented flows 501
Gaussian distribution profile 554
GC-MS interface devices 551
generalized standard addition method (GSAM) 219

glass disk   138
glass electrodes   621
glassy carbon   115
glucose   67, 117
– chemiluminescence-based determination
    67
glucosidase bioreactor   413
glutamic acid   607
gradient dilution calibration   216
Griess reaction   56, 155, 647, 649
Griess–Saltzman reaction   144
Grignard reaction   159

h

Hagen–Poiseuille equation   128
hemoglobin   117, 118ff.
high-performance liquid chromatography
    (HPLC)   241, 483, 490, 496, 501
– analysis. 483
– instruments   241
– methodologies   496
high temperature liquid chromatography
    (HTLC)   501
high throughput screening (HTS)
    techniques   483
homogeneous systems   602, 603
horseradish peroxidase (HRP)   67
hydride generators   555
hydrobromic acid   610
hydrodynamic injection   53, 61, 142
hydrodynamics   109ff., 110ff.
hydrolysis reactions   350
hyphenated techniques   529

i

immobilized biomacromolecules   237
– nucleic acids   237
– proteins   237
immunoassay system   162
– semi-automated heterogeneous   162
immunochemical determinations   95
in-line distillation   639
in-line UV digestion   639
– industrial and environmental
    applications   639
in situ spectrophotometric detection   51
index analysis   654
– flow diagram   654
indirect determinations   519
indophenol blue methods   580
inductively coupled plasma-atomic emission
    spectrometry (ICP-AES)   244
inductively coupled plasma-mass spectrometry
    (ICP-MS)   244

– detection   93
infrared absorption   607
inhibition   116
– enzyme-catalyzed hydrolysis   116
injection   47, 48
– merged   48, 70
– process   68
– proportional   48
– rotary valves   47
– techniques   43
– valve   49, 58
inorganic fillers   237
inorganic oxidants Ce(IV)   603
interpolative standard addition method   211
ion carrier   428
ion-exchange membranes   50
ion exchanger beads   139
ion-exchange resins   374
ion exchanger, *see* ionophore
ion-pair extraction reaction   152, 154
ion-selective electrode(s)   415, 425, 426, 619
– direct determinations   619
– indirect determinations   620
ion-selective ionophores   153, 428, 430, 433
IR spectrophotometry   607
IR systems   607
isotope dilution (ID)   556

j

jet ring cell   85, 87
Journal of Flow Injection Analysis (JFIA)   327

l

lab-on-a-chip, *see* microfluidic devices
lab-on-valve (LOV)   60, 149, 521, 601, 634
– bead injection   635
– microconduits   69
– systems   61
lab-on-valve to capillary electrophoresis
    equipment (LOV-CE)   61
laser ablation   550
LED-based absorbance detector   135
ligand-binding flow assays   493
– on-line   493
limit of detection (LOD)   578
linear sweep voltammetry   109
lipid regulators   69
lipophilic pH indicator   154
liquid chromatographs   271ff.
– post-column arrangements   271
– pre-column arrangements   271
liquid flow techniques   425
liquid handling system   127
liquid-liquid extraction   606, 610

– systems 609
luminescence phenomena 346ff., 610
– chemiluminescence 346
– fluorescence 346
– phosphorescence 346
luminescence detection 53
– bioprocess monitoring 53
– immunoassay 53
luminescence, solid-phase 366
luminol chemiluminescence 617
lytic agent 156

m
magnetic microparticles 82
magnetic nanoparticles 99
mass spectrometry-based biochemical
   assays 488
– requirements 488
matrix-assisted laser desorption/ionization
   (MALDI) 485
matrix effects 555
measurement of dispersion 15
– approaches 21
– axial dispersion, degree and intensity 19
– "D" coefficient 16
– peak variance and theoretical plate
   height 18
– peak width and time of appearance 17
membrane-based systems 607
membrane liquid/liquid extraction 270
– hollow-fiber membranes 270ff.
merging-zones technique 219
metal-ligand interaction
   (metalloporphyrins) 430
methylene blue active substances (MBAS)
   640
Michaelis–Menten theory 412
micro-column reactors 237
micro flow systems 127
micro-total analysis systems (μTAS) 142, 220
micro-unit operations 149
– phase separation 150
– solvent extraction 150
microbead suspensions 82
microchannels 159
microdialysed samples 524
microelectrode arrays 116
microemulsion 526
microfluidics 94
– channels 137
– T-shape 137
microfluidic chip(s) 95, 138
– platforms 138
microfluidic devices 149, 162, 163

microfluidic systems 374
micropipette tip 110
microporous materials 141
– monolithic column materials 141
– porous glass frits 141
microprocessor-controlled electronic
   pipette 206
microsensor array 244
microvolumes 107
miniaturization 95, 265
modalities of flow analysis 181
models of transport 25
– axially dispersed plug flow model 26
– descriptive "Black Boxes" 25
– random walk 28
– tanks-in-series model 27
– uniform dispersion model 26
molecular absorption spectrophotometry
   602
molecular imprinted polymers (MIPs) 374,
   587
molecular imprinting processes 618
molybdenum blue method 584
monoenzymatic electrodes 397, 402
MS ionization chamber 551, 552
multi-collector (MC) spectrometer 559
multi-commutated flow techniques 72
multi-commutation 167, 169ff., 171
– addition/removal of manifold
   components 171
– concentration-oriented feedback
   mechanisms 169
– concept 181
– definition 169ff., 171ff.
– intermittent stream 169
– merging zones 170
– multi-commuted flow system 171
– sample stopping 170
multi-dissolution vessels 628
multi-injection systems 241
multi-insertion principle 62
multi-ion sensing system 152, 154
– integration of 152
multi-point calibration graphs 205
multi-port valves 241
multi-pulse amperometry 456
multi-pumping flow injection analysis 346
multi(bio) sensor arrays 244
multicollector mass spectrometer (MCMS)
   548
multicommutated flow injection analysis
   (MCFIA) 61
– manifold 61
– systems 61

multicomponent analysis   229
– principal   229
– strategies   229
multicomponent determinations   239, 602
multiparametric analysis   57
multipumping flow systems (MPFS)   69
multisyringe burettes   66
multisyringe flow injection analysis
    (MSFIA)   63
– LOV system   69
– MPFS system   72
multivariate calibration   249
multivariate measurement   219
multivariate optimization   231, 232
– simplex algorithm   232
multivariate responses   219

n
Nafion membranes   410
nanofluidics   100
– devices   164
nanoparticles   98
nanotechnology   98
nebulizer-flame devices   244
Nikolskii–Eisenmann equation   427
nitrate ions   50, 56, 57
– reduction   57
– SIA determination   56
nitrite   56, 117, 144
– SIA analysis   144
– SIA determination   56
non-chromatographic separation
    techniques   274
– dialysis   274
– gas diffusion   274
non-chromatographic separation
    technique   268, 280
– liquid/liquid extraction   270
– liquid-phase microextraction   280
– membrane liquid/liquid extraction   270
– microwave-assisted extraction   269
– pervaporation   270
– pressurized hot water extraction   270
– solid-phase microextraction   280
– sonication-assisted solvent extraction   270
non-steroidal antiinflammatory drugs   69
non spectroscopic interferences   555
– matrix effects   555
nuclear magnetic resonance (NMR)   157,
    158, 244
– instrument   157, 158
– measurement   158
– microscale   157
nucleoside hydrolysates   242

o
off-scale dilutions   659
on-chip ion sensing   154
– system   152
on line   277
– chemical derivatization reactions   293
– immobilized enzymes   293, 296
– chemical vapor generation   513
– digestion
    – microwave radiation   292
    – ultrasound radiation   292
    – UV radiation   292
– dilution   526
– gas sampling   303
– membrane-based separation   302
    – diffusion scrubber   304
    – Donnan dialysis   310
    – gas diffusion   292
    – microdialysis   304
    – passive dialysis   304
    – pervaporation   292
    – supported-liquid membrane   292
– photodegradation tests   615
– sample processing   291
    – cold vapour   309
    – co-precipitation   300
    – dilution   292
    – gas-liquid separation   301
    – hydride generation   301
    – precipitation   300
– solid samples   306
    – dissolution   306
    – extraction   306
    – leaching   306
– solvent extraction   294
    – back-extraction   294
    – cloud-point extraction   296
    – extraction chromatography   296
    – flow-batch extraction   292
    – liquid membranes   292
    – micelle-mediated extraction   292, 296
    – wetting-film extraction   297
– sorbent extraction   297
    – bead injection   300
    – extractive membranes   297
    – imprinted polymers   299
    – knotted reactors   292
    – packed columns   299
    – solid-phase optosensing   299
– split-flow interface   277ff.
on-tube detectors   127
optosensing, RTP-based   371
organic compounds   586
– toxic/hazardous   586

organic/aqueous interface   150
organized media, fluorescence   349
organophosphoric insecticides   415
– spectrophotometric assay   415

p
p-aminobenzoic acid   604
packed-bed reactor   67
packed columns   522
parallel configurations   241
parallel factor analysis (PARAFAC)   248
partial least square regression (PLS)   248
passively operated device   171
– sample stopping   172ff.
penicilloic acid   621
pentoxyverine citrate   616
– chemiluminescent reaction   616
peristaltic pump   49, 54, 63, 128, 209
pH indicator dye   152
pharmaceutical compounds   117
pharmaceutical formulations   602
– analysis   602
phenyl barbituric acid   615
phosphate   117
phospho-molybdic acid complex   649
phosphonium salts   429
phosphorescent light   614
photoluminescence detection   87
physico-chemical processes   601
– heterogeneous   601
– homogeneous   601
piezoelectric pumps   128
piston pumps   55, 129
Plackett–Burman designs   232
plant and soil analysis   188
plant digests   188
poly(methylene blue)   117
polymer films   115
polymeric chip   127
polymeric ion exchangers   139
polymeric substrate (PDMS)   136
– membrane   159
– valves   159
polytetra-fluoroethylene (PTFE)   160
– membrane diffusion cell   619
– tubes   127
porous borosilicate glass   139
post-column reaction   67
potentiometric biosensors   402
potentiometric detection   91, 244
precipitation reactions   520
preconcentration   514, 527
– on minicolumns   513
preconcentration/separation systems   512

preparative techniques   511
principal component regression (PCR)   248
properties of fluids and transport
    phenomena   7
– diffusion   8
– diffusivity   8
– thermal conductivity   8
– viscosity   7
protein–ligand interactions   483, 485
protein–protein interactions   483
pulse-based injection   70
pulsed sample introduction interface
    (PSI)   553
PVC membranes polymer   435
– matrix   430

q
quadrupole mass spectrometer (QMS)   547
quantitative structure-activity relationship
    (QSAR)   433
– approach   433
quantum dots   375
quasi-independent techniques   440

r
radial dispersion, intensity   20
radio frequency (RF) power supply   554
radiometric detections   91
reaction-based calibration   217
rearguard configuration   266 ff.
renewable column   83, 84, 91
reporter-free assays   489
retaining microbeads   85
room temperature phosphorescence (RTP)
    355ff.
rotary hexagonal valve   47
rotary injection valve   47
rotary valves   47, 64
RTD waves and generation of signals   12
– controlled dispersion   13
– dispersion process   12
– transient profile   14

s
sample dilution   519
sample incubation   178, 179
– zone trapping   179
sample inserted volume   177
sample introduction   175, 176
– hydrodynamic injection   175
– introduced sample aliquot   175
– loop-based injection   175
– multiple injections   176
– time-based injection   175

sandwich technique   55
screen-printed (disposable) electrodes   465
screen-printing technology   435
second-order calibration   252
second-order data, *see* two-way data
sector field mass spectrometers (SFMS)
    547
segmentation-retention process   55
segmented flow analysis technique   45
segmented flow analyzers (SFA)   639
selection valve (SV) ports   59
semi-automated flow/bead-injection system
    413
sensors   366
sequential determination   240
sequential injection   71
– multisyringe flow injection systems   71
sequential-injection analysis (SIA)   53, 57,
    346, 436, 565, 577
– amperometric immunosensing   238
– based colorimetric analysis   138
– bead injection system   634
– electroanalytical systems   622
– flow cell   604
– liquid drivers   55
– method   238, 611
– optimization of   143
– sandwich systems   56
– switching valves   71
– system   54, 59, 71, 43, 239, 565, 611, 612,
    628, 630, 632
sequential spectrometers   559
set of standards method (SSM)   203, 204, 207,
    211, 214, 215
Shinn's reagent   49
silica-based pumps   140
simplification   265ff.
single-channel FIA assemblies   603
single-line FIA system   134
single-syringe burette   71
six-way valve, *see* rotary valve
sliding bar commuter   170
– intermittent stream   170
small-sized thin solid phase (STSP)
    columns   590
Smoluchowski equation   130
soft ionization techniques   485
soil extracts   188
solenoid piston micropumps   69
solenoid pumping system   579
solenoid valves   559
solid-liquid extraction   618
solid-liquid heterogeneous systems   613
solid-phase absorptiometry   86

solid-phase enzymatic reactors   627
solid-phase extraction   55, 179, 269, 272, 468,
    514, 586
solid-phase reactor   609, 612, 613, 621
solid-state electrodes   435
solid particles   79, 80, 81
solution additions to the sample   178
– flow titrations   178
– standard addition method   178
spectrophotometric method   582, 603
spectroscopic detection methods   587
– atomic absorption   587
– ICPAES   587
– ICP-MS   587
square-shaped signal   45
square wave voltammetry (SWV)   458
stand-alone analytical techniques   127
standard addition method (SAM)   556
– application   556
– methods   205, 207, 211, 214, 215
stopped-flow technique   219
strict-sense flow   463
stripping voltammetry   115
– anodic   115
sugar analysis   655
– flow diagram   655
sulfur dioxide   241
– binary determination of   241
surface-enhanced resonance Raman
    spectroscopy (SERRS)   470
surfactant-coated alumina   610
synchronous injection   67
syringe micropumps   128
system operating mode   172, 173
– aspirating   173
– hydrodynamic pressure   172ff., 173ff.
– propelling   172

t

tandem streams   62, 174
– binary sampling   174
– injection   62, 174
– liquid interfaces   174
– mixing conditions   174
tannic acid system   617
Teflon membrane   139
tetrabutylammonium cation   134
thermal/calorimetric   114ff. 119
– optical   114, 119
thick-film urea biosensors   412
thin-layered flow cells   466
three-channel solenoid valve   554
time-based injection   61, 62, 65
– techniques   43

time-based injector   206
time of flight mass spectrometers
      (TOFMS)   549
time window   167
– external timing   167
TiO$_2$-mediated photodegradation reactor   466
total analysis systems (mTAS)   149
trace elements   515
trace metal ions   115, 116ff.
transient-sampling technology, *see* laser ablation
transient signal   441
tubular reactors   237
two-point multi-line graphs   213
two way data   246
two-way liquid drivers   55

u
univariate calibration, *see* zero-order
      calibration   248
univariate optimization   231
UV digester   647
UV light source   603
UV–vis detection   496, 602
UV–vis heterogeneous systems   606
UV–vis spectrophotometry   53, 208, 243, 607
– instruments   242
– measurements   59

v
vanguard configuration   266ff.
volatile reaction   52

voltage drop   138
voltammetric curve   112
– pseudo-steady-state   112
voltammetric method   624
voltammetric microcells   465
voltammetric studies   624
voltammetric techniques   244, 458
voltammetric/amperometric flow
      analysis   456
– principles   456
– techniques   456
voltammetry   112
– cyclic   112
– square wave curve   112
volume-based injection   43

w
wall-jet   109ff., 110, 111
– detection mode   434
water analysis   184
weak acid dissociable cyanide   651
widening dynamical concentration range
      178
Wittig reaction   159

z
zero-order calibration   248
zone penetration technique   219
zone sampling   168, 169, 178ff.
– external timing   169ff.
– technique   219